ANNALS OF THE NEW YORK ACADEMY OF SCIENCES
Volume 320

HEALTH EFFECTS OF HALOGENATED AROMATIC HYDROCARBONS

Edited by William J. Nicholson and John A. Moore

The New York Academy of Sciences
New York, New York
1979

Library of Congress Cataloging in Publication Data

Main entry under title:

Health effects of halogenated aromatic hydrocarbons.

(Annals of the New York Academy of Sciences; v. 320)
1. Halocarbons—Toxicology—Congresses. 2. Aromatic compounds—Toxicology—Congresses. I. Nicholson, William Jamieson, 1930– II. Moore, John Arthur, 1939– III. New York Academy of Sciences. IV. Series: New York Academy of Sciences. Annals; v. 320.
[DNLM: 1. Hydrocarbons, Halogenated—Toxicity—Congresses. 2. Hydrocarbons, Halogenated—Metabolism—Congresses. 3. Environmental pollutants—Toxicity—Congresses. 4. Immunity—Congresses. W1 AN626YL v. 320 / QD341.H9 H434 1978]
Q11.N5 vol. 320 [RA1242.H35] 508'.1s 79-12253
ISBN 0-89766-008-0 [615.9'51'1]

SP
Printed in the United States of America
ISBN 0-89766-008-0

ANNALS OF THE NEW YORK ACADEMY OF SCIENCES

VOLUME 320

May 31, 1979

HEALTH EFFECTS OF HALOGENATED AROMATIC HYDROCARBONS*

Editors and Conference Chairmen

WILLIAM J. NICHOLSON AND JOHN A. MOORE

Advisory Committee

KARIM AHMED, JAMES R. ALLEN, PERRY J. GEHRING, JAMES GIBSON, LEO J. HETLING, RENATE KIMBROUGH, JAMES D. MCKINNEY, ALAN POLAND, J.G. VOS, AND MARCUS WASSERMANN

CONTENTS

Part I. Production, Chemistry, and Distribution

*This volume is the result of an international conference entitled *Health Effects of Halogenated Aromatic Hydrocarbons* and several related workshops, all of which were held 24–27 June, 1978 by The New York Academy of Sciences as part of its Science Week program.

Part II. Animal Toxicity and Metabolism

Papers from the Workshop Entitled Body Clearance of Halogenated Hydrocarbons

Part III. General Human Health Effects

Part IX. Surveillance for Future Environmental Contaminants

Late Papers

Part II.

Part VII.

Paper Presented at the Conference but not Submitted for Publication:

Part VIII.

This volume is one of three resulting from the **SCIENCE WEEK** sponsored by The New York Academy of Sciences on June 21–June 30, 1978. The title of the overall program was: **THE SCIENTIFIC BASIS FOR THE PUBLIC CONTROL OF ENVIRONMENTAL HEALTH HAZARDS.**

Chairman of Science Week: Herbert J. Kayden

Science Week Conferences were organized by:

E. Cuyler Hammond
John A. Moore
William J. Nicholson
Irving J. Selikoff

Science Week Administrative Coordinator: Paul Milvy

Financial Assistance was Received from:

- ABBOTT LABORATORIES
- ALCOA FOUNDATION
- ALLIED CHEMICAL CORPORATION
- AMALGAMATED CLOTHING AND TEXTILE WORKERS UNION
- THE B.F. GOODRICH COMPANY
- CELANESE CORPORATION
- E.I. DU PONT DE NEMOURS & COMPANY
- EXXON CORPORATION
- FANNIE E. RIPPEL FOUNDATION
- THE FORD FOUNDATION
- THE GOODYEAR TIRE & RUBBER COMPANY
- INSULATORS HEALTH HAZARD FUND
- INTERNATIONAL CHEMICAL WORKERS UNION
- JOHNS-MANVILLE CORPORATION
- NATIONAL BUREAU OF STANDARDS
- NATIONAL CANCER INSTITUTE (1 R13 CA 24515-01)
- NATIONAL INSTITUTE FOR OCCUPATIONAL SAFETY & HEALTH
- NATIONAL INSTITUTE OF ENVIRONMENTAL HEALTH SCIENCES
- OCCUPATIONAL SAFETY AND HEALTH ADMINISTRATION
- OFFICE OF NAVAL RESEARCH
- OIL, CHEMICAL AND ATOMIC WORKERS INTERNATIONAL UNION
- UNITED STATES ARMY MEDICAL RESEARCH AND DEVELOPMENT COMMAND
- UNITED STATES BUREAU OF MINES
- UNITED STATES CONSUMER PRODUCT SAFETY COMMISSION
- UNITED STATES DEPARTMENT OF ENERGY
- UNITED STATES ENVIRONMENTAL PROTECTION AGENCY
- UNITED STEELWORKERS OF AMERICA

HEALTH EFFECTS OF HALOGENATED AROMATIC HYDROCARBONS

Part I. Production, Chemistry, and Distribution

DIOXINS, DIBENZOFURANS AND OTHER POLYHALOGENATED AROMATICS: PRODUCTION, USE, FORMATION, AND DESTRUCTION

Christoffer Rappe

Department of Organic Chemistry
University of Umeå
S-901 87 Umeå, Sweden

Hans Rudolf Buser and Hans-Paul Bosshardt

Swiss Federal Research Station
CH-8820 Wädenswil, Switzerland

The polyhalogenated aromatic compounds of technical use are the polyhalogenated benzenes and toluenes, biphenyls (PCBs), terphenyls (PCTs), naphthalenes (PCNs), chlorinated phenols, phenoxy acids, and diphenyl ethers. There is no known technical use for the polyhalogenated dibenzodioxins (PCDDs) and dibenzofurans (PCDFs).

Polyhalogenated Benzenes and Alkylbenzenes

Figure 1 shows a number of polyhalogenated benzenes used as additives in various products or as intermediates in the production of other halogenated aromatics. They can all by prepared by the chlorination of benzene. The bromine analogues are used as flame retardants. A series of polyhalocymenes has recently been identified in spent bleach liquors from a sulfite pulp mill.[1]

PCBs, PCTs, and PCNs

PCBs PCTs, and PCNs (Figure 2) can all be prepared by chlorination of the parent hydrocarbon. The PCBs especially have been widely used in a long series of various applications.[2,3] The total world production of PCBs exceeds 500,000 tons. The three groups of compounds are all very stable to biodegradation and they have shown a pronounced tendency for bioaccumulation. For these reasons the use of PCBs is now, in several countries, restricted to "closed systems" such as transformers and capacitors. A mixture of PBBs, mainly the 2,2',4,4',5,5'-hexabromo isomer,[4] has been used as a flame retardant under the trade name Fire Master BP 6 or FF-1.

Chlorinated Phenols, Phenoxy Acids and Diphenyl Ethers

The annual world production of chlorinated phenols (Figure 3) is estimated to be about 150,000 tons. The compounds 2,4-dichloro-, 2,4,6-trichloro, 2,3,4,6-tetrachloro- and pentachlorophenol are prepared by the chlorination of phenol and *p*-chloro-*o*-cresol is prepared by chlorination of *o*-cresol. Another method for the preparation of

0077–8923/79/0320–0001 $01.75/0 © 1979, NYAS

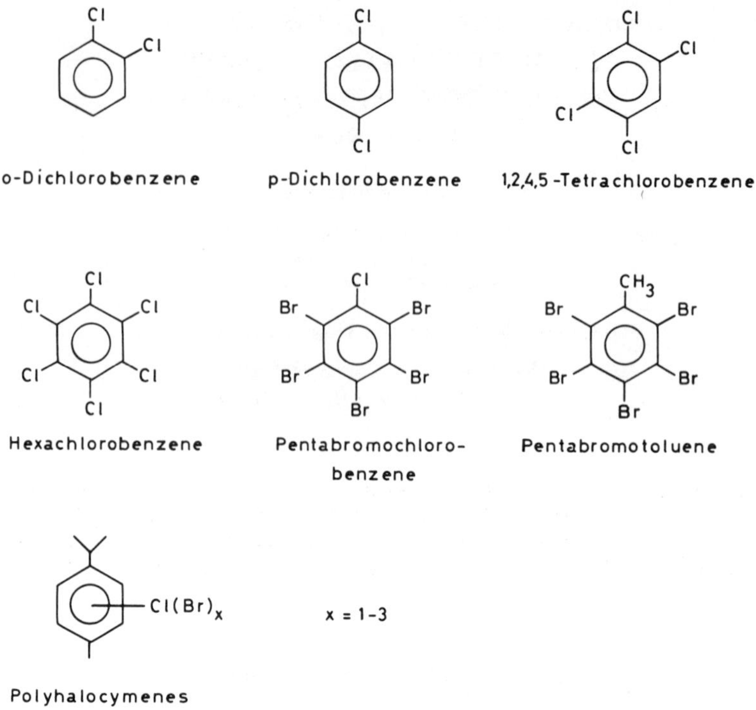

FIGURE 1. Formulae of polyhalogenated benzenes and alkylbenzenes.

chlorophenols is the alkaline hydrolysis of the corresponding halobenzene (2,4,5-trichloro- and pentachlorophenol). The chlorophenols are used as fungicides, herbicides, slimacides, and bactericides, and they are also key intermediates in the production of the chlorinated phenoxyacetic acids (2,4-D, 2,4,5-T and MCPA), which are widely used herbicides in agriculture and forestry. The bactericide, hexachlorophene, is also prepared from 2,4,5-trichlorophenol.[5]

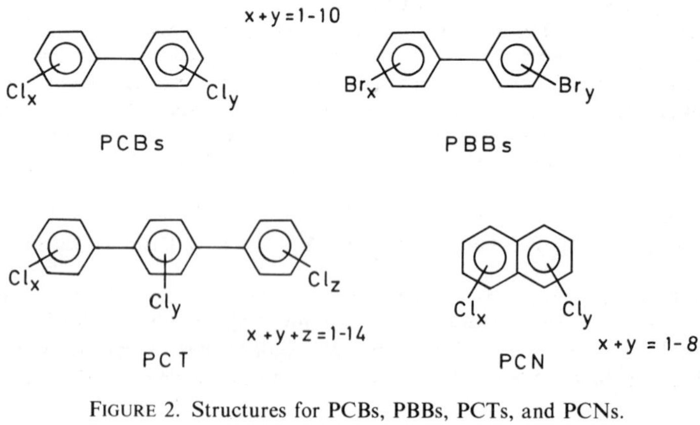

FIGURE 2. Structures for PCBs, PBBs, PCTs, and PCNs.

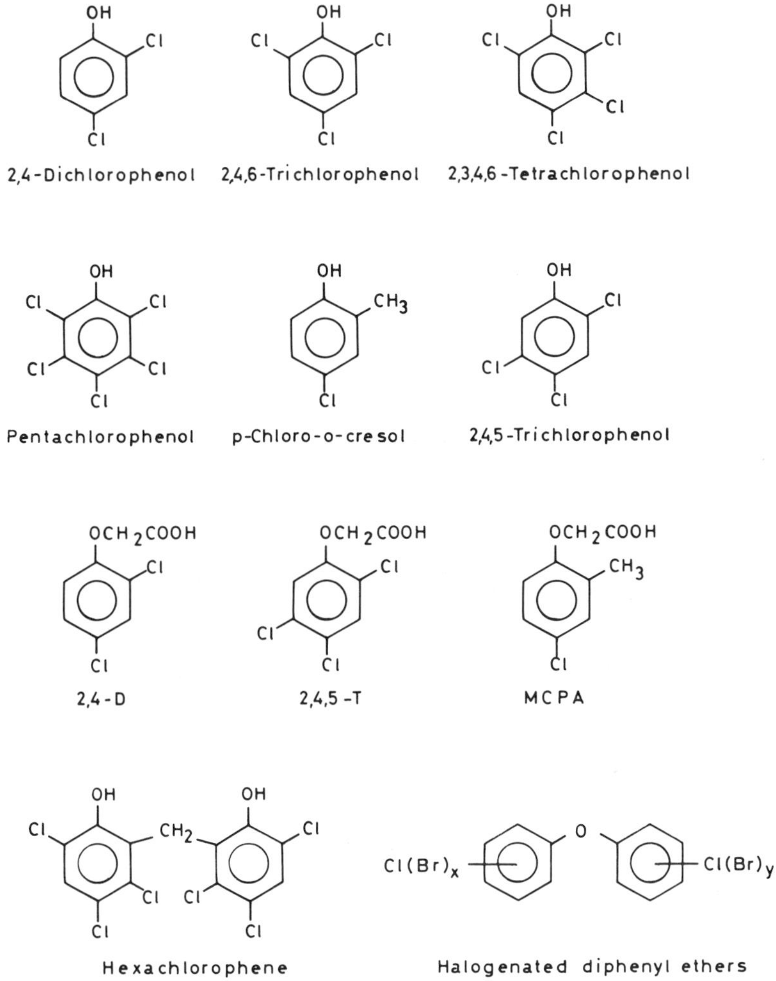

FIGURE 3. Structures for chlorinated phenols, phenoxy acids, and diphenyl ethers.

The chlorinated diphenyl ethers have been used as thermostable heat exchange liquids, while the bromine analogues have found use as flame retardants.

POLYCHLORINATED DIOXINS (PCDDs) AND DIBENZOFURANS (PCDFs)

The PCDDs and PCDFs are two series of tricyclic aromatic compounds that exhibit similar physical, chemical, and biologic properties. The basic structures and numbering are given in FIGURE 4. The number of chlorine atoms can vary between one and eight, and the number of positional isomers is quite large. In all there are 75 PCDDs and 135 PCDFs, and the isomers possible for a certain number of chlorine atoms are given in TABLE 1.

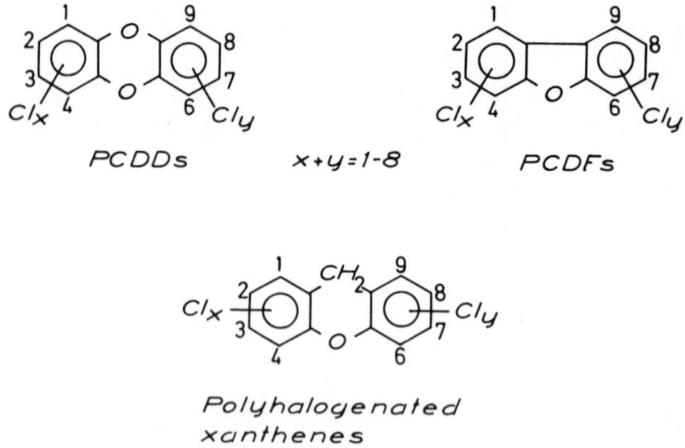

FIGURE 4. Structures and numbering of PCDDs, PCDFs, and polyhalogenated xanthenes.

The isomers with the highest acute toxicity are 2,3,7,8-tetra-CDD, 1,2,3,7,8-penta-CDD, 1,2,3,6,7,8- and 1,2,3,7,8,9-hexa-CDD, 2,3,7,8-tetra-CDF and 2,3,4,7,8-penta-CDF (see FIGURE 5) which all have LD_{50} values in the range 1–100 μg/kg for the most sensitive animal species.[6] The bromine analogoues like 2,3,7,8-tetra-BDD and 2,3,7,8-tetra-BDF seem to have the same toxicity as the corresponding chlorine compounds.[7]

Recent work by McConnell *et al.*, Poland *et al.* and Bradlaw have shown that the positional isomers of PCDDs and PCDFs vary highly in their a acute toxicity and biologic activity.[6–8] A factor of 1000–10,000 can be found for so closely related isomers as 2,3,7,8- and 1,2,3,8-tetra-CDD.

Because of their extreme toxicity, as well as the large variation in toxicity between different isomers, there is a special need for an unusually sensitive and very specific analytic method for PCDDs and PCDFs, particularly for monitoring levels in environmental samples. A desirable detection limit for the most toxic PCDDs and PCDFs would be the part per trillion level, which means that 1 pg could be detected in a 1 g sample.

Essential requirements for such an analytic method are

1. efficient and selective clean up;
2. good separation;

TABLE 1

NUMBER OF PCDD AND PCDF ISOMERS FOR A CERTAIN NUMBER OF CHLORINE ATOMS

	PCDDs	PCDFs
Cl_1	2	4
Cl_2	10	16
Cl_3	14	28
Cl_4	22	38
Cl_5	14	28
Cl_6	10	16
Cl_7	2	4
Cl_8	1	1

3. selective and ultra-sensitive quantification; and
4. when possible, confirmation

A good technique for analyzing individual PCDDs and PCDFs has recently been described.[9] It involves one or two steps of column chromatography clean-up followed by high-resolution gas-chromatography (GC), separation using a glass capillary column and quantification using mass-spectrometry (MS) (mass chromatography and/or mass fragmentography). Artifacts usually disturb the analytic work at these extreme low concentration levels, but this risk can be minimized by a careful inspection of the complete mass spectrum.[10] For a correct structure assignment of the PCDDs, the low mass ions can be useful.[11] Additional information can be obtained by a comparison of the retention times using high-resolution GC with authentic standards, but the number of synthetic standards is limited, especially for the PCDFs where only about 20–25 out of 135 isomers are available.

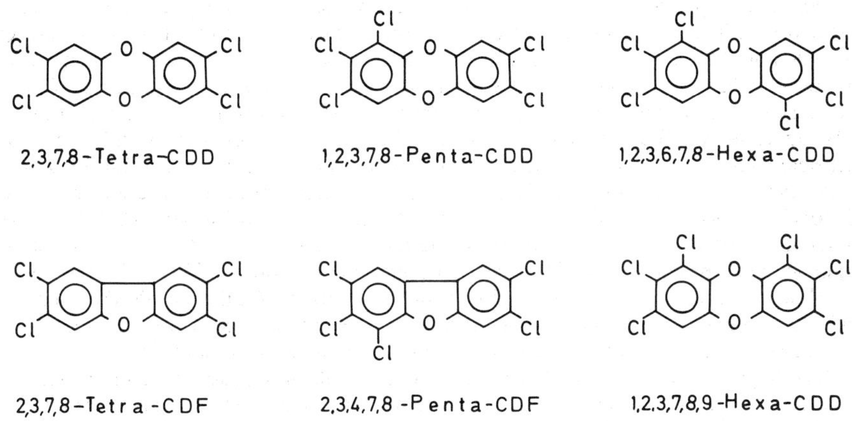

FIGURE 5. Structures for the most toxic of the PCDD and PCDF isomers.

LEVELS OF PCDDS AND PCDFS

The dioxin problem was first observed in connection with teratogenic effects found for the phenoxy acid 2,4,5-T. These effects were shown to be caused by 30 μg/ of 2,3,7,8-tetra-CDD present in this particular sample.[12] The levels of 2,3,7,8-tetra-CDD in drums of Herbicide Orange placed in storage in the USA (or in the Pacific) before 1970 have been found to vary between 0.1 and 47 μg/g.[13] Since Herbicide Orange was formulated as a 1:1 mixture of the butyl esters of 2,4-D and 2,4,5-T, the levels of 2,3,7,8-tetra-CDD in individual 2,4,5-T preparations used in the 1960s could be as high as 100 μg/g. As a result of governmental regulations, efforts were made during the 1970s to control and to minimize the formation of 2,3,7,8-tetra-CDD, and now all producers claim that their products contain less than 0.1 μg/g of 2,3,7,8-tetra-CDD.

In addition, it has also been reported that samples of Herbicide Orange, as well as 2,4,5-T samples presently on the market, contain penta-CDDs, tetra-CDFs, and penta-CDFs at levels corresponding to that found for the 2,3,7,8-tetra-CDD.[14,15] However, the analytic technique used in these investigations (low-resolution GC/MS) do not allow the identification or quantification of the individual PCDD and PCDF isomers. In analyses using high-resolution GC/MS (and MS confirmation) we have reported that in other samples of Herbicide Orange, as well as in European 2,4,5-T formulations from the 1960s, the 2,3,7,8-tetra-CDD was the dominating PCDD and PCDF, and only minor amounts of other PCDDs and PCDFs could be found, mainly lower chlorinated PCDDs in the samples of Herbicide Orange.[10]

Up to now most of the interest from scientists, regulatory agencies and the public has been concentrated on 2,4,5-T as the only source of PCDDs and PCDFs. However, using the analytic technique briefly described above, it has been possible to identify other important sources of these hazardous products.

The bactericide hexachlorophene (see FIGURE 3) is prepared from the same starting material as 2,4,5-T. Because of additional purification, however, the level of 2,3,7,8-tetra-CDD in this product is 0.03 mg/kg or less. However, hexachlorophene also contains about 100 mg/kg of a hexachloroxanthene (FIGURE 4), the 1,2,4,6,8,9-isomer.[16,17]

Pentachlorophenol and other chlorinated phenols are widely used for several applications. They have been found to contain a variety of contaminants, including PCDDs and PCDFs, and TABLE 2 collects a number of relevant analyses of these products. It has been reported that several positional isomers of PCDDs and PCDFs are present in these products. A series of pentachlorophenols from commerical sources in Switzerland was found to have a ratio for PCDFs/PCDDs less than 1 and an almost identical pattern of hexa- and hepta-CDDs. The main isomers were 1,2,4,6,7,9- (or 1,2,4,6,8,9-), 1,2,3,6,8,9- (or 1,2,3,6,7,9-) and 1,2,3,6,7,8-hexa-CDD present in a ratio of 10:40:50. Both hepta-CDDs were present with 1,2,3,4,6,7,8-hepta-CDD as the major constituent.[21] In commerical Scandinavian 2,4,6-tri- and 2,3,4,6-tetrachlorophenol both prepared by chlorination of phenol, the ration PCDFs:PCDDs was found to be greater than 20. The products contained about 10 different PCDF isomers, and the isomeric distribution was about the same. However, very little information is yet available on the identity and quantity of the individual isomers.[10,20]

The hepatic effects in female rats of technical and pure grade pentachlorophenol were investigated by Goldstein *et al.*[18] The technical product, which was contaminated by PCDDs and PCDFs (TABLE 2), produced a number of hepatic effects that cannot be attributed to the pentachlorophenol itself, but are consistent with effects associated with PCDDs and PCDFs.

In 1970, Vos *et al.*, identified PCDFs as toxic impurities in European PCBs at the μg/g level. The toxic effects of these PCBs were found to be parallel to the levels of PCDFs.[22] The same impurities have also been reported in American and Japanese PCBs.[23] Using packed column/MS, Bowes *et al.*, found that the most abundant PCDFs had the same retention times as 2,3,7,8-tetra-CDF and 2,3,4,7,8-penta-CDF.[24]

After being used for two years a heat exchange system containing a PCB was found to have a fourfold increase of PCDFs (15–20) μg/g). More than 40 different PCDF isomers were found, and the dominating isomer could be identified as 2,3,7,8-tetra-CDF, at a level of 1.25 μg/kg (peak 44, FIGURE 6).[25]

In 1968, more than 1200 persons in southwestern Japan were intoxicated by consuming a commerical rice oil contaminated by 1000 μg/g of PCBs. Analysis of the rice oil (Yusho oil) showed that it contained 5 μg/g of PCDFs. Consequently, the levels of PCDFs in the PCB were dramatically increased.[26] The major PCDF constituents have been shown to be the toxic 2,3,7,8-tetra-CDF (0.45 μg/g) and 2,3,4,7,8-penta-CDF (peaks 44 and 68, FIGURE 7), and there is a striking similarity between the PCDF isomers present in the Yusho oil and in the used Japanese PCB, see above (FIGURE 6).[25]

TABLE 2

LEVELS OF PCDDS AND PCDFS FOUND IN COMMERCIAL SAMPLES OF CHLORINATED PHENOLS

Phenol	Ref.	PCDDs (mg/kg)					PCDFs (mg/kg)				
		Cl_4	Cl_5	Cl_6	Cl_7	Cl_8	Cl_4	Cl_5	Cl_6	Cl_7	Cl_8
Cl_5–	18	<0.1	<0.1	8	520	1380	≤4	40	90	400	260
Cl_5–	19	<0.02	<0.03	10	130	210	0.20	0.20	13	70	55
Cl_4–	20	0.7	5.2	9.6	5.6	0.7	10	10	60	60	10
Cl_3–	21	—	<0.3	<0.5	<0.2	—	1.5	17.5	36	4.8	—

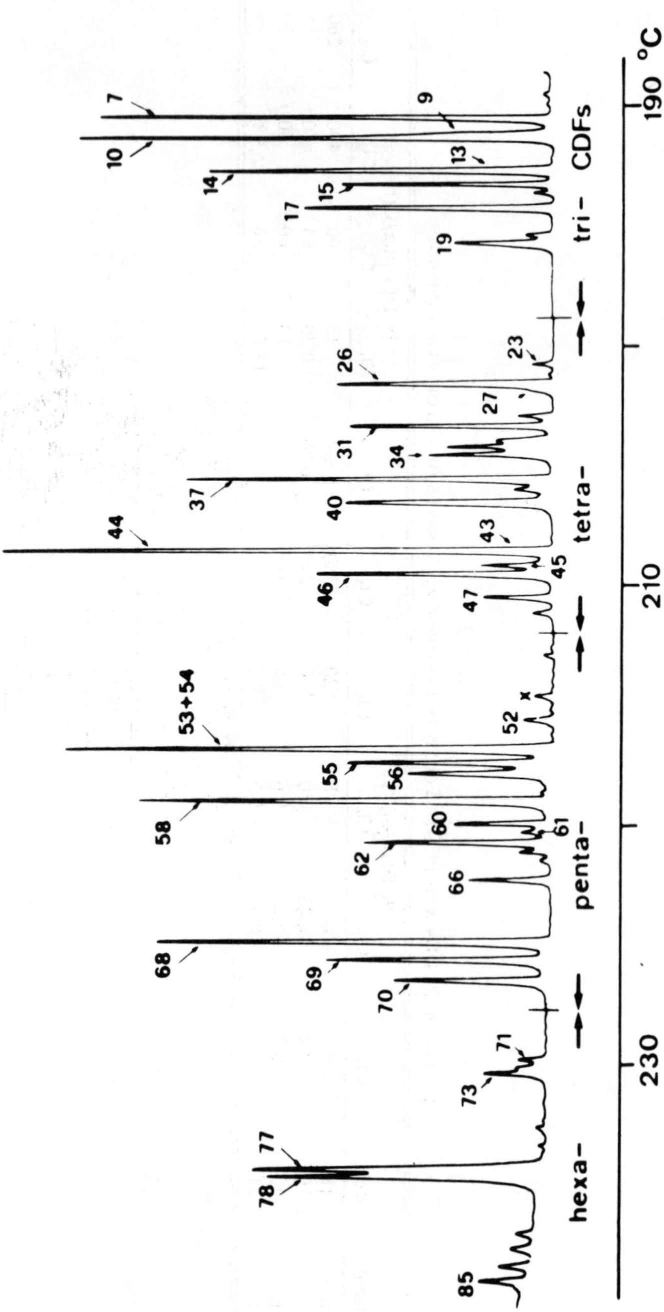

FIGURE 6. Mass fragmentograms showing elution of PCDFs in a used Japanese PCB.

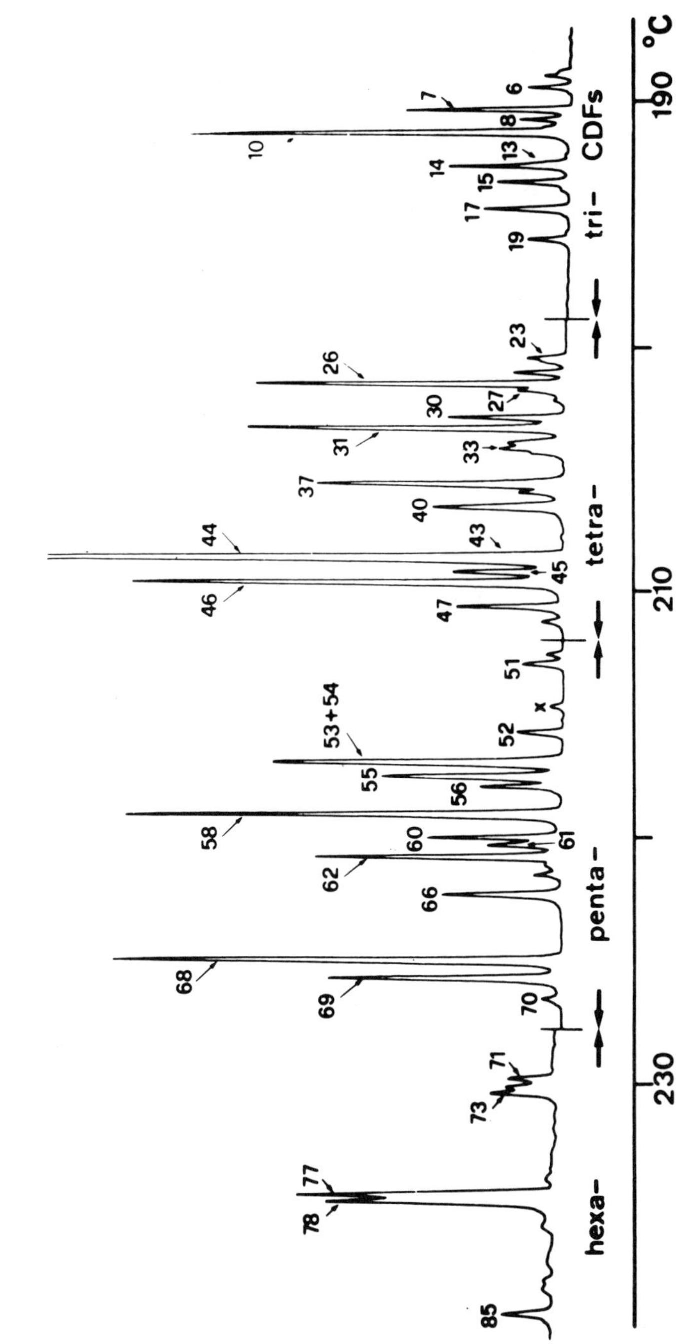

FIGURE 7. Mass fragmentograms showing elution of PCDFs in Yusho oil.

Recently Olie *et al.*, reported the occurrence of PCDDs and PCDFs in fly ash and flue gases of municipal incinerators in the Netherlands.[27] No quantitative data were given in this report, but more recently Buser and Bosshardt made a quantification that the total amounts of PCDDs and PCDFs in fly ash from a municipal incinerator were 0.2 μg/g and 0.1 μg/g, respectively, and the fly ash from an industrial heating facility, also in Switzerland, contained 0.6 μg/g and 0.3 μg/g, respectively.[28] It has also been found that the number of individual isomers was quite large; as many as 30–40 isomers of PCDDs as well as PCDFs were found, and in other studies it has been shown that in fly ash the highly toxic PCDD-isomers 2,3,7,8-tetra- and 1,2,3,7,8-

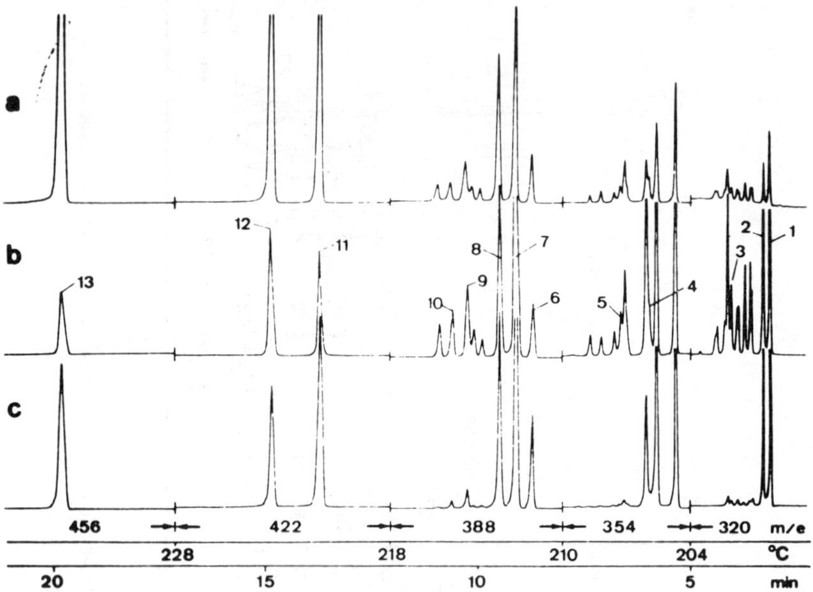

FIGURE 8. Mass fragmentograms showing elution of PCDDs in (a) fly ash of a municipal incinerator, (b) fly ash of an industrial heating facility, and (c) micropyrolysis of combined 2,4,6-tri, 2,3,4,6-tetra- and pentachlorophenate.

Peak identifications:

1 = 1,3,6,8-, 2 = 1,3,7,9- and 3 = 2,3,7,8- tetra-CDD; 4 = 1,2,4,7,8- and 5 = 1,2,3,7,8-penta-CDD; 6 = 1,2,4,6,7,9-, 7 = 1,2,3,4,6,8-, 8 = 1,2,3,6,8,9-, 9 = 1,2,3,6,7,8- and 10 = 1,2,3,7,8,9,-hexa-CDD; 11 = 1,2,3,4,6,7,9- and 12 = 1,2,3,4,6,7,8-hepta-CDD; 13 = octa-CDD.

penta-CDDs are only minor constituents (FIGURE 8), while the known highly toxic PCDF-isomers are the major constituents. Peak 44 FIGURE 9, is 2,3,7,8-tetra-CDF and peak 68 is 2,3,4,7,8-penta-CDF.[29,30]

FORMATION OF PCDDS AND PCDFS

The photochemical dimerization of chlorophenols to PCDDs has been studied by Crosby *et al.*[31] The only PCDD formed in this study was the octa-CDD. Other PCDDs

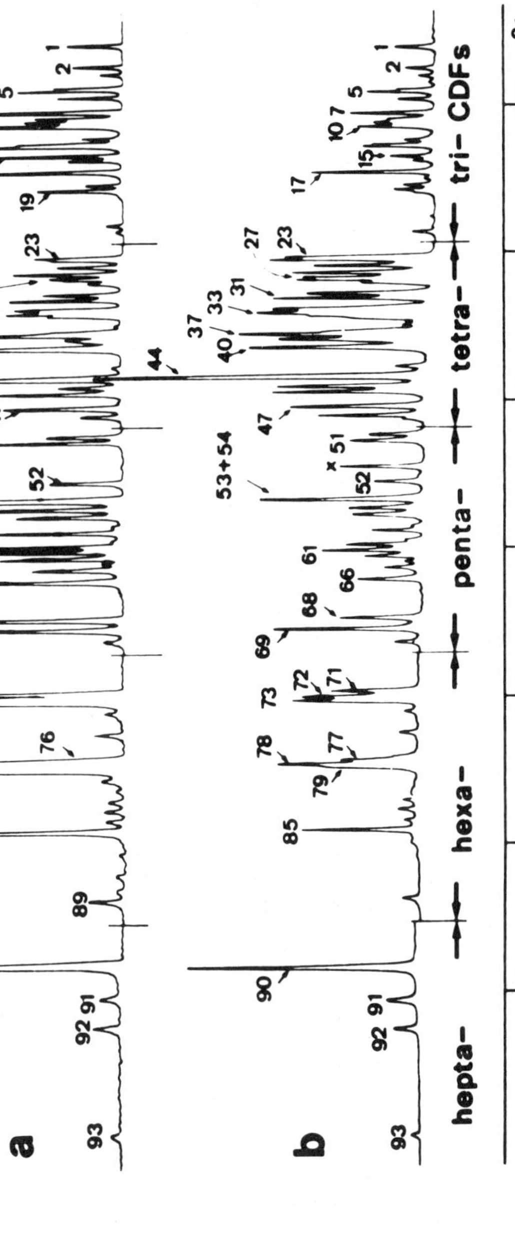

FIGURE 9. Mass fragmentograms showing elution of tri- to hepta-CDFs in fly ash of (a) an industrial heating facility and (b) a municipal incinerator.

can be formed by photochemical cyclizations of chlorinated *o*-phenoxyphenols, the so-called predioxins.[32] These predioxins are very common impurities (1–5%) in commerical chlorophenols.[5]

Another photochemical process of potential environmental importance by which the highly toxic penta- and tetrachloro compounds might be formed is dechlorination of the higher chlorinated PCDDs and PCDFs, octa-CDD and octa CDF.[33] The products formed in the photolysis of octa-CDD have now been identified. By comparison with authentic standards it was found that the main tetrachloro isomer was the 1,4,6,9-tetra-CDD, the major pentachloro compound is expected to be the 1,2,4,6,9-isomer and the main hexachloro compound was the 1,2,4,6,7,8-(or 1,2,4,6,8,9-) isomer. The major heptachloro compound was the 1,2,3,4,6,7,9-isomer. The reaction scheme deduced from the data above shows that the chlorine atoms are removed preferably from the lateral positions on the carbon ring. Consequently the most toxic isomers, 1,2,3,7,8-penta and 2,3,7,8-tetra-CDD, are not likely to be formed from the photolysis of the higher PCDDs.[11] The preferential positions for the photochemical loss of chlorine atoms in the dibenzofuran system has not been studied.

The observation that PCDDs and PCDFs can be found in fly ash and flue gases indicates that these hazardous compounds can be formed in pyrolytic processes or by burning. Salts of chlorophenols have been found to undergo a pyrolytic dimerization yielding PCDDs. This reaction has been used for the preparation of a large number of PCDD standards. The main PCDDs found in the fly ash are the same as those found in a laboratory pyrolysis of a mixture of 2,4,6-tri-, 2,3,4,6-tetra- and pentachlorophenol, the most common used commerical chlorophenols, this indicates that these products are the precursors to PCDDs in fly ash, see FIGURE 8.[29]

The burning of materials impregnated with commercial 2,3,4,6-tetrachlorophenates yielded 150–1000 μg of mixed PCDDs/g chlorophenate. Although only found as minor constituents, 2,3,7,8-tetra (peak 3, FIGURE 10 a) and 2,3,4,7,8-penta-CDD (peak 13) were both quantified at levels exceeding 10 μg/g chlorophenate.[20] It can be pointed out that these hazardous compounds are not the expected dimerization products of any chlorophenate present in these commerical mixtures. It is suggested that they are formed in cyclization reactions of impurities present in the commerical formulation.[20]

Curves b and c, FIGURE 10 are from the burning of purified penta- and 2,4,6-trichlorophenate. In the case of the 2,4,6-trichlorophenate, the only tetra-CDDs were the 1,3,6,8- and 1,3,7,9-isomers (peaks 1 and 2). In the case of the pentachlorophenate, the octa-CDD was the main product, but in addition, both hepta-CDDS, eight hexa-CDDs and several penta- and tetra-CDDs were identified. In the formation of the lower chlorinated PCDDs a nonspecific dechlorination process must be involved.[30]

It has recently been found that pyrolysis of technical PCB-mixtures yielded about 30 major, and more than 30 minor, PCDFs (see FIGURE 11). One of the main constituents was 2,3,7,8-tetra-CDF, the most toxic of the PCDFs (peak 44). Taking into consideration the amount of PCB recovered after the pyrolysis, the yield of PCDFs was between 3–25% calculated on the amount of PCB decomposed.[30,34] Consequently, uncontrolled burning of PCBs can be an important environmental source of the hazardous PCDFs. Moreover, a comparison showed a striking similarity between the PCDF pattern in the fly ash (FIGURE 8) and the pyrolysis of the commerical PCBs suggesting PCBs to be precursors of the PCDFs found in fly ash. Because of this, it is recommended that all destruction of PCB-containing wastes using incinerators must be carefully controlled. Pyrolysis of individual synthetic PCB-isomers showed that the formation of PCDFs can follow three different reaction

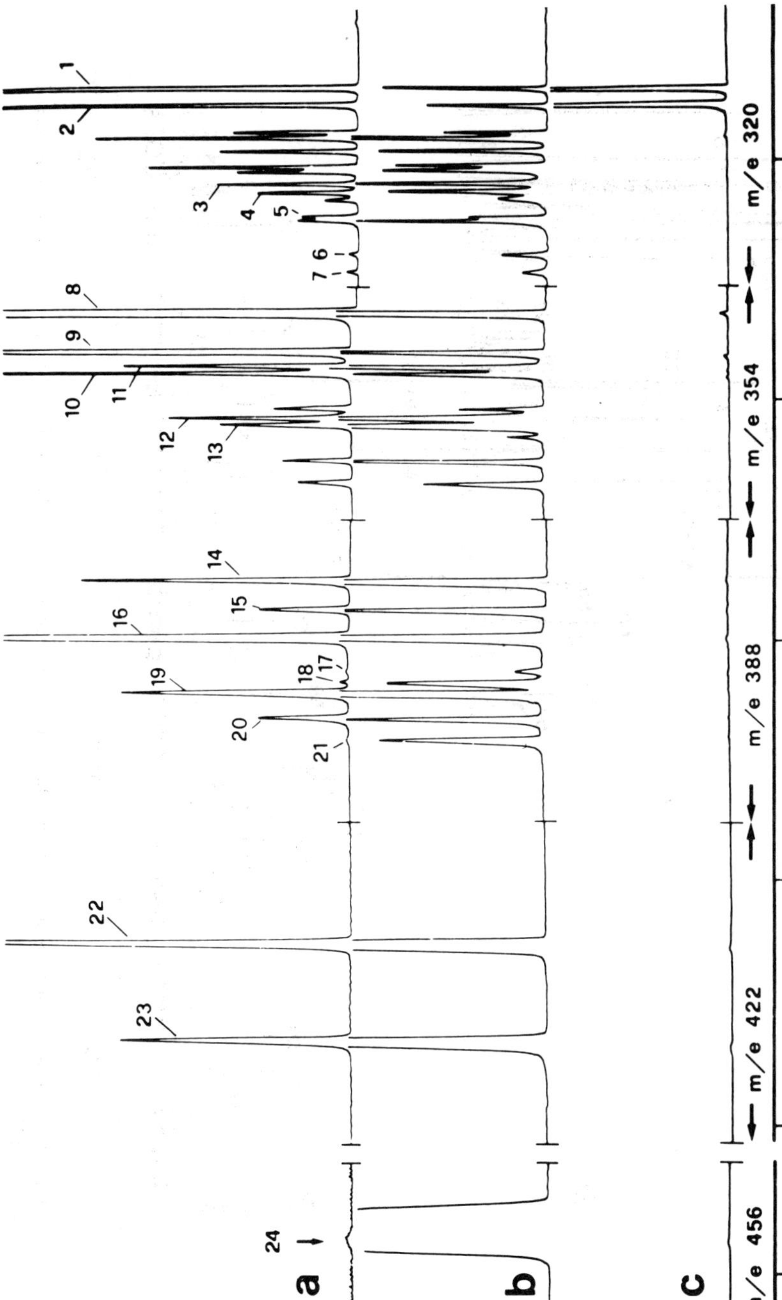

FIGURE 10. Mass fragmentograms showing elution of PCDDs from the burnings on birch leaves of (a) Servarex, (b) purified pentachlorophenate, and (c) purified 2,4,6-trichlorophenate. Peak identification, see TABLE 3.

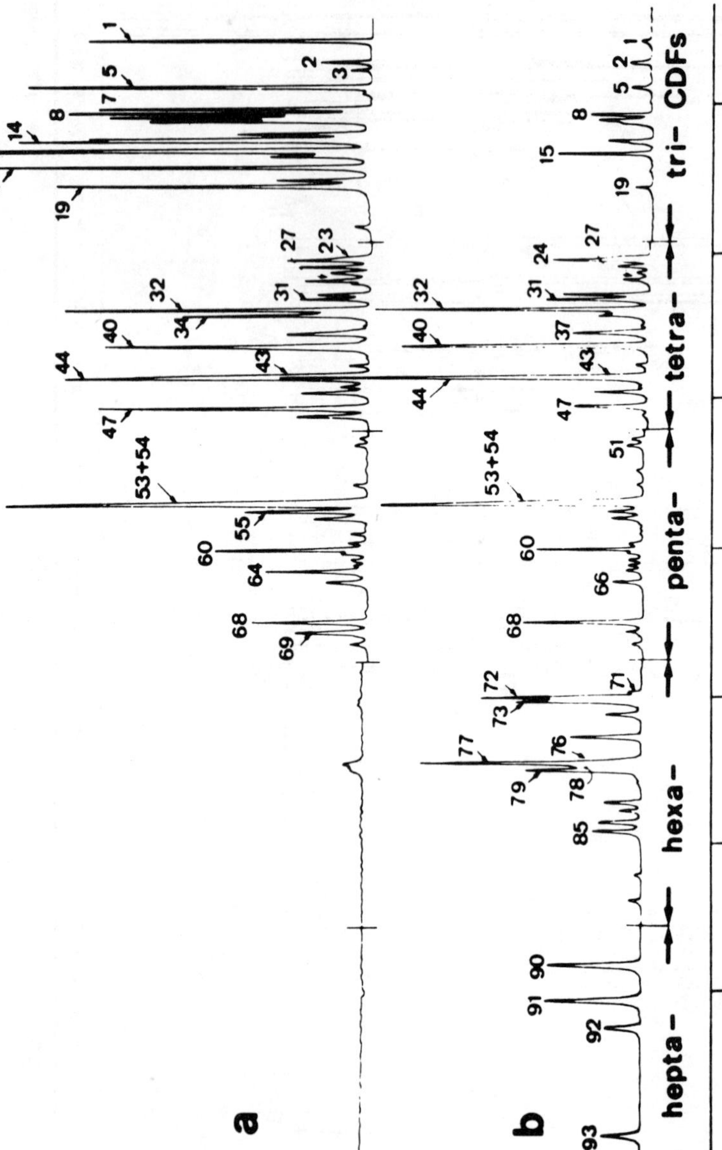

FIGURE 11. Mass fragmentograms showing elution of tri- to hepta-CDFs in (a) pyrolyzed Aroclor 1254 and (b) pyrolyzed Aroclor 1260.

TABLE 3

IDENTIFICATION OF PCDDS IN EXTRACTS (SEE FIGURE 10)

Peak	PCDD isomer	Peak	PCDD isomer
1	1,3,6,7-tetra-CDD	13	1,2,3,7,8-penta-CDD
2	1,3,7,9-	14	1,2,4,6,7,8,9-hexa-CDD†
3	2,3,7,8-	15	1,2,3,4,6,8-
4	1,2,3,8- *	16	1,2,3,6,8,9- ‡
5	1,2,6,9,-	17	1,2,3,4,6,9,-
6	1,2,6,7-	18	1,2,3,4,7,8-
7	1,2,8,9-tetra-CDD	19	1,2,3,6,7,8-
8	penta-CDDs from pyro-	20	1,2,3,7,8,9-
9	lysis of 2,4,6- tri-	21	1,2,3,4,6,7-hexa-CDD
10	and 2,3,4,6-tetrachlorophenate	22	1,2,3,4,6,7,9-hepta-CDD
11	1,2,4,7,8-penta-CDD	23	1,2,3,4,6,7,8-hepta-CDD
12	1,2,3,4,7-penta-CDD	24	octa-CDD

*Or the Smiles product 1,2,3,7-tetra-CDD.
†Or the Smiles product 1,2,4,6,8,9-hexa-CDD.
‡Or the Smiles product 1,2,3,6,7,9-hexa-CDD.

pathways; an example (FIGURE 12) is pyrolysis of 2,2′,4,4′5,5′-hexachlorobiphenyl, one of the main constituents in commercial PCBs.[30] A pyrolysis of the commercial flame retardant Fire Master BP 6, 2,2′,4,4′,5,5′-hexabromobiphenyl, yielded the very toxic 2,3,7,8-tetra-BDF.[34]

In addition to PCBs, polychlorinated diphenylethers could be considered as possible precursors of PCDFs. It has been shown in preliminary experiments that the thermal conversion of the ethers into PCDFs is of the same order as that from PCBs.[30] Although polychlorinated diphenylethers have found use as heat exchange media and hydraulic liquids, the PCBs must be considered as a more important source for the

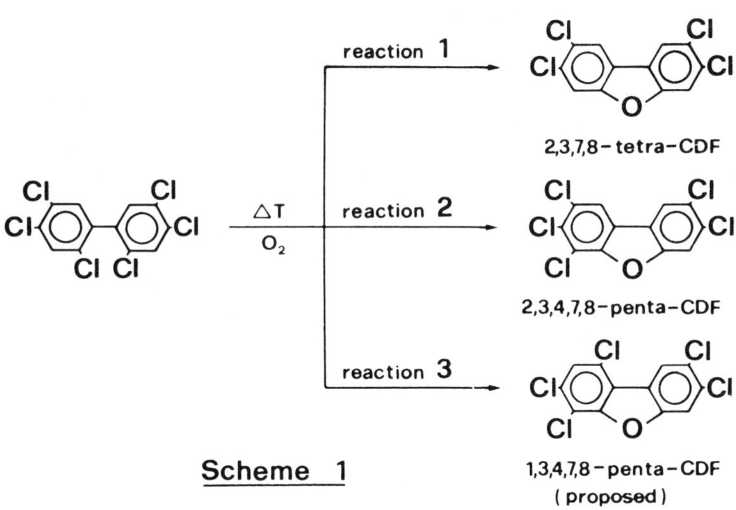

FIGURE 12. Reaction scheme for the pyrolysis of 2,2′,4,4′,5,5′-hexachlorobiphenyl.

PCDFs because of the much larger production volume of the PCBs as compared to the polychlorinated diphenylethers.

The possible formation of 2,3,7,8-tetra-CDD and other PCDFs as the result of thermal reactions of the esters of the phenoxy acid 2,4,5-T has been the subject of much controversy. However, recent investigations have shown that this formation is only of very limited importance, if it takes place at all.[15,35,36]

Exposure and Risks

Human exposure to PCDDs and PCDFs may be the result of specific exposure, mainly of occupational origin, or be caused by general exposure of the public.

Occupational exposure may occur:

- in chemical plants producing chlorinated phenols or PCBs;
- in factories utilizing these chemicals for the production of other substances, e.g. 2,4,5-T, phosphorous insecticides or hexachlorophene from 2,4,5-trichlorophenol;
- in the use process of these chemicals under various occupational conditions, such as spraying of herbicides, using chlorinated phenols for a variety of applications; especially as a wood preservative, slimacide, bactericide, in cutting fluids, or hide preservation;
- in the use of hexachlorophene by workers in sanitation applications;
- in factories manufacturing or repairing transformers or capacitors;
- in factories having heat-exchange systems containing PCBs;
- in factories utilizing casting waxes containing PCBs; or
- in various incinerators where material containing chlorinated phenols, PCBs, PBBs, polychlorinated or polybrominated diphenyl ethers are incinerated.

A general exposure to the public by 2,3,7,8-tetra-CDD has previously been discussed in relation to the herbicide spraying program in Vietnam in the 1960s and the explosion in a chemical plant near Seveso in Italy, July 1976. The Yusho oil accident in Japan 1968 resulted in an exposure of the public to various PCDFs.

In addition to these known and widely discussed exposures, our recent results indicate other possible situations where exposure to the general public may occur. One such exposure is in the vicinity of a saw-mill, where impregnated wood-shavings have been incinerated for the production of heat. This incineration can result in an exposure of workers to various PCDDs including the most toxic isomers. Continuous inhalation of fly ash or flue gases from municipal incinerators and industrial heating facilities may also be discussed because of the low levels of PCDDs and PCDFs found in these products, especially when material containing chlorinated phenols or PCBs are being burned. This risk is well illustrated by a recent fire in downtown Toronto of a transformer containing PCBs. This accident released soot and PCBs in the air. Analyses of the soot have indicated an increased level of tetra-CDF (unknown isomer).[37] The general intoxication in the state of Michigan by 2,2′,4,4′,5,5′-hexabromobiphenyl and the use of this compound as a flame retardant might possibly be associated with an exposure of the highly toxic 2,3,7,8-tetra-BDF, which is being formed in pyrolytic reactions.[34]

A limited number of fish samples from sprayed areas in Vietnam have responded positively on analyses for 2,3,7,8-tetra-CDD.[38] The accident near Seveso in Italy resulted in contamination of soil, buildings, vegetation, and animals by the same toxic compound. The US Environmental Protection Agency (EPA) in its dioxin monitoring program arranged a collaborative study on beef-fat samples taken from cattle that had grazed on rangelands which were known to have been treated with 2,4,5-T. Of the 85 beef-fat samples analyzed, one was positive at a level of 60 ppt 2,3,7,8-tetra-CDD, two were reported to have 20 ppt and five may have had levels of 2,3,7,8-tetra-CDD within a range of 5–10 ppt.[39]

References

1. Eklund, G., B. Josefsson & A. Bjöseth. 1978. Determination of chlorinated and brominated lipophilic compounds in spent bleach liquors from a sulphite pulp mill. J. Chromatogr. **150:** 161–169.
2. Hutzinger, O., S. Safe & V. Zitko. 1974. The chemistry of PCB's.:7–39. CRC Press, Cleveland, Ohio.
3. International Agency for Research on Cancer. 1978. IARC monographs on the evaluation of carcinogenic risk of chemicals to man. **19:** IARC. Lyon, France. In Press.
4. Andersson, K., Å. Norström, C. Rappe, B. Rasmuson & H. Swahlin. 1975. Photochemical degradation of PCB, PBB and other flame retardants. Environ. Qual. Safety Supply. Vol. **III:** 798–802.
5. Nilsson, C.-A., Å. Norström, K. Andersson & C. Rappe. 1978. Impurities in commercial products related to pentachlorophenol in Pentachlorophenol, Chemistry, Pharmacology, and Environmental Toxicology. K. Ranga Rao, Ed.:313–323. Plenum Press. New York.
6. McConnell, E. J., J. Moore, J. Haseman & M. Harris. 1978. The comparative toxicity of chlorinated dibenzo-*p*-dioxin isomers in mice and guinea-pigs. Toxicol. Appl. Pharmacol. **44:** 335–356.
7. Poland, A. & E. Glover. 1977. Chlorinated Biphenyl Induction of aryl hydrocarbon hydroxylase activity: a study of the structure-activity relationship. Mol. Pharmacol. **13:** 924–938.
8. Bradlaw, J., FDA Washington. Personal communication.
9. Buser, H. R. 1977. Determination of 2,3,7,8-tetrachlordibenzo-*p*-dioxin in environmental samples using high-resolution gas chromatography and low resolution mass spectrometry. Anal. Chem. **49:** 918–922.
10. Rappe, C., H. R. Buser & H.-P. Bosshardt. 1978. Identification and quantification of polychlorinated dibenzo-*p*-dioxins (PCDDs) and dibenzofurans (PCDFs) in 2,4,5-T-ester formulations and Herbicide Orange. Chemosphere. **7:** 431–438.
11. Buser, H. R. & C. Rappe. 1978. Identification of substitution patterns in polychlorinated dibenzo-*p*-dioxins (PCDDs) by mass spectroscopy. Chemosphere **7:** 199–211.
12. Courtney, K. D. & J. A. Moore, 1971. Teratology studies with 2,4,5-trichlorophenoxyacetic acid and 2,3,7,8-tetrachlorodibenzo-*p*-dioxin. Toxicol. Appl. Pharmacol. **20:** 396–403.
13. Young, A. L., C. E. Thalken, E. L. Arnold, J. H. Cupello & L. G. Cocerham. 1976. Fate of 2,3,7,8-tetrachlorodibenzo-*p*-dioxin (TCDD) in the environment: Summary and decontamination recommendations. USAFA-TR-76-18 Boulder, Colo. :1–41.
14. Huckins, J. N., D. L. Stalling & W. A. Smith. 1978. Foam-charcoal chromatography for analyses of polychlorinated dibenzodioxins in Herbicide Orange. J. Assoc. Off. Anal. Chem. **61:** 32–38.
15. Ahling, B., A. Lindskog, B. Jansson & G. Sundström. 1977. Formation of polychlorinated dibenzo-*p*-dioxins and dibenzofurans during combusion of a 2,4,5-T formulation. Chemosphere **6:** 461–466.
16. Baughman, R. W. 1974. Tetrachlorodibenzo-*p*-dioxins in the environment. High resolution mass spectrometry at the picogram level. Thesis Harvard University. Appendices C and D.
17. Göthe, R. & C. A. Wachtmeister. 1972. Synthesis of 1,2,4,5,7,8-hexachloroxanthene. Acta Chem. Scand. **26:** 2523.
18. Goldstein, J. A., M. Friesen, R. E. Linder, P. Hickman, J. R. Hass & H. Bergman. 1977. Effects of pentachlorophenol on hepatic drug metabolizing enzymes and porphyria

related to contamination with chlorinated dibenzo-*p*-dioxins and dibenzofurans. Biochem. Pharmacol. **26:** 1549–1557.

19. Buser, H. R. & H.-P. Bosshardt. 1976. Determination of polychlorinated dibenzo-*p*-dioxins and dibenzofurans in commerical pentachlorophenols by combined gas chromatography—mass spectrometry. J. Assoc. Off. Anal. Chem. **59:** 562–569.
20. Rappe, C., S. Marklund, H. R. Buser & H.-P. Bosshardt. 1978. Formation of polychlorinated dibenzo-*p*-dioxins (PCDDs) and dibenzofurans (PCDFs) by burning or heating chlorophenates. Chemosphere **7:** 269–281.
21. Rappe, C., A. Garå & H. R. Buser. 1978. Identification of polychlorinated dibenzofurans (PCDFs) in commercial chlorophenal formulations. Chemosphere. In press.
22. Vos, J. G., J. H. Koeman, H. L. van der Maas, M. C. ten Noever de Brauw & R. H. de Vos. 1970. Identification and toxicological evaluation of chlorinated dibenzofuran and chlorinated naphthalene in two commerical polychlorinated biphenyls. Food Cosmet. Toxicol. **8:** 625–633.
23. Roach, J. A. G. & I. H. Pomerantz. 1974. The finding of chlorinated dibenzofurans in a Japanese polychlorinated biphenyl sample. Bull. Environ. Contam. Toxicol. **12:** 338–342.
24. Bowes, G. W., M. J. Mulvihill, M. R. De Campe & A. S. Kende. 1975. Gas Chromatographic characteristics of authentic chlorinated dibenzofurans; Identification of two isomers in American and Japanese polychlorinated biphenyls. J. Agric. Food Chem. **23:** 1222–1223.
25. Buser, H. R., C. Rappe & A. Garå. 1978. Polychlorinated dibenzofurnas (PCDFs) found in Yusho oil and in used Japanese PCB. Chemosphere. **7:** 439–449.
26. Nagayama, J., M. Kuratsune & Y. Masuda. 1976. Determination of chlorinated dibenzofurans in Kanechlor and "Yusho Oil." Bull Environ. Contam. Toxicol. **15:** 9–13.
27. Olie, K., P. L. Vermeulen & O. Hutzinger. 1977. Chlorodibenzo-*p*-dioxins and chlorodibenzofurans are trace components of fly ash and flue gas of some municipal incinerators in the Netherlands. Chemosphere **6:** 455–459.
28. Buser, H. R. & H.-P. Bosshardt. 1978. Polychlorierte Dibenzo-*p*-dioxine, Dibenzofurane und Benzole in der Asche kommunaler und industrieller Verbrennungsanlagen. Mitt. Geb. Lebensmittelunter. Hyg. **69:** 191–199.
29. Buser, H. R. H.-P. Bosshardt & C. Rappe. 1978. Identification of polychlorinated dibenzo-*p*-dioxin isomers found in fly ash. Chemosphere **7:** 165–172.
30. H. R. Buser, H.-P. Bosshardt, C. Rappe & R. Lindahl. 1978. Identification of polychlorinated dibenzofuran isomers in fly ash and PCB pyrolyses. Chemosphere **7:** 419–429.
31. Crosby, D. G. & A. S. Wong. 1976. Photochemical generation of chlorinated dioxins. Chemosphere **5:** 327–332.
32. Nilsson, C.-A., K. Andersson, C. Rappe & S.-O. Westermark. 1974. Chromotographic evidence for the formation of chlorodioxins from chloro-2-phenoxyphenols. J. Chromatogr. **96:** 137–147.
33. Pimmer, J. R., U. I. Klingebiel, D. G. Crosby & A. S. Wong. 1973. Photochemistry of dibenzo-*p*-dioxins. Adv. Chem. Ser. **120:** 44-54.
34. Buser, H. R., H.-P. Bosshardt & C. Rappe. 1978. Formation of polychlorinated dibenzofurans (PCDFs) from the pyrolysis of PCBs. Chemosphere **7:** 109–119.
35. Andersson, K., H.-P. Bosshardt, H. R. Buser, S. Marklund & C. Rappe. 1978. Chlorinated phenoxy acids and their dioxins. Ecol. Bull. (Stockholm) **27:** 26.
36. Stehl, R. H. & L. L. Lampraski, 1977. Combustion of several 2,4,5-trichlorophenoxy compounds: Formation of 2,3,7,8-tetrachlorodibenzo-*p*-dioxin. Science **197:** 1008–1009.
37. Ministry of the Environment, Ontario, Canada. News Release, February 13, 1978.
38. Baughman, R. & M. Meselson. 1973. An analytical method for detecting TCDD (Dioxion): Levels of TCDD in samples from Vietnam. Environ. Health Perspect. **5:** 27–35.
39. McKinney, J. D. 1978. Analysis of 2,3,7,8-tetrachlorodibenzo-*para*-dioxin in environmental samples. Chlorinated phenoxy acids and their dioxins. Ecol. Bull. (Stockholm) **27:** 55–68.

PROBLEMS IN ANALYTIC METHODOLOGY: SAMPLE HANDLING, EXTRACTION, AND CLEANUP

Phillip W. Albro

Environmental Biology and Chemistry Branch
National Institute of Environmental Health Sciences
Research Triangle Park, North Carolina 27709

The field of residue analysis has advanced greatly in the past 20 years, and may be said to have passed the "how to do it" into the "how to do it *best*" state. For almost any type of pollutant residue there exists a variety of effective analytic procedures, such that choosing between them has become to a certain extent a matter of convenience. That different methods give comparable results has not been shown in all cases, so to insure uniformity several laboratory groups have established collections of "standard methods," such as the Pesticides Analytical Manual of the Food and Drug Administration. As analysts encounter new sources of pollution, become interested in new types of pollutants, and extend their studies to new samples and sample matrices, they continue to experience some of the old problems in new forms. The establishment of a collection of standard methods tends to inhibit the development of new methods, and as a result, there are relatively few laboratories currently engaged in methods development for the common pesticides and pollutants.

Virtually every modern method of residue analysis starts with some form of extraction, to separate the compounds of interest from the bulk of the sample matrix. In most cases, extraction is followed by some sort of cleanup procedure to eliminate or at least minimize interfering substances. In both of these preliminary treatments there is a need to know the fractional recoveries of the compounds of interest, and to have a meaningful way of relating the amounts found in the subsequent assay to the concentrations originally in the sample matrix. These parameters are influenced significantly by the manner in which the sample was collected, preserved, stored, shipped, and otherwise processed prior to extraction. This paper is an attempt to document the extent to which commonly used methods of sample handling, extraction, and cleanup either cause or fail to eliminate problems in the subsequent analysis of pollutant residues. As our laboratory is mainly concerned with the "accidental" pollutants, polychlorinated or polybrominated biphenyls (PCBs and PBBs), chlorinated dibenzo-*p*-dioxins and dibenzofurans, most of the examples will involve these classes of compounds.

Sample Handling

The problems our laboratory most often encounters relative to sample handling prior to extraction can be grouped into four categories: contamination, loss, desiccation, and alteration. The first two, contamination and loss, are closely related problems and will be discussed together.

Samples received for analysis in our laboratory from outside the Institute are most often in the form of animal tissues. When we have no input into the selection of storage conditions, we frequently receive the tissues either in glass jars with screw caps or wrapped in aluminum foil. Both containment media can cause serious problems for the analyst. Figure 1a shows a gas chromatogram of pentane rinse of the shiny side of

0077–8923/79/0320–0019 $01.75/0 © 1979, NYAS

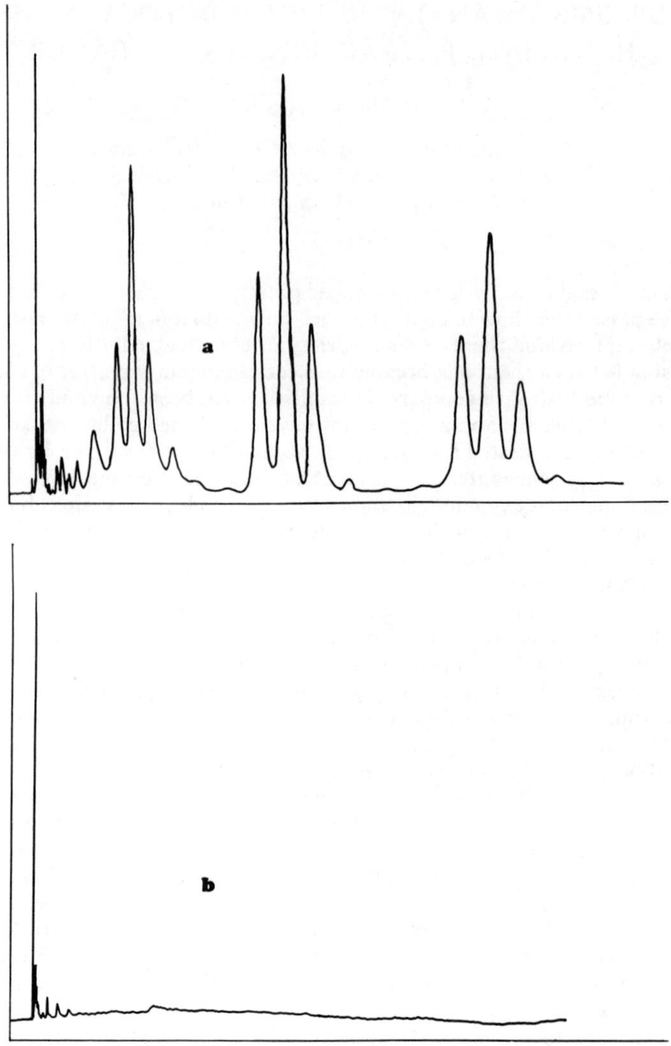

FIGURE 1. GLC of pentane rinse of aluminum foil on OV-210 column with ^{63}Ni-electron capture detector. Amount injected correspnded to 2 cm^2 of foil. a = shiny side of foil; b = dull side.

commercially available aluminum foil. The gas chromatography (GLC) conditions, involving a column packed with 10% OV-210 on Gas Chrom Q and the use of a ^{63}Ni-electron capture detector, are typical of those used for analysis of PCBs. The amount of pentane rinse injected corresponded to 2 cm^2 of foil. FIGURE 1b shows the corresponding rinse of the dull side of the same foil. In general, the amount of interfering material varies with the brand and lot of foil, but the risk of contamination from this source is sufficiently serious that foil should either not be used for packaging of analytic samples, or it should be thoroughly rinsed with acetone before use.

We occasionally receive samples frozen in polyethylene bottles (e.g. liquid scintillation vials, linear polyethylene). In most cases, these bottles do not introduce interfering materials into the samples, and have the advantage that they will not break in the freezer. However, we stored liver samples containing ^{14}C-labeled DDT in linear polyethylene bottles at 4°C overnight, and found that up to 2% of the ^{14}C had been "extracted" into the plastic such that it could not be washed out. This loss of hydrophobic compounds from tissues during storage in plastic containers, while serious in itself, can also constitute a source of contamination if the (washed) bottles are subsequently used to store other samples. Similar loss/contamination phenomena are seen if tissue samples are stored in screw-cap glass bottles having rubber liners in the caps.

A common type of screw-cap bottle has a foil-covered cardboard liner in the cap. FIGURE 2 shows a gas chromatogram of a hexane extract of the cardboard liner from one such bottle, and is typical of chromatograms obtained from many such cap liners over the years. The peaks, representing a hydrogen flame ionization detector response in this case, were found by GC-mass spectrometry[1] to be due to butyl phthaloyl butyl glycolate, butyl benzyl phthalate, and other related phthalates, benzoates and salicylates. We are not certain whether the source of materials is the liner itself or the glue used to secure it to the cap. These plasticizers, while of no great concern if analysis is to involve GLC with electron capture or halogen-specific detection, do potentially interfere with analyses by GC-mass spectrometry.

Sample desiccation and alteration are also related problems. When tissues are stored in a "self-defrosting" freezer in unsealed containers, the results shown in TABLE 1 can be observed. If the tissues were not weighed prior to freezing, or if they are to be subdivided at a later time, this desiccation may make it impossible to relate the

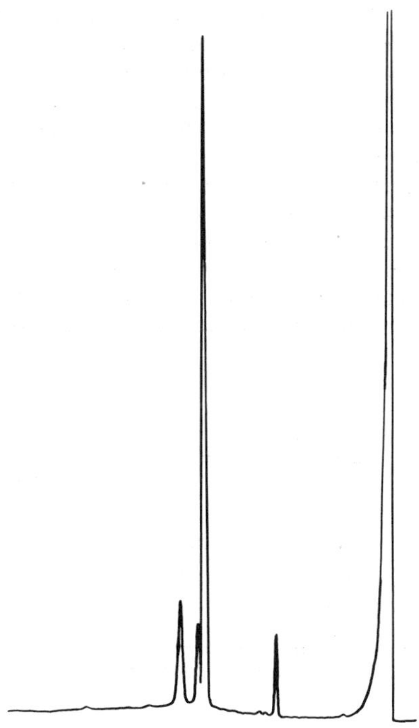

FIGURE 2. GLC of hexane extract of cardboard bottle cap liner. 10% OV-3 column with hydrogen flame ionization detector at 16 × 10^{-11} amp/mV sensitivity. Amount injected represents 1/100 of the extract from one 3/4″ diameter liner.

TABLE 1

DESICCATION OF RAT LIVER AT −20° C IN A FROST-FREE FREEZER*

Days at −20°	Weight g	Percentage of Day 0
0	1.2263	100
7	1.0175	83
14	0.9339	76
21	0.8527	70
28	0.7295	59
35	0.6724	55

*A solid piece of rat liver was stored in an unsealed container at −20° C and weighed weekly.

amount of substance determined analytically to its original concentration in the tissue. A related problem occurs when samples experience repeated freezing and thawing: adipose tissue in particular has a tendency to "leak" lipid when the cell membranes are disrupted by a freeze-thaw cycle, and in a series of experiments in which such cycles were deliberately applied to a collection of samples of adipose tissue from a rat, the apparent lipid content of the tissue (mg per g tissue) decreased by an average of 10% after three freezings. This loss was only apparent, and was not observed if the tissue was extracted in the original storage container.

EXTRACTION

Although it has been shown in some cases that recovery of pesticides from tissues during extraction does not necessarily correlate with recovery of lipids,[2] it is usually desirable to quantitatively extract the lipids with the environmental pollutants for reporting purposes.[3] It is not, however, common for residue analysts to use traditional, extensively validated lipid extraction procedures involving such well-known solvents as chloroform-methanol[4,5] or ethanol-ether.[6] A recent study of the recovery of lipid from human adipose tissue for purposes of pesticide determination, for example, compared nine different solvent mixtures none of which were among the traditional systems mentioned above.[2] The absence of an extensive experiential background for the use of the solvents that are favored in residue work generates the requirement that a given extraction procedure be validated for each type of sample matrix to which it is applied. Moreover, an extraction procedure validated for one class of compounds in a given

TABLE 2

RECOVERY OF [^{14}C]HEXABROMOBIPHENYL* FROM VARIOUS RAT TISSUES

	Recovered, Percentage of Spike				
Tissue	Tissue Residue	Flask Rinses	Alumina, 3% CH_2Cl_2	Alumina, 20% CH_2Cl_2	Lost
Liver	2.2	2.8	91.0	2.0	2.0
Adipose	0.6	0.7	95.6	3.0	0.1
Blood	12.1	1.8	84.6	1.4	0.1

*1g Samples were spiked with 60 ppb (60 ng) of [U-^{14}C]-2,4,5,2,′,4′,5′-hexachlorobiphenyl in 0.1 ml of DMSO.

tissue may not be suitable for the extraction of a different but closely related class of compounds from the same tissue.

We have described an extraction and cleanup procedure that we found to be essentially quantitiative for the recovery of PCBs, dibenzofurans, and dibenzo-*p*-dioxins from animal tissues.[7] When this procedure was applied to PBBs in those same tissues, however, the results shown in TABLE 2 were obtained. The method used for extraction of blood[8] gave a poor recovery of hexabromobiphenyl, and this compound "tailed" badly during the chromatographic cleanup, resulting in further losses.

In general, the nature of the sample matrix influences the effectiveness of given extraction procedures through three matrix properties. The *toughness* of the matrix determines the relative ease or difficulty in obtaining a sufficiently finely divided preparation to allow high extraction efficiencies. Thus, while it may be possible to extract liver in a teflon-glass homogenizer, brain tissue may require a blender, whereas skin will require an extremely powerful disintegrator such as the Polytron.

The *water content* of the sample matrix must be considered. Whereas acetone or acetonitrile are excellent solvents for the extraction of PCBs from soil,[9] the admixture of a small amount of water so decreases the solubility of pesticides in these solvents that they can not be used directly on undried animal tissues. The solubility of the toxic

TABLE 3

EFFECT OF WATER ON THE SOLUBILITY OF 3,3′,4,4′-TETRACHLOROBIPHENYL IN ACETONE

Water (%)	Solubility,* mg/ml
<0.1	8.60
1	8.15
2	7.67
5	4.90
10	3.16

*At 22°C.

3,4,3′,4′-tetrachlorobiphenyl in acetone as a function of the water content is shown in TABLE 3. Mixtures of polar and nonpolar solvents such as benzene-methanol may break into two phases in the presence of a small amount (2%) of water, resulting in nonreproducible extraction.

Lipid contents of different tissues influence both the amount of solvent needed for extraction and the proportion of nonpolar component in a mixed solvent. For example, a chloroform-methanol extraction procedure[8] that is very effective for serum or plasma gives completely unsatisfacfory results when applied to milk with its much greater lipid content. Whereas a single liquid phase containing a filterable protein precipitate occurs with serum or plasma, milk gives a lipid-solvent phase separated by a proteinaceous interface from a solvent-water-water solubles phase, and the combination does not lend itself to filtration. This parameter will be discussed in more detail later in terms of its effect on cleanup.

The most serious problem in the area of extraction has been discussed previously relative to analysis of PCBs.[3] How can one be sure that recovery of a spiked compound is the same as the recovery of the corresponding endogenous compound? The most important factor in generating a valid spiked sample may be the solvent in which the spike is dissolved. TABLE 4 shows the effect of different spiking solvents on the recovery of the pesticide Mirex from fish muscle. The tissue was extracted the same

TABLE 4

EFFECT OF SPIKING SOLVENT ON RECOVERY OF MIREX FROM FISH MUSCLE

Solvent	Typical Recovery (%)
n-Hexane	89
Benzene	88
Chloroform	85
DMSO	41

way in all four cases, but recovery varied from 89% to 41%. Unfortunately, we do not know which answer is more nearly correspondant to the recovery of endogenous Mirex.

TABLE 5 shows the recovery of a PCB mixture, Aroclor 1254, from rat skin by two different extraction procedures. One rat was fed ^{14}C-labeled PCBs by gavage in cottonseed oil 24 hours prior to sacrifice. Control rat skin was spiked with ^{14}C-PCBs using either *n*-hexane or DMSO as spiking solvent. In each case, DMSO gave recoveries more closely approximating the recovery of endogenous PCB, but an exact match could not be achieved. This experiment was designed to allow variations to be seen, by the use of extraction procedures that do not quantitatively extract the lipids from skin. In general, then, comments in the literature to the effect that "recovery from spiked samples was X%" may not provide any valid information about the recovery of endogenous material. Details of the spiking procedure should always be given, and some attempt should be made, when possible, to determine the actual recovery of endogenous compounds. The whole aspect of spiking needs additional study.

CLEANUP

Early attempts to avoid a need for cleanup (e.g. Dale *et al.*[10]) by using a relatively poor extracting solvent and an electron capture detector to minimize interference gave unreliable results because of the extremely poor reproducibility of extraction, as well as the generally low recoveries. For example, the recovery of aldrin from serum by extraction with hexane averaged only 26%, with a relative standard error of 28%.[10] Present cleanup procedures utilize one or more of four principles: (1) chemical destruction of interfering substances with acid, alkali or oxidizing agents; (2) liquid-liquid partition; (3) absorption chromatography or (4) gel filtration (permeation).

TABLE 5

RECOVERY OF ENDOGENOUS AND SPIKED AROCLOR 1254 FROM RAT SKIN

Spiking Solvent	Percentage Recovered ± S.D.	
	Method A	Method B
None*	20.3 ± 5.6 (3)	95.6 ± 24.6 (7)
Hexane	98.3 ± 2.8 (3)	81.8 ± 4.2 (3)
DMSO	52.8 ± 6.1 (3)	92.4 ± 4.0 (3)

*Endogenous PCBs extracted 24 hours after feeding ^{14}C-labeled 1254 to male CD rat. Method A = boiling briefly in benzene. Method B = Polytron homogenization in benzene:methanol 3:1 by vol. Number of replicates in parentheses.

Chemical destruction methods are extremely useful, but caution must be used in applying them to cleanup of compounds other than those for which they have specifically been validated. For example, 2,3,7,8,-tetrachlorodibenzo-*p*-dioxin is commonly determined following alkaline hydrolysis of tissue or extracted lipids[11,12] This treatment completely destroys octachlorodibenzo-*p*-dioxin.[7] The degree of cleanup achieved by partitioning a lipid extract in a nonpolar solvent (hexane or carbon tetrachloride) against sulfuric acid is usually excellent when the lipid consists primarily of triglycerides and/or phospholipids,[7] but we observe up to 10% of cholesteryl esters remaining in the organic phase. These compounds elute from adsorption columns in the same fraction as dibenzofurans and dibenzo-*p*-dioxins, thus, may interfere with subsequent determinations.

The lipid content of a particular sample has a significant effect on the recovery of pesticides in solvent partition procedures. The effect of fat on the partitioning of DDT between hexane and polar solvents is shown in TABLE 6. Thus, if a cleanup procedure involved, for example, partitioning a hexane solution containing pesticide against acetonitrile, three extractions with acetonitrile would recover 95% of the DDT in the absence of lipid. If the partitioning system contained 5% fat, only 90% of the DDT would be recovered in the combined acetonitrile extracts. This factor, then, contributes to the variability of recovery in cleanup procedures involving partitioning methods. An amimal exposure experiment involving an agent that produces a fatty liver might result in a 3- to 5-fold difference in fat content of the livers between treated and control animals. This would result in a difference in percentage recovery of pesticides from the livers of the two groups. Since hexane-acetonitrile partitioning is a very commonly used cleanup step,[13] this limitation must be taken into consideration.

TABLE 6

PARTITION OF DDT BETWEEN HEXANE AND A POLAR SOLVENT IN THE PRESENCE OR ABSENCE OF LIPID (COTTONSEED OIL)

	Partition Coefficient	
	Without Lipid	With Lipid*
Acetonitrile	0.486	0.727
85% DMF	0.684	1.062

*We added 90 mg of cottonseed oil per 2 ml of 1:1 partitioning mixture. The partition coefficients were corrected for change in volume due to the lipid.

Griffitt and Craun[14] made a comparison of the recoveries of various pesticides using the multiresidue procedure outlined in the Pesticide Analytical Manual (PAM method) *versus* gel permeation chromatography (GPC). Some of their results are summarized in Table 7. This table illustrates the fact that the procedure giving the highest mean recovery is not necessarily the best procedure to use. Thus, while the recovery of DDE was 97% by GPC and 91% by PAM, the former had a relative standard deviation (R.S.D.) of ± 19.9%, the latter only 6.0%. The GPC method recovered *o,p*-DDT with only a ± 4.6% R.S.D., while the PAM method gave an R.S.D. of ± 12.0%. Recoveries, otherwise acceptable, that show high variation from sample to sample, are probably indications that the cleanup procedure has an unexpected source of interference and efforts to compensate for that source would probably be worthwhile. The assay of several replicates for *p,p*-DDE by a procedure involving GPC cleanup can not, according to the data in TABLE 7, give an analytic result with less uncertainty than ± 19.9%, making statistical comparison of two sample means unnecessarily difficult. In general, then, a cleanup procedure giving a moderate but highly reproducible recovery may be a better choice than one giving a high but variable recovery.

The nature of the lipids in a sample extract again becomes of importance in adsorption chromatography. For example, it has been reported[15] that dibenzo-*p*-dioxins may be cleaned up on alumina columns. The separation of 2,3.7,8-tetrachlorodibenzo-*p*-dioxin from PCBs was accomplished with equal facility on acidic, neutral, and basic alumina, and no reason for choosing one over the other was given. However, if lipid extracts are chromatographed on alumina without a preliminary cleanup, we have seen that the capacity of alumina to hold back lipids decreases in the order acidic>basic>neutral. In particular, lecithin is poorly retained on neutral alumina and overloading occurs at 9 mg of phospholipid per gram of alumina. Acidic alumina was also observed to retard sterol esters to a greater extent than did either neutral or basic alumina, when all three were activated at 130°C overnight. Under identical conditions, loading the columns with a mixture of cottonseed oil (90%) and cholesteryl oleate (10%), the sterol ester was quantitatively

TABLE 7

RECOVERY OF PESTICIDES FROM MILK FAT BY TWO DIFFERENT METHODS*

Pesticide	Percentage Recovered GPC	Percentage Recovered PAM	Relative S.D. (%) GPC	Relative S.D. (%) PAM
Aldrin	78	92	9.5	6.3
p,p'-DDE	97	91	19.9	6.0
o,p'-DDT	113	88	4.6	12.0
Dieldrin	113	96	8.3	3.2
Lindane	95	101	8.6	1.8
Heptachlor	98	73	10.0	4.0
Aroclor 1254	100	77	1.0	3.5

*N = 3. GPC = Gel Permeation Chromatography; PAM = Standard Method from Pesticide Analytical Manual, involving partitioning with acetonitrile and chromatography on Florisil. Modified from Griffitt and Craun.[14]

eluted from neutral or basic alumina with 10 ml of 15% methylene chloride in hexane per gram of alumina. Elution from acidic alumina required 25% methylene chloride in hexane. So refinements in even the established cleanup procedures are still possible.

In summary, the recent trends toward analytic procedures other than GLC-electron capture detector measurements have increased the demands on the separation scientist relative to the degrees of cleanup needed. Assumptions about acceptable means of sample storage, extraction procedures, and the use of spiked samples for determination of recovery efficiencies are now being questioned. A great deal of caution is needed to validate the use of a given work-up procedure for different classes of compounds and different sample matrices.

REFERENCES

1. OSWALD, E. O., P. W. ALBRO & J. D. MCKINNEY. 1974. Utilization of gas-liquid chromatography coupled with chemical ionization and electron impact mass spectrometry

for the investigation of potentially hazardous environmental agents and their metabolites. J. Chromatogr. **98:** 363.

2. MES, J. & D. S. CAMPBELL. 1976. Extraction efficiency of polychlorinated biphenyl, organochlorine pesticides and phthalate esters from human adipose tissue. Bull. Environ. Contam. Toxicol. **16:** 53.
3. JENSEN, S., L. RENBERG and R. VAZ 1973. Problems in the quantification of PCB in biological material. Coll. Lect. PCB Conf. II, Nat. Swed. Environ. Prot. Board, Stockholm, Swed. pp. 7–14.
4. FOLCH, J., M. LEES & G. H. SLOANE-STANLEY. 1957. A simple method for the isolation and purification of total lipides from animal tissues. J. Biol. Chem. **236:** 497.
5. BLIGH, E. C. & W. J. DYER. 1959. A rapid method of total lipid extraction and purification. Can. J. Biochem. Physiol. **37:** 911.
6. ENTEMAN, C. 1961. The preparation of tissue lipid extracts. J. Am. Oil Chemists' Soc. **38:** 534.
7. ALBRO, P. W. & B. J. CORBETT. 1977. Extraction and clean-up of animal tissues for subsequent determination of mixtures of chlorinated dibenzo-*p*-dioxins and dibenzofurans. Chemosphere **7:** 381.
8. KATES, M. 1972. Techniques of Lipidology. American Elsevier. New York, pp. 351–352
9. SEIDL, G. & K. BALLSCHMITER. 1976. Isolation of PCB's from soil. Recovery rates using different solvent systems. Chemosphere **5:** 373.
10. DALE, W. E., A. CURLEY & C. CUETO, JR. 1966. Hexane extractable chlorinated insecticides in human blood. Life Sci. **5:** 47.
11. BAUGHMAN, R. & M. MESELSON. 1973. An analytical method for detecting TCDD (dioxin): Levels of TCDD in samples from Vietnam. Environ. Health Perspect., Exp. Issue No. **5:** 27.
12. FRIES, G. F. & G. S. MARROW. 1975. Retention and excretion of 2,3,7,8,-tetrachlorodibenzo-*p*-dioxin by rats. J. Agric. Food Chem. **23:** 265.
13. PORTER, M. O., S. J. V. YOUNG & J. A. BURKE, 1970. A method for the analysis of fish, animal, and poultry tissue for chlorinated pesticide residues. J. Assoc. Off. Anal. Chem. **53:** 1300.
14. GRIFFITT, K. R. & J. C. CRAUN. 1974. Gel permeation chromatographic system: An evaluation. J. Assoc. Off. Anal. Chem. **57:** 168.
15. PORTER, M. L. & J. A. BURKE. 1971. Separation of three chlorodibenzo-*p*-dioxins from some polychlorinated biphenyls by chromatography on an aluminum oxide column. J. Assoc. Off. Anal. Chem. **54:** 1426

QUALITATIVE AND QUANTITATIVE METHODS FOR DIOXIN ANALYSIS

J. Ronald Hass and Marlin D. Friesen

Environmental Biology and Chemistry Branch
National Institute of Environmental Health Sciences
Research Triangle Park, North Carolina 27709

INTRODUCTION

The cause for public health concern over the widespread use of chemicals contaminated with polychlorinated dibenzo-*p*–dioxins (PyCDDs) is well documented both in papers presented at this conference and earlier literature.[1] The exquisite toxicity exhibited by certain of the PyCDDs in combination with their tendency to accumulate through the food chain and the economic importance of the chemicals in which they are found have contributed to making their specific detection and determination one of the more challenging problems currently facing analytic chemists. Achieving biologically significant detection limits in the analysis of most samples requires careful attention to detail in all aspects of the work from sampling through data interpretation. In fact, because of the potential public interest in this area, care should even extend to the reporting of results. Since these analyses are frequently being performed at or near the capabilities of available instrumentation, the interdependence of various steps in the analytic process are accentuated. However, for purposes of this presentation, we will restrict our attention to the analytic step in which the PyCDD content of a suitable extract of the sample is to be measured.

As in any analytic problem, useful criteria by which various methods may be judged include: sensitivity, specificity, dynamic range, reproducibility and convenience. In electronic devices, sensitivity limitations generally arise from amplifier noise. Thus, sensitivity generally is expressed as the quantity of material necessary to give a specified ratio of signal to amplifier noise (S:N ratio). The sensitivity of one's measurement device defines the minimum quantity of sample necessary for a given detection limit under optimal conditions. This detection limit is rarely achieved in practice because of the nonspecific nature of most detectors. This response to sample components which are not the compound of interest is the factor limiting the detection limits for most analysis. That is, one can not decrease detection limits idenfinitely by simply using larger samples, even if they are available. Generally, the overall specificity of the analysis is some combination of the specificities of each step so that care in experimental design can lead to improvement in detection limits with a detector of any given specificity and sensitivity. The dynamic range of the detector is important since it sets the limit on concentration ranges that can be handled in a given analysis. Convenience includes such factors as frequency and difficulty of maintenance, frequency of calibration, skill required for proper operation, etc.

In the design of analytic methods for environmental screening, the toxicity of the compound(s) of interest and any tendency to accumulate in the food chain are important considerations in establishing the desired detection limit. For example, with 2,3,7,8,-TCDD, a dose of ca. 0.9 μg/kg fed orally to preganant mice over a seven day period led to malformation in the developing fetus and fetal death.[2] Furthermore, there is evidence to suggest that toxic effects associated with TCDD exposure are

0077–8923/79/0320–0028 $01.75/0 © 1979, NYAS

cumulative.[3] Assuming equivalent toxicology for humans and a 50 kg woman, a female of child-bearing age should limit TCDD exposure to a total of ca. 10^{-5} g. If the sole route of exposure is beef and there are no synergistic effects and allowing a safety factor of 100, this would require the consumption of ca. 45 kg of beef contaminated with 1 ppt TCDD. Of course, there is not firm scientific basis at present to assess the validity of any of the above assumptions. However, this does serve to illustrate the orders of magnitude one should consider in setting the desired detection limits for this compound. Thus, limits of 0.1–1 ppt are justified. Further assuming a 1 g sample, the instrument sensitivity must be such that 10^{-13}–10^{-12} g of TCDD give a measurable signal and all interferring substances have been removed from the sample. Since some TCDD losses during work-up are inevitable, this is generally not possible so that an analytic sensitivity of 10^{-14} g for 2,3,7,8-TCDD is indicated.

Fortunately, other PyCDDs tested have been found to be less toxic by factors of 100 or more, a least in terms of LD_{50}. The teratological effects of most of PyCDDs remain to be investigated, however. If one assumes that the variation in all toxic effects is reflected in relative LD_{50}s, then detection limit of 1–10 ppt for, say, the 1, 2, 3, 6, 7, 8-hexachlorodibenzo-*p*-dioxin (HxCDD) are adequate. For this detection limit, an analytic sensitivity of 10^{-13}–10^{-12} g will suffice.

Early Methods for the Detection and Determination of PyCDD

The earliest methods for the analysis for PyCDDs were directed toward detection of an unidentified "chick edema factor." These methods involved either a biologic assay[4] or the detection of peaks with unusually large retention times in the gas chromatographic (GC) analysis of the appropriate fraction from the sample in question.[5] These assays were nonspecific and, at best, semiquantitative. A number of laboratories[6–10] improved the GC analysis by preparation of suitable reference compounds so that tentative identification of the particular PyCDD present could be made. The certainty of these identifications rested on the similarity of solubility properties required by the sample work-up as well as identical retention indices for the reference compound(s) and the unknown(s) on one or more GC columns. The likelihood of interferences was reduced by use of gas chromatographic detectors that have a poor response toward most biologic compounds relative to their response to PyCDDs; however, these electron capture detectors do have finite response to most compounds. Examination of typical data reported for the analysis of even relatively simple samples show many spurious peaks in the chromatograms.[6] This limits the confidence one can place in the identification.

If one is willing to assume the unknown has been identified, quantitative analysis is straightforward, if tedious. The tedium results from properties of the electron capture detector. These are: limited dynamic range, susceptibility to contamination, and a wide range of sensitivities for reasonably similar compounds. Thus, samples whose content of individual PyCDDs covers a wide range may require multiple analyses in order to have the response of each PyCDD fall within the linear region of its calibration curve. Furthermore, the overall performance of the detector is markedly dependent upon its degree of contamination so that frequent analysis of standard solutions is required. Also, dependence of detector response upon chemical structure requires one to measure and monitor the response for each compound of interest. Finally, and perhaps most important, the sensitivity of the device is inadequate for many analyses. The introduction of pulsed electron capture detectors is reported to increase the dynamic range and reduce the contamination effects.[11] However, I am unaware of any reported analyses for PyCDDs using this improved detector. In our hands, this pulsed detector has a lower sensitivity for 2,3,7,8-TCDD than the more traditional designs.[12]

Then, in summary, the electron capture detector was initially employed for PyCDD analysis, but presently is not widely used mainly because of specificity and sensitivity limitations.

Methods Involving Mass Spectrometry

The early contributions of mass spectrometry to the analysis for PyCDDs was restricted mainly to confirming their presence in samples previous assayed by electron capture GC.[8,9] This can likely be attributed to equipment costs and experimental difficulties associated with combined gas chromatography/mass spectrometry (GC/MS). The first attempt at quantitative analysis by mass spectrometry was the high resolution technique reported by Baughman and Meselson,[13] in which they adapted a method first reported by Biros.[14] These workers take advantage of the relatively large negative mass defect 2,3,7,8,-TCDD (319.8956 for TCDD compared to 320.2674 for $C_{16}H_{36}N_2O_4$, for example). By operating a mass spectrometer at high resolution, the TCDD signal can be cleanly separated from most biologic interferences. Thus, a mass spectrometer was modified to permit rapid, repetitive high resolution scans over a narrow mass range. The use of a signal averaging computer resulted in a sensitivity of such that ca. 1×10^{-12} g of 2,3,7,8-TCDD gave a peak detectable above electronic noise where the sample was introduced by means of the direct insertion probe and a 0.3 AMU window centered at ca. m/z 322 was scanned with 0.25 sec scan cycles. Since the measurement step involves no chromatographic separation, the detection limits depend upon how well the sample clean-up removes interfering substances. As noted previously, the mass spectrometer will resolve most biologic interferences. It is, however, still necessary to remove most of the sample matrix since the source sensitivity is adversely affected when samples larger than a few μg total material are analyzed. Assuming a sensitivity of 10^{-12} g for TCDD, a 1 g sample and a desired detection limit of 10^{-12} g/g, this requires a 200,000-fold concentration of the TCDD relative to the general matrix if the ion source loading is to be maintained ≤ 5 μg. Furthermore, there are endogenous compounds, not of biologic origin, which are not as well resolved from TCDD. These include members of the DDT family and certain polychlorinated biphenyls (PCBs). Since these compounds are typically present at the 0.1–1 part-per-million (ppm) in biologic samples, it is necessary for the concentration of the TCDD relative to the DDT family and PCBs to be increased by as much as 10,000-fold if the detection limits is to approach the desired sensitivity. The effect of excess PCB (FIGURE 1) illustrates potential difficulties from even moderate amounts of these compounds on quantitative analysis at TCDD levels ca. 10-fold above the instrumental limit of sensitivity.

A perhaps more serious limitation of this particular high resolution technique arises from the combination of the sample work-up and introduction techniques. The mass spectrometer will give a positive response for any compound whose elemental composition is $C_{12}H_4Cl_4O_2$. Obviously, this includes, all TCDDs, not just 2,3,7, 8-TCDD. Unfortunately, there are also other possible interferences. In our laboratory, we have isolated a material from chicken excretia from chicks fed PCBs that gave an exact mass indicative of an elemental composition of $C_{12}H_4Cl_4O_2$. This material could be reduced to give a material that reacted with CH_3I/K_2CO_3 to give a dimethyl derivative. We assume this to be a *para* dihydroxyl metabolite of the PCB, which was oxidized to the quinone during treatment of the sample with NaOH. Needless-to-say, the presence of these dihydroxyl-PCB metabolites can result in serious positive interference if the sample is subjected to caustic conditions during preparation.

The Baughman/Meselson method was the first analytic method for 2,3,7,8-TCDD which was reported to have adequate sensitivity for application to environmental samples. Their work represents a very important step in the development of methodology for TCDD analysis in that the feasibility for reaching environmentally meaningful detection limits was established and many of the more subtle problems in these analyses were indicated. In answer to some of these problems and in conjunction with the steadily improving technology for GC/MS, the next developments involve application of this technique to the PyCDD problem.

In fact, Baughman and Meselson recognized the advantages offered by GC and soon published a modification of their procedure to include a preparative GC step.[15] The additional separation step substantially reduced the PCB and DDT family interferences making detection limits of a few ppt possible. Unfortunately, the recovery of the preparative GC step was rather low. This is complicated by the potential hazards of working with gaseous 2,3,7,8-TCDD. At approximately the same time, Crummett and Stehl reported a combined GC/MS method for PyCDD determination in various chlorinated phenols and their derivatives.[16] Using a low

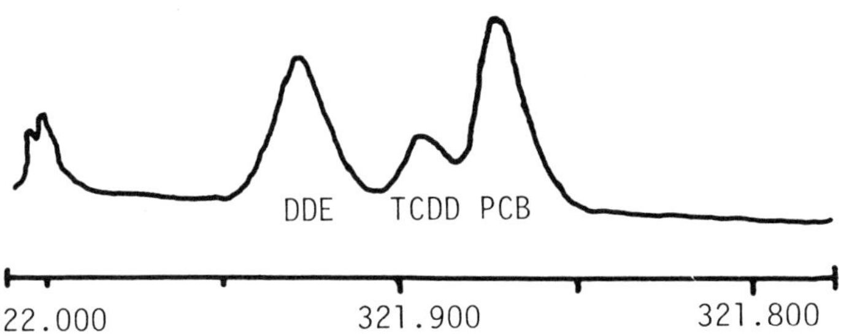

FIGURE 1. Partial high resolution mass spectrum obtained from 2.5 x 10^{-10} g TCDD plus matrix from 10 g human milk.[13]

resolution mass spectrometer, they depended upon the combination of clean-up methodology, GC retention time, and conformity of the signal to the expected mass spectrum for their specificity. By monitoring three ions from the mass spectrum, they reported a sensitivity adequate to give an S:N ratio of 10:1 from 6 $\times$ 10^{-12} g of 2,3,7,8-TCDD injected on the gas chromatograph. These authors report a detection limit of only .01 ppm for a 5 g sample. Perhaps a lower detection could have been achieved by injection of a larger fraction of the sample. In any event, Buser and Bosshardt developed a technique based upon monitoring m/z 320 (molecular ion of TCDD) using a quadrupole mass spectrometer with a sensitivity of 5 $\times$ 10^{-10} g injected to give a detected signal, which they reported to give a detection limit of 0.001 ppm based upon a 20 g sample of chlorinated phenol or derivative.[17]

Buser modified the low resolution selected ion monitoring (SIM) GC/MS analysis to accommodate dioxins with differing chlorine content[18] reporting of detection limit of 0.2 ppm PyCDD in pentachlorophenol. He indicated this could have been lowered to 0.01 ppm by appropriate concentration of the sample. Blaser *et al.* reported a similar method for PyCDD in PCP which makes use of a simplified sample clean-up and resulted in a detection limited of 0.5 ppm.[19] Meanwhile, Buser prepared and separated 11 of the 13 possible PyCDDs that contain 6 or more chlorines.[20] Using a capillary column with ca. 35,000 theoretical plates and a 2°C/min temperature program, the 8 HxCDD he prepared eluted within a 10° temperature span. Similar work with the TCDDs[21] demonstrated that two different capillary columns are necessary to separate 2,3,7,8-TCDD from the 9 available of the 22 possible TCDD. Obviously, GC methods with lower resolution are even more likely to suffer interference in the analysis for a given PyCDD. In conjunction with high chromatographic resolution, capillary columns offer the additional advantage of high sensitivity as a result of greater instantaneous sample concentrations in the ion source. Thus, Buser was able to improve his sensitivity to approximately 5×10^{-12} g injected for a detectable response. A detection limit of 3–5 ppt in 5 g soil and grass samples was reported.

Meanwhile in the USA, the Environmental Protection Agency (EPA) became concerned over possible 2,3,7,8-TCDD accumulation through the use of 2,4,5-T as a herbicide on rangeland. This interest stimulated the development of a combined GC-high resolution mass spectrometric technique in which two ions (m/z 320 and 322) are monitored at high resolution,[22–24] using a signal averaging computer to improve S:N ratio. This technique was further improved by incorporation of capillary columns in the GC.[25] By spiking the sample with $^{37}Cl_4$-2,3,7,8-TCDD prior to extraction, analytic precisions on the order of 10% were achieved. Positive identification of 2,3,7,8-TCDD required that the intensity ratios of m/z 319.8965 and 321.8936 match those of TCDD, the compound elute at the proper retention time, and the signals by $\geq$2.5 times background noise. Typical detection limits of a few ppt are attained.

Negative Chemical Ionization Mass Spectrometry

Although the improvements in instrumental sensitivity and specificity over the past several years have been impressive, the actual improvements in detection limits have been less impressive. This can be attributed to a combination of a number of effects, all of which fall under the general classification of matrix effects. The first of these is a result of the finite abundance sensitivity of any mass spectrometer. The abundance sensitivity is a measurement of the contribution of a peak at m/z = M to a peak at m/z = M+ΔM. A typical value for a modern double focusing instrument operated at a resolution of 1000 with ΔM=1 dalton is ca. 10^6:1. This is a result of the tail of the approximately gaussian mass spectral peak. Clearly, the abundance sensitivity of the mass spectrometer determines the ability of the instrument to give accurate intensity measurements when intense peaks are present within 0.2 to 0.3 daltons. This sets the practical limit for the concentration of the PyCDD in the sample extract indicating the importance of the extraction and clean-up efficiency in analyses being performed at maximum sensitivity. A more subtle problem can arise for mass spectrometers whose ion optical design is such that mass analysis occurs immediately prior to ion detection. If some parent ion should fragment in the field-free region immediately prior to the magnetic sector to give a daughter ion (a so-called metastable ion), this ion will be transmitted through the magnetic sector as a peak several daltons wide centered, in general, at some non-integer m/z. As the total quantity of material introduced into the ion source increases, the probability that one of these diffuse peaks will interfere with any given m/z increases. This again acts to limit the ratio of matrix/PyCDD that can be tolerated. So again, we find that decreases in detection limit will be the result of matrix removal. It is worth noting that instruments whose ion optical arrangement is such that energy analysis is performed between mass analysis and ion detection are not subject to this problem.

As discussed in the paper by Phillip W. Albro in this volume, extraction and clean-up methodologies are now highly refined. Any future improvements can be anticipated to be generally expressed as percentages rather than factors. Furthermore, the specificity of the mass analyzer has already reached its theoretical limits, except possibly for improvements in abundance sensitivity, which will likely be small and costly. Thus, we decided to investigate the improvements possible through the use of more specific ionization techniques. Dougherty and co-workers[26,27] have noted the selectivity of negative chemical ionization (NCI) techniques for chlorinated aromatic compounds in biologic matrices. Hunt *et al.*[28] published an oxygen NCI spectrum of 2,3,7,8-TCDD in which the most abundant ion resulted nominally from the cleavage of two oxygen-carbon bonds to form a dichloro-quinoxide anion. These authors noted this ion is well removed in mass from the previously reported interferences for TCDD measurement. The reported sensitivity was 50:1 S:N ratio for 2×10^{-12} g of 2,3,7,8-TCDD introduced by means of direct probe, making this technique at least competitive with electron impact. These results stimulated us to undertake a study of the NCI spectra of the available PyCDD and to investigate the suitability of this technique for the determination of these compounds in biologic matrices.

Negative Chemical Ionization Mass Spectrometry of PyCDD

General Considerations

Sample ionization by electron impact involves transfer of kinetic energy from an energetic electron (typically 70–80 eV) to the internal energy of a gaseous molecule. When the amount of this energy exceeds the ionization potential of the molecule, an electron may be ejected, forming a positive molecular ion. If sufficient energy is added (as is generally the case) various fragmentation reactions from this positive ion may occur. Relative sensitivities depend upon the cross section of the energy transfer reaction, the number of fragment ions produced and, to a lesser extent, the difficulty of forming positive ions from the sample. Thus, most compounds have relative sensitivities within a factor of 10. Chemical ionization is a modification of the electron impact process in which a large excess of a reagent gas introduced into the ion source. The reagent gas is ionized by electron impact and the ions thus formed react with neutral molecules to form a reagent plasma. The ions in the reagent plasma may then react with neutral sample molecules. In general, some chemical species (say, a proton) is transferred from the reagent ion to the sample molecule. Thus, for sample ionization to occur, the affinity of the sample molecule for the ionizing specie must exceed that of the reagent. This can be exploited to achieve a degree of selective ionization of mixture components by using, for example, a base such as NH_4^+ as the reagent ion to selectively ionize more basic compounds (e.g. amines) in the presence of more acidic compounds. The term "negative chemical ionization" as presently used describes the experiment in which the mass spectrometer is modified to analyze the negative ions emulating from a chemical ionization source. There are at least two fundamentally different processes leading to the formation of sample negative ions under these conditions. First, one can have ion-molecule reactions analogous to the positive ion case. Typical examples include proton abstraction by OH^- to form $(M\text{-}1)^-$ ions when M indicates the molecular weight of the compound,[29] and transfer of a halide ion to give $(M+X)^-$ ions.[30] Again, sample ionization occurs when the relative affinities of the sample and reagent ion are proper for chemical reaction to occur. The other process involves sample ionization by means of resonant electron capture. This type of ionization will occur only if the molecular anion is more stable than the molecule and separated electron. This is a condition generally met by halogenated aromatic compounds and rarely met by aliphatic compounds. The stability of an anion relative to the molecule plus an electron is measured by the electron affinity of the molecule. When materials of significantly differing electron affinities are introduced into the ion source simultaneously, the material with the higher electron affinity is preferentially ionized (FIGURE 2). Furthermore, certain molecules apparently can have very high cross-sections for electron capture and thus are capable of giving negative ion beams one or two orders of magnitude more intense than positive ion beams from similar quantities of material. Thus, in terms of achieving the maximum sensitivity and selectivity in the ionization process, NCI techniques involving electron capture ionization of the sample appear the most promising.

ELECTRON CAPTURE NCI SPECTRA OF PYCDD

The electron capture NCI spectra of the 14 available PyCDDs (FIGURE 3) were obtained using CH_4 as the chemical ionization reagent gas.[31] Under these circumstances, the CH_4 serves mainly to reduce the energy of the source electrons to near thermal energies so that electron capture probabilities are increased and to act as an energy sink for newly formed negative molecular ions. These spectra are dominated by $M^{-\cdot}$, $(M\text{-}Cl)^{-}$ and $(M\text{-}Cl_2)^{-\cdot}$ ions. In some cases, intense ions resulting from the reaction of $M^{-\cdot}$ with residual oxygen in the vacuum system are observed. These ions are discussed below. Examination of these spectra indicates that $M^{-\cdot}$ and $(M\text{-}Cl)^{-}$ are generally the ions providing the greatest sensitivity for analysis. The exception is 2,3,7,8-TCDD in which the $(M\text{-}Cl_2)^{-\cdot}$ ion is the most intense.

The use of electron capture NCI results in an increase in specificity for 2,3,7,8-TCDD analysis in that the interferences from penta- and heptachlorobiphenyls and the DDT family are reduced by virtue of a reduction in ion current at m/z 320 and 322 from these compounds by at least three orders of magnitude.[32] In view of the ability of present clean-up methods to separate these compounds from PyCDDs, it is more significant that the fragment ion current from endogenous biologic materials is reduced. This reduction results from two factors. First, these compounds generally have low electron affinities and hence tend not to be ionized. Second, the ionization process necessarily involves little transfer of energy so that fragmentation reactions are unlikely. Thus, two benefits are derived. The reduction in primary ion current from extraneous materials reduces the requirements on the abundance sensitivity of the mass spectrometer and the reduction in fragmentation reactions lowers the contribution of diffuse peaks due to fragmentation reactions within field-free regions or the mass analyzer. The one situation in which the specificity is not increased is the PCB metabolite derivative discussed earlier. This can be minimized by avoiding basic conditions during sample work-up. However, if this is undersirable for some reason, further specificity can be gained by exploitation of a highly specific anion-molecule reaction of the PyCDD.

EFFECT OF OXYGEN ADDED TO THE ION SOURCE ON THE NCI SPECTRA OF THE PYCDD

The observations by Hunt *et al.*[28] that 2,3,7,8-TCDD underwent an ion-molecule reaction under NCI conditions when O_2 was used as the reagent gas led us to investigate the reactivity of the PyCDD negative molecular ions by adding 5–10% O_2 to the CH_4 in our ion source.[31] These spectra (FIGURE 4) show peaks corresponding to

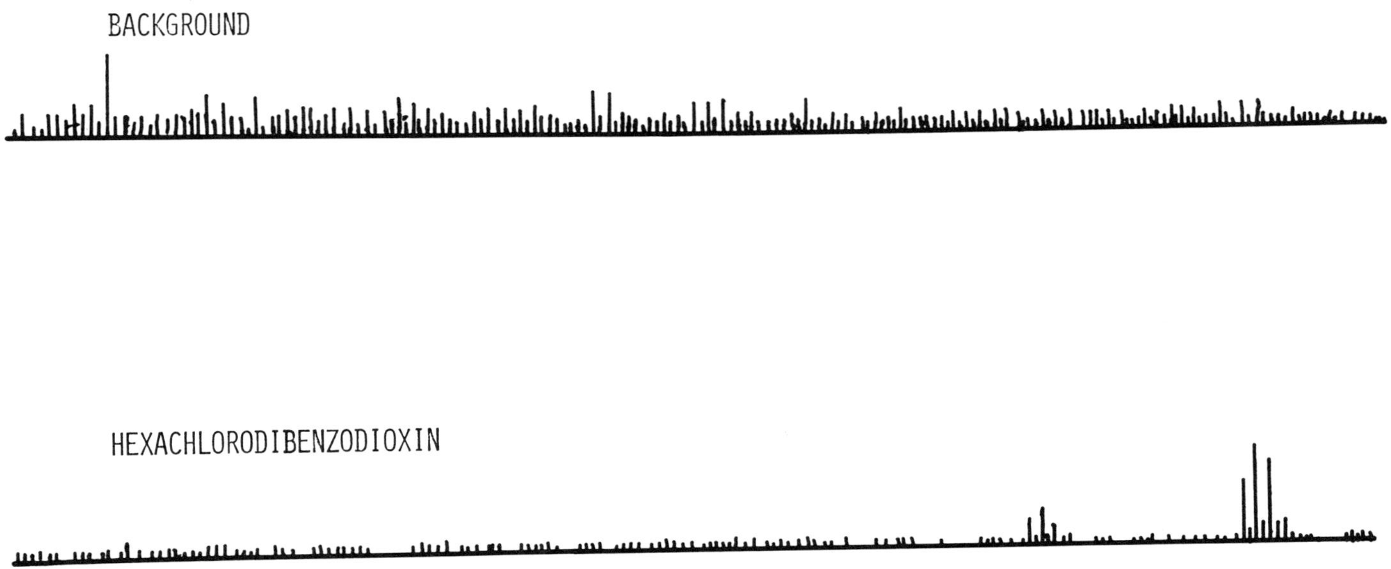

FIGURE 2. Upper trace is background ion current in the mass spectrometer. The lower trace resulted from the addition of 10^{-9} g of a hexachlorodibenzo-*p*-dioxin. This material has a higher electron affinity than the instrumental contaminants and is thus preferentially ionized.

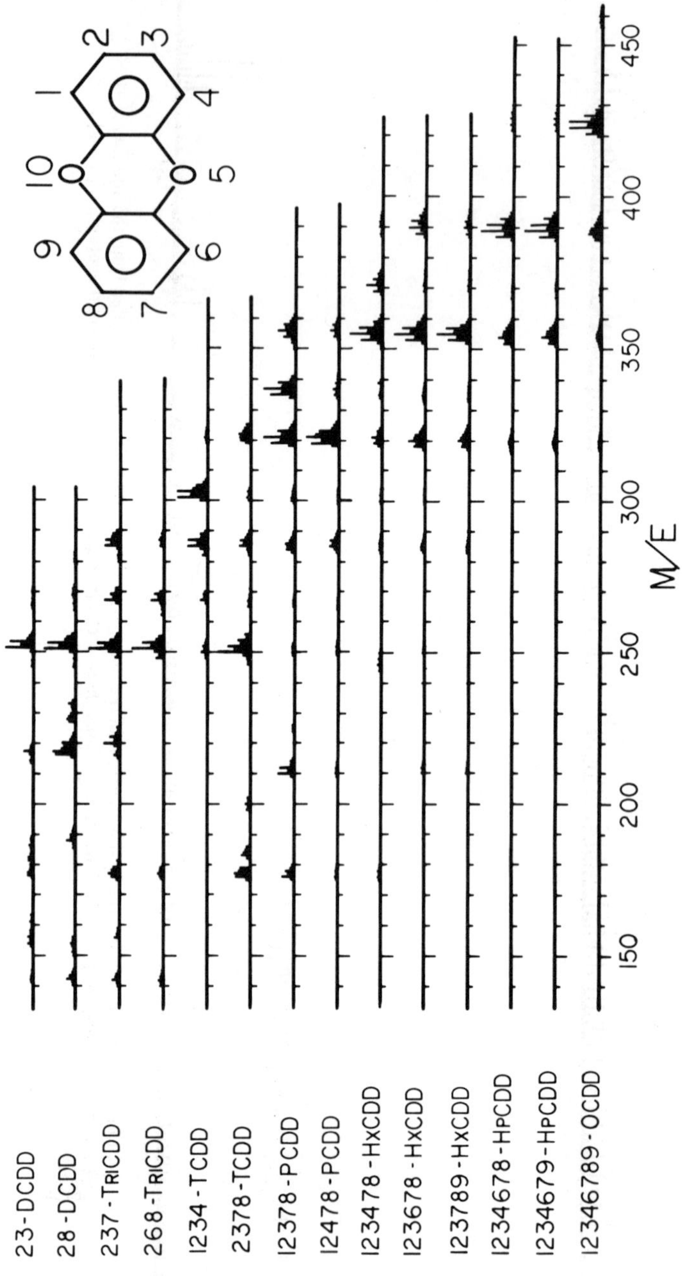

FIGURE 3. Negative chemical ionization mass spectra of the 14 available PyCDDs obtained using methane as the reagent gas.

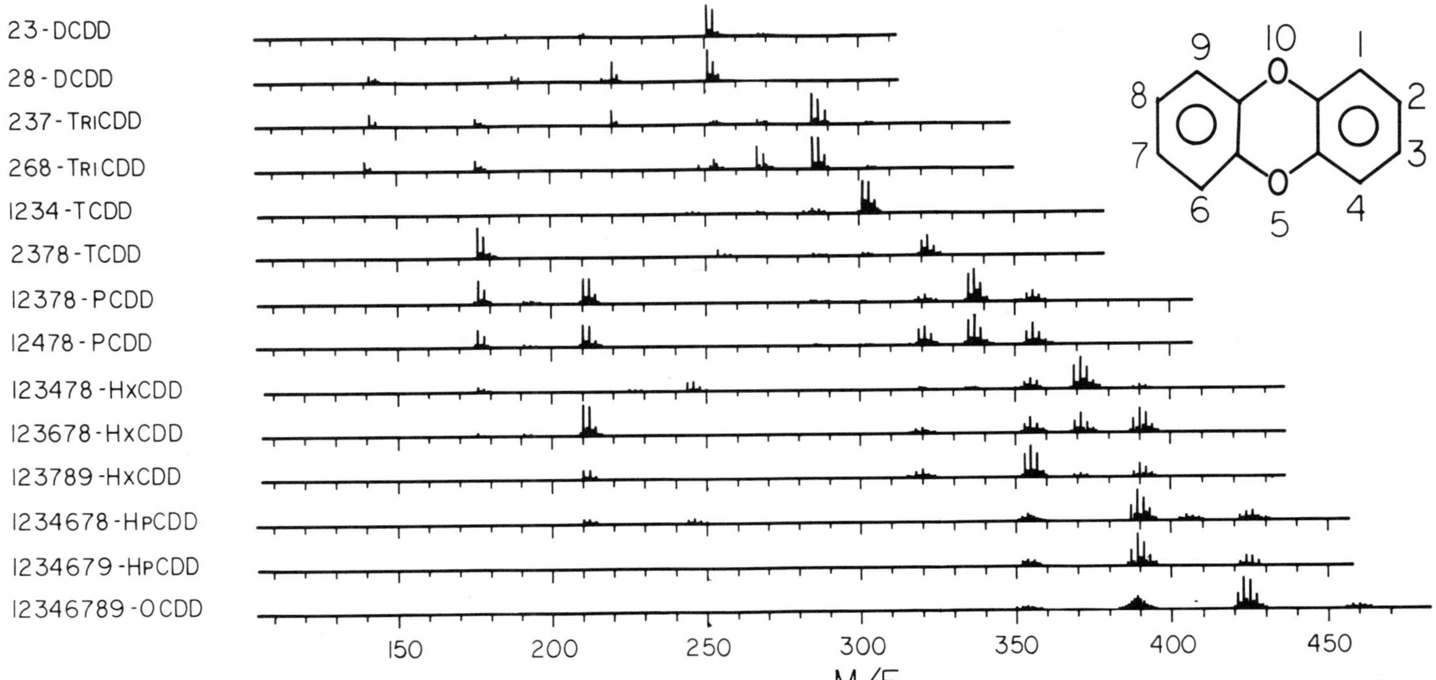

FIGURE 4. Negative chemical ionization mass spectra of the 14 available PyCDDs obtained using 10% oxygen in methane as the reagent gas.

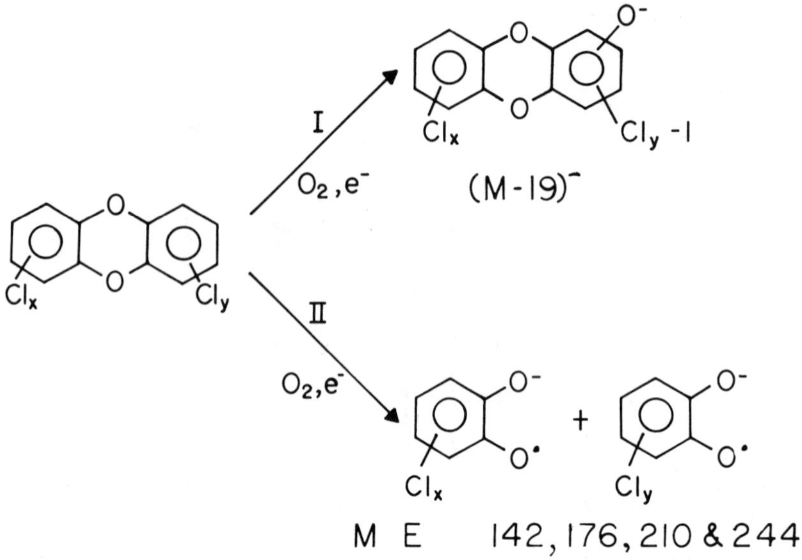

FIGURE 5. Ion molecule reactions observed between PyCDD negative molecular ions and oxygen.

two ion-molecule reactions (FIGURE 5) in addition to the peaks corresponding to electron capture processes. Reaction pathway I corresponds to reaction with O_2 followed by elimination of ClO. Reaction pathway II is of greater interest since it serves to define the number of chlorines on each ring. More importantly, a diquinone formed from a dihydroxyl-PCB will not give an interfering peak since insufficient energy is available to break a carbon-carbon bond. An unexpected benefit of adding O_2 to the reagent mixture became apparent from the dependence of relative intensities from pathway I/pathway II upon chlorine distribution in the dioxin molecule. In particular, pathway I is favored for those PyCDDs with *peri* chlorines whereas pathway II is favored in the absence of *peri* chlorines. In the case of the TCDDs, the 2,3,7,8-TCDD is the only isomer with no *peri* chlorines. Thus, the methane/oxygen NCI spectrum of this compound in combination with gas chromatography offers the most specific means currently available to identify this compound when $\leq 10^{-9}$ g of material is available.

ANALYSIS OF BIOLOGIC SAMPLES FOR PYCDD BY NCI MASS SPECTROMETRY

The appropriate test for a new detection method is to compare its performance to that of an established method for the analysis of typical samples, keeping as many other factors as possible constant. With this objective in mind, we undertook the analysis of a series of bovine liver and fat samples for PyCDD.[33] The samples were subjected to an efficient extraction and clean-up method, then the PyCDD fraction was analyzed on mass spectrometers with similar mass analyzers and ion detectors (Finnigan Model 3300). One instrument is dedicated for electron impact work and is interfaced to the gas chromatograph with a single stage glass jet separator (~40–50% yield). The other instrument is designed for chemical ionization and has been modified to permit the analysis of negative ions.[31,32] In this case, the gas chromatograph is coupled directly to the mass spectrometer. The same data acquisition system (INCOS 2300) was used with both instruments for the actual sample analyses. By systematically adjusting the gain of each ion detection system, differences in detector sensitivity were eliminated. Assuming equal sensitivity of the mass analyzers, the differences in ion source sensitivity should be the only factor affecting relative analytic performance except for the factor of approximately two resulting from separator losses in the electron impact instrument.

The results of the analysis of the liver are summarized in TABLE 1. Of the 63 separate determinations made by each technique, 49 resulted in measurable quantities

TABLE 1

DIOXINS IN BOVINE TISSUE SAMPLES
(ng/g wet tissue)

SAMPLE		OCDD EI	OCDD MNCI	HpCDD EI	HpCDD MNCI	HxCDD EI	HxCDD MNCI
154892	LIVER	ND	0.23	ND	0.07	ND	0.06
154893		ND	0.27	ND	0.03	ND	0.01
COW 201		2.3	3.1	0.4	0.63	ND	0.03
SM. Y. H.		3.9	3.9	D	0.93	ND	0.03
LG. Y. H.		1.1	1.3	D	0.31	ND	0.03
BULL		3.0	2.1	0.3	0.61	ND	D
COW 555		5.0	7.7	1.2	2.4	1.4	1.3
7745		1.9	1.7	D	0.37	ND	0.01
401		5.7	5.7	0.8	0.81	ND	ND
154889		4.6	4.9	1.5	1.3	ND	ND
COW 180		3.6	5.0	D	0.33	ND	ND
COW 359		6.0	3.9	ND	0.44	ND	0.11
4 MO. CALF		23	36	8.0	12	ND	ND
152812		4.2	4.3	ND	3.5	ND	ND
154881		ND	0.15	ND	ND	ND	ND
4 MO. CALF	FAT	D	5.1	ND	0.39	ND	ND
152449		26	37	ND	ND	ND	ND
154892		18	10	ND	ND	ND	ND
154893		50	47	12	0.33	ND	ND
154881		37	41	ND	0.44	ND	ND
154889		ND	4.4	ND	2.1	ND	0.39

of PyCDD. Of these, only 24 or about one-half of the determinations, could be made by EI. Thus, under similar conditions, NCI offers a very practical advantage over EI for the analysis of bovine liver and fat. In FIGURE 6 is plotted the EI and NCI results for those samples with sufficiently high PyCDD levels for EI determination. The 45° correspondence line has been drawn in to show that the agreement between the two techniques is quite reasonable. The least squares regression line is given by [EI] = 1.058 [NCI] + 0.01. That is, within experimental error, the slope is the predicted unity and the intercept is (0,0).

A sample of wood from the barn used to house the cows was also available for analysis. Wood shavings were obtained from this sample. These were subjected to the same sample clean-up and the PyCDD fraction again analyzed by both EI and NCI

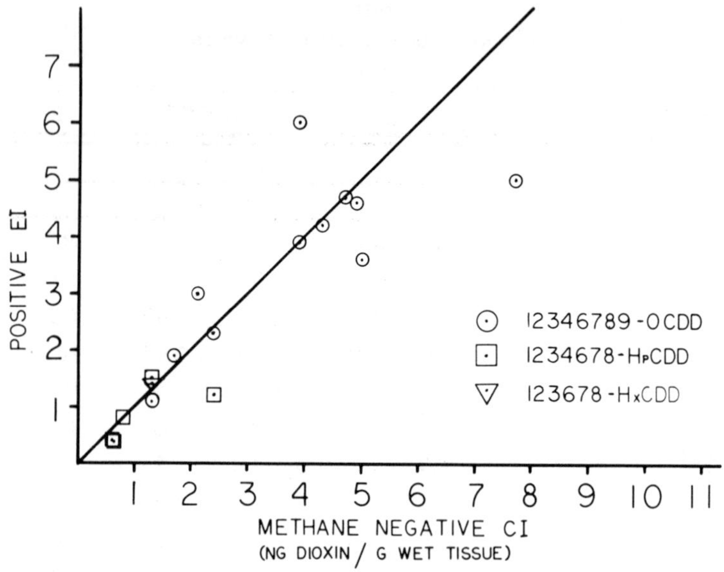

FIGURE 6. Comparison of EI and NCI results from fat and liver samples. The unity slope line has been drawn to show expected correspondence.

mass spectrometry. The clean-up method failed to remove sufficient extraneous material to permit EI analysis. However, the octa-and heptachlorodibenzo-*p*-dioxin content could be estimated by NCI. The sample complexity was too great to allow an estimate of HxCDD.

This work was performed using relatively inefficient mass analyzers and packed columns in the gas chromatographs. A two-three order of magnitude improvement in detection limit should be readily attained through the use of capillary columns and a higher performance mass analyzer.

CONCLUSION

In this paper, the methodology currently available for the determination of PyCDD in biologic/environmental samples has been reviewed. Although electron capture gas chromatography is by far the least expensive technique available, it does not have adequate sensitivity or specificity for PyCDD analysis in biologic matrices. Mass spectrometry in combination with gas chromatography offers considerable improvements in sensitivity and specificity compared to the electron capture methods. However, in the important Cl_4 to Cl_6 range of PyCDD, the gas chromatography is not presently able to separate all isomeric compounds. Since the electron impact mass spectra of all PyCDD of a given chlorine content are very similar, interferences are likely. By use of such selective ionization techniques as NCI, even greater improvement in the ratio of signal from PyCDD to signal from matrix can be attained. Furthermore, the mass spectra produced have a higher information content than those produced by EI ionization. In summary, the most sensitive and specific analysis for PyCDD in biologic matrices would involve an efficient sample clean-up with PyCDD determination on a large double-focusing mass spectrometer operated at a resolution of *ca.* 3000 to separate any extraneous hydrocarbon signal, with sample introduction by means of a glass-capillary-column-equipped gas chromatograph. By careful attention to detail and the use of suitable carrier compounds to compensate for sample losses, detection limits of *ca.* 10^{-13} g PyCDD/g sample should be possible from one g samples. To the best of my knowledge, no single laboratory has all these techniques available at the present time.

Acknowledgments

We would like to thank Dr. C. E. Parker, Mr. D. J. Harvan and Ms. Juaria Hines for their assistance in sample work-up and Dr. P. W. Albro for many helpful discussions.

References

1. For a recent review see: Environ. Health Perspect., September, 1973.
2. Sparschu, G. L., F. L. Dunn & V. K. Rowe. 1971. Food Cosmet. Toxicol. **9:** 405.
3. Allen, J. R. & L. A. Carstens. 1967. Am J. Vet. Res. **28:** 1513.
4. Verrett, M. J., J. Marliac & J. McLaughlin. 1964 J. Assoc. Off. Anal. Chem. **47:** 1003.
5. Neal, P. 1967. J. Assoc. Off. Anal. Chem. **50:** 1338.
6. Elvidge, D. A. 1971. Analyst **96:** 721.
7. Williams, D. T. & B. J. Blanchfield. 1972. J. Assoc. Off. Anal. Chem. **55:** 93.
8. Firestone, D., J. Ress, N. L. Brown, R. P. Barron & J. N. Damico. 1972. J. Assoc. Off. Anal. Chem. **55:** 85.
9. Woolson, E. A., R. F. Thomas & P. D. J. Ensor. 1972. J. Agric. Food Chem. **20:** 351.
10. Hutzinger, O., S. Safe & V. Zitko. 1972. Int. J. Environ. Anal. Chem. **2:** 95.
11. Lovelock, J. E. 1961. Nature **189:** 729.
12. Hass, J. R. Unpublished data.
13. Baughman, R. & M. Meselson. 1973. In Advances in Chemistry Ser. 120. E. Blair., Ed. American Chemical Society. Washington, D.C., p. 92.
14. Biros, F. J. 1970. Anal. Chem. **42:** 537.
15. Baughman, R. & M. Meselson. 1973. Environ. Health Perspect. **5:** 27.
16. Crummett, W. B. & R. H. Stehl. 1973. Environ. Health Perspect. **5:** 15.
17. Buser, H. -R. & H. -P. Bosshardt. 1974. J. Chromatog. **90:** 71.
18. Buser, H. -R. 1975. J. Chromatog. **107:** 295.
19. Blaser, W. W., R. A. Bredeweg, L. A. Shadoff & R. H. Stehl. 1976. Anal. Chem. **48:** 984.

20. BUSER, H.-R. 1975. J. Chromatog. **114:** 95.
21. BUSER, H.-R. 1977. Anal. Chem. **49:** 918.
22. HARLESS, R. L. & E. O. OSWALD. 1977. Presented at Twenty-fourth American Society for Mass Spectrometry Meeting, San Diego, Ca., Paper No. TP-10.
23. TAYLOR, M. L., R. L. C. WU, C. D. MILLER & T. O. TIERNAN. Presented at Twenty-fourth American Society for Mass Spectrometry Meeting, San Diego, Ca. Paper No. H5.
24. SHADOFF, L. A. & R. A. HUMMEL. 1978. Biomed. Mass Spectrom. **5:** 7.
25. HARLESS, R. L. & E. O. OSWALD. 1978. Presented at the Twenty-fifth American Society for Mass Spectrometry Meeting, St. Louis, Mo., Paper No. RP-7.
26. DOUGHERTY, R. C. & K. PIOTROWSKA. 1976. Proc. Nat. Acad. Sci. USA **73:** 1977.
27. DOUGHERTY, R. C. & K. PIOTROWSKA. 1976. J. Assoc. Off. Anal. Chem. **59:** 1023.
28. HUNT, D. F., T. M. HARVEY & J. W. RUSSELL. 1975. J. Chem. Soc. Chem. Commun. 1975: 152.
29. SMIT, A. L. C. & F. H. FIELD. 1977. J. Am. Chem. Soc.: 6471.
30. DOUGHERTY, R. C., J. DALTON & F. J. BIROS. 1972. Org. Mass Spectrom. **6:** 1171.
31. HASS, J. R., M. D. FRIESEN, & M. K. HOFFMAN. 1978. Org. Mass Spectrom. In press.
32. FRIESEN, M. D. 1977. Ph.D. Thesis, Kansas State University.
33. HASS, J. R., M. D. FRIESEN, D. J. HARVAN & C. E. PARKER. 1978. Anal. Chem. In press.

THE PROBLEM OF MEASUREMENTS NEAR THE LIMIT OF DETECTION

Warren B. Crummett

Analytical Laboratories
Dow Chemical, U.S.A.
Midland, Michigan 48640

Concern about potential adverse health effects of chemicals in the environment has placed great pressure on the analytic chemist to provide instant and unequivocal solutions to problems relating to the detection and determination of

1. trace impurities in products;
2. trace contaminants in the workplace; and
3. trace contaminants in the environment.

Problems arise because the results obtained in these analyses are very uncertain because of difficulties in validating methods near the limit of detection, which is often at 1 part per billion or lower. The limit of detection itself is defined and determined by a wide variety of approaches,[1] which are generally subjective and often designed to serve the investigators purpose. The interpretation of the data is further hindered because the *amount* of uncertainty is very seldom known.

Abuse of analytic systems and data in these situations is widespread among those who are relatively inexperienced in the application of analytic chemistry. However, even highly trained experienced professional chemists can be misled by results, either positive or negative, generated near the "limit of detection." The examples that follow were generated by "state-of-the-art" technology.

The measurement of trace impurities in products is the most important of the various industrial analyses because it provides data from which processes can be changed or modified to minimize the presence of an impurity and thus eliminate it in the workplace and the environment.

One of the most urgent and highly visible problems in product quality is the reduction of the dioxin content in chlorinated phenolic products. The presence of TCDD (2,3,7,8-tetrachlorodibenzo-*p*-dioxin) in 2,4,5-trichlorophenol and its derivatives is generally thought to be unavoidable.[2] Yet, as analytic methodology has improved the ability of the manufacturing plant to monitor and subsequently produce material of lower TCDD content has also improved. TABLE 1 provides an excellent example of this. As can be seen, this manufacturing-analytic interaction produces a "push-pull" effect to improve both sensitivity and certainty of methods as well as quality of the product.

There is an unfortunate tendency among some to assume that contamination of similar substances is predictably consistent. Such assumptions are dangerous.

Although Ronnel®* brand insecticide is a derivative of 2,4,5-trichlorophenol, TCDD has never been conclusively shown to be present in this product. In the early 1970s, everyone assumed that a signal due to it was buried under the background noise and interferences on the mass spectrogram (MS) and was roughly equal to the contribution of TCDD from 2,4,5-trichlorophenol. Nevertheless, as clean-up procedures became more effective, TCDD could still not be found—even with improved

*Trademark of the Dow Chemical Company.

0077–8923/79/0320–0043 $01.75/0 © 1979, NYAS

TABLE 1

2,4,5-TRICHLOROPHENOL IMPROVEMENT CONCURRENT WITH ANALYTIC METHODS IMPROVEMENT

Analytic Method		TCDD Concentration (ppm)		
Type	Date Developed	Limit of Detection	Product Specification	Date of Specification
GLC*	1964	1.0	1.0	1966
GLC	1970	0.5	0.5	1970
GC-MS	1971	0.05	0.1	1971
GC-MS	1977	0.001	0.02	1977
LC	1978	0.001	0.02	1977

GLC = gas-liquid chromatography; GC-MS = gas chromatography-mass spectrometry; LC = liquid chromatography.

recovery data at low levels. The history of the analytic method development is shown in TABLE 2.

These are just two instances which show that widely held assumptions by even the most knowledgeable are only speculation unless supported by valid data. In both examples considerable care and judgment had to be exercised to test the methods for accuracy, precision, and reliability.

Workers must be protected from exposure to hazardous materials in the workplace. In practice this can best be achieved if specific methods are available to detect the materials at appropriate levels. Problems often result when methods are sensitive but not sufficiently specific.

Concern about the carcinogenic properties of bis-chloromethyl ether led our laboratory to develop two highly specific methods[3,4] for its determination in air with detection limits of 1 part per billion to assure the safety of our workers in plants where chloromethylmethyl ether (which contains bischloromethyl ether as an impurity) is used. A laboratory in another company also developed a method thought by some to be equally specific.[5] Studies conducted using this method produced results which indicated that bis-chloromethyl ether could be formed at unsafe levels whenever formaldehyde and hydrogen chloride were present together in air.[6,7] Further, the study purported to show that bis-chloromethyl ether could be found wherever formaldehyde is in contact with ionic chloride. In spite of data refuting the above reports[8,9] and additional work[10] which produced few positive results, the National Institute for Occupational Safety and Health initiated an elaborate study to measure bis-chloromethyl ether formation in the textile, dye, fertilizer, and foundry industries, and hospital laboratories. Using a more specific method,[8] the conclusion was reached that bis-chloromethyl ether is not formed regularly or in sufficient quantity to be detected at tenths of a part per billion level in the environments studied.

TABLE 2

RONNEL ANALYTIC METHODS IMPROVEMENT

Analytic Method		TCDD Concentration (ppm)
Type	Date Developed	Limit of Detection
GC-MS*	1971	0.1
GC-MS	1975	0.02
GC-MS	1977	0.005
LC	1978	0.001

*GC-MS = gas chromatography-mass spectrometry; LC = Liquid chromatography.

False alarms of environmental hazards lead to wasteful expenditures of time, scientific resources and public funds. Fortunately, the independent analytic methods development and investigation were revealingly successful.

Analytic methods designed to determine if a substance accumulates or concentrates in the environment must be extraordinarily specific as well as sensitive.

The search for 2,3,7,8-tetrachlorodibenzo-*p*-dioxin (TCDD) in environmental samples is probably the most difficult and thorough one ever undertaken. The early work of Baughman and Meselson,[11] was viewed as not only the most sensitive (one part per trillion in favorable cases), but the most specific (said by some to be unequivocal). However, the limit of detection (2.5 times noise) of their method as well as other methods involving mass spectrometry has been found to vary with each sample and several interferences have since been reported.

We have improved this methodology[12] and applied it to beef fat,[13] bovine milk,[14] fish,[15] human milk,[15] and a host of other environmental samples. We have, also, participated in the United States Environmental Protection agencies Dioxin Implementation Plan in which EPA, USDA, EDF, and Dow cooperated in examining beef fat from cattle grazed on rangeland areas sprayed with the 2,4,5-T brand herbicide in Texas, Kansas, Oklahoma, and Missouri. Samples were prepared at EPA's Mississippi Test Facility, and identical portions examined by GC-MS at Wright State University, the University of Utah, and Dow Chemical Co. A selected few of these were also examined at Harvard University. Although the finest equipment and the best protocol were used, and the most able scientists involved, the results were difficult to interpret. The limit of detection varied extensively with the sample and between laboratories. Many results were identical with the limit of detection making it impossible to say whether TCDD *was* or *was not* present. In those cases, all that could be concluded was that TCDD was not present above that level.

A new Dioxin Implementation Plan study is now underway. This study will further define the capability of the analytic methodology by statistically analyzing a series of control samples of beef fat and human milk and the same samples fortified with very small quantities of TCDD. The results of this study should be known soon.

We have cited the above examples to show that:

1. results obtained with the finest equipment and the best protocol can sometimes be misleading;
2. the best scientists always test their conclusions by experiments intentionally designed to prove themselves wrong; and
3. coping with the problems of products, the workplace and the environment require the continuous application of the best analytic science. Less than this produces not inadequate data but incorrect data.

How then can we evaluate work that reports the discovery of hazardous materials in environmental samples? We believe that a minimum set of criteria can be developed for assuring the integrity of the analytic methodology supporting these reports. The concepts have already been developed and used by such groups as the American Society for Testing Materials and The Association of Official Analytical Chemists. However, their action takes too long to aid in effective decision making for individual cases. But the work reported here today as well as that reported in scientific journals, and in newspapers can be evaluated. The following questions on a number of topics can help with the evaluations.

Sampling

Were the samples taken by statistical design to answer a specific scientific question or were a few (that happened to be on hand) examined to see if a sensational discovery could be made? Were adequate control samples obtained? Can contamination and degradation in the sample during storage and analysis be ruled out?

Method Validation

Was the 10-10-10 rule[18] or equivalent applied to determine recovery and make a comparison with controls? Do the recovered concentrations bracket the results reported? Was the calibration or recovery curve extrapolated to lower concentrations to be able to report results? Has the uncertainty of the method been measured or estimated?

Collaboration

Has a second non-allied laboratory used the method with equivalent results?

Limit of Detection

Is the limit of detection defined? Is it considered to be the same as limit of determination? Are indefinite terms such as "trace" reported without numerical values assigned? Are results reported with values below the detection limit?

Proof of Identity

Has identity been corroborated by more than one highly specific analytic technique?

The answers to these questions are critical to the evaluation of any data collected near the limit of detection.

If, when all the results are in and no signals above the limit of detection have been found, we still cannot say that the sample is absolutely free of the contaminant. Professor Frederick D. Rossini, Chairman, Environmental Measurements Advisory Committee, Science Advisory Board, United States Environmental Protection Agency put it best in a recent letter to members of the committee[16]: "If we can't measure it," he said, "we know nothing about it, and, when we do measure it, the resulting figure has associated with it a specified range of uncertainty." To this we can add that as the measurement approaches the level of the limit of detection, the uncertainty increases exponentially so that the uncertainty is more difficult to measure than the desired signal.

Professor Rossini goes on to say: "There is no such thing as absolute zero pollution." We can be assured that all the material (including the halogenated aromatic compounds) ever made in any quantity are present at some level in all environmental samples[17] and we should not be surprised or alarmed when they are found at part per trillion or lower levels. We need to educate both the scientific community and the general public of these facts. The Environmental Measurements Advisory Committee is considering how best to accomplish this.

Until some authoritative scientific body establishes precise criteria for assuring the integrity of analytic methodology near the detection limit, we must ask the questions raised above. Doing so will not only enhance the credibility of the scientific community, but may avoid having scientific questions decided in the courts.

REFERENCES

1. CURRIE, L. A., 1978. Limits for Qualitative Detection and Quantitative Determination. Anal. Chem. **50:** 586.
2. DIOXIN WORKING GROUP, US ENVIRONMENTAL PROTECTION AGENCY. Dioxin: Position Document. Draft, April 2, 1977.
3. SHADOFF, L. A., G. J. KALLOS & J. S. WOODS. 1973. Determination of Bis (Chloromethyl) Ether in Air by Gas Chromatography-Mass Spectrometry. Anal. Chem. **45:** 2341.
4. SOLOMON, R. A. & G. J. KALLOS. 1975. Determination of Chloromethyl Methyl Ether and Bis-Chloromethyl Ether in Air at the Part Per Billion Level by Gas Liquid Chromatography. Anal. Chem. **47:** 955.
5. COLLIER, L. 1972. Determination of Bis-Chloromethyl Ether at the PPB Level in Air Samples by High-Resolutions Mass Spectroscopy. Environ. Sci. Technol. **6:** 930.
6. FRANKEL, L. S., K. S. MCCALLUM & L. COLLIER. 1974. Formation of Bis (Chloromethyl) Ether from Formaldehyde and Hydrogen Chloride. Environ. Sci. Technol. **8:** 356.
7. ROHM & HASS COMPANY, *News Release,* Dec. 27, 1972.
8. KALLOS, G. J. & R. A. SOLOMON. 1973. Investigation of the Formation of Bis-Chloromethyl. Ether in Simulated Hydrogen Chloride-Formaldehyde Atmospheric Environments, Am. Ind. Hyg. Assoc. J.: 469.
9. TOU, J. C. & G. J. KALLOS. 1976. Possible Formation of Bis (Chloromethyl) Ether from the Reactions of Formaldehyde and Chloride Ion. Anal. Chem. **48:** 958.
10. MARCELENO, P. E. & P. J. BIERBAUM. A Preliminary Report on the Formation and Detection of Bis-Chloromethyl Ether in the Industrial and Medical Environment, Industrial Hygiene Association Conference, Miami, FL, May 12–17, 1974.
11. BAUGHMAN, R. & MESSELSON, M. 1973. An Analytical Method for Detecting TCDD (DIOXIN): Levels of TCDD in Samples from Vietnam. Environ. Health Perspect. **5:** 27.
12. SHADOFF, L. A. & R. A. HUMMEL. 1978. The Determination of 2,3,7,8-Tetrachlorodibenzo-*p*-Dioxin in Biological Extracts by Gas Chromatography-Mass Spectrometry. Biomed. Mass Spectrom. **5:** 7.
13. KOCHER, C. W., N. H. MAHLE, R. A. HUMMEL, L. A. SHADOFF & M. E. GETZENDANER. 1978. A Search for the Presence of 2,3,7,8-Tetrachlorodibenzo-*p*-Dioxin in Beef Fat. Bull. Environ. Contam. Toxicol. **19:** 229.
14. MAHLE, N. H., H. S. HIGGINS & M. E. GETZENDANER. 1977. Search for the Presence of 2,3,7,8-Tetrachlorodibenzo-*p*-Dioxin in Bovine Milk. Bull. Environ. Contam. Toxicol. **18:** 123.
15. SHADOFF, L. A., R. A. HUMMEL & L. LAMPARSKI. 1977. A Search for 2,3,7,8-Tetrachlorodibenzo-p-Dioxin (TCDD) in an Environment Exposed Annually to 2,4,5-Trichlorophenoxy Acid Ester (2,4,5-T) Herbicides. Bull. Environ. Contam. Toxicol. **18:** 478.
16. ROSSINI, F. D. 1978. Letter to Environmental Measurements Advisory Committee of the US Environmental Protection Agency, May 31, 1978.
17. DONALDSON, W. T. 1977. Trace Organics in Water. Environ. Sci. Technol. **11:** 348.
18. HARRIS, T. H. & J. G. CUMMINGS. 1964. Enforcement of the Federal Insecticide, Fungicide and Rodenticide Act in the United States. Residue Reviews **6:** 104.

AN EXPANDED APPROACH TO THE STUDY AND MEASUREMENT OF PCBs AND SELECTED PLANAR HALOGENATED AROMATIC ENVIRONMENTAL POLLUTANTS*

D. L. Stalling, J. N. Huckins, J. D. Petty
J. L. Johnson, and H. O. Sanders

United States Department of the Interior
Fish and Wildlife Service
Columbia National Fishery Research Laboratory
Columbia, Missouri 65201

Analysis of environmental polychlorinated biphenyl (PCB) residues is complicated by the presence of a large number of congeners in the various commercial mixtures. Currently, environmental levels of PCBs are often reported in terms of total PCB concentration or Aroclor® content. Marked differences exist in the toxicity of individuals chlorobiphenyls and impurities in technical PCB mixtures.[1,2] To predict adequately the environmental hazards associated with PCB pollution, detailed information is required on the composition of technical PCB mixtures and environmental PCB residues. Assessment of the biologic significance of environmental PCB residues necessitates the establishment of definitive relationships between PCB structure and toxicity. Recent studies by Goldstein *et al*[1] and Poland *et al.*[2] indicate that PCB components lacking *o,o′*-Chlorine (Cl) substitution and having four or more Cl atoms, may account for a significant portion of the toxicologic activity resulting from acute exposures to PCB mixtures.

To date, use of high resolution (HR) wall-coated open tubular (WCOT) glass capillary gas chromatographic (GC) columns has provided the most complete separation of PCB congeners in technical mixtures.[3,4] Earlier, we found that 28M WCOT glass capillary C_{87},[5] Apiezon-L and SE-30 columns failed to resolve PCBs lacking *o,o′*-Cl substitution from several other components present in Aroclors 1248 and 1254.[4] However, application of PX-21 carbon or PX-21 carbon/foam column chromatography[6] provided separations adequate for the quantitation of non-*o,o′*Cl-substituted PCBs and trace planar impurities in technical mixtures.[4] The retention volumes of PCB components on carbon columns generally increased with decreasing *o,o′*-Cl substitution and with *meta* and *para* Cl substitution. Large volumes of toluene were required to elute planar non-*o,o′*-substituted PCBs, chlorinated dibenzofurans, and dioxins from granular (>325 mesh) PX-21 carbon columns.[4] Use of small amounts of PX-21 carbon on foam as an adsorbent for the isolation of these compounds permitted a significant reduction in the solvent required.

We report the carbon separation of Aroclor 1248 into four fractions to determine the toxicity of structurally related PCB components to fish. PCB components in each fraction were characterized using HRGC 28M WCOT glass capillary columns of Apiezon-L, and C_{87} hydrocarbon. Rainbow trout (*Salmo gairdneri*) were exposed to Aroclor 1248 fractions containing 1–4 *o,o′*-Cl for 28 days with no visible effects at

*Reference to trade names does not imply government endorsement of commercial products.

0077–8923/79/0320–0048 $01.75/0 © 1979, NYAS

concentrations equal to or less than 8 μg/1. However, all fish exposed to 0.1–1.3 μg/1 of non *o,o′*-Cl substituted 3,4,3,′,4′-tetrachlorobiphenyl (TCB) exhibited poor growth and impaired equilibrium after 34 days exposure.

EXPERIMENTAL

Reagents

Solvents were nanograde (distilled in glass) from Burdick and Jackson, Muskegon, MI. AMOCO carbon grade PX-21 was obtained from AMOCO Research Corporation, 200 East Randolph Drive, P.O. Box 5910-A, Chicago, IL. Polyurethane plugs (Gaymar Identi Plugs #L800) were obtained from Arthur H. Thomas Company, Philadelphia, PA. Silica gel was CC-7 Silicar 70-230 mesh from Mallinckrodt, P.O. Box 5439, St. Louis, MO. Chlorobiphenyl standards were obtained from RFR Inc., 1 Main Street, Hope, RI; Analabs Inc., 80 Republic Drive, North Haven, CT.; and as a gift of Göran Sundström, National Swedish Environment Protection Board, Wallenberg Laboratory, Stockholm, Sweden.

Apparatus

Apiezon-L and C_{87} WCOT glass capillary columns (0.22 mm × 28 m) were purchased from the Quadrex Corporation, P.O. Box 3881 Amity Station, New Haven, CT. Aroclor 1248 and carbon fractions were characterized using a Varian model 3700 gas chromatograph equipped with small volume pulsed ^{63}Ni electron capture detectors and a capillary injection system. A 2 mm id × 4.6 m glass GC column with 2.5% purified Apiezon-L[7] on Chromasorb W HP (w/w) was employed for the analysis of most fish extracts. The gel permeation chromatograph (GPC) utilized for sample cleanup was an Autoprep 1001, Analytical Biochemistry, P.O. Box 1097, Columbia, MO. A flow-through proportional diluter system as described by McAllister *et al.*[8] was used for rainbow trout exposures.

Method

PX-21 carbon was sieved into batches >325 mesh and <325 mesh. Particles <325 mesh were used in the preparation of carbon/foam adsorbent as described by Huckins *et al.*[6] and carbon particles >325 mesh were used without modification. Chromatographic columns were 1 cm id glass, filled to 10 cm bed depth with carbon or carbon/foam. A stepwise gradient of toluene in cyclohexane (Figures 1–3) was utilized for the fractionation of PCBs and PCB impurities on columns of carbon and carbon/foam. Column flow rate was maintained at 3 ml/min and the recovery of non-*o,o′*-Cl PCBs from Aroclor 1248 was determined using carbon purified Aroclor 1248 spiked with known quantities of non-*o,o′*-Cl PCBs. The sample capacity of a 1.75 g (>325 mesh) PX-21 carbon column for 1-4 *o,o′*-Cl substituted PCB mixtures was 200 mg. Similar sized carbon columns removed >95% of non-*o,o′*-Cl-substituted PCBs from 1 g quantities of Aroclor 1248. Columns of PX-21 carbon (>325 mesh) were used for the fractionation of 1–4 *o,o′*-Cl Aroclor 1248 components for toxicity studies. Structural assignments of PCB congeners present in 0–4 *o,o′*-Cl carbon fractions of Aroclor 1248 (TABLE 1) were based on HRGC comparison with PCB standards and previously published data by Jensen and Sundström[7] and Sisson and Welti.[9]

Rainbow trout fingerlings (0.3–1.0g) reared at the Columbia National Fisheries Research Laboratory were exposed to analytic grade (>95 pure) 3,4,3′,4′-TCB and 1–4 *o,o′*-Cl carbon fractions of Aroclor 1248 in a flow-through diluter system.[8] Exposure times were 28 or 50 days and selected fish were maintained in fresh water for a period of 28 days after cessation of exposure (TABLES 2 & 3). Analytic grade 3,4,3′,4′-TCB was used for the non-*o,o′*-Cl study because isolation of sufficient

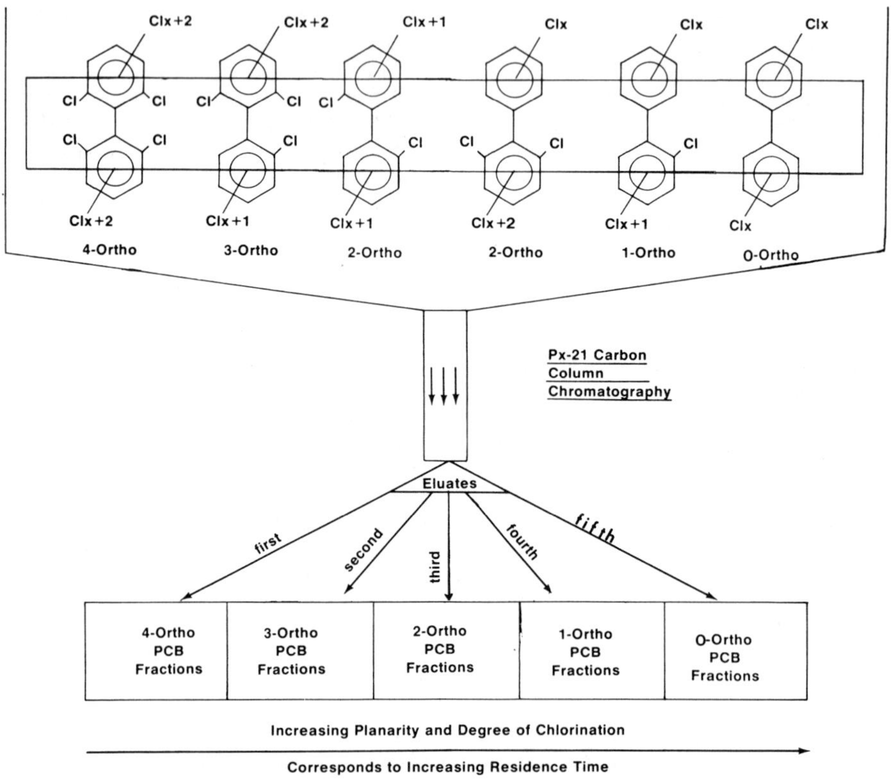

FIGURE 1. Elution order of the six possible *o,o′*-Cl substitution patterns of PCBs from PX-21 carbon or carbon/foam columns.[14] See FIGURE 2 for solvent system.

quantities of 3,4,3′4′-TCB for a flow-through exposure would have required the carbon fractionation of 25 g of Aroclor 1248.

Whole body fish samples (<5 g) were ground with four times their weight of anhydrous sodium sulfate and placed in 1 cm id × 10 cm glass columns fitted with Teflon® stopcocks.[10] PCB residues (neutrals) were extracted with 150 ml of 5% diethyl ether in petroleum ether using a column flow rate of 2–3 ml/min. Sample cleanup was accomplished using automated GPC[11] and Florisil chromatography. The GPC column (2.5 cm id × 30 cm.) was packed with 35 g of Bio Beads SX-3 and eluted with 25%

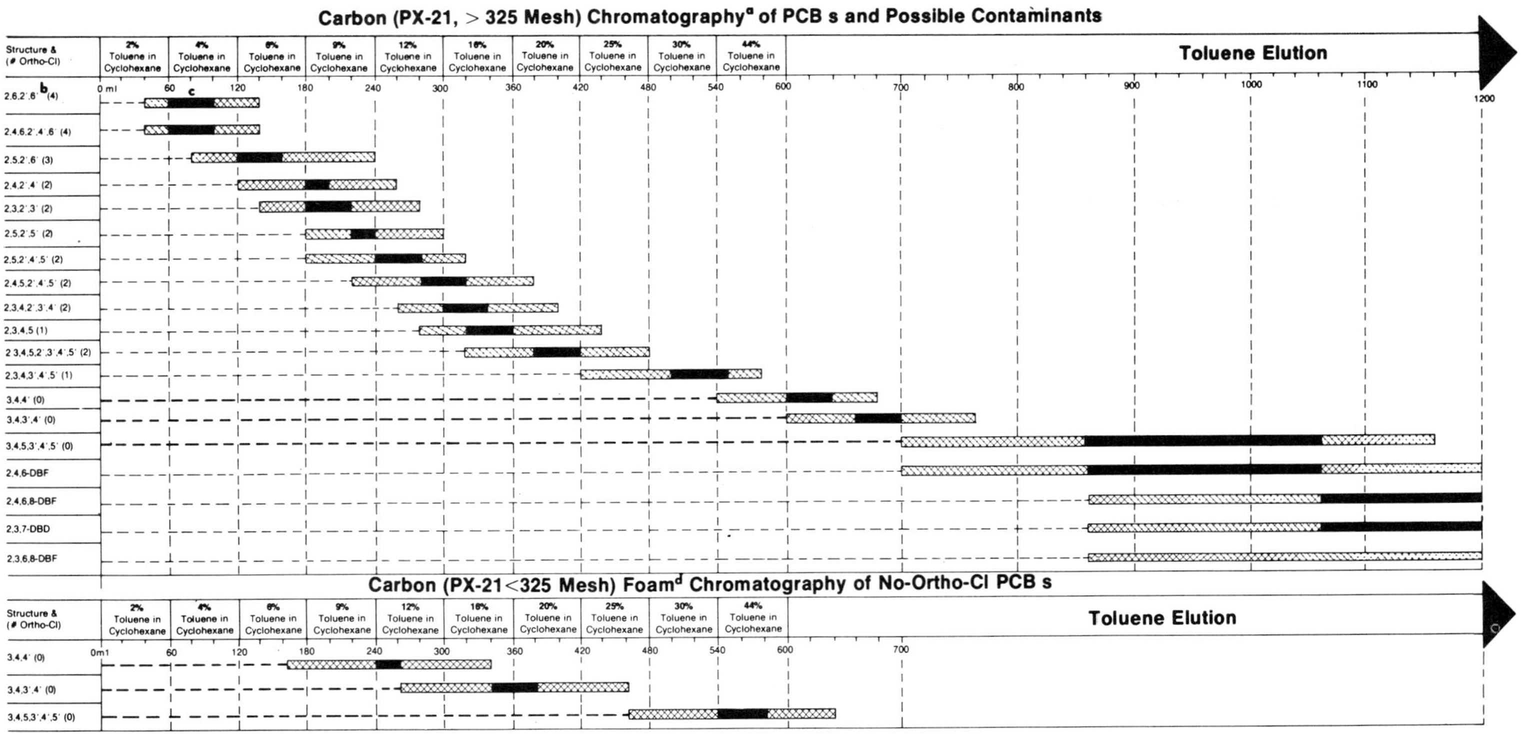

FIGURE 2. PX-21 carbon and carbon/foam chromatography of selected PCBs, chlorinated dibenzofurans (DBF), and a dibenzo-*p*-dioxin (DBD). *a*, Stepwise elution of two 1 cm id × 10 cm carbon (1.75g) columns; *b*, carbon columns were spiked with 20 μg of each PCB and 5 μg of DBF and DBD, *c*, dark portion of bar represents maximum concentration; *d*, stepwise elution of a 1 cm id × 10 cm carbon/foam (1.1g, 15% carbon w/w) column.

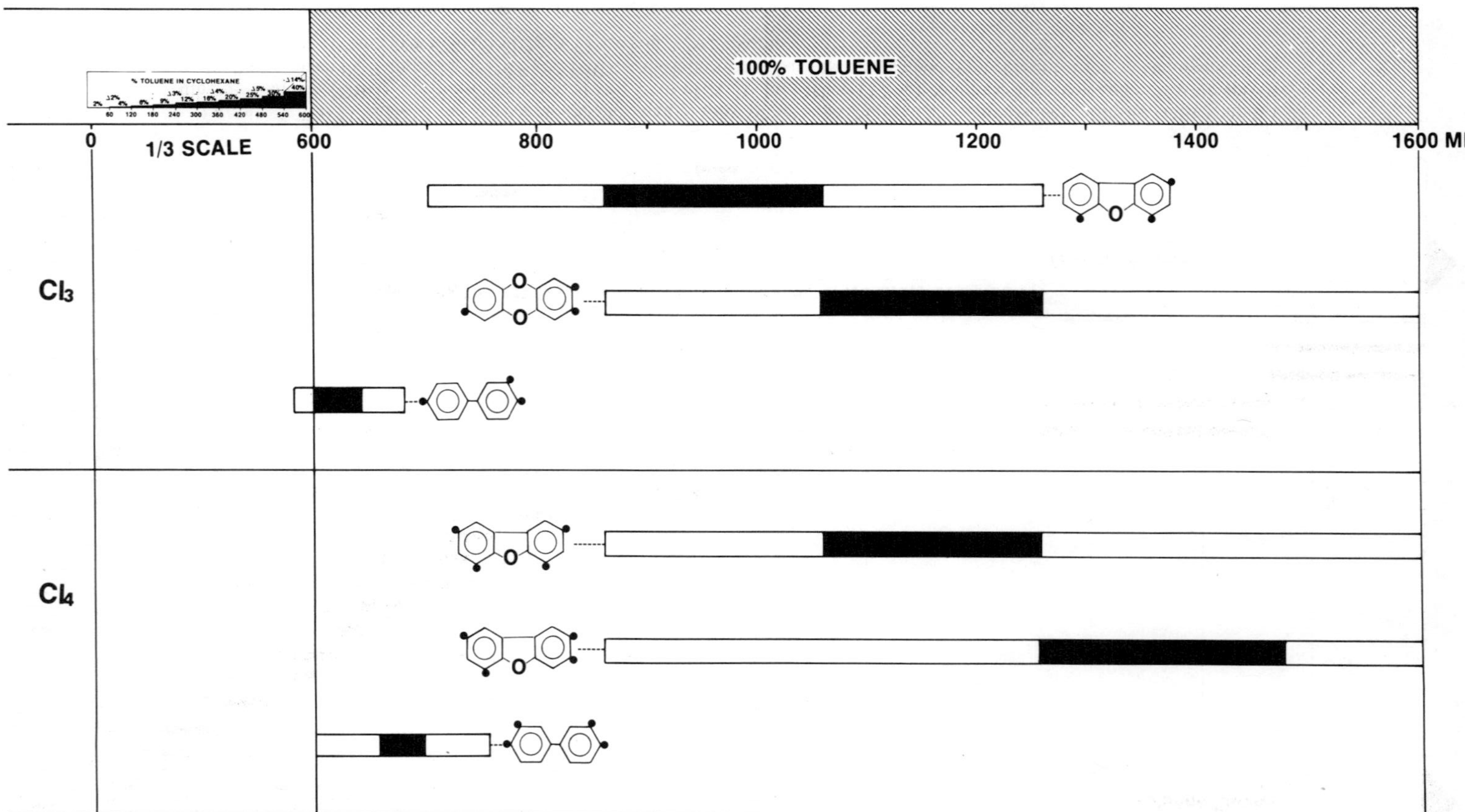

FIGURE 3. Elution profiles of planar PCBs, chlorinated DBFs, and a DBD from PX-21 carbon (>325 mesh) columns (1 cm id × 10 cm).

TABLE 1

MAJOR AROCLOR® 1248 COMPONENTS USED IN FOUR PCB EXPOSURES TO RAINBOW TROUT

40–180 ml* 2–4 *o,o'*-Cl (GC Peak #)	140–320 ml 1&2 *o,o'*-Cl (GC Peak #)	340–520 ml 1 *o,o'*-Cl (GC Peak #)	540–780 ml 0 *o,o'*-Cl (GC Peak #)
2,4' (8̇)†	2,3',5' (15)	3,4,2',5' (8)	3,4,3',4'
2,2',5' (9̇)	4,2'5' (1̇6)	2,3',4',5' ‡	
2,2',4' (1̇0)	3,2',4' (1̇7)	3,2',3',4' ‡	
2,2',3' (11)	2,3',4' (2̇0)	2,4,3',4' ‡	
2,6,2',6'§	2,3,4' (2̇0)	2,3,3',4' ‡	
	2,5,2', 5' (1)		
2,2',4',6' (14)	2,4,2',5' (2)	3,4,2',4',5' (25)	
2,5,2',5' (1)	2,3,2',5' (3)	3,4,2',3',4' (28)	
2,4,2',5' (2)	2,2',3',4' (4)		
2,3,2',5' (3)	2,6,3',4'‡		
2,4,2',4'	4,2',3',6' ¶		
2,2',3',4' (4)	2,3,2',4' ¶		
2,5,2',3',6' (6)	3,4,2',5' (8)		
2,3,2',3',6' (7)	2,5,2',4',5' (12)		
	2,4,2',4',5' (13)		
	2,3,2',4',5' (14)		
	2,5,2',3',4' (15)		
	3,4,2',3',6' (17)		
	2,3,2',3',4' (16)		

*Carbon fraction containing maximum concentration of the 1248 components listed.
†A 1 *o,o'*-Cl PCB; numbers with dots represent different GC peaks, see FIGURE 4.
‡Identities are tentative.
§A 4 *o,o'*-Cl PCB; coelutes with GC peak 11.
¶Two possible structures for a single GC peak.

TABLE 2

WHOLE BODY RESIDUES (μg/g) IN RAINBOW TROUT* EXPOSED TO 3,4,3',4'-TCB AND 1 *o,o'*-Cl 1248 COMPONENTS IN A FLOW-THROUGH DILUTER SYSTEM

Toxicant (# of *o,o'*-Cl)	Mean† H_2O Concentration (μg/l)	Days Exposure 16	29	50	Days Elimination 7	14	28
	0.1			0.8	0.4	0.2	ND‡
3,4,3',4'	0.4			1.9	—	—	—
(0 *o,o'*-Cl)	0.7			5.1	4.7	3.8	1.9
	1.4			12.0	—	—	—
(1 *o,o'*-Cl)§	3.8		82.2		62.2	34.6	26.9
	7.8	60.2¶	—		—	—	—

*Control fish contained <0.1 μg/g total PCB residue and each sample was a composite of 2 to 4 fish.

†Water samples were analyzed weekly.

‡None detected; minimum detection limit of 3,4,3',4' was 5 ng/g.

§ See TABLE 1 for identity of 1248 components in the 1 *o,o'*-Cl carbon fraction.

¶Complete mortality due to a diluter malfunction was observed at 16 days in the 7.8 μg/l concentration.

toluene in ethyl acetate. The flow rate of the column was maintained at 4–5 ml/min. The first 105 ml of eluate from the column was discarded and the 105–200 ml fraction containing the PCBs was collected. Subsequent purification with Florisil was necessary for samples with less than 5 μg/g of PCBs. GPC fractions containing PCBs were concentrated to 5 ml volume in isooctane and chromatographed using a 1 cm id column containing 4 g of activated Florisil. PCBs were eluted with 40 ml of 10% diethyl ether in petroleum ether at a flow rate of 3–5 ml/min. A second column wash of 40 ml of 20% diethyl ether in petroleum ether was also collected to monitor PCB recovery from Florisil columns.

Residues of PCBs accumulated by fish were measured by GC using a electron capture detector. GC conditions are given in apparatus section. The peak heights of the major components in each fraction were summed and the totals of these peak heights were used to prepare calibration curves. These curves were used to quantitate the PCBs accumulated by fish during the exposure periods.

TABLE 3

WHOLE BODY RESIDUES (μg/g) IN RAINBOW TROUT* EXPOSED TO 1 AND 2 *o,o′*-Cl AND 2–4 *o,o′*-Cl 1248 COMPONENTS IN A FLOW-THROUGH DILUTER SYSTEM

Toxicant	Mean† H_2O Concentration	Days of Exposure		Days Elimination		
(# of *o,o′*-Cl)	(μg/l)	10	29	7	14	28
(1&2 *o,o′*-Cl)‡	4.5		36.5	30.0	14.9	—
	19.4		72.1§	27.5	13.2	7.1¶
(2–4 *o,o′*-Cl)‡	3.9		46.5	19.8	17.4	8.2
	16.0		144.7	74.0	42.9	27.0

*Control fish contained <0.1 μg/g total PCB residue and each sample analyzed was a composite of 2 to 4 fish.

†Water samples were analyzed weekly.

‡See TABLE 1 for identity of 1248 components.

§Represents mortalities.

¶One fish sampled.

Results and Discussion

Goldstein[1] and Poland[2] have indicated that adverse physiologic activity associated with technical PCB mixtures may be due to the presence of small quantities of non-*o,o′*-Cl-substituted PCBs. The hazards posed to aquatic organisms by the presence of trace levels of toxic chlorinated dibenzofurans in PCBs[12,13] are still unclear.

The major objective of this study was to determine if differences in PCB toxicity relate to *o,o′*-Cl substitution. Assignment of Aroclor toxicity to the various PCB classes of *o,o′*-Cl substitution necessitated that structurally related subgroups be available for testing. In addition, an effective means was required to minimize any influence of toxic PCB impurities such as chlorinated dibenzofurans. Preparative carbon column chromatography as described in the method section was used to

FIGURE 4. HRGC of 2–4 *o,o′*-Cl, 1 and 2 *o,o′*-Cl, and 1 *o,o′*-Cl carbon fractions. Capillary column was 28m Apiezon-L and column temperature was set at 240°C. See TABLE 1 for identities of numbered peaks.

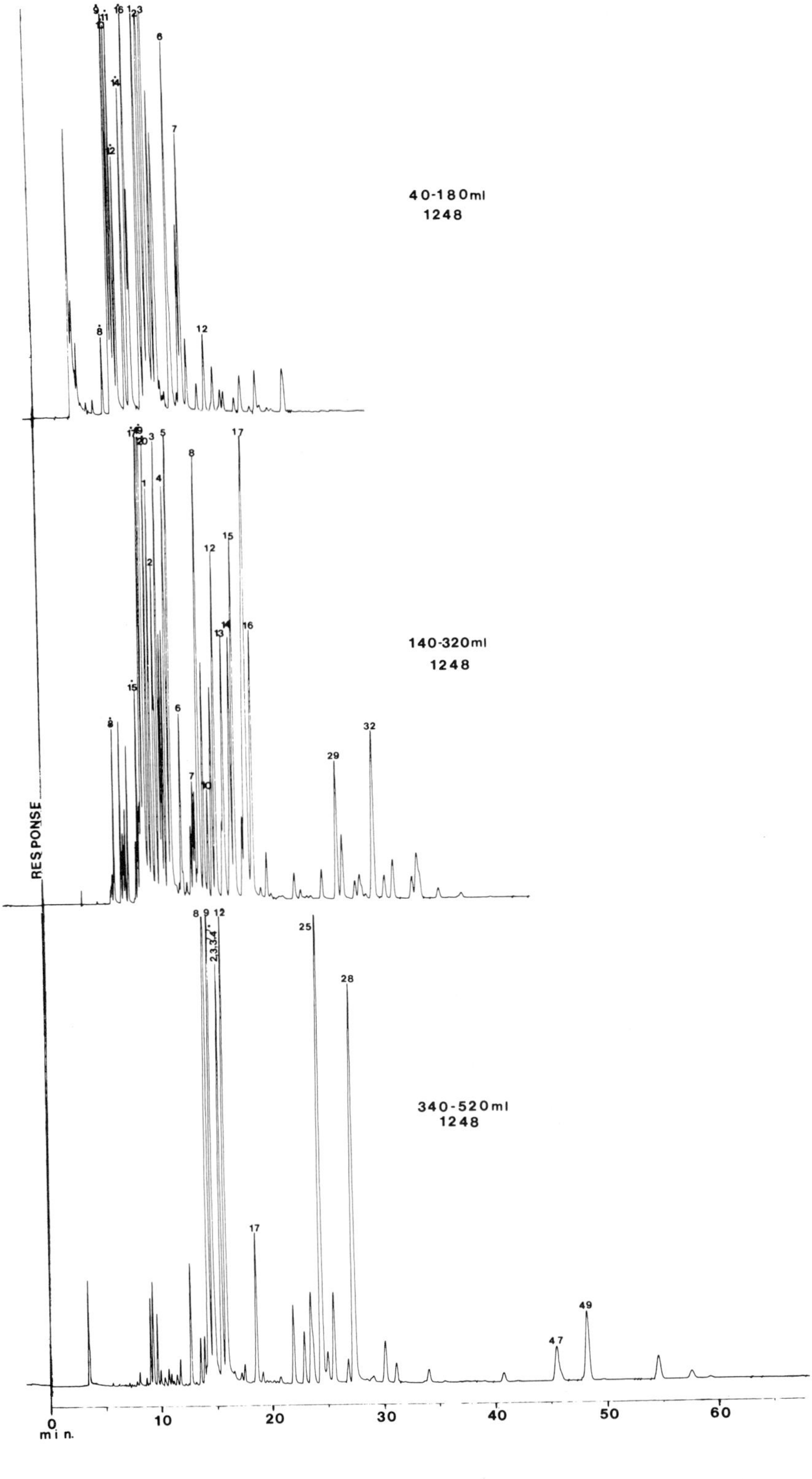
40-180ml
1248
140-320ml
1248
340-520ml
1248
RESPONSE
min.

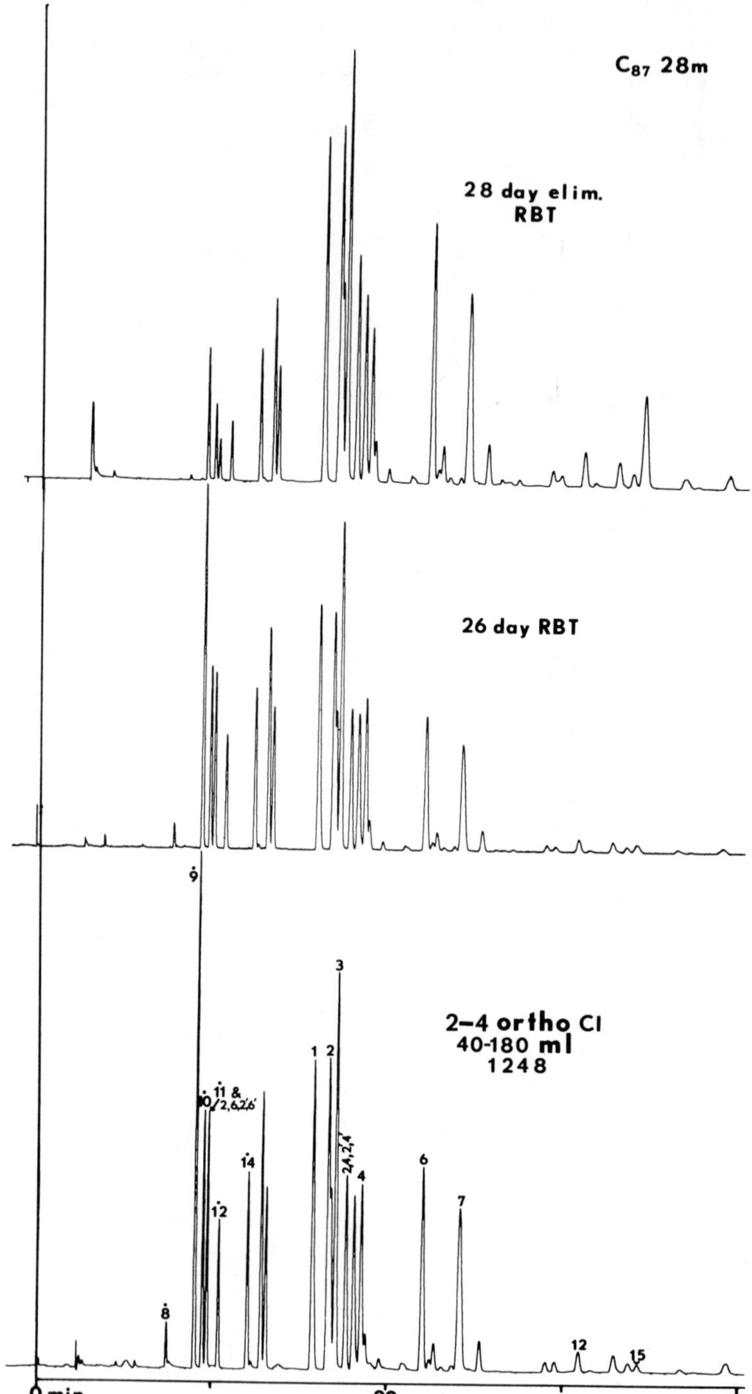
C87 28m
28 day elim.
RBT
26 day RBT
2–4 ortho Cl
40-180 ml
1248
0 min.
20
40

provide 1–4 *o,o'*-Cl fractions for toxicity testing. The 1–4 *o,o'*-Cl PCB fractions were assumed to be free of chlorinated dibenzofurans because of the large volumes of toluene required for the elution of these planar compounds from carbon columns (Figure 2).

We found that PCB components are more strongly retained by PX-21 carbon as: (1) substitution of the *o,o'* position decreases, and (2) total Cl substitution increases.[4] Elution volume of non-*o,o'*-Cl-substituted PCB components examined (3,4,4'-trichlorobiphenyl, 3,4,3',4'-TCB, and 3,4,5,3',4',5'-hexachlorobiphenyl) increased with their degree of chlorination (Figure 2). The presence of a 1 *o,o'*-Cl substituted PCB (2,4'dichlorobiphenyl) in the fraction containing the 2–4 *o,o'*-Cl substituted PCBs illustrates the decreased interaction of carbon with less chlorinated components (TABLE 1).

HR Apiezon-L and C_{87} glass capillary GC was used to characterize 1–4 *o,o'*-Cl fractions (TABLE 1 & FIGURES 4 & 5). PCB components without *o,o'*-Cl substitution and those with 4 *o,o'*-Cl substitution represented <1.0% (w/w) of the PCBs present in Aroclor 1248. Earlier, we quantitated the non-*o,o'*-Cl substituted PCBs present in several Aroclor mixtures.[4] We found that Aroclor 1248 contained the largest amount of non-*o,o'*-Cl-substituted PCBs with 4 or more Cl atoms. The major non-*o,o'*-Cl isomer (3,4,3',4'-TCB) represented 0.22% (w/w) of the PCBs in Aroclor 1248.

We exposed rainbow trout to several concentrations of each Aroclor 1248 fraction (TABLE 1), and analytic grade non-*o,o'*-Cl-substituted 3,4,3'4'-TCB. Residue data from whole body fish samples are summarized in TABLES 2 and 3. Unfortunately, a malfunction of the diluter system used for the 1 *o,o'*-Cl exposure (TABLE 2) resulted in a temporary rise in the water concentration of toxicant and may have significantly increased residues in fish sampled during the experiment. The half-life of total PCB residues in fish exposed to the various concentrations of the four PCB fractions did not differ greatly, and ranged from 7 to 14 days. However, comparison of the PCBs accumulated in fish to the components in the original fractions revealed major differences in the elimination rates of individual PCBs in the 1 and 2 *o,o'*-Cl, and 2-4 *o,o'*-Cl fractions (FIGURE 5). Elimination rates of chlorobiphenyls appeared to relate directly to the degree of chlorination and number and ring position of vicinal unsubstituted carbon atoms.

No major difference in toxicity was observed in rainbow trout exposed to similar concentrations of 1–4 *o,o'*-Cl fractions of Aroclor 1248. Fish exposed to less than 8 μg/l were not visibly affected (TABLE 4). However, exposure to all concentrations (0.1–1.4 μg/l) of the non-*o,o'*-Cl-substituted 3,4,3',4'-TCB resulted in poor growth, reduced food consumption, and disorientation in test fish. Mortalities were not observed until 28 days following cessation of the 1.4 μg/l 3,4,3',4'-TCB exposure. Rainbow trout used for the 3,4,3',4'-TCB study had an average weight of 0.6 g at the start of the dilutor exposure. After 50 days fish maintained in untreated water weighed an average of 3.1 g. However, fish exposed to 0.1 and 1.4 μg/l of 3,4,3',4'-TCB for 50 days weighted an average of 2.2 and 1.2 g respectively. Fish sampled during the 3,4,3',4'-TCB study were pooled for residue analysis and each of the above average weights represented 5 to 14 individuals.

The recent detection of 3,4,3',4'-TCB in a composite fish sample (33% largemouth bass and 67% goldfish, w/w) from the Hudson River[4] underscores the need for

FIGURE 5. Bottom chromatogram, C_{87} HRGC of 2–4 *o,o'*-Cl PCB components from carbon fractionation of Aroclor 1248. *Middle chromatogram,* extract from 26-day exposure of rainbow trout (RBT) to same PCB components. *Top chromatogram,* extract of rainbow trout exposed to same PCBs for 26 days and then maintained in fresh water for 28 days. HRGC conditions similar in all chromatograms; see text.

TABLE 4

INCIDENCE OF ADVERSE EFFECTS (AE) FROM FLOW-THROUGH EXPOSURES OF RAINBOW TROUT* TO AROCLOR® 1248 COMPONENTS

Toxicant (# of *o,o'*-Cl)	Mean H_2O Concentration (μg/l)	Days of Exposure 29	Days of Exposure 50	Days Elimination 28
3,4,3',4' (0 *o,o'*-Cl)	0.1		AE†	AE
	0.4		AE	AE
	1.4		AE	AE (25%m)‡
(1 *o,o'*-Cl)§	0.75	NE¶		NE
	3.8	NE		NE
	7.8	NE		NE
(1&2 *o,o'*-Cl)§	2.0	NE		NE
	4.5	NE		NE
	11.4	AE (16%m)		NE
(2–4 *o,o'*-Cl)§	2.4	NE		NE
	3.9	NE		NE
	16.0	AE (25%m)		NE

*Control fish appeared to be healthy during the studies and no mortalities were observed.

†Adverse effects consisted of poor growth, reduced food consumption and disorientation; these effects were first observed at day 16 in the 0.4 and 1.4 μg/l 3,4,3',4' concentrations.

‡Percent mortalities of beginning population.

§See TABLE 1 for identify of 1248 components in these fractions.

¶No effect observed.

additional research on non-*o,o'*-Cl-substituted PCBs. Routine detection of non-*o,o'*-Cl-substituted PCBs, chlorinated dibenzofurans, and dioxins in environmental extracts containing technical PCB mixtures will require delineation of their behavior through current multiresidue analytic procedures and the use of carbon or carbon/foam chromatography. Carbon chromatography using PX-21 will permit fractionation of chlorinated aromatic mixtures into structurally related subgroups for future structure/toxicity studies. Correlation of environmental residues of non-*o,o'*-Cl-substituted PCBs and planar contaminants, such as chlorinated dibenzofurans and dioxins, to similar concentrations in laboratory studies should provide a better perspective on their environmental impact.

ACKNOWLEDGMENTS

We thank staff members G. Gibson, B. Faerber, and T. Durham for their technical assistance.

REFERENCES

1. GOLDSTEIN, J. A., P. HICKMAN, H. BERGMAN, J. MCKINNEY & M. WALKER. 1977. Chem. Biol. Interact. **17:** 69–87.
2. POLAND, A. & E. GLOVER. 1977. Mol. Pharmacol. **13:** 924–938.
3. KRUPCIK, J., P. A. LECLERCQ, A. SIMOUA, P. SUCHANEK, M. COLLAK & J. HRIVNAK. 1976. J. Chromatogr. **119:** 271–283.
4. STALLING, D. L., J. N. HUCKINS & J. D. PETTY. 1978. *In* Proceeding of the International

Symposium on the Analysis of Hydrocarbons and Halogenated Hydrocarbons in the Aquatic Environment. Plenum Publishing Corp. New York, N.Y. In press.
5. RIEDO, F., D. FRITZ, G. TARJAN & E. KOVATS. 1976. J. Chromatogr. **126:** 63–83.
6. HUCKINS J. N., D. L. STALLING & W. A. SMITH. 1978. J. Assoc. Off. Anal. Chem. **61:** 32–38.
7. JENSEN, S. & G. SUNDSTRÖM. 1974. Ambio **3:** 70–76.
8. MCALLISTER, W. A., JR., W. L. MAUCK & F. L. MAYER, JR. 1972. Trans. Am. Fish. Soc. **101:** 555–557.
9. SISSONS, D. & D. WELTI. 1971. J. Chromatogr. **60:** 15–21.
10. HESSELBERG & J. L. JOHNSON. 1972. Bull. Environ. Contam. Toxicol. **7:** 115–120.
11. TINDLE, R. C. & D. L. STALLING. 1972. Anal. Chem. **44:** 1768–1773.
12. BOWES, G. W., M. J. MULVIHILL, B. R. T. SIMONEIT, A. L. BURLINGAME & R. W. RISEBROUGH. 1975. Nature **256:** 305–307.
13. MORITA, M. J. NAKAGAWA, K. AKIYAMA, S. MIMURA & N. ISONO. 1977. Bull Environ. Contam. Toxicol. **18:** 67–73.
14. STALLING, D. L., L. M. SMITH & J. D. PETTY. 1978. Proceedings of the Symposium on Measurement of Organic Pollutants in Water and Wastewater, American Society for Testing and Materials (ASTM) Committee D–19 on water. To be published as a special technical publication by the ASTM.

SURVEY OF SELECTED ORGANOCHLORINE PESTICIDES IN THE GENERAL POPULATION OF THE UNITED STATES: FISCAL YEARS 1970–1975

F. W. Kutz, S. C. Strassman, and J. F. Sperling

Ecological Monitoring Branch (TS-768)
United States Environmental Protection Agency
Washington, D.C. 20460

In decisions regarding the uses of aldrin, dieldrin, heptachlor, and chlordane, the Administrator of the U.S. Environmental Protection Agency cited and considered the findings of these chemicals in human tissues. Findings of pesticide residues and their metabolites in representative samples from the general population provide a major element in pesticide regulatory decision-making. These residues demonstrate the extent of the environmental distribution of particular pesticides and, when coupled with laboratory animal data showing adverse biologic effects, signal a potential public health hazard.

One program that examines pesticide residues and associated chemicals in human tissues and fluids is the National Human Monitoring Program for Pesticides. This program, operated by the U.S. Environmental Protection Agency, monitors on a national scale, the prevalence and level of exposure to selected pesticides experienced by the general population. These data are used to identify trends in this exposure and to assess the effects of regulatory actions.

Residues of pesticides and their metabolites in various human tissues and body fluids represent the total body burden of these pesticides and reflect past and present exposure to them. Pesticides may gain entrance into the human body through the gastrointestinal tract subsequent to ingestion; through the respiratory tract as a result of inhalation of airborne pesticide-laden dusts, vapors, and aerosols; by penetration through the intact skin and mucus membranes; and (rarely) by absorption directly into the bloodstream through broken skin.

This paper reports findings of selected organochlorine pesticide residues and their metabolites in human adipose tissue collected annually on a national basis from fiscal years 1970 through 1975. During this period, a fiscal year began on July 1 and extended to the end of June of the following year. For example, fiscal year 1970 began on July 1, 1969, and ended on June 30, 1970.

Materials and Methods

Samples of human adipose tissue were obtained in geographic locations selected on the basis of an experimental survey design covering the conterminous 48 states. A proportionate, stratified-random sampling design was followed for selecting cities in which samples were collected.

The sampling design for the surveys conducted during fiscal years 1970, 1971, and 1972 was based on the population distribution in the four census regions as provided by the 1960 Census. For the surveys conducted during fiscal years 1973, 1974, and 1975, the population strata reflected the nine census divisions as defined by the 1970 Census. In each case, the sampling units were cities with populations in excess of 25,000 people. The number of cities sampled in each region or division was determined

0077–8923/79/0320–0060 $01.75/0 © 1979, NYAS

by the proportion of the population residing in that unit. Cities which actually served as collecting sites were selected by randomization (see FIGURES 1 and 2).

This design provided for samples statistically representative of the pesticide residue levels in the general population. For each collection site, an annual sample quota was established and allocated proportional to the demographic distribution found in the Census. Age, sex, and race were the major demographic elements recognized in the human monitoring survey.

Human adipose tissues were collected by cooperating medical pathologists and medical examiners from unembalmed cadavers during postmortem examinations and from specimens previously excised surgically for therapeutic or elective reasons. Thus, tissues were received from patients having pathologic conditions as well as those killed by traumatic injury. Sampling performed during autopsy was accomplished as quickly

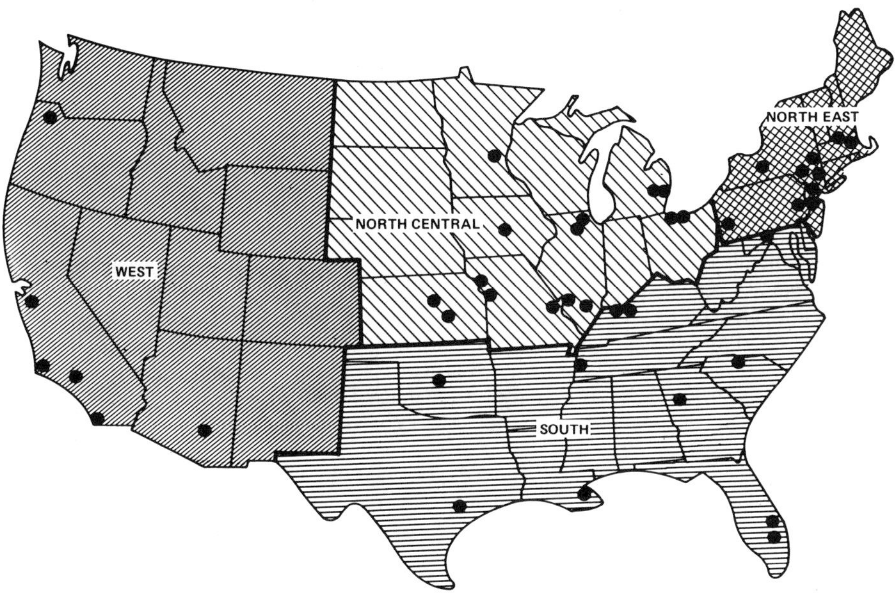

FIGURE 1. Sampling sites for Fiscal Years 1970, 1971, and 1972 human monitoring survey design.

as feasible after death, and in all instances within 24 hours. Then the specimens were placed directly into clear pesticide-free glass bottles having aluminum foil-lined caps. No fixatives or preservatives were used. The samples were frozen immediately and maintained in that state until analyzed. After analysis, any remaining tissues were archived in frozen storage.

Information recorded for each individual sample included age, sex, race, and pathologic diagnosis. Geographic residence was assumed to be in the general vicinity of the hospital. Since the objective of the program was to reflect the pesticide burden in the general population, no samples were collected from patients with known or suspected pesticide poisoning, from cachectic patients, or from patients institutionalized for extended periods. Further details of the program have been presented by Yobs.[1]

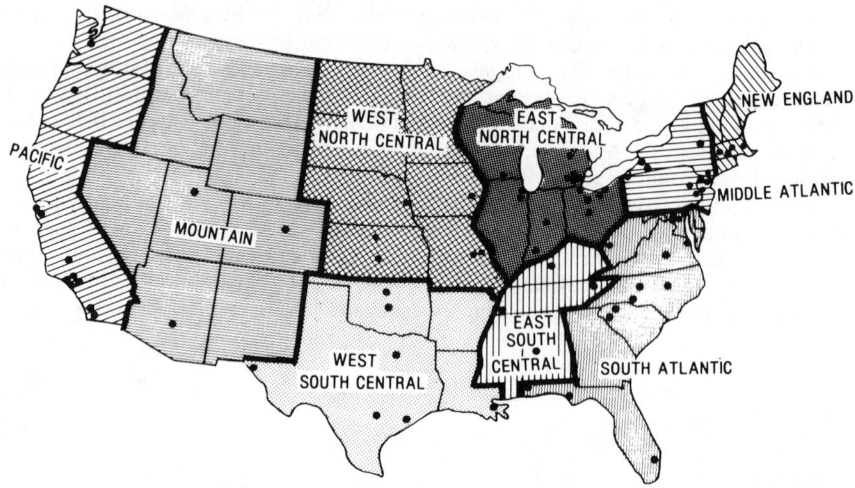

FIGURE 2. Sampling sites for Fiscal Years 1973, 1974, and 1975 human monitoring survey design.

All chemical analyses were conducted by contract laboratories using only methodologies specified by the program. These laboratories were equipped with gas-liquid chromatographs with electron capture detectors. Laboratories were required to maintain acceptable performance levels as evidenced by satisfactory participation in the interlaboratory quality assurance program established and moderated by the EPA Environmental Toxicology Division, Research Triangle Park, N.C. This laboratory also served as technical consultants for the analytic portion of the program.

Samples were analyzed for selected chlorinated hydrocarbon insecticides and polychlorinated biphenyls (TABLE 1) using a modified Mills-Onley-Gaither procedure. Thin layer chromatography, Coulson electrolytic conductivity detectors, microcoulometry, and in some cases, combined gas chromatography-mass spectrometry were employed for confirmation. The exact analytic methods used in this program have been published in manual form by Thompson.[2]

TABLE 1

CHEMICALS DETECTABLE IN HUMAN ADIPOSE TISSUE
(Modified Mills-Onley-Gaither Method)

o,p'-DDT	Aldrin
p,p'-DDT	Endrin
o,p'-DDE	Dieldrin
p,p'-DDE	Heptachlor
o,p'-DDD	Heptachlor Epoxide
p,p'-DDD	Oxychlordane
α-BHC	*trans*-Nonachlor
β-BHC	Mirex
Lindane	Hexachlorobenzene
δ-BHC	Polychlorinated biphenyls

In calculating the results for this report, only quantifiable levels of pesticide residues were considered; reports of trace amounts were regarded as zero. Since these residue data tended not to be distributed normally statistically, the geometric mean was determined to be the most reliable measure of central tendency. All residue levels were calculated on a percent lipid basis. This conversion was computed by dividing the whole tissue (wet-weight) value by the proportion of lipid extractable material for each specimen. This technique reduced the inherent variation in residue levels attributable to differences in lipid content of individual tissue specimens.

Results and Discussions

The results of analyses for selected organochlorine insecticides in adipose tissue follow. Residues of δ-BHC, aldrin, heptachlor, and endrin, although detectable in this analytic approach, were not found in these general population samples.

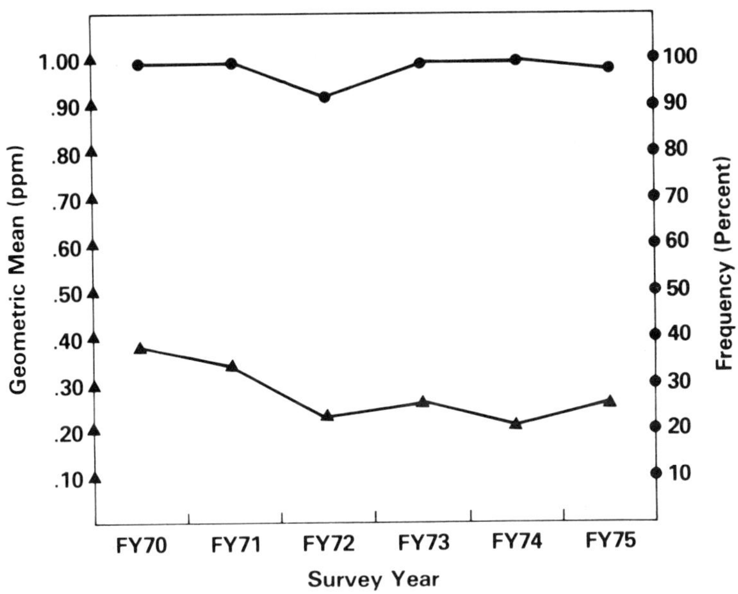

FIGURE 3. Annual frequencies and geometric mean levels of β-benzene hexachloride in human adipose tissue.

β-Benzene Hexachloride (FIGURE 3)

This residue was representative of exposure to the insecticide benzene hexachloride (BHC). These data showed that this chemical was widely distributed at low levels in the human population during all 6 years. Although there seemed to be a trend toward reduction in the quantity of residue, there was no reduction in the frequency of detection, which remained high during all years. It has been hypothesized that certain laboratory animals convert the various isomers of BHC, including lindane, to the *beta*

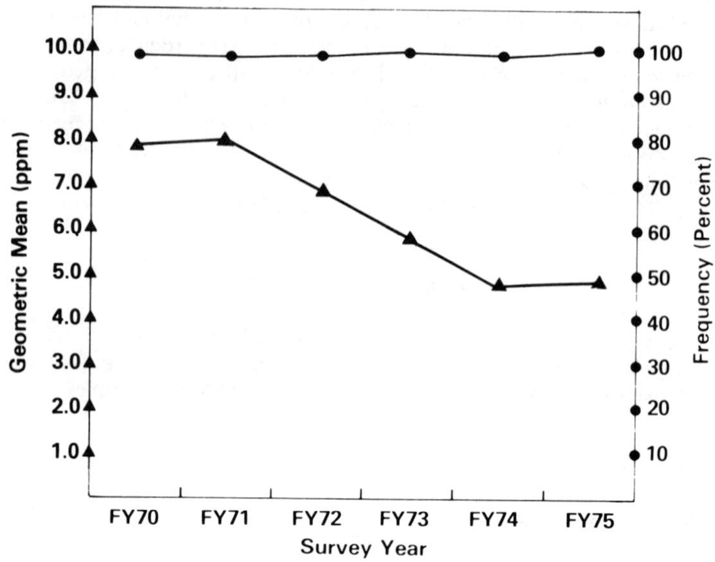

FIGURE 4. Annual frequencies and geometric mean levels of the total DDT equivalent in human adipose tissue.

configuration for storage.[3] If this is true in the human system, then storage of the *beta* isomers may be representative of exposure to all configurations of BHC.

Total DDT Equivalent (FIGURE 4)

Biochemically, DDT is dechlorinated in the human body to DDD, and then either metabolized to the water-soluble and excretable DDA, or excreted directly as DDD. Another DDT analog, DDE, is not appreciably derived from ingested DDT, but rather from ingestion of DDE previously formed from DDT in the environment. Since DDE is not effectively eliminated from the body, the result is a gradual increase in the body burden of this chemical. On the other hand, DDT is broken down and excreted more rapidly than DDE.[4,5]

The total DDT equivalent is a conglomerate figure reflecting the burden of DDT and its analogs found in human tissue.* DDT and certain of its transformation products had the distinction of being stored in greater quantities than any of the other organochlorine pesticides detected in these human adipose tissues. Although the frequency of detecting these residues remained at practically 100% during the 6 survey years, a trend toward reduction in concentration, apparent in the geometric mean levels, was observed. The agricultural uses of this chemical were cancelled by order of the Administrator of the Environmental Protection Agency in late 1972. Further details of this reduction have been presented by Kutz *et al.*[6–8]

*The total DDT equivalent [*o,p′*-DDT + *p,p′*-DDT + 1.114 (*o,p′*-DDE + *p,p′*-DDE + *o,p′*-DDD + *p,p′*-DDD)] is calculated by adjusting the DDE and DDD transformation products of DDT by a molecular weight based constant to convert them to an equivalent of DDT.

Dieldrin (FIGURE 5)

Dieldrin residues were indicative of exposure to the insecticides aldrin and dieldrin. After exposure, aldrin residues are rapidly converted by the human body to dieldrin. Dieldrin residues were found in almost every tissue collected; these low levels remained appreciably unchanged during the 6 survey years. Most of the agricultural uses of both these insecticides were terminated in late 1974.

Heptachlor Epoxide (FIGURE 6)

Residues of this chemical were representative of exposure to two pesticides, heptachlor and chlordane. This chemical was also found at low levels in most of the tissues submitted to the program.

Oxychlordane (FIGURE 7)

Oxychlordane is a mammalian metabolite of the chloro-isomers of the insecticides chlordane and heptachlor. The capability to detect oxychlordane was introduced into this multiresidue analytic approach during the mid-1971 fiscal year survey; the fiscal year 1972 survey was the first complete annual sampling for the detection of this chemical. As with its close relative heptachlor epoxide, oxychlordane residues were found with high frequency at levels of about 0.10 ppm in adipose tissue.

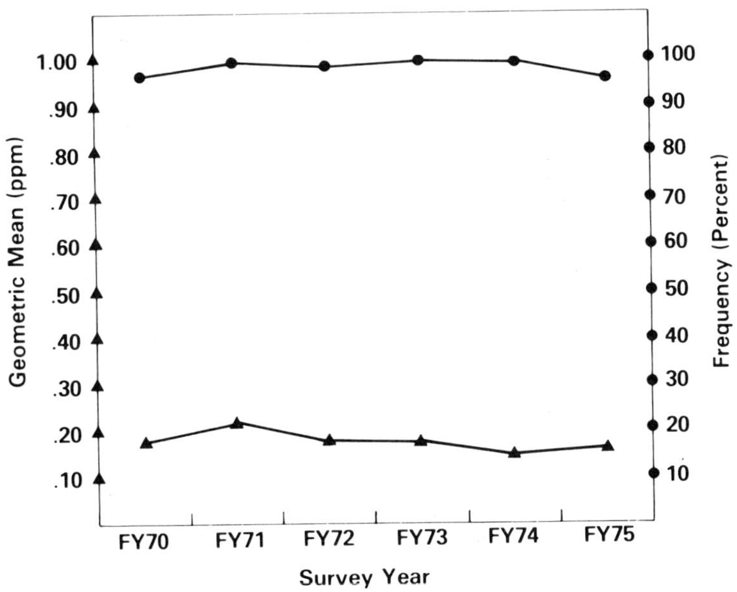

FIGURE 5. Annual frequencies and geometric mean levels of dieldrin in human adipose tissue.

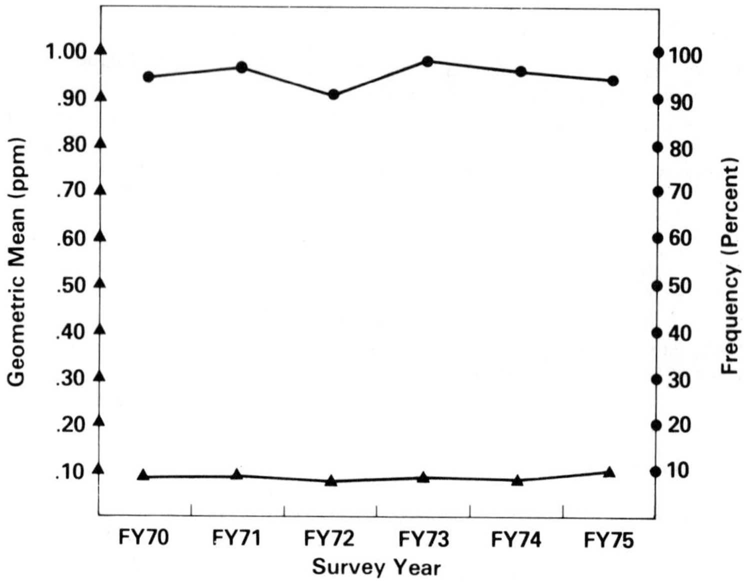

FIGURE 6. Annual frequencies and geometric mean levels of heptachlor epoxide in human adipose tissue.

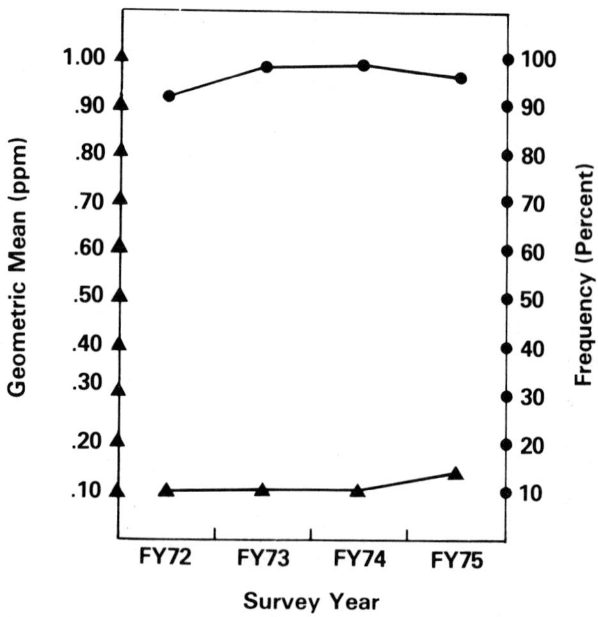

FIGURE 7. Annual frequencies and geometric mean levels of oxychlordane in human adipose tissue.

TABLE 2

ANNUAL FREQUENCIES AND GEOMETRIC MEAN LEVELS OF *TRANS*-NONACHLOR IN HUMAN ADIPOSE TISSUE

Survey Year	Sample Size	Frequency (Percent)	Geometric Mean (ppm)
FY74	47	95.74	0.10
FY75	683	96.78	0.15

trans-*Nonachlor (TABLE 2)*

Another indication of human exposure to technical chlordane and heptachlor was the presence of residues of *trans*-nonachlor. *trans*-Nonachlor is not a metabolite of chlordane or heptachlor, but rather a constituent of the commercially available formulation. The technique to detect this chemical in human adipose tissue was developed in 1974[9] and introduced into the multiresidue method during the analysis of samples collected for both fiscal years 1974 and 1975. Consequently, reduced sample sizes for this residue occurred during this period. Data from the 1975 fiscal year survey indicated that this substance also was found at low levels in almost every tissue analyzed. However, *trans*-nonachlor was present at levels slightly higher than the two metabolites of chlordane and heptachlor.

Hexachlorobenzene (TABLE 3)

Hexachlorobenzene is a fungicide used to protect wheat seed, and also may be a contaminant of some other pesticides as well as an industrial pollutant from chlorinated benzene manufacture. Since the analytic methodology to detect hexachlorobenzene was introduced into the existing multiresidue technique during the analysis of fiscal year 1973 samples, data is presented for the last 3 survey years. These data demonstrated that this was another residue ubiquitously distributed in the human population.

Other Residues

Residues of α-benzene hexachloride, lindane (γ-benzene hexachloride) and mirex have also been detected in human adipose tissue from the general population. These chemicals were found at very low frequencies during most survey years. Additionally,

TABLE 3

ANNUAL FREQUENCIES AND GEOMETRIC MEAN LEVELS OF HEXACHLOROBENZENE IN HUMAN ADIPOSE TISSUE

Survey Year	Sample Size	Frequency (Percent)	Geometric Mean (ppm)
FY73	1095	70.9	0.02
FY74	898	93.9	0.04
FY75	779	91.5	0.05

analytic results have indicated that certain polychlorobiphenyls are commonly detected in human adipose tissue.

The National Human Monitoring Program is a continuing function of the U.S. Environmental Protection Agency. As refinements and new methodologies for the analyses of human tissues and fluids have permitted, the human monitoring program has expanded to include additional pesticides and toxic substances. Reports of pesticide residues in human milk and urine have been previously published.[10-12]

References

1. Yobs, A. R. 1971. The national human monitoring program for pesticides. Pest. Monit. J. **5**(1): 44–46.
2. Thompson, J. F., ed. 1977. Manual of Analytical Methods for the Analysis of Pesticide Residues in Human and Environmental Samples. Prepared by the Environmental Protection Agency, Environmental Toxicology Division, Health Effects Research Laboratory, Research Triangle Park, N.C.
3. Kamasa, T. 1971. Hygenic studies of pesticide residues. Report No. 1. Accumulation of the BHC isomers (α, β, γ and δ) in the body of the rat and urinary excretion following oral administration. Japan. Hyg. **26:** 358–364 (in Japanese).
4. Morgan, D. P. & C. C. Roan. 1971. Absorption, storage, and metabolic conversion of ingested DDT and DDT metabolites in man. Arch. Environ. Health **22**(3): 301–308.
5. Roan, C. C., D. P. Morgan & E. H. Pashal. 1971. Urinary excretion of DDA following ingestion of DDT and DDT metabolites in man. Arch. Environ. Health **22**(3): 309–315.
6. Kutz, F. W., A. R. Yobs, S. C. Strassman & J. F. Viar, Jr. 1977. Effects of reducing DDT usage on total DDT storage in humans. Pest. Monit. J. **11**(2): 61–63.
7. Kutz, F. W., A. R. Yobs & S. C. Strassman. 1977. Racial stratification of organochlorine insecticide residues in human adipose tissue. J. Occup. Med. **19**(9): 619–622.
8. Kutz, F. W., S. C. Strassman & A. R. Yobs. 1977. Survey of pesticide residues and their metabolites in humans. *In* Pesticide Management and Insecticide Resistance. D. L. Watson & A. W. A. Brown, Eds. pp. 523–539. Academic Press. New York.
9. Kutz, F. W., G. W. Sovocool, S. C. Strassman & R. G. Lewis. 1976. *trans*-Nonachlor residues in human adipose tissue. Bull. Environ. Contam. Toxicol. **16**(1): 9–14.
10. Strassman, S. C. & F. W. Kutz. 1977. Insecticide residues in human milk from Arkansas and Mississippi, 1973–74. Pest. Monit. J. **10**(4): 130–133.
11. Kutz, F. W. & S. C. Strassman. 1977. Human urinary metabolites of oragnophosphate insecticides following mosquito adulticiding. Mosquito News **37**(2): 211–218.
12. Kutz, F. W., R. S. Murphy & S. C. Strassman. 1978. Survey of pesticide residues and their metabolites in urine from the general population. *In* Pentachlorophenol. K. R. Rao, Ed. pp. 363–369. Plenum Pub. Corp. New York.

WORLD PCBs MAP: STORAGE AND EFFECTS IN MAN AND HIS BIOLOGIC ENVIRONMENT IN THE 1970s

Marcus Wassermann, Dora Wassermann, Similica Cucos, and Howard J. Miller

Department of Occupational Health
Hebrew University-Hadassah Medical School
Jerusalem, Israel

Early in the 1960s L. Goldwater planned and conducted a study on "normal" values for toxic substances in humans.[1] Under the aegis of the World Health Organization and Columbia University, this study investigated populations of 15 areas of the world and provided the first map of mercury, lead, and arsenic levels in human blood and urine. In 1965 when this study's panel of investigators was convened in Geneva, I suggested that a world-wide assessment of DDT storage in humans be done. I felt motivated to do this because intensive studies had already been conceived and undertaken by Wayland J. Hayes, Jr., under the auspices of the Communicable Disease Center of Atlanta, and by other investigators of North America and Europe, to explore this new and interesting topic. Because of the kind and valuable support of the Communicable Disease Center, Atlanta, and of Professor Hayes, the Department of Occupational Health of the Hebrew University of Jerusalem was enabled to explore the storage of DDT in humans of Israel. Later, with the kind support of the World Health Organization and particularly of J. W. Wright, Head, Vector Biology and Control, and the International Agency for Research on Cancer, Lyon, we studied the same topic in populations of Australia, Sri Lanka, Thailand, Nigeria, Kenya, Uganda, South Africa, Argentina, Brazil, and Spain. In this context we stated in 1966, that "DDT is a current constituent of the human body."[2] Eight years later we presented, in collaboration with L. Tomatis of the International Agency for Research on Cancer, Lyon, a world map of the storage of organochlorine insecticides in humans and described the natural history of this phenomenon.[3]

During the same period as we were reporting that DDT was a current constituent of the human body, S. Jensen of Karolinska University in Stockholm was opening a new pathway into the field of environmental toxicology by means of identifying the family of PCBs as pollutants of the environment and constituents of animals.[4] This discovery stimulated further studies aimed at assessing the level of PCBs in the environment and living beings, and the biologic effects of these already ubiquitous pollutants.

This report uses the scientific data available to us in an attempt to describe the epidemiologic features of the PCBs stored in man and his biologic environment and some of the effects of PCBs.

PCBs first found their use in industry in 1929. Since then they have been progressively manufactured and marketed. Except for sporadic reports in the 1930s and 1940s of the toxic effects exerted by these compounds on people occupationally exposed, they were not generally regarded as chemicals of high toxicity or as potential environmental contaminants. The large-scale contamination by PCBs of cooking oil in Japan (1968), as well as other toxic episodes, such as occurred in 1968 when 400,000 chickens died in Japan because of contaminated chicken feed[5] and in North Carolina where 75,000 eggs were destroyed (the fish-meal chicken-feed was shown to contain 400 ppm PCBs),[6] led to the restriction in the use of these compounds. They are now

0077-8923/79/0320-0069 $01.75/0 © 1979, NYAS

used mainly for closed electrical apparatuses, and consequently, beginning with 1971, their production has decreased.

At some time or other in the last 50 years, PCBs have become current constituents of the human body and the biologic environment. This condition is illustrated by the following data:

PCB Residues in Marine Organisms (Table 1)

PCBs were found in nearly all plants and animals from the Atlantic Ocean.[7] The lowest PCB residue levels in plankton, 0.03 ppm, were reported in the neighborhood of Scotland[8] and a relatively high level, 1.5 ppm, in the Atlantic.[7]

Atlantic hagfish collected in the Gullmar fjord had 5.0 ppm PCB liver residue levels, whereas the open-sea hagfish had only 0.2 ppm liver PCB residues.[9] Mussels from polluted areas had higher PCB residue levels, 0.017–1.70 ppm, as compared to those of unpolluted areas, 0.01–0.67 ppm.[10] Pink shrimp from the Oregon coast had very low PCB residue levels, 0.008 ppm,[11] as compared to shrimp from Escambia Bay (FL), accidentally polluted with PCBs, who contained up to 7.0 ppm.[12] Mussels from the Mediterranean coast of Spain had higher PCB residue levels than those of the Atlantic coast.[13]

In larger marine animals (seals, whales, sharks) PCB residue levels varied from several ppb to tens and hundreds of ppm. Grey seals from the Gulf of St. Lawrence (Canada) had a mean blubber PCB level of 14.5 ppm.[14] Seals from English waters had 212 ppm PCBs in the blubber.[15] Whales and dolphins from Canada had 0.7–147.0 ppm PCB blubber residues,[16] whereas in the Lesser Antilles these animals had only 2.7–5.0 ppm blubber PCB residues.[17]

PCB residues in male marine mammals increased with age. Female seals showed no increase of PCB residues with age.[14] Parturition and lactation may be the cause of this difference between male and female,[18] since during pregnancy the PCBs body burden is lowered by placental transfer and during lactation it is lowered through excretion in milk.

The present residue levels in marine organisms are high enough to produce toxic effects *quo ad vitam* and *quo ad futurum* and justify the efforts to reduce the contamination of oceans and seas by PCBs.[16]

PCB Residues in Fish (Table 2)

Fish constitute an important link in the food chain.

Fresh-water fish, especially fish caught in industrial countries, in rivers and their estuaries where industrial waste is disposed of, have higher PCB residue levels than those caught in nonindustrialized areas. Fish from the rivers of the eastern USA had up to 140 ppm PCBs in their flesh.[30]

Lake fish showed intermediary values for PCB residues, between those of fresh-water fish and deep-sea fish: Lake Erie fish contained up to 4.40 ppm,[31] Lake Ontario fish, 5.15 ppm,[32] Lake Michigan, 2.7–15.0 ppm,[33] Lake Superior, 1.0–5.0 ppm,[20] and Lake Superior trout 7.0 ppm.[34]

Sardines from the Mediterranean coast of Spain had 0.68–4.72 ppm PCBs, while those from the Atlantic coast had only 0.34–0.40 ppm.[13] Sardines, from the Adriatic Sea, had 0.37–1.06 ppm PCBs,[35] while sardines from the Bay of Marseilles had 0.7–5.1 ppm PCBs[36] and those from Tokyo Bay 1.0–11.0 ppm.[37]

TABLE 1

PCB RESIDUES IN MARINE ORGANISMS

Ref.	Authors	Year	Country	Species	Tissue	Mean	Range
7	Anonymous	1972	? Atlantic Ocean	Zooplankton		1.50	
19	Risebrough *et al.*	1972	USA, Atlantic Ocean	Zooplankton	Lipid extract	40.00	(median)
20	Veith *et al.*	1977	USA, W. Lake Superior	Zooplankton		0.10	
21	Giam *et al.*	1973	Mexico, Gulf of Mexico & N. Caribbean	Zooplankton			<0.003–1.055
8	Holden,	1970	Scotland	Zooplankton		0.03	
22	Linko *et al.*	1974	Finland, Turku archipelago		Lipid extract	25.00	
23	Ware & Addison	1973	Gulf of St. Lawrence	Plankton		0.19	0.090–3.00
9	Falkmer *et al.*	1977	Sweden, Gullmar fjord	Atlantic hagfish	Liver	5.00	
		1977		Open sea hagfish	Liver	0.20	
10	Holden	1973	Scotland ('69–'71)	Mussels from unpolluted and polluted areas			0.01–0.067
							0.017–1.70
13	Baluja *et al.*	1973	Spain				0.58–1.36
11	Claeys *et al.*	1975	USA, Oregon coast estuaries	Bivalve mollusks		<0.003	
			Oregon-Washington coast	Pink shrimp		0.008	
			Oregon-Washington coast	Euphausiids		0.025	
12	Duke *et al.*	1970	USA, Escambia Bay, FL.	Shrimp and crab			up to 7.00
24	Bowes *et al.*	1973	USA, San Miguel Island, CA.	Sea lions	Blubber	62.00	
					Liver	3.00	
25	Buhler *et al.*	1975	USA, The central Oregon coast	Sea lions	Blubber		21.20–34.10
18	Addison *et al.*	1973	Canada, Gulf St. Lawrence	Harp seals			3.0–22.0
14	Addison & Brodie	1977	Canada, Gulf St. Lawrence	Grey seals	Blubber lipid	14.50	
					Milk lipid	60.00	
15	Holden	1972	England, The Baltic, North & Irish Seas	Seals	Blubber	212.00	
26	Helle *et al.*	1976	Finland	Ringed seals—pregnant	Blubber	56.00	
				Ringed nonpregnant	Blubber	77.00	
27	Saschenbrecker	1973	Canada, Atlantic coast	N. Am. Fin-back whales	Blubber		0.012–0.185

TABLE 1 (*continued*)
PCB RESIDUES IN MARINE ORGANISMS

Ref.	Authors	Year	Country	Species	Tissue	Mean	Range
17	Gaskin *et al.*	1974	Lesser Antilles	Cetaceans: Short-finned pilot whales and long snouted dolphin	Blubber		2.69–5.00
16	Taruski *et al.*	1975	Canada	Cetaceans and dolphins	Blubber		0.70–147.00
28	Zitko & Hutzinger	1972	Canada & USA, Gulf of Maine, Bay of Fundy	Basking shark	Muscle	0.07	
				White shark	Liver	218.00	
29	Jensen *et al.*	1969	Sweden	11 marine species			0.019–15.0

TABLE 2
PCB RESIDUES IN FISH (ppm)

Ref.	Author	Year	Country	Species	Mean	Range
40	Anonymous	1971	USA, Alabama	Fish		up to 360.0
31	Carr	1972	USA, The Ohio shore of Lake Erie	Different species of fish		0.08–4.40
30	Buhler & Claeys	1972	USA, Eastern rivers	Different species of fish		up to 140.0
39	Bache *et al.*	1972	USA, Caynga lake, Ithaca	A 12-year-old trout	26.0	
41	Smith *et al.*	1974	USA, Utah	Different species of fish		0.05–0.97
32	Haile *et al.*	1975	USA, Lake Ontario	Different species of fish	5.15	
11	Claeys *et al.*	1975	USA, Columbia river, Rogue river	Selected fish species	0.480	
				Summer-run steel head	0.125	
34	Parejko *et al.*	1975	USA, Lake Superior ('71)	Lake trout	7.00	
33	Veith	1975	USA, Lake Michigan ('71)	13 Species-rainbow smelt	2.7	lowest average
				-Lake trout	15.0	highest average

20	Veith *et al.*	1977	USA, West Lake superior ('72–'73)	Fish	0.1	
				Larger fish	5.0	
42	Baird *et al.*	1975	Mexico, Gulf	17 Species of mesopelagic fishes		up to 0.02
28	Zitko & Hutzinger	1972	Canada, Nova Scotia & New Brunswick	Marine fish		0.2–0.5
			USA	Fresh-water fish		2.1–10.0
43	Zitko	1972	Canada, Nova Scotia & New Brunswick	American eel, herring, Atlantic Salmon		0.3–0.7
44	Sims *et al.*	1975	Canada, Atlantic coast ('71)	Cod liver		3.4–23.3
37	Doguchi *et al.*	1971	Japan, Tokyo Bay	Sardines		1.0–10.0
45	Bjerk & Sundby	1970	Norway	Dogfish	0.03	
				Pike	0.04	
46	Bjerk	1972	Norway, Coastal waters	Sprat		up to 3.4
			Oslo harbor	Herring		up to 16.0 (fat basis)
38	Hauge	1972	Norway	Sea fishes		0.01–3.0
47	Linko *et al.*	1974	Finland, the Turku archipelago	Baltic herring	0.31	
				Pike	0.05	
				Pike (Ovaries)	190.0	
48	Jensen *et al.*	1973	Finland	Herring	16.0	1.8–110.0 (fat basis)
8	Holden	1970	Scotland	Clupeoid fish	2.60	
				Fresh water fish	0.00	
36	Monod *et al.*	1971	France, Bay of Marseilles	Sardines		0.7–5.1
49	Eichner	1973	Germany, Lake Constance,	White fish perch		0.03–1.2
			Tribut. of the Rhine	Roak, trout		0.05–3.9
50	Huschenbeth	1973	Germany, North Sea	Lean fish	0.2	
				Fat fish	1.3	
35	Viviani *et al.*	1971	Adriatic Sea	Anchovies		0.18–0.96
				Sardines		0.37–1.06
				Sprats		0.62–0.92
13	Baluja *et al.*	1973	Spain, Mediterranean coast	Sardines		0.34–0.40
			Atlantic	Sardines		0.34–0.40

Deep-sea fish have relatively low PCB residue levels; 0.2–0.5 ppm PCBs were found in marine fish in Canada[28] and 0.1–3.0 ppm in fish from Norwegian seas.[38]

PCBs residue values in fish varied with the species. Different species of fish caught off the Ohio shore of Lake Erie (yellow perch, coho salmon, carp, channel catfish, fresh water drum, and white brass) showed a range with average PCB residues of 0.08 to 4.40 ppm. Significantly higher levels were found in channel catfish.[31] From 13 species of fish sampled from Lake Michigan, the rainbow smelts had the lowest average of 2.7 ppm and the lake trout the highest, 15.0 ppm.[33]

PCB residue levels were higher in larger fish. Most trout and salmon from Lake Michigan longer than 12 inches, contained PCB levels higher than 5.0 ppm, which is the tolerance level established by FDA.[33] Larger fish from Lake Superior had a residue level of PCBs of 5.0 ppm, in comparison to smaller fish whose PCB residue level was 1.0 ppm.[20] A 12-year-old trout was reported with 26.0 ppm PCBs.[39]

PCB Residues in Birds (Tables 3 and 4)

The avian species accumulated the highest PCB residues. They constitute the top of the most affected and therefore the most studied food chain.

PCB residue levels in birds are very high in those regions of industrialized countries where industrial wastes are disposed of via fresh waters and coastal waters.

PCB residues in the adipose tissue of birds vary from several ppm to hundreds and even thousands of ppm. Amongst the birds of the USA, White-tailed eagles were reported to have a PCB residue level of 14,000.0 ppm, peregrine falcons 2,000.0 ppm[51] and herons 900.0 ppm.[52] Common murres, found dead, had 4,000.0 ppm in the brain.[53] Chacalacas fed on unsprayed native fruits had comparatively very low adipose-tissue, PCB-residue levels of 0.17 ppm.[54]

English fresh-water fish-feeding birds had liver PCB residue levels of up to 900.0 ppm,[55] and gannets of the eastern and western coasts of England had a liver PCBs residue range of 4,720.0–9,590.0 ppm,[56] while terrestrial predatory birds had liver PCB residues in the range of only 1.3–272.0 ppm.[57] High levels of PCBs in the birds of Germany, 24.0–740.0 ppm,[58] and much lower levels, 5.0–6.2 ppm, in terrestrial and aquatic avian species of Spain[59] were reported.

Adult merlins were shown to have a residue level of 196.0 ppm as against immature merlins with only 28.6 ppm.[60]

Guillemots found dead had higher liver residue levels, 50.0 ppm, than shot guillemots, 0.5 ppm.[61] Experimentally in cormorants, death occurred at a total body burden similar to that found in cormorants found dead naturally.[62]

Reduction of avian populations, as an effect of PCBs, occurs because of the excessively high residue levels in birds and their eggs, as a result of neurologic changes induced by PCBs (changes that reduce the defensive fitness, the birds becoming easy prey for predators), of immunologic changes that render the birds more susceptible to infectious diseases, and because of a reduction of reproduction success. Some of these negative effects of PCBs may be potentiated by one or another of the xenobiotics that abound in the current environment.

PCB Residues in Avian Eggs (Table 5)

The eggs of fish-eating birds have the highest residue levels of PCBs. The average for PCB residues in egg-extracted lipids of 13 species of fish-eating birds of the Upper Great Lake States was 339.7 ppm. Partial averages for the PCB residues of the eggs of

TABLE 3

PCB RESIDUES IN BIRDS OF NORTH AMERICA (ppm)

Ref.	Author	Year	Country	Species	Tissue	Mean	Range
51	Gustafson	1970	USA	Peregrine falcons		2000.00	
				White-tailed eagles		14000.00	
52	Anonymous	1970	?	Herons		900.00	
60	Risebrough *et al.*	1970	USA, Western shore of Lake Michigan ('68–'69)	Young peregrine falcons	Adipose	52.50	
				Adult merlins		196.00	
				Immature merlins		28.60	
63	Anonymous	1971	USA, Long Island Sound	Terns	Breast muscle		5.00–175.00
64	Hays & Risebrough	1972	USA, Great Gull Island, NY ('69–'70)	Abnormal young terns	Breast muscle	25.00	
				Fish brought by adult birds		1.00	
65	Belisle *et al.*	1972	USA	Bald eagles (one)	Brain	235.0	
66	Mclane *et al.*	1973	USA, (11 states)	Woodcock wings	Whole bird extr. lipid		4.27–8.63
67	Greichus *et al.*	1973	USA, S. Dakota (Lake Poinset)	Pelicans	Adipose	22.40	
68	Fisher	1973	USA, N. Pacific ('69)	Birds	Adipose	15.30	
69	Dacre	1974	USA	Mutton birds of New Zealand	Adipose	1.83	
70	Kreitzer	1974	USA, several states ('70–'71)	Mourning dove	Breast muscle	0.12	
53	Scott *et al.*	1975	USA, Oregon coast ('69)	Common murres	Brain	4000.00	dead birds
			('69)			3700.00	during
			('70)			1100.00	die-off
71	Cromartie *et al.*	1975	USA, (18 states) ('71)	Bald eagles	Carcasses		0.30–290.00
					Brain		0.10–150.00
			('72)		Carcasses		0.60–1200.00
					Brain		0.65–190.00
54	Marion	1976	USA, South Texas, Rio Grande	Chacalacas	Adipose	0.17	
72	Zitko *et al.*	1972	Canada, Bay of Fundy	Herring gulls	Adipose	1.40	Aroclor 5460
28	Zitko and Hutzinger	1972	Canada and USA	Double crested cormorant	Breast muscle	3.38	

TABLE 4

PCB RESIDUES IN BIRDS OF EUROPE (ppm)

Ref.	Authors	Year	Country	Species	Tissue	Mean, range
45	Bjerk and Sundby	1970	Norway	Starling	Breast muscle	0.014
38	Hauge	1972	Norway	Carnivorous birds	Breast muscle	3.00
73	Odsjö	1973	Sweden, S.E. Coast	Eagle owls (found dead)	Brain	260.00
					Breast muscle	110.00
74	Bogan and Bourne	1972	Scotland-Arctic	Large gulls and skuas	Muscle & liver	535.00
55	Prestt *et al.*	1970	England ('68)	Fresh water fish-feeding birds	Liver	up to 900.00
				Bird-feeding raptors	Liver	up to 70.00
61	Parslow & Jefferies	1973	England, Irish Sea area	Guillemots I dead	Carcasses	3.00
					Liver	50.00
				II shot	Carcasses	1.00
					Liver	0.53
56	Parslow *et al.*	1973	England, E. & W. coasts	Gannets	Liver	4720.00–9570.00
57	Karlog *et al.*	1971	Denmark & Greenland	Terrestrial predatory birds	Liver	1.30–272.00
75	Braestrup *et al.*	1974	Denmark & W. coast of Greenland	Ducks and purple sandpipers	Adipose	1.20–3.30
				Ptarmigans and Bunnick's guillemots		10.10–12.90
				Cormorants and ravents		26.30–37.10
76	Koeman *et al.*	1972	The Netherlands	Dead cormorants	Brain	190.00
					Liver	319.00
62	Koeman *et al.*	1973	The Netherlands	Dead cormorants; Total of	Carcasses	0.50–1.0 g.
				Nestling cormorants	Brain	0.69
58	Acker & Schulte	1971	Germany	Long eared owls, sparrows, hawks, falcons	Adipose	24.00–740.00
77	Figge *et al.*	1976	Germany	Birds from N. Germany		2.20 (T. DDT + PCBs)
59	Baluja *et al.*	1974	Spain	Terrestrial and aquatic species		5.00–6.20

these birds species showed species differences. The eggs of herring gulls had a PCB residue average of 2224.5 ppm (on a fat basis) and the red-necked grebe eggs an average of 744.6 ppm (on a fat basis). The average residue level of PCBs for egg-extracted lipids from fish-eating birds of Louisiana was only 35.0 ppm. Signs of eggshell thinning were found in 9 out of 13 species in the Upper Great Lake States and in 2 out of 7 species in Louisiana.[78] Eggs of the bald eagle had the lowest PCB residue levels in Alaska, 2.2 and 1.1 ppm (wet weight) and the highest in the state of Michigan, 300 ppm, in a survey made during 1969–1970.[79] In 1973 in Ontario, a level of 434 ppm was reported.[80]

PCB levels of eggs of the brown pelican were 1.0–11.0 ppm (wet weight), during 1970.[81] These levels showed a continuous decrease over the following years, reaching an average of 4.7 ppm in 1973, when for the first time in many years the brown pelican reproduced well.[82] Wet-weight, egg PCB-residue levels of the double crested cormorant averaged 45.6[83] and 43.5 ppm.[84]

Birds of industrialized areas have higher egg PCB residue levels in comparison to birds of nonindustrialized areas, 3.34 ppm versus 1.02 ppm.[85] Another example of this situation is a study carried out in Germany during 1974–1976 on eggs of 19 native species. A range of 2.0 to 847.0 ppm PCB residue levels was reported, with higher contamination in eggs from industrialized areas of the Ruhr and Nürenberg.[86]

White-tailed eagles of the Gulf of Bothnia showed high PCBs contamination with egg residue levels averaging 157.0 ppm. Reproduction failures in these birds have been reported.[87] In Schleswig-Holstein, Germany, PCB residue levels in eggs from white-tailed eagles ranged from 6.1 ppm to 97.0 ppm. Here, the PCB levels have not been definitely correlated with a low reproductive rate.[88]

In northern Europe, guillemots,[29] ospreys,[87] eiders,[89] pochards,[85] and sparrow hawks[90] showed relatively high egg PCB residue levels, 16.0, 25.0, 35.0, 40.7, and 47.0 ppm, respectively. Eiders from a protected island of Denmark, were reported to have substantially decreased their PCB residue levels, from 35.0 ppm to 14.0 ppm, in a two-year period.[89]

The effort to reduce the contamination of rivers, lakes and costal waters by PCBs and other organochlorine pollutants, will result in lower egg xenobiotic residues and will prevent the upsetting of the biologic equilibrium in nature.

PCB Residues in Terrestrial Animals (Table 6).

PCB residues were found in terrestrial living beings in different concentrations, from the smallest, at the limit of detection, 0.01 ppm, in adipose tissue of herbivorous mammals in Sweden[73] to thousands of ppm in fish-eating birds.[51]

PCBs are current constituents of living beings in our epoch. They have been reported in the polar bear at less than 400 miles from the North Pole[96] and in every species sampled throughout the earth.

In nursing cows and newborn pups, in the USA (PA), 5.8 and 5.5 ppm PCBs respectively, were found in the adipose tissue. In 2-month-old pups an average of 33.0 ppm was reported, perhaps due to the milk content of PCBs.[97] In minks, the PCB residues were among the highest found in terrestrial mammals, 45.0 ppm in the adipose tissue. The high levels were presumably due to a fish diet.[73] Dead minks found in 1969–1970 in New England, had 73 times the level of PCBs found in healthy minks in 1974. The authors of the report do not think that the high level of PCBs found in the carcasses were responsible for the death of the animals.[98] In hares and foxes in Germany, relatively small PCB residue levels were found in their adipose tissue, an

TABLE 5

PCB RESIDUES IN EGGS (MEAN, RANGE-ppm)

In North America							
Ref.	Authors	Oear	Country	Bird species	Wet weight		Lipids
91	Gress *et al.*	1971	USA, Farallon islands	Common murre			168.0
81	Blus *et al.*	1971	USA	Brown pelican		<1.0–11.0	
19	Wiemeyer *et al.*	1972	USA, Alaska-Kodiak ('69)	Bald eagle	2.2		
			"Admiral. Isl. ('70)		1.1		
			Maine ('69)		9.9		
			Minnesota ('69–'70)		7.7		
			Florida ('69)		12.2		
			Michigan ('69)		27.7		
24	Browes *et al.*	1973	USA	Herring gull	300.0		
78	Faber & Hickey	1973	USA, Upper Great Lake States	Fish-eating birds			339.7
				Red necked grebe			744.6
				Common merganser			260.2
				Red crested merganser			489.0
				Herring gull			2224.5
			Louisiana	Fish-eating birds			35.0
82	Blus *et al.*	1977	USA, South Carolina ('70)	Brown pelican	6.1	5.0–74.5	
			('71)		5.2	3.9–70.4	
			('71)		6.5	1.5–36.5	
			('72)		7.5	2.6–32.3	
			('73)		4.7	0.9–19.3	
80	Grier	1977	USA, N. East & Ontario ('66–'73)	Bald eagle	434.0		
92	Longcore & Mulhern	1973	USA, N. East & Canada ('71)	Black duck	3.30	1.0–6.90	
					<0.05	0.02–0.05	
83	Zitko	1972	Canada, New Brunswick	Double crested cormorant	45.6		
				Gull	7.4		
84	Zitko & Choi	1972	Canada, New Brunswick	Double crested cormorant	43.5		
72	Zitko *et al.*	1972	Canada, Bay of Fundy	Herring gull	0.1	(Aroclor 5460)	
93	Mes *et al.*	1974	Canada, Bay of Fundy	Chicken	<0.010		

In Europe						
94	Bjerk and Holt	1971	Norway	Herring gull		0.2–38.0
				Common gull		trace–0.8
29	Jensen *et al.*	1969	Sweden, Baltic	Guillemot	16.0	
87	Koivusaari *et al.*	1972	Finland, Quarken area	Osprey	25.0	
			Gulf of Bothnia ('70–'71)	White-tailed eagle	157.0	
89	Fog & Kraul	1973	Denmark (from a protec- island)	Eider '70	35.0	
				'72	14.0	
85	Block & Kraul	1972	Denmark (collected from two localities)	Swan	1.02	
					3.34	Industrial areas
				Pochard	40.70	
90	Koeman *et al.*	1972	The Netherlands (east)	Sparrow hawk '69–'71	47.0	
88	Koeman *et al.*	1972	Germany-Schleswig-Holstein	White-tailed eagle		6.1–97.0
86	Conrad	1977	Germany-Schleswig-Holstein	19 Native bird species '74–'76		2.0–847.0
95	Mestres & Illes	1973	France	Gull	0.8	23.0 PCTs

TABLE 6

PCB RESIDUES IN TERRESTRIAL ANIMALS (ppm)

Ref.	Authors	Year	Country	Species	Tissue	Mean	Range
96	Anomymous	1972	?	Polar bears, present (Less than 400 miles from North Pole)			
98	Friedman *et al.*	1977	USA (New England)	Minks (found dead) '69–'70		73-fold higher levels than in healthy minks in 1974	
97	Kurts & Kim	1977	USA	Cows (nursing)	Adipose	5.8	
				(newborn)		5.5	
				(2-month-old pups)		33.0 (3.4 ppm plasma level was found in 2 month-old pups)	
73	Odsjö	1973	Sweden	Herbivorous mammals		0.01 (at the limit of detection)	
				American minks	Muscle	0.58	
					Adipose	45.00	
99	Bruggemann	1974	Germany	Hares	Adipose	0.22 (in 20 out of 72 samples)	
				Foxes	Adipose	2.50 (in 1 out of 5 samples)	
100	Hidaka *et al.*	1972	Japan, Lake Biwa	Corbiculae '63	(preserved)		0.04–2.4
				'71			0.06–1.1

average of 0.22 ppm PCBs in 20 positive out of 72 samples in hares and 2.50 ppm in the adipose tissue of 1 positive out of 5 samples of adipose tissue of foxes.[99]

PCB Residues in the Adipose Tissue of the General Population (Table 7)

PCB average residues in the adipose tissue of groups of the general population from different countries of Europe, range from about 1.0 ppm to several ppm. The lowest levels are reported from the Northern countries: Norway 0.90 ppm, the Oslo area 0.80 ppm,[101] England[102] and The Netherlands[103] up to 1.0 ppm and Finland with an average of 1.97 ppm.[104] Austria had a higher average for the adipose tissue PCB residue levels, 3.5 ppm, with a range of 0.3–7.3 ppm.[105] Germany had the highest level of PCB residues in the general population with a range of averages from 6.8 ppm in Cologne to 10.0 ppm in Munich.[106]

From the USA, several surveys[96,107,108] reported in 1972 showed that the majority of cases (57.6–68.9% of the cases studied) had an adipose tissue PCBs residue level of less than 1.0 ppm, about a third (25.9–36.7%) had PCB residue levels ranging from 1.0 to 2.0 ppm and only 5.2 to 7.6% of cases had residue levels of over 2.0 ppm. A study of the EPA's Perrine laboratory, FL., found PCBs residue levels ranging from 1.0 to 6.0 ppm, but as much as 600 ppm has been found.[40]

In Japan where the storage of PCBs in man has been extensively studied, averages of less than 1.0 ppm, 0.6 ppm,[109] 0.8 ppm,[110] and 0.9 ppm[111] were reported during 1973 and 1974, but at the same time, averages of several ppm were also found: 5.0 ppm,[112] 2.86 ppm,[113] and 2.6 ppm.[114]

The studies carried out in Germany show a tendency for PCB residues of the general population to increase during the years 1970 to 1974: 5.7 ppm for 1970,[115] 6.4 ppm for 1971,[58] and 8.3 ppm for 1974[116] having been reported. Such a tendency cannot be refuted for other countries since annual reports are very few: Inoue *et al.*, reported in Japan 1.60 ppm for 1974 and 2.60 ppm for 1975.[114,117]

PCB levels according to sex, showed higher values in males.[118]

PCB Plasma Residues in Man (Table 8).

Plasma PCB residues were studied especially in Japan because of the accidental poisoning of humans that occurred in 1968. Plasma PCB residue levels ranged from several ppb to tens of ppb, in the general population of those countries where the sampling was performed. The Yusho patients presented high adipose tissue and plasma residue levels during the 1–3 years after the alimentary poisoning took place. In recent years PCB plasma levels have dropped to almost the levels found in the general population. Yet, there remains for those Yusho cases with PCB levels towards the upper range of plasma values, a specific gas chromatographic pattern: peak 17 was higher and peak 13 was lower as compared with the controls.[124–126] Besides, 7 years after the accident, very high levels may still be found in several Yusho patients, with 92.0 and 230.0 ppb being reported in two patients.[114]

Occupationally exposed people have rather high plasma PCB residue levels: 75.0–1900.0 ppb plasma PCBs were reported in persons working in a capacitor plant in Finland,[127] and 100.0–650.0 ppb in a capacitor plant in Japan. Plasma PCB levels of 90.0–740.0 ppb were found in these workers one month after exposure.[128] Skin lesions were reported.[128]

Increased plasma levels of PCBs occurred also in cachectic patients and in starved animals.

TABLE 7

PCB RESIDUES IN THE GENERAL POPULATION (ADIPOSE TISSUE, ppm)

Ref.	Authors	Year	Country	Wet tissue	(mean, range)
101	Bjerk	1972	Norway	0.90	
			Oslo area	0.80	
102	Abbott *et al.*	1972	England		≤1.0
104	Hattula *et al.*	1976	Finland, Pori Iyvaskyla	1.97	
103	Freudenthal & Greve	1973	The Netherlands		≤1.0
115	Acker & Schulte	1970	Germany	5.7	
58	Acker & Schulte	1971	Germany	6.4	4.3–9.3
116	Acker & Schulte	1974	Germany	8.3	
106	Acker & Schulte	1974	Germany, Cologne	6.8	
			Germany, Götigen	7.5	
			Germany, Hamburg	7.9	
			Germany, Berlin	9.1	
			Germany, Munich	10.0	
105	Pesendorfer *et al.*	1973	Austria, Vienna	3.5	0.3–7.3
40	EPA's Perrine Lab (FL)	1971	Austria, Vienna		1.0–6.0
96	Anonymous	1972	Austria, Vienna	29.7% popn.	1.0–2.0
			Austria, Vienna	7.6% popn.	2.0
107	Price and Welch	1972	Austria, Vienna	57.6% popn.	≤1.0
			Austria, Vienna	36.7% popn.	1.0–2.0
			Austria, Vienna	5.6% popn.	>2.0
108	Yobs	1972	USA	68.9% popn.	≤1.0
			USA	25.9% popn.	1.0–2.0
			USA	5.2% popn.	2.0

TABLE 7 (*continued*)

PCB RESIDUES IN THE GENERAL POPULATION (ADIPOSE TISSUE, ppm)

Ref.	Authors	Year	Country	Wet tissue	(mean, range)
119	Wassermann *et al.*	1973	Israel	2.75	
112	Ui	1972	Japan	5.0	
113	Nishimoto *et al.*	1972	Japan, Kochi, Kyoto	2.86	≤7.5
120	Mizutani	1972	Japan	4.7	
110	Curley *et al.*	1973	Japan	0.8	0.3–1.48
121	Doguchi	1973	Japan	5.1	(M 70–79 year)
			Japan	2.4	(F 60–69 year)
109	Doguchi *et al.*	1974	Japan	0.6	0.1–2.1
111	Masuda & Kagawa,	1974	Japan, Saga ('70)	0.9	0.3–1.9
117	Inoue *et al.*	1974	Japan	1.6	
114	Inoue *et al.*	1975	Japan	2.6	
118	Fukano & Doguchi	1977	Japan	2.01 (M)	1.1 PCTs
				0.59 (F)	1.04 PCTs
122	Solly & Shanks	1974	New Zealand	0.82	
123	Dyment *et al.*	1971	New Guinea	Absent	

TABLE 8

PCB RESIDUES IN MAN (mean, range)

Ref.	Authors	Year	Country	General Population (plasma levels-ppb)			Cases	
127	Karppanen & Kolho	1973	Finland		3.1– 12.0			
132	Finklea *et al.*	1972	USA, Charleston		– 29.0		723	Volunteers
133	Polishuk *et al.*	1977	Israel	19.3				Pregnant women
124	Inoue *et al.*	1973	Japan	3.3		4.5 (urban)	42	
						2.7 (rural)		
117	Inoue *et al.*	1974	Japan	4.9				
125	Masuda *et al.*	1974	Japan	3.0			37	
126	Okumura *et al.*	1974	Japan	2.8			37	
134	Yakamatsu *et al.*	1974	Japan		1.0– 7.0		82	
135	Doguchi & Fukano	1975	Japan	3.2		5.0 PCTs		
118	Fukano & Doguchi	1977	Japan ('73)		2.2– 5.1	0.7–19.6 PCTs		
			Japan ('75)		1.8– 3.8	1.1– 9.4 PCTs		
				Occupationally Exposed Persons (plasma levels-ppb)				
127	Karppanen & Kolho	1973	Finland		36.0– 63.0	(in an analytic lab)		
					75.0–1900.0	(in a capacitor plant)		
128	Hasegawa *et al.*	1972	Japan		100.0– 650.0	(in a capacitor plant)		
					90.0– 740.0	(1 month after exposure)		
136	Kitamura *et al.*	1973	Japan	800.0		(in an electric condenser plant, 1 month after exposure)		
				300.0		(in an electric condenser plant, 3 months after exposure)		
				Yusho Patients				
137	Yamaguchi *et al.*	1971	Japan ('69)	75.5		(adipose tissue-ppm)		
112	Ui	1972	Japan	13.0		(adipose tissue-ppm)		
138	Masuda & Kagawa	1974	Japan ('69–'72)	2.5	0.7– 4.3	(adipose tissue-ppm)		
124	Inoue *et al.*	1973	Japan ('73)	7.0	2.0– 15.0	(plasma-ppb)		
125	Masuda *et al.*	1974	Japan ('73)	7.0		(plasma-ppb)	41	
126	Okumura *et al.*	1974	Japan	8.6		(plasma-ppb)		Yusho LGC pattern
				3.8		(plasma-ppb)		Intermediate pattern
134	Takamatsu *et al.*	1974	Japan		2.0– 15.0	(plasma-ppb)	32	
114	Inoue *et al.*	1975	Japan ('75)	6.1		(plasma-ppb)		
				Severe Wasting Diseases (blood-ppm)				
129	Hesselberg and Scherr	1974	USA	47.0				

In patients with severe wasting diseases, 47.0 ppm PCBs was reported in venous blood.[129] An increase in the death of female eiders, at the end of the physiologic starvation that occurs during the breeding cycle, was explained by the release of OCC residues during the using up of stored body fat.[130] In rats, starved during the recovery from the effects of Kanechlor-400 or Kanechlor-500 given previously, there was a temporary but definite shortening of hexobarbital sleeping time, suggesting the possibility of PCB release from the adipose tissue deposits.[131]

PCB Residues in Human Milk (Table 9)

The average PCB residue levels in human milk have been reported for several countries of Europe, North America and Asia. These averages vary from ten ppb in Norway,[101] to about a hundred ppb in Germany.[115]

PCB residues in human milk were higher than those found in colostrum, 74.5 ppb[139] versus 44.3 ppb.[133]

For milk and colostrum the individual chlorinated biphenyl compounds showed different percentages of the total PCB value found, peaks 9, 13, and 14 representing higher percentages in colostrum, whereas peaks 2, 3, 4, and 6 showed higher percentages in milk collected during the third through the fifth month of nursing.[139]

Smaller concentrations of PCBs were excreted in milk in overweight women and in women over 30 years old as compared to normal or underweight women and to women under 30 years of age.[133]

Although the excretion of PCBs and other xenobiotics in milk represent an efficient means for their removal from the female body it should be remembered that these compounds are transferred with the milk to the infant.

Because of the great range of milk PCB residues, due to differences in exposure and biochemical individuality, it is obvious that special care for the nutrition and environment of potential mothers will lower the xenobiotic content of the human female body and consequently, the placental and milk xenobiotics transfer to the fetus and infant will be decreased.

Effects of PCBs on Reproduction in Aquatic Organisms (Table 10)

PCBs are toxic to some aquatic organisms at concentrations existing presently in sea water. Phytotoxicity occurred at levels less than 5 ppb.[153] Selective toxicity of PCBs for phytoplankton communities was reported. Netplankton production was inhibited by more than 50% by Aroclor 1242 at levels of 10 and 25 ppb, while nannoplankton (which contributes about 73% to the entire phytoplankton production) was not affected.[153] Reduction of phytoplankton division in media containing PCBs was also reported.[154,155] This selective toxicity for phytoplankton species may affect marine herbivorous populations.

At a concentration of 21 ppb Aroclor 1254 in sea water, a decrease in mating occurred in *Gammarus oceanicus*. The decrease seems to be the result of toxic effects in males.[156]

Premature hatching and decrease in hatchability as well as decreased survival and growth in the young fish were observed when coho salmon eggs were kept in Aroclor-containing media.[157]

A diminished percentage of pregnant females was noted in marine mammals. Pregnant Ringed seals, caught in the Bosnian Bay, Choska Inlet, and Ochotskan Sea,

TABLE 9

PCB RESIDUES IN HUMAN MILK (mean, range)

Ref.	Author	Year	Country	Whole milk (ppb)		Milk-extracted Lipids (ppm)	
115	Acker & Schulte	1970	Germany	103.0		3.5	
58	Acker & Schulte	1971	Germany			5.3	
140	Knoll & Jayaraman	1973	Germany	90.0			
141	Pesendorfer	1975	Austria, Vienna			1.54	
			Mistelbach			1.29	
142	Juszkiewicz *et al.*	1975	Poland	9.0			0.01–1.40
143	Falandysz	1977	Poland	390.0			
144	Westöö *et al.*	1970	Sweden	16.0			
145	Westöö & Norén	1972	Sweden (1967)	14.0			
			Sweden (1971–1972)	25.0			
146	Westöö	1974	Sweden	Increased			
101	Bjerk	1972	Norway	10.0			
147	Risebrough & Brodine	1969	USA, California	60.0			
123	Dyment *et al.*	1971	USA, Texas	Absent			
			New Guinea	Absent			
148	Savage *et al.*	1973	USA, Colorado, rural		40.0–100.0		
149	Strassman & Kutz	1975	Arkansas & Mississippi		Trace		
150	Musial *et al.*	1974	Canada, New Brunswick	22.0		15.3	
			Canada, Nova Scotia	18.0		18.6	
133	Polishuk *et al.*	1977	Israel	44.3	(colostrum)	3.10	
139	Miller *et al.*	1978	Israel	74.5		2.29	
151	Oura *et al.*	1972	Japan, Toyama	30.0		1.10	
152	Hidaka *et al.*	1972	Japan, Kyoto	50.0			
113	Nishimoto	1972	Japan, Kochi & Nangoku	30.0			
111	Masuda & Kagawa	1974	Japan, Fukuoka (Jan.–Mar. '72)	30.0	10.0– 60.0	1.2	0.30– 5.6
			Japan, Ishigaki (Mar. '72)	14.0	10.0– 20.0	0.4	0.10– 0.7

TABLE 10

EFFECTS OF PCBs ON REPRODUCTION IN AQUATIC ORGANISMS

Ref.	Authors	Year	Species	Compound, dosage	Effect
155	Morgan	1972	Chlamidomonas	Aroclor	Decreased cell density. Dose-related decrease in uptake of ^{14}C.
153	Moore	1973	Phytoplankton	Aroclor 1242 10 & 25 ppb	Net plankton production was inhibited > 50%. Nannoplankton was not inhibited.
154	Fisher	1975	Phytoplankton	PCBs	Reduction of cell division rate.
159	French	1976	Flagellated protozoan	Aroclor 10.5 ppm	50% depression of cell population growth.
				DDT 425.0	50% depression of cell population growth.
160	Maki & Johnson	1975	Daphnia magna Strauss	PCBs 19.0 ppb	50% reduction in the number of young.
156	Wildish	1972	*Gammarus oceanicus*	Aroclor 1254 10 & 100 ppm in sea water/10 hr.	Survival and development were hindered in the embryos.
				Aroclor 1254 21.0 ppb	Decrease in mating.
157	Halter & Johnson	1974	Eggs of Coho Salmon	Aroclor 1254 4.4 ppb until 4 w. after hatch. 15.0 ppb until 2 d. before hatching	Decrease of egg hatchability, premature hatching and decreased survival and growth in the young fishes.
161	De Lang *et al.*	1973	Sea lions of S. California	Naturally exposed	A high rate of premature births in those with higher levels of PCBs.
26	Helle *et al.*	1976	Ringed seals from: Bosnian bay, Choska inlet, Ochotskan Sea	Naturally exposed	Pregnant seals had lower PCB and DDT residue levels. About half of the non-pregnant population showed evidence of implantations followed by resorption or abortion.
158	Helle *et al.*	1976	Seals, Gulf of Bothnia	Naturally exposed	Only 30% were pregnant. The nonpregnant seals had higher PCB and DDT residues. They presented uterine changes.

Finland, were shown to have lower organochlorine residues (75 ppm DDT and 56 ppm PCBs) as compared with a nonpregnant population (130 ppm DDT and 77 ppm PCBs). About half of the nonpregnant population showed signs of implantation.[26] Seals from the Gulf of Bothnia presented a 30% pregnancy rate (which is considered as a low rate). The nonpregnant seals had higher organochlorine residues.[158]

Effects of PCBs on Reproduction in Birds (Table 11)

Decrease in the reproduction success in birds is the most striking effect of the ever-increasing environmental chemical pollution of the last decades. It is the result of diminished egg production, eggshell thinning and increased rate of embryonic mortality. The reproductive success for sparrow-hawk eggs collected in east Netherlands during 1969–1971 was approximately 25% lower than in the period 1931–1943. PCBs and DDT residues were 47 and 26 ppm respectively.[162]

Impaired egg production and decline in eggshell parameters were correlated with increased xenobiotic residues (OCC: DDT, dieldrin, PCBs and such heavy metals as mercury and cadmium) in several avian species.[78]

Inhibition of ovulation was reproduced experimentally in birds given PCBs.[163]

Eggshell thinning was reproduced experimentally with PCBs in several avian species. The eggs of Japanese quail given 50 ppm PCBs in their diet for 15 weeks presented a decreased eggshell thickness (4.5%) and an increased percentage of cracked eggs.[164,165] High levels of DDT and DDE (100 and 200 ppm) given to Japanese quail (about 500 and 600 ppm residues in their eggs) produced a 5% decrease in the shell thickness and consequently an increased shell breakage in the eggs of these birds. At this level, the activity of carbonic anhydrase in the shell glands was decreased by 12–15%.[164] Eggshell thinning was also reproduced in chickens given Aroclor 1254 or 1242[166] but not in ring doves given 100 ppm Aroclor 1254 for 3 months,[167,168] or in white Leghorn chickens and Japanese quail given 10 and 20 ppm PCBs.[169] These negative results seem to be due at least in some cases to inadequate dosage, since the Japanese quail egg thickness was affected by larger doses of PCBs as mentioned above.

Eggshell thinning, occurring in nature, has often been attributed to other environmental pollutants rather than to PCBs. The fact that eggshell thinning was reproduced experimentally with PCBs suggests that the responsibility for such eggshell thinning in nature may be shared by many xenobiotics among which the most frequently cited were DDT, dieldrin, PCBs, and mercury.

Normal eggshell production depends on the plasma estrogen levels and the activity of carbonic anhydrase in the shell gland. Chlorinated hydrocarbons activate normally dormant hydroxylases in the liver, which enhance the hydroxylation of the steroid estrogen rings, resulting in lower estrogen levels.[170] PCBs are more potent activators of estrogen degradation than are DDT and DDE.[51] The lower estrogen levels result in a lower supply of calcium to the oviduct and contribute thus to eggshell thinning.[170] Calcium carbonate is produced from bicarbonate and calcium in the presence of carbonic anydrase.

Reduction in hatchability due to embryonic mortality was reproduced experimentally in mallards,[171] in Leghorn chickens,[172] in ring doves,[167] in pheasants,[173,174] in chickens[166] and in hens.[175]

PCBs residues in birds and bird eggs reached levels higher than in any other species.

TABLE 11

EFFECTS OF PCBs ON REPRODUCTION IN BIRDS

Ref.	Authors	Year	Species	Compound	Effects
80	Grier	1977	Bald eagles, N-W Ontario	Natural Contamination	Eggshell thinning and lowering of reprod. rate (434 ppm PCB residues in eggs).
175	Kolbye	1972	Hens	Accidental Contamination	Reduced hatchability of eggs.
176	Cecil *et al.*	1974	Hens	Aroclor 1232 & 1954 20 mg/kg	Reduction of hatchability in fertile eggs.
177	Haegele & Tucker	1974	Mallards	Aroclor 1254	Short-term eggshell thinning (single dose)
171	Risebrough & Anderson	1975	Mallards	PCBs 40 ppm & DDE 40 ppm/7 wk	Reduced egg production and hatchability. Extr. lip. residues: 344 ppm DDE, 364 ppm PCBs
165	Call & Hanell	1974	Japanese quail	Aroclor ppm 1242 312.5 & 5000* 1254 78.1 & 1250 1260 62.5* & 1000*	Significantly lower egg production and eggshell thinning in all the groups. (*Lower egg weights in marked groups.)
164	Chang & Stokstad	1975	Japanese quail	PCBs 50 ppm/15 wk	Eggshell thinning (4.5%). Increase in the percentage of cracked eggs.
178	Hill *et al.*	1976	Japanese quail 5-month old hens	Aroclor 1242 10 ppm/45 days	The shell of eggs was 4.7% (for the 1st egg) and 5.2% (for the 1st 5 eggs) thinner.
179	Cecil *et al.*	1973	Young J. quail	Aroclor 1242 100 ppm/2 months	Depression of egg production. Egg weight and eggshell thickness not affected.
166	Dustman *et al.*	1971	Chickens	Aroclor 1242 & 1254	Impaired egg production, egg hatchability and eggshell thickness.
172	Scott *et al.*	1975	White Leghorn chickens	PCBs	Some decrease in egg production. Drastic reduction in hatchability.
169	Scott	1977	White Leghorn chickens	PCBs 10–20–100 ppm	Reduction in hatchability.
			Japanese quail	PCBs 10–20–100 ppm	No effect on eggshell quality in either species.
168	Peakall *et al.*	1971	Ring doves	Aroclor 1254 10 ppm/35 days	No effect on eggshell quality.
167	Peakall *et al.*	1972	Ring doves	Aroclor 1254/ 2 generations	Low hatching of eggs in the second generation.
173	Dahlgren *et al.*	1971	Pheasants	Aroclor 1254 50 mg/wk./4 months	Lowered egg production and hatchability. The hatched chicks were significantly lighter.
174	Dahlgren *et al.*	1972	Penned pheasants	Aroclor 1254 12.5 & 50 mg/wk./4 m.	Egg production and hatchability were lowest in the 50 mg. group.
163	Velle	1971	Birds	PCBs & other OC	Inhibition of ovulation.

Effects of PCBs on Reproduction in Terrestrial Mammals During Experimental or Accidental Exposure to PCBs (Table 12)

Lengthening of estrus cycle in mice[181–183] and Rhesus monkeys[184] and changes in the menstrual cycle in Yusho patients[185] were reported. Also, low mating indices in rats,[186] lower fertility in monkeys,[187] decrease in the frequency of implanted ova in mice[182,183,188] and in dogs,[189] fewer offspring per litter in rats,[190] complete failure to produce offspring in mink[191] and in dogs[189] at a certain dosage level of PCBs were reported. Fetotoxicity and abortion in rabbits,[192] embryonic mortality in mink,[193] diminished weight of whelps[189] and monkeys[187] at birth, small-for-date babies in Yusho mothers[137,194,195] stillbirths in mink,[196] dogs,[189] and man,[195] decreased survival of offspring in rats,[186] deaths of newborn kits in minks[193] and decreased percentage of whelps living more than 5 days after birth[189] have also been reported.

The effects of PCBs on reproduction vary in intensity with dosage. They occurred in a large variety of living organisms under conditions of natural exposure or laboratory experiments.

These effects were largely explained as being caused by the influence of PCBs on naturally occurring biochemical processes, by upsetting quantitatively the activity of enzymatic systems (on which they depend), in mother and fetus.

The placental transfer of PCBs was reported in marine and terrestrial mammals. In seals from the Gulf of Bothnia,[158] mice,[197,198] rats,[110] rabbits,[199,200] cows,[201] and in man (general population[202]) and in Yusho patients.[203–206]

The concentration of PCBs in rabbit fetal tissues was dose dependent.[199]

In seals, fetuses contained somewhat lower levels of PCBs and DDT than did their mothers.[158] PCB residues in rabbit fetal livers were higher than in maternal livers,[207] and the residues in the kidneys of cow fetuses were higher than in the maternal kidneys.[201]

Effect of PCBs on Development and Growth (Table 13)

Chromosomal abnormalities were observed in birds that received PCBs in their diets.[7] 100 ppm Aroclor 1254 in the diet of ring doves produced a significant increase in chromosome abnormalities. Increased aberration rates were obtained in embryos following administration of 10.0 ppm Aroclor 1254 in the diet for two generations.[167]

No evidence of mutagenic effects was observed in *Drosophila melanogaster* with Clophen A30 or Clophen A50,[209] nor in rat bone marrow and spermatogonia after single Aroclor 1242 doses of 1250, 2500 and 5000 mg/kg or repeated doses of 500 mg/kg/day, for 4 days,[210] nor in human lymphocytes after treatment in media containing 100.0 ppm Aroclor 1254 for 24 hours and the last 8 hours of a 52-hour experiment.[211]

Teratogenic effects were reported in birds experimentally exposed to PCBs.[7] Beak abnormalities in chicks were obtained after injection of Aroclor 1242 into the yolk sac.[212]

Teratogenic effects in offspring were produced also by feeding PCBs to hens.[176]

Among terns from Great Gull Island, NY., naturally exposed to PCBs, having a muscle PCB residue level of 5.0–175.0 ppm[63] and 25.0 ppm,[64] a high number of deformities of the eye, bill, and leg plus loss of flight feathers and failure to hatch were observed.[63,64,213]

No teratogenic effects were obtained in rats given Aroclor 1242, 10 or 30 mg/kg[186] or in rabbits given 10, 50 or 100 mg Aroclor 1254 per kg.[192]

TABLE 12

EFFECT OF PCBs ON REPRODUCTION IN TERRESTRIAL MAMMALS

Ref.	Authors	Year	Species	Compound, dosage	Effects
181	Johansson *et al.*	1972	NMRI mice	PCBs 20 mg/kg (1 inj.)	Lengthening of the estrus cycle by one d./3 cycles.
182	Kihlström *et al.*	1973	NMRI mice	Clophen A60 25 μg/62 d. to the nursing mothers.	Increase in the length of the estrus cycle. Decrease in the frequency of implanted ova/offspring.
183	Orberg & Kihlström	1973	NMRI mice	Clophen A60 25 μg/d. Chronic exposure	Lengthening of estrus cycle. Decrease in the frequency of implanted ova.
188	Kihlström *et al.*	1975	NMRI mice	PCBs 50 mg/kg/wk/4 wk beginning the day of deliv.	The frequency of implantation in the young decreased from 94% (controls) to 75%.
186	Keplinger *et al.*	1971	Rats	Aroclor 1242 & 1254 100 ppm in diet	Low mating indices. Decreased survival in pups.
				Aroclor 1260 100 ppm in diet	No effect.
190	Linder *et al.*	1974	Rats	Aroclor 1254 20 & 100 ppm in diet	Fewer offspring per litter.
				Aroclor 1260 20 & 100 ppm in diet	No effect.
208	Jonsson *et al.*	1976	PCBs		High doses markedly reduced reproduction.
192	Villeneuve *et al.*	1971	Rabbits (pregn.)	Aroclor 1254 12.5 mg/kg/d. first 28 d.	Abortions. Fetotoxicity.
193	Aulerich *et al.*	1971	Mink	Coho salmons and other Great Lake fishes	Embryonic mortality. Death of newborn kits.
191	Ringer *et al.*	1972	Mink	Aroclor 1254 5–10 ppm	Complete failure to produce offspring.
196	Aulerich *et al.*	1973	Mink	Coho salmons of Lake Michigan	58% of whelped 55% of the young were stillborn. Toxicity varied inversely with chlorine content.
189	Jensen *et al.*	1977	Dogs	PCBs 22, 11 mg/d. doses	Decrease in the number of whelps per bitch.
				PCBs 3.3 ppm + DDT 3.3 ppm.	The number of delivering bitches, whelps per bitch, whelps living after 5 days and weight of whelps was decreased. The number of stillborns was increased.
				PCBs 11 mg/d/66 d. (beginning 1 m. before mating).	No whelp was born in this group.
187	Allen	1975	Monkeys	Aroclor 1248 2.5 ppm in diet	Lower fertility. Diminished weight of young at birth.
195	Nishimura	1973	Man	Yusho patients	Stillbirths.

TABLE 13

EFFECTS OF PCBs ON DEVELOPMENT AND GROWTH

Ref.	Authors	Year	Species	Compound, dosage	Effect
209	Nilsson & Ramel	1974	*Drosophila melanogaster*	Clophen A30	No chromosomal damage was observed.
				Clophen A50	
7	Anonymous	1972	Birds	PCBs in diet	Some evidence of chromosomal damage
167	Peakall *et al.*	1972	Ring doves	Aroclor 1254 100 ppm	Significant increase in chromosomal abnormalities in embryos.
				10 ppm/2 generations	13/17 embryos had aberration rates exceeding the mean control rate and 4/17 exceeding the highest control rate. One chromosomal rearrangement was observed.
210	Green *et al.*	1975	Rats	Aroclor 1242 1250, 250 or 5000 mg per kg in single dose	No evidence of mutagenic effects in bone marrow and spermatogonia.
				500 mg/kg/d/4 d	Idem.
211	Hoopingarner *et al.*	1972	Man	Aroclor 1254 100 ppm/24 hr + last 8 hr of 52 hr	No effect on chromosomes of lymphocytes exposed "in vitro" was observed in a 52 hours experiment.
212	Mc Laughlin *et al.*	1963	Chick eggs	Aroclor 1242 inj. into the yolk sac.	Beak abnormalities in young.
7	Anonymous	1972	Birds	PCBs in diet.	Some evidence of teratogenesis.
176	Cecil *et al.*	1974	Hens	Aroclor 1232 and 1254 20 mg/kg	Teratogenic effects.
63	Anonymous	1972	Terns of Long Island Sound	Natural exposure	A high incidence of deformities. Muscle PCB residues range was 5.0–175.0 ppm.
64	Hays & Risebrough	1972	Common and roseate terns of Great Gull Island, NY ('69–'70)	Natural exposure	Eye, bill and foot deformities, flight feathers loss and failure of hatch. Muscle PCB residues in young terns, 25 ppm.
213	Hays & Risebrough	1972	Terns from Great Gull Island, NY	Natural exposure	Bill and leg deformities and missing flight feathers.
187	Allen	1975	Monkeys	PCBs	Pregnancies produced small infants.
137	Yamaguchi *et al.*	1971	Man	Yusho mothers	Small babies

TABLE 13 (*continued*)

EFFECTS OF PCBs ON DEVELOPMENT AND GROWTH

Ref.	Authors	Year	Species	Compound, dosage	Effect
194	Funatsu	1972	Man	Yusho mothers	Small babies
195	Nishimura	1973	Man	Yusho mothers	Substandard body weight in new born.
214	Fraumeni	1974	Man	Yusho babies	Growth retardation, early teeth eruption.
215	Kuratsune *et al.*	1971	Man	Yusho children	Diminished growth, in boys
216	Yoshimura	1971	Man	Yusho children	Diminished growth, in boys

TABLE 14

EFFECTS OF PCBs ON LIVER MORPHOLOGY

Ref.	Authors	Year	Species	Compound, dosage	Effects
223	Ito *et al.*	1973	Mice	Kanechlor-500, 100 & 250 ppm Kanechlor-400 & -300 300, 400 & 500 ppm	Amyloid degeneration.
217	Treon *et al.*	1956	Rats	Aroclor 1242 & 1254, vapors	Liver enlargement.
218	Grant *et al.*	1971	Rats	Aroclor 1254, 500 mg/kg/1d.	Significantly enlarged livers. Increased liver lipids.
219	Keplinger *et al.*	1971	Rats	Aroclor 1242 & 1254 100 mg/kg	Increased liver weight.
220	Kimbrough *et al.*	1972	Rats	Aroclor 1260	Increase in fat droplets and smooth endoplasmic reticulum (SER). SER cisternae were condensed in hyaline inclusions.
221	Norbach & Allen	1972	Rats	Aroclor 5460, 10 ppm. Aroclor 1260 0.2 ppm/21d.	Hypertrophy of the liver. Proliferation of SER and multilayered concentric membrane arrays.

222	Vos & Nottenboom-Rom	1972	Rabbits	Hexachlorobiphenyl	Lipid accumulation, pigment deposits, nuclear changes and necrosis of liver parenchymatous cells.
224	Rogers *et al.*	1976	Rat liver lysosomes *in vitro*	Aroclor 1254	Labilization of lysosomal membranes and consequent release of hydrolytic enzymes into the cytoplasm, cellular damage.
228	Nishihara & Yamamoto	1972	Guinea pigs (M)	Kanechlor-400 100 mg/kg single dose	Proliferation of SER which persisted 2 months
			Rats	Kanechlor-400 100 mg/kg single dose + cycloheximide	Slight inhibition of SER proliferation.
				Kanechlor-400 200 mg/kg	The glycogen deposits disappeared completely.
225	Iatropoulos *et al.*	1977	Rhesus monkeys (F)	Clophen A30	Dose dependent degenerative changes in the liver.
				2,5,4′-trichlorobiphenyl	Reversible changes in the endothelial cells of small vessels with secondary hypoxia and degenerative changes in the liver, the kidney and central nervous system.
			Lactating Rhesus monkeys (F)	Chlophen A30	Slight to moderate degenerative changes in infant monkeys. The severity of these changes increased with age.
227	Hirayama *et al.*	1969	Man	Yusho patients	Increase in SER (liver biopsy).

Monkeys fed diets containing Aroclor 1248 produced small infants.[187]

Substandard body weight in new born Yusho babies,[137,194,195] growth retardation and early teeth eruption[214] and diminished growth in boys[215,216] were reported after the PCBs poisoning in Japan.

Effects of PCBs on Liver Morphology (Table 14)

In animal experiments, exposure to PCBs brought about the following: significantly enlarged livers,[217–219] lipid accumulation (increased number of lipid droplets), proliferation of the smooth endoplasmic reticulum (SER) in the cytoplasm of centrilobular parenchymatous cells,[220–222] adenomas, carcinomas, adenofibrosis (proliferation of bile ducts within a connective tissue rich in collagen fibers—(see Table 25), amyloid degeneration[223] and degenerative changes, and necrosis due either to labilization of lysosomal membranes and consequent release of hydrolytic enzymes into the cytoplasmic matrix[224] or to reversible changes in the endothelial cells of small vessels, with secondary hypoxia.[225]

The morphologic changes were more pronounced in females than in males.[226] The size and quality of the morphologic changes were correlated to the dosage.[226]

Increase in SER was reported in a Yusho patient (liver biopsy) in 1969.[227]

The proliferation of SER, of the tubulo-vesicular type, as being the morphologic expression of increased enzymatic activity was considered an adaptive phenomenon, used by the cell as a resource (of defense) to get rid of xenobiotics.

The endoplasmic reticulum may also proliferate in the form of smooth cisternae, arranged in whorls, which may be very large or multiple and occupy most of the available cytoplasmic space.

An increased phospholipid to protein ratio and a decreased ratio of cholesterol to proteins in the concentric membrane arrays within the liver cells of chlorinated-aromatic-hydrocarbon-treated rats, suggested to Norbach and Allen,[221] that these structures are a distinct entity of the endoplasmic reticulum which is more suited to appropriating foreign hydrophobic compounds to the enzymes of the membranes. In accord with this concept, the smooth cisternae whorls may also be seen as an adaptive response of the liver cell to the exposure to increased amounts of xenobiotics.

These morphologic changes in the SER may be also the morphologic substrate for the induction of enzymatic systems that upset normal metabolic pathways in the cell and so explain most of the biologic effects of xenobiotics.

Effects of PCBs on Enzymatic Activity (Tables 15 and 16)

PCBs were shown to induce the activity of a number of enzymes and to inhibit the activity of others.

Highly chlorinated PCB mixtures were more potent inducers of enzymatic activity.[162,229–232] Aroclor 1254 is a more potent inducer than Aroclor 1242[229] and than 1216.[233] Pentachlorobiphenyl was however more effective than hexachlorobiphenyl as an enzyme inducer in the rat,[230] and Aroclors 1242 through 1254 (in the intermediary range of chlorination) were more effective inhibitors than Aroclors 1221 and 1268 of oligomycin-insensitive $Mg^{2\pm}$ ATPase, in fish.[234] On the other hand, it seems that the degree of chlorination is of minor significance for the depression of NADH cytochrome c reductase and succinate cytochrome c reductase in the beef heart heavy microsomal fraction.[235]

N-demethylase is more sensitive to the activating effect of Aroclor 1260 than to that of Aroclor 1254, while *O*-demethylase is more sensitive to the activating effect of

TABLE 15

EFFECTS OF PCBs ON ENZYMATIC ACTIVITY IN FISH AND BIRDS

Ref.	Author	Year	Species	Compound, dosage	Tissue	Effects
242	Cutkomp *et al.*	1972	Blue-gill fish	PCBs	Muscle	Inhibition of oligomycin-insensitive Mg^{2+}-ATPase. (*in vitro*) Less effective in liver, kidney, and brain.
234	Desaiah *et al.*	1972	Fish	PCBs	4 tissues	Inhibition of oligomycin-insensitive Mg^{2+}-ATPase.
				Aroclor 1242	Brain	Inhibition of $(Na^+ + K^+)$-ATPases.
243	Davis *et al.*	1972	Rainbow trout	Aroclor 1242 & 1254 400 ppm	Gill	Inhibition of ATPases.
244	Risebrough *et al.*	1968	Pigeons	PCBs	Liver	Activation of microsomal hydroxylating enzymes.
245	Lincer & Peakall	1970	American kestrels	PCBs	Liver	Activation of microsomal hydroxylating enzymes.
163	Velle	1971	Birds	PCBs	Liver	Increased steroid hydroxylase activity.
229	Bailey & Bunyan	1972	Pigeons & quails	Aroclor 1254 Aroclor 1242	Liver	Increase in microsomal proteins. Induction of cytochrome P-450 activity. Aroclor 1254 is a more potent inducer.
241	Bitman *et al.*	1972	Japanese quail	PCBs & PCTs		Initial inhibition followed by stimulation of pentobarbital metabolism.
242	Cutcomp *et al.*	1972	Fat head minnows	Aroclor 1242/4m	Brain	75% inhibition on mitochondrial Mg^{2+}-ATPase and 35% in $(Na^+ + K^+)$ ATPase.
246	Koch *et al.*	1972	Fat head minnows (reared in aquaria)	Aroclor 1242 Aroclor 1254/4m		Inhibition of mitochondrial Mg^{2+}-ATPase. in brain. in brain.
247	Goldstein *et al.*	1975	Chickens	Hexachlorobiphenyl 400 ppm/3 wk		Increase of *p*-nitrophenol glucuronyl transferase, *d*-aminolevulinic acid synthetase and cytochrome P-450 activity.
248	Dieter	1974	Coturnix quail	Aroclor 1254/12 wk		Increased activity of plasma enzymes.
249	Dieter	1974	Coturnix quail	Aroclor 1254		Inhibition of biochemical adaptation to motion stress. The enzyme responses to stress were 2–3 times greater.
250	Dieter	1975	Wild starlings	PCB 5–100 ppm/ 12 wk		Induction of lactate dehydrogenase and creatinine kinase activity.
169	Scott	1977	White Leghorn Chickens	PCBs	Plasma Liver	Decreased glutathione peroxidase activity. Induction of microsomal benzpyrene hydroxylase.

TABLE 16

EFFECTS OF PCBs ON ENZYMATIC PRODUCTION IN MAMMALS

Ref.	Authors	Year	Species	Compound, dosage	Effects in Liver
238	Konat & Clausen	1973	Mice	Aroclor 1254 1500 ppm/7 d.	Induction in cytochrome P-450 and mixed function oxidase activities.
251	Abraham	1974	Charles River CD-1 mice	PCBs 500 ppm/7 d	Increased demethylase activity. Decreased glucose-6-phosphatase activity.
230	Fujita *et al.*	1971	Rats	Di- tetra -penta -hexachloro biphenyls	2,4,5,3′,4′-pentachlorobiphenyl was the most potent inducer.
252	Benthe *et al.*	1972	Rats	Di- and tetra-chlorobiphenyls (CB)	Stimulation of microsomal drug oxidizing enzymes is effective with tetra-CB
233	Bickers *et al.*	1972	Rats	Aroclor 1254 25 mg/kg/6 d.	Two fold increase in aniline hydroxylase, ethyl morphine *N*-demethylase and cytochrome P-450 and b_5 activity. Increased microsomal proteins.
				Aroclor 1216	1.4 fold increase in ethyl morphine *N*-demethylase and and cytochrome P-450 activity.
253	Letterst & Van Loon	1972	Rats	PCBs 150 ppm/kg/30d	Induction of nitroreductase and nitrohydroxylase activity. More active than DDT.
236	Letterst *et al.*	1972	Rats	Aroclors 1240–1260	Induction of demethylase, nitroreductase and cytochrome P-450 activity. Decreased activity of glucose-6-phosphatase. Aroclor 1260 is the more effective inducer.
221	Norbach & Allen	1972	Rats	Aroclor 1254, 200 ppb	Increased microsomal proteins/g of liver.
				Aroclor 5460 1 ppm	Increased esterase activity.
239	Bickers *et al.*	1973	Rats	PCBs skin application 25 mg/kg/d.6 d.	Induction of cytochrome P-450, ethylmorphine *N*-demethylase, aniline hydroxylase and benzpyrene hydroxylase. More than DDT.
162	Chen & DuBois	1973	Rats	PCBs	Increased activity of *N*-demethylase, most sensitive to Arochlor 1260, and of *O*-demethylase, most sensitive to Arochlor 1254.

240	Bickers *et al.*	1974	Rats	PCBs	3-fold induction of cytochrome P-448 and 10-fold induction of Benzo (a)-pyrene hydroxylase activity. More potent inducers than DDT.
254	Den Tonkelaar & Van Esch	1974	Rats	PCBs 2–200 ppm/2 wk	Increased aniline hydroxylase, aminopirine demethylase and hexobarbital activity.
218	Grant *et al.*	1974	Rats	Aroclor 1254 20 mg/kg/246d	Increase in microsomal enzyme activity.
				2 mg/kg	No effect.
231	Johnstone *et al.*	1974	Rats	Pure PCB compounds	Increase in *N*-demethylase, *O*-demethylase, aniline hydroxylase and nitroreductase activity.
255	Alvares & Kappas	1975	Rats (pregnant)	PCBs	Induction of aryl hydrocarbon hydroxylase in fetus, placenta and in neonatal livers, during lactation.
256	Cappon & Nicholls	1975	Rats	Aroclor 1254 i.p. 200 mg/kg/2 d	Increased incorporation of ^{14}C-phenylalanine-tRNA into polypeptide.
257	Fulfs & Abraham	1976	Rats	Aroclor 1221 25 mg/kg/7 m	Inhibition of glucose-6-phosphatase activity.
237	Marniemi *et al.*	1976	Rats	PCBs Single oral dose of 100 mg/kg	Induction of cytochrome P-450, NADPH cytochrome c reductase, *p*-nitroanisole *O*-demethylase, benzpyrene hydroxylase, glutathione S-transferase and UDP-glucuronuyl transferase activities.
200	Villeneuve *et al.*	1971	Rabbits (pregnant)	Aroclor 1254 10 mg/kg/28 d below 10 mg/kg	Induction of aniline hydroxylase and aminopirine *N*-demetylase activities. No-effect level.
258	Sharp *et al.*	1974	Rabbits (kidney)	PCBs	Inhibition of ($Na^+ + K^+$) ATPase activity.
			Beef (brain)		Inhibition of ($Na^+ + K^+$) ATPase activity.
235	Pardini	1971	Beef (heart)	PCBs	Inhibition of NADH-cytochrome c reductase and succinate cytochrome c reductase activities by 80%.
259	Wilson & Sharp	1972	Beef (brain)	Aroclor 1251 Aroclor 1254	Inhibition ($Na^+ + K^+$) ATPase activity by 50%.
189	Jensen *et al.*	1977	Dogs	PCBs 3.3 & 11.0 ppm in diet	Increased concentrations of liver cytochrome P-450 and renal cadmium.

Aroclor 1254.[162] That is, one PCB mixture may be a more potent inducer for one enzyme and a poorer inducer for another within a given animal species.

A relatively low concentration of PCBs is enough to activate some enzymes, while for other enzymes the concentration of PCBs has to be very high and/or the PCBs mixture used has to be highly chlorinated. Aroclors 1240 through 1260, at a dosage of 0.5 mg/kg for four weeks increase the activity of nitroreductase, but only a dosage of 50 mg will induce cytochrome P450, in rats.[236]

The degree of induction of enzymatic activity may differ in the organs of the same animal. The inhibitory effect of PCBs on the activity of oligomycin-insensitive $Mg^{2\pm}$ ATPase (present in endoplasmic reticulum) is more effective in muscle homogenates as compared to kidney, brain, and liver in Blue gill fish.[234]

The peak activity of several enzymes occurred between 2 weeks and 2 months in rats, after a single oral dose of 100 mg PCBs per kg.[232]

The induction of *N*-demethylase in rats was more pronounced in males than in females, and more in young than in adults.[162]

Organochlorine compounds have in common the biologic effect of upsetting the homeostasis of enzymatic activity. Generally, PCBs seem to be more potent inducers of enzymatic activities than DDT[238] and the effect is more persistent.[239] PCBs are more potent inducers of the hepatic mixed function oxydase system than DDT in rats[240] and mice.[238] Aroclor 1254 caused a 360% induction of cytochrome P450 activity, but lindane and DDT only induced 93% and 76%, respectively.[238] Aroclor 1260 has an estradiol-degrading potential four to five times that of *p,p'*-DDE or technical grade DDT.[51] After a single skin application of PCBs to rats, the enzyme induction was greater and persisted longer than with DDT.[239] However, PCBs were less effective than DDT in their inhibitory effect on mitochondrial $Mg^{2\pm}$ ATPase, in fish muscle homogenate.[234]

Stimulation of the enzymatic activity of pentobarbital metabolism by PCBs in the Japanese quail, follows an initial inhibition that did not occur with DDT and its analogues.[241]

Differences in the effects on enzymatic systems may explain differences in the biologic effects of organochlorine compounds.

Effects of PCBs on Drug Metabolism (Table 17)

PCBs were shown to reduce the sleeping time induced by several hypnotic drugs like phenobarbital, pentobarbital, and hexobarbital, the time of zoxazolamine-generated paralysis, and the time of action of other drugs. The reduction of drug-induced sleeping time was proportional to the chlorine content of the PCBs mixture used[232] and to the dose of PCBs,[260] and it was inversely proportional to the time elapsed between the administration of the PCBs and the performance of the test.[261] Divided doses were less active than a comparable total single dose.[131]

The reduction of phenobarbital sleeping time was greater in male quails than in females.[232] The reduction of hexobarbital sleeping time was equal for male, female, castrated, and sex-hormone-treated rats.[131] The effect of PCBs + DDT on the hexobarbital sleeping time was greater than that obtained with a 2.5 times higher dosage of PCBs when administered alone, and the effect of PCBs + β-BHC was that obtained with a ten times higher dosage of PCBs when given alone. The doses of DDT and β-BHC were respectively about twice and eight times lower than those which would have given the same effect when given alone.[261]

The paralysis following administration of zoxazolamine to rats, which had previously received Aroclor 1216, was reduced by about 50% and in rats which

TABLE 17

EFFECTS OF PCBs ON DRUG METABOLISM

Ref.	Authors	Year	Species	Compound	Effects
232	Cecil *et al.*	1975	Japanese quail	Aroclor 1221 through 5460 Single dose of 100 mg/kg	Prolonged phenobarbital sleeping time after 2 hours, similar to controls after 24 hours and reduced by 50% after 48 hours. The reduction was greater in males and was correlated with the percentage of chlorine in the PCBs. The effect was lower with PCTs.
260	Arachi & Tanaka	1974	Mice (F)	PCBs	Reduction of hexobarbital sleeping time, parallel to the dose of PCBs.
261	Arachi & Tanaka	1974	Mice (F)	PCBs 20 mg/kg, DDT 200 mg/kg and β-BHC 100 mg/kg.	Reduction of hexobarbital sleeping time (minimum effective doses).
				PCBs 20 mg/kg + DDT 200 mg/kg	Reduced to 44% after 2 days; 63% 7; 89% 14; as controls 21
				PCBs 5 mg/kg + β-BHC 25 mg/kg	65% 2; 70% 7; 92% after 14 days.
263	Kimura & Taniguchi	1975	Mice	PCBs	Shortened hypnotic effect of hexobarbital sodium.
			Rats	PCBs	Shortened hypnotic effect of hexobarbital sodium.
264	Street *et al.*	1969	Rats (F)	Aroclor 1221 50 mg/kg	Reduction of hexobarbital sleeping time by 11%.
				Aroclor 1248 50 mg/kg	Reduction of hexobarbital sleeping time by 35%.
				Aroclor 1268 50 mg/kg	Reduction of hexobarbital sleeping time by 48%.

TABLE 17 (*continued*)

EFFECTS OF PCBs ON DRUG METABOLISM

Ref.	Authors	Year	Species	Compound	Effects
233	Bickers *et al.*	1972	Rats	Aroclor 1254 25 mg/kg/6 d	Abolishment of zoxazolamine paralysis.
				Aroclor 1216 mg/kg/6 d	Reduction of zoxazolamine paralysis by about 50%.
				Aroclor 1254 25 mg/kg/6 d	Reduction of hexobarbital sleeping time by about 75%.
				Aroclor 1216 25 mg/kg/6d	by about 50%.
131	Tanaka & Komatsu	1972	Rats	Kanechlor-500 2 mg/kg/d/3d	to 49%.
				Kanechlor-0.4 mg/kg/d/15d	to 100%.
				30d	to 82%.
				/45d	to 87%.
				/53d	of hexobarbital sleeping time to 88%.
				Kanechlor-400 100 mg/kg/3d	Reduction equal for female, male, castrated and sex hormones treated rats.
192	Villeneuve *et al.*	1972	Rats	Aroclor 1254 chronic exp.	Reduction of pentabarbital sleeping time
				Aroclor 1260 chronic exp.	Reduction of pentabarbital sleeping time
				Aroclor 1221 chronic exp.	No effect.
231	Johnstone *et al.*	1974	Rats	PCBs	Reduction of phenobarbital sleeping time.

received Aroclor 1254, was abolished.[233] PCBs were the most potent inducers of zoxazolamine-metabolizing enzymes.[262]

There are quantitative and qualitative differences between xenobiotics so far as their influence on enzymatic activity is concerned. The resultant of the action of the xenobiotics of the internal milieu added to the biochemical individuality of drug-metabolizing enzymes will decide the plasma level of drugs and in turn their efficacy.

Effects of PCBs on Lipid Metabolism (Table 18)

Increase in plasma cholesterol in rats given 1000 and 2000 ppm PCBs in the diet[265] and 0.5, 2.5, 5.0, and 50.0 mg Kanechlor-400 per day during 2–6 weeks[266] and PCTs[267] were reported. A higher dose of Kanechlor-400, 50 mg per day for 2–6 weeks, was necessary in order to obtain an increase in plasma triglycerides in rats, whereas the plasma cholesterol and phospholipids increased with only 0.5 mg Kanechlor-400 under the same conditions.[266] A tenfold increase in serum triglyceride level was produced in rats given 100 mg Kanechlor-400 per kg during 4 weeks.[268] In rabbits given 8 ccs PCBs 1% in olive oil per day for 3 days, a twofold increase of total liver lipids and triglyceride levels was noted. When the treatment was extended to 11 days, an abnormal increase in serum triglyceride levels occurred.[269] In *Macaca mulatta* monkeys given 100 or 300 ppm Aroclor 1248 in the diet for 2–3 months, total serum lipids, cholesterol and triglycerides were reduced.[187] In Yusho patients increased serum triglyceride levels were reported.[126,270–272]

Effects of PCBs on Liver Porphyrins (Table 19)

PCBs produced hepatic porphyria in mice,[274] in rats, rabbits, quail, and chickens.[275] The increase in liver porphyrins was proportional to the dosage. Female rats[276] and female Japanese quail[277] were more susceptible than males to hepatic porphyria. Porphyrin granules were associated with lysosomes in the centri-lobular hepatocytes near the bile canaliculi, in mice and rats given Aroclor 1254, 25 ppm in diet[274] or Aroclor 1221, 25 ppm.[257] A 20-fold increase in the activity of mitochondrial ALA-synthetase occurred in male Japanese quail given 500 ppm Aroclor 1260 during 7 days and in female Japanese quail given only 100 ppm. The no-effect level was considered to be 0.1 ppm.[277]

Marked increase in the urinary excretion of coproporphyrin and fecal excretion of protoporphyrin was observed: in rats that received 500 ppm Kanechlor-300 or Kanechlor-500 in the diet,[278] in rabbits following 27 dermal applications of 118 mg PCBs for 38 days[279] or subcutaneous injections of Kanechlor-400 for 55 days[280] and in chickens given Phenchlor PD60, Clophen A60 or Aroclor 1260, 400 ppm.[281] PCBs were claimed to be nonporphyrinogenic in the guinea pig.[282]

A rapid accumulation of uroporphyrin III occurred in liver cell culture of chicken embryos treated with PCBs—by inhibition of uroporphyrinogen decarboxylase.[283]

The porphyrinogenic effect of PCBs was attributed to compounds that were detected in very small quantities in the technical PCBs. Pure tetrachlorobiphenyl did not produce porphyria in rats,[284] whereas some chickens from a group given hexachlorobiphenyl became porphyric.[247] These two experiments may suggest that higher chlorinated biphenyl compounds are responsible for the porphyrinogenic action or the tetrachlorobiphenyl was ineffective because of an inadequate dosage.

TABLE 18

EFFECTS OF PCBs ON LIPID METABOLISM

Ref.	Authors	Year	Species	Compound, dosage	Effects
265	Kiriyama *et al.*	1972	Rats	PCBs in the diet 1000 and 2000 ppm	Increase in plasma cholesterol (no effect if konjac mannan—a nonnutritive polysaccharide was administered simultaneously).
267	Kiriyama *et al.*	1974	Rats	PCBs	Significant increase in plasma and liver cholesterol.
				PCTs	Increase in plasma cholesterol.
				PCB substitutes	No effect, with the exception of kanetherm-T.
266	Nagai *et al.*	1971	Rats	Kanechlor-400 0.5, 2.5, 5, 50 mg/d/2–6 wk	Increase in plasma triglyceride in the 50 mg group. Plasma cholesterol and phospholipids increased in all the groups with the dosage.
268	Tanaka *et al.*	1969	Rats	Kanechlor-400 100 mg/kg 4 wk	Tenfold increase of serum triglyceride level.
236	Letterst *et al.*	1972	Rats	Aroclor 1242–1260 0.5, 5, 50, 500 mg/kg/4 wk	Increase in liver triglyceride level. The greatest increase was obtained with Arochlor 1248.
273	Nagai *et al.*	1972	Rats	Kanechlor-400 1 mg/kg/d/30 d	The rate of incorporation of ^{14}C-acetate into skin fatty acids of sterol esters was enhanced, suggesting that the chain elongation and desaturation of those fatty acids were activated.
269	Ito *et al.*	1971	Rabbits	PCBs 1% in olive oil 8 cc/d/3 d	Twofold increase of liver total lipid and triglyceride levels.
				11 d	Abnormal increase in serum triglyceride levels.
187	Allen	1975	Monkeys	Aroclor 1248 100 and 300 ppm/1 m.	Reduced total serum lipids, cholesterol and phospholipids.
270 271	Uzawa *et al.*	1969, 1971	Man	Yusho patients	Serum triglyceride levels higher than 300 mg/100 ml in 60 out of 396 patients.
272	Uzawa *et al.*	1972	Man	Yusho patients	Increase in serum triglyceride levels: 136 mg/100 ml (up to four times the normal values).
126	Okumura *et al.*	1974	Man	Yusho patients	Increase of plasma triglycerides: 134 mg/100 ml.

TABLE 19

EFFECT OF PCBs ON LIVER PORPHYRINS

Ref.	Authors	Year	Species	Compound, dosage	Effects
274	Fulfs	1976	Mice Rats	Aroclor 1254 20 mg/kg	Porphyrin granules associated with lysosomes in centrilobular hepatocytes.
220	Kimbrough *et al.*	1972	Mice Rats	PCBs	Hepatic porphyria.
285	Goldstein *et al.*	1973 1973	Mice Rats	Aroclor 1254 10 mg/kg/d/4 m	Hepatic porphyria.
278	Kawanishi *et al.*	Mice Rats		Kanechlor-300 500 ppm Kanechlor-500 500 ppm	Marked increase in urinary excretion of coproporphyrin and fecal excretion of protoporphyrin.
276	Goldstein *et al.*	1974	Mice Rats	Aroclor 1254 100 ppm Aroclor 1242 10, 100 ppm/21 d	Increased liver porphyrins (correlated to dose) Increased liver porphyrins (correlated to dose) Female rats more sensitive.
284	Kawanishi *et al.*	1974	Mice Rats	Tetrachlorobiphenyl (pure)	It did not produce porphyria.
257	Fulfs & Abraham	1976 Rats	Mice Rats	Aroclor 1221 25 ppm/7 m	Porphyrin granules in the centrilobular hepatocytes, near the bile canaliculi (first noted at 6 months).
282	Vos & Van Driel-Grootenhuis	1972	Guinea pigs	PCBs	Nonporphyrinogenic.
279	Vos & Beems	1971	Rabbits	PCBs dermal application	Increased excretion of copro- and protoporphyrins.
280	Miura *et al.*	1973	Rabbits	Kanechlor-400 s.c. inj. 55d	Porphyrinogenic.
275	Vos	1972	Rabbits Rats Quail Chickens	PCBs	Hepatic porphyria
247	Goldstein *et al.*	1974	Chickens	Hexachlorobiphenyl in diet 400 ppm/3 w.	Some chickens become porphyric.
281	Vos & Koeman	1970	Chickens	Phenchlor PD60, Clophen A60, Aroclor 1260, 400 ppm	Increased excretion of copro- and protoporphyrins.
283	Sinclair & Granick	1974	Chick embryos	PCBs	Rapid accumulation of uroporphyrin III in liver cells in culture. (Inhibition of uroporphyrinogen decarboxylase).

TABLE 19 (*continued*)

EFFECT OF PCBs ON LIVER PORPHYRINS

Ref.	Authors	Year	Species	Compound, dosage	Effects
277	Vos *et al.*	1971	Japanese quails (males)	Aroclor 1260 50–1000 mg/kg 7 d	Porphyrin fluorescence. 20-fold increase in mitochondrial ALA-synthetase at 500 mg/kg level. 10-fold increase at the 50 mg/kg level. The no-effect level was 0.1 mg/kg (liver residue 10.5 ppm.)
			(females)		20-fold increase of ALA-synthetase activity in the 100 mg/kg group.

TABLE 20

OTHER METABOLIC EFFECTS OF PCBs

Ref.	Authors	Year	Species	Compound, dosage	Effects
241	Bitman *et al.*	1972	Rats (M) & (F)	Aroclor 1242	Decrease of about 50% of the liver vitamin A
			Japanese quail (M) and females kept in the dark	Aroclor 1242	Decrease of about 50% of the liver vitamin A
179	Cecil *et al.*	1973	White Leghorn chickens	PCBs in diet	Potentiation of a vitamin E-selenium deficiency
289	Greichus *et al.*	1975	Penned white pelicans	100 mg PCBs/d/10 d.	Significant lowering of potassium, calcium and protein serum levels.
169	Scott	1977	Young Japanese quail	Aroclor 1242 100 ppm 2 m	Increase in liver weight Decrease in liver vitamin A level by about 50%
286	Bastomsky *et al.*	1975	Rats	PCBs	Decrease in serum bilirubin level by almost 50%
			Rats	Aroclor 1254 25 mg/kg d/4 d/in corn oil	Hyperbilirubinemia—no effective increase in UDP-glucuronosyntransferase (1.4.1.7.).
			Gunn rats		Serum bilirubin levels were significantly decreased.
287	Hirayama *et al.*	1974	Man	Yusho patients	Significant decrease of serum bilirubin. The levels were negatively correlated with PCBs and triglyceride blood levels.

OTHER METABOLIC EFFECTS OF PCBS (TABLE 20)

Besides these metabolic actions of PCBs, several others were reported:

A decrease of nearly 50% in serum bilirubin was described in Gunn rats, genetically deficient in UDP glucuronosyntransferase (2.4.1.7), given 25 mg Aroclor 1254 per kg per day for 4 days in corn oil.[286] In Yusho patients, a significant decrease was noted in serum bilirubin, negatively correlated with PCB and triglyceride blood levels.[287]

A decrease of about 50% was found in the liver vitamin A in rats and Japanese quail given Aroclor 1242 for 2 months.[179,288] A potentiation of a vitamin E-selenium deficiency was reported in white Leghorn chickens given PCBs in the diet.[286]

A significant lowering of serum potassium, calcium and of serum proteins was observed in penned white pelicans that received 100 mg of PCBs per day for 10 days.[289]

HORMONAL EFFECTS OF PCBs (TABLE 21)

Increases in the metabolism of gonadal steroid hormones in pigeons[244] of testosterone, estradiol 17 β and 4-androstene-3, 17 dione in one-week-old white Leghorn cockerels,[290] of estradiol in the American kestrel,[245] of estrogen in ring doves,[170] of progesterone in rats,[208] of estrogens and dehydroepiandrosterone in the adult boar,[291] and of androsterone, etiocolanolone and dehydroepiandrosterone in Yusho patients[292] were reported. Inhibition of 17 β [$^{-3}$H] estradiol binding, to the uterine receptor, by PCBs given to rats[293] and pathologic changes in the uterine horns (stenosis and occlusions) in seals with high PCBs and DDT residue levels were also described. An estrogenic activity of PCBs and PCTs, evaluated by the 18-hour glycogen response of the immature rat uterus test, was shown.[241,288]

One of the consequences of the low sexual hormones levels was a decrease in primary and secondary sexual characteristics. A decrease in comb and wattles and of the testicular weight in white Leghorn cockerels,[294–297] and a decrease of seminal vesicles in mice[298] occurred following PCBs administration. Another consequence of the changes induced by PCBs, in the metabolism of gonadal steroid hormones and other hormones like those of the thyroid and adrenal glands, was an impairement of the reproduction success at several stages of this process.

The effects of PCBs on the thyroid hormones were studied experimentally in animals. A decreased PBI serum level was found in rats given PCBs.[299,300] A reduced binding of thyroid hormones to serum proteins was reported, as shown by enhanced resin uptake of ^{125}I-triiodothyronine from serum. Also, an increased biliary excretion of ^{125}I-thyroxine and a larger proportion of biliary ^{125}I present as thyroxine-glucuronide were reported. In support of an increased degradation of thyroid hormones rather than a secretion failure by the thyroid gland, as the cause of the lowered thyroid hormones serum levels, is the reduced binding of thyroid hormones to proteins (that is, a higher percentage of free hormones allowed for their more active degradation in the liver).[299,300] Further, the negative feed back of the adenohypophysis-thyroid axis has been demonstrated by the following findings: enlarged thyroid glands in bobwhite quail,[301] enhanced uptake by the thyroid gland of ^{131}I in rats,[300] and ultrastructural features of hyperfunction in the thyroid cells that secrete thyroxine and triiodothyronine.[299]

The decreased levels of circulating gonadal and thyroid hormones seem therefore to be the result of their increased degradation by liver enzymes activated by the presence of PCBs in the animal body.

TABLE 21

HORMONAL EFFECTS OF PCBs IN BIRDS AND MAMMALS

Ref.	Authors	Year	Species	Compound	Effects on the Gonads
244	Risebrough *et al.*	1968	Pigeons	PCBs	Increased steroid metab. in liver homogenates.
287	Rehfield *et al.*	1971	Chickens	Aroclor 1248	Depression of secondary sexual characteristics.
294	Platonow & Funnell	1971	White Leghorn cockerels	Aroclor 1254, 500 ppm	Decreased development of comb and wattles.
				"250 ppm/13 wk	Testicles decreased to 50%.
295	Platonow & Funnell	1972	White Leghorn cockerels	PCBs, 500 ppm/4 wk	Lower comb and testicular weight.
290	Nowicki and Norman	1972	1 wk old	Aroclor 1254 & 1260	Increase in the metab. of: testosterone, estradiol-17β and 4-androstene-3,17 dione.
296	Iturri	1974	Single comb white Leghorn cockerels	Various PCBs, Chronic 50–200 ppm oral adm.	Depression of secondary sexual characteristics. Decreased testes weight.
245	Lincer & Peakall	1970	American kestrels	Cockerels inj. with Aroclor 1254 & 1262	Dose-dependent microsomal breakdown of estradiol to a more polar metabolite in liver.
170	Anonymous	1970	Ring doves (F)	PCBs 10 ppm/3 wk	Lowered estrogen levels.
145	Oerberg & Lundberg	1974	Mice, castrated (M)	Clophen A60 0.25 mg/d. during 28 days	Decreased seminal vesicles in testosterone propionate treated animals.
288	Bitman & Cecil	1970	Rats (immature)	PCBs 1221–1248 One inj.	Low estrogenic activity.
				PCTs	Low estrogenic activity.
306	Second report for Yusho	1971	Rats (castrated)	PCBs	Potentiation of uterotropic action of estradiol (it did not increase the uterine weight).
			Rats (immature)	PCBs	Increased uterine weight and uterine glycogen.
241	Bitman *et al.*	1972	Rats (immature)	PCBs	Low estrogenic activity.
293	Nelson	1974	Rats (immature)	PCBs	Inhibition of 17 β (-^{3}H) estradiol binding to uterine receptor.

207	Jonsson *et al.*	1976	Rats (F)	PCBs 75 ppm/8–36 wk. 150 ppm/8–36 wk.	Reduction of plasma progesterone. Histologic changes in ovarian stroma.
291	Platonow *et al.*	1972	Boar (adult)	Aroclor 1254 10 & 100 mg	Reduced urinary estrogen and dehydroepiandrosterone (with the higher dose).
292	Nagai *et al.*	1969	Man	Yusho patients	40% showed elevated urinary androsterone, etiocolanolone and dehydroepiandrosterone.
302	Wassermann & Wassermann	1972	Rats	PCBs 1221	Ultrastructural feature of adrenal zona fasciculata hyperfunction.
303	Wassermann & Wassermann	1973	Rats	PCBs 1221	Increased plasma corticosterone levels.
304	Flick *et al.*	1965	Rats	Aroclor 1242 200 ppm	Enlarged adrenals.
305	Wassermann *et al.*	1976	Rats	Aroclor 1254	Ultrastructural features of hyperfunction of β-cells of the endocrine pancreas.
299	Wassermann & Wassermann	1973	Rats	Aroclor 1221 Aroclor 1254	Decreased PBI serum levels. Ultrastructural features of hyperfunction of the thyroid cells.
300	Bastomsky	1974	Rats	Aroclor 1254 i.p. 25 mg/kg/d/4 d	4–5 fold increase in biliary excretion of ^{125}I-thyroxine. Increased biliary ^{125}I-thyroxine glucuronide. Enhanced uptake of ^{131}I. Reduced serum PBI levels. Reduced binding of thyroid hormones to serum proteins.
301	Hurst *et al.*	1974	Bobwhite quail	PCBs 500 ppm	Enlarged thyroid glands. They took up more iodide.

In contrast, corticosterone was found to be increased in the plasma of rats treated with PCBs,[302] and the adrenal zona fasciculata in these rats showed ultrastructural features of hyperactivity.[303] Enlarged adrenal glands were reported in cockerels given Arochlor 1242 in the diet.[304] This effect on glucocorticoids may be a consequence of a need for higher levels of glucosteroids as a defense against the stressor character of PCBs and perhaps also of the need for the catatoxic activity of corticosteroids.

The increased level of glucosteroids may produce in its turn, other biologic effects as a chain reaction to persistent heterostasis. Such may be the case of the immunosuppressive effect of PCBs and the hyperactivity of the β-cells of the endocrine pancreas. Ultrastructural features of hyperactivity were reported in the β-cells of the endocrine pancreas of rats given PCBs.[305] These findings were considered to be a result of solicitation of β-cells for carbohydrate homeostasis.

Effects of PCBs on the Immunologic Defense System (Table 22)

Decreased weight of lymphoid tissues in chickens[275,281,304] and in rabbits,[275] marked decrease in spleen weight in penned pheasants,[174] in single-comb white Leghorn cockerels,[296] and in rats,[281] atrophy of the thymus cortex and reduction of lymph nodes in rabbits,[279] and decrease in the bursa of Fabricius and spleen weight in progeny of chickens[307] were reported after PCB administration.

Decrease in total proteins and IgG were obtained in rats during subacute exposure to Aroclor 1221. The effect was higher than with DDT.[308] Decreased antibody response in chickens,[307] decreased serum antitoxin titers to tetanus toxoid and delayed hypersensitivity to tuberculin in guinea pigs[282] and a lower number of antitoxin-producing cells in the popliteal lymph nodes in guinea pigs given tetanus toxoid[309] were obtained with commercial mixtures of PCB.

In Yusho patients, a decrease in IgA and IgM and an increase in IgG were reported 2 years after the accidental alimentary poisoning.[310] Lower IgM levels were found in Yusho patients showing chloracne.[311]

Increased susceptibility to fungal disease in esturarine fish,[312] an increased susceptibility to hepatitis virus in Mallard ducklings,[313] an increased incidence of pneumonia and lung and intracranial abscesses in rats,[314] and an increased incidence of pneumonia and diarrhea in monkeys[315] were obtained experimentally with PCBs.

Chronic airway infections that persisted for several years were found in the majority of Yusho patients.[316]

From the experimental work and from the accidental poisoning which happened in Japan, it appears that the decreased immunologic response produced by PCBs may result in increased susceptibility to viral, bacterial, or fungal diseases.

The increased level of glucosteroids in animals given PCBs[303] may partially explain the reduced activity of the immunologic system. The fact that DDT also has been shown in animal experiments to reduce the immunologic response, though to a lesser degree[308] and in the absence of an increased glucosteroid plasma level, suggests the existence of more than one mechanism for the decrease in the immunologic response by means of organochlorine compounds.

Effects of PCBs on the Nervous System (Table 23)

An uncoordinated locomotor activity was reported in the weil-tailed goldfish kept in water containing 5 ppm Clophen A50.[318] An increased locomotor activity was

TABLE 22.

EFFECTS OF PCBs ON THE IMMUNOLOGIC DEFENSE SYSTEM

Ref.	Authors	Year	Species	Compound, Dosage	Biologic Effects
312	Hansen *et al.*	1971	Estuarine fish	Aroclor 1254	Increased susceptibility to fungal disease.
304	Flick *et al.*	1965	Chickens	PCBs	Atrophy of lymphoid tissues.
313	Friend & Trainer	1970	Mallard ducklings 10 d. old.	Aroclor 1254, 0, 0, 25, 50, 100 ppm	Increased susceptibility to duck hepatitis virus; 14, 35, 65, 94% mortality respectively.
281	Vos & Koeman	1970	Chickens	PCBs	Atrophy of lymphoid tissues.
275	Vos	1972	Chickens	PCBs	Atrophy of lymphoid tissues.
174	Dahlgren *et al.*	1972	Penned pheasants	Aroclor 1254 1:9 in corn oil	Marked splenic atrophy in the dead birds. (300–400 ppm PCBs levels in brain tissue).
296	Iturri	1974	Single comb white Leghorn cockerels	Aroclor 1254 chronic oral adm.	Decreased spleen weight.
307	Harris	1976	Chickens	PCBs	Decreased antibody response and decrease in Bursa of Fabricius and spleen weight in progeny.
279	Vos & Beems	1971	Rabbits	PCBs	Atrophy of the thymus cortex. Reduction of lymph nodes.
275	Vos	1972	Rabbits	PCBs	Atrophy of lymphoid tissue.
308	Wassermann *et al.*	1973	Rabbits	Aroclor 1221 250 ppm/3 m.	Decrease in total proteins and IgG. (More than with *p,p′*-DDT).
281	Vos & Koeman	1970	Rats	Phenchlor DP6 2‰	Small spleens.
314	Kimura & Baba	1973	Rats	Kanechlor-400 in diet.	Increased incidence of pneumonia. Lung and intra-cranial abscesses.
316	Shigematsu *et al.*	1974	Rats	PCBs	Increased number of lysosomes and large vacuoles in alveolar macrophages.
282	Vos & Van Driel-Grootenhuis	1972	Guinea pigs	Aroclor 1260 Clophen A60 50 mg/kg	Decreased serum antitoxin titres, delayed hyper-sensitivity to tuberculin, atrophy of the thymus.
309	Vos & Roij	1972	Tetanus toxoid	Aroclor 1260 0, 10, and 500 ppm for 8 wk	Significant reduction of immunoglobulins. Lower number of antitoxin-producing cells in the popliteal lymph nodes.
315	Nishizumi	1970	Monkeys	PCBs	Pneumonia or diarrhea were the main causes of death.
317	Allen & Norbach	1973	Rhesus monkeys	PCBs	Lymphopenia
310	Shigematsu *et al.*	1971	Man	Yusho patients	Decrease in IgA and IgM and increase in IgG, two years after the alimentary poisoning.
311	Saito *et al.*	1972	Man	Yusho patients	Lower IgM levels in patients showing chlorance.
316	Shigematsu *et al.*	1974	Man	Yusho patients	Chronic airway infections that persisted several years.

TABLE 23

EFFECTS OF PCBs ON THE NERVOUS SYSTEM

Ref.	Authors	Year	Species	Compound, Dosage	Effects
318	Hattula & Karlog	1972	Weil-tailed goldfish	Clophen A50 5 ppm in water	Uncoordinated movements.
319	Ulfstand *et al.*	1971	Robins	Clophen A50, 5 μg inj. into mealworms	Increased locomotor activity. PCB residues level in the breast muscle was 4 times that in controls.
320	Karlsson *et al.*	1974	Red starts	PCBs 11 μg into meal-worms/d/12 d	Increased locomotor activity during the early night hours, at the peak of normal migratory activity. The residue levels were as those reported for wild birds.
174	Dahlgren *et al.*	1972	Penned pheasants	Aroclor 1254 in corn oil, 12.5 & 50 mg/wk	The ability to avoid hand capture was less efficient in the offspring.
321	Kreitzer & Heinz	1974	Coturnix quail	Aroclor 1254 in diet 200 ppm/8 d at 7 d of age followed by 6 d of normal diet.	Group avoidance response significantly suppressed. No recovery was observed. Increased mortality from predators may occur.
322	Fukuda	1972	Rats	Kanechlor-400 100 mg/kg/d/12 d 300 mg/kg/d/6 d	Changes in cortical evoked responses elicited by stimulation of sciatic nerve: high N_2 waves and lack of N_2 waves.
324	Fukuda *et al.*	1974	Rats & mice	Kanechlor-400	No effect on the maximal electroshock seisure in mice.
225	Iatropoulos *et al.*	1977	Rhesus monkeys	Clophen A30 4, 16, 64 mg/kg/28 d	Central nervous system degenerative changes due to secondary hypoxia (reversible changes in the endothelium of small vessels).
323	Murai & Kuroiwa	1971	Man	Yusho patients	Numbness or pains in arms and legs. Decrease in the conduction velocity in peripheral sensory nerves (in the radial or sural nerves or both).

obtained in robins fed Clophen A50-injected mealworms,[319] and in redstarts given 11 μg PCBs (in mealworms) per day, for 12 days.[320] The increased locomotor activity in redstarts began during the early night hours at the peak of normal migratory activity.

Offspring of penned pheasants given Aroclor 1254 in corn oil, 12.5 and 50 mg per week were less able to avoid hand capture-.[253] The group-avoidance response was significantly suppressed in coturnix quail treated with 200 ppm Aroclor 1254 for 8 days, beginning at the age of seven days and followed by 6 days of normal diet.[321] Such effects in birds naturally exposed to PCBs may cause increased mortality from predators.

Changes in cortical-evoked responses elicited by stimulation of the sciatic nerve, high N_2 waves and lack of N_2' waves were noted in rats given Kanechlor-400.[322] Decrease in the conduction velocity in peripheral sensory nerves and numbness and pains in arms and legs were reported in Yusho patients.[323]

Degenerative changes in the myelin sheath of nerve fibers and in the brain were described in Rhesus monkeys given Clophen A30 for 28 days. The brain lesions were attributed to reversible changes, produced by PCBs, in the endothelium of small vessels, which lead to secondary hypoxia.[324]

Dermatologic Effects of PCBs (Table 24)

Hyperplasia and hyperkeratosis of the epidermal and follicular epithelium were obtained with Phenchlor DP6 and Clophen A60 in rabbits.[279] Aroclor 1260 in dermal application gave less severe skin lesions, as compared with the above mentioned compound,[222,279] whilst 2,4,5,2′,4′,5′-hexachlorobiphenyl produced moderate skin lesions.[222]

Other authors found that the skin lesions produced by technical compounds in rabbits were identical to those produced by pure chlorinated biphenyls (hyperkeratosis, dilatation of hair follicles and cysts filled with keratinous material).[225]

In Rhesus monkeys given 25 ppm Aroclor 1248 in the diet, loss of hair and acne occurred after one month.[226] *Macaca mulatta* monkeys given 100 or 300 ppm Aroclor 1248 in the diet had lost, within 3 weeks to 1 month, most of their hair and eyelashes. The eyelids and the mouth were edematous. Small comedones appeared on the cheeks and the neck. Erythema and acne of the face and neck developed within 1–2 months. The lesions were severe in females and moderate in males. The major microscopic changes were of large intrafollicular keratin cysts and epithelial hyperplasia of the hair follicles, particularly of the face.[187]

In occupationally exposed persons, skin lesions were noted.[127] Cases of chloracne were reported among workers exposed to these compounds for 19 months.[327] Three cases of malignant melanomas, in persons occupationally exposed to PCBs, were ascribed to the carcinogenic effects of PCBs,[328] although, other compounds to which those patients were exposed concomitantly could be charged with the responsibility.[329]

In Yusho patients, pigmentation of the skin, nails, and gingiva, and chloracne eruptions appeared.[330,331] Hypersecretion of Meibomian glands was also reported.[331] The severely affected patients showed no spontaneous improvement in their cutaneous manifestations.[332] In Yusho babies, chloracne and a dark discoloration of the skin and mucous membranes, resembling that of Addison's disease was found to occur.[204,214]

TABLE 24

DERMATOLOGIC EFFECTS OF PCBs

Ref.	Authors	Year	Species	Compound, Dosage	Effects
318	Hattula & Karlog	1972	Weil-tailed goldfish	Clophen A50 5 ppm in water	Change in skin color (orange to pale yellow).
279	Vos & Beems	1971	Rabbits	Phenchlor DP6 Clophen A60 Aroclor 1260 (dermal application)	Reddening of skin, transverse wrinkling, hyperplasia, and hyperkeratosis of the epidermal and follicular epithelium Less marked effects and minimal hyperplasia and hyperkeratosis of the follicular epithelium.
222	Vos & Notenboom	1972	Rabbits	Aroclor 1260 2,4,5,2′,4′,5′-hexachlorobiphenyl	Severe skin lesions. Less severe skin lesions.
325	Komatsu & Kikuki	1972	Rabbits	Kanechlor-400 Kanechlor-500 3,4,3′,4′-tetrachlorobiphenyl	The skin lesions produced by technical compounds were identical to those produced by pure chlorinated biphenyls (hyperkeratosis, dilatation of hair follicles and cysts filled with keratinous material).
326	Allen *et al.*	1974	Rhesus monkey	Aroclor 1248 25 ppm/2 months	Loss of hair and acne occurred after one month.

187	Allen	1975	Macaca mulatta monkeys	Aroclor 1248 100 or 300 ppm/2–3 m	Within 3 wk-1 m. the animals had lost most of their hair (face, head and neck), the mouth and eyelids were edematous and small comedones appeared on the cheeks, the neck and around the mouth. The major microscopic changes were of large intrafollicular keratin cysts, epithelial hyperplasia of the hair follicles, particularly of the face.
				Aroclor 1248 2.5 or 5.0 ppm	Facial edema, alopecia, erythema and acne that involved the face and neck developed in 1–2 months. The lesions were severe in females and moderate in males.
128	Hasegawa *et al.*	1972	Man	Occupational exposure	Skin lesions were noted.
327	Meighs *et al.*	1954	Man	Occupational exposure	Cases of chloracne among the workers exposed for 19 months.
330	Toshitani	1972	Man	Yusho patients	Acneform eruptions in covered areas. Improvement in exposed areas.
331	Urabe	1974	Man	Yusho patients	Muco-cutaneous symptoms. Hyperpigmentation of gingiva. Hypersecretion of Meibomian glands.
332	Khoda *et al.*	1974	Man	Yusho patients	The severely affected patients showed no spontaneous improvement of cutaneous manifestations.
204	Miller	1974	Man	Yusho babies	Chloracne in 10 cases.
214	Fraumeni	1974	Man	Yusho babies	Peculiar dark color of skin and mucous membranes resembling those found in Adisson disease.

TABLE 25

TUMORIGENIC EFFECTS OF PCBs

Ref.	Authors	Year	Species	Compound, Dosage	Effects
9	Falkmer *et al.*	1977	Hagfish of fjord of open sea	Naturally occurring	Hepatomas 5.8% in 1972 and 0.6% in 1974–1976 Hepatomas 2.8% in 1972 and 0.9% in 1974–1976 Islet cell neoplasms and hamartomas 0.5% and less than 0.1% for the surveys of 1972 and 1974–1976 for the fjord and open sea hagfish.
328	Bahn *et al.*	1976	Man	Occupational exposure	Melanoma
223	Ito *et al.*	1973	Mice	Kanechlor-500 in diet 500 ppm/12 m	Nodular hyperplasias and well differentiated hepatocellular carcinomas in the liver Promotion of carcinogenesis by α-BHC and β-BHC.
333	Kimbrough *et al.*	1974	Mice	Aroclor 1254	Hepatomas and adenofibrosis
334	Nagasaki *et al.*	1974	Mice	Kanechlor-500 500 ppm/ α-BHC 250 ppm/24 wk Kanechlor-500 100 ppm + α-BHC 50 ppm	32 wk Neoplastic changes in liver 32 wk Neoplastic changes in liver 32 wk Neoplastic changes in liver
337	Nagasaki *et al.*	1975	Mice	Penta- tetra- or trichloro-biphenyls, 100, 250, 500 ppm for 24 wk Penta- and tetra-chloro-biphenyl 100 ppm + 50 ppm α-BHC	Liver tumorigenesis in higher incidence at higher dose rates. Liver tumorigenesis of α-BHC was promoted also by pure chlorinated biphenyls.
314	Kimura & Baba	1973	Rats	Kanechlor-400 (1000–1500 mg—the total for 400 days)	Tumoral liver change in 6/10 females and 0/10 males.
336	Makiura *et al.*	1974	Rats	Kanechlor-500	Inhibited the induction of liver tumors by several liver carcinogens.
337	Kimbrough *et al.*	1975	Rats (F)	Aroclor 1260 100 ppm/21 m	Hepatocellular carcinomas 26/184 Hyperplastic nodules 146/184 Adenofibrosis
338	Wassermann *et al.*	1978	Rats (F)	Aroclor 1254 200 ppm/25 m	Hepatocellular adenomas.
339	Kimura *et al.*	1976	Rats	PCBs	Promotors in experimental carcinogenesis.
340	Allen & Norbach	1973	Rhesus monkeys	Aroclor 1248 300 mg/kg PCTs 5000 mg/kg	Hyperplasia and dysplasia of gastric mucosa with invasion of adjacent tissues.

Tumorigenic Effects of PCBs (Table 25)

Carcinogenicity in animal experimentation is one of the tests for toxicity of environmental chemical pollutants. Experimental studies of the carcinogenicity of PCBs were carried out in rodents. Hepatocellular adenomas, carcinomas, and adenofibrosis were obtained in mice with commercial mixtures of PCBs[223,333,334] and with pure biphenyl compounds[335] and in rats with technical mixtures of PCBs.[314,336–338] The incidence of the tumors increased with the dosage.[335] Very high doses were needed to produce adenomas in male rats.[336] The administration of technical PCBs or of pure chlorinated biphenyls (tetra- and penta-chlorobiphenyls) together with α-BHC, in doses which given separately would have no carcinogenic effect, induced liver tumors.[335,339] These facts suggest that PCBs may act as promotors of experimental liver carcinogenicity. For other hepatocarcinogens like 3′- methyl - *p* - dimethylazobenzene, *N*- 2 - fluosenyl acetamide and diethylnitrosamine, PCBs proved to be inhibitors of carcinogenesis.[336] Hyperplasia of the epithelium of the hair follicles, particularly of the face and severe hypertrophic and hyperplastic gastritis were produced experimentally in adult monkeys.[339]

Hepatomas and islet-cell hamartomas and neoplasms occurred in cyclostomes, particularly in hagfish in the Gullmar fjord, Sweden, and in deep sea waters. The incidence of these tumors was 5.8% for hepatomas in the Gullmar fjord hagfish and only 2.8% for those collected at open sea, some 12 km from the coast, in 1972. In 1974–1976 these percentages became 0.6 and 0.9% respectively. Islet cell hamartomas and frank neoplasms, which were found in 0.5% of cases in 1972, decreased to less than 0.1% in 1974–1976. Subcutaneous and mesenterial neoplasms were occasionally observed. PCB levels of 5.0 ppm and 0.1–0.4 total DDT were found in fjord hagfish livers and only 0.2 ppm PCBs in open sea hagfish.[9]

Two malignant melanomas were found in a group of 31 persons occupationally exposed to PCBs. An incidence rate of only 0.04 malignant melanomas would be expected for this group. Another case, out of a group of 41 persons occupationally exposed to PCBs, was also reported.[328] There is no general agreement concerning the induction of melanomas by PCBs, since it seems that the group of workers mentioned was exposed concomitantly to other carcinogens, especially epoxides.[329]

Summary

PCB residues are found in biota all over the world. A biologic magnification of PCBs has been demonstrated in the food chain: plankton-fish-fish eating birds.

A world map of the PCBs residues in biota and some of their biologic effects are given in this study.

The biologic effects of PCBs are varied and may generally be explained by the induction or inhibition of the activity of a large number of enzymes, which upsets quantitatively, normal biochemical processes. Harm to reproduction, growth, development, defense systems, tissues, and organs appeared in susceptible organisms as a result of such changes or as a chain reaction to heterostases. The adverse health effects, observed in persons occupationally exposed and in those accidentally poisoned, point to the risk for the general population of an ever-increasing environmental pollution by PCBs.

There is need for an integrated approach, consisting of epidemiologic studies on high risk groups in the general population and in persons occupationally exposed, as well as periodic assessment of PCB residues in man, his food, and feed supplies.

References

1. Goldwater, L. 1967. WHO Chronicle **21:** 191–192.
2. Wassermann, M. & D. Wassermann. 1966. 15th Int. Congr. Occup. Health. Vienna, Abstracts, Part 2, **6:** 954.
3. Wassermann, M., L. Tomatis & D. Wassermann. 1975. Pure Appl. Chem. **42:** 189–208.
4. Jensen, S. 1966. New Sci. **32:** 612.
5. Kohanawa, M., S. Shoya, Y. Ogura, M. Moriwaki & M. Kawasaki. 1969. Nat. Inst. Anim. Health Quart. **9:** 213–219.
6. Nimmo, D. K., R. R. Blackman, A. J. Wilson Jr. & J. Forester. 1971. Mar. Biol. **11:** 191–197.
7. Anonymous. 1972. Chemistry **45:** 22.
8. Holden, A. V. 1970. Nature **228:** 1220–1221.
9. Falkmer, S., S. Marklund, P. E. Mattson & C. Rappe. 1977. Ann. N.Y. Acad. Sci. **298:** 342–355.
10. Holden, A. V. 1973. Pestic. Monit. J. **7:** 37–52.
11. Claeys, R. R., R. S. Caldwell, N. H. Cutshall & R. Holton. 1975. Pestic. Monit. J. **9:** 2–10.
12. Duke, T. W., J. I. Lowe & J. I. Wilson Jr. 1970. Bull. Environ. Contam. Toxicol. **5:** 171–180.
13. Baluja, G., J. M. Franco & M. A. Murado. 1973. Invest. Pesq. **37:** 593–620.
14. Addison, R. R. & P. F. Brodie. 1977. J. Fish. Res. Board. Can. **34:** 937–941.
15. Holden, A. V. 1972. Monitoring organochlorine contamination of the marine environment by the analysis of residues in seals. *In* Marine Pollution and Sea Life. pp. 266–272. Fishing News (Books) Ltd. London.
16. Taruski A. G., C. E. Olney & H. E. Winn. 1975. J. Fish. Res. Board Can. **32:** 2205–2209.
17. Gaskin, D. E., G. J. D. Smith, P. W. Arnold, M. V. Louisy, R. Frank, M. Holdrinet, J. W. Mc Wade. 1974. J. Fish. Res. Board. Can. **31:** 1235–1239.
18. Addison, R. F., S. R. Kerr & J. Dale. 1973. J. Fish Res. Board. Can. **30:** 595–600.
19. Risebrough, R. W., V. Wreeland, G. R. Harvey, H. P. Miklas & G. M. Carmignani. 1972. Bull. Environ. Contam. Toxicol. **8:** 345–355.
20. Veith, G. D., D. W. Kuehl, F. A. Puglisi, G. E. Glass & J. G. Eaton. 1977. Arch. Environ. Contam. Toxicol. **5:** 487–499.
21. Giam, C. S. M. K. Wong, A. R. Hanks, W. M. Sackett & R. L. Richardson. 1973. Bull. Environ. Contam. Toxicol. **9:** 376–382.
22. Linko, R. R., P. Rantimaki & K. Urpo. 1974. Bull. Environ. Contam. Toxicol. **12:** 733–738.
23. Ware, D. M. & R. F. Addison 1973. Nature **246:** 519–521.
24. Bowes, G. W., B. R. Simoneit, A. L. Burlingam, B. W. De Lappe & R. W. Risebrough. 1973. Environ. Health Perspect. **5:** 191–198.
25. Buhler, D. R., R. R. Claeys & B. R. Mate. 1975. J. Fish. Res. Board Can. **32:** 2391–2397.
26. Helle, E., M. Olsson & S. Jensen. 1976. Ambio **5:** 188–189.
27. Saschenbrecker, P. W. 1973. Can. J. Comp. Med. **37:** 203–206.
28. Zitko, V. & O. Hutzinger. 1972. Schriftenr. Ver. Wasser Boden Lufthyg. Berlin Dahlem **37:** 121–128.
29. Jensen, S., A. G. Johnels, M. Olsson & G. Otterlind. 1969. Nature **224:** 247–250.
30. Buhler, D. R. & R. R. Claeys. 1972. Chlorinated hydrocarbon pesticide and mercury content of freshwater fishes. In Oreg. State Univ. Environ. Health Sci. Cent. Annual Progr. Rep. p. 215.
31. Carr, R. L. 1972. Pestic. Monit. J. **6:** 23–26.
32. Haile, G. L., G. D. Veith, W. C. Boyle & G. F. Lee. 1975. Nat. Tech. Inform. Serv. PB-243, 364, 35p.
33. Veith, G. D. 1975. Pestic. Monit. J. **9:** 21–29.

34. Parejko, R., R. Johnston & R. Keller 1975. Bull. Environ. Contam. Toxicol. **14:** 480–488.
35. Viviani, R., G. Crisetig, V. Petruzzi & P. Cortesi. 1971. Atti. Soc. Ital. Sci. Vet. **25:** 429–433.
36. Monod, J. L., H. Raybaud, C. Venot & G. Carrara. 1971. Bull. Soc. Pharm. Marseille **20:** 155–160.
37. Doguchi, M., F. Ushio & M. Abe. 1971. Annu. Rep. Tokyo Metrop. Res. Lab. Pub. Health **23:** 271–276.
38. Hauge, S. 1972. Nor. Skogind. **26:** 181–182.
39. Bache, C. A., J. W. Serum, W. D. Youngs & D. J. Lisk. 1972. Science **177:** 1191–1192.
40. Anonymous 1971. Med. World News **12:** 24.
41. Smith, F. A., R. P. Sharma, R. J. Lynn & J. B. Low. 1974. Bull. Environ. Contam. Toxicol. **12:** 218–223.
42. Baird, R. C., N. P. Thompson, T. L. Hopkins & W. R. Weiss. 1975. Bull. Mar. Sci. **25:** 473–481.
43. Zitko, V. 1972. Int. J. Environ. Anal. Chem. **1:** 221–231.
44. Sims, G. G., C. E. Cosham, J. R. Campbell, M. C. Murray. 1975. Bull. Environ. Contam. Toxicol. **14:** 505–512.
45. Bjerk, J. E. & R. Sundby. 1970. Nor. Vet. Tidsskr. **82:** 241–246.
46. Bjerk, J. E. 1972. Nord. Veterinar. med. **24:** 451–457.
47. Linko, R. R., J. Kaitaranta, P. Rantamaki & L. Eronen. 1974. Environ. Pollut. **7:** 193–207.
48. Jensen, S., A. G. Johnels, M. Olsson & G. Otterlind. 1973. Oikos **15:** 230–231.
49. Eichner, M. 1973. Z. Lebensm. Unters. Forsch. **151:** 376–383.
50. Huschenbeth, E. 1973. Arch. Fischereiwiss. **24:** 105–116.
51. Gustafson, C. G. 1970. Environ. Sci. Technol. **4:** 814–819.
52. Anonymous 1970. Nature **228:** 103.
53. Scott, J. M., J. A. Wiens & R. R. Claeys. 1975. J. Wildl. Manage. **39:** 310–320.
54. Marion, W. R. 1976. Pestic. Monit. J. **10:** 84–86.
55. Prestt, J., J. Jefferies & N. W. Moore. 1970. Environ. Pollution **1:** 3–26.
56. Parslow, J. L. F., D. J. Jefferies & H. M. Hanson. 1973. Mar. Pollut. Bull. **4:** 41–43.
57. Karlog, O., J. Kraul & S. Dalgaard-Mikkelsen. 1971. Acta Vet. Scand. **12:** 310–312.
58. Acker, L. & E. Schulte. 1971. Ernaehrungsforschung **16:** 559–567.
59. Baluja, G., J. M. Franco & M. A. Murado. 1974. Rev. Agroquim. Technol. Aliment. **14:** 229–245.
60. Risebrough, R. W., G. L. Florant & D. D. Berger. 1970. Can. Field Natur. **84:** 247–253.
61. Parslow, J. L. F. & D. J. Jefferies. 1973. Environ. Pollut. **5:** 87–101.
62. Koeman, J. H., H. C. W. Van Velzen-Blad, R. De Vries & J. G. Vos. 1973. J. Reprod. Fert. **19:** 353–364.
63. Anonymous. 1971. Chem. Eng. News. **49:** 32–33.
64. Hays, H. & R. W. Risebrough. 1972. Auk **89:** 19–35.
65. Belisle, A. A., W. L. Reichel, L. N. Locke, T. G. Lamont, B. M. Mulhern, R. M. Prouty, R. B. De Wolf & E. Cromartie. 1972. Pestic. Monit. J. **6:** 133–138.
66. Mc Lane, M. A. R., L. F. Stickel, E. R. Clark & D. L. Hughes. 1973. Pestic. Monit. J. **7:** 100–103.
67. Greichus, Y. A., A. Greichus & R. J. Emerick. 1973. Bull. Environ. Contam. Toxicol. **9:** 321–328.
68. Fisher, H. J. 1973. Pacific Sci. **27:** 220–225.
69. Dacre, J. C. 1974. Bull. Environ. Contam. Toxicol. **11:** 517–522.
70. Kreitzer, J. F. 1974. Pestic. Monit. J. **7:** 195–199.
71. Cromartie, E., W. L. Reichel, L. N. Locke, A. A. Belisle, T. E. Kaiser, T. G. Lamont, B. Mulhern, R. M. Prouty & D. M. Swineford. 1975. Pestic. Monit. J. **9:** 11–14.

72. Zitko, V., O. Hutzinger, W. D. Jamieson & P. M. K. Choi. 1972. Bull. Environ. Contam. Toxicol. **7:** 200–201.
73. Odsjö, R. 1973. PCB in some Swedish terrestrial organisms. *In* PCB Conference II Nat. Swedish Environ. Protection Board Publications: 4E. pp. 45-58.
74. Bogan, J. A. & W. R. P. Bourne. 1972. Nature (London) **240:** 358.
75. Braestrup, L., J. Clausen & O. Berg. 1974. Bull. Environ. Contam. Toxicol. **11:** 326–332.
76. Koeman, J. H., T. Bothof, R. De Vries, H. Van Velzen-Blad & J. G. Vos. 1972. TNO-Nieuws **27:** 561–569.
77. Figge, K., H. Hoerschelmann & K. Polzhofer. 1976. Hosp. Hyg. Gesundheitswes. Desinfekt. **68:** 354–360.
78. Faber, R. A. & J. J. Hickey. 1973. Pestic. Monit. J. **27:** 36.
79. Wiemeyer, S. N., B. M. Mulhern, F. J. Ligas, R. J. Hensel, J. E. Mathiven, F. C. Robards & S. Postupalsky. 1972. Pestic. Monit. J. **6:** 50–55.
80. Grier, J. W. 1977. Can. Field Nature **8:** 469–475.
81. Blus, L. J., R. G. Heath, C. D. Gish, A. A. Belisle & R. M. Prouty. 1971. Bioscience **21:** 1213–1215.
82. Blus, L. J., B. S. Neely Jr., T. G. Lamont & B. Mulhern. 1977. Pestic. Monit. J. **11:** 40–53.
83. Zitko, V. 1972. Bull. Environ. Contam. Toxicol. **7:** 105–110.
84. Zitko, V. & P. M. K. Choi. 1972. Bull. Environ. Contam. Toxicol. **7:** 63–64.
85. Bloch, D. & I. Kraul. 1972. Acta Vet. Scand. **13:** 588–590.
86. Conrad, B. 1977. Naturwissenschaften **64:** 43–44.
87. Koivusaari, J., A. Laamanen, J. Nuuja, R. Palokangas & V. Vihko. 1972. Work Environ. Health **9:** 44–45.
88. Koeman, J. H., R. H. Hadderingh & M. F. J. J. Bijleveld. 1972. Biol. Conserv. **4:** 373–377.
89. Fog, M. & J. Kraul. 1973. Acta Vet. Scand. **14:** 350–352.
90. Koeman, J. H., C. F. van Bensekom & J. J. M. de Greif. 1972. TNO-Nieus **27:** 542–550.
91. Gress, F., R. W. Risebrough & F. C. Sibley. 1971. Condor **73:** 368–369.
92. Longcore, J. R. & B. M. Mulhern. 1973 Pestic. Monit. J. **7:** 62–66.
93. Mes, J., D. E. Coffin & D. Campbell. 1974. Pestic. Monit. J. **8:** 8–11.
94. Bjerk, J. E. & G. Holt. 1971. Acta Vet. Scand. **12:** 429–441.
95. Mestres, R. & S. Illes. 1973. Trav. Soc. Pharm. Montpellier **33:** 201–208.
96. Anonymous. 1972. Chem. Eng. News **50:** 28–30.
97. Kurtz, D. A. & K. C. Kim. 1977. Pestic. Monit. J. **10:** 79–83.
98. Friedman, M. A., F. D. Griffith & S. Woods. 1977. Arch. Environ. Contam. Toxicol. **5:** 457–469.
99. Bruggeman, Von J., L. Busch, U. Drescher-Kadan, W. Eiseli & P. Hoppe. 1974. Z. Jagdwiss. **20:** 70–74.
100. Hidaka, K. T. Ohe & K. Fujiwara. 1972. J. Food Hyg. Soc. Jpn. **13:** 523–530.
101. Bjerk, J. E. 1972. Tidsskr. Nor. Laegeforen. **92:** 15–19.
102. Abbot, D. C., G. B. Collins & R. Goulding. 1972. Br. Med. J.: 553–556.
103. Freudenthal, J. & P. A. Greve. 1973. Bull. Environ. Contam. Toxicol. **10:** 108–111.
104. Hattula, M. L., J. Jkkala, M. Isomaki, K. Maatta & A. V. Arstila. 1976. Acta Pharmacol. Toxicol. **39:** 545–554.
105. Pesendorfer, von H., I. Eichler & E. Glofke. 1973. Wien. Klin. Wochenschr. **85:** 218–222.
106. Acker, L. & E. Schulte. 1974. Naturwissenschaften **61:** 32.
107. Price, H. A., & R. L. Welch. 1972. Env. Health. Pespect. **1:** 73–78.
108. Yobs, A. R. 1972. Environ. Health. Perspect. **1:** 79–81.
109. Doguchi, M., S. Fukano & F. Ushio. 1974. Bull. Environ. Contam. Toxicol. **11:** 157–158.
110. Curley, A., V. W. Burse & M. E. Grim. 1973. Food Cosmet. Toxicol. **11:** 471–476.
111. Masuda, Y. & R. Kagawa. 1974. Bull. Environ. Contam. Toxicol. **11:** 213–216.
112. Ui, J. 1972. Laekartidningen **69:** 2789–2796.

113. NISHIMOTO, T., M. UYETA, S. TAUE, K. OKIKAZAWA, J. NISHIUCHI & K. KONDO. 1972. Proc. Med. Sci. **82:** 515–516.
114. INOUE, Y., S. ABE, M. TAKAMATSU, N. AOKI, S. MIKI & K. FUJIWARA. 1975. Fukuoka Igakkai Zasshi **66:** 610–616.
115. ACKER, L. & E. SCHULTE. 1970. Naturwissenschaften **57:** 497.
116. ACKER, L. & E. SCHULTE. 1974. Naturwissenschaften **61:** 1–4.
117. INOUE, Y., J. ABE, M. TAKAMATSU & N. AAKI. 1974. Jpn . J. Hyg. **29:** 92.
118. FUKANO, S. & M. DOGUCHI. 1977. Bull. Environ. Contam. Toxicol. **17:** 613–617.
119. WASSERMANN, M., L. TOMATIS, D. WASSERMANN, N. E. DAY, Y. GRONER, D. LAZAROVICI & D. ROSENFELD. 1974. Pestic. Monit. J. **8:** 1–7.
120. MIZUTARU, T., M. MATSUMOTO & K. FUJIWARA. 1972. Progr. Med. **81:** 666.
121. DOGUCHI, M. 1973. Chlorinated hydrocarbon in the environment in the Kanto Plain and Tokyo Bay, as reflected in fishes, birds and man. *In* New methods in environmental chemistry and toxicology. F., Coulston, F. Korte M. Goto, Eds. Tokyo Acad. Sci. Book Inc. pp 269–289.
122. SOLLY, S. R. B. & V. SHANKS. 1974. N. Z. J. SCI. **17:** 535–544.
123. DYMENT, P. G., L. M. HEBERTSON, E. D. GOMES, J. S. WISEMAN & R. W. HORNABROOK. 1971. Bull. Environ. Contam. Toxicol. **6:** 532.
124. INOUE, Y., S. ABE, H. ESAKI & M. TADAMATSU. 1973. Kurume Med. J. **20:** 83–86.
125. SHIMAMUZA, K., M. TAKADA & M. KURATSUNE. 1974. Fukuoka Acta Med. **65:** 25–27.
126. OKUMURA, M., Y. MASUDA & S. NAKAMUTA. 1974. Fukuoka Acta Med. **65:** 84–87.
127. KARPPANEN, E. & L. KOLHO. 1973. PCB Conference II Nat. Swedish Environ. Protection Board Publication 4E. pp. 124–127.
128. HASEGAWA, H., M. SATO & H. TSURUTA. 1972. Occup. Health **10:** 50–55.
129. HESSELBERG, R. J. & D. D. SCHERR. 1974. Bull. Environ. Contam. Toxicol. **11:** 202–205.
130. ROOTH, J. & D. A. JONKERS. 1972. TNO Nieuws **27:** 551–555.
131. TANAKA, K. & F. KOMATSU. 1972. Fukuoka Acta Med. **63:** 360–366.
132. FINKLEA, J., L. E. PRIESTER, J. P. CREASON, T. HAUSER, T. HINNERS & D. J. HAMMER. 1972. Am. J. Pub. Health **62:** 645–651.
133. POLISHUK, Z. W., M. RON, M. WASSERMANN, S. CUCOS, D. WASSERMANN & C. LEMESCH. 1977. Pest. Monit. J. **10:** 121–129.
134. TAKAMATSU, M., Y. INOUE & S. ABE. 1974. Fukuoka Acta Med. **65:** 28–31.
135. DOGUCHI, M. & S. FUKANO. 1975. Bull. Environ. Contam. Toxicol. **13:** 57-63.
136. KITAMURA, M., T. TSUKAMOTO, K. SUMINO, K. HAYAKAWA, T. SHIBOTA & J. HIRANO. 1973. Jpn. J. Indust. Health **47:** 354–355.
137. YAMAGUCHI, A., T. YOSHIMURA & M. KURATSUNE. 1971. Fukuoka Acta Med. **62:** 112–117.
138. MASUDA, Y. & R. KAGAWA. 1974. Fukuoka Acta Med. **65:** 17–24.
139. MILLER, H. J., S. CUCOS, D. WASSERMANN & M. WASSERMANN. Presented at the 10th Inter-American conference on Toxicology and Occupational Medicine. October 22–25, 1978.
140. KNOLL, W. & S. JAYARAMAN. 1973. Nahrung **17:** 599–615.
141. PESENDORFER, H. 1975. Wien. Klin. Wochenschr. **87:** 732–736.
142. JUSZKIEWICZ, T., A. NIEWIADOMSKA, R. RADOMANSKI. 1975 Pol. Tyg. Lek. **30:** 1821–1824.
143. FALANDYSZ, J. 1977. Porzem. Spozyw. **31:** 307–309.
144. WESTÖÖ, G., K. NORÉN & M. ANDERSSON. 1970. Var Föda **22:** 10–31.
145. WESTÖÖ, G. & K. NORÉN. 1972. Var Föda **24:** 41–54.
146. WESTÖÖ, G. 1974. Ambio **3:** 79–83.
147. RISEBROUGH, R. & V. BRODINE. 1969. Environment **12:** 16–27.
148. SAVAGE, E. P., J. D. TESSARI, J. W. MALBERG, H. W. WHEELER & J. R. BAGBY. 1973. Pestic. Monit. J. **7:** 1–5.
149. STRASSMAN, S. C. & F. W. KUTZ. 1977. Pestic. Monit. J. **10:** 130–133.
150. MUSIAL, C. Y., O. HUTZINGER, V. ZITKO & J. CROCKER. 1974. Bull. Environ. Contam. Toxicol. **12:** 258–267.

151. Oura, H., H. Kobayashi, T. Oura, I. Senda & K. Kubota. 1972. J. Jpn. Assoc. Rural Med. **21:** 300–301.
152. Hidaka, K., T. Ohe & K. Fujiwara. 1972. Prog. Med. **82:** 519–520.
153. Moore, S. A. Jr. 1973. Nat. Tech. Inform. Serv. PB-231:618.
154. Fisher, N. S. 1975. Science **189:** 463–464.
155. Morgan, J. R. 1972. Bull. Environ. Contam. Toxicol. **8:** 129–137.
156. Wildish, D. J. 1972. Bull. Environ. Contam. Toxicol. **7:** 182–187.
157. Halter, M. T. & H. E. Johnson. 1974. J. Fish. Res. Board Can. **31:** 1543–1548.
158. Helle, E., M. Olsson & S. Jenson. 1976. Ambio **5:** 261–263.
159. French, J. E. 1976. Diss. Abstr. Int. B. **37:** 96–97.
160. Maki, A. W. & H. E. Johnson. 1975. Bull. Environ. Contam. Toxicol. **13:** 412–416.
161. De Lang, R. L., W. J. Gilmartin & J. G. Simpson. 1973. Science **181:** 1168–1169.
162. Chen, T. S. & K. P. Dubois. 1973. Toxicol. Appl. Pharmacol. **26:** 504–512.
163. Welle, W. 1971. Tidsskr. Nor. Laegeforen. **91:** 2299–2302.
164. Chang, E. S. & E. L. R. Stokstad. 1975. Poultry Sci. **54:** 3–10.
165. Call, D. J. 1974. Bull. Environ. Contam. Toxicol. **11:** 70–77.
166. Dustman, E. H., L. F. Stickel, L. J. Blus, W. L. Reichel & S. N. Wiemayer. 1971. Trans. North Am. Wildl. Nat. Resour. Conf. **36:** 118–133.
167. Peakall, D. B., J. L. Limiar & S. E. Bloom. 1972. Environ. Health. Perspect. **1:** 103–104.
168. Peakall, D. B. 1971. Bull. Environ. Contam. Toxicol. **6:** 100–101.
169. Scott, M. L. 1972. Fed. Proc. Fed. Am. Soc. Exp. Biol. **36:** 1888–1893.
170. Anonymous. 1970. Chem. Eng. News **48:** 31.
171. Risebrough, R. W. & D. W. Anderson. 1975. J. Wildl. Manage. **39:** 508–513.
172. Scott, M. L., J. R. Zimmermann, S. Marinsky, P. A. Mullenhoff, G. L. Rumsey & R. W. Rice. 1975. Poultry Sci. **54:** 350–368.
173. Dahlgren, R. B. & R. L. Linder. 1971. J. Wildl. Manage. **32:** 315–319.
174. Dahlgren, R. B., R. L. Linder & C. W. Carlson. 1972. Environ. Health Perspect. **1:** 89–101.
175. Kolbye, A. C., Jr. 1972. Environ. Health Perspect. **1:** 85–88.
176. Cecil, H. C., J. Bitman, R. J. Lillie, G. F. Fries & J. Verrett. 1974. Bull. Environ. Contam. Toxicol. **11:** 489–495.
177. Heagelle, M. A. & R. K. Tucker. 1974. Bull. Environ. Contam. Toxicol. **11:** 98–102.
178. Hill, E. F., R. G. Heath & J. D. Williams. 1976. Bull. Environ. Contam. Toxicol. **16:** 445–453.
179. Cecil, H. C., S. J. Harris, J. Bitman & G. F. Fries. 1973. Bull. Environ. Contam. Toxicol. **9:** 179–185.
180. Deleted.
181. Johansson, J. O., J. E. Kihlström & C. Lundberg. 1972. Ambio **1:** 148–149.
182. Kihlström, J. E., J. Oerberg, C. Lundberg. P. O. Danielsson & J. Sydhoff. 1973. *In* PCB Conference II Nat. Swedish Environ. Protection Board Publications 4E:109–111.
183. Orberg, J. & J. E. Kihlström. 1973. Environ. Res. **6:** 176–179.
184. Barsotti, D. A. & J. R. Allen. 1975. Fed. Proc. **34:** 338.
185. Kusuda, M. 1971 Sankato Fujinka **4:** 1063–1072.
186. Keplinger, M. L., O. E. Fancher & J. C. Calandra. 1971. Toxicol. Appl. Pharmacol. **19:** 402–403.
187. Allen, J. R. 1975. Fed. Proc. **34:** 1675–1679.
188. Kihlström, J. E., C. Lundberg, J. Orberg, P. O. Danielsson & J. Sydhoff. 1975. Environ. Physiol. Biochem. **5:** 54–57.
189. Jensen, S., J. E. Kihlström, M. Olsson, C. Lundberg & J. Orberg. 1977. Ambio **6:** 239.
190. Linder, R. E., T. V. Gaines & R. D. Kimbrough. 1974. Food. Cosmet. Toxicol. **12:** 63-77.
191. Ringer, R. K., R. J. Aulerich & M. Zabik. 1972. Proc. Am. Chem. Soc. 164th A. C. S. Meeting New York pp. 149–154.

192. Villeneuve, D. C., D. L. Grant, K. Khera, D. J. Clegg, H. Baer & W. E. Y. Phillips. 1971. Environ. Physiol. **1:** 67–71.
193. Aulerich, R. J., R. K. Ringer, H. L. Seagran & W. G. Youatt. 1971 Can. J. Zool. **49:** 611–616.
194. Funatsu, I. 1972. Kurume, Med. J. **19:** 43–51.
195. Nishimura, H. 1973. Int. Chem. Eng. **13:** 774–780.
196. Aulerich, R. J., R. K. Ringer & S. Iwamoto. 1973. J. Reprod. Fertil. **19:** 365–376.
197. Berlin, M., J. Gage & S. Holm. 1975. Arch. Environ. Health. **30:** 141–147.
198. Melvas, B. & I. Brandt. 1973. Nat. Swedish Environ. Protection Board Publication.
199. Grant, D. L. 1971. Environ. Physiol. **1:** 61–66.
200. Villeneuve D. C., D. L. Grant, W. E. J. Phillips, M. L. Clark & D. J. Clegg. 1971. Bull. Environ. Contam. Toxicol. **6:** 120–128.
201. Platonow, N. & N. Y. Chen. 1973. Vet. Rec. **92:** 69–70.
202. Polishuk, Z. W., D. Wassermann, M. Wassermann, S. Cucos & M. Ron. 1977. Environ. Res. **13:** 278–284.
203. Kuratsune, M., T. Yoshimura, J. Matsuzaka & A. Yamaguchi. 1972. Environ. Health Perspect. **1:** 119–128.
204. Miller, R. W. 1971. Teratology **4:** 211–212.
205. Taki, I., M. Kuratsune & Y. Masuda. 1973. Research coordination bureau, Science and Technology Agency, pp. 87–95.
206. Tsukamoto, H. *et al.* 1969. Fukuoka Acta Med. **60:** 496–512.
207. Grant, D. L., D. C. Villeneuve, K. Khera, D. J. Clegg & W. E. J. Phillips. 1971. Proc. Can. Fed. Biol. Soc. **14:** 52.
208. Jonsson, H. T. Jr., J. E. Keil, R. G. Gaddy, C. B. Loadholt, G. R. Henniger & E. M. Walker, Jr. 1976. Arch. Environ. Contam. Toxicol. **3:** 479–490.
209. Nilsson, B. & C. Ramel. 1974. Hereditas **77:** 319–322.
210. Green, S. G., J. V. Carr, K. A. Palmer & E. J. Oswald. 1975. Bull. Environ. Contam. Toxicol. **13:** 14–22.
211. Hoopingarner, R., A. Samuel & D. Krause. 1972. Environ. Health Perspect. **1:** 155–158.
212. McLanghlin, Y. Jr., G. P. Marliac, M. J. Verrett, M. K. Mutchler & O. G. Fitzhugh. 1963. Toxicol. Appl. Pharmacol. **5:** 760–771.
213. Hays, H. & R. W. Risebrough. 1972. Natr. Hist. **80:**39–47.
214. Fraumeni, J. F. 1974. Pediatrics **53:** 807–812.
215. Kuratsune, M., T. Yoshimura, J. Matsuzuka & A. Yamaguchi. 1971. Health Rep. **86:** 1083–1091.
216. Yoshimura, T. 1971. Fukuoka Acta Med. **62:** 109–116.
217. Treon, J. F., F. P. Cleveland, J. W. Cappel & R. W. Atchley. 1956. Am. Ind. Hyg. Assoc. Quart. **17:** 204–213.
218. Grant, D. L. & W. E. J. Phillips. 1974. Bull. Environ. Contam. Toxicol. **12:** 145–152.
219. Keplinger, M. L., O. E. Fancher & J. C. Calandra. 1971. Toxicol. Appl. Pharmacol. **19:** 402–403.
220. Kimbrough, R. D., R. E. Linder & T. B. Gaines. 1972. Arch. Environ. Health. **25:** 354–364.
221. Norbach, D. H. & J. R. Allen. 1972. Environ. Health. Perspect. **1:** 137–143.
222. Vos., J. G. & E. Notenboom-Ram. 1972. Toxicol. Appl. Pharmacol. **23:** 563–578.
223. Ito, N., H. Nagasaki, M. Arai, S. Makiura, S. Sugihara & K. Hirao. 1973. J. Nat. Cancer Int. **51:** 1637–1646.
224. Rogers, A., A. Mellors & S. Safe. 1976. Res. Commun. Chem. Pathol. Pharmacol. **13:** 341–344.
225. Iatropoulos, M. J., R. Felt, F. Coulston & F. Korte. 1977. Toxicol. Appl. Pharmacol. **41:** 173.
226. Wassermann, D. Unpublished data.
227. Hirayama, C., T. Irisa & T. Yamamoto. 1969. Fukuoka Acta Med. **60:** 455–461.
228. Nishihara, H. & T. Yamamoto. 1972. Fukuoka Acta Med. **63:** 352–359.
229. Bailey, S. & P. J. Bunyan. 1972. Nature **236:** 34–36.

230. Fujita, S., M. Tsuji, K. Kato, S. Saeki & H. Tsukamoto. 1971. Fukuoka Acta Med. **62:** 30–34.
231. Johnstone, G. J., D. J. Ecobichon & O. Hutzinger. 1974. Toxicol. Appl. Pharmacol. **28:** 66-81.
232. Cecil, H. C., S. J. Harris & J. Bitman. 1975. Arch. Environ. Contam. Toxicol. **3:** 183–192.
233. Bickers, D. R., L. C. Harber, A. Kappas & A. P. Alvares. 1972. Res. Commun. Chem. Pathol. Pharmacol. **3:** 505–512.
234. Desaiah, D., L. K. Cutkomp, H. H. Yap & R. B. Koch. 1972. Biochem. Pharmacol. **21:** 857–865.
235. Pardini, R. S. 1971. Bull. Environ. Contam. Toxicol. **6:** 539–545.
236. Letterst, C. L., T. M. Farber, A. M. Baker & E. J. Van Loon. 1972. Toxicol. Appl. Pharmacol. **23:** 112–122.
237. Marniemi, J., M. G. Parkki & H. Vainio. 1976. Acta Physiol. Scand. Suppl. **440:** 139.
238. Konat, G. & J. Clausen 1973. Environ. Physiol. Biochem. **3:** 139–147.
239. Bickers, DR., A. Kappas & A. P. Alvares. 1973. Fed. Proc. Fed. Am. Soc. Exp. Biol. **32:** 235.
240. Bickers, DR., A. Kappas & A. P. Alvares. 1974. J. Pharmacol. Exp. Therap. **188:** 300–309.
241. Bitman, J., H. C. Cecil & S. J. Harris. 1972. Environ. Health Perspect. **1:** 145–149.
242. Cutkomp, L. K., H. H. Yap, D. Desaiah & R. Y. Koch. 1972. Environ. Health. Perspect. **1:** 165–168.
243. Davis, P. W., J. M. Friedhoff & G. A. Wedemeyer. 1972. Bull. Environ. Contam. Toxicol. **8:** 69–72.
244. Risebrough, R. W., P. Rieche, C. B. Peckall, S. G. Kerman & M. N. Kirven. 1968. Nature **220:** 1098–1102.
245. Lincer, J. L. & D. B. Peakall. 1970. Nature **228:** 783–784.
246. Koch, R. B., D. Desaiah, H. H. Yap & L. K. Cutkomp. 1972. Bull. Environ. Contam. Toxicol. **7:** 87–92.
247. Goldstein, J. A., J. D. McKinney, G. W. Lucier, J. A. Moore, P. Hickman H. Bergman. 1975. Pharmacologist **16:** 239.
248. Dieter, M. P. 1974. Toxicol. Appl. Pharmacol. **27:** 86–98.
249. Dieter, M. P. 1974. Toxicol. Appl. Pharmacol. **29:** 110–111.
250. Dieter, M. P. 1975. Arch. Environ. Contam. Toxicol. **3:** 142–150.
251. Abraham, R., U. C. Koepke, L. Goldberg & F. Coulston. 1974. Toxicol. Appl. Pharmacol. **29:** 128–129.
252. Benthe, H. F., A. Schmoldt & H. Schmidt. 1972. Arch. Toxicol. **29:** 97–106.
253. Letterst, C. L. & E. J. Van Loon. 1972. Proc. Soc. Exp. Biol. Med. **141:** 765–768.
254. Den Tonkelaar, E. M. & G. J. Van Esch. 1974. Toxicology **2:** 371–380.
255. Alvares, A. P. & A. Kappas. 1975. FEBS Lett. **50:** 172–174.
256. Cappon, I. D. & D. M. Nicholls. 1975. Pestic. Biochem. Physiol. **5:** 109–118.
257. Fulfs, J. C. & R. Abraham. 1976. toxicol. Appl. Pharmacol. **37:** 119–120.
258. Sharp, C. W., D. G. Hunt, S. T. Clements & W. E. Wilson. 1974. Mol. Pharmacol. **10:** 119–129.
259. Wilson, W. W. & C. W. Sharp. 1972. Fed. Proc. Fed. Am. Soc. Exp. Biol. **31:** 591.
260. Araki, Y. & K. Tanaka. 1974. Fukuoka Acta Med. **65:** 47–52.
261. Araki, Y. & K. Tanaka. 1974. Fukuoka Acta Med. **65:** 65–68.
262. Truhaut, R., D. P. Hien & P. L. Nguyen. 1974. C. R. Acad Sci. Paris Ser. D. **278:** 3003–3006.
263. Kimura, A. & H. Taniguchi. 1975. Tokushima J. Exp. Med. **22:** 11–19.
264. Street, J. C., F. M. Urry, D. J. Wagstaff & A. D. Blau. 1969. Abstr. Papers Am. Chem. Soc. No. 158.
265. Kiriyama, S., S. Inoue, N. Machinaka & A. Yoshida. 1972. Jpn. Soc. Food. Nutr. **25:** 205.
266. Nagai, J., M. Furukawa, Y. Yae & K. Higuchi, 1971. Fukuoka Acta Med. **62:** 42–47.

267. KIRIYAMA, S., M. BANJO & H. MATSUSHIMA. 1974. Nutr. Rep. Int. **10:** 79–88.
268. TANAKA, K., F. SETSUHARU, F. KOMATSU & N. TAMURA. 1969. Fukuoka Acta Med. **60:** 544–547.
269. ITO, Y., H. UZAWA & A. NOTOMI. 1971. Fukuoka Acta Med. **62:** 48–50.
270. UZAWA, H., Y. ITO, A. NOTOMI & S. KATSUMI. 1969. Fukuoka Acta Med. **60:** 499.
271. UZAWA, H. Y. ITO, A. NOTOMI, S. HORI, Y. IKEURA & S. KATSUDI. 1971. Fukuoka Acta Med. **62:** 66.
272. UZAWA, H., A. NOTOMI, S. NAKAMUTA & Y. IKEURA. 1972. Fukuoka Acta Med. **63:** 401–404.
273. NAGAI, J., K. HIGUCHI & Y. YAE. 1972. Fukuoka Acta Med. **63:** 367–373.
274. FULFS, J. C. 1976. Diss Abstr. Int. B. **37:** 2751.
275. VOS, J. G. 1972. Environ. Health Perspect. 1:105–117.
276. GOLDSTEIN, J. A., P. HIKMAN & D. L. JUE. 1974. Toxicol Appl. Pharmacol. **27:** 437–448.
277. VOS, J. G., J. J. T. W. A. STRIK, CATHERINA W. M. VAN HOLSTEŸN & J. H. PENNINGS. 1971. Toxic Appl. Pharmacol. **20:** 232–240.
278. KWANISHI, S., S. SANO, T. MIZUTANI & M. MASTUMOTO. 1973. Jpn. J. Hyg. **28:** 84.
279. VOS, J. G. & R. B. BEEMS. 1971. Toxicol. Appl. Pharmacol. **19:** 617–633.
280. MIURA, H., S. OMORI & M. KATOH. 1973. Jpn. J. Hyg. **28:** 83.
281. VOS, J. G. & J. H. KOEMAN. 1970. Toxicol. Appl. Pharmacol. **17:** 656–668.
282. VOS, J. G. & L. VAN DRIEL GROOTENHUIS. 1972. Sci. Total Environ. **1:** 289–302.
283. SINCLAIR, P. R. & S. GRANICK. 1974. Biochem. Biophys. Res. Commun. **61:** 124–133.
284. KAWANISHI, S., S. SANO, T. MIZUTANI & M. MATSUMOTO. 1974. Jpn. J. Hyg. **29:** 81.
285. GOLDSTEIN, J. A., P. HICKMAN & D. JUE. 1973. Fed. Proc. **32:** 702.
286. BASTOMSKY, CH., B. SALYMOSS, G. SZIGMOND & J. M. WYSE. 1975. Clin. Chim. Acta **61:** 171–174.
287. HIRAYAMA, C., M. OKUMURA, J. NAGAI & Y. MASUDA. 1974. Clin. Chim. Acta **55:** 97–100.
288. BITMAN, J. & H. C. CECIL. 1970. J. Agric. Food Chem. **18:** 1108–1112.
289. GRAICHUS, Y. A., D. J. CALL, B. M. AMMANN, A. GREICHUS & H. SHAVE. 1975. Arch. Environ. Contam. Toxicol. **3:** 330–343.
290. NOWICKI, H. G. & A. W. NORMAN. 1972. Steroids **19:** 85–99.
291. PLATONOW, N. S., R. M. LIPTOP & H. D. GEISSINGER. 1972 Bull. Environ. Contam. Toxicol. **7:** 358–365.
292. NAGAI, J., M. FURUKAWA, Y. YAE & Y. IKEDA. 1969. Fukuoka Acta Med. **60:** 475–488.
293. NELSON, J. A. 1974. Biochem. Pharmacol. **23:** 447–451.
294. PLATONOW, N. S. & H. S. FUNNELL. 1971. Vet. Rec. **88:** 109–110.
295. PLATONOW, N. S. & H. S. FUNNELL. 1972. Can. J. Comp. Med. Vet. Sci. **36:** 89–93.
296. ITURRI, S. J. 1974. Diss. Abstr. Int. **35:** 1371 B.
297. REHFIELD, B. M., R. L. BRADLEY JR. & M. L. SUNDE. 1971. Poultry. Sci. **50:** 1090–1096.
298. ORBERG, J. & C. LUNDBERG. 1974. Environ. Physiol. Biochem. **4:** 116–120.
299. WASSERMANN, D., M. WASSERMAN, SIMI CUCOS, & M. DJAVAHERIAN. 1973. Thyroid Gland Changes in Rats Receiving Polychlorinated Biphenyls. *In* Pesticides and The Environment: A Continuing Controversy **2:** 151–161. W. B. Deichmann, Ed. Intercontinental Medical Book Corp. New York.
300. BASTOMSKY, C. H. 1974. Endocrinology **95:** 1150.
301. HURST, J. G., W. S. NEWCOMER & J. A. MORRISON. 1974. Poultry Sci. **53:** 125–133.
302. WASSERMANN, D., M. WASSERMANN, S. CUCOS & M. DJAVAHERIAN. 1973. Environ. Res. **6:** 334–338.
303. WASSERMANN, D. & M. WASSERMANN. 1972. Ultrastructure of adrenal zona fasciculata in polychlorinated biphenyls (PCBs) receiving rats. 17th International Congress of Occupational Health. Buenos Aires, 1972.
304. FLCIK, D. F., R. G. O'DELL & V. A. CHILDS. 1965. Poultry Sci. **44:** 1460–1465.
305. WASSERMANN, D., M. WASSERMANN & C. LEMESCH. 1975. Environ. Physiol. Biochem. **5:** 332–340.

306. Second report of the study for "Yusho" (chlorobiphenyls poisoning). 1971. Kyushu University, Fukuoka, Japan.
307. HARRIS, S. J. 1976. Poultry Sci. **55:** 1933–1940.
308. WASSERMANN, M., D. WASSERMANN & E. KEDAR. 1973. Bull. Environ. Contam. Toxicol. **10:** 42–50.
309. VOS, J. G. & TH. DE ROIJ. 1972. Toxicol. Appl. Pharmacol. **21:** 549–555.
310. SHIGEMATSU, N., Y. NORIMATSU, T. ISHIBASHI, M. YOSHIDA, S. SUETSUGU, T. KAWATSU, T. IKEDA, R. SAITO, S. ISHIMARU, T. SHIRAKISA, M. KIDO, K. EMORI & H. TOSHIMITSU. 1971. Fukuoka Acta Med. **62:** 150–156.
311. SAITO, R., N. SHIGEMATSU & S. ISHIMARU. 1972. Fukuoka Acta Med. **63:** 408–411.
312. HANSEN, D. J., P. R. PARRISH, J. I. LOWE, A. J. WILSON JR. & P. D. WILSON. 1971. Bull. Environ. Contam. Toxicol. **6:** 113–119.
313. FRIEND, MILTON & D. O. TRAINER. 1970. Science **170:** 1314–1316.
314. KIMURA, N. T. & T. BABA. 1973. Gann **64:** 105–108.
315. NISHIZUMI, M. 1970. Arch. Environ. Health **21:** 620-632.
316. SHIGEMATSU, N., S. ISHIMARU, T. HIROSE, T. IKEDA, K. EMORI & N. MIYAZAKI. 1974. Fukuoka Acta Med. **65:** 88-95.
317. ALLEN, J. R. & D. A. NORBACH. 1973. Science **179:** 498–499.
318. HATTULA, M. L. & O. KARLOG. 1972. Acta Pharmacol. Toxicol. **31:** 238–240.
319. ULFSTRAND, S., A. SØDERGREN & J. KAHOEL. 1971 Nature **231:** 467–468.
320. KARLSSON, B., B. PERSSON, A. SODERGREN & S. ULFSTRAND. 1974. Environ. Pollut. **7:** 53–63.
321. KREITZER, J. F. & G. H. HEINZ. 1974. Environ. Pollut. **6:** 21–29.
322. FUKUDA, T. 1972. Fukuoka Acta Med. **63:** 378–381.
323. MURAI, Y. & Y. KUROIWA. 1971. Neurology **21:** 1173–1176.
324. FUKUDA, T., J. MORI 6 Y. ARAKI. 1974. Fukuoka Acta Med. **65:** 69–73.
325. KOMATSU, F. & M. KIKUCHI. 1972. Fukuoka Acta Med. **63:** 384-386.
326. ALLEN, J. R., L. A. CARSTENS & D. A. BARSOLLI. 1974. Toxicol. Appl. Pharmacol. **30:** 440–451.
327. MEIGHS, J. W., J. J. ALBOM & B. L. KARTIN. 1954. J. Am. M. **154:** 1417-1418.
328. BAHN, A. K., J. R. ROSENWAIKE, N. HERRMANN, P. GROVER, J. STELLMAN & K. O'LEARY. 1976. New Engl. J. Med. **295:** 450.
329. LAWRENCE, Ch. 1977. N. Engl. J. Med. **296:** 108.
330. TOSHITANI, S. 1972. Fukuoka Acta Med. **63:** 392–395.
331. URABE, H. 1974. Fukuoka Acta Med. **65:** 1–4.
332. KOHDA, H., S. ASAHI & S. TOSHITANI. 1974. Fukuoka Acta Med. **65:** 81–83.
333. KIMBROUGH, R. D. & R. E. LINDER. 1974. J. Nat. Cancer Inst. **53:** 547–552.
334. NAGASAKI, H., S. TOMII, T. MEGA, S. SUGIHARA, S. MIYATA & N. ITO. 1974. J. Nara Med. Assoc. **25:** 635–648.
335. NAGASAKI, H., S. TOMII & T. TSUMASHIKA. 1975. Jpn. J. Hyg. **30:** 134.
336. MAKIURA, S., H. AOE & S. SUGIHARA. 1974. J. Nat. Cancer Inst. **53:** 1253–1257.
337. KIMBROUGH, R. D., R. A. SQUIRE, R. E. LINDER, J. D. STRANDBERG, R. J. MONTALI & V. W. BURSE. 1975. J. Nat. Cancer Inst. **55:** 1453–1459.
338. WASSERMANN, D., H. MILLER & M. WASSERMANN. 1978. Toxicological European Research, Vol. 1. In press.
339. KUMURA, N. T. *et. al.* 1976. Z. Krebsforsch. **87:** 257–266.
340. ALLEN, J. R. & D. H. NORBACK. 1973. Science **179:** 498–499.

SUMMARY OF THE WORKSHOP ON RECENT ADVANCES IN ANALYTIC TECHNIQUES FOR HALOGENATED AROMATIC COMPOUNDS

Chairpersons:
Phillip W. Albro and J. Ronald Hass

Environmental Biology and Chemistry Branch
National Institute of Environmental Health Sciences
Research Triangle Park, North Carolina 27709

Warren B. Crummett

Analytical Laboratories
Dow Chemical, U.S.A.
Midland, Michigan 48640

This workshop was intended to be a forum for the informal exchange of views and ideas concerning problems in the analysis of environmental and biologic matrices for chlorinated aromatic compounds. The following is a summary of the discussions of various selected topics.

DETERMINATION OF POLYCHLORINATED BIPHENYLS IN ENVIRONMENTAL AND BIOLOGIC SAMPLES

Effects of Sample Distortion on PCB Measurement Using an Electron Capture Detector

The majority of laboratories performing assays for PCBs use a gas chromatograph (GC) equipped with an electron capture detector (ECD) for the isolation/measurement step. In general, the composition of a PCB mixture changes when it is exposed to environmental/biologic systems. It is well known that the response of an ECD varies greatly depending upon such structural features as halogen content of the molecule. However, the variation of ECD response with structure for isomeric PCBs is apparently less widely appreciated. For example, the ECD response for the symmetrical hexachlorobiphenyl isomers spans a range of 20.[1] Thus, even alteration of the composition of a GC peak consisting of coeluting isomeric compounds can introduce serious errors in the analysis.

As an illustration of the practical results of these effects, the following experiments were undertaken using Aroclor® 1260 so that evaporative losses would be minimized.

A sample of Aroclor 1260 was fractionated over Florisil using hexane as the solvent. The tail of the PCB peak was collected so as to give a distorted PCB mixture. A weighed aliquot of this material was spiked into cottonseed oil. A separate sample of cottonseed oil was spiked with a weighed portion of the original Aroclor 1260. Both samples were then subjected to an extraction/clean-up procedure for PCB analysis.[2] The PCB content was then assayed using GC-ECD and Aroclor 1260 as the reference material. The Aroclor 1260 spike gave the correct answer with 95% recovery. The

0077-8923/79/0320-0125 $01.75/ © 1979, NYAS

other spike gave a GC pattern similar to Aroclor 1260 but the analysis was in error by a factor of 3. Quantitation was based either on total area measured over the elution time of Aroclor 1260 or by comparison of the signals from prominent peaks. The latter method gave much worse results.

A second study involved the fractionation of Aroclor 1260 by reverse-phase high pressure liquid chromatography to give six fractions separated more-or-less on the basis of chlorine content. When these fractions were used as spikes in assays similar to those described above, the results ranged from $\frac{1}{7}$ to 7 × the correct answer. Thus, very serious errors may result from the use of GC-ECD for the analysis of samples containing PCB mixtures whose composition has been altered.

Other Approaches to PCB Analysis

Gas Chromatography/Mass Spectrometry

From the earlier discussion it is clear that variation in detector response can and probably does result in serious errors in the measurement of PCBs in biologic/environmental samples. One obvious approach to this problem is to use a detector that retains the selectivity of the ECD but exhibits a more uniform response to various individual PCBs. The mass spectrometer (MS) is a detector offering these characteristics. The most straight-forward experiment is to perform a GC/MS analysis either scanning complete spectra continuously or, if greater sensitivity is desired, monitoring a limited number of ions which are "characteristic" of PCBs. The results of such a selected ion monitoring (SIM) analysis of Aroclor 1248 (FIGURE 1) indicate a ratio for the total area from m/z 326 (pentachlorobiphenyls) to that from m/z 292 (tetrachlorobiphenyls) of approximately 0.5. Similar data for Aroclor 1254 give a 326/292 ratio of *ca.* 3.0 (see FIGURE 2). Thus, one should be able to measure the 326/292 ratio for an unknown, then select the appropriate Aroclor for standardization. Unfortunately, the situation is not quite that simple. The 326/292 ratio for a quail feed extract (FIGURE 3) was found to be approximately 1.6. Clearly this is another case of a PCB mixture that has been altered. While it is tempting to assume that a mixture of approximately equal concentration of Aroclor 1248 and Aroclor 1254 should be used for standardization, this cannot be justified unless the cause of the alteration can be determined. In addition to the mixture of different Aroclors, the pattern could result from selective loss of the more volatile tetrachlorobiphenyls from Aroclor 1254 during work-up or their loss as a result of their more facile microbial metabolism. This again serves to illustrate the potential hazards in attempting to assign a single concentration for what is a mixture of components.

Complete Fractionation and Determination of Each Individual PCB

The crux of the problem in PCB analysis is the fact that we are dealing with a mixture of individual components whose relative concentrations within the organism/environment will be changing from the time of exposure until, at least, the time of sample collection. Thus, the approach most likely to give a meaningful result is to measure each PCB individually rather than attempt to assign a concentration for a mixture such as an Aroclor.

A fractionation scheme that shows promise for this separation is based upon the use of charcoal on a polyurethane foam substrate to initially separate the PCB mixture on the basis of the number of chlorines in the 2,2′,6,6′ positions[3] followed by capillary column GC analysis of each fraction. Each PCB present is identified automatically by a computer. The program makes the identification on the basis of the combined chromatography information. Accurate quantitative analysis can then be achieved for all PCBs which are available as standards. The advantages of this method are obvious. The major disadvantage is the time required per analysis. The GC analysis of each of the 4 fractions requires about 4 hours. Thus one obtains a great deal of information about the sample with a commensurate expenditure of effort. Clearly careful thought must go into experimental design so that the usefulness of the analytic data justifies the work necessary for their generation.

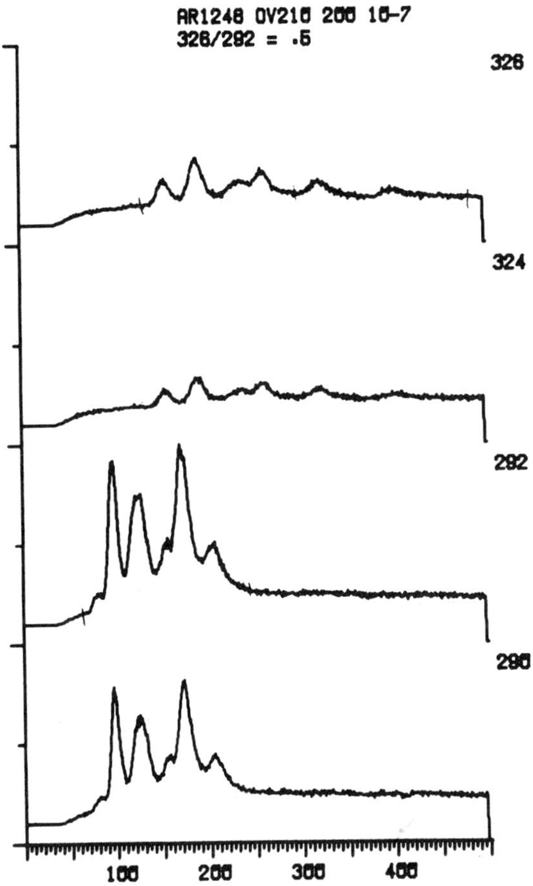

FIGURE 1. Selected ion monitoring analysis of Aroclor 1248.

METHODS PRESENTLY BEING DEVELOPED

Radioisotope Dilution Assay (RIDA)

In this method the sample is spiked with uniformly labeled (^{14}C) PCBs similar in composition to Aroclor 1254. After extraction and clean-up the PCB fraction is perchlorinated using $SbCl_5$. The amount of decachlorobiphenyl is determined by

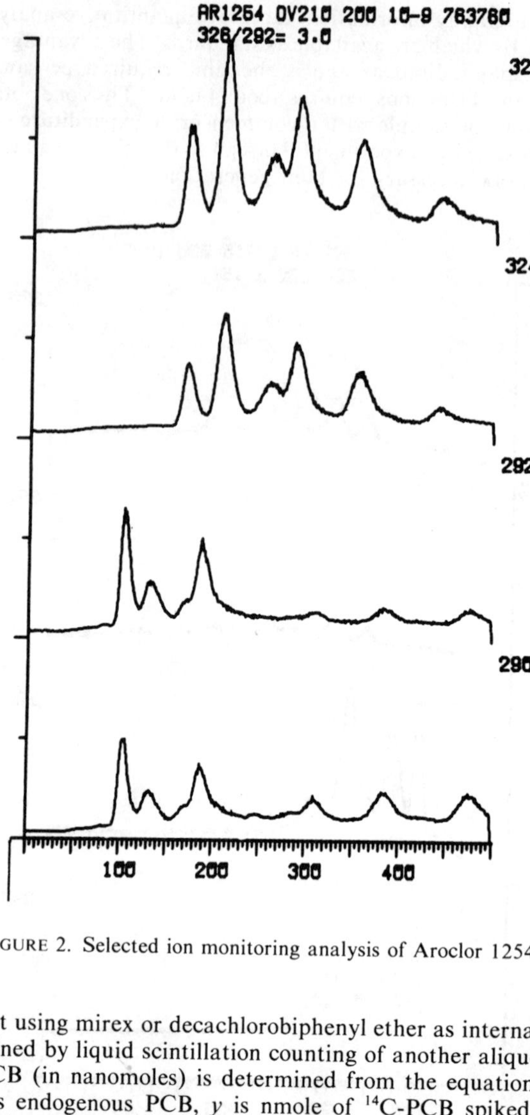

FIGURE 2. Selected ion monitoring analysis of Aroclor 1254.

GLC of an aliquot using mirex or decachlorobiphenyl ether as internal standard, and the ^{14}C is determined by liquid scintillation counting of another aliquot. The amount of endogenous PCB (in nanomoles) is determined from the equation $G = y(So - S)/S$, where G is endogenous PCB, y is nmole of ^{14}C-PCB spiked, So is specific radioactivity of ^{14}C-PCB spiked and S is specific activity determined for the decachlorobiphenyl. The assay gives approx. 95% accountability in samples of cottonseed oil or rat liver containing Aroclor 1248, and gives "total PCB content" independent of the PCB composition.

Radioimmunoassay

Antisera have been produced in rabbits with varying selectivities to PCB isomers. No other classes of compounds have been found to interfere. At present, using a double antibody technique and ^{125}I-labeled PCB derivative as assay reagent, the assay requires about 5 ng of an Aroclor as limit of measurement. Unfortunately, extraction and clean-up are still required.

A Modified Procedure for the Cleanup of TCDD Residues in Human Milk[4]

A neutral cleanup procedure, involving liquid chromatography with successive columns of MgO/celite 545, alumina (two columns) and Florisil, was successfully developed in the Harvard laboratory for the cleanup of TCDD residues in beef fat.[5,6] In the case of human milk samples, it was necessary to perform an initial extraction with a diethyl ether/hexane mixture to remove TCDD together with the milk lipids. It was found, in the subsequent cleanup of the extract, that the first two columns of the

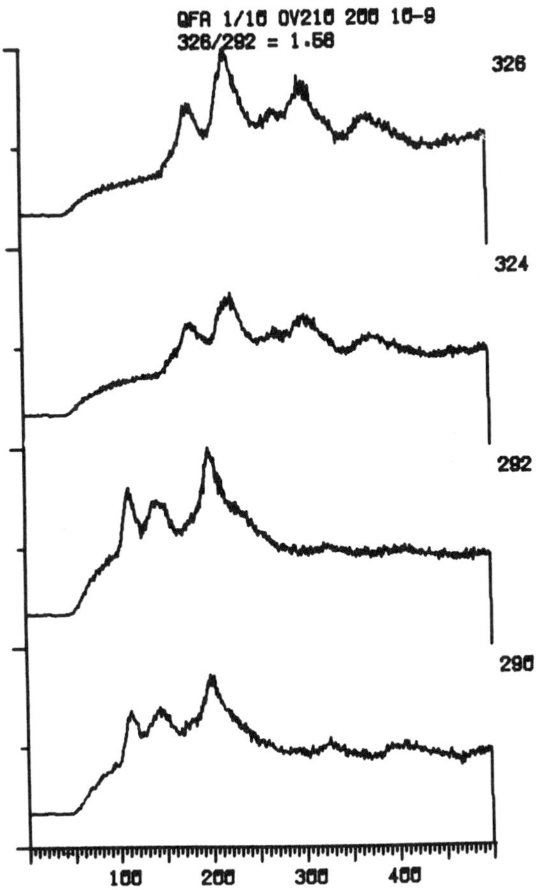

FIGURE 3. Selected ion monitoring analysis of a non-polar fraction of quail feed.

original procedure for beef fat could be combined together by using the technique of reverse flow chromatography. This modification involved the addition of alumina to the top of the MgO/celite 545 column after sample loading. The bulk of the lipid was eluted as usual with diethyl ether/hexane. The adsorbents were then moved to the opposite end of the column prior to elution of TCDD with benzene. The major portion of the remaining lipid, which was displaced by benzene from the MgO, was retained to a considerable extent by alumina. An alumina column of 2 mm i.d. rather than the original 6 mm i.d. column was used for the next step, in order to achieve a more complete separation of TCDD from interfering substances. The final Florisil chromatographic step was unchanged from the original procedure.

As a part of a continuing effort by several laboratories to determine TCDD levels in the environment, a series of human milk samples were received from NIEHS for cleanup by the modified neutral procedure, followed by analysis by direct probe high resolution mass spectrometry. Many of the samples came from population groups which had a potential for TCDD ingestion, either from proximity to a trichlorophenol industrial accident or from the use of 2,4,5-T on rangeland. Other samples were collected from mothers living in an urban area where there was no known use of TCDD-contaminated products. Several of these control samples had known amounts of TCDD added to them by NIEHS. All samples had been suitably coded to preclude any knowledge of their origin.

When a 20 ml sample size was used for cleanup it was found that subsample equivalent to 5 ml could be analyzed with a limit of detection of approximately 0.2 parts per trillion (ppt). In certain samples there was a signal at m/e 320, equivalent to 0.2 ppt TCDD, which was inseparable in mass from TCDD m/e 320 ion at a resolution of 10,000. Additional signals were noted in the m/e 320 and 322 mass regions but they did not significantly interfere with the TCDD ions. Results from the study are presently being confirmed by another laboratory.

References

1. Kohli, K. K., P. W. Albro & J. D. McKinney. 1979. J Anal. Toxicol. In press.
2. Albro, P. W. & B. J. Corbett. 1977. Chemosphere **7:** 381–385.
3. Huckins, J. N., D. L. Stalling & W. A. Smith. 1978. J. Assoc. Off. Anal. Chem. **61:** 32–38.
4. O'Keefe, P. W. 1978. *In* Dioxin, Toxicology and Chemical Aspects. F. Cattabeni, A. Cavalaro & G. Galli, Eds. Spectrum Publications. New York, pp. 59–78.
5. O'Keefe, P. W., M. S. Meselson & R. W. Baughman. 1978. J. Assoc. Off. Anal. Chem. **61:** 621–626.

THE METABOLISM AND DISPOSITION OF HALOGENATED AROMATICS

H. B. Matthews and S. Kato

Laboratory of Pharmacology
National Institute of Environmental Health Sciences
Research Triangle Park, North Carolina 27709

To describe the disposition of all halogenated aromatic hydrocarbons in a short paper such as this would be an impossible task. This group of chemicals, which has found some of the most diverse industrial and agricultural uses of all classes of synthetic chemicals, includes compounds that have a variety of chemical structures and substitutions. The disposition of halogenated aromatics is determined by a number of factors the importance of which may vary according to the species exposed. However, almost all halogenated aromatics are sufficiently lipophilic to be readily absorbed from the gastrointestinal tract, and in most mammalian species polarity and the rate at which they are metabolized to polar compounds are the major determinants of the disposition of halogenated aromatics once they enter the body. The rate of metabolism of halogenated aromatics to compounds that can be readily excreted may vary greatly within a single type of compound, but types of halogenated aromatics can be broadly classified as Type I, II, or III compounds according to their decreasing polarity.

TYPE I HALOGENATED AROMATICS

Type I compounds, the most polar of the halogenated aromatics, contain a polar moiety that facilitates their excretion as the parent compound or conjugates of the parent compound. Excretion of Type I compounds is independent of metabolism, and the rate of removal of the major portion of a dose of these compounds from the body can usually be described by a single-component exponential decay curve. Chlorinated phenoxy herbicides and chlorinated phenols are two types of halogenated aromatics that could be classified as Type I compounds.

The hydrophilic moiety of chlorinated phenoxy herbicides is the organic acid. These compounds, 2,4-D, 2,4,5-T, and silvex, are readily absorbed from the intestine, but because of the presence of the acidic moiety they are rapidly excreted, primarily in the urine, as the parent compounds or as conjugates of the parent compounds.[1–4] The half-lives of these compounds in tissue are usually only about 5 to 10 hours when animals are exposed to low or moderate doses, but the half-lives may be increased as much as threefold for high to very high doses. This effect is probably an indication of the active excretion of these compounds and may be due to a saturation of the renal organic acid secretory mechanism. The highest tissue concentrations of the chlorinated phenoxy herbicides are found in the kidney followed in a dose-dependent manner by the plasma and liver.[3,4]

The hydrophilic moiety of the chlorinated phenols is the hydroxyl group. These compounds are also readily absorbed from the gastrointestinal tract and are excreted primarily as the parent compounds or conjugates of the parent compounds.[5–7] Biologic half-lives of chlorinated phenols may vary with the degree of chlorination and species exposed but usually do not exceed 24 hours.[5–7] The chlorinated phenols are subject to

0077–8923/79/0320–0131 $01.75/0 © 1979, NYAS

more metabolism than the phenoxy herbicides, and both the tetrachlorophenols and pentachlorophenols have been shown to be further hydroxylated in the *para* position.[7,8] Pentachlorophenol dechlorination and hydroxylation to tetrachloro-*p*-hydroquinone has been shown to be mediated by the hepatic microsomal enzymes of laboratory animals,[9] and this metabolite of pentachlorophenol has been detected in the urine of men who were occupationally exposed to pentachlorophenol.[8] In studies with laboratory animals the highest concentrations of pentachlorophenol were found in liver, kidney, and blood, respectively.[6]

TYPE II HALOGENATED AROMATICS

Type II halogenated aromatics have an intermediate polarity due to substitution(s) other than organic halogens. Type II compounds may be polar enough to be excreted in the bile prior to their metabolic alteration, but since they are readily absorbed from the intestine, most of the unaltered compound that is excreted in the bile will be reabsorbed from the intestine and returned to the liver. Therefore, the elimination of Type II compounds is largely dependent upon their rate of metabolism, and since the metabolism and the greatest tissue concentration of Type II compounds occurs in the liver, the decrease in the body burden of Type II compounds can usually be described by a single exponential component. Examples of compounds that could be classified as Type II halogenated aromatics are the halogenated dibenzodioxins, dibenzofurans, and Kepone, although Kepone is not a true aromatic.

The dioxins are formed as unwanted contaminants in the synthesis of halogenated phenols. The extremely toxic 2,3,7,8-tetrachlorodibenzo-*p*-dioxin (TCDD) is formed in the manufacture of 2,4,5-trichlorophenol which is, in turn, used to synthesize the herbicide 2,4,5-T. Other dioxins have been reported as contaminants of commercial preparations of pentachlorophenol. TCDD has been reported to concentrate in the liver and to a lesser extent in the adipose tissue of the laboratory rat.[10,11] The half-life of TCDD in the laboratory rat following a single dose is 3 to 4 weeks, and with chronic exposure a steady state is reached in about 7 weeks.[11] Because of its extreme toxicity and the very low doses which must be administered, very little is known about the metabolism of TCDD. However, recent work in our laboratory has shown that TCDD is slowly excreted in the bile and that approximately 90% of the TCDD-derived material in bile does not cochromatograph with the parent compound. We believe that this material may represent one or more TCDD metabolites, but as yet we have not isolated enough of this material to allow structural confirmation.

Polychlorinated dibenzofurans (PCDFs) are formed in the commercial synthesis of polychlorinated biphenyls (PCBs)[12] and have been isolated from the PCB-contaminated rice oil that was involved in the Yusho incident in Japan.[13] PCDFs are structurally and toxicologically similar to the chlorinated dioxins and have likewise been shown to be concentrated in the liver and to a lesser degree in the adipose tissue of laboratory animals.[14] Because of its similarity to TCDD, 2,4,7,8-tetrachlorodibenzofuran (TCDF) is thought to be the most toxic of the PCDFs. TCDF has a half-life of approximately one week in the liver of the mouse, but the half-life of TCDF in other tissues may be longer because the liver appears to have a greater ability to eliminate TCDF.[14]

Kepone and mirex are not aromatic hydrocarbons, they are perchlorinated, polycyclic carbon cage structures, but they serve very well to illustrate the effect a single substitution may have on the disposition of a halogenated xenobiotic. Mirex is completely chlorinated, virtually inert to biologic degradation, and is concentrated in adipose tissue and skin.[15] The disposition of mirex is similar to that of persistent Type

III halogenated hydrocarbons. The biologic half-life of mirex has not been predicted, but it is probably greater than the life time of most laboratory animals. Ivie *et al.*[16] fed a chronic dose of mirex to rats, quail, and fish for 16 months without reaching a point where the tissue levels ceased to increase, thus demonstrating the great potential of mirex for bioaccumulation with chronic exposure. On the other hand, Kepone, which is structurally similar to mirex except for a single carbonyl group that becomes hydrated under biologic conditions,[17] is much more polar than mirex and concentrates in the liver and to a lesser extent in the adipose tissue. Kepone is slowly metabolized, and the metabolites along with significant amounts of the parent compound are excreted in the bile. The unaltered Kepone that is excreted in the bile is subject to enterohepatic circulation, and the interruption of this enterohepatic cycle has been the basis for the therapeutic treatment of Kepone intoxication with cholestyramine.[18,19] The half-life of Kepone is four to five weeks in the laboratory rat,[20] and varied from 97 to 177 days in men who were occupationally exposed to Kepone.[18] Treatment with cholestyramine significantly reduced the half-life of Kepone in these men.[18,19]

Type III Halogenated Aromatics

The third class of halogenated aromatic hydrocarbons, Type III compounds, contains those compounds that have no substitutions other than halogens. All of these compounds are quite non-polar and are not excreted at an appreciable rate prior to their metabolism to more polar compounds. All but the very highly halogenated Type III halogenated aromatics are readily absorbed from the intestine when administered at low to moderate doses. Following absorption, the initial distribution of Type III compounds is to the more highly perfused tissues, particularly liver, where they may be metabolized and excreted or redistributed to the more slowly perfused tissues, adipose tissue, and skin, which have a higher affinity for these very lipophilic compounds. Thus, the Type III halogenated aromatics have what I refer to as an initial half-life, which includes the initial absorption, distribution, and redistribution phases. Since a larger portion of the dose is available for metabolism and excretion during the initial half-life, it is invariably shorter than the subsequent half-lives. Following equilibration with adipose tissue, Type III compounds are eliminated at a rate which is the product of the blood:tissue ratio, the rate of tissue perfusion, and the rate at which the individual compounds are metabolized to more polar compounds that can be excreted. Therefore, depending upon a number of variations, the initial half-life of a Type III compound may be only one-fifth to one-third as long as its subsequent half-lives.

Most of the metabolism of Type III halogenated aromatics is oxidative, results in the addition of one or more hydroxyl groups to the aromatic ring, and is the product of hepatic mixed-function oxidase enzymes. Hydroxylated metabolites of Type III compounds are rapidly conjugated and thus have short biologic half-lives. Most of the tissue residues of Type III compounds represent the parent compound.

The most simple Type III halogenated aromatics are the halogenated benzenes, and the fundamental work on the metabolism and excretion of halogenated benzenes was done in the laboratory of R. T. Williams during the 1950's. Some of the most dramatic effects of chlorination on the metabolism and excretion of chlorinated benzenes were reported by Jondorf, Parke, and Williams from their work with a series of tri- and tetrachlorobenzenes.[21,22] Their work demonstrated that there is very little excretion of chlorinated benzenes prior to metabolism to more polar compounds and that the position of the chlorine atoms on the benzene ring exerted a profound effect on the rate of metabolism and, consequently, the rate of excretion. The ratios for the rate of excretion of trichlorobenzene by the rabbit were approximately 3 to 2 to 1 for the 1,2,3,-; 1,2,4-; and 1,3,5-trichlorobenzene isomers, respectively.[21] For the tetrachlorobenzene isomers the effect of chlorine position on metabolism and excretion was even more marked, and the relative rates of excretion were approximately 20 to 5 to 1 for the 1,2,3,4-; 1,2,3,5-; and 1,2,4,5-tetrachlorobenzene isomers, respectively.[22] Thus, it was demonstrated that the presence of two adjacent, unsubstituted carbon atoms facilitates the metabolism and excretion of chlorinated benzenes.

More recent work on the halogenated benzenes by Safe and his co-workers has provided evidence that the presence of two adjacent unsubstituted carbon atoms in chlorinated[23] and brominated[24] benzenes facilitate the formation of an arene oxide as a metabolic intermediate. The arene oxide intermediate is a precursor to the hydroxylated metabolites which are eventually conjugated and excreted. Halogenated benzenes without adjacent unsubstituted carbon atoms, i.e., benzenes with halogen atoms in at least the 1, 3, and 5 positions, may still be metabolized via an arene oxide intermediate but at a reduced rate, and the metabolites should show evidence of the "NIH-shift" as predicted by Daly and Jerina.[25,26] The shift of both chlorine and bromine atoms were observed in the studies of halogenated benzene metabolism by Safe's group.[23,24] This group has also demonstrated "NIH-shifts" of chlorine atoms in the metabolsim of chlorinated naphthalenes.[27]

The unsubstituted biphenyl molecule, like benzene, is readily metabolized by the hepatic mixed-function oxidases. The two phenyl rings do, however, have a directing effect on the enzymatic attack on the biphenyl molecule. The major metabolite of this aromatic is the 4-hydroxy derivative, and the secondary metabolite is 2-hydroxy biphenyl.[28] Furthermore, it appears that these two metabolites may be formed by two separate enzymes, since the formation of 4-hydroxy biphenyl is induced to a greater extent by pretreatment of the animal with phenobarbital whereas 3-methylcholanthrene is a better inducer of 2-hydroxy biphenyl formation.[28] Formation of a third and minor metabolite of biphenyl, 3-hydroxy biphenyl, is thought to be the product of a completely different oxidative mechanism, the direct insertion of a hydroxyl group which does not involve an arene oxide intermediate.[28]

Halogenation of the biphenyl molecule can, depending upon the degree and position of halogenation, have marked and varied effects on its metabolism and disposition. The addition of a single chlorine atom directs the hydroxylation of the biphenyl ring to the unhalogenated ring but decreases its rate of metabolism only slightly.[29]

Since the major metabolite of biphenyl, 4-hydroxy biphenyl, is thought to result from an arene oxide intermediate between the meta and para positions of the unhalogenated ring, it follows that halogenation of both para positions of biphenyl may have the greatest effect on PCB metabolism. Therefore, with the addition of two or more halogen atoms to the biphenyl molecule the position of halogenation becomes a critical determinant of the rate of metabolism and thus of the biologic half-life of the halogenated biphenyl. The initial half-life of 4,4′-dichlorobiphenyl in the rat is approximately one day, twice as long as that of 4-chloro biphenyl.[30] The major metabolite of 4,4′-dichlorobiphenyl excreted by the rat, 4-hydroxy-3,4′-dichlorobiphenyl, provides evidence for an "NIH-shift" of a chlorine atom and thus implies metabolism via an arene oxide intermediate that would have to be formed between a halogenated and an unhalogenated carbon atom.[31]

There is evidence, however, that not all species can metabolize a para halogenated biphenyl equally well. For example, the dog and rat metabolize an excrete 4,4′-dichlorobiphenyl at a rate sufficient to result in an initial half-life of about one day in each species. On the other hand, the monkey metabolizes and excretes this same PCB much more slowly and the initial half-life of 4,4′-dichlorbiphenyl in the monkey is two to three weeks. Therefore, this PCB has a potential for bioaccumulation in the tissues of monkeys, but not dogs and rats.

The degree of metabolic hindrance provided by halogenation may vary considerably with a single species depending upon the position of halogenation. For example, the initial half-life of 2,4,5,2′,5′-pentachlorobiphenyl in rats is only about twice that of 4,4′-dichlorobiphenyl.[33] The major metabolites of this pentachlorobiphenyl are 3-hydroxy and 3,4-dihydroxy 2,5,2′,4′,5′-pentachlorobiphenyl.[34] However, the addition of a single chlorine atom to the 4′ position of the pentachlorobiphenyl to form 2,4,5,2′,4′,5′-hexachlorobiphenyl inhibits the metabolism of this PCB to such an extent that computer extrapolation of the rate of excretion indicates that the initial half-life of this PCB in the rat is greater than the lifetime of the animal.[30]

On the other hand, the rat readily metabolizes 2,3,6,2′,3′,6′-hexachlorobiphenyl to the 4-hydroxy derivative and the initial half-life of this hexachlorobiphenyl in the rat is only about one day.[35] There is also less species variation in rate of metabolism of 2,3,6,2′,3′,6′-hexachlorobiphenyl than was observed with 4,4′-dichlorobiphenyl and the half-lives of this PCB in the dog and monkey are approximately 1 and 5 days, respectively.[32] Thus, it appears as though a relatively high degree of chlorination does not greatly prolong the biologic half-life of a PCB in either of these species so long as a meta-para pair of adjacent unsubstituted carbon atoms are open for enzymatic oxidation.

In addition to the effect of the degree and position of halogenation on biphenyl metabolism the type of halogenation may also have an effect on the rate of metabolism. For example, 2,4,5,2′,4′,5′-hexachlorobiphenyl, which has one *ortho* and one *meta* position open on each phenyl ring, is slowly metabolized by the rat to 3-hydroxy and 6-hydroxy-2,4,5,2′,4′,5′-hexachlorobiphenyl, 4-hydroxy-2,3,5,2′,4′,5′-hexachlorobiphenyl and 4-hydroxy-2,5,2′,4′,5′-pentachlorobiphenyl.[35] The 3- and 6-hydroxy metabolites could have been formed either as a result of an arene oxide intermediate between a chlorinated and an unchlorinated carbon atom or by the direct insertion of a hydroxyl group. The 4-hydroxy-2,3,5,2′,4′,5′-hexachlorobiphenyl and 4-hydroxy-2,5,2′,4′,5′-pentachlorobiphenyl metabolites show evidence of an "NIH-shift" of a chlorine atom and oxidative dechlorination, strong indications that they are products of an arene oxide intermediate which was formed between the unchlorinated meta and the chlorinated *para* positions of one phenyl ring.[25,26,35] The major component of the PBB mixture that was involved in the Michigan incident, 2,4,5,2′,4′,5′-hexabromobiphenyl, has the same pattern of substitution as this PCB; however, a similar investigation of the distribution, metabolism, and excretion of this PBB in the rat failed to reveal evidence of its metabolism.[36] Furthermore, the dog, which appears to be particularly adept at metabolizing PCBs that do not have two adjacent unsubstituted carbon atoms and metabolizes and excretes 2,4,5,2′,4′,5′-hexachlorobiphenyl with an initial half-life of approximately 8 days, is apparently unable to metabolize the PBB at an appreciable rate and eliminates it quite slowly.[32] The reasons for the slower rate of PBB metabolism are not yet understood. However, two possible explanations for the slower metabolism of the PBB would be either that the larger bromine atoms more effectively hinder the approach of the enzyme to the carbon skeleton or that electrochemical differences cause the bromine atoms to inhibit more effectively the formation of an arene oxide intermediate.

And finally, it should be pointed out that a short biologic half-life for the Type III halogenated aromatic is not necessarily a positive aspect. Type III compounds that have short half-lives are most likely to be metabolized via arene oxide intermediates and the toxicity and/or carcinogenicity of a number of compounds have been attributed to their metabolism via arene oxide intermediates that covalently bind to subcellular macromolecules.[26] Work in Safe's laboratory has demonstrated that

4-chlorobiphenyl is metabolized via an arene oxide intermediate and that this halogenated biphenyl binds covalently to subcellular macromolecules *in vitro*.[37] Work in our laboratory indicates that 2,3,6,2′,3′,6′-hexachlorobiphenyl is also metabolized via an arene oxide intermediate[35] and was covalently bound to the nucleic acids and protein of a number of tissues of mice that were exposed to this PCB *in vivo*.[38] However, the toxicologic significance of covalent binding of halogenated biphenyls to subcellular macromolecules remains to be determined.

References

1. Erne, K. 1966. Distribution and elimination of chlorinated phenoxyacetic acid in animals. Acta Vet. Scand. **7:** 240–256.
2. Courtney, D. K. 1970. 2,4,5-T in the rat: Excretion pattern, serum levels, placental transport and metabolism. *In* Pesticides Symposia, pp. 277–283. Hales and Associates. Florida.
3. Sauerhoff, M. W., W. H. Braun, G. E. Blau & P. J. Gehring. 1976. The dose-dependent pharmacokinetic profile of 2,4,5-trichlorophenoxyacetic acid following intravenous administration to rats. Toxicol. Appl. Pharmacol. **26:** 352–361.
4. Sauerhoff, M. W., W. H. Braun & J. E. Lebeau. 1977. Dose-dependent pharmacokinetic profile of silvex following intravenous administration in rats. J. Toxicol. Environ. Health **2:** 605–618.
5. Jakobson, I. & S. Yllner. 1971. Metabolism of ^{14}C-pentachlorophenol in the mouse. Acta Pharmacol. Toxicol. **29:** 513–524.
6. Larsen, R. V., L. E. Kirsch, S. M. Shaw, J. E. Christian & G. S. Born. Excretion and tissue distribution of uniformly labeled ^{14}C-pentachlorophenol in rats. J. Pharm. Sci. **61:** 2004–2006.
7. Ahlborg, U. G. & K. Larsson. 1978. Metabolism of tetrachlorophenols in the rat. Arch. Toxicol. **40:** 63–74.
8. Ahlborg, U. G., J. E. Lingren & M. Mercier. 1974. Metabolism of pentachlorophenol. Arch. Toxicol. **32:** 271–281.
9. Ahlborg, U. G., K. Larsson & T. Thunberg. 1978. Metabolism of pentachlorophenol in vivo and in vitro. Arch. Toxicol. **40:** 45–53.
10. Allen, J. R., J. P. Van Miller & D. H. Norback. 1975. Tissue distribution excretion and biological effects of ^{14}C-tetrachlorodibenzo-p-dioxin in rats. Food Cosmet. Toxicol. **13:** 501–505.
11. Rose, J. Q., J. C. Ramsey, T. H. Wentzler, R. A. Hummel & P. J. Gehring. 1976. The fate of 2,3,7,8-tetrachlorodibenzo-p-dioxin following single and repeated oral doses to the rat. Toxicol. Appl. Pharmacol. **36:** 209–226.
12. Vos, J. G., J. H. Koeman, H. L. Vander Maas, M. C. ten Noever de Braun & R. H. de Vos. 1970. Identification and toxicological evaluation of chlorinated dibenzofuran and chlorinated naphthalene in two commercial polychlorinated biphenyls. Food Cosmet. Toxicol. **8:** 625–633.
13. Morita, M., J. Nakagawa, K. Akiyama, S. Mimura & N. Isono. 1977. Detailed examination of polychlorinated dibenzofurans in PCB preparations and Kanemi Yusho oil. Bull. Environ. Contam. Toxicol. **18:** 67–73.
14. Morita, M. & S. Oishi. 1977. Clearance and tissue distribution of polychlorinated dibenzofurans in mice. Bull. Environ. Contam. Toxicol. **18:** 61–66.
15. Mehendale, H. M., L. Fishbein, M. Fields & H. B. Matthews. 1972. Fate of mirex ^{14}C in the rat and plants. Bull. Environ. Contam. Toxicol. **8:** 200–207.
16. Ivie, G. W., J. R. Gibson, H. E. Bryant, J. J. Begin, J. R. Burnett & H. W. Dorough. 1974. Accumulation, distribution and excretion of mirex-^{14}C in animals exposed for long periods to the insecticide in the diet. J. Agric. Food Chem. **22:** 626–653.
17. Blanke, R. V., M. W. Fariss & F. D. Griffith. 1977. Analysis of chlordecone (Kepone®) in biological specimens. J. Anal. Toxicol. **1:** 57–62.

18. COHN, W. J., J. J. BOYLAN, R. V. BLANKE, M. W. FARISS, J. R. HOWELL & P. S. GUZELIAN. 1978. Treatment of chlordecone (kepone) toxicity with chloestyramine. New Engl. J. Med. **298:** 243–248.
19. BLANKE, R. V., M. W. FARISS & P. S. GUZELIAN. 1978. Identification of a reduced form of chlordecone (Kepone®) in human stool. Bull. Environ. Contam. Toxicol. In press.
20. EGLE, J. L., S. B. FERNANDEZ, P. S. GUZELIAN & J. F. BORZELLECA. 1978. Distribution and excretion of chlordecone (Kepone) in the rat. Drug Metab. Dispos. **6:** 91–95.
21. JONDORF, W. R., D. V. PARKE & R. T. WILLIAMS. 1955. Studies in detoxication 66. The metabolism of halogenobenzenes. 1:2:3-, 1:2:4- and 1:3:5-trichlorobenzenes. Biochem. J. **61:** 512–521.
22. JONDORF, W. R., D. V. PARKE & R. T. WILLIAMS. 1958. Studies in detoxication 76. The metabolism of halogenobenzenes. 1:2:3:4-, 1:2:3:5- and 1:2:4:5-tetrachlorobenzenes. Biochem. J. **69:** 181–189.
23. KOHLI, J., D. JONES & S. SAFE. 1976. The metabolism of higher chlorinated benzene isomers. Can. J. Biochem. **54:** 203–208.
24. RUZO, L. O., S. SAFE & O. HUTZINGER. 1976. Metabolism of bromobenzenes in the rabbit. J. Agric. Food Chem. **24:** 291–293.
25. DALY, J. W., D. M. JERINA & B. WITKOP. 1972. Arene oxides and the NIH shift: The metabolism, toxicity and carcinogenicity of aromatic compounds. Experientia **28:** 1129–1264.
26. JERINA, D. M. & J. W. DALY. 1974. Arene oxides: A new aspect of drug metabolism. Science **185:** 573–582.
27. RUZO, L., D. JONES, S. SAFE & O. HUTZINGER. 1976. Metabolism of chlorinated naphthalenes. J. Agric. Food Chem. **24:** 581–583.
28. BILLINGS, R. E. & R. E. MCMAHON. 1978. Microsomal biphenyl hydroxylation: The formation of 3-hydroxybiphenyl and biphenyl catechol. Mol. Pharmacol. **14:** 145–154.
29. BLOCK, W. D. & H. H. CORNISH. 1959. Metabolism of biphenyl and 4-chlorobiphenyl in the rabbit. J. Biol. Chem. **234:** 3301–3302.
30. MATTHEWS, H. B. & M. W. ANDERSON. 1975. Effect of chlorination on the distribution and excretion of polychlorinated biphenyls. Drug Metab. Dispos. **3:** 371–380.
31. HASS, J. R., L. T. JAO, N. K. WILSON & H. B. MATTHEWS. 1977. Metabolism of 4-chlorobiphenyl and 4,4′-dichlorobiphenyl in the rat: Quantitative and qualitative aspects. J. Agric. Food Chem. **25:** 1330–1333.
32. SIPES, I. G. Personal communication.
33. MATTHEWS, H. B. & M. W. ANDERSON. 1975. The distribution and excretion of 2,4,5,2′,5′-pentachlorobiphenyl in the rat. Drug Metab. Dispos. **3:** 211–219.
34. CHEN, P. R., J. D. MCKINNEY & H. B. MATTHEWS. 1976. Metabolism of 2,4,5,2′,5′-pentachlorobiphenyl in the rat. Drug Metab. Dispos. **4:** 362–367.
35. KATO, S. & H. B. MATTHEWS. 1978. The effect of chlorine position on the metabolism of hexachlorobiphenyl isomers by the rat. Abstract No. 270: Toxicol. Appl. Pharmacol. **45:** 334.
36. MATTHEWS, H. B., S. KATO, N. M. MORALES & D. B. TUEY. 1977. Distribution and excretion of 2,4,5,2′,4′,5′-hexabromobiphenyl, the major component of Firemaster BP-6. J. Toxicol. Environ. Health **3:** 599–605.
37. WYNDHAM. C. & S. SAFE. 1978. *In vitro* metabolism of 4-chlorobiphenyl by control and induced rat liver microsomes. Biochemistry **17:** 208–215.
38. MORALES, N. M. & H. B. MATTHEWS. 1978. *In vivo* binding of PCBs to hepatic macromolecules of mice. Abstract No. 269: Toxicol. Appl. Pharmacol. **45:** 334.

TOXICOPATHOLOGY CHARACTERISTICS OF THE HALOGENATED AROMATICS

E. E. McConnell and J. A. Moore

Environmental Biology and Chemistry Branch
National Institute of Environmental Health Sciences
Research Triangle Park, North Carolina 27709

INTRODUCTION

Halogenated dibenzo-*p*-dioxins, dibenzofurans, biphenyls and napthalenes are structurally related compounds (FIGURE 1) of considerable environmental interest because of their ubiquitous occurrence, persistance in the environment, possible magnification in the food chain and relatively high toxicity of the more potent isomers. Numerous reports can be found that describe the lesions in various species of animals caused by these chemicals using various dosing regimens given over various periods of time. The purpose of this paper is to review these lesions in the common laboratory animal species including nonhuman primates, and discuss their apparent commonality.

CLINICAL SYNDROME

The clinical syndrome in nonhuman primates is similar for all these classes of compounds.[1–7] In addition, the differences between acute high-level and low-level chronic exposures are mainly quantitative. The first indication of toxicity is loss of body weight, which is often unrelated to food intake.[6] In lethal exposures, the weight loss is progressive, with monkeys losing over one-third their original weight.[2] The next most characteristic feature of early toxicity is a cutaneous syndrome manifested by alopecia and edema, particularly on the face and eyelids (FIGURES 2 and 3). Later acneform eruptions (chloracne) develop in hair follicles of the face and lips and blepharitis due to impaction of the Meibomian glands is striking (FIGURE 4). A similar lesion in the ceruminous glands causes a keratinaceous/waxy filling of the external auditory canal. The skin on the remainder of the body, especially on the extremities, becomes dry and scaly and the finger- and toe-nails are often lost with little evidence of pain (FIGURE 5). Terminally, there is progressive weakness, but even at this stage the monkeys remain fairly alert to external stimuli.

Chronic, low-level, nonlethal exposures of female monkeys cause reproductive difficulties.[8,9] Even when conception occurs, there is often early abortion and those offspring born are weak and usually die within a few weeks, although death may be caused by an additional exposure through their mother's milk.

Overt signs of intoxication with these compounds are much less conspicuous in rodents than monkeys. Rabbits are the only common laboratory species that show acne-like changes, which have been demonstrated with both skin painting and systemic studies.[10] Other than weight loss, very few clinical signs are observed in rats and guinea pigs.[11–13] Weight loss is usually immediate, dose-related, and progressive.[14] Just before death, mice (FIGURE 6) and chickens may show a distended abdomen due to ascites.[15,16]

It is difficult to determine the relative toxicity (dose required to produce a given effect) of these classes of compounds because they normally occur as mixtures of

0077–8923/79/0320–0138 $01.75/0 © 1979, NYAS

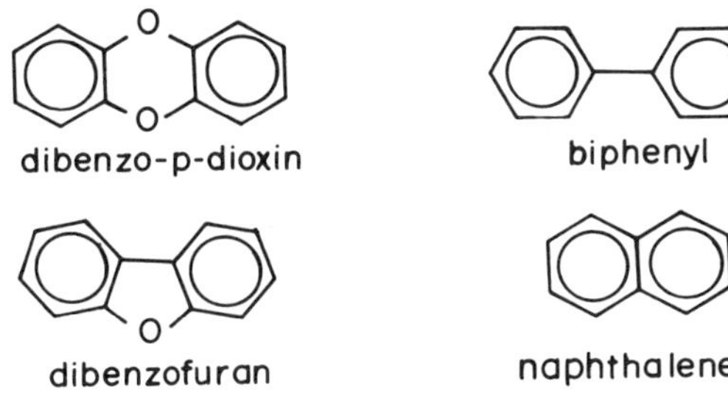

FIGURE 1. Structural relationship of toxic compounds.

various isomers,[17,18] which differ in toxicity.[13,19–21] In addition, it has been shown that some PCBs may be contaminated with furans, which adds to the complexity.[22–24] However, based on a study in guinea pigs using the most toxic isomer of each class of compound, the relative toxicity was dioxin > furan ≪ PCB ≪ napthalene.[25]

At the acute LD_{50} dose of these compounds, the median time to death in most species of animals is at least a few weeks. Importantly, higher (superlethal) doses do not markedly reduce this time.[13] Females and neonates of all species tend to be more sensitive to the effects of these compounds. In regard to relative species susceptibility to these compounds, guinea pigs[13] and chickens[26,27] appear to be the most susceptible animals to intoxication. Long-term, low-level exposures of monkeys suggest that they are also highly sensitive.[3] Rats and rabbits[28] are less sensitive and mice are the most resistant.[7,13]

POSTMORTEM AND HISTOPATHOLOGIC OBSERVATIONS

The lesions produced by the four different classes of compounds (at least the most toxic isomers) are similar within a given species but vary to some degree between species.[13] The only consistent findings in all species studied are almost complete loss of

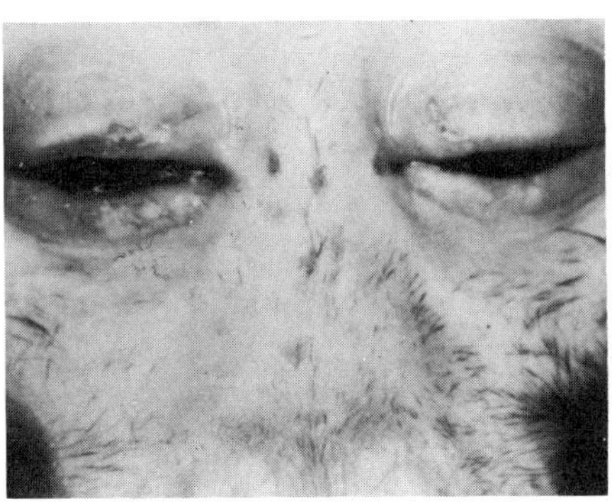

FIGURE 2

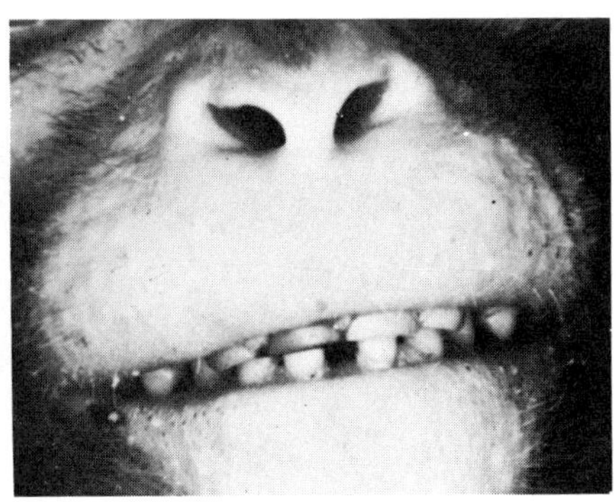

FIGURE 3

FIGURE 2. Facial appearance of monkey 47 days postexposure to 70 μg/kg 2,3,7,8-tetrachlorodibenzo-*p*-dioxin (TCDD). Note narrow opening, thickened eyelids, and loss of eyelashes and facial hair. (Toxicol. Appl. Pharmacol. **43:** 175–187, 1978).

FIGURE 3. Same monkey as FIGURE 2. Note the edematous appearance of the lips and accentuated hair follicles. (Toxicol. Appl. Pharmacol. **43:** 175–187, 1978).

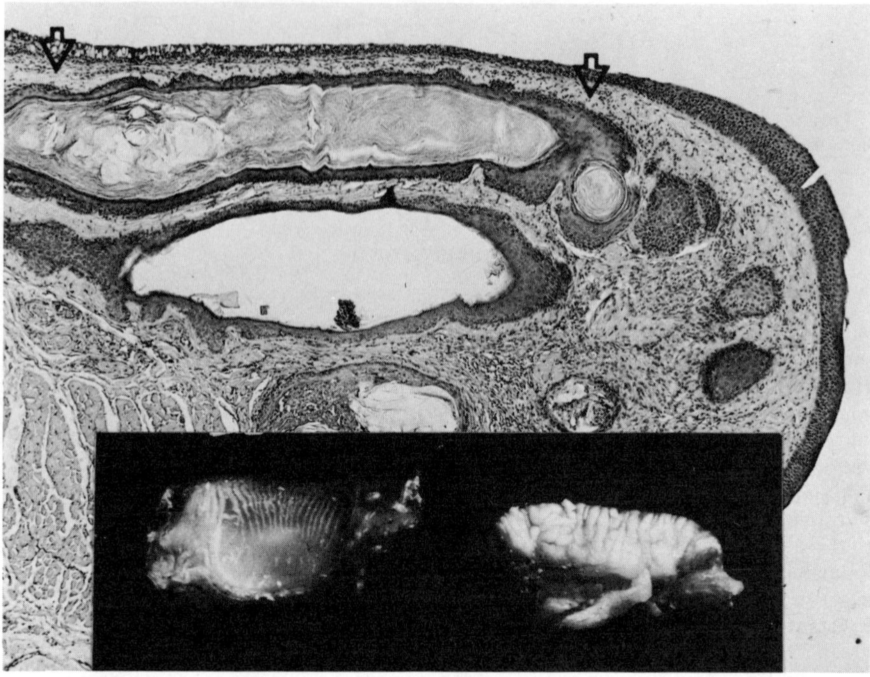

FIGURE 4. Insert—Ocular surface (inside) of eyelid from above monkey (right) compared to that of a nonexposed monkey (left). The Meibomian glands in the treated animal are thick and tortuous as compared to the normal fine regular pattern. Photomicrograph shows a longitudinal section of abnormal Meibomian gland (arrows) dilated with keratinaceous debris and loss of surrounding sebaceous glands.

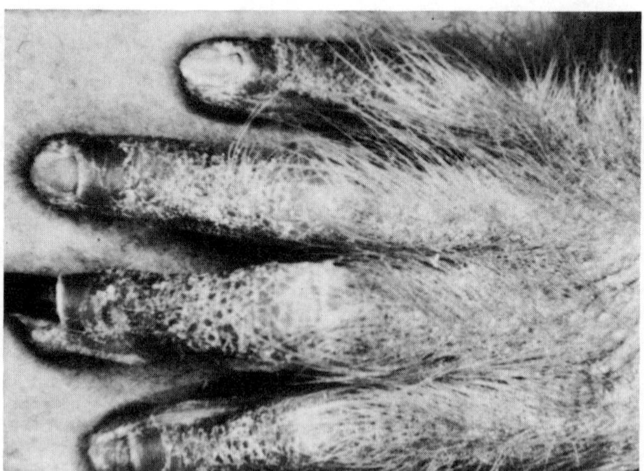

FIGURE 5. Hand of monkey showing loss of three of four fingernails and scaly appearance of skin (Toxicol. Appl. Pharmacol. **43:** 175–187, 1978).

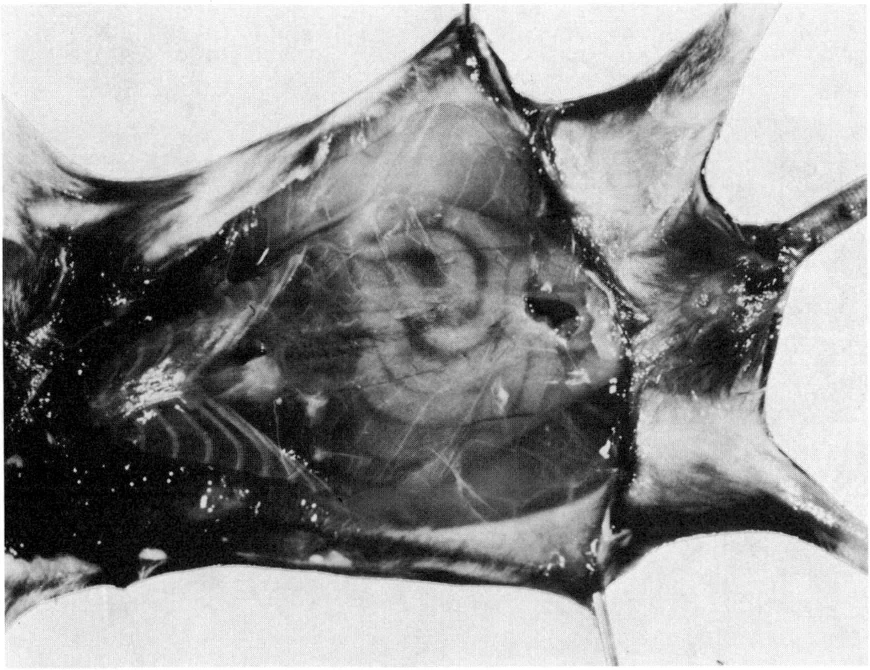

FIGURE 6. Mouse which died after exposure to 3,3′,4, 4′,5,5′-hexachlorobiphenyl. The skin has been reflected showing subcutaneous edema and ascites.

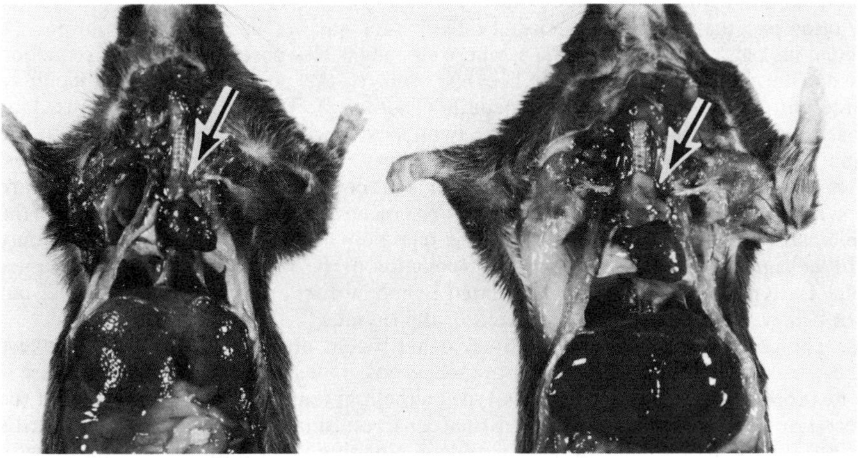

FIGURE 7. Dioxin treated mouse on left and normal mouse on right. Note the marked reduction in the size of the thymus as compared to the normal which covers the base of the heart (arrows). There is also much less fat in the axillary region and the liver is mottled in the treated mouse.

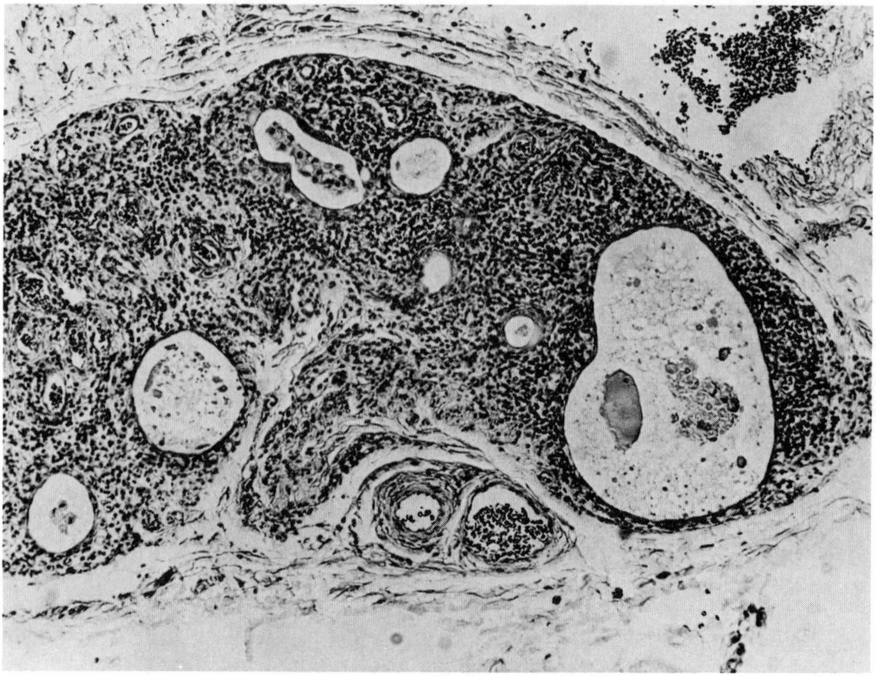

FIGURE 8. Photomicrograph of thymus of a monkey exposed to polychlorinated biphenyls (PCBs). There is a loss of the normal architecture and it is impossible to differentiate the cortex from the medulla.

body fat and involution of the thymus (FIGURE 7).[16,27,29] At the lethal dose the thymus is often less than one-fourth normal size;[30] much smaller than the reduction in size found in "sick" or starved animals, where atrophy is also observed. The size reduction is due to a loss of cortical lymphocytes, often to the point where it is difficult to differentiate the cortex from the medulla (FIGURE 8). This appears to be caused, at least to some degree by necrosis of the lymphocytes within a few days after exposure and is characterized by numerous macrophages phagocytizing the necrotic debris. After the involution has occurred (<10 days in most species), the necrotic events are less conspicuous.[13] Additionally, if an animal receives a less than lethal exposure, the microscopic appearance of the thymus is relatively normal although the weight may still be significantly less than normal. Reduction in the size of other lymphoid tissues (spleen, lymph nodes, and gut-associated lymphoid tissue) may also be observed, but not nearly to the degree as that noted in the thymus.[31]

The cutaneous, eyelid and auditory canal lesions observed clinically in monkeys are caused primarily by changes in the sebaceous glands within these tissues. There is a metaplastic change to a squamous-type epithelium that changes the character of the secretions from normal oily to keratinaceous resulting in dry skin and blepharitis (FIGURE 4). In addition, this often results in plugging and rupture of the glands with subsequent inflammatory changes (acne). Hyperkeratosis of the associated hair follicles often occurs concurrently to the metaplastic lesion. The chloracne-like lesions observed in rabbits have a similar pathogenesis.[32]

Fluid in the abdominal cavity (ascites) is observed with variable frequency in three animal species (chicken,[15] mice,[16] and monkeys[6]) exposed to these compounds. It has been related to hypoproteinemia or direct vascular damage. It, along with hydropericardium, is such a characteristic feature in fowl that it assumes diagnostic significance in this species.[33]

Proliferative lesions of the gastrointestinal tract are a feature of exposure (particularly chronic) to these compounds by nonhuman primates (FIGURE 9).[2,34] The initial response in the gastric mucosa is a loss of parietal cells with replacement by mucous-secreting cells. This is accompanied by hyperplasia and subsequent downgrowth of the glandular epithelium, which often breaks through the subjacent lamina propria (FIGURE 10). Ulceration and cystlike dilation with invasion by these glands

FIGURE 9

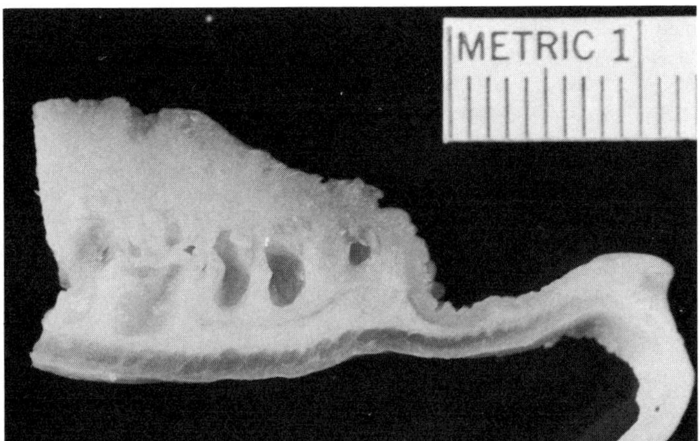

FIGURE 10

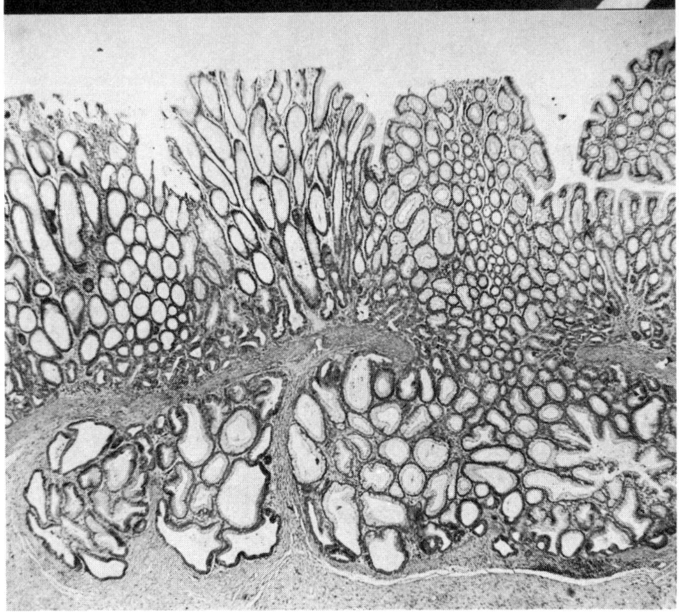

FIGURE 9. Section from the fundic region of the stomach of a monkey exposed to PCBs showing hyperplastic appearance of mucosa compared to normal thickness on right.

FIGURE 10. Histologic appearance of lesion similar to that in FIGURE 9 showing dilated crypts and invasive downgrowth into subjacent tissues.

results in a neoplastic-appearing lesion, but metastasis has not been recorded. This lesion has been reported as "Simian gastropathy."[35] A similar but less severe lesion is found in the duodenum and colon. In the latter organ the lesion has been referred to as "colitis cystica profunda."[36]

Gastrointestinal lesions in other animal species are of less diagnostic importance and are found only terminally. For the most part this consists of hemorrhage, usually in the stomach and proximal small intestine. Microscopically, the hemorrhage appears to result from loss of integrity of the small vessels (diapedesis) of the lamina propria.[18]

Pathology of the liver is interesting in that it is so conspicuous in some species (mice, rats, rabbits, and chickens[18,26]) and yet so minimal in others (monkeys and guinea pigs.[6,13]) However, hepatomegaly (both absolute and percent of body weight) is consistently observed in all species, and at exposures considerably below the lethal dose. The increased size and weight of the liver has been ascribed to hepatocellular enlargement, which ultrastructurally has been related to an increase in endoplasmic reticulum.[37–40] The hepatic lesions in rats and even more so in mice consist of focal necrosis of hepatocytes with inflammatory infiltrates, cytomegaly and fatty change.[13,16,41,42] Hepatocellular necrosis with hemorrhage is the hallmark of acute toxicity in rabbits.[43] In more chronic exposures, bile-duct hyperplasia is found to varying degrees in all animals species. This may extend into the extrahepatic biliary system and macroscopic enlargement of the common bile duct is often striking in nonhuman primates (FIGURES 11 and 12).[6]

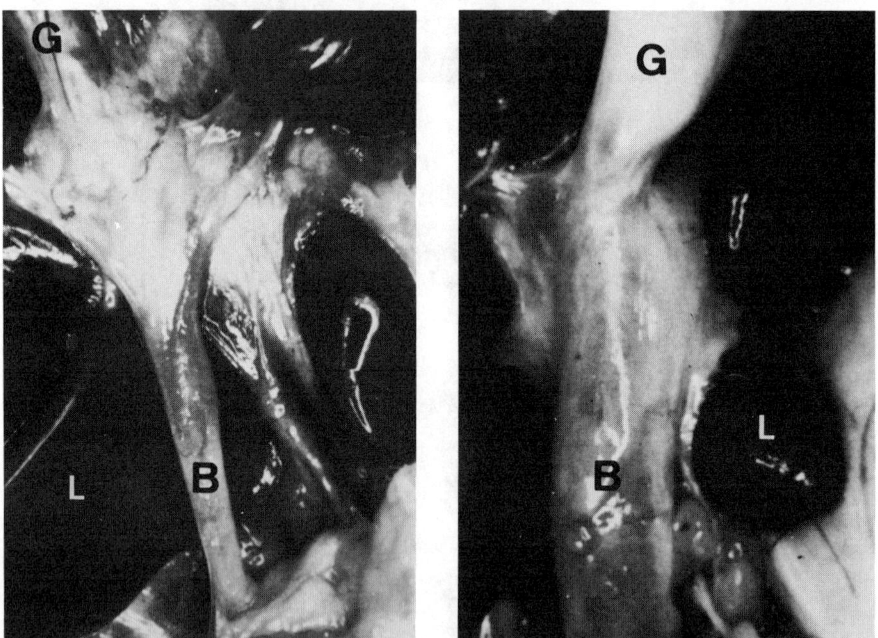

FIGURES 11 and 12. Bile duct (B) from normal monkey (FIGURE 11, left) compared to a monkey exposed to 2,3,7,8–TCDD. Note the thickened appearance which extends into the gall bladder (G), Liver (L).

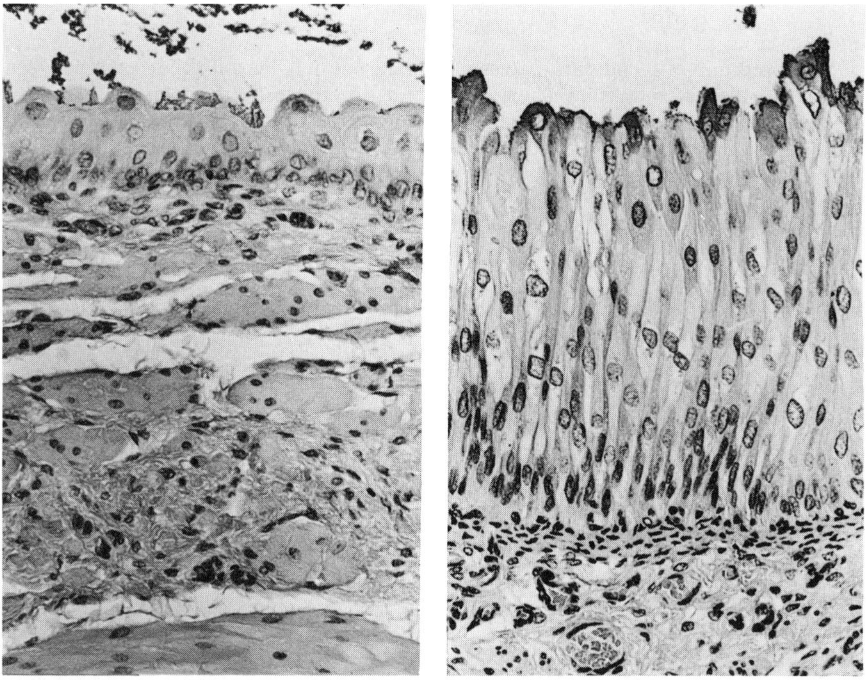

FIGURES 13 and 14. Urinary bladder from guinea pig exposed to 2,3,7,8-TCDD showing marked epithelial hyperplasia. (FIGURE 13, left, shows the normal bladder and FIGURE 14, the effects of treatment.) (Toxicol. Appl. Pharmacol. **44:** 335–356, 1978).

Significant urinary tract lesions have been found only in guinea pigs[13,16] and monkeys.[6] These consist of hyperplasia of the transitional epithelium of the renal pelvis, ureter, and urinary bladder. In these cases, the epithelial lining may be 2 to 3 times normal thickness (FIGURES 13 and 14).

Testicular degeneration has been observed in mice,[13] rats,[11] guinea pigs[13] and monkeys.[27] It is restricted to degenerative changes in the seminiferous components and may be so severe that only a single layer of germinal tissue remains (FIGURES 15 and 16). However, in most cases this appears to be a terminal event and it is unclear as to whether this is a direct response to toxicity.

Varying degrees of hypocellularity of the bone marrow have been found in all species of animals in which it was studied.[3,13] Bone marrow changes are more apparent in animals that die but may also be found to some degree in animals that receive a sublethal dose. While in the more severe cases it is obviously pancytopenic in nature, depression of erythroid precursors may be more pronounced as evidenced by an increase in the myeloid:erythroid ratio.

The adrenal gland of mice may show micro-foci of abscessation and atrophy of the zona glomerulosa has been observed in guinea pigs.[13] Hemorrhage in the adrenal has been observed in many species. However, adrenal lesions appear to be restricted to animals that die or have been killed in a moribund condition. Experimental evidence (incomplete for some chemicals) suggests that these compounds are weakly to moderately carcinogenic in mice and rats.[12,44–46]

Hematology and Clinical Chemistry

Interpretation of blood parameters is difficult not only because they vary between species but also because they show significant differences within the same species depending on whether exposures are acute or chronic. In acute studies with rats and monkeys the packed cell volume is normal[4,6] or only mildly reduced as compared to chronic studies[47] where anemia is more pronounced, especially in monkeys.[4] Anemia has also been described in mice and chickens.[48] The anemic state is probably related to bone marrow depression as described before.

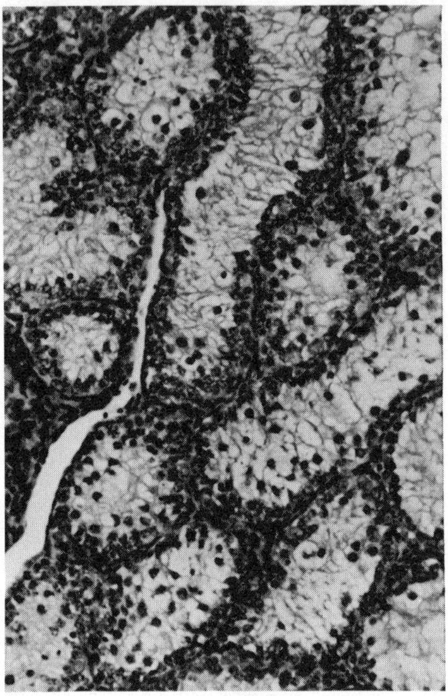

Figure 15

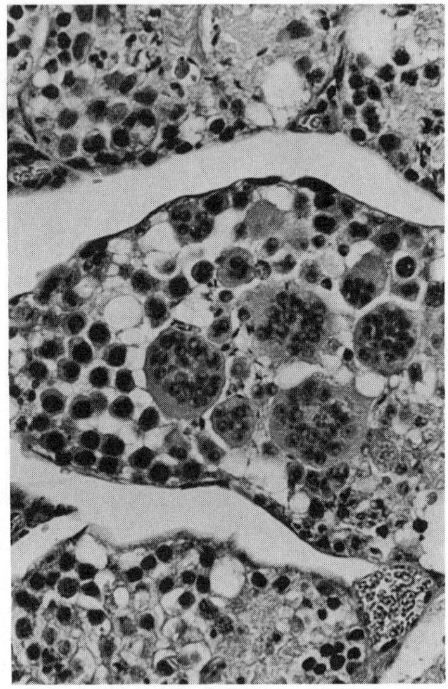

Figure 16

Figure 15. Testicle from guinea pig exposed to 2,3,6,7-tetrabrominated napthalene showing loss of germinal cells in the seminiferous tubules.

Figure 16. Testicle from a mouse exposed to 2,3,7,8-TCDD showing abnormal seminiferous tubule containing several multinucleated giant cells and necrotic spermatocytes. (Toxicol. Appl. Pharmacol. **44:** 335–356, 1978).

Leukocytosis has been reported in monkeys after acute exposures,[6] but leukopenia is more characteristic of chronic exposures.[4] The decrease in the total number of circulating leukocytes is caused primarily by a decrease in the number of lymphocytes. Similarly, leukopenia has been observed in mice, rats and guinea pigs.[49] In contrast, leukocytosis due to an increase in neutrophiles is the usual response in rats to the toxic action of these compounds, especially in long-term studies. A decrease in the number of circulating platelets has been reported in rat and monkey studies,[3,49] but information is lacking on this parameter in other species. In fact, thrombocytopenia with resulting hemorrhage has been incriminated as a contributing factor to the death of monkeys chronically exposed to dioxin.[3] Again, hematologic dyscrasias are more apparent in female animals, but this may be related to their overall increased susceptibility to intoxication with these chemicals.

A reduction in total serum protein is found routinely in monkeys,[2] mice[6] and chickens[48] but not guinea pigs[6] and rats.[12] The reduction is primarily the result of a decrease in the albumin fraction, but specific globulin fractions may be also altered.

Alterations in serum enzymes and other blood-chemistry parameters reflect alterations in various organs observed histopathologically. Increased levels of glutamic-oxaloacetic transaminase (GOT),[47,50] glutamic-pyruvic transaminase (GPT)[4] and bilirubin[11] all reflect liver lesions and the former two parameters may be related to overall cachexia. Serum cholesterol levels are increased in rats[49] but decreased in monkeys.[4] In contrast serum triglycerides increase in monkeys[6] but remain normal in rats. These two parameters have not been studied in other species. Glucose levels are decreased terminally in all species studied and are felt to be related to debilitation. Indicators of renal damage such as blood urea nitrogen, creatinine, and blood electrolytes are usually within normal limits.

Porphrin levels (particularly in the liver) are dramatically increased in mice[51] and chickens[52] and less so in rats, rabbits, and monkeys. This also may occur at dose levels far below the lethal range. In fact, one of the dioxins (2,3,7,8-tetrachlorodibenzo-*p*-dioxin) is thought to be the most porphrinagenic compound ever produced by man.[53] Interestingly, increases in porphrin have not been demonstrated in monkeys or guinea pigs.

General Discussion

The overall appearance of the toxic syndrome produced by these compounds suggests that there are certain organs that show a commonality of lesions: *i.e.* (1) thymus and to a lesser degree other lymphoid tissues, (2) bone marrow and the hematologic dyscrasias derived from it. In contrast, the highly diagnostic combination of cutaneous and gastrointestinal lesions in monkeys appear to be restricted to this species, although rabbits may also show acne-like lesions of lesser severity. Several studies have shown differences in tissue residues of these compounds. The monkey[54] shows a much higher concentration of the retained dose in the skin than most other species,[55,56] which may explain the cutaneous response. A similar difference in tissue concentrations may explain the more limited hepatic pathology in monkeys and guinea pigs. A great deal of kinetic work remains before this aspect can be elucidated. Excretion rates via the kidney are also higher in monkeys[57] than rats[58] and this may explain the hyperplastic changes in the urinary epithelium in monkeys.

In most cases, the degree of tissue destruction is insufficient to define the cause of death. The pathologic appearance gives one the impression of a starvation-like syndrome albeit other species-specific manifestations. In addition, the lesions of the immune system may result in secondary disease syndromes, particularly in long-term, low-level exposure studies.

An excellent review of the toxicity of polychlorinated polycyclic compounds, particularly as they relate to human health, has been compiled by Kimbrough.[29]

References

1. Allen, J. R. 1975. Response of the nonhuman primate to polychlorinated biphenyl exposure. Fed. Proc. **34:** 1675–1679.
2. Allen, J. R., L. J. Abrahamson & D. H. Norback. 1973. Biological effects of

polychlorinated biphenyls and triphenyls on the subhuman primate. Environ. Res. **6:** 344–354.
3. Allen, J. R., D. A. Barsotti, J. P. Van Miller, L. J. Abrahamson & J. J. Lalich. 1977. Morphological changes in monkeys consuming a diet containing five hundred parts per trillion of 2,3,7,8-tetrachlorodibenzo-*p*-dioxin. Food Cosmet. Toxicol. **15:** 401–410.
4. Allen, J. R., L. A. Carstens & D. A. Barsotti. 1974. Residual effects of short-term, low-level exposure of nonhuman primates to polychlorinated biphenyls. Toxicol. Appl. Pharmacol. **30:** 440–451.
5. Brinkman, U. A., Th & H. G. M. Reymer. 1976. Polychlorinated napthalenes. J. Chromatog. **127:** 203–243.
6. McConnell, E. E., J. A. Moore & D. W. Dalgard. 1978. Toxicity of 2,3,7,8-tetrachlorodibenzo-*p*-dioxin in Rhesus monkeys *(Macaca mulatta)* following a single oral dose. Toxicol. Appl. Pharmacol. **43:** 175–187.
7. Moore, J. A. & E. E. McConnell. Comparative toxicity of halogenated dibenzofurans. Ann. N.Y. Acad. Sci. In press. This volume.
8. Allen, J. R., D. A. Barsotti & J. P. Van Miller. 1978. Reproductive dysfunction in nonhuman primates exposed to dioxins. 17th Annual Meeting Society of Toxicology, San Francisco, CA.
9. Barsotti, D. A., R. J. Marlar & J. R. Allen. 1976. Reproductive dysfunctions in Rhesus monkeys exposed to low levels of polychlorinated biphenyls (Aroclor 1248). Food Cosmet. Toxicol. **14:** 99–103.
10. Jones, E. L. & H. Krizek. 1962. A technic for testing acnegenic potency in rabbits, applied to the potent acnegen, 2,3,7,8-tetrachlorodibenzo-*p*-dioxin. J. Invest. Dermatol. **39:** 511–517.
11. Kociba, R. J., P. A. Keeler, C. N. Park & P. J. Gehring. 1976. 2,3,7,8-tetrachlorodibenzo-*p*-dioxin (TCDD): Results of a 13-week oral toxicity study in rats. Toxicol. Appl. Pharmacol. **35:** 553–574.
12. Kociba, R. J., D. G. Keyes, J. E. Beyer, R. M. Carreon, C. E. Wade, D. A. Dittenber, R. P. Kalnins, L. E. Frauson, C. N. Park, S. D. Barnard, R. A. Hummel & C. G. Humiston. Results of a two year chronic toxicity and oncogenicity study of 2,3,7,8-tetrachlorodibenzo-*p*-dioxin (TCDD) in rats. Toxicol. Appl. Pharmacol. In press.
13. McConnell, E. E., J. A. Moore, J. K. Haseman & M. W. Harris. 1978. The comparative toxicity of chlorinated dibenzo-*p*-dioxins in mice and guinea pigs. Toxicol. Appl. Pharmacol. **44:** 335–356.
14. Harris, M. W., J. A. Moore, J. G. Vos & B. N. Gupta. 1973. General biologic effects of TCDD in laboratory animals. Environ. Health Perspect. **5:** 101–109.
15. Allen, J. R. & J. J. Lalich. 1962. Response of chickens to prolonged feeding of crude "toxic fat." Proc. Soc. Exp. Biol. Med. **109:** 48–51.
16. Gupta, B. N., J. G. Vos, J. A. Moore, J. G. Zinkl & B. C. Bullock. 1973. Pathologic effects of 2,3,7,8-tetrachlorodibenzo-*p*-dioxin in laboratory animals. Environ. Health Perspect. **5:** 125–140.
17. Lillie, R. J., H. C. Cecil, J. Bitman & G. F. Fries. 1974. Differences in response of caged white Leghorn layers to various polychlorinated biphenyls (PCBs) in the diet. Poultry Sci. **53:** 726–732.
18. Vos, J. G. & J. H. Koeman. 1970. Comparative toxicologic study with polychlorinated biphenyls in chickens with special reference to porphyria, edema formation, liver necrosis, and tissue residues. Toxicol. Appl. Pharmacol. **17:** 656–668.
19. Bell, W. B. 1953. The relative toxicity of the chlorinated napthalenes in experimentally produced bovine hyperkeratosis (X-Disease). Vet. Med. **48:** 135–146.
20. Biocca, M., J. A. Moore, B. N. Gupta & J. D. McKinney. 1975. Toxicology of selected symmetrical hexachlorobiphenyl isomers: I. Biological responses in chicks and mice. Proceed. Natl. Conf. Polychlorinated Biphenyls EPA-560/6-75-004. pp. 67–72.
21. Moore, J. A., B. N. Gupta & J. G. Vos. 1975. Toxicity of 2,3,7,8-tetrachlorodibenzofuran—Preliminary results. Proc. Nat. Conf. PCBs. EPA 560/6-75-004. pp. 77–80.
22. Bowes, G. W., M. J. Mulvihill, R. B. Simoneit, A. L. Burlingame & R. W. Risebrough. 1975. Identification of chlorinated dibenzofurans in American polychlorinated biphenyls. Nature **256:** 305–307.
23. Roach, J. A. G. & I. H. Pomerantz. 1974. The finding of chlorinated dibenzofurans in a

Japanese polychlorinated biphenyl sample. Bull. Environ. Contam. Toxicol. **12:** 338–342.

24. Vos, J. G., J. H. Koeman, H. L. van der Maas, M. C. ten Noever de Brauw & R. H. De Vos. 1970. Identification and toxicological evaluation of chlorinated dibenzofuran and chlorinated napthalene in two commercial polychlorinated biphenyls. Food Cosmet. Toxicol. **8:** 625–633.
25. McConnell, E. E. & J. D. McKinney. 1978. Exquisite toxicity in the guinea pig to structurally similar halogenated dioxins, furans, biphenyls and napthalenes. 17th Annual Meeting, Society of Toxicology, San Francisco, CA.
26. McKinney, J. D., K. Chae, B. N. Gupta, J. A. Moore & J. A. Goldstein. 1976. Toxicological assessment of hexachlorobiphenyl isomers and 2,3,7,8-tetrachlorodibenzofuran in chicks. Toxicol. Appl. Pharmacol. **36:** 65–80.
27. Norback, D. H. & J. R. Allen. 1973. Biological response of the nonhuman primate, chicken, and rat to chlorinated dibenzo-*p*-dioxin ingestion. Environ. Health Perspect. **5:** 233–240.
28. Schwetz, B. A., J. M. Norris, G. L. Sparschu, V. K. Rowe, P. J. Gehring, L. L. Emerson & C. G. Gerbig. 1973. Toxicology of chlorinated dibenzo-*p*-dioxins. Environ. Health Perspect. **5:** 87–100.
29. Kimbrough, R. D. 1974. The toxicity of polychlorinated polycyclic compounds and related chemicals. CRC Critical Reviews in Toxicology. **2:** 445–499.
30. Vos, J. G., J. A. Moore & J. G. Zinkl. 1973. Effect of 2,3,7,8-tetrachlorodibenzo-*p*-dioxin on the immune system of laboratory animals. Environ. Health Perspect. **5:** 149–162.
31. Vos, J. G., J. A. Moore & J. G. Zinkl. 1974. Toxicity of 2,3,7,8-Tetrachlorodibenzo-*p*-dioxin (TCDD) in C57B1/6 mice. Toxicol. Appl. Pharmacol. **29:** 229–241.
32. Vos, J. G. & R. B. Beems. 1971. Dermal toxicity studies of technical polychlorinated biphenyls and fractions thereof in rabbits. Toxicol. Appl. Pharmacol. **19:** 617–633.
33. McClure, E. L., J. E. Savage & B. L. O'Dell. 1962. Hydropericardium and ascites in chicks fed chlorinated hydrocarbon. Poultry Sci. **41:** 295–299.
34. Allen, J. R. & D. H. Norback. 1973. Polychlorinated biphenyl- and triphenyl-induced gastric mucosal hyperplasia in primates. Science **179:** 498–499.
35. Scotti, T. M. 1973. Simian gastropathy with submucosal glands and cysts. Arch. Pathol. **96:** 403–408.
36. Scotti, T. M. 1975. Colitis cystica profunda in rhesus monkeys. Lab. Animal Sci. **25:** 55–61.
37. Fowler, B. A., G. W. Lucier, H. W. Brown & O. S. McDaniel. 1973. Ultrastructural changes in rat liver cells following a single oral dose of TCDD. Environ. Health Perspect. **5:** 141–148.
38. Hansell, M. M. & D. J. Ecobichon. 1974. Effects of chemically pure chlorobiphenyls on the morphology of rat liver. Toxicol. Appl. Pharmacol. **28:** 418–427.
39. Jones, G. & W. H. Butler. 1974. A morphological study of the liver lesion induced by 2,3,7,8-tetrachlorodibenzo-*p*-dioxin in rats. J. Pathol. **112:** 93–97.
40. Kasza, L., M. A. Weinberger, C. Carter, D. E. Hinton, B. F. Trump & E. A. Brouwer. 1976. Acute, subacute, and residual effects of polychlorinated biphenyl (PCB) in rats. II. Pathology and electron microscopy of liver and serum enzyme study. J Toxicol. Environ. Health. **1:** 689–703.
41. Allen, J. R. & L. J. Abrahamson. 1973. Morphological and biochemical changes in the liver of rats fed polychlorinated biphenyls. Arch. Environ. Contam. Toxicol. **1:** 265–280.
42. Kimbrough, R. D., R. E. Linder & T. B. Gaines. 1972. Morphological changes in livers of rats fed polychlorinated biphenyls. Arch. Environ. Health **25:** 354–364.
43. Koller, L. D. & J. G. Zinkl. 1973. Pathology of polychlorinated biphenyls in rabbits. Am. J. Pathol. **70:** 363–378.
44. Di Giovanni, J., A. Viaje, D. L. Berry, T. J. Slaga & M. R. Juchau. 1977. Tumor-initiating ability of 2,3,7,8,-tetrachlorodibenzo-*p*–dioxin (TCDD) and Arochlor 1254 in the two-stage system of mouse skin carcinogenesis. Bull. Environ. Contam. Toxicol. **18:** 552–556.
45. Kimbrough, R. D., R. A. Squire, R. E. Linder, J. D. Strandberg, R. J. Montali &

V. W. BURSE. 1975. Induction of liver tumors in Sherman strain female rats by polychlorinated biphenyl Arochor 1260. J. Nat. Cancer Inst. **55:** 1453–1459.

46. VAN MILLER, J. P., J. A. LALICH & J. R. ALLEN. 1977. Increased incidence of neoplasms in rats exposed to low levels of 2,3,7,8-tetrachlorodibenzo-*p*-dioxin. Chemosphere **10:** 625–632.
47. BRUCKNER, J. V., K. L. KHANNA & H. H. CORNISH. 1973. Biological responses of the rat to polychlorinated biphenyls. Toxicol. Appl. Pharmacol. **24:** 434-448.
48. ALLAN, J. R. & L. A. CARSTENS. 1966. Electron microscopic alterations in the liver of chickens fed toxic fat. Lab. Invest. **15:** 970-979.
49. ZINKL, J. G., J. G. VOS, J. A. MOORE & B. N. GUPTA. 1973. Hematologic and clinical chemistry effects of 2,3,7,8-tetrachlorodibenzo-*p*-dioxin in laboratory animals. Environ. Health Perspect. **5:** 11-118.
50. BUU-HOI, N. P., C. PHAM-HUU, G. SESQUE, M. C. AZUM-GELADE & G. SAINTRUF. 1972. Enzymatic functions as targets of toxicity of "dioxin" (2,3,7,8-tetrachlorodibenzo-*p*-dioxin). Naturwissenschaften **59:** 173–174.
51. GOLDSTEIN, J. A., P. HICKMAN, H. BERGMAN & J. G. VOS. 1973. Hepatic porphria induced by 2,3,7,8-tetrachlorodibenzo-p-dioxin. Res. Commun. Chem. Pathol. Pharmacol. **6:** 919–928.
52. GOLDSTEIN, J. A., J. D. MCKINNEY, G. W. LUCIER, P. HICKMAN, H. BERGMAN & J. A. MOORE. 1976. Toxicological assessment of hexachlorobiphenyl isomers and 2,3,7,8-tetrachlorodibenzofuran in chicks. Toxicol. Appl. Pharmacol. **36:** 81–92.
53. POLAND, A. & E. GLOVER. 1973. Chlorinated dibenzo-*p*-dioxins: Potent inducers of δ-aminolevulinic acid synthetase and aryl hydrocarbon hydroxylase. II. A study of the structure-activity relationship. Mol. Pharmacol. **9:** 736-747.
54. VAN MILLER, J. P., I. C. HSU & J. R. ALLAN. 1975. Distribution and metabolism of H^3-2,5,2′,5′-tetrachlorobiphenyl in rats (38610). Proc. Soc. Exp. Biol. Med. **148:** 682-687.
55. ALLEN, J. R., J. P. VAN MILLER & D. H. NORBACK. 1975. Tissue distribution, excretion and biological effects of [^{14}C] tetrachlorodibenzo-*p*-dioxin in rats. Food Cosmet. Toxicol. **13:** 501–515.
56. BRANDT, I. 1975. The distribution of 2,2′,3,4,4′,6′ and 2,3′,4,4′,5′,6-hexachlorobiphenyl in mice studied by whole-body autoradiography. Toxicology **4:** 275-287.
57. VAN MILLER, J. P., R. J. MARLAR & J. R. ALLAN. 1976. Tissue distribution and excretion of tritiated tetrachlorodibenzo-*p*-dioxin in non-human primates and rats. Food Cosmet. Toxicol. **14:** 31–34.
58. BURSE, V. W., R. D. KIMBROUGH, E. C. VILLANUEVA, R. W. JENNINGS, R. D. LINDER & G. W. SOVOCOOL. 1974. Polychlorinated biophenyls. Arch. Environ. Health **29:** 301–307.

COMPARATIVE TOXICITY OF THREE HALOGENATED DIBENZOFURANS IN GUINEA PIGS, MICE, AND RHESUS MONKEYS

J. A. Moore, E. E. McConnell, D. W. Dalgard,* and M. W. Harris

Environmental Biology and Chemistry Branch
National Institute of Environmental Health Sciences
Research Triangle Park, North Carolina 27709

**Hazleton Research Laboratories of America, Inc.*
Vienna, Virginia 22180

INTRODUCTION

The chlorinated dibenzofurans have been reported to be present in a variety of polychlorinated biphenyls (PCBs)[1-3] and in pentachlorophenol.[4] They were also detected in the PCB-contaminated rice oil that was responsible for the human intoxication referred to as 'Yusho'.[5,6] Polychlorinated dibenzofurans were also detected in tissues of patients with 'Yusho'.[7] A polychlorinated dibenzofuran fraction extracted from PCB's was found to be highly toxic in chickens.[8] A mixture of polychlorinated dibenzofurans also caused toxic effects when fed to rats.[9] 2,3,7,8-Tetrachlorodibenzofuran (TCDF) has been evaluated for toxicity in chicks,[10] guinea pigs,[11] and mice.[11]

The results reported in this manuscript extend earlier studies with TCDF in guinea pigs and mice, report the toxic effects in rhesus monkeys, and describe the toxic effects of 2,3,4,7,8-pentachlorodibenzofuran (PCDF) and 2,3,7,8-tetrabromodibenzofuran (TBDF) in guinea pigs.

MATERIALS AND METHODS

TCDF was obtained from the Illinois Institute of Technology Research Institute, Chicago, Illinois; PCDF was obtained from the Food and Drug Administration, Washington, D.C.; and TBDF was obtained from Dr. A. Kende, University of Rochester, Rochester, New York. All chemicals were determined to be of greater than 98% purity using gas chromatographic and mass spectral techniques. Each chemical was dissolved in reagent grade acetone and subsequently diluted in corn oil, after which the acetone was removed *in vacuo*. Corn oil was added so that each guinea pig received a volume of 0.1 ml/100 gram body weight; each mouse received 0.01 ml/1 gram body weight; each monkey received a total dose volume of 3.5 ml/kg body weight. All animals were dosed by gastric intubation.

A dose group consisted of 6 guinea pigs, 8 mice, or 2 monkeys. For guinea pigs and mice, a minimum of 9 animals served as controls; for monkeys, 2 animals served as controls.

Mice were of the C57 Bl/6fh(J67) strain (Jackson Laboratories, Bar Harbor, Maine), male, 6 weeks old at time of dosing; they were housed singly in polycarbonate plastic cages, 18.4 × 27.9 cm, which contained autoclaved corn cob bedding, autoclaved NIH-31 food (Zeigler Brothers, Gardner, Pa.) and water. Guinea pigs were of the Hartley strain (Charles River Breeding Laboratories, Wilmington, Ma.),

0077-8923/79/0320-0151 $01.75/0 © 1979, NYAS

male, 3–4 weeks old at time of dosing; they were housed singly in suspended stainless steel wire cages, 41.3 × 28.3 cm and fed NIH Feed A (Zeigler Brothers, Gardner, Pa.) and water *ad libitum*.

Rhesus monkeys (*Macaca mulatta*) were young, sexually immature females weighing between 2.0 and 3.7 kg, fully quarantined and conditioned, and housed individually in stainless steel cages equipped with automatic water founts an additional 3 weeks prior to dosing; they were fed Purina Monkey Chow (25% protein) twice daily.

Each animal was weighed 3 times weekly throughout the experiment, which was 30 days for mice and guinea pigs and 60 days for monkeys.

In studies with guinea pigs and mice, the first animal in each dose group was allowed to die, after which most moribund animals were killed by inducing deep anesthesia and then exsanguinated by transecting blood vessels leading to a limb. All monkeys were killed in a similar manner when they reached a moribund condition. Necropsies were performed on all monkeys and in guinea pigs and mice dying or killed during the observation period. Survivors in dosage groups in which deaths occurred and from all animals in the highest dosage group in which death was not observed were also necropsied at the end of the experiment. Tissues selected for histopathology were esophagus, trachea, heart, liver with gall bladder, lung, stomach, small and large intestines, spleen, thymus, kidney, urinary bladder, testes, skin (ear), brain, eye, thyroid, pancreas, adrenal, sternum (bone marrow), and skeletal muscle. Additional tissues selected in monkeys were spinal cord (thoracic and lumbar), eyelids, salivary gland, mesenteric lymph nodes, mandilular lymph nodes, tracheal lymph nodes, mandibular lymph nodes, ovaries, uterus, ear canal, lips, foot, and hand. Liver was examined under long wavelength ultraviolet light for red fluorescence as an indication of excess porphyrins. Tissue specimens were fixed in buffered neutral 10% formalin, embedded in paraffin, sectioned at 6μ, and stained with hematoxylin and eosin (HE); cryostat sections of formalin fixed liver were stained with Oil Red 0 for neutral lipid evaluation.

The following hematology and clinical chemistry studies were performed at the termination of the experiment on mice that received 22 doses of 300 mg/kg TCDF *per os* and on guinea pigs that received a single oral dose of TCDF: hematocrit, hemoglobin total leukocyte count, platelet, total erythrocytes, differential leukocyte count, total serum proteins by the buiret method, and cellulose acetate electrophoresis (microzone® electrophoresis system, Beckman, Fullerton, California) for serum fractionations. In the monkey experiment, the following clinical chemistry studies were performed in addition to those outlined above: sedimentation rate, serum glucose, blood urea nitrogen, total bilirubin, sodium, potassium, chloride, calcium, alkalin phosphatase, glutamine oxalecetic transaminase (SGOT), cholesterol, triglycerides, aldolase, and porphyrin analysis. These determinations were performed twice prior to dosing; on the day of dosing and on days 7, 14, 28, and 56 post-dosing.

The dosing protocol for the various halogenated dibenzofurans is depicted in TABLE 1.

RESULTS

Guinea Pigs

A lethal effect was observed with certain doses of TCDF, PCDF, and TBDF (TABLE 2). The pattern of toxic symptoms was similar for the three chemicals. Guinea pigs that were destined to die typically demonstrated an immediate and

TABLE 1

DOSING PROTOCOL FOR HALOGENATED DIBENZOFURANS

	Dose (μg/kg)			Dose		Observation
	Guinea Pig	Mice	Monkey	Frequency	Route	Period (days)
TCDF	0, 1, 5, 10, 15	0, 400, 600, 800, 1200, 1500, 2500, 4000, 6000	0, 500, 1000, 1500	single	oral	30 60 (monkey)
	ND*	6000	ND	single	SQ	30
	ND*	0, 30, 100, 300	ND	5/week 22 total	oral	30
PCDF	0, 1, 3, 10, 30	ND	ND	single	oral	30
TBDF	0, .47, 1.58, 4.74, 15.81†	ND	ND	single	oral	30

*ND - not done.
†Molar equivalent to 0., 0.3, 1.0, 3, 10 μg/kg TCDF.

TABLE 2

GENERAL TOXIC EFFECTS OF HALOGENATED DIBENZOFURANS IN GUINEA PIGS*

Chemical	Dose (μg/kg)	Mortality	Mean Time to Death (days)	Comments
TCDF	0	0/9†	—	Normal growth
	1	0/6	—	Decreased body weight gain
	5	0/6	—	Body weight loss in 2; other decreased weight gain
	10	6/6	14.6 (12–20)	Immediate and progressive weight loss
	15	6/6	11.8 (9–16)‡	Immediate and progressive weight loss
PCDF	0	0/9	—	Normal growth
	1	0/6	—	Decreased body weight gain
	3	0/6	—	Decreased body weight gain
	10	6/6	15.8 (13–19)	Immediate and progressive weight loss
	30	6/6	9.2 (8–11)	Immediate and progressive weight loss
TBDF	0	0/9	—	Normal growth
	.47	0/6	—	No effect
	1.58	0/6		No effect
	4.74	1/6	26	Decreased body weight gain; progressive weight loss in animal that died
	15.84	6/6	11.7 (10–13)	Immediate and progressive weight loss

*All animals received single oral dose in corn oil, controls received only corn oil.
†Numerator indicates number dead, denominator indicates number tested.
‡Mean time to death; number in parenthesis is range.

progressive body weight loss with body weights at time of death 50% of those at time of dosing. The impression was that food consumption only decreased during the several days preceding death. These guinea pigs developed a spectrum of progressive signs characterized by rough, soiled hair coat; listlessness; dehydration; and wetness of hair under the chin and neck.

All guinea pigs that received 10 μg/kg or 15 μg/kg TCDF died 9 to 20 days after dosing. The mean time to death was earlier (11.8 days) at the 15 μg/kg dose than that recorded for the 10 μg/kg dose (14.6 days). No deaths occurred at the 5 or 1 μg/kg level, although toxic effects were induced, as denoted by decreased body weight gain or absolute weight loss.

PCDF killed all guinea pigs a mean of 15.8 or 9.2 days after administration of the 10 and 15 μg/kg doses, respectively. All guinea pigs survived the 1 or 3 μg/kg doses, although a decreased body weight gain during the 30-day observation period was recorded.

All guinea pigs died at a mean of 11.7 days after receiving the 15.84 μg/kg dose of TBDF. One of 6 guinea pigs died 26 days after receiving a 4.74 μg/kg dose. No deaths or body weight effects were observed in guinea pigs that received the 0.47 or 1.58 μg/kg of TBDF. Decreased body weight gain was recorded in the 5 guinea pigs that survived the 4.74 μg/kg dose.

At necropsy, there was no difference in the pattern of lesions present in guinea pigs that received TCDF, PCDF, or TBDF. There was marked reduction in the size of the thymus, lack of body fat, and reduction of muscle mass. Histopathologic examination of tissues revealed a loss of lymphoid cells in the thymic cortex, and hyperplasia of epithelial cells in the renal pelvis, ureter, and urinary bladder. These lesions along with lack of liver pathology, hypocellularity of bone marrow, seminiferous tubules, and lymphoid elements in spleen and Peyers patches, plus adrenal hemorrhage, were similar to those described for guinea pigs that developed intoxications from chlorinated dibenzodioxins.[12,13] Mild evidence of thymus lymphoid hypoplasia was the only histologic finding in guinea pigs that survived the 30-day observation period. This impression, which was sporadic, was observed only in guinea pigs that received the highest nonlethal dose. No hematologic or serum protein alterations were apparent in guinea pigs that survived the 30-day observation period.

Mice

Previous experiments failed to establish a lethal dose for TCDF in mice that received a single 6,000 μg/kg dose administered by gavage or subcutaneous injection. Mice that received the TCDF by subcutaneous injection had a transient depression in body weight gain, thymic involution, and mild hepatotoxicity.[11]

In a subsequent experiment, mice that received 22 oral doses over a 30-day period of 30, 100, or 300 μg/kg TCDF did not develop clinical signs of toxicity. Terminal body weights were similar in all groups (TABLE 3). Livers from mice in the 300 μg/kg group did not fluoresce when examined under long-wavelength ultraviolet light, a technique routinely used in our laboratories to detect the presence of porphyria. Liver weight and liver to body weight ratios were elevated, whereas thymus weight and thymus to body weight ratios were markedly reduced. Hematologic determinations failed to reveal any differences between controls and the 300 μg/kg group.

Total erythrocyte count, hemoglobin, packed cell volume, and platelet counts were similar in the treated and 300 μg/kg dose group. Total leukocyte counts were decreased in the 300 μg/kg TCDF group (2.91×10^3, S.E. 0.30) when compared to

TABLE 3

BODY WEIGHT AND SELECTED ORGAN WEIGHTS AT NECROPSY IN C57Bl/6J MICE TREATED WITH TCDF*

Dose (mg/kg)	No. Mice	Body Weight (grams)	Body Weight Gain (grams)	Liver		Thymus	
				Weight (grams)	L/BW Ratio	Weight (grams)	T/BW Ratio
0	8	27.6	2.4	1.56	5.62	.045	.163
30	8	26.9	2.4	1.91	7.08	ND†	ND
100	8	27.1	2.1	2.01	7.37	ND	ND
300	8	26.1	1.7	1.83	7.01	.007	.027

†ND = not done.

*Mice received oral dose of TCDF or vehicle (.1 ml/10 gm body weight) on 22 days over a 30-day period.

the controls (4.60×10^3, S.E. 0.71). No significant differences in the differential leukocyte counts were found. Total serum protein was moderately elevated in treated mice (7.02 gm % versus 6.31 gm %) due entirely to increased serum globulin (2.66 gm % versus 1.93 gm %). The albumin/globulin ratio was decreased proportionately (1.64 versus 2.38).

Rhesus Monkeys

A series of single oral exposure experiments was performed in female Rhesus monkeys to establish the toxic dose for TCDF and characterize the toxicopathologic effects. The two monkeys that received 1500 mg/kg died 14 and 23 days after TCDF administration (TABLE 4). Two of four monkeys died 25 and 31 days subsequent to a dose of 1000 μg/kg TCDF; the remaining two monkeys survived the 60-day period

TABLE 4

TOXIC EFFECTS OF A SINGLE ORAL TCDF DOSE IN RHESUS MONKEYS

Dose (μg/kg)	Mortality	Time of Death (days)	Clinical Pathology*
0	0/2	—	
500	0/2	—	Mild anemia at Day 28; reduced serum cholesterol at Day 14; neutrophilia and mild lymphopenia at Days 14 and 28
1000	2/4	25; 31	Mild anemia at Day 28; persistent, reduced serum cholesterol; moderate decrease in serum protein and albumin; neutrophilia and mild lymphopenia at Days 14 and 28; terminal increase in SGOT and BUN
1500	2/2	14; 23	Persistent reduced serum cholesterol; moderate decrease in serum protein and albumin; neutrophilia and mild lymphopenia at Days 14 and 28; terminal increase in SGOT and BUN

*Performed at days 14, 28, 42, and 60.

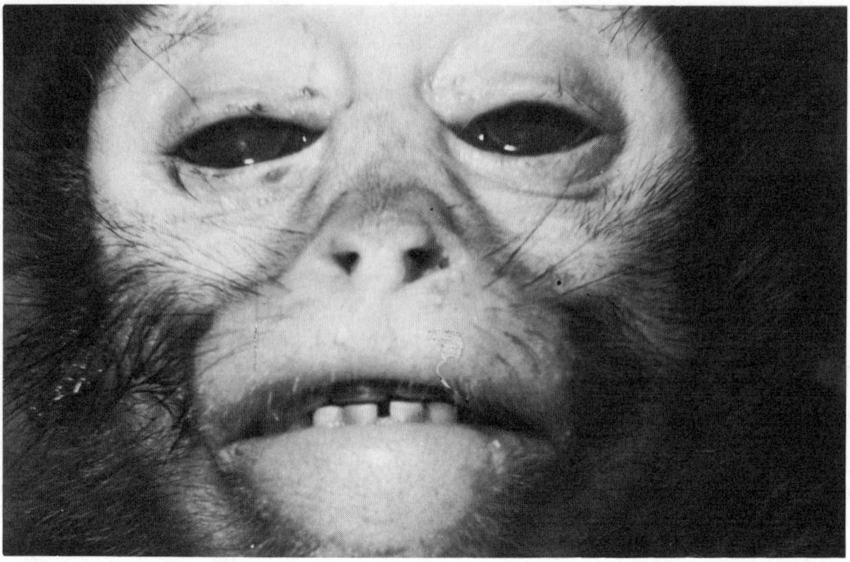

FIGURE 1. Monkey exposed to TCDF. Note edema of the lips and eyelids, loss of facial hair and especially the eyelashes.

TABLE 5
SUMMARY OF PATHOLOGIC EFFECTS OF TCDF IN RHESUS MONKEYS

Tissue	Gross	Microscopic
Skin and adnexa	Facial edema, dry skin, loss of eye lashes, loss of finger and toe nails, streaks on ocular surface of eyelid, and occlusion of external ear canal.	Hyperkeratosis Squamous metaplasia Cystic dilation of glands
Fat	Moderate to severe loss	None
Lymphoid	None	Thymic atrophy due to lymphoid cell loss Moderate atrophy of lymphoid cells in spleen and lymph nodes.
Liver	Increased weight	No lesions observed
Bile Duct	Markedly thickened	Hypertrophy of biliary duct and gall bladder epithelium Increased mucous secretion
Gastrointestinal	Hyperemia, petechial hemorrhage, focal ulceration and mucosal cysts	Loss of parietal cells, increased mucous cells, microcystic dilation of crypts

after exposure, but experienced weight loss and a progression of toxic symptoms. The two monkeys that received 500 μg/kg TCDF survived, remained alert and active, but did develop skin lesions and had a reduced weight gain.

The toxic response was dose related and did not begin to manifest itself clinically until 7–10 days after dosing. At this time, monkeys began developing evidence of slight facial edema in tissue around the eye socket, anorexia, occasional vomiting, and depression. Frequent rubbing of the eyes was common, and was believed to be due to insufficient lacrimation. Most lost all of their eyelashes (FIGURE 1); loss of fingernails and toenails also occurred, particularly at the two higher doses. Most animals

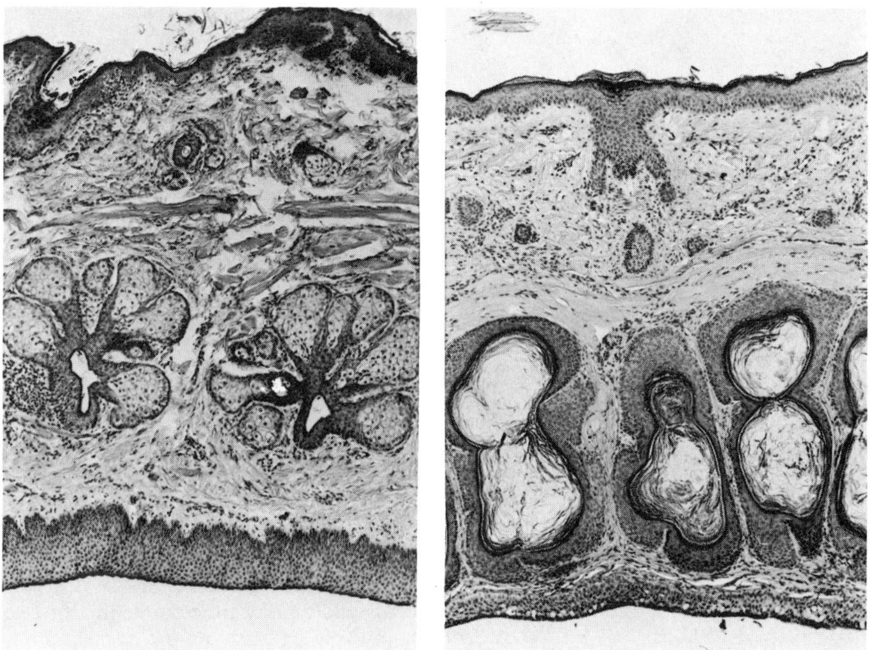

FIGURE 2. Cross section through the eyelid of a control (left) and TCDF-exposed monkey (right). The normal grape-like clusters of sebaceous glands (meibomian glands) around the central duct are missing in the treated animal. In addition, the ducts are dilated with keratinaceous debris.

accumulated a white, wax-like exudate in the ear canal; prominent white streaks were present in the ocular surface of the eyelid. A dry, leathery texture to the skin also developed. Monkeys that survived usually showed some improvement in condition by the end of the 60-day observation period.

The clinical pathology findings are summarized in TABLE 5. A mild anemia was present 28 days after dosing in all treated monkeys surviving to this time. In all treated groups, a marked relative and absolute neutrophilia was observed at the 14- and 28-day hematologic examination, as was a consistent relative lymphopenia and less striking absolute lymphopenia. Serum cholesterol was decreased by 33–50% in all treated groups on day 14, and persisted until death or sacrifice in monkeys that

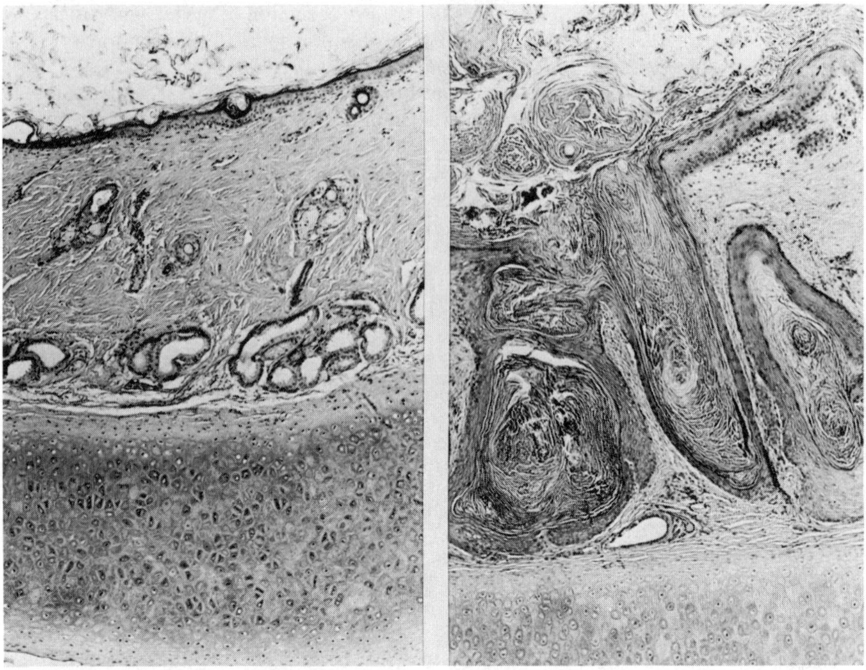

FIGURE 3. Section of ear canal from control monkey (left) and one exposed to TCDF (right). The sebaceous (ceruminous) glands have undergone a change similar to that observed in FIGURE 2. In addition, the sweat glands have been compressed.

received 1000 or 1500 μg/kg TCDF. SGOT and BUN values were increased only at the period that immediately preceded death. Moderate decreases in total serum protein and serum albumin were noted at the 14- and 28-day time points in monkeys that received 1000 or 1500 μg/kg. A decrease in these serum protein parameters was observed only at the 14-day time point in monkeys that received 500 μg/kg TCDF.

Post-mortem examination of the treated monkeys revealed a spectrum of lesions, which for the most part reflected the signs described above. The cutaneous, eyelid, external ear canal and toe nail lesions were apparent in all animals, but were more severe at the higher two doses and in animals that died. In fact, some improvement in these abnormalities in the two 500 μg/kg animals was noted at necropsy (60 days post exposure). All treated monkeys showed a moderate to severe loss of body fat. Although the livers were relatively normal, 3 of 6 treated monkeys had distinctly enlarged bile ducts. Livers did not fluoresce under long-wavelength ultraviolet light. No consistent differences in organ weight or organ to body weight ratios were found for liver, spleen, or kidney. Thymus weights were not recorded. Hyperemia, scattered areas of petechial hemorrhage, focal ulceration, and small bubble-like cysts were found in the fundic and duodenal area of the stomach and proximal small intestine in 3 of 6 animals. No other lesions were noted at necropsy.

Microscopic lesions were confined to the eyelid and ear canal in monkeys that received 500 μg/kg TCDF. In these tissues, squamous metaplasia of the meibomian and ceruminous glands, and hyperkeratosis of the epidermis was present along with cystic dilatations of the glandular ducts. Similar, but more severe, changes were observed in all animals at the higher two doses (FIGURES 2 and 3), as was a follicular hyperkeratosis of the skin in monkeys that received 1500 μg/kg (FIGURE 4). Thymus atrophy, the severity of which was dose-related, was observed, as denoted by loss of cortical lymphocytes and necrotic debris within Hassal's corpuscles (FIGURE 5). Spleen lymphoid elements showed moderate atrophy of germinal centers at the 1500 μg/kg doses; atrophic changes were less marked at the 1000 μg/kg dose, but more significant in the two monkeys that died. Similar, but mild, morpholic changes in mesenteric, tracheobronchial, and cervical lymph nodes were found at the 1500 μg/kg dose. No lesions were observed in the liver. Four of 6 animals had hyperplastic changes of the bile duct and gall bladder epithelium. The mucosal surface was nodular as a result of cystic hyperplasia, and there was an impression of exaggerated mucous secretion (FIGURE 6). In the four monkeys that were sacrificed in a moribund state due to TCDF toxicity, there was degranulation of exocrine pancreatic cells. A definite indication of bone marrow hypoplasia was also observed. In the glandular portion of the stomach, there was a loss of parietal cells and replacement by mucous cells. The crypt epithelium was hyperplastic with down growth into the underlying mucosa with microcystic dilation (FIGURE 7). Similar, but less severe, changes were observed in the proximal duodenum and colon.

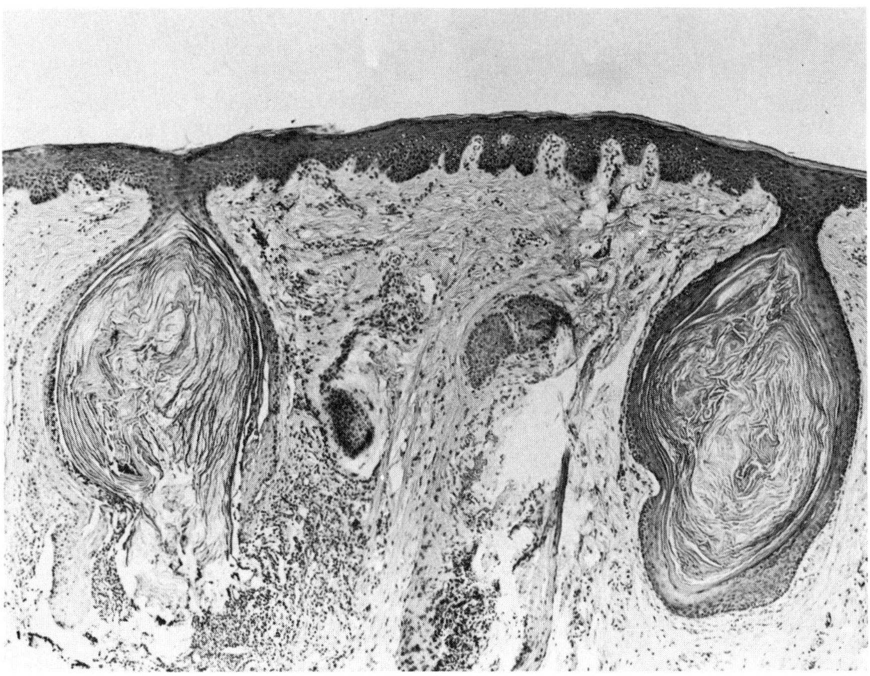

FIGURE 4. Section of lip from a monkey exposed to TCDF. The hair follicles are dilated with keratinaceous debris. The hair follicle on the left has ruptured, causing adjacent inflammation.

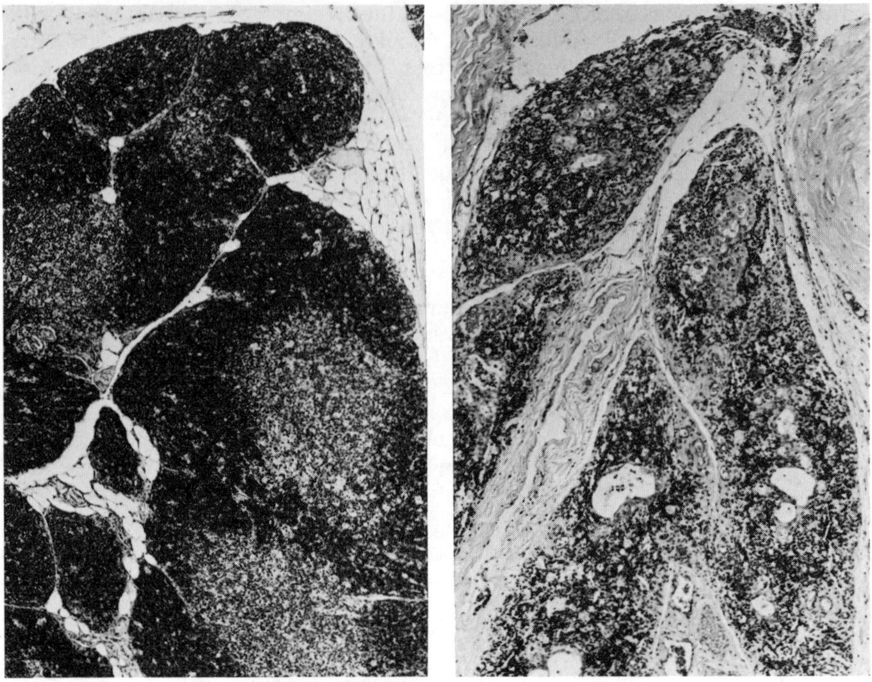

FIGURE 5. Section from the thymus of a control monkey (left) and one treated with TCDF (right). The cortex (right) is markedly reduced in size and there is necrotic debris in the medulla.

DISCUSSION

TCDF

The results of this study confirm the preliminary report of toxicity in guinea pigs.[11] The single oral $LD_{50/30}$ in guinea pigs is $>5<10$ $\mu g/kg$ body weight, and the toxicity is characterized by immediate and progressive weight loss with death at lethal doses occurring an average of 12–15 days after exposure. Histologic effects are principally associated with depletion of lymphoid cells within tissue. The effects are indistinguishable from those reported for the halogenated dibenzodioxins[13] and other halogenated aromatic hydrocarbons.[14] The $LD_{50/30}$ for 2,3,7,8-TCDF is 2 to 4-fold higher than reported for 2,3,7,8-tetrachlorodibenzo-*p*-dioxin (TCDD).

These studies established that TCDF is toxic at low dose levels to a primate species. The single oral $LD_{50/60}$ for TCDF is established as 1,000 $\mu g/kg$, a dose that is about 20-fold higher than that estimated for TCDD.[15] The toxicity is characterized by onset of the following symptoms 7–10 days after dosing: progressive weight loss, facial edema, loss of eyelashes and nails, development of white streaks in eyelids, thickening of skin, and death within 2–4 weeks. Histologic features are lymphoid hypoplasia, hyperkeratosis of skin, squamous metaplasia and cystic dilatation of components of the epidermal adnexa. Decreases in serum cholesterol and serum albumin were also consistently observed. The disease pattern produced cannot be distinguished from that produced by polychlorinated biphenyls or TCDD. The similarity of disease process mandates that experiments with both TCDF and PCBs be performed in order to better assess the respective roles of these chemicals in 'Yusho.'

TCDF at a total dose of up to 6,600 μg/kg administered over a 30-day period failed to cause clinical signs of toxicity in mice. These results and those reported earlier indicate that doses of ± 6 mg/kg TCDF given orally (either as a single or divided dose) or subcutaneously do not cause overt toxicity in mice. This dose is 23-fold higher than the TCDD $LD_{50/30}$ in the same strain of mouse and 30 to 33-fold higher than the TCDD dose required to produce equivalent organ-weight effects.[13] Thymus-weight decrease and liver hypertrophy provide evidence of some absorption and biologic response. It has been reported that 2,3,7,8-TCDF, when administered orally to mice with a number of other chlorinated dibenzofurans, was absorbed and had an approximate half-life of one week in liver.[16] The apparent decreased susceptibility of the mouse to TCDF when compared to toxicity due to TCDD needs to be explored further. Experiments on absorption, distribution, and metabolism of TCDF are planned.

The toxicity of halogenated dibenzofurans indicates that their presence as contaminants in a number of commercial products may confer toxic potential that is in excess of that established for the primary ingredient. Halogenated dibenzodioxins are

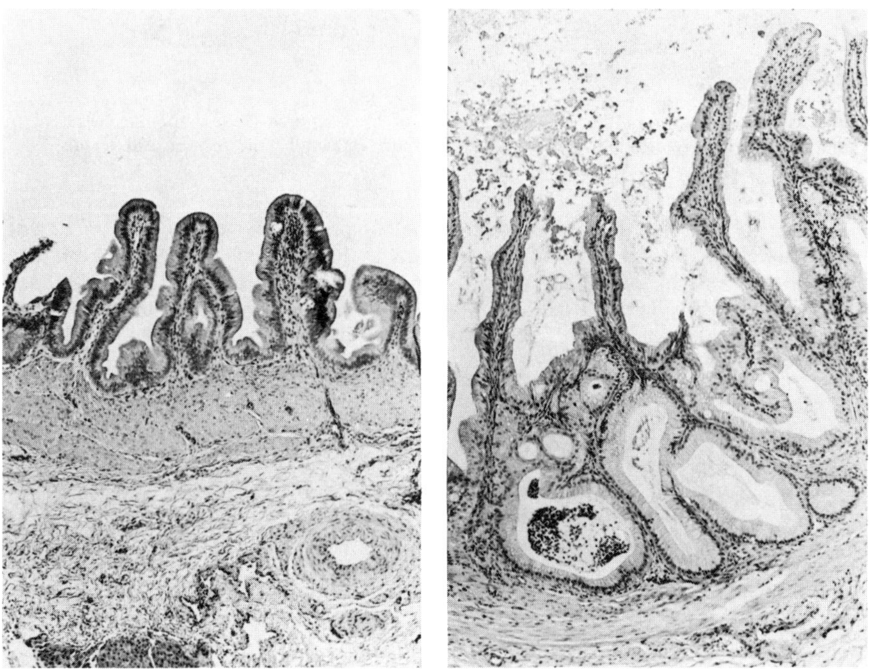

FIGURE 6. Mucosa of normal bile duct (left) and that from a monkey that died after a single exposure of 1500 μg/kg TCDF (right). The mucosa (right) is extremely hyperplastic with cystic dilation and inflammation at the base.

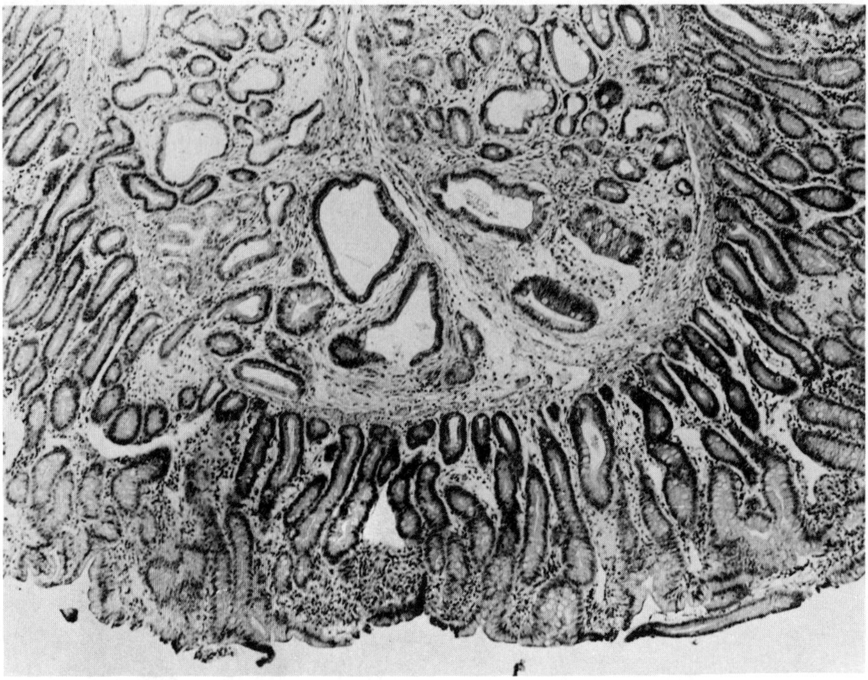

FIGURE 7. Section from the duodenal portion of the stomach from a monkey exposed to TCDF. There is early dilation of the crypts, and more especially, the submucosal glands.

also impurities in some of these products. The comparable pattern of toxic effects of the dibenzofurans and dibenzodioxins suggests a similar mechanism of action. The plausibility of this hypothesis is supported by their similar capability to induce aryl hydrocarbon hydroxylase.[17] It is therefore prudent to consider the combined levels of these contaminants in assessing the toxic potential of low-level contaminants in commercial products.

REFERENCES

1. BOWES, G. W., M. J. MULVIHILL, B. R. T. SIMONET, A. L. BURLINGAME & R. W. RISEBROUGH. 1975. Nature **256:** 305–307.
2. BOWES, G. W., M. J. MULVIHILL, M. R. DECAMP & A. S. Kende. 1975. J. Agric. Food Chem. **23 (6):** 1222–1223.
3. ROACH, J. A. G. & I. H. POMERANTZ. 1974. Bull. Environ. Contam. Toxicol. **12:** 330.
4. GOLDSTEIN, J. A., M. FRIESEN, R. E. LINDER, P. HICKMAN, J. R. HASS & H. BERGMAN. 1977. Biochem. Pharmacol. **26:** 1549–1557.
5. NAGAYAMA, J., M. KURATSUNE & Y. MASUDA. 1976. Bull. Environ. Contam. Toxicol. **15 (1):** 9–13.
6. MORITA, M., J. NAKAGAWA, K. AKIYAMA, S. MIMURA & N. I. SONO. 1977. Bull. Environ. Contam. Toxicol. **18 (1):** 67–73.
7. NAGAYAMA, J., Y. MASUDA & J. KURATSUNE. 1977. Food Cosmet. Toxicol. **15:** 195–198.

8. Vos, J. G., J. H. Koeman, F. L. van der Maas, M. C. ten Noever de Brauw & R. H. Vos. 1970. Food Cosmet. Toxicol. **8:** 625–633.
9. Oishi, S., M. Morita & H. Fukuda. 1978. Toxicol. Appl. Pharmacol. **43:**13–22.
10. McKinney, J., K. Chae, B. N. Gupta, J. A. Moore & J. A. Goldstein. 1976. Toxicol. Appl. Pharmacol. **36:** 65–80.
11. Moore, J. A., B. N. Gupta & J. G. Vos. 1976. Proceedings of the National Conference on Polychlorinated Biphenyls, November 1976: 77–79.
12. Gupta, B. N., J. G. Vos, J. A. Moore, J. G. Zinkl & B. C. Bullock. 1973. Environ. Health Perspect. **5:** 125–140.
13. McConnell, E. E., J. A. Moore, J. Haseman & M. Harris. 1978. Toxicol. Appl. Pharmacol., **44:** . In press.
14. McConnell, E. E. & J. A. Moore. 1978. Ann. N.Y. Acad. Sci. This volume.
15. McConnell, E. E., J. A. Moore & D. W. Dalgard. 1978. Toxicol. Appl. Pharmacol. **43:** 175–187.
16. Morita, M. & S. Oishi. 1977. Bull. Environ. Contam. Toxicol. **18** (**1**): 61–67.
17. Goldstein, J. A., J. R. Hass, P. Linko & D. J. Harvan. 1978. Drug Metab. Dispos. In press.

THE STRUCTURE-ACTIVITY RELATIONSHIPS OF HALOGENATED BIPHENYLS AS ENZYME INDUCERS

Joyce A. Goldstein

National Institute of Environmental Health Sciences
National Institutes of Health
Research Triangle Park, North Carolina 27709

Polychlorinated biphenyls (PCBs) are industrial chemicals that have become widespread environmental pollutants. Commercial mixtures of PCBs have been manufactured in the United States, Germany, and Japan under the trade names Aroclor, Clophen, and Kaneclor. These include a number of commercial mixtures with various degrees of chlorination. Recently, polybrominated biphenyls (PBBs) were accidently introduced into livestock feed and were thus introduced into the food chain in the state of Michigan.[1] The toxicity of PCBs has been studied extensively.[2] Although PBBs have received less attention, recent work indicates their toxicity is similar to that of PCBs.[3]

PCBs and PBBs produce a number of hepatic effects including an increase in liver weight, fatty liver, porphyria, and even necrosis at high doses.[2–4] PCBs are also potent inducers of the hepatic microsomal monoxygenase system, which is responsible for the oxidation of xenobiotics and steroids.[5,6] Cytochrome P-450 is the terminal oxidase of this system. PCBs induce the hepatic monoxygenase system in man as well as in experimental animals.[7]

Most inducers of hepatic monoxygenases can be divided into two groups.[8] The prototypes of these groups are phenobarbital and 3-methylcholanthrene (3-MC). In the rat, phenobarbital induces cytochrome P-450 and increases the metabolism of a variety of substrates. 3-MC induces a distinct subspecies of cytochrome P-450, often termed cytochrome P-448, which differs in its spectral properties and substrate specificity from the cytochrome present in microsomes from control or phenobarbital-treated animals.[9,10] Actually, there is considerable evidence for multiple forms of cytochrome P-450 in purified hemeproteins from control, phenobarbital-, and 3-MC-induced microsomes.[11,12] However, the major form of cytochrome P-450 induced by phenobarbital differs from the major form induced by 3-MC. For purposes of this paper, the cytochrome(s) induced by 3-MC will be referred to as cytochrome P-448. Cytochrome P-450 will be used as a more general term to include the constituitive and phenobarbital-induced cytochromes.

The inductive properties of the commercial PCB mixtures are somewhat unusual. Their effects are relatively long-lasting, and they induce a number of extrahepatic tissues[13,14] including kidneys, adrenals, lung, gut, skin, and testes. They are effective transplacental inducers, induce the newborn by means of excretion in the milk, and induce hepatic enzymes after cutaneous applications.[13] Moreover, Alvares *et al.*[14,15] have shown that Aroclor® 1254, a commercial mixture of PCBs containing 54% chlorine, exhibits properties of both classes of inducers. It induces ethylmorphine *N*-demethylase, an enzyme preferentially induced by phenobarbital, and aryl hydrocarbon hydroxylase (AHH), an enzyme preferentially induced by 3-MC (TABLE 1). In addition, the effects of PCBs on the ratio of the 455 to 430 nm peaks of the ethyl isocyanide difference spectra and on the absorption maximum of the CO-difference spectra are intermediate between those of phenobarbital and 3-MC-induced microsomes. The absorbance maximum of the CO-hemeprotein complex of partially

0077–8923/79/0320–0164 $01.75/0 © 1979, NYAS

TABLE 1

A COMPARISON OF CHANGES IN DRUG-METABOLIZING ENZYMES INDUCED BY PHENOBARBITAL, 3-METHYLCHOLANTHRENE, AND AROCLOR 1254

	AHH (nmol/mg/hr)	Ethylmorphine *N*-Demethylase (nmol/mg/hr)	Cytochrome P-450 (nmol/mg)	EtNC Ratios (^{A}455-490/$^{\Delta A}$430-490)	CO λmax
Controls	4.2 ± 0.7	340 ± 20	0.73 ± 0.02	0.6	450
Phenobarbital	16.4 ± 0.3*	1130 ± 30*	2.32 ± 0.03*	0.8	450
3-MC	55.8 ± 1.5*	350 ± 30	1.70 ± 0.05*	1.7	448
Aroclor 1254	42.1 ± 3.5*	1230 ± 40*	2.01 ± 0.10*	1.4	448–449

From Alvares and Kappas.[14] Male rats were given 25 mg/kg/day × 6 days of Aroclor 1254, 75 mg/kg/day × 4 days of phenobarbital or 25 mg/kg/day × 4 days of 3-MC. Values are means ± S.E..

*Significantly different from controls, $p < 0.05$.

TABLE 2

THE EFFECTS OF SYNTHETICALLY PURE AND COMMERCIALLY AVAILABLE GRADES OF 2,4,5,2′,4′,5′-HCB, AND TCDF ON DRUG-METABOLIZING ENZYMES

Treatment	AHH (nmol/g/min)	Aminopyrine *N*-Demethylase (μmol/g/30 min)	Cytochrome P-450 (nmol/g)	EtNC ratios (455/430 nm)
Control	5.4 ± 0.6^{a}	6.0 ± 0.3^{a}	18.4 ± 0.6^{a}	0.60 ± 0.05^{a}
Pure HCB	29.1 ± 2.5^{b}	12.6 ± 1.4^{b}	57.1 ± 10.4^{bc}	0.59 ± 0.02^{a}
Analabs HCB	169.8 ± 10.3^{c}	11.3 ± 1.2^{b}	57.8 ± 6.8^{c}	0.85 ± 0.02^{b}
TCDF	207.4 ± 16.9^{bcd}	4.4 ± 0.8^{a}	40.3 ± 1.7^{b}	1.54 ± 0.10^{c}
Pure HCB & TCDF	235.9 ± 4.5^{d}	11.0 ± 1.7^{b}	79.5 ± 11.5^{c}	0.99 ± 0.04^{b}
Aroclor 1254	177.7 ± 19.8^{c}	8.9 ± 0.9^{b}	37.4 ± 4.5^{b}	0.85 ± 0.08^{b}

Female weanling rats were given 3 daily i.p. doses of corn oil (control), 50mg/kg of pure 2,4,5,2′,4′,5′-HCB, 50 mg/kg of Analab 2,4,5,2′,4′,5′-HCB, 2.5 μg/kg TCDF + 50 mg/kg of pure 2,4,5,2′,4′,5′-HCB, or 50 mg/kg Aroclor 1254.

$^{a-d}$Values that do not show a superscript are significantly different ($p < 0.05$).

purified cytochrome P-450 from phenobarbital-, 3-MC-and Aroclor 1254-treated rats are 450, 447.5, and 448 nm, respectively.[14] Ryan *et al.*[12,16] isolated three forms of cytochrome P-450 from Aroclor 1254-treated rats with approximate molecular weights of 48,000 (P-450_a), 52,000 (P-450_b), and 56,000 (P-450_c) (FIGURE 1). Cytochromes P-450_a and P-450_b have also been found in phenobarbital-treated rats, whereas cytochromes P-450_a and P-450_c are found in 3-MC-treated rats. Cytochrome P-450_b is the major form of cytochrome P-450 in phenobarbital-treated rats while

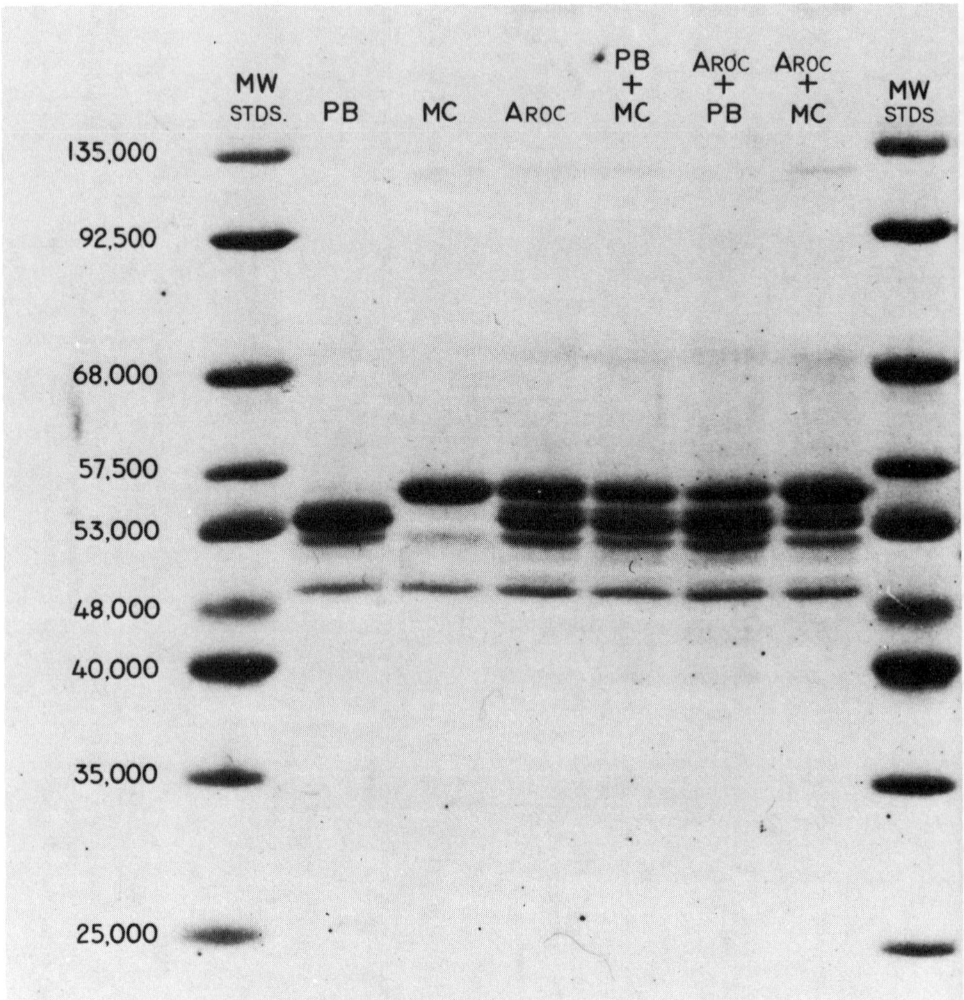

FIGURE 1. SDS-polyacrylamide slab gels of cytochromes P-450 from Aroclor 1254-, phenobarbital, and 3-MC-treated rats. The hemeproteins were treated with SDS, heated, and subjected to electrophoresis. One μg of each hemeprotein or mixtures of 0.5 μg of each hemeprotein were applied. From Ryan *et al.*,[12] with permission of the author.

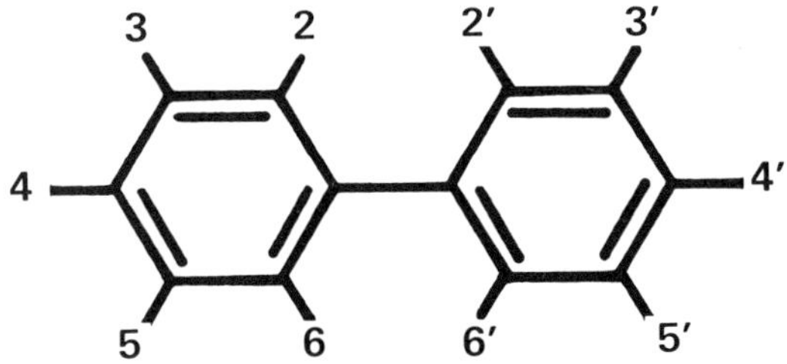

FIGURE 2. The biphenyl molecule.

cytochrome P-450$_c$ is the major form in 3-MC treated rats. The cytochromes from Aroclor 1254-treated rats are indistinguishable from a mixture of hemeproteins from phenobarbital- and 3-MC-treated rats by spectral, electrophoretic, immunologic, or catalytic properties. All of these data suggest that PCBs induce several forms of cytochrome P-450 that are identical to a mixture of cytochromes from the phenobarbital- and 3-MC-induced animals. The induction of multiple pathways by Aroclor 1254 has led to the use of this commercial mixture as an inducer of the rat liver enzymes used for activation of chemicals to possibly mutagenic metabolites in the *S. typhimurium* mutagenesis test system.[17]

STRUCTURE-ACTIVITY RELATIONSHIPS OF PCB ISOMERS

Commercial mixtures of PCBs contain a number of isomers with various chlorination patterns. The numbering system of the biphenyl molecule is shown in FIGURE 2. Theoretically, there are 210 possible isomers. At least 60 different PCB isomers have been identified in Aroclor 1254.[18] Commercial mixtures are also contaminated by chlorinated dibenzofurans.[19] Certain of the chlorinated dibenzo-*p*-dioxins and dibenzofurans, particularly those chlorinated in the 2,3,7,8-positions, are potent inducers of cytochrome P-448 and AHH, resembling 3-MC but 30,000 times more potent.[20–22]

Early work with the commercial mixtures indicated that induction by PCBs is related to the degree of chlorination.[5] Subsequent work with various isomers has shown that both the position and degree of chlorination are important.[23–27] Highly chlorinated isomers are generally more potent. This observation is consistent with the lower rate of metabolism of these isomers.[28]

It now seems clear that PCB isomers can be divided into the classical groups of phenobarbital or 3-MC types of inducers.[26,29] FIGURE 3 shows dose-response curves for the effects of two such isomers on hepatic microsomal enzymes in female rats. 2,4,5,2′,4′,5′-Hexachlorobiphenyl resembles phenobarbital as an inducer. It increases cytochrome P-450 and aminopyrine *N*-demethylase (a cytochrome P-450-mediated enzyme) but produces only a slight increase in AHH (an enzyme mediated preferentially by cytochrome P-448). In contrast, 3,4,5,3′,4′,5′-hexachlorobiphenyl resembled 3-MC as an inducer. Like 3-MC, it increases AHH dramatically, but decreases aminopyrine *N*-demethylase. The cytochrome P-450 induced by 2,4,5,2′,4′,5′-hexachlorobiphenyl has a CO absorption maximum at 450 nm, whereas that induced by 3,4,5,3′,4′,5′-hexachlorobiphenyl has a maximum at 448.5 nm (FIGURE 4). In addition, the ratio of the 455:430 nm peaks of the ethyl isocyanide difference spectrum is reversed by 3,4,5,3′,4′,5′-hexachlorobiphenyl, but not altered appreciably by 2,4,5,2′,4′,5′-hexachlorobiphenyl. Neither 2,4,5,2′,4′,5′- or 3,4,5,3′,4′,5′-hexachlorobiphenyl produces a mixed response at any dose.

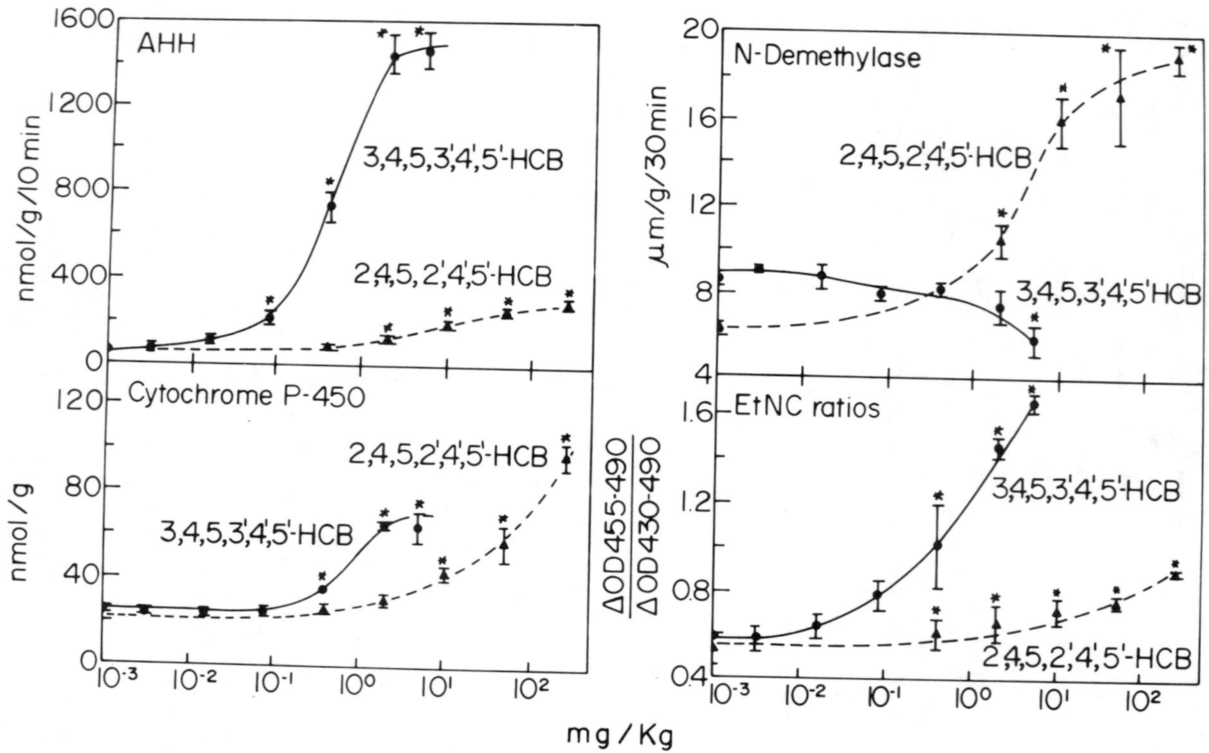

FIGURE 3. Comparison of the log dose-response curves for induction of aryl hydrocarbon hydroxylase (AHH), aminopyrene *N*-demethylase, cytochrome P-450, and the ethyl isocyanide difference spectra by 2,4,5,2′,4′,5′-hexachlorobiphenyl (HCB) versus 3,4,5,3′,4′,5′-HCB. Rats were injected i.p. daily for 3 days at the designated dose and sacrificed 24 hours after the last dose. Values represent the mean ± S.E. from four rats. From Goldstein *et al.*[26] Reproduced by permission of the publishers.

FIGURES 5–7 compare the effects of a number of hexa-, tetra- and dichloro-biphenyl isomers at relatively high doses.[26] Only two PCB isomers can be classified as 3-MC type inducers. These are 3,4,5,3′,4′,5′-hexa- and 3,4,3′,4′-tetrachlorobiphenyl. 2-Hydroxy-3,4,5,3′,4′,5′-hexachlorobiphenyl, the potential metabolite of the most active and toxic hexachlorobiphenyl isomer (3,4,5,3′,4′,5′-), is much less potent than the parent compound, suggesting that the metabolite is not the active intermediate. The inactivity of 3,5,3′,5′-tetrachlorobiphenyl (FIGURE 6) as an inducer relative to

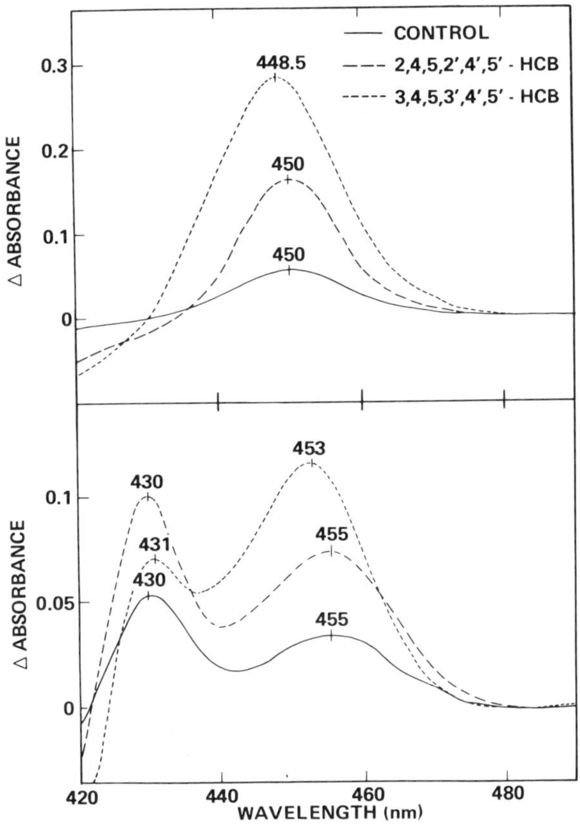

FIGURE 4. Carbon monoxide and ethyl isocyanide difference spectra of liver microsomes from female treated corn oil (control) or 10 mg/kg (30 μmol/kg) of 2,4,5,2′,4′,5′- and 3,4,5,3′,4′,5′-hexachlorobiphenyls (HCBs) i.p. daily for 3 days and sacrificed 24 hours after the last dose. From Goldstein *et al.*[26] Reproduced by permission of the publishers.

3,4,3′,4′-tetrachlorobiphenyl is not related to differences in metabolism, since 3,5,3′,5′-tetrachlorobiphenyl has a longer half-life (10 days)[30] than 3,4,3′,4′-tetrachlorobiphenyl (1 day).[31]

FIGURE 8 summarizes the structures of various types of halogenated hydrocarbons which induce cytochrome P-448.[26,29] The structural similarities between these biphenyls and the active chlorinated dibenzo-*p*-dioxins and dibenzofurans is striking, although PCB isomers are 10^3 times less potent than 2,3,7,8–tetrachlorodibenzo-*p*-dioxin (TCDD). The structural requirements appear to be the presence of at least two adjacent halogens in the lateral positions of the biphenyl molecule and an absence of halogens in the *ortho* positions. The addition of one or two *ortho* chlorines to the biphenyl molecule abolishes this type of induction (e.g. 2,3,4,5,3′,4′,5′-hepta-, 2,3,4,2′,3′,4′-hexa- and 2,4,3′,4′-tetrachlorobiphenyls are inactive).

It seems likely that planarity is required for interaction at the site of action, and that chlorination of the *ortho* positions of the biphenyl molecule does not favor planarity. This would explain the fact that certain PCBs appear to mimic the effects of the chlorinated dibenzo-*p*-dioxins but are less potent. The dibenzo-*p*-dioxin, biphenylene, and dibenzofuran molecules are locked into a rigid, relatively planar structure and are thus more potent[29] than 3,4,5,3′,4′,5′-hexa- or 3,4,3′,4′-tetrachlorobiphenyls,

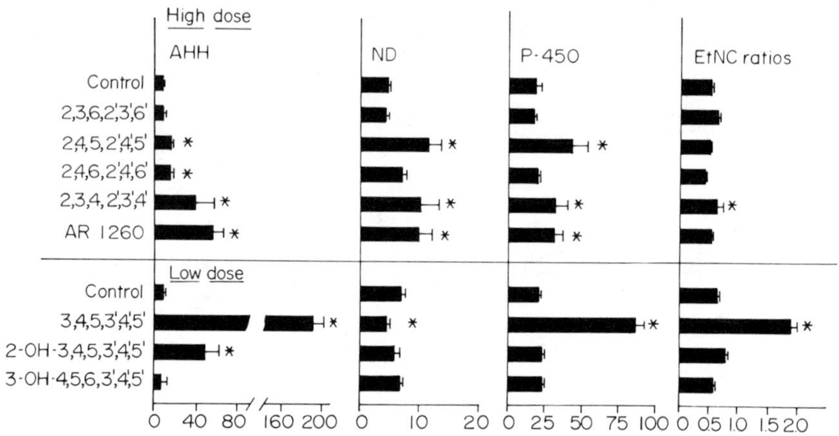

Figure 5. Comparative effects of hexachlorobiphenyl isomers on hepatic drug metabolizing enzymes. Rats were given hexachlorobiphenl isomers or Aroclor 1260 i.p. for 3 days at a dose of 140 μmol/kg (50 mg/kg) (high dose) or 30 μmol/kg (low dose) and sacrificed 25 hours after the last dose. Values represent mean ± S.E. AHH is expressed as μmol/g, liver/min, *N*-demethylase as μmol/g liver/30 min, cytochrome P-450 as μmol/g, ethyl isocyanide (EtNC) rations refer to the Δ Absorbance 455–490/Δ Absorbance 430–490. *Significantly different from controls, $p < 0.05$.

which are free to rotate. Biphenyls with an *ortho* substitution may not be able to assume a planar configuration. Some substantiation of this hypothesis is provided by Poland and Glover.[29] They have shown that 3,4,5,3′,4′,5′-hexa- and 3,4,3′,4′-tetrachlorobiphenyl inhibit the binding of TCDD to a binding species in hepatic cytosol believed to be the receptor for induction of cytochrome P-448. Compounds that do not induce this form of the cytochrome, e.g. 2,4,5,2′,4′,5′-hexachlorobiphenyl, do not inhibit this binding of TCDD. Poland[29] has postulated that the required structure is a rectangle, 3 × 10 Å, with halogens in three of the four corners.

Most PCB isomers are phenobarbital-type inducers, or are inactive (FIGURES 5–7)[26] These structure-activity relationships are summarized in FIGURE 9. In general, the more highly chlorinated isomers are more potent than the lower chlorinated congeners. Most of the hexachlorobiphenyls (2,3,4,2′,3′,4′-; 2,4,5,2′,4′,5′-; and 2,3,5,2′,3′,5′-hexachlorobiphenyl[32]) are relatively potent phenobarbital-type induc-

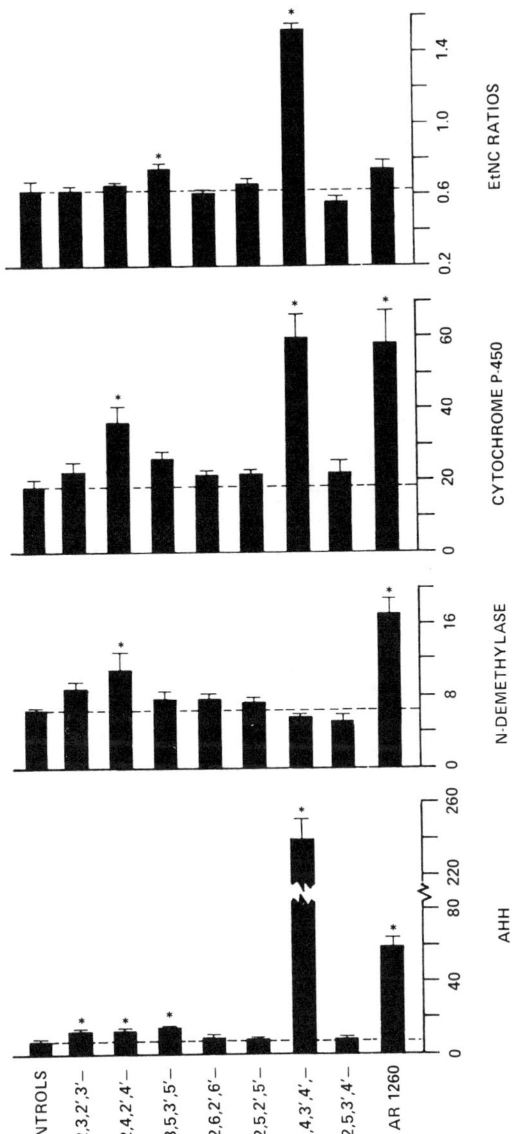

FIGURE 6. Comparative effects of tetrachlorobiphenyl isomers on hepatic drug-metabolizing enzymes. Rats were given 140 μmol/kg of Aroclor (AR) 1260 or tetrachloro isomers. Units of enzyme activity same as in FIGURE 4. Values represent mean ± S.E. *Significantly different from controls, $p < 0.05$.

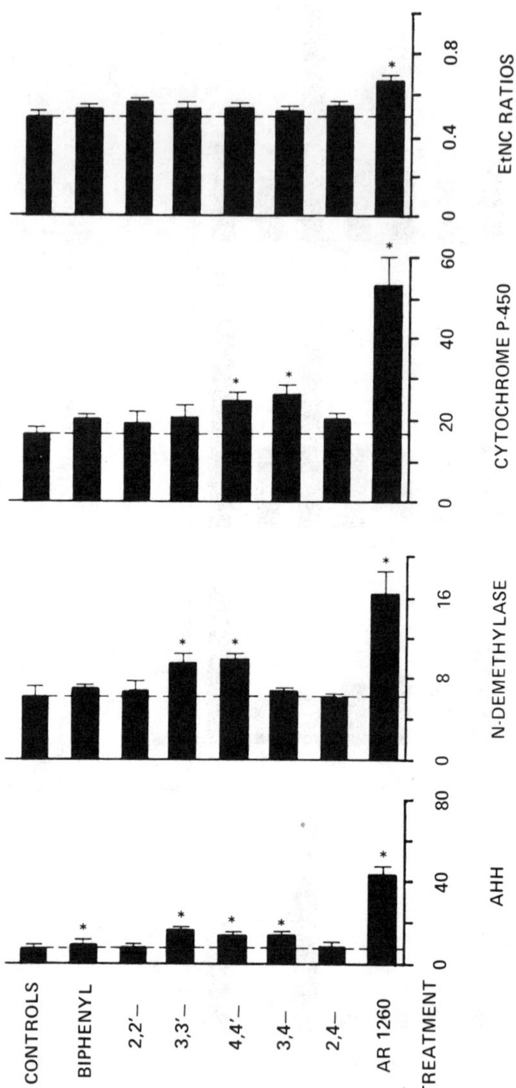

FIGURE 7. Comparative effects of dichlorobiphenyl isomers, Units of enzyme activity same as in FIGURE 4. Values represent mean ± S.E. *Significantly different from controls, $p < 0.05$.

ers.[26] 2,4,6,2′,4′,6′- And 2,3,6,2′,3′,6′-hexachlorobiphenyls are relatively weak phenobarbital-type inducers.[25,26] Most tetra-isomers are weak or inactive with the exception of 2,4,2′,4′-tetrachlorobiphenyl. Two dichloribiphenyls, 4,4′-and 3,3′-, are weak inducers at high doses. 2,3,4,5,2′,3′,4′,5′-Octachlorobiphenyl is also an inducer, probably belonging to the phenobarbital class,[23] but decachlorobiphenyl is relatively weak.[29] The weak activity of 2,3,6,2′,3′,6′-hexachlorobiphenyl and many of the di- and tetra- isomers could be related to their relatively rapid metabolism. Hansel *et al.*[27] have found that the relative potency of 2,4,5,2′,4′,5′-hexa-; 2,4,2′,4′- and 2,5,2′,5′-tetra-; and 4,4′-dichlorobiphenyls is proportional to the disappearance of these compounds from the liver. Unfortunately, most investigators did not determine tissue levels of the various isomers. The structural requirements for induction of cytochrome P-450 are not clear. However, many chlorinated hydrocarbons with diverse structures are inducers (e.g. chlordane, DDT, dieldrin).

The question of whether there are any PCB isomers that are mixed inducers has

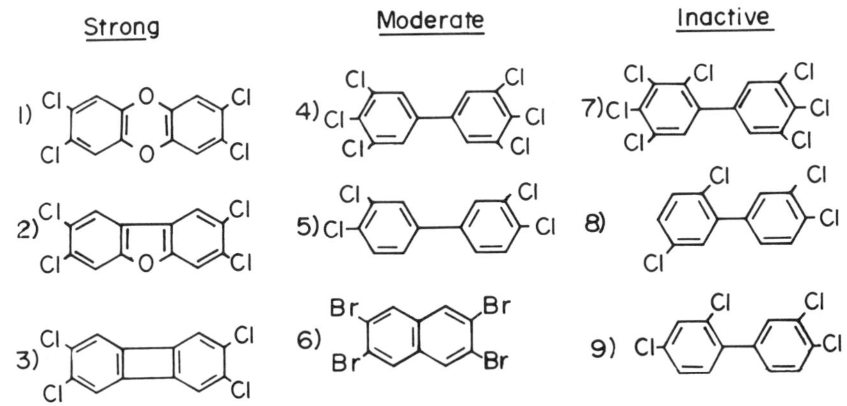

FIGURE 8. Summary of PCB isomers and related halogenated aromatics which induce cytochrome P-448. (1) 2,3,7,8-tetrachloro-*p*-dibenzodioxin; (2) 2,3,7,8-tetrachlorodibenzofuran; (3) 2,3,6,7-tetrachlorobiphenylene; (4) 3,4,5,3′,4′,5′–hexachlorobiphenyl; (5) 3,4,3′,4′-tetrachlorobiphenyl; (6) 2,3,6,7-tetrabromonapthalene; (7) 2,3,4,5,3′,4′5′-heptachlorobiphenyl, 2,5,2′,4′-tetrachlorobiphenyl, 2,4,3′,4′-tetrachlorobiphenyl. Data compiled from Poland and Glover[29] and Goldstein *et al.*[26]

been a controversial one. Early work indicated that 2,4,5,2′,4′,5′- and 2,3,4,2′,3′,4′-hexachlorobiphenyls might be mixed inducers,[33,34] while our work[26] and that of Poland and Glover[29] indicated that they were phenobarbital-type inducers. Moreover, synthesis of 2,4,5,2′,4′,5′-hexachlorobiphenyl by the Ullman reaction gives a product contaminated by as much as 3% of 2,3,7,8-tetrachlorodibenzofuran.[35] We found the commercially available 2,4,5,2′,4′,5′-isomer used in the previous study to be contaminated with 44 ppm of 2,3,7,8-tetrachlorodibenzofuran.[36] The "mixed induction" seen with this isomer (TABLE 2) can be totally accounted for by the presence of the dibenzofuran. The "mixed induction" reported for 2,3,4,2′,3′,4′- and 2,3,4,2′,4′,5′-hexachlorobiphenyls[34] remains controversial at this point. However, synthesis of 2,3,4,2′,3′,4′-hexachlorobiphenyl by the Ullman reaction could produce a chlorinated dibenzofuran as a byproduct by a mechanism similar to that reported for 2,4,5,2′,4′,5′-hexachlorobiphenyl. If an isomer is proposed as a mixed inducer, it should be the

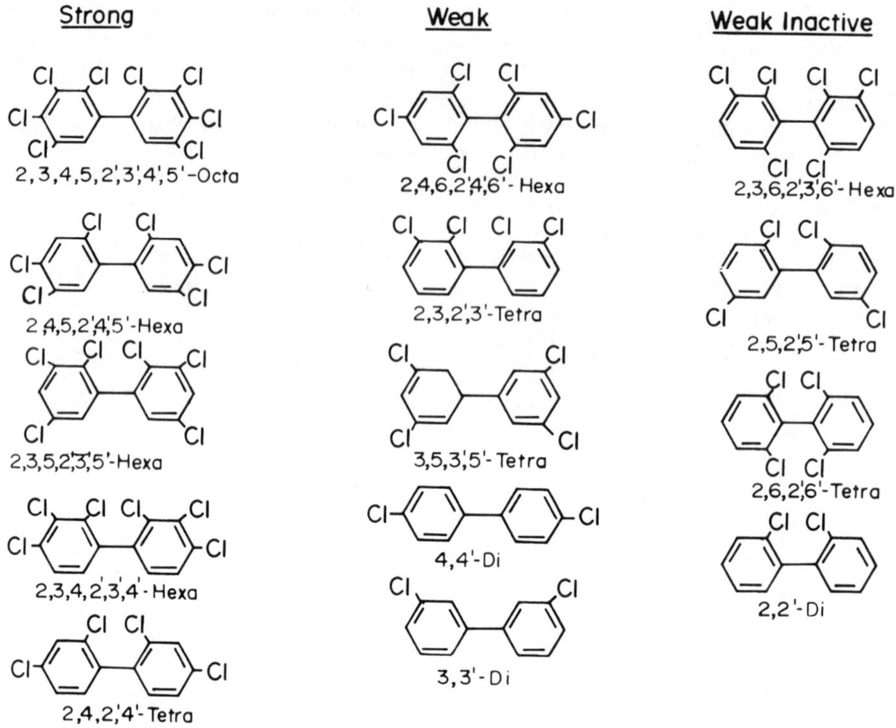

FIGURE 9. Summary of PCB isomers that induce cytochrome P-450. Compiled from Goldstein *et al.*,[26] Johnstone *et al.*[23] (2,3,4,5,2′,3′,4′,5′-octa) and Kohli *et al.*[32] (2,3,5,2′,3′,5′-hexa).

responsibility of the investigator to prove the identity and purity of the chemical used, particularly with respect to possible contamination by chlorinated dibenzofurans. At present, no PCB isomer has been proven to be a mixed inducer. Addition of a single *ortho* chlorine to 3,4,5,3′,4′,5′-hexachlorobiphenyl abolishes induction of cytochrome P-448.[26] Moreover, a biphenyl that contains one "P-450 inducing ring" and one "P-448 inducing ring" (2,4,3′,4′-tetrachlorobiphenyl), is inactive as an inducer of AHH.[29]

POLYBROMINATED BIPHENYLS

Firemaster BP6, the commercial polybrominated biphenyl mixture involved in the accident in Michigan, appears to be a "mixed inducer" resembling Aroclor 1254 in potency.[37] The structure-activity relationships for the PBB isomers have not been studied as thoroughly as those of the PCB isomers, but the structure-activity relationships will undoubtedly be similar. 2,4,5,2′,4′,5′-Hexa- and 2,3,4,5,2′,4′,5′-heptabromobiphenyl, the predominant isomers in the commercial mixture, are cytochrome P-450 (phenobarbital-type) inducers.[38–40] Interestingly, the 2,4,5,2′,4′,5′-hexa- isomer is relatively nontoxic compared to Firemaster,[3] despite its extremely long half-life.[41] 3,4,5,3′,4′,5′-Hexabromobiphenyl, like its hexachloro- analogue, is a

3-MC-type inducer;[29] however, it has not been detected in the commercial mixture. All of the components that have been identified in Firemaster BP6 thus far[42] would probably be phenobarbital type inducers by analogy with PCB structure activity relationships.[26] 2,3,6,7-Tetrabromonapthalene is an inducer of cytochrome P-448, but is so weak (ED_{50} = 20 mg/kg)[32] that the brominated napthalene content of Firemaster appears to be biologically unimportant.[42] If the TCDD receptor is a rectangle 3 × 10 Å,[29] the distance between the lateral bromines in the tetrabromonapthalene molecule is probably not sufficient for a good fit.

Toxicity and Potency of PCB Isomers as Inducers

PCB isomers that induce cytochrome P-450 are much less toxic than those which induce cytochrome P-448 in the mouse,[4] rat,[26] and guinea pig.[43] In fact, it is not clear whether the classic symptoms of PCB poisoning can be reproduced by the

	Induction		Relative potency	
	Type	ED_{50} µmol/kg	ED_{50}	LD_{50}
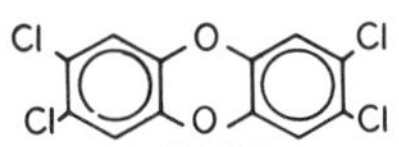 2,3,7,8-TCDD	P-448	0.0006	4000	250
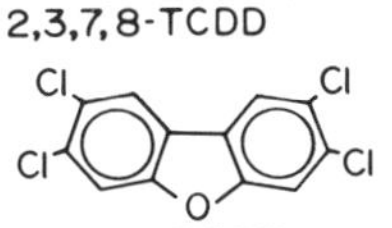 2,3,7,8-TCDF	P-448	0.005	700	65
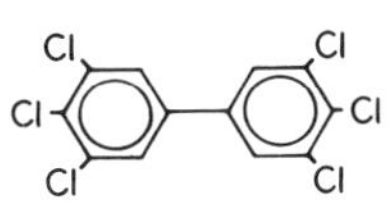 3,4,5,3',4',5'-HCB	P-448	3.4	1*	1*
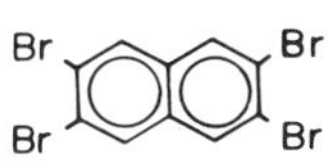 2,3,6,7-TBN	P-448	42	0.1	3
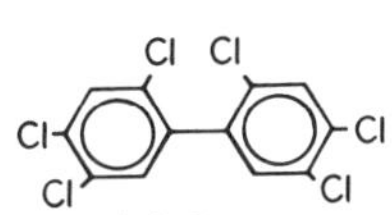 2,4,5,2',4',5'-HCB	P-450	42	0.1	$<$0.05
2,4,5,2',4',5'-HBB	P-450	45	0.1	—

FIGURE 10. Relative potency of some PCB isomers and related halogenated hydrocarbons and enzyme induction in the rat and toxicity to the guinea pig. The ED_{50} for induction of AHH by TCDD in the male rat is from Poland and Glover,[22] that of 2,3,7,8-tetrachlorodibenzofuran (TCDF), 3,4,5,3′,4′,5′-hexachlorobiphenyl (HCB), and 2,3,6,7-tetrabromonaphthalene in the female rat are from Goldstein *et al.*[26,36,38] The ED_{50} for 2,4,5,2′,4′,5′-hexachloro and hexabromobiphenyls (HBB) represent the ED_{50} for the induction of aminopyrine *N*-demethylase in the female rat.[26,38] All compounds were given i.p., and the dose shown is the total dose administered. The dosing schedule varied from a single[22,38] to three daily divided doses.[26] The single dose LD_{50} in the guinea pig is from McConnell and McKinney.[43]

2,4,5,2′,4′,5′-type of inducer, despite the long half-life of this compound. It is certainly apparent that these compounds are less potent. Atrophy of the thymus and lymphoid organs, porphyrin accumulation, and fatty liver have been reported in mice treated with 3,4,5,3′,4′,5′-hexachlorobiphenyl, but not with the 2,4,5,2′,4′,5′-isomer, even at a 30-fold higher dose.[4] In the chicken, 3,4,5,3′,4′,5′-hexachlorobiphenyl produces thymic atrophy and edema, responses characteristic of poisoning with TCDD or 2,3,7,8-tetrachlorodibenzofuran poisoning, while all other hexachlorobiphenyls (2,3,6,2′,3′,6′- 2,4,6,2′,4′,6′-; 2,3,4,2′,3′,4′-; 2,4,5,2′,4′,5′-; and 2,3,5,2′,3′,5′-) were inactive.[44] Yamamoto *et al.*[45] have reported toxicity with 2,3,4,3′,4′-pentachlorobiphenyl similar to that reported for the 3,4,5,3′,4′,5′-hexachloro-isomer.[4,6] It would be desirable to have additional information on the identification and purity of this isomer, since this type of toxicity would not be predicted for a biphenyl with an *ortho* halogen. In our hands, synthesis of 2,3,4,3′,4′-pentachlorobiphenyl by procedures identical to those used by Yamamoto *et al.*[45] yields a number of products including two pentachlorobiphenyls, 2,3,4,3′,4′- and 3,4,5,3′,4′- with a 40% yield of the latter (McKinney, unpublished data). We would predict this type of toxicity and 3-MC-type of induction for the 3,4,5,3′,4′-isomer.

For the TCDD-type inducers, relative toxicity can be roughly predicted by the potency as an inducer.[29] FIGURE 10 compares the relative toxicity of several PCB and PBB isomers, TCDD, 2,3,7,8-tetrachlorodibenzofuran and 2,3,6,7-tetrabromonapthalene in the guinea pig with their potency as inducers of rat liver microsomal enzymes. Unfortunately, these data are not available in the same species; therefore this comparison is a crude one. However, despite possible variations between species, there does appear to be some correlation between potency as inducers, the type of induction (cytochrome P-450 versus P-448) and toxicity. TCDD and 2,3,7,8-tetrachlorodibenzofuran are 10^3 times more potent than 3,4,5,3′,4′,5′-hexachlorobiphenyl or 2,3,6,7-tetrabromonapthalene as inducers,[26,29] and are 10^2 times more toxic to the guinea pig.[43] On the other hand, 3,4,5,3′,4′,5′-hexachlorobiphenyl is at least 10–25 times more potent as an inducer than 2,4,5,2′,4′,5′-hexachlorobiphenyl,[26] and is much more toxic.[44] The structural requirements for induction of cytochrome(s) P-448 and for the syndrome of toxic effects associated with TCDD poisoning appear similar within this series of halogenated aromatic compounds, which includes chlorinated dibenzo-*p*-dioxins and PCBs.

SUMMARY

Polyhalogenated biphenyls induce microsomal mixed-function oxygenases in many species including man. They can be divided into two classes of inducers. 3,4,5,3′,4′,5′-Hexa- and 3,4,3′,4′-tetrachlorobiphenyls appear to interact with the TCDD receptor, albeit at 100 to 1,000 times the dose. The structure required for 3-MC (TCDD)-type induction seems to be the presence of at least two adjacent halogens in the lateral positions (*meta* and *para*) of the benzene rings, and an absence of halogens in the *ortho* positions adjacent to the biphenyl bridge. It is likely that halogenation of the *ortho* positions prevents binding by blocking planar configuration. These isomers, although less toxic than TCDD, are much more toxic than other PCB isomers. Most halogenated biphenyl isomers are phenobarbital type inducers or are inactive. The structure-activity relationship for this type of induction is less clear, but active isomers include 2,4,2′,4′-tetra-; 2,4,5,2′,4′,5′-; 2,3,4,2′,3′,4′-; 2,3,5,2′,3′.5′-and 2,4,6,2′,4′,6′-hexachlorobiphenyls and 2,4,5,2′,4′,5′-hexabromobiphenyl. The inactivity of many of the lower chlorinated isomers as inducers (mono-through tetrachloro-) may be related to their rapid metabolism. The cytochrome which is induced by commercial PCB and PBB mixtures appears to be identical to a mixture of cytochromes from phenobarbital- and 3-MC-treated rats. These structure-activity relationships suggest that the "mixed inductive" effects and much of the toxicity of PCB and PBB mixtures are due to the presence of compounds related structurally to the active chlorinated dibenzofurans or 3,4,3′,4′-tetrachlorobiphenyl.

References

1. Dunckel, A. E. 1975. J. Am. Vet. Med. Assoc. **167:** 838–841.
2. Kimbrough, R. D. 1974. CRC Crit. Rev. Toxicol. **2:** 445–498.
3. Moore, J. E., M. I. Luster, B. N. Gupta & E. E. McConnell. 1978. Presented at the 17th Annual Meeting of the Society of Toxicology, San Francisco, CA, March 12–16.
4. Biocca, M., J. A. Moore, B. N. Gupta & J. D. McKinney: 1976. Conference Proceedings of the National Conference on Polychlorinated Biphenyls, Chicago, IL, November 1975, pp. 67–72.
5. Litterst, C. L., M. T. Farber, A. M. Baker & E. J. Van Loon. 1972. Toxicol. Appl. Pharmacol. **23:** 112–122.
6. Risebrough, R. W., P. Rieche, D. B. Peakall, S. G. Herman & M. N. Kirven. 1968. Nature **220:** 1098–1102.
7. Alvares, A. P., A. Fishbein, K. E. Anderson & A. Kappas. 1977. Clin. Pharmacol. Therap. **22:** 140–146.
8. Conney, A. H. 1967. Pharmacol. Rev. **19:** 317–366.
9. Sladek, N. E. & G. J. Mannering. 1966. Biochem. Biophys. Res. Commun. **23:** 668–674.
10. Lu, A. Y. H. & W. Levin. 1974. Biochem. Biophys. Acta. **334:** 205–240.
11. Thomas, P. E., A. Y. H. Lu, D. Ryan, S. B. West, J. Kawalek & W. Levin. 1976. J. Biol. Chem. **251:** 1385–1391.
12. Ryan, D. E., P. E. Thomas, D. Korzeniowski & W. Levin. 1978. Fed. Proc. **37:** 766 (Abstract).
13. Alvares, A. P. & A. Kappas. 1977. Clin. Pharmacol. Therap. **22:** 807–816.
14. Alvares, A. P. & A. Kappas. 1977. J. Biol. Chem. **252:** 6373–6378.
15. Alvares, A. P., D. R. Bickers & A. Kappas. 1973. Proc. Nat. Acad. Sci. USA **70:** 1321–1325.
16. Ryan, D. E., P. E. Thomas & W. Levin. 1977. Mol. Pharmacol. **13:** 521–532.
17. Ames, B. N., J. McCann & E. Yamasaki. 1975. Mutation Res. **31:** 347–364.
18. Sissons, D. & D. Welti. 1971. J. Chromatogr. **60:** 15–32.
19. Bowes, G. W., M. J. Mulvihill, B. R. T. Simonett, A. L. Burlingame & R. W. Risebrough. 1975. Nature **256:** 305–307.
20. Poland, A. & E. Glover. 1973. Mol. Pharmacol. **9:** 736–747.
21. Lucier, G. S., O. S. McDaniel, G. E. R. Hook, B. A. Fowler, B. R. Sonawane & E. Faeder. 1973. Environ. Health Perspect. **5:** 199–209.
22. Poland, A. & E. Glover. 1974. Mol. Pharmacol. **10:** 349–359.
23. Johnstone, G. J., D. J. Ecobichon & O. Hutzinger. 1974. Toxicol. Appl. Pharmacol. **28:** 66–81.
24. Ecobichon, D. J. & A. M. Comeau. 1975. Toxicol. Appl. Pharmacol. **33:** 94–105.
25. Goldstein, J. A. J. D. McKinney, G. W. Lucier, P. Hickman. H. Bergman & J. A. Moore. 1976. Toxicol. Appl. Pharmacol. **36:** 81–92.
26. Goldstein, J. A., P. Hickman, H. Bergman, J. D. McKinney & M. P. Walker. 1977. Chem. Biol. Interact. **17:** 69–87.
27. Hansell, M. M., D. J. Ecobichon, A. M. Comeau & P. H. Cameron. 1977. Exp. Mol. Pathol. **26:** 75–84.
28. Matthews, H. B. & M. W. Anderson. 1975. Drug Metab. Dispos. **3:** 371–380.
29. Poland, A. & E. Glover. 1977. Mol. Pharmacol. **13:** 924–938.
30. Tuey, D. B. & H. B. Matthews. 1977. Drug Metab. Dispos. **5:** 444–450.
31. Matthews. H. B. Personal communication.
32. Kohli, K. K., H. Mukhtar, J. R. Bend, P. W. Albro & J. D. McKinney. 1978.

Presented at the Joint Meeting of the American Society of Pharmacology and Toxicology and The Society of Toxicology, Houston, TX, August 13–17 (Abstract).
33. ALVARES, A. P. & A. KAPPAS. 1976. Hoppe-Seyler's Z. Physiol. Chem. **357:** 1019–1020.
34. STONARD, M. D. & J. B. GREIG. 1976. Chem. Biol. Interact. **15:** 365–379.
35. MORON, M. G., G. SUNDSTROM & C. A. WACHTMEISTER. 1973. Acta Chem. Scand. **27:** 1321–1322.
36. GOLDSTEIN, J. A., J. R. HASS, P. LINKO & D. J. HARVAN. 1978. Drug Metab. Dispos. **6:** 258–264.
37. DENT, J. G., K. J. NETTER & J. E. GIBSON. 1976. Toxicol. Appl. Pharmacol. **38:** 237–249.
38. GOLDSTEIN, J. A. & P. HICKMAN. 1978. Presented at the 17th Annual Meeting of the Society of Toxicology, San Francisco, CA, March 1216 (Abstract).
39. MOORE, R. W. & S. D. AUST. 1977. Pharmacologist **19(2):** 162 (Abstract).
40. MOORE, R. W. & S. D. AUST. 1978. Presented at the 17th Annual Meeting of the Society of Toxicology, San Francisco, CA, March 12–16 (Abstract).
41. MATTHEWS, H. B., S. KATO, N. M. MORALES, T. B. TUEY. 1977. J. Toxicol. Environ. Health. **3:** 599–605.
42. HASS, J. R., E. E. MCCONNELL & D. J. HARVAN. 1978. Agric. Food Chem. **26:** 94–99.
43. MCCONNELL, E. E. & J. D. MCKINNEY. 1978. Presented at the 17th Annual Meeting of the Society of Toxicology, San Francisco, CA, March 12–16 (Abstract).
44. MCKINNEY, J. D. K. CHAE, B. N. GUPTA, J. A. MOORE & J. A. GOLDSTEIN. 1976. Toxicol. Appl. Pharmacol. **36:** 65–80.
45. YAMAMOTO, H. A., H. YOSHIMURA, M. FUGITA & T. YAMAMOTO. 1976. Chem. Pharm. Bull. **24:** 2168–2174.

POSSIBLE CORRELATION BETWEEN INDUCTION MODES OF HEPATIC ENZYMES BY PCBs AND THEIR TOXICITY IN RATS

Hidetoshi Yoshimura, Shin'ichi Yoshihara, Naoki Ozawa, and Masanobu Miki

Faculty of Pharmaceutical Sciences
Kyushu University
Fukuoka 812, Japan

PCBs are well-known as a widespread environmental contaminant[1,2] and also as a potent and long-lasting inducer of hepatic microsomal monooxygenase systems.[3–8] However, since commercial PCB preparations are complex mixtures of various chlorobiphenyl congeners, the important question has arisen as to whether these congeners exert their biologic effects in the same manner. To answer this question, an increasing number of studies has been undertaken comparing the biologic effects of individual PCBs.[9–12] Several workers have shown that the position as well as the number of chlorines substituted is important for the biologic effects of PCBs.[3,13–15]

Recently, Goldstein *et al.*[16] and Yoshimura *et al.*[17] reported that PCB congeners fell into two distinct groups of inducers, that is, the phenobarbital (PB)-type and the 3-methylcholanthrene (MC)-type. These workers also pointed out that chlorination at the 3,4,3′ and 4′-positions of the biphenyl is a minimum requirement for exhibiting MC-type of inductions. Poland *et al.* demonstrated that most of the halogenated aromatic compounds, such as dibenzo-*p*-dioxins,[18] dibenzofurans,[19] azoxybenzenes,[20] and biphenyls,[21] to induce aromatic hydrocarbon hydroxylase (AHH) were planar and bound to the hepatic cytosol protein which appeared to be the receptor for induction of cytochrome P448. Goldstein and her co-workers compared the toxicologic effects in the chicken of five symmetrical hexachlorobiphenyls (HCB) and 2,3,7,8-tetrachlorodibenzofuran.[15,22]

In the present study, acute toxicity of several individual PCBs, which are categorized as either the PB- or the MC-type, has been examined in adult and weanling male rats, in order to find some correlations with their modes and potencies to induce hepatic enzymes.

Materials and Methods

Chemicals

We purchased 2,5,2′,5′- and 3,4,3′,4′-tetrachlorobiphenyl (TCB) from Analabs, Inc., North Haven, Conn. 2,4,3′,4′-TCB,[23] 2,3,4,3′,4′- and 3,4,5,3′,4′-pentachlorobiphenyl (PenCB),[24] and 3,4,5,3′,4′,5′-HCB[17] were prepared according to the method described previously. These individual PCBs synthesized were exhaustively purified by silica gel and alumina column chromatography followed by recrystallization, and their high purities were proved by ECD-gas chromatography (GC) and GC-mass spectrometry. 3-Methylcholanthrene (MC) and sodium phenobarbital (PB) were purchased from Wako Pure Chemical Ind., Ltd., Osaka and Yoneyama Chemical Ind., Ltd., Osaka, respectively. Benzo[a]pyrene (BP) and 3-hydroxy-benzo[a]pyrene (3-OH-BP) were generously provided by Dr. Kinoshita, School of Health Sciences,

0077-8923/79/0320-0179 $01.75/0 © 1979, NYAS

Kyushu University, and benzphetamine (BZ) hydrochloride was kindly donated by Dr. Neal, Department of Biochemistry, Vanderbilt University. NADP, NADPH, NADH, glucose-6-phosphate (sodium salt, G-6-P), G-6-P dehydrogenase (from Bakers yeast, type V) and HEPES were purchased from Sigma Chemical Company, St. Louis, Mo.

Animal Treatments

Adult (about 200 g, 8 wk old) and weanling (about 100 g, 4 wk old) male Wistar rats were purchased from Kyudo Company Ltd., Tosu, Japan. All PCBs were dissolved in salad oil and injected into the rats i.p. at designated single doses. They were sacrificed 5 days after the injection. MC dissolved in salad oil and PB dissolved in 0.9% NaCl were injected i.p. once a day for 5 consecutive days, and animals were sacrificed 24 hours after the last injection. Control animals were injected with salad oil or 0.9% NaCl at a dose equivalent to treated animals. A group of 3–4 rats was submitted to each treatment and housed in stainless steel cages without bedding. They were allowed water and chow (Nippon Clea Co., Ltd., Tokyo) *ad libitum.*

Cell Fractionation

The rats were sacrificed by decapitation after fasting for 18 hours. The livers were removed, weighed, and homogenized with 3 vol. of ice-cold 1.15% KCl in a Potter-Elvehjem-type glass-teflon homogenizer after perfusion with 0.9% NaCl. The homogenate was centrifuged at 9000 × *g* for 20 minutes and the supernatant fraction was used for the following enzyme assays. In the enzyme distribution experiment, the pellet obtained by the first centrifugation was resuspended in the same vol. of 1.15% KCl and centrifuged again at 9000 × *g* for 20 minutes. The resultant supernatant was used as the second 9000 × *g* supernatant, and the pellet was suspended in a half of the initial vol. of 1.15% KCl.

Enzyme Assays

Benzo[a]pyrene 3-hydroxylase (BP hydroxylase) in a liver 9000 × *g* supernatant was assayed by the method of Nebert *et al.*[25] with a minor modification. Assay of benzphetamine *N*-demethylase (BZ demethylase) in the supernatant was conducted in a final vol. of 1.0 ml containing 50 μmoles of HEPES (pH 7.6), 0.1 μmole of EDTA, 6 μmoles of $MgCl_2$, 1 μmole of BZ, 0.5 ml of 9000 × *g* supernatant, 0.33 μmole of NADP, 8 μmoles of G-6-P and 0.1 unit of G-6-P dehydrogenase. The reaction was started by addition of NADPH-generating system, continuing for 15 minutes at 37°C, and terminated by addition of 0.4 ml of 20% TCA. The amount of formaldehyde formed was determined by the method of Nash.[26] DT diaphorase activity in the supernatant was measured by the method of Ernster *et al.*[27] The incubation mixture (3 ml) consisted of 150 μmoles of potassium phosphate (pH 7.5), 1 μmole of KCN, 0.24 μmole of 2,6-dichlorophenol-indophenol (DCPIP), 0.45 μmole of NADH, and 5 or 10 μl of 9000 × *g* supernatant, added to start the reaction. The reaction was carried out at 25°C for 2 minutes with recording the decrease of absorbance of DCPIP at 600 nm by a Hitachi Spectrophotometer, model 124, and then the reaction associated with DT diaphorase was inhibited by addition of 30 μl of 10 mM dicoumarol in 0.1 N NaOH. The activity of DT diaphorase was calculated from the change of absorbance inhibited by dicoumarol, using the extinction coefficient of 21 $cm^{-1}mM^{-1}$. Content of cytochrome P450(448) in the liver homogenate or 9000 × *g* supernatant was determined by the dithionite-reduced difference spectrum of CO-bubbled sample according to the method of Matsubara *et al.*,[28] with a slight modification. Glucose-6-phosphatase (G-6-Pase) was assayed by determination of inorganic phosphate released from G-6-P according to the method of Swanson.[29]

Analytic Methods

Protein was determined by the method of Lowry *et al.*[30] using bovine serum albumin as a standard. Inorganic phosphate was determined by the method of Fiske and Subbarow.[31] Total liver lipids were extracted by the method of Folch *et al.*[32] and weighed after drying. Triglycerides, total cholesterol and phospholipids in the extracted lipids of liver and in serum were determined by the method of Flecher,[33] Zurkowski,[34] and Bertlett,[35] respectively.

Toxicologic Studies

Thymus and spleen as well as liver of the rat were removed immediately after killing, washed in a cold 1.15% KCl and weighed as a wet weight.

Statistical Methods

Statistical anaylsis of significance of the various parameters measured in treated and control groups was assessed by two-sided Mann-Whitney U tests.[36]

RESULTS

Inductive Effect and Toxicity of Several Individual PCBs in Adult Rats

In order to estimate the ability of several individual PCBs to induce hepatic enzymes and also to confirm their characteristics as PB- or MC-type inducers, activities of BZ demethylase, BP hydroxylase, DT diaphorase and a content of cytochrome P450(448) were determined in the liver 9000 × *g* supernatant from adult rats pretreated with each PCBs, together with CO absorption maximum. As shown in FIGURE 1 (right), BZ *N*-demethylation, a reaction predominantly mediated by PB-induced cytochrome P450,[37] was considerably increased by pretreatment with 2,5,2′,5′- and 2,4,3′,4′-TCB at a dose of 100 mg/kg, and also with 2,3,4,3′,4′-PenCB at a dose of 10 mg/kg, but strongly decreased by pretreatment with 3,4,5,3′,4′-PenCB and 3,4,5,3′,4′,5′-HCB at a dose of 10 mg/kg.

On the contrary, BP 3-hydroxylation which is mainly catalyzed by MC-induced cytochrome P448,[38] was markedly enhanced by pretreatment with 3,4,5,3′4′-PenCB and 3,4,5,3′,4′,5′-HCB. The specific activity per nmole of cytochrome P448 was slightly higher in 3,4,5,3′,4′-PenCB pretreatment than in 3,4,5,3′,4′,5′-HCB pretreatment. This activity to hydroxylate BP was also increased by pretreatment with 2,4,3′,4′-TCB, 3,4,3′,4′-TCB and 2,3,4,3′,4′-PenCB, but their magnifications were much less as compared to those in 3,4,5,3′,4′-PenCB and 3,4,5,3′,4′,5′-HCB.

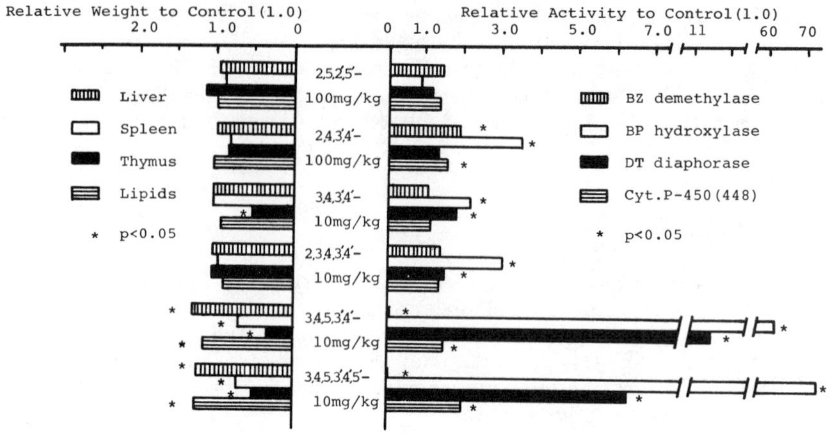

FIGURE 1. Comparative effects of pretreatment with pure PCB congeners on liver enzyme activity, organ weight, and liver lipid content in adult rats. The activities in the control (mean ± S.D.) were: BZ demethylase 12.93 ± 1.52 nmoles HCHO formed/mg protein/15 min, BP hydroxylase 1.35 ± 0.06 nmoles 3-OH-BP formed/mg protein/5 min, DT diaphorase 0.583 ± 0.072 nmoles DCPIP reduced/mg protein/min, cytochrome P450 0.144 ± 0.008 nmoles/mg protein. The weight of organs (g/100g bw) and liver lipids in the control were: liver 3.60 ± 0.25, spleen 0.558 ± 0.091, thymus 0.252 ± 0.058, lipids 44.79 ± 4.98 mg/g liver.

This evidence indicates that, while 2,5,2′,5′- and 2,4,3′4′-TCB are categorized as PB-type inducers, 3,4,5,3′,4′-PenCB and 3,4,5,3′,4′,5′-HCB are MC-type inducers, as reported previously.[17] This was further supported by a CO absorption maximum of cytochromes induced by respective PCBs (TABLE 2). Although 3,4,3′,4′-TCB is known as a prototype of MC-type PCBs,[16,17,21] a dose level of 10 mg/kg of this PCB seemed insufficient to induce significant amounts of cytochrome P448, judging from the results shown in FIGURE 1 and TABLE 2. At a low dosage of 10 mg/kg, the induction mode of 2,3,4,3′,4′-PenCB seemed rather to resemble that of the PB-type than that of the MC-type. It was very interesting to note that DT diaphorase was greatly induced only by MC-type PCBs. This activity was about 11 and 6 times as high as the control values after pretreatment with 3,4,5,3′,4′-PenCB and 3,4,5,3′,4′,5′-HCB, respectively, whereas by other PCBs the activity was enhanced less than twofold.

Acute toxicity of these six PCBs was tested by measuring the changes of organ and body weights in the same animals, on which the inductive effects were examined. As shown in FIGURE 1 (left), the PB-type PCBs, 2,5,2′,5′- and 2,4,3′,4′-TCB, did not affect the weight of the liver, although this did slightly affect the weights of the spleen and thymus. Body weight gain was also unaffected by 2,5,2′5′-TCB, but somewhat suppressed by 2,4,3′,4′-TCB (data not shown). 3,4,3′,4′-TCB and 2,3,4,3′,4′-PenCB, which could not induce significant amounts of cytochrome P448 at a dosage of 10 mg/kg, did not show any toxic signs, except that thymus atrophy was observed after 3,4,3′4′-TCB pretreatment. In contrast, the potent MC-type inducers, 3,4,5,3′,4′-PenCB and 3,4,5,3′,4′,5′-HCB, caused a marked loss of body weight (data not shown) and severe atrophies of the thymus and spleen after a single i.p. dose of 10 mg/kg. In addition, liver enlargement accompanied by fatty liver was observed by pretreatment with both MC-type PCBs. These results clearly indicate that MC-type PCBs are much more toxic than PB-type ones in adult rats.

Inductive Effect and Toxicity of Several Individual PCBs in Weanling Rats

Since a dose of 10 mg/kg of 3,4,3′,4′-TCB seemed insufficient as an effective inducer of cytochrome P448, a relatively high dose of 50 mg/kg was given to weanling rats. In this experiment, 3,4,5,3′,4′-PenCB was used as a positive marker.

As indicated in FIGURE 2 (right), a dose of 50 mg/kg of 3,4,3′,4′-TCB markedly increased the activities of BP hydroxylase and DT diaphorase, but this treatment did not appear to induce the metabolism of BZ. The potency of 3,4,3′,4′-TCB as a MC-type inducer was, however, much less than that of 3,4,5,3′,4′-PenCB, since the latter PCB could induce BP hydroxylase and DT diaphorase 3 and 6 times, respectively, as high as the former at fivefold higher dose. The cytochrome P450 induced by 3,4,3′,4′-TCB gave a CO absorption maximum at 449 nm, whereas 3,4,5,3′,4′-PenCB-induced hemoprotein gave the maximum at 448 nm (TABLE 2). Pretreatment with 2,3,4,3′,4′-PenCB at 50 mg/kg showed unique inductive effects. Although this PCB significantly increased the content of cytrochrome P448 and DT diaphorase activity like 3,4,3′,4′-TCB, BP hydroxylase was enhanced to a lesser extent than by 3,4,3′,4′-TCB.

The results from FIGURE 1 and 2 suggest that the responsiveness to the induction with 3,4,5,3′,4′-PenCB is almost the same in adult and weanling rats. As shown in FIGURE 2 (left), however, this potent MC-type PCB exhibited slightly higher toxicity in all parameters tested in the weanlings than in the adults, suggesting that weanling rats were more sensitive to the toxic effects of MC-type PCBs than adult rats. Unlike at lower dosages, at a higher dosage (50 mg/kg) 3,4,3′,4′-TCB exhibited toxic signs similar to those of 3,4,5,3′,4′-PenCB, causing liver enlargement, atrophies of the thymus and spleen as well as lipid accumulation in the liver. Its toxic potency was, however, much less than that of 3,4,5,3′,4′-PenCB, in parallel with its weak MC-type inducing ability as MC-type. On the other hand, 2,3,4,3′,4′-PenCB pretreatment at a dose of 50 mg/kg induced liver enlargement and slight atrophy of the thymus, but did not cause any lipid accumulation in the liver or atrophy of the spleen. Body weight

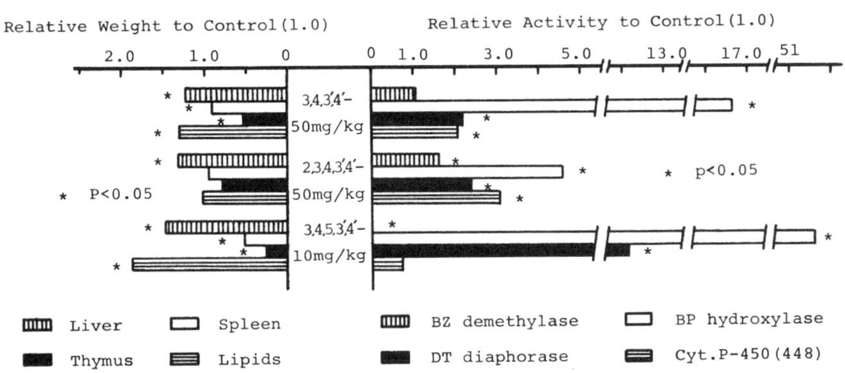

FIGURE 2. Comparative effects of pretreatment with pure PCB congeners on liver enzyme activity, organ weight, and liver lipid content in weanling rats. The activities in the control (mean ± S.D.) were: BZ demethylase 13.98 ± 1.76 nmoles HCHO formed/mg protein/15 min, BP hydroxylase 0.344 ± 0.010 nmoles 3-OH-BP formed/mg protein/5 min, DT diaphorase 0.469 ± 0.050 nmoles DCPIP reduced/mg protein/min, cytochrome P450 0.172 ± 0.018 nmoles/mg protein. The weight of organs (g/100g bw) and liver lipids in the control were: liver 3.80 ± 0.30, spleen 0.477 ± 0.014, thymus 0.478 ± 0.023, lipids 43.5 ± 2.5 mg/g liver.

TABLE 1

COMPARATIVE EFFECT OF PRETREATMENT WITH PURE PCB CONGENERS ON LIPID COMPOSITION OF LIVER AND SERUM IN WEANLING RATS

Treatment	Dose (mg/kg)	Total Lipid	Triglyceride	Cholesterol	Phospholipid
Liver		(mg/g liver)	(mg/g liver)	(mg/g liver)	(μmoles/g liver)
Control		43.5 ± 2.5*	3.43 ± 0.52	3.09 ± 0.39	33.8 ± 1.2
		(1.00)	(1.00)	(1.00)	(1.00)
3,4,3′,4′,-TCB	50	57.5 ± 6.0†	5.77 ± 1.94†	2.96 ± 0.19	34.1 ± 1.6
		(1.32)	(1.68)	(0.96)	(1.01)
2,3,4,3′,4′-PenCB	50	47.8 ± 1.6	3.41 ± 0.58	2.93 ± 0.23	36.1 ± 1.3†
		(1.05)	(0.99)	(0.95)	(1.07)
3,4,5,3′,4′-PenCB	10	81.6 ± 8.7†	25.81 ± 9.48†	6.74 ± 0.28†	40.4 ± 0.8†
		(1.88)	(7.53)	(2.18)	(1.20)
Serum			(mg/100 ml)	(mg/100 ml)	
Control			72.8 ± 7.0	181.6 ± 9.1	
			(1.00)	(1.00)	
3,4,3′,4′-TCB			67.6 ± 3.6	231.9 ± 11.6†	
			(0.93)	(1.28)	
2,3,4,3′,4′-PenCB			73.9 ± 10.5	181.2 ± 26.4	
			(1.02)	(1.00)	
3,4,5,3′,4′-PenCB			65.4 ± 4.1	98.0 ± 5.0†	
			(0.90)	(0.54)	

*Each value represents mean ± S.D. of 4 rats and that in parenthesis represents relative value to that of the control (salad oil).

†$p < 0.05$.

gain was almost completely suppressed by 3,4,5,3′4′-PenCB; however, neither 2,3,4,3′,4′-PenCB nor 3,4,3′,4′-TCB affected growth rate during the 4 days after the injection (FIGURE 4).

TABLE 1 shows the comparative effects of three PCBs on lipid composition of liver and serum in weanling rats. Severe fatty liver caused by 3,4,5,3′,4′-PenCB pretreatment was evidenced by this remarkable increase of liver triglycerides. In this case, a slight increase of cholesterol and phospholipids was also observed. 3,4,3′,4′-TCB and 2,3,4,3′,4′-PenCB treatments increased only triglycerides and phospholipids, respectively, but not other two components. On the contrary, serum cholesterol and triglyceride levels were decreased by 3,4,5,3′,4′-PenCB treatment. 3,4,3′,4′-TCB showed a similar tendency to decrease serum triglycerides, but increased cholesterol.

Toxic Effects of Phenobarbital (PB) and 3-Methylcholanthrene (MC) in Weanling Rats

Toxic potencies of MC-type PCBs, 3,4,3′,4′-TCB, 3,4,5,3′,4′-PenCB and 3,4,5,3′,4′,5′-HCB, appeared to be correlated with their inducing potencies in rats. If this induction *per se* is concerned with the toxicity of MC-type PCBs, MC should also cause similar toxicity. Therefore, the toxic effect of MC was examined in weanling rats and compared with that of 3,4,5,3′,4′,5′-HCB, a potent MC-type PCB. Effect of PB pretreatment was tested simultaneously.

As shown in FIGURE 3 (right), pretreatment with MC for 5 consecutive days at 20 mg/kg/day exhibited inductive effects on hepatic enzymes almost like those of 3,4,5,3′,4′,5′-HCB at a single dose of 10 mg/kg. In addition, MC and HCB gave similar CO absorption maximum at 448 nm. MC also enhanced DT diaphorase at the same extent by HCB, whereas PB did not increase this activity so much.

On the contrary, certain differences were observed between toxicities induced by MC and 3,4,5,3′,4′,5′-HCB (FIGURES 3 and 4). MC caused liver hypertrophy and reduced the weight of the thymus; however, it did not affect the weight of the spleen or the growth rate. Furthermore, the lipid content of the liver in MC-treated rats did not change at all, while HCB increased it by 30%. These results suggest that the toxic potency of PCBs is correlated with their MC-type inducing ability, but the toxic characteristics are different from those of MC itself.

Effect of 3,4,5,3′,4′-PenCB Pretreatment on Fractionation of the Hepatic Cell

When the livers from rats pretreated with potent MC-type PCBs were homogenized with 1.15% KCl, increased amounts of pellet were obtained after centrifugation at 9000 × *g* for 20 minutes. This is not the case with control and PB-type PCBs pretreatment. These observations might indicate that liver microsomes of rats pretreated with potent MC-type PCBs tended to aggregate and precipitate easily on centrifugation at 9000 × *g*. Therefore, the distribution of G-6-Pase and cytochrome P450(448) in the liver 9000 × *g* supernatant and pellet of normal and 3,4,5,3′,4′-PenCB pretreated rats were compared by repeated centrifugations of 20% homogenate of liver with 1.15% KCl.

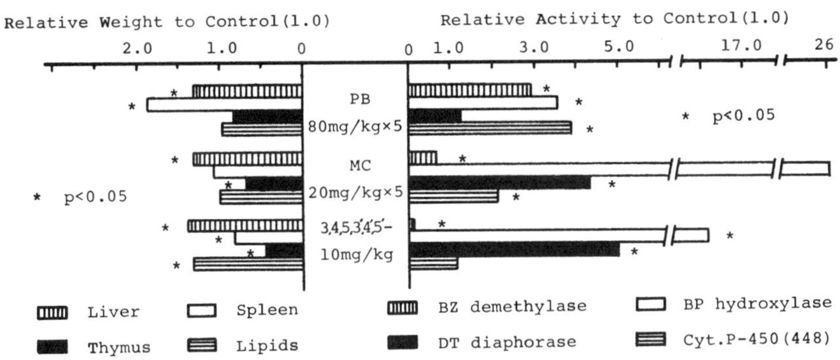

FIGURE 3. Comparative effects of pretreatment with 3,4,5,3′,4′,5′-HCB, PB and MC on liver enzyme activity, organ weight, and liver lipid content in weanling rats. Rats treated with HCB were further injected i.p. with salad oil (2 ml/kg) for the 4 following days. Control groups were injected i.p. with 2 ml/kg of saline or salad oil for 5 days. The enzyme activities in the control for saline and salad oil treatment were given, respectively, as follows (mean ± S.D.): BZ demethylase 13.69 ± 1.58, 12.29 ± 1.88 nmoles HCHO formed/mg protein/15 min; BP hydroxylase 0.139 ± 0.020, 0.127 ± 0.036 nmoles 3-OH-BP formed/mg protein/5 min; DT diaphorase 0.206 ± 0.057, 0.216 ± 0.014 nmoles DCPIP reduced/mg protein/min; cytochrome P450 0.151 ± 0.028, 0.145 ± 0.012 nmoles/mg protein. The weights of organs (g/100g bw) and liver lipids in the respective controls were: liver 3.69 ± 0.47, 3.60 ± 0.21; spleen 0.307 ± 0.015, 0.365 ± 0.018; thymus 0.417 ± 0.083, 0.344 ± 0.009; lipids 51.71 ± 1.81, 48.53 ± 1.75 mg/g liver.

TABLE 2

ABSORPTION MAXIMUM OF CO COMPLEX OF HEMOPROTEINS INDUCED BY PURE PCB CONGENERS

Treatment	Dose (mg/kg)	n	Maximum (nm)
Control (salad oil)		8	450.0 ± 0.1
2,5,2′,5′-TCB	100	4	449.9 ± 0.4
2,4,3′,4′-TCB	100	4	449.6 ± 0.5
3,4,3′,4′-TCB	10	4	449.5 ± 0.4
	50	4	448.8 ± 0.1*
2,3,4,3′,4′-PenCB	10	4	449.7 ± 0.2
	50	4	448.8 ± 0.1*
3,4,5,3′,4′-PenCB	10	7	448.0 ± 0.2*,†
3,4,5,3′,4′,5′-HCB	10	7	448.5 ± 0.2*
Control (saline)		4	449.5 ± 0.0
PB	80 × 5	4	450.4 ± 0.1
MC	20 × 5	4	448.5 ± 0.2

Values represent mean ± S.D. of 4 to 8 rats.
*Significantly different from control, $p < 0.05$.
†Significantly different from MC-treated, $p < 0.05$.

The results were shown in TABLE 3. Protein distributed in the pellet was increased and that in the supernatant was decreased significantly by 3,4,5,3′,4′-PenCB pretreatment as compared with the control values. Furthermore, a marked difference in the distribution of G-6-Pase, a marker enzyme of microsomes, was observed

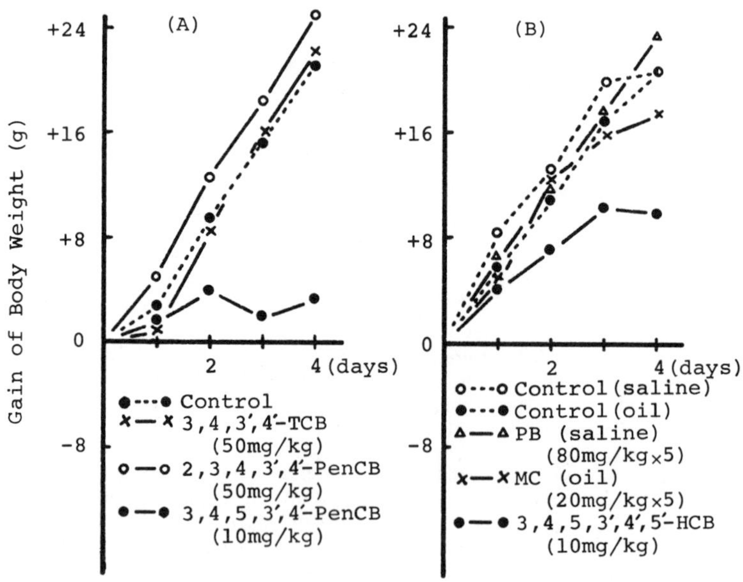

FIGURE 4. Effects of pretreatment with pure PCB congeners, PB and MC on daily body weight gain in weanling rats.

between the control and 3,4,5,3′,4′-PenCB pretreatment. Approximately 40% of the total activity was retained in the first supernatant of the control liver. In this same fraction from PenCB pretreatment, however, only 7% of G-6-Pase was retained. Most activity was precipitated into the pellet after the second centrifugation.

Such patterns observed in the distribution of G-6-Pase were more emphasized in the distribution of cytochrome P450(448). The content of cytochrome P450(448) in whole liver was increased more than 10 times by PenCB pretreatment as compared with the control value. While approximately 70% of cytochrome P450 of control liver was found in the first 9000 × *g* supernatant, but only 8% was recovered into this fraction from PenCB-pretreated rats. Most of cytochrome P448 precipitated after the first centrifugation could not be recovered into the supernatant by recentrifugation in isotonic KCl, similarly as in the case of G-6-Pase. These results strongly suggest that the liver microsomes from rats pretreated with potent MC-type PCBs were altered not only quantitatively but also qualitatively.

TABLE 3

EFFECTS OF PRETREATMENT WITH 3,4,5,3′,4′-PenCB ON CELL FRACTIONATION OF RAT LIVER

Fraction	Protein (mg/total liver) Control	PenCB	G-6-Pase (μmoles Pi formed/ total liver/min) Control	PenCB	Cyt. P-450 (448) (nmoles/ total liver) Control	PenCB
Whole homogenate	1763.3*	1650.3	111.7	133.4	148.2	1537.0
	(100)	(100)	(100)	(100)	(100)	(100)
9000 × g	560.2	420.3	43.4	9.2	102.1	126.7
supernatant-I	(31.8)	(25.5)	(38.8)	(6.9)	(69.8)	(8.2)
9000 × g	94.5	87.0	3.3	1.5	9.8	3.5
supernatant-II	(5.4)	(5.3)	(3.0)	(1.1)	(6.6)	(0.2)
Pellet	735.7	1036.5	60.3	100.0	34.0	1237.9
	(41.7)	(62.8)	(54.0)	(75.0)	(22.9)	(80.5)

*Data are expressed as the values of pooled liver (control, 7.7 g, PenCB, 11.5 g) from 2 rats per group. Values in parentheses represent % of the values of whole homogenate.

DISCUSSION

3,4,5,3′,4′,5′-HCB has been reported as the most potent MC-type PCB of inducers in rats[16,21] and mice.[21] We have examined the ability of six individual PCBs to induce hepatic enzymes in adult and weanling rats, and found that these can be classified into three types as inducers (FIGURE 5), and that 3,4,5,3′,4′-PenCB is most potent as an inducer of the MC-type. As indicated in FIGURE 2, 3,4,5,3′,4′-PenCB enhanced BP hydroxylase in weanling rats by 50 times the control value, 5 days after a single i.p. injection of 10 mg/kg. In contrast, the enhancement by 3,4,5,3′4′,5′-HCB was 16 times the control value at the same dose. DT diaphorase is an enzyme that is located mainly in the hepatic soluble fraction and partly in microsomes, and inducible by MC[39] and 2,3,7,8-tetrachlorodibenzo-*p*-dioxin,[40] but not by PB.[39] 3,4,5,3′,4′-PenCB and 3,4,5,3′,4′,5′-HCB could induce this enzyme more than 12 times and 5 times over control, respectively, in weanling rats (FIGURES 2 and 3).

Although 3,4,3′,4′-TCB is known as a prototype of MC-type PCBs, the ability of this PCB to induce BP hydroxylase and DT diaphorase was much less than that of

3,4,5,3′,4′-PenCB and 3,4,5,3′,4′,5′-HCB (FIGURES 1 and 2). From these results, an increasing order of induction potencies as MC-type PCBs is considered to be 3,4,3′,4′-TCB ≪ 3,4,5,3′,4′,5′-HCB ≦ 3,4,5,3′,4′-PenCB. It is recognized that the induction potencies of MC-type PCBs for DT diaphorase are quite parallel to those for BP hydroxylase, suggesting that the induction mechanism of BP hydroxylase and DT diaphorase by MC-type inducers may be closely correlated with each other.

2,3,4,3′,4′-PenCB exhibited a unique inductive nature. As reported previously, a minimum requirement for the structure to induce cytochrome P448 and AHH is chlorination of the 3,4,3′ and 4′-positions of the biphenyl ring while chlorination of the ortho-positions (2,2′ or 6,6′) abolishes this ability.[16,17,21] 2,3,4,3″,4′-PenCB, which is chlorinated at one of the ortho-positions of 3,4,3′,4′-TCB, had a markedly decreased induction potency for AHH, although it still retained an ability to induce cytochrome

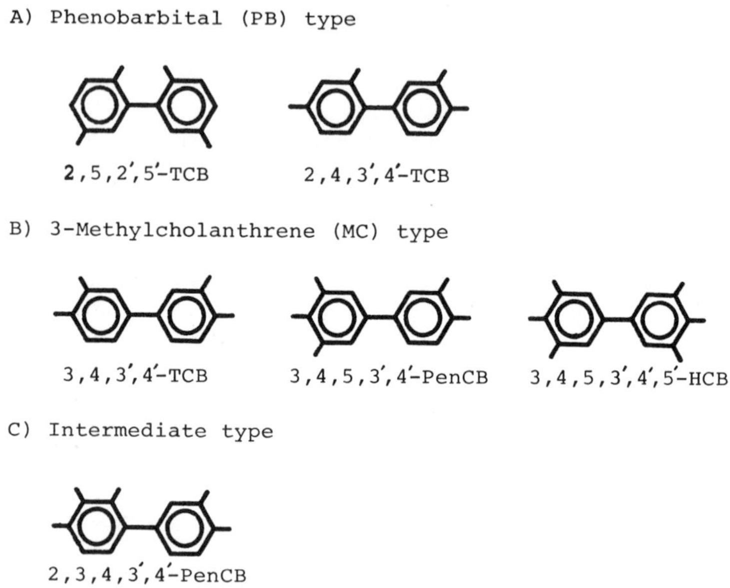

FIGURE 5. Classification of pure PCB congeners used in this study into three types of inducers.

P448 (FIGURE 2). This may suggest that 2,3,4,3′,4′-PenCB induces a different type of cytochrome P448 which is not associated with BP hydroxlation.

A number of studies on the toxicity of commerical PCBs and related compounds were made in this decade, and these findings have been excellently reviewed by Kimbrough.[41] However, only very limited numbers of reports on the general toxicity of individual PCBs are available. In the present study, toxicologic evaluation of six PCB congeners was made by measuring the gravimetric changes of the body and of several organs and also the lipid content of the liver, in comparison with their inducing potency described above. At a dose of 100 mg/kg, 2,5,2′,5′-TCB did not show any significant effects on all parameters tested. Allen *et al.* also reported that 2,5,2′,5′-TCB caused only a slight increase of aminopyrine *N*-demethylase activity and did not appear to cause any gross abnormality in rats even at a high dose of 1.0 g/kg.[42]

2,4,3′,4′-TCB also did not affect the liver weight, but did somewhat suppress the body and thymus weights. As previously reported, 2,4,3′,4′-TCB was mainly metabolized in rats to 5-hydroxy-2,4,3′,4′-TCB, which showed five-fold higher toxicity than the parent PCB.[43] Therefore, these partial toxic effects observed for 2,4,3′,4′-TCB may be due to the formation of this toxic metabolite.

As opposed to PB-type PCBs, MC-type PCBs were much more toxic. The thymus and spleen showed high sensitivites to PCBs and related compounds.[41] Severe atrophy was induced in both organs by 3,4,5,3′,4-PenCB and 3,4,5,3′,4′,5′-HCB at a dose of 10 mg/kg, and by 3,4,3′,4′-TCB at a dose of 50 mg/kg 5 days after a single injection, respectively. Information on the comparative toxicity of individual PCBs is limited, but McKinney *et al.* have reported that of the five symmetric hexachlorobiphenyls, 2,3,6,2′,3′,6′-, 2,4,5,2′,4′,5′-, 2,3,4,2′,3′,4′-, 2,4,6,2′,4′,6′- and 3,4,5,3′,4′,5′-HCB, the last one was the only one to cause significant edema and thymus atrophy in the chicken.[22] The results obtained here in rats indicated that 3,4,5,3′,4′-PenCB exhibited more profound toxicity to the thymus and spleen than did 3,4,5,3′,4′,5′-HCB. 3,4,3′,4′-TCB started to exhibit the toxic effect in rats at 50 mg/kg. 3,4,5,3′,4′-PenCB caused severe atrophies of the adrenals and kidneys; these effects were caused to a lesser extent by 3,4,5,3′,4′,5′-HCB and least by 3,4,3′,4′-TCB (data not shown).

Liver hypertrophy and increased hepatic lipid content are the most common observations in rats[11,44] and chickens[22] after pretreatment with commercial PCB preparations and also after certain individual PCBs. In our experiments, liver enlargement was observed not only by the three MC-types PCBs, but also by 2,3,4,3′,4′-PenCB as well as PB and MC. It is interesting to note, however, that only the liver enlargements caused by MC-types PCBs were accompanied by a marked increase of liver lipid content, without increase of serum lipids (TABLE 1). These results suggest that fatty liver produced by commercial PCB preparations is mainly due to MC-type PCBs in them and the mechanism appears to be a disturbance of lipid clearance from the liver.

Body weight gains during 4 days after PCBs injection were suppressed strongly by 3,4,5,3′,4′-PenCB and moderately by 3,4,5,3′,4′,5′-HCB (FIGURE 4). However, no significant effects were observed with 2,3,4,3′,4′-PenCB and 3,4,3′,4′-TCB at a dose of 50 mg/kg. Food consumption was only slightly reduced by potent MC-type PCBs.

Although the numbers of PCBs tested are so limited that we cannot make a conclusive statement, the results obtained strongly suggest that MC-type PCBs are much more toxic than those of the PB-type and that their toxic potency is well correlated with their MC-type inducing ability (FIGURE 6). Pretreatment with MC for 5 consecutive days at 20 mg/kg/day caused a similar liver enlargement and a lesser atrophy of the thymus, compared with MC-type PCBs. But no effect was observed on the spleen weight and liver lipid content, the changes in which are typical toxic syndromes with MC-types PCBs, and body weight gain was also unaffected by MC pretreatment.

At present we are unable to explain why the toxic potency of PCBs is parallel with their MC-type inducing ability, while their toxic characteristics differ from that of MC itself. On the structure-activity relationship in toxicity of halogenated aromatic compounds, such as polyhalogenated dibenzo-*p*-dioxins, dibenzofurans, azoxybenzenes, and biphenyls, Poland *et al.* have speculated that the induction of AHH activity *per se* does not cause toxicity, but toxic syndromes observed by potent MC-type PCBs may be mediated by one (or more) of the other members of the gene battery which is coordinately expressed by halogenated aromatic compounds.[21]

As shown in TABLE 3, the liver microsomes from 3,4,5,3′,4′-PenCB-pretreated rats became markedly aggregative as compared with those from untreated rats. Similar properties were observed with 3,4,5,3′,4′,5′-HCB, but not with MC (data not shown).

The low recovery of microsomes pretreated with potent MC-type PCBs was considerably improved by using 0.25 M sucrose as a homogenizing or suspending medium instead of 1.15% KCl (data not shown). Therefore, an inducing potency of MC-type PCBs should not be judged only from cytochrome P450(448) content in microsomes obtained by differential centrifugation. Pretreatment with 3,4,5,3′,4′-PenCB has increased specifically cytochrome P448, and resulted in a marked decrease of cytochrome P450 relative to cytochrome P448 whereas a considerable amount of cytochrome P450 was retained in MC-treatment.[25] Furthermore, cytochrome P448 purified from 3,4,5,3′,4′-PenCB-treated rats has shown different properties from

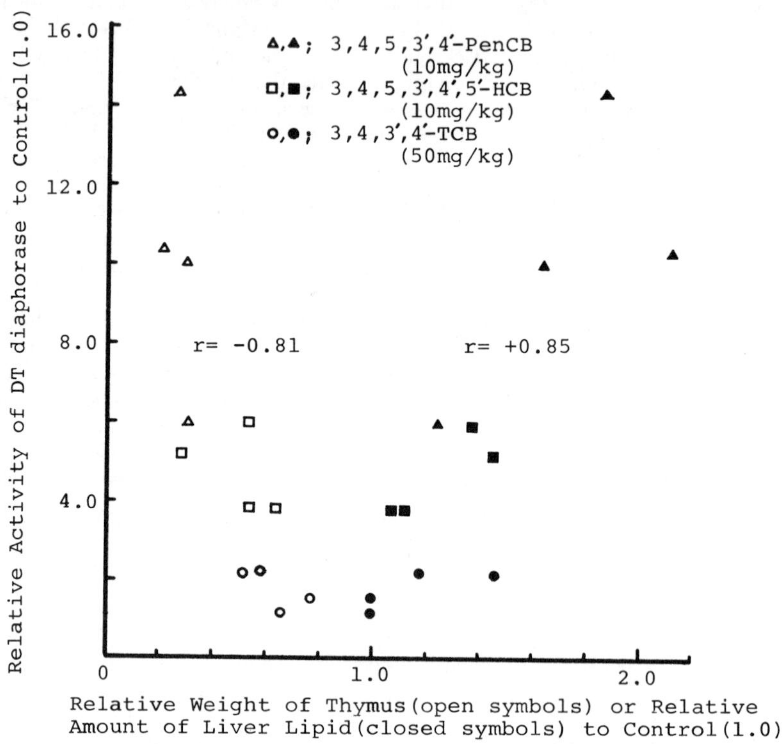

FIGURE 6. Correlation between DT diaphorase activity and toxicity induced by MC-type PCBs in weanling rats.

cytochrome P448 purified from MC-treated rats.[46] Therefore, a clearer understanding of these different characteristics as inducers might assist in clarifying the discrepancy between the inductive nature and the toxicity observed with MC-type PCBs and MC itself.

SUMMARY

Acute toxicity of individual PCBs, which were categorized as either phenobarbital (PB)- or 3-methylcholanthrene (MC)-type inducers, was examined in young male

Wistar rats, comparing their effects on growth rate, organ weight and liver lipid content, 5 days after a single i.p. injection. PB-type PCBs (2,4,3′,4′- and 2,5,2′5′-tetrachlorobiphenyl), which slightly increased a content of cytochrome P450, did not show any significant toxicity at a dose of 100 mg/kg. On the contrary, MC-type PCBs (3,4,5,3′,4′-pentachloro- and 3,4,5,3′,4′,5′-hexachlorobiphenyl), which markedly increased a content of cytochrome P448, strongly reduced growth rate and weights of thymus and spleen at a dose of 10 mg/kg. Liver enlargement accompanied by fatty liver was also observed only with MC-type PCBs. 3,4,3′,4′-Tetrachlorobiphenyl was also toxic at a dose of 50 mg/kg, in keeping with its weak MC-type-inducing ability. Pretreatment with MC affected neither growth rate, spleen weight, nor liver lipid content. These results suggest that the toxic potency of PCBs is related to their MC-type inducing ability, but the toxic characteristics are different from those of MC itself.

References

1. Risebrough, R. W., P. Rieche, D. B. Peakall, S. G. Herman & M. N. Kirven. 1968. Nature (London) **220:** 1098–1102.
2. Biros, F. J., A. C. Walker & A. Medberg. 1970. Bull. Environ. Contam. Toxicol. **5:** 317–323.
3. Fujita, S., H. Tsuji, K. Kato, S. Saeki & H. Tsukamoto. 1971. Fukuoka Acta Med. **62:** 30–34.
4. Villeneuve, D. C., D. L. Grant, W. E. J. Phillips, M. L. Clark & D. J. Clegg. 1971. Bull. Environ. Contam. Toxicol. **6:** 120–128.
5. Litterst, C. L., T. M. Farber, A. M. Baker & E. J. Van Loon. 1972. Toxicol. Appl. Pharmacol. **23:** 112–122.
6. Bickers, D. R., L. C. Harber, A. Kappas & A. P. Alvares. 1972. Res. Commun. Pathol. Pharmacol. **3:** 505–512.
7. Alvares, A. P., D. R. Bickers & A. Kappas. 1973. Proc. Nat. Acad. Sci. USA **70:** 1321–1325.
8. Ecobichon, D. J. & A. M. Comeau. 1974. Chem.-Biol. Interact. **9:** 341–350.
9. Vos, J. G. & E. Notenboom-Ram. 1972. Toxicol. Appl. Pharmacol. **23:** 563–577.
10. Chen, P. R., H. M. Mehendale & L. Fishbein. 1973. Arch. Environ. Contam. Toxicol. **1:** 36–47.
11. Hansell, M. M. & D. J. Ecobichon. 1974. Toxicol. Appl. Pharmacol. **28:** 418–427.
12. Stonard, M. D. & J. B. Greig. 1976. Chem.-Biol. Interact. **15:** 365–379.
13. Johnstone, G. J., D. J. Ecobichon & O. Hutzinger. 1974. Toxicol. Appl. Pharmacol. **28:** 66–81.
14. Ecobichon, D. J. & A. M. Comeau. 1975. Toxicol. Appl. Pharmacol. **33:** 94–105.
15. Goldstein, J. A., J. D. McKinney, G. W. Lucier, P. Hickman, H. Bergman & J. A. Moore. 1976. Toxicol. Appl. Pharmacol. **36:** 81–92.
16. Goldstein, J. A., P. Hickman, H. Bergman, J. D. McKinney & M. P. Walker. 1977. Chem.-Biol. Interact. **17:** 69–87.
17. Yoshimura, H., N. Ozawa & S. Saeki. 1978. Chem. Pharm. Bull. (Tokyo) **26:** 1215–1221.
18. Poland, A. & E. Glover. 1973. Mol. Pharmacol. **9:** 736–747.
19. Poland, A. & E. Glover. 1976. J. Biol. Chem. **251:** 4936–4946.
20. Poland, A., E. Glover, A. S. Kende, M. DeCamp & C. M. Giandomenico. 1976. Science **194:** 627–630.
21. Poland, A. & E. Glover. 1977. Mol. Pharmacol. **13:** 924–938.
22. McKinney, J. D., K. Chae, B. N. Gupta, J. A. Moore & J. A. Goldstein. 1976. Toxicol. Appl. Pharmacol. **36:** 65–80.
23. Saeki, S., A. Tsutsui, K. Oguri, H. Yoshimura & H. Hamana. 1971. Fukuoka Acta Med. **62:** 20–24.

24. YAMAMOTO, H., H. YOSHIMURA, M. FUJITA & T. YAMAMOTO. 1976. Chem. Pharm. Bull. (Tokyo) **24:** 2168–2174.
25. NEBERT, D. W. & H. V. GELBOIN. 1968. J. Biol. Chem. **243:** 6242–6249.
26. NASH, T. 1953. Biochem. J. **55:** 416–421.
27. ERNSTER, L., L. DANIELSON & M. LJUNGGREN. 1962. Biochem. Biophys. Acta. **58:** 171–188.
28. MATSUBARA, T., M. KOIKE, A. TOUCHI, Y. TOCHINO & K. SUGENO. 1976. Anal. Biochem. **75:** 596–603.
29. SWANSON, M. A. 1955. Methods in Enzymology **2:** 541–543.
30. LOWRY, O. H., N. J. ROSEBROUGH, A. L. FARR & R. J. RANDALL. 1951. J. Biol. Chem. **193:** 265–275.
31. FISKE, C. H. & Y. SUBBAROW. 1925. J. Biol. Chem. **66:** 375–400.
32. FOLCH, J., M. LEES & G. H. S. STANLEY. 1957. J. Biol. Chem. **226:** 497–509.
33. FLETCHER, M. J. 1968. Clin. Chim. Acta **22:** 393–397.
34. ZURKOWSKI, P. 1964. Clin. Chem. **10:** 451–453.
35. BARTLETT, G. R. 1959. J. Biol. Chem. **293:** 466–468.
36. HOLLANDER, M. & D. A. WOLFE. 1973. Nonparametric Statistical Methods. John Wiley & Sons, Inc. New York.
37. LU, A. Y. H., W. LEVIN, S. B. WEST, M. JACOBSON, D. RYAN, R. KUNTZMAN & A. H. CONNEY. 1973. J. Biol. Chem. **248:** 456–460.
38. RYAN, D., A. Y. H. LU, J. KWALEK, S. B. WEST & W. LEVIN. 1975. Biochem. Biophys. Res. Commun. **64:** 1134–1141.
39. LIND, C. & L. ERNSTER. 1974. Biochem. Biophys. Res. Commun. **56:** 392–400.
40. BEATTY, P. & R. A. NEAL. 1976. Biochem. Biophys. Res. Commun. **68:** 197–204.
41. KIMBROUGH, R. D. 1974. CRC Crit. Rev. Toxicol. **2:** 445–498.
42. ALLEN, J. R., L. A. CARTENS, L. J. ABRAHAMSON & R. J. MARLAR. 1975. Environ. Res. **9:** 265–273.
43. YAMAMOTO, H. & H. YOSHIMURA. 1973. Chem. Pharm. Bull. (Tokyo) **21:** 2237–2242.
44. KIMBROUGH, R. D., R. E. LINDER & T. B. GAINES. 1972. Arch. Environ. Health **25:** 354–364.
45. OZAWA, N., S. YOSHIHARA & H. YOSHIMURA. 1978. In preparation.
46. OZAWA, N., S. YOSHIHARA, K. KAWANO, Y. OKADA & H. YOSHIMURA. 1978. In preparation.

INDUCTION OF MICROSOMAL HEMOPROTEIN(S) P-450 IN THE RAT AND RAINBOW TROUT BY POLYHALOGENATED BIPHENYLS*

Clifford R. Elcombe, Ronald B. Franklin, and John J. Lech

Department of Pharmacology
Medical College of Wisconsin
Milwaukee, Wisconsin 53226

INTRODUCTION

The polyhalogenated biphenyls are widespread and persistent environmental contaminants,[1-6] and both the polychlorinated and polybrominated biphenyls have been found to accumulate in fish tissues.[7,8] A pure polychlorinated biphenyl isomer, 2,2′,5,5′-tetrachlorobiphenyl, was recently shown to possess a half-life of approximately 2 years in the rainbow trout,[9] and certain higher chlorinated biphenyls are expected to have even longer tissue half-lives. The polychlorinated and polybrominated biphenyls are known to be potent inducers of mixed-function oxidase activity in several mammalian species.[10-19] Crude mixtures of polychlorinated biphenyls, such as Aroclors® 1254 and 1242, exhibit enzyme-inducing properties that have been termed "mixed." That is, they appear to induce both cytochromes P-450 and P_1-450 (P-448); these are classically associated with phenobarbital-and 3-methylcholanthrene-induction, respectively. Crude polybrominated biphenyls also appear to induce forms of cytochrome(s) P-450 that exhibit "mixed" properties.

Much work has demonstrated the presence of multiple forms of hemoproteins P-450 in hepatic microsomes of rats, mice, and rabbits. It has further been shown that the relative proportions of these isoenzymes may be altered considerably by various chemical agents.[20-26] These variations in hemoprotein(s) P-450 subpopulations may lead to distinct changes in the metabolic pathways of chemicals.[27,28]

More recently, attention has been directed to the potential of other species to metabolize foreign compounds. Of particular concern has been the ability of fish to metabolize xenobiotics. Early work suggested a lack of drug-metabolizing enzymes in aquatic species; however, it is now recognized that fish do indeed possess an array of drug-metabolizing enzymes comparable in many respects to those found in mammals.[29-32] Induction of hemoprotein(s) P-450-mediated monooxygenase activity has been demonstrated in fish by using agents such as the polycyclic aromatic hydrocarbons,[33,34] crude petroleum,[35] and the polyhalogenated biphenyls.[36-38] Compounds such as 2,3-benzanthracene and β-naphthoflavone not only increase the metabolism of model mixed-function oxidase substrates *in vitro* but also increase the *in vivo* metabolism and disposition of various chemicals.[34,39,40] Recently, reports have appeared indicating that fish possess multiple forms of hepatic microsomal hemoprotein(s) P-450 and that treatment of fish with polycyclic aromatic hydrocarbons or polychlorinated biphenyls results in an alteration of the hemoprotein(s) P-450 subpopulations.[39,41]

The present paper compares the effect of polychlorinated-and polybrominated biphenyls on the hepatic microsomal hemoprotein(s) P-450 and the monooxygenase systems of rainbow trout and rats.

*This paper is based upon research supported by National Institute of Environmental Health Sciences Grant 01080 and NIEHS Aquatic Biomedical Research Core Grant ES01985.

0077–8923/79/0320–0193 $01.75/0 © 1979, NYAS

Materials and Methods

Rainbow trout (*Salmo gairdneri*) weighing 70–100 g, were obtained from Kettle Moraine Springs Trout Hatchery (Adell, Wisconsin), and were held in flowing, charcoal-filtered water at 12°C for a minimum of two weeks prior to use. A 12-hour (0600–1800) photocycle was in use.

$NADP^+$, NADPH, glucose-6-phosphate, glucose-6-phosphate dehydrogenase, and 7-hydroxycoumarin (umbelliferone) were obtained from Sigma Chemical Company (St. Louis, Missouri). 7-Ethoxycoumarin was synthesized as described by Ullrich and Weber.[42] Ethoxyresorufin was most kindly synthesized by Dr. S. R. Challend, (Wellcome Research Laboratories, Beckenham, Kent, England) from resorufin obtained from Eastman Kodak, Ltd, Rochester, New York. Aroclors® 1254 and 1242 were generous gifts from Monsanto Chemical Company. Dr. D. Rickert (Chemical Industry Institute of Toxicology, Raleigh, North Carolina) generously donated Firemaster BP6. [7,10^{14}C] Benzo(a)pyrene (51 mCi/mmol) was purchased from Amersham-Searle (Des Plaines, Illinois). All other reagents were of the highest purity commercially available.

Pretreatment of Fish and Preparation of Microsomes

Aroclor 1254, Aroclor 1242 and Firemaster BP6 were administered to rainbow trout by intraperitoneal injection as solutions in corn oil (1 ml/kg). Routinely, fish received 150 mg/kg of these compounds. Control fish were injected with corn oil alone. After injection, fish were maintained in 50-liter tanks (4–6 fish per tank) until sacrifice. The fish were killed by a blow to the head and the livers rapidly excised into ice-cold 0.154 M KCl after carefully removing the gall bladders. Washed microsomes were prepared as described previously.[38,41]

Male Sprague-Dawley rats (approximately 200 g) were administered one dose of 150 mg/kg intraperitoneally of either Aroclor 1242, Aroclor 1254, or Firemaster BP6. Animals pretreated with 3-methylcholanthrene were injected intraperitoneally for 4 consecutive days at 20 mg/kg. All animals were sacrificed 96 hours after initiation of the pretreatment regimes. Hepatic microsomes were prepared by ultracentrifugation.

Microsomal Enzyme Assays

Arylhydrocarbon (benzo(a)pyrene) hydroxylation was determined by the method of Hansen and Fouts[43] as modified by Statham *et al.*[44] Ethoxyresorufin-*0*-deethylation, ethoxycoumarin-*0*-deethylation and ethylmorphine-*N*-demethylation were measured using published procedures.[45,52,46] Temperature optima were 29–30°C for the two deethylations and 25–26°C for arylhydrocarbon hydroxylation and ethylmorphine-*N*-demethylation. All enzymic activities were measured as initial velocities from the linear portion of the reaction time course. Protein dependency for these reactions in rainbow trout hepatic microsomes have been described previously.[41] Microsomal protein concentrations were estimated by the method of Ross and Schatz[47] using bovine serum albumin standards.

Spectral Measurements

Hemoprotein(s) P-450 were estimated by the method of Estabrook *et al.*[48] using an extinction coefficient of 100 mM^{-1} cm^{-1} between 450 and 510 nm. This method consists of bubbling both cuvettes with carbon monoxide and reducing the contents of the sample cuvette with $Na_2S_2O_4$. In this manner spectral interference due to hemoglobin is eliminated.

Sodium Dodecylsulphate-Polyacrylamide Gel Electrophoresis (SDS-PAGE)

SDS-PAGE was performed using slab gels (1.75 mm thick) essentially by the procedure of Laemmli and Favre.[49] Total acrylamide concentrations were 3% (w/v) and 7.5% (w/v) in the stacking and resolving gels, respectively. The resolving gel was approximately 15 cm long. Coomassie Blue R-250 was used to visualize proteins and the method of Thomas *et al.*[50] was used to detect peroxidase activity.

Individual electrophoretograms were analyzed densitometrically at 550 nm using

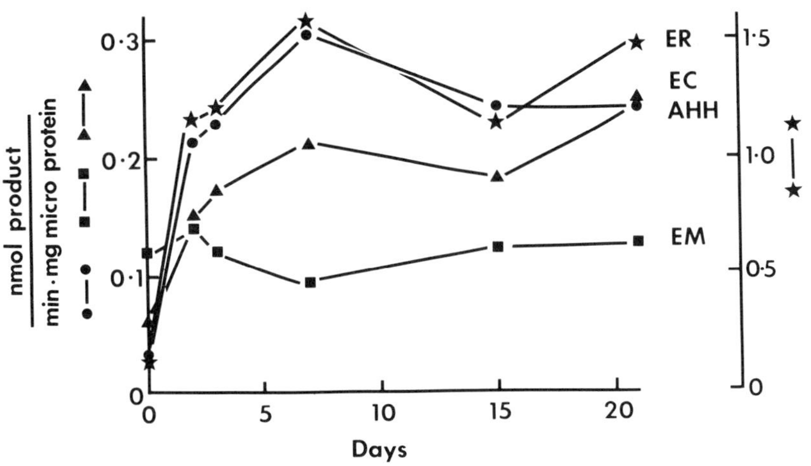

FIGURE 1. Time course of induction by Aroclor 1254 in rainbow trout: *—*, ethoxyresorufin-0-deethylation: ▲—▲, ethoxycoumarin-*0*-deethylation; ●—●, arylhydrocarbon (benzo(a)pyrene) hydroxylation; ■—■, ethylmorphine-*N*-demethylation. Each point represents value from microsomes obtained from the pooled livers of six fish. Aroclor 1254 was administered i.p. in corn oil at a dose 150 mg/kg. (Reproduced with permission from Elcombe & Lech.[38])

an ISCO Model 1319 gel scanner in combination with an ISCO Type 6 optical unit and Model UA-5 absorbance monitor.

RESULTS

Injection of rainbow trout with Aroclor 1254 resulted in a time-dependent increase in various hemoprotein P-450-mediated monooxygenase activities. FIGURE 1 illustrates that arylhydrocarbon (benzo(a)pyrene) hydroxylation, ethoxycoumarin-*0*-deethylation and ethoxyresorufin-*0*-deethylation were all increased 2 days after treatment of the fish. These activities attained plateau values between 4 and 7 days after injection of Aroclor 1254 and remained elevated for at least 21 days. It is of considerable interest that ethylmorphine-*N*-demethylation was unaffected by Aroclor

1254 pretreatment. FIGURE 2 documents a similar experiment utilizing Firemaster BP6 as an inducing agent in fish. Monooxygenase activities were observed to increase more slowly after Firemaster BP6 treatment than after Aroclor 1254, and a relatively rapid decline in the activities was seen. Fifteen days after pretreatment with Firemaster BP6, ethoxyresorufin-*0*-deethylation and arylhydrocarbon (benzo(a)pyrene) hydroxylation approximated control values. Although not illustrated, ethylmorphine-*N*-demethylation was unaffected by Firemaster BP6 treatment.

Total hepatic microsomal hemoprotein P-450 content was not significantly increased by Aroclors 1254 or 1242, or Firemaster BP6. Furthermore, the Soret absorbance maxima of carboxyferrocytochrome P-450 was at 449 nm in all fish microsomal preparations examined (TABLE 1).

Using male rats, large increases in hemoprotein content were observed after

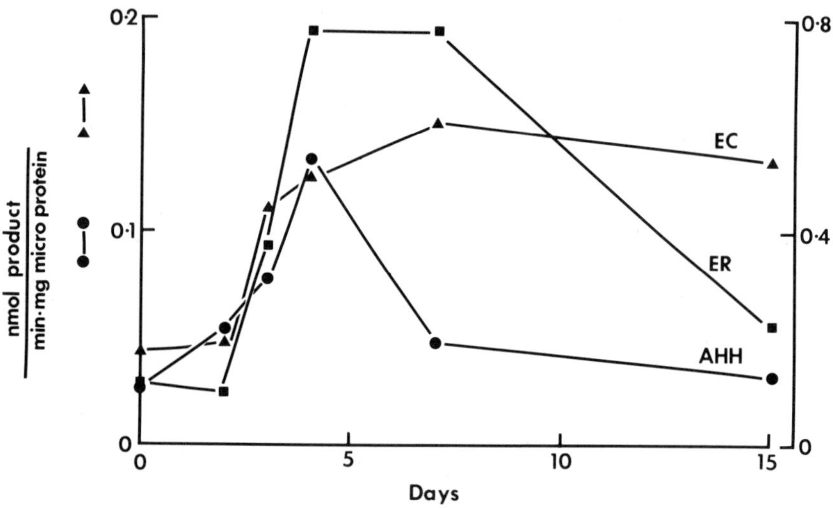

FIGURE 2. Time course for induction by Firemaster BP6 in rainbow trout: ■—■, ethoxyresorufin-0-deethylation; ▲—▲, ethoxycoumarin-*0*-deethylation; ●—●, arylhydrocarbon (benzo(a)pyrene) hydroxylase. Each point represents value from microsomes obtained from the pooled livers of six fish. Firemaster BP6 was administered i.p. at a dose of 150 mg/kg. (Reproduced with permission from Elcombe & Lech.[38])

treatment with Firemaster BP6 or Aroclors 1254 and 1242 and the Soret absorbance maxima of the carbon monoxide-ferrocytochrome complex underwent a hyposochromic shift (TABLE 1).

Although total hepatic microsomal hemoprotein(s) P-450 of the trout was not increased in this experiment, monooxygenase activities were greatly stimulated by Firemaster BP6, Aroclor 1242, and Aroclor 1254. TABLE 2 demonstrates that ethoxycoumarin-*0*-deethylation was increased 6- and 9-fold by Firemaster BP6 and Aroclor 1242, respectively, while ethoxyresorufin-*0*-deethylation was increased 19.5-, 30- and 18-fold by Firemaster BP6, Aroclor 1242, and Aroclor 1254 respectively.

Sodium dodecylsulphate-polyacrylamide gel electrophoresis of hepatic micro-

TABLE 1

HEMOPROTEIN P-450 CONTENT AND SORET MAXIMA OF RAINBOW TROUT AND MALE RAT LIVER MICROSOMES

	Trout		Rat	
Pretreatment*	P-450 Content (nmol/mg†)	λmax (nm)	P-450 Content (nmol/mg†)	λmax (nm)
Corn oil	0.11 ± 0.01‡	449	0.76 ± 0.02‡	450.03 ± 0.03‡
Firemaster BP6	0.13 ± 0.01	449	2.2 ± 0.1	449.60 ± 0.08
Aroclor 1242	0.13 ± 0.00	449	ND‡	ND‡
Aroclor 1254	0.14 ± 0.02	449	2.1 ± 0.1	448.60 ± 0.10

*Fish and rats were injected intraperitoneally with the compounds (150 mg/kg) dissolved in corn oil (1 ml/kg) and sacrificed 4 days later.
†Nmol hemoprotein per mg microsomal protein.
‡Values are mean ± S.E. (n = 4–6); ND = not determined.

somes from variously induced trout indicated the presence of several Coomassie blue-staining polypeptide bands in the 40,000–60,000 molecular weight region. FIGURE 3 illustrates densitometric scans of such polyacrylamide gels. Staining for peroxidase activity suggested that bands A, B, C, and D were hemoproteins. Changes in polypeptide patterns may be observed in microsomes obtained from β-naphthoflavone- and Aroclor 1254-pretreated trout. By far the most striking alteration was the appearance of an intense Coomassie blue-staining band (A) at 57,000 daltons in microsomes from β-naphthoflavone- or Aroclor 1254-pretreated trout (FIGURE 3). Also, it appeared that the relative intensities of bands B, C, and D (molecular weights 51,000 48,000 and 45,000 respectively) were different in the three preparations examined. The polypeptide at 57,000 daltons, induced by β-naphthoflavone and Aroclor 1254 was barely detectable in control microsomes.

FIGURE 4 illustrates densitometric scans of electrophoretograms obtained using male rat liver microsomes. The vertical line indicates the 53,000 dalton band (cytochrome P_1-450) induced by 3-methylcholanthrene and Aroclor 1254. It should be noted that this band has a different molecular weight from the comparable band induced in rainbow trout hepatic microsomes by Aroclor 1254 and β-naphthoflavone.

TABLE 2

ETHOXYCOUMARIN- AND ETHOXYRESORUFIN-*O*-DEETHYLATIONS BY RAINBOW TROUT HEPATIC MICROSOMES

	Ethoxycoumarin-*O*-Deethylation		Ethoxyresorufin-*O*-Deethylation	
Pretreatment*	nmol/min/mg†	% Control	nmol/min/mg	% Control
Corn oil	0.02 ± 0.00‡	100	0.02 ± 0.00‡	100
Firemaster BP6	0.12 ± 0.04	600	0.39 ± 0.18	1950
Aroclor 1242	0.18 ± 0.05	900	0.60 ± 0.37	3000
Aroclor 1254	ND‡	—	0.36 ± 0.10	1800

*Fish were injected intraperitoneally with the compounds (150 mg/kg) dissolved in corn oil (1 ml/kg) and were sacrificed 4 days later.
†Activity expressed as nmol product formed per min per mg microsomal protein.
‡Values are Mean ± S.E. (n = 4–6); ND = not determined.

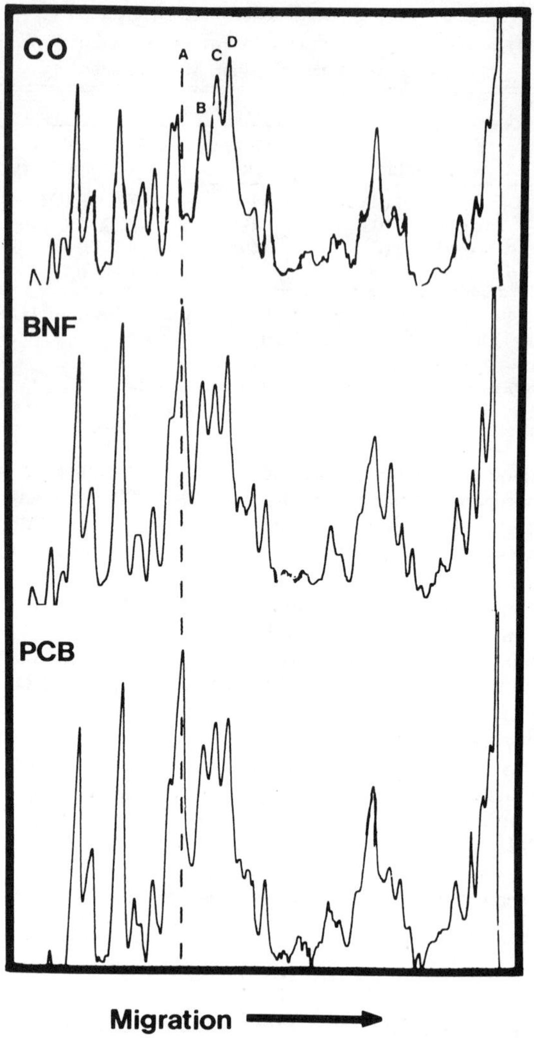

FIGURE 3. Densitometric scans of electrophoretograms of trout hepatic microsomes. CO, corn oil-microsomes; BNF, microsomes from fish treated with β-naphthoflavone (100 mg/kg): PCB, microsomes from fish treated with Aroclor 1254 (150 mg/kg). 90 μg of microsomal protein were applied to gel in all cases. Staining was with Coomassie Blue R-250. Vertical dashed line is at 57,000 dalton, Bands A (57,000), B (51,000), C (48,000) and D (45,000) all appeared to exhibit peroxidase activity.

Discussion

It is apparent that trout hepatic microsomal monooxygenase activities remained elevated longer after a single injection of Aroclor 1254 than after Firemaster BP6. Several explanations for this effect are plausible. It is possible that the constituents of

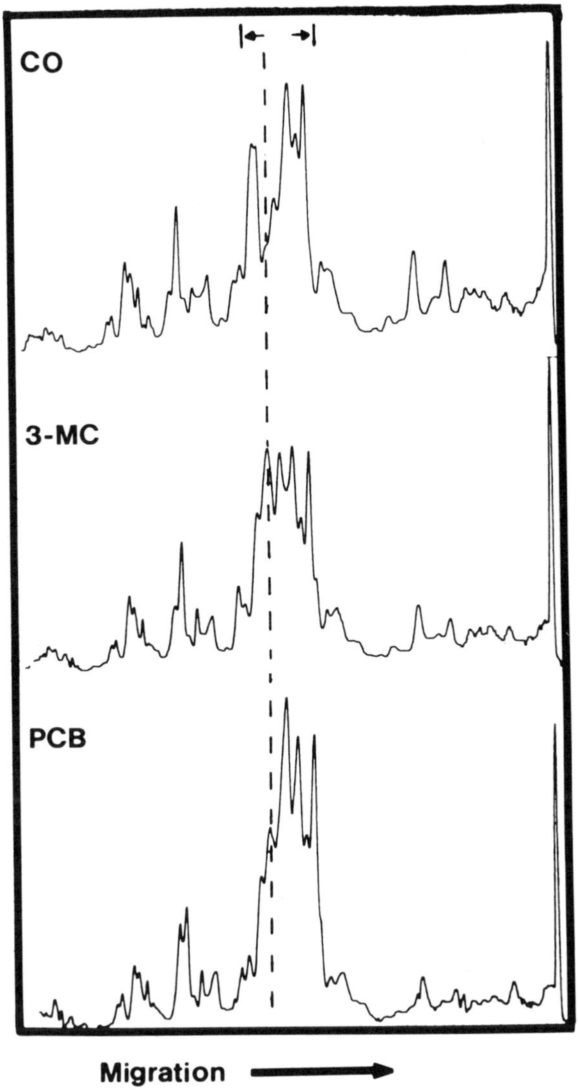

FIGURE 4. Densitometric scans of electrophoretograms of male rat hepatic microsomes. CO, corn oil-microsomes; 3-MC, microsomes from rats treated with 3-methylcholanthrene (4 daily doses, 20 mg/kg); PCB, microsomes from rats treated with Aroclor 1254 (150 mg/kg). 45 μg of microsomal protein was applied to gel in all cases. Staining was with Coomassie Blue R-250. Vertical dashed line is at 53,000 daltons. Several bands in the marked ⊢□⊣ region appeared to exhibit peroxidase activity.

Aroclor 1254 (either polychlorinated biphenyls or impurities such as chlorinated dibenzofurans) are metabolized slower and are therefore more persistent; or they are more potent as inducers than the components of Firemaster BP6. Alternatively, it may be that polychlorinated biphenyls and polybrominated biphenyls may stimulate monooxygenation by different mechanisms.

It is of note that neither Aroclor 1254 nor Firemaster BP6 increased ethylmorphine-*N*-demethylation. This enzymic activity is greatly stimulated by polyhalogenated biphenyl mixtures in rodents[10,11,16–18] and is associated with cytochrome P-450 (phenobarbital-inducible) rather than cytochrome P_1-450 (3-methylcholanthrene-inducible). Other reports[33,41,51] have also indicated the refractive nature of fish to induction of monooxygenation by phenobarbital-like inducers (e.g. phenobarbital, DDT, phenylbutazone).

Initial studies suggested that polychlorinated biphenyls were "mixed" inducers of drug metabolism of rodents.[12,52] However, the elegant studies of Poland and Grover[53] and Goldstein *et al.*[54] have illustrated that the type of induction is dependent upon the chlorine substitution pattern. Biphenyl isomers chlorinated in the 2, 6, 2′, or 6′ positions are non-coplanar and induce cytochrome P-450, whereas isomers lacking these substituents are cytochrome P_1-450 inducers.

Preliminary studies in our laboratories have suggested that rainbow trout hepatic microsomal monooxygenase activity can be induced by planar polychlorinated biphenyl isomers (e.g. 3,3′,4,4′-tetrachlorobiphenyl) but not by nonplanar isomers (e.g. 2,2′,4,4′-tetrachlorobiphenyl or 2,2′,4,4′,5,5′-hexachlorobiphenyl). In addition to stimulating monooxygenase activity the planar isomers appeared to increase the amount of the hemoprotein found at 57,000 daltons using SDS-PAGE (Elcombe, Lech, and Franklin, unpublished observations).

Although the polychlorinated and polybrominated biphenyls induce monooxygenation in the rainbow trout they only induce hemoprotein(s) P-450, which catalytically appears to be similar to rat cytochrome P_1-450. However, despite the catalytic similarity, polychlorinated biphenyl- or β-naphthoflavone-induced trout hemoprotein(s) P-450 is not indentical to rat cytochrome P_1-450. The absorbance maxima of the trout microsomal carboxyferrocytochrome complex did not exhibit a hyposochromic shift after induction with polychlorinated biphenyls or β-naphthoflavone. Also the ethylisocyanide-ferrocytochrome complex was effectively unchanged after induction.[38] Finally, comparative SDS-PAGE (FIGURES 3 and 4) demonstrated marked differences in polypeptide patterns of trout and rat microsomes in the 40,000–60,000 dalton regions. The polycyclic hydrocarbon- or polychlorinated biphenyl-induced hemoprotein(s) P-450 of the trout has a molecular weight of 57,000 as opposed to 53,000 in the rat (FIGURES 3 and 4). Of interest if that the polychlorinated biphenyls and β-naphthoflavone also appear to give rise to alterations in the relative populations of other peroxidase staining hands (FIGURE 3, Bands B, C, and D).

It is apparent that inducing agents such as the polyhalogenated biphenyls and polycyclic aromatic hydrocarbons may alter the subpopulations of hepatic microsomal hemoproteins P-450 of rainbow trout. These alterations may lead to changes in the metabolism and toxicity of other chemical agents. Several reports[34,39,40] have demonstrated *in vivo* changes in the metabolic profile of chemicals due to induction of monooxygenation in rainbow trout. Furthermore, it has been shown that treatment of rainbow trout by Aroclor 1254 protects against the hepatocarcinogenicity of aflatoxin B_1.[55] This phenomenon appears to be due to alterations in aflatoxin B_1 metabolism after Aroclor 1254 treatment, because the substitution of Aroclor 1254-induced trout microsomes (as postmitochondrial supernatant) for control trout microsomes resulted in decreased mutagenicity of aflatoxin B_1 in an Ames test.[56] Whether this decrease in aflatoxin B_1 carcinogenicity and mutagenicity is due to polychlorinated biphenyl-induced alterations in mixed function oxidation, epoxide hydratase, glutathione-*S*-epoxide transferase, or other conjugating enzymes remains to be elucidated.

References

1. Risebrough, R. W., P. Rieche, D. B. Peakall, S. G. Herman & M. N. Kirven. 1968. Polychlorinated biphenyls in the global ecosystem. Nature **220:** 1098–1102.
2. Koeman, J. H., M. C. Ten Noever De Brauw & R. H. De Vos. 1969. Chlorinated Biphenyls in fish, mussels and birds from the river Rhine and the Netherlands coastal area. Nature **221:** 1126–1128.
3. Karatsune, M., T. Yoshimura, J. Matsuzaka & A. Yamaguchi. 1972. Epidemiologic study on Yusho, a poisoning caused by ingestion of rise oil contaminated with a commercial brand of polychlorinated biphenyls. Environ. Health Perspect. **1:** 119–128.
4. Yobs, A. R. 1972. Levels of polychlorinated biphenyl in adipose tissue of the general population of the nation. Environ. Health Perspect. **1:** 79–81.
5. Fishbein, L. 1974. Toxicity of chlorinated biphenyls. Ann. Rev. Pharmacol. **14:** 139–155.
6. Kay, K. 1977. Polybrominated biphenyls (PBB) environmental contamination in Michigan, 1973–1976. Environ. Res. **13:** 74–93.
7. Panel on Hazardous Trace Substances. 1972. PCBs-environmental impact. Environ. Res. **5:** 249–362.
8. Zitko, V. 1977. The accumulation of polybrominated biphenyls by fish. Bull. Environ. Contam. Toxicol. **17:** 285–292.
9. Guiney, P. D., R. E. Peterson, M. J. Melancon & J. J. Lech. 1977. The distribution and elimination of 2,5,2′,5′-[^{14}C] tetrachlorobiphenyl in rainbow trout (*Salmo gairdneri*). Toxicol. Appl. Pharmacol. **39:** 329–338.
10. Bickers, D. R., L. C. Harber, A. Kappas & A. P. Alvares. 1972. Polychlorinated biphenyls: Comparative effects of high and low chlorine containing Aroclors on hepatic mixed function oxidase. Res. Commun. Chem. Pathol. Pharmacol. **3:** 505–512.
11. Litterst, C. L., T. M. Farber, A. M. Baker & E. J. Van Loon. 1972. Effect of polychlorinated biphenyl on hepatic microsomal enzymes in the rat. Toxicol. Appl. Pharmacol. **23:** 112–122.
12. Alvares, A. P., D. R. Bickers & A. Kappas. 1973. Polychlorinated biphenyls: A new type of inducer of cytochrome P-448 in the liver. Proc. Nat. Acad. Sci. USA. **70:** 1321–1325.
13. Ecobichon, D. J. & A. M. Comeau. 1974. Comparative effects of commercial Aroclors on rat liver enzyme activities. Chem. Biol. Interact. **9:** 341–350.
14. Bickers, D. R., J. Eiseman, A. Kappas & A. P. Alvares. 1975. Microscope immersion oils: effects on skin application on cutaneous and hepatic drug-metabolizing enzymes. Biochem. Pharmacol. **24:** 779–783.
15. Goldstein, J. A., P. Hickman, V. W. Burse & H. Bergman. 1975. A comparative study of two polychlorinated biphenyl mixtures (Aroclors 1242 and 1016) containing 42% chlorine on induction of hepatic porphyria and drug metabolizing enzymes. Toxicol. Appl. Pharmacol. **32:** 461–473.
16. Dent, J. G., K. J. Netter & J. E. Gibson. 1976. Effects of chronic administration of polybrominated biphenyls on parameters associated with hepatic drug metabolism. Res. Commun. Chem. Pathol. Pharmacol. **13:** 75–82.
17. Dent, J. G., K. J. Netter & J. E. Gibson. 1976. The induction of hepatic microsomal metabolism in rats following acute administration of a mixture of polybrominated biphenyls. Toxicol. Appl. Pharmacol. **38:** 237–249.
18. Dent, J. G., U. Roes, K. J. Netter & J. E. Gibson. 1977. Stimulation of hepatic microsomal metabolism in mice by a mixture of polybrominated biphenyls. J. Toxicol. Environ. Health **3:** 651–661.
19. Dent, J. G. 1978. Characteristics of cytochrome P-450 mixed function oxidase enzymes following treatment with PBBs. Environ. Health Perspect. **23:** 301–308.

20. COMAI, K. & J. L. GAYLOR. 1973. Existence and separation of three forms of cytochrome P-450 from rat liver microsomes. J. Biol. Chem. **248:** 4947–4955.
21. WELTON, A. F. & S. D. AUST. 1974. Multiplicity of cytochrome P-450 hemoproteins in rat liver microsomes. Biochem. Biophys. Res. Commun. **56:** 898–906.
22. HAUGEN, D. A., T. A. VAN DER HOEVEN & M. J. COON. 1975. Purified liver microsomal cytochrome P-450. J. Biol. Chem. **250:** 3567–3570.
23. HAUGEN, D. A., M. J. COON & D. W. NEBERT. 1976. Induction of multiple forms of mouse liver cytochrome P-450. J. Biol. Chem. **251:** 1817–1827.
24. THOMAS, P. E., A. Y. H. LU, S. B. WEST, D. RYAN, G. T. MIWA & W. LEVIN. 1977. Accessibility of cytochrome P-450 in microsomal membranes: Inhibition of metabolism by antibodies to cytochrome P-450. Mol. Pharmacol. **13:** 819–831.
25. BIRNBAUM, L. S., M. B. BAIRD & H. R. MASSIE. 1976. Pregnenolone-16α-carbonitrile-inducible cytochrome P-450 in rat liver. Res. Commun. Chem. Pathol. Pharmacol. **15:** 553–562.
26. DICKINS, M., J. W. BRIDGES, C. R. ELCOMBE & K. J. NETTER. 1978. A novel hemoprotein induced by isosafrole pretreatment in the rat. Biochem. Biophys. Res. Commun. **80:** 89–96.
27. NEBERT, D. W. & J. S. FELTON. 1976. Importance of genetic factors influencing the metabolism of foreign compounds. Fed. Proc. **35:** 1133–1141.
28. GILLETTE, J. R. 1976. Environmental factors in drug metabolism. Fed. Proc. **35:** 1142–1147.
29. ADAMSON, R. H. 1967. Drug metabolism in marine vertebrates. Fed. Proc. **26:** 1047–1054.
30. BUHLER, D. R. & M. E. RASMUSSON. 1968. The oxidation of drugs by fishes. Comp. Biochem. Physiol. **25:** 233–239.
31. DEWAIDE, J. H. & P. TH. HENDERSON. 1968. Hepatic-*N*-demethylation of aminopyrine in rat and trout. Biochem. Pharmacol. **17:** 1901–1907.
32. POHL, R. J., J. R. BEND, A. M. GUARINO & J. R. FOUTS. 1974. Hepatic microsomal mixed-function oxidase activity of several marine species from coastal Maine. Drug. Metab. Disp. **2:** 545–555.
33. BEND, J. R., R. J. POHL & J. R. FOUTS. 1973. Further studies of the microsomal mixed-function oxidase system of the little skate, *Raja erinacea,* including its response to some xenobiotics. Bull. Mt. Desert Island. Biol. Lab. **13:** 9–13.
34. STATHAM, C. N., C. R. ELCOMBE, S. P. SZYJKA & J. J. LECH. 1978. Effect of polycyclic aromatic hydrocarbons on hepatic microsomal enzymes and disposition of methylnaphthalene in rainbow trout *in vivo.* Xenobiotica **8:** 65–71.
35. PAYNE, J. F. & W. R. PENROSE. 1975. Induction of arylhydrocarbon (benzo[a]pyrene) hydroxylase in fish by petroleum. Bull. Environ. Contam. Toxicol. **14:** 112–116.
36. HILL, D. W., E. HEJTMANCIK & B. J. CAMP. 1976. Induction of hepatic microsomal enzymes by Aroclor 1254 in *Ictalurus punctatus* (Channel Catfish). Bull. Environ. Contam. Toxicol. **16:** 495–502.
37. LIDMAN, U., L. FÖRLIN, O. MOLANDER & G. AXELSON. 1976. Induction of the drug metabolizing system in rainbow trout (*Salmo gairdneri*) liver by polychlorinated biphenyls (PCBs). Acta Pharmacol. Toxicol. **39:** 262–272.
38. ELCOMBE, C. R. & J. J. LECH. 1978. Induction of monooxygenation in rainbow trout by polybrominated biphenyls: A comparative study. Environ. Health Perspect. **23:** 309–314.
39. ELCOMBE, C. R., M. J. MELANCON & J. J. LECH. 1978. Induction of cytochrome(s) P-450 mediated monooxygenation in rainbow trout *in vivo* and *in vitro.* Presented at 7th Int. Congress Pharmacology, Paris, France.
40. MELANCON, M. J. & J. J. LECH. 1978. Uptake, biotransformation, disposition and elimination of 2-methylnaphthelene and naphthelene in several fish species. Proc. 2nd Annual Symp. on Aquatic toxicology. ASTM, Philadelphia. In press.
41. ELCOMBE, C. R. & J. J. LECH. 1978. Induction and characterization of cytochrome(s) P-450-mediated monooxygenation in rainbow trout. Toxicol. Appl. Pharmacol. In press.
42. ULLRICH, V. & P. WEBER. 1972. The *O*-dealkylation of 7-ethoxycoumarin by liver microsomes. Hoppe-Seyler's Z. Physiol. Chem. **353:** 1171–1177.

43. HANSEN, A. R. & J. R. FOUTS. 1972. Some problems in Michaelis-Menten kinetic analysis of benzpyrene hydroxylase in hepatic microsomes from polycyclic hydrocarbon-pretreated animals. Chem. Biol. Interact. **5:** 167–182.
44. STATHAM, C. N., S. P. SZYJKA, L. A. MENAHAN & J. J. LECH. 1977. Fractionation and subcellular localization of marker enzymes in rainbow trout liver. Biochem. Pharmacol. **26:** 1395–1400.
45. BURKE, M. A. & R. T. MAYER. 1974. Ethoxyresorufin: Direct flurometric assay of a microsomal deethylation which is preferentially inducible by 3-methylcholanthrene. Drug Metab. Disp. **2:** 583–588.
46. ANDERS, M. W. & G. J. MANNERING. 1966. Kinetics of the inhibition of the *N*-demethylations of ethylmorphine by 2-diethylamino-ethyl 2,2-diphenyl valerate HCl (SKF 525-A) and related compounds. Mol. Pharmacol. **2:** 319–327.
47. ROSS, E. & G. SCHATZ. 1973. Assay of protein in the presence of high concentrations of sulfhydryl compounds. Anal. Biochem. **54:** 304–306.
48. ESTABROOK, R. W., J. PETERSON, J. BARON & A. G. HILDEBRANDT. 1972. The spectrophotometric measurement of turbid suspensions of cytochromes associated with drug metabolism. *In* Methods in Pharmacology. C. F. Chignell, Ed. Vol 2: 303–350. Appleton-Century-Crofts. New York.
49. LAEMMLI, U. K. & M. FAVRE. 1973. Maturation of the head of bacteriophage T4. I. DNA Packaging Events. J. Mol. Biol. **80:** 575–599.
50. THOMAS, P. E., D. RYAN & W. LEVIN. 1976. An improved staining procedure for the detection of the peroxidase activity of cytochrome P-450 on sodium dodecyl sulphate polyacrylamide gels. Anal. Biochem. **75:** 168–176.
51. ADDISON, R. F., M. E. ZINCK & D. E. WILLIS. 1977. Mixed function oxidase enzymes of trout (*Salvelinus fontinalis*) liver: absence of induction following feeding of p,p′-DDT or p,p′-DDE. Comp. Biochem. Physiol. **57C:** 39–43.
52. ALVARES, A. P. & P. SIEKEVITZ. 1973. Gel electrophoresis of partially purified cytochromes P-450 from liver microsomes of variously-treated rats. Biochem. Biophys. Res. Commun. **54:** 923–929.
53. POLAND, A. & E. GLOVER. 1977. Chlorinated biphenyl induction of aryl hydrocarbon hydroxylase activity: a study of the structure-activity relationship. Mol. Pharmacol. **13:** 924–938.
54. GOLDSTEIN, J. A., P. HICKMAN, H. BERGMAN, J. D. MCKINNEY & M. P. WALKER. 1977. Separation of pure polychlorinated biphenyl isomers into two types of inducers on the basis of induction of cytochrome P-450 or P-448. Chem. Biol. Interact. **17:** 69–87.
55. HENDRICKS, J. D., T. P. PUTNAM, D. B. BILLS & R. O. SINNHUBER. 1977. Inhibitory effect of a polychlorinated biphenyl (Aroclor 1254) on aflatoxin B_1 carcinogenesis on rainbow trout (Salmo gairdneri). J. Nat. Cancer Inst. **59:** 1545–1551.
56. STOTT, W. T. & R. O. SINNHUBER. 1978. Trout hepatic enzyme activation of aflatoxin B_1 in a mutagen assay system and the inhibitory effects of PCB. Bull. Environ. Contam. Toxicol. **19:** 35–41.

STUDIES OF THE MECHANISMS OF TOXICITY OF 2,3,7,8-TETRACHLORODIBENZO-*p*-DIOXIN (TCDD)

Robert A. Neal, Patrick W. Beatty, and Thomas A. Gasiewicz

Center in Environmental Toxicology
Department of Biochemistry
Vanderbilt University School of Medicine
Nashville, Tennessee, 37232

The major interest of our laboratory relative to TCDD has been the biochemical mechanism or mechanisms of acute lethality of this compound. As has been noted by other speakers at this conference, the major feature of a single lethal dose of TCDD across all animal species is a progressive weight loss and general debilitation of the animal with death occurring in a few weeks or a few months. Gross and histopathology have not, to date, provided information that has been particularly useful in focusing biochemical investigations. For example thymic involution is a consistent finding in all animal species. However, numerous investigations have yet to provide information as to the mechanism of thymic involution. Also the major pathologic effect of TCDD in mice and rats is degeneration of the liver. However, the degree of injury to the liver does not appear to be sufficient to be the cause of death. In addition, in the guinea pig, the species most sensitive to TCDD, little damage to the liver is seen.

The extreme lethality of TCDD, particularly in the guinea pig, suggests that TCDD is probably affecting some fundamental process in animal cells leading to a general disfunction of those cells. Studies of the biochemical effects of TCDD intoxication have focused on some of these fundamental processes. Thus, Greig and coworkers[1] have shown that liver DNA synthesis stimulated by partial hepatectomy of rats does not appear to be inhibited by TCDD. Also Lucier *et al.*[2] have shown that state 3 and state 4 respiration of liver mitochondria, the process by which ATP is synthesized, and ADP:0 are not affected in TCDD-treated rats. In our laboratory we have also examined state 3 and state 4 respiration of liver mitochondria from terminally ill rats treated with TCDD and find state 3 and state 4 respiration is progressing normally. Work by Cunningham and Williams[3] indicates that protein and lipid synthesis also appear to be continuing normally in TCDD-treated rats. Work by Vos *et al.*[4-6] has indicated that TCDD has a marked effect on the immunocompetence of young rodents. However, work by Grieg *et al.*[7] has shown that specific pathogen-free and germ-free rats are as susceptible to the lethal effects of TCDD as normal rats. Thus the reduced immunocompetence is not likely the cause of the acute lethality of TCDD.

We have carried out a number of studies examining for effects of TCDD on other fundamental biochemical processes within mammalian cells. It must be stated at the outset that these and other studies have not yet provided us with information on which to base a logical hypothesis for the mechanism or mechanisms of the acute lethal effects of TCDD. However, the data that have been forthcoming from the various studies, in conjunction with data from future experiments, may eventually allow investigators in this field to postulate a mechanism for the acute lethal effects of TCDD.

One of the initial studies in our laboratory was based on an article by Jackson[8] who reported that TCDD at a concentration of 0.1 μg/liter in the culture medium caused a significant inhibition of all phases of mitosis of dividing endosperm cells of the African

0077-8923/79/0320-0204 $01.75/0 © 1979, NYAS

Blood Lily. These findings suggested that TCDD could function as an inhibitor of cell division. To test the hypothesis that TCDD could also inhibit mitosis and effect the growth of mammalian cells we examined the effect of TCDD on the growth of several mammalian cells in culture.[9] The cell types used in these studies were HeLa cells, Balb-3T3 cells, SV101 cells, human diploid fibroblasts from human foreskin, and normal human lymphocytes. The results of these studies indicated there was no significant effect of 1×10^{-6}M TCDD on the growth of these cells. These results suggest that TCDD does not have an inhibitory effect on the growth of mammalian cells in culture. It should be noted, however, that Niwa and coworkers[10] have reported some toxicity of TCDD to established hepatic cell lines.

Bostomsky[11] has reported that TCDD increased fourfold the biliary excretion of thyroxine (T_4) as its glucuronide. This resulted in a reduction of serum T_4 concentration to one-half normal. TCDD also elevated serum thyrotropin concentrations, produced goiters, and increased thyroid ^{131}I uptake. These data coupled with the fact that some of the symptomatology of TCDD toxicity resembles the hypothroid state prompted us to examine for an effect of TCDD on thyroid hormone function. TABLE 1 shows the results of an experiment in which we examined the effect of daily injections of triiodothyronine (T_3), the active form of the thyroid hormones, on the

TABLE 1

EFFECT OF 3,5,3′-TRIIODOTHYRONINE (T_3) ON TCDD-MEDIATED TOXICITY IN MALE MICE

TCDD μg/kg	T_3 μg/kg/day	Time for 50% Mortality (days)	Mean time* to Death (days)	Mortality at 21 Days (%)
200	0	12	11.0 ± 2.8 (10)	100.0
200	50	15	13.1 ± 2.9 (7)	87.5
200	100	15	15.0 ± 3.2 (7)	77.8
200	200	16	16.3 ± 2.4 (7)	70.0

*The number of animals is indicated in parentheses.

survival of mice administered a lethal dose of TCDD. As can be seen there was some increase in survival time; however, all animals eventually died. Subsequent experiments suggested the increased survival times were the result of increased food intake in the T_3 treated animals.

We also examined the ability of TCDD to interfere with the metamorphosis of *Rana catesbeiana* tadpoles.[12] If TCDD inhibits thyroid function by a basic mechanism such as competition for tissue receptor sites, then administration of TCDD to bullfrog tadpoles should interfere with T_3- or T_4-dependent processes such as metamorphosis. In the experiment with *Rana catesbeiana* tadpoles, there was no mortality attributable to TCDD at any of the dosages examined (25–1000 μg/kg, i.p.). A mortality rate of approximately 20% was seen in all groups including controls. Those deaths that did occur may be attributed to the injection procedure or to infection by polyhedral cytoplasmic deoxyribovirus (tadpole edema virus). All surviving tadpoles were observed to successfully complete metamorphosis with no observable morphologic abnormalities. In addition, histopathologic examination of the liver, heart, kidney, lung, and reproductive organs of these animals shortly after the completion of metamorphosis revealed no lesions other than occasional parasites. TCDD is a potent embryotoxic agent in rats and mice. Therefore, it is of interest that TCDD appears to have no effect on amphibian metamorphosis, a somewhat analogous developmental process.

In light of these results, the toxicity of TCDD in adult bullfrogs was also examined.[12] There are biochemical and physiologic differences in the larval as compared to the adult forms of *Rana catesbeiana*. It was reasoned, therefore, if TCDD was toxic to the adult bullfrog but not the larval form, that this animal might be a good model for studying the mechanism of toxicity of TCDD. Adult bullfrogs were administered TCDD at levels of 0, 50, 100, 250, and 500 μg/kg, i.p. in a single dose and observed for a period of 35 days. During this period, there was no mortality in any treatment group. The only observable difference was somewhat lessened food intake during the initial part of the experiment in the frogs receiving 500 μg/kg TCDD. However, at the termination of the experiment, the food consumption of the high-dose animals was not different from that of the controls. Histopathologic examination of liver, kidney, lung, heart, and gonads of these animals revealed no significant lesions at any of the doses of TCDD administered.

The observation that T_3 has no direct protective effect other than the stimulation of food intake in adult male mice receiving TCDD indicates that there is no causal role for a hypothyroid state in TCDD-mediated toxicity. The experiments which showed that TCDD did not adversely affect metamorphosis in *Rana catesbeiana* tadpoles strongly suggests that TCDD does not interfere with the binding of thyroid hormones to tissue receptors.

Although the state 3 and state 4 respiration of hepatic mitochondria are not affected by TCDD, we postulated that TCDD may be acting to change the oxidation-reduction state of the cell or to decrease the level of intracellular ATP by an increased activity of various ATPases or flavoenzymes involved in oxidation of reduced pyridine nucleotides. We therefore examined for the oxidation-reduction state of the liver using the freeze-clamp method. We also examined for ATP levels by this same technique.

In the initial experiments we examined the content of pyridine nucleotides as a function of time in the livers of male rats which had received 90 μg/kg TCDD or pair-fed controls. The results are presented in TABLE 2. TCDD had no significant effect on the liver content of reduced pyridine nucleotides during a 15-day time period when compared to pair-fed controls. The only exception was a reduction in NADH (36%) at day 15 after TCDD administration. Similarly there was no consistent effect of TCDD treatment upon liver NADP concentrations. However, beginning with day 3 after administration there was a consistently lower NAD content in livers from TCDD-treated rats. Although pair-fed rats showed a trend toward decreased NAD content compared to day one values, the TCDD-treated rats at each time point had even lower values. It must be noted that values for total tissue content of pyridine nucleotides has limited significance in that both free and protein bound nucleotides are measured, thereby giving no indication of the free pool sizes. Additionally there are contributions from the blood content of the tissue assayed since the organ must be frozen immediately upon excision.

The intrinsic errors in measuring pyridine nucleotide levels can be circumvented by measurement of the redox state of the tissue in question. The measurement of the concentrations of oxidized and reduced substrates of enzyme systems considered to be at equilibrium can be used to determine the redox state of their respective subcellular compartments. Accordingly, the cytoplasmic NAD/NADH ratio of cytosol and mitochondrial compartments may be determined from measurements of the substrates involved in the lactate dehydrogenase and β-hydroxybutyrate dehydrogenase systems, respectively.

The results of the determination of the NAD/NADH ratio and ATP concentra-

TABLE 2

EFFECT ON TCDD ON PYRIDINE NUCLEOTIDE CONTENT IN RAT LIVER*

Days Post-TCDD	NAD		NADP		NADH		NADPH	
	Control	TCDD	Control	TCDD	Control	TCDD	Control	TCDD
1 (30 hrs)	.676	.537	.080	.094	.067	.045	.381	.363
	± .089	± .057	± .018	± .012	± .023	± .006	± .091	± .220
3	.522	.276†	.097	.088	.069	.057	.372	.625
	± .159	± .043	± .044	± .017	± .024	± .034	± .188	± .242
7	.604	.344†	.114	.087	.057	.038	.294	.215
	± .072	± .110	± .012	± .023	± .014	± .004	± .050	± .150
11	.349	.281†	.113	.057†	.059	.064	.539	.624
	± .036	± .093	± .012	± .020	± .007	± .041	± .178	± .405
15	.491	.282†	.130	.087	.067	.043†	.329	.292
	± .110	± .033	± .023	± .033	± .003	± .013	± .255	± .214

*Pyridine nucleotide content is expressed as μmoles/g liver (frozen weight) and are the mean ± S.D. of values from four rats. TCDD was administered i.p. 90 μg/kg in olive oil. Control rats were pair fed against TCDD-treated rats.

†$p < .05$ versus pair-fed controls; Student's t-test.

tion of livers of rats treated with 50 μg/kg TCDD and both pair-fed and *ad lib* controls are contained in TABLE 3. The rats were sacrificed 14 days after administration of TCDD. As can be seen there is a significant difference between TCDD-treated rats and pair-fed controls in all parameters measured. However, in all cases the values in TCDD-treated rats are very nearly equivalent to those in the control rats fed *ad libitum*. An examination of TABLE 3 indicates that the cytosol and mitochondria compartments of the liver cells of TCDD-treated rats are in a more oxidized state relative to pair-fed control rats. However, the observation that the values in TCDD-treated rats are similar to those of untreated, fed rats would seem to contradict the hypothesis that there is a pathologic alteration of the cellular energy state in TCDD-treated rats.

One of the most striking and consistent effects of the administration of TCDD to laboratory animals is depletion of lymphoid tissues and especially the thymus. Glucocorticoids are known to exert dramatic catabolic effects on lymphoid tissues including the thymus, spleen, and lymph nodes.[13] Thus, TCDD might exert its toxic effects on lymphoid tissues by causing increased glucocorticoid production. Other effects of overproduction of glucocorticoids that mimic the symptoms of TCDD

TABLE 3

EFFECT OF TCDD ON RAT LIVER ENERGY STATE*

Treatment	NAD/NADH		ATP content†
	Cytosol	Mitochondria	
Ad lib control	625 ± 111	15.2 ± 10	1.00 ± 0.6
Pair-fed control	826 ± 10	3.98 ± 0.3	0.476 ± 0.05
TCDD	419 ± 94‡	13.04 ± 2.4‡	0.944 ± 0.28‡

*Male Sprague-Dawley rats, 150 gm, were injected i.p. with 50 μg/kg TCDD 14 days prior to sacrifice. Values shown are the mean ± S.D. of values from four rats.

†μmoles ATP/gm liver (frozen weight).

‡$p < .025$ versus pair-fed controls; Student's t-test.

toxicity to varying degrees are muscle atrophy due to protein catabolism, peptic ulceration, psychopathologic changes similar to those seen in some humans with chloracne,[14] edema due to the mineralocorticoid activity of glucocorticoids, and mobilization of fatty acids from adipose tissue.[13]

There appear to be two possible mechanisms by which TCDD could cause an increase in glucocorticoid activity. The biosynthesis of adrenal steroids involves several hydroxylation reactions which are carried out by adrenal steroid hydroxylases. These hydroxylases are cytochrome P-450-containing enzymes similar to the mixed function oxidase (MFO) enzyme system in liver. The liver also contains steroid hydroxylases which are responsible for the metabolism of various steroids. These enzymes can be induced by compounds that induce hepatic MFO enzyme activities. Although the adrenal steroid hydroxylases have not been reported to be induced by common MFO inducers, TCDD has been reported in many animal species to be a powerful inducer of hepatic aryl hydrocarbon hydroxylase. Thus, TCDD might be capable of direct stimulation of the activity of the adrenal steroid hydroxylases even though this property has not been reported for other compounds.

Such a mechanism operating at the level of cholesterol 20 α-hydroxylase, the rate-limiting step of adrenal steroid biosynthesis, would effectively remove steroid synthesis from feedback control by means of ACTH. Alternatively induction of the adrenal 17 α-hydroxylase could lead to the formation of large amounts of cortisol. This would be of importance in laboratory animals such as the rat in which corticosterone is the predominant glucocorticoid, since cortisol has greater physiologic potency than corticosterone.

The second hypothesis is that TCDD could mimic the effects of glucocorticoids by binding to cytosol or nuclear glucocorticoid receptors causing an irreversible or long-acting stimulation of glucocorticoid-like effects in the tissues in which this mimicry occurred. This mechanism is attractive for several reasons. There is precedent for specific binding of TCDD to cytosol proteins with the discovery of the cytosol receptor for AHH induction in rat and mouse liver.[15] Further, it has been demonstrated[16] that *in vivo* administration of carcinogenic polycyclic aromatic hydrocarbons to rats can modify binding of the steroid hormone—cytosol receptor complex to nuclear binding sites for this complex.

In addition, there are other possible mechanisms by which TCDD could cause increased glucocorticoid activity, which are less easily explainable. These include decreased catabolism of steroid hormones, blockage of pituitary or hypothalamic glucocorticoid receptors leading to loss of feedback control of ACTH production, and hypersensitization of certain tissues to endogenous levels of circulating glucocorticoids.

In order to investigate this possible modulation of glucocorticoid function and metabolism, the effect of TCDD on blood glucocorticoid levels and toxicity in adrenalectomized rats was examined. TCDD was further examined as a possible competitor for cytosol glucocorticoid receptor sites in rat liver. Finally, the ability of TCDD to cause a functional interference with the peripheral actions of glucocorticoid hormones was examined.

Measurement of the fluorescence of a steroid extract in ethanolic sulfuric acid reveals the presence of 11-hydroxycorticosteroids which in the rat is represented primarily by corticosterone, the major glucorticoid in that species. An examination of plasma "F" (fluorescence) values expressed as μg of corticosterone per 100 ml of plasma (TABLE 4) reveals that an approximate LD_{50} dose of TCDD (50 μg/kg) slightly depresses the level of circulating glucocorticoids during the first four days after administration. However, at 7 and 14 days after administration of TCDD the plasma F value of TCDD-treated animals has risen to a value nearly two and one-half times that of pair-fed controls. It is interesting that gross signs of TCDD toxicity in rats are usually observed to begin with anorexia at about four days after administration and outward symptoms appear within two to four days later. It is at this time that plasma corticosteroid levels exhibited a marked increase.

It appears reasonable to posit that if TCDD exerts a toxic effect mediated by adrenal hyperfunction, then prior adrenalectomy should result in a substantial protective effect against subsequent TCDD administration. The 20 day LD_{50} of TCDD in adult male rats is 50 μg/kg. However, administration of TCDD to adrenalectomized rats at doses of 10, 20, 40, and 80 μg/kg resulted in 100% mortality within six days in all TCDD-treated, adrenalectomized rats. The lack of mortality in vehicle-injected, adrenalectomized controls indicates that death was not due to the trauma of injection. Sham-operated rats receiving TCDD exhibited typical dose-related mortality, with the average time required for death ranging from 15 to 18 days depending on the dose of TCDD.

FIGURE 1 graphically presents the results of a competitive binding assay in which the ability of TCDD to compete with dexamethasone, a potent synthetic glucocorticoid, is examined. The upper line on the graph indicates that TCDD is unable to displace ^{3}H-labeled dexamethasone from the cytosol receptor from normal rats even when TCDD is present in approximately a 200-fold molar excess to ^{3}H-dexamethasone. The lower curve indicates the displacement of ^{3}H-dexamethasone by increasing concentrations of unlabeled dexamethasone.

TABLE 4

PLASMA "F" VALUES OF RATS RECEIVING TCDD*

	Time After TCDD Injection (Days)					
Control†	0.5	1	2	4	7§	14§
18.68 ± 6.9	24.80 ± 0.89	17.14 ± 5.22	11.53 ± 4.06	11.84 ± 3.35	50.94 ± 23.7	53.35 ± 11.2

*Adult male Sprague-Dawley rats, 125–150 gm, received 50 μg/kg TCDD in olive oil vehicle and were pair fed to controls receiving olive oil only. Plasma fluorescence due to 11-hydroxycorticosteroids is expressed as the mean ± S.D. in μg equivalents of corticosterone/100 ml plasma.

†The control value reported is the mean ± S.D. of control values determined at each time point. Experiments values are the means ± S.D. of determinations from four rats and were normalized by comparison of control values at each time point to the mean control value.

‡$p < .05$ versus control values; Student's t-test.

§Mean ± S.D. of determinations from three rats.

It is well established that the activity of rat liver tyrosine aminotransferase (TAT) is markedly induced as a specific response to the administration of glucocorticoids.[17] Therefore, the induction of TAT activity in rat liver cytosol should provide a parameter by which the ability of TCDD to interfere with the physiologic response of an animal to glucocorticoids could be estimated. TABLE 5 illustrates the ability of a challenge dose of dexamethasone (50 μg/kg) to cause induction of TAT activity within 5 hours in the livers of TCDD-treated (50 μg/kg) and pair-fed control rats. The induction of TAT activity was studied as a function of time after TCDD administration in order to discern any possible progressive alteration of the response. From TABLE 5 it can be seen that TCDD pretreatment has no significant inhibitory effect on TAT induction by dexamethasone. Four days after TCDD administration, both

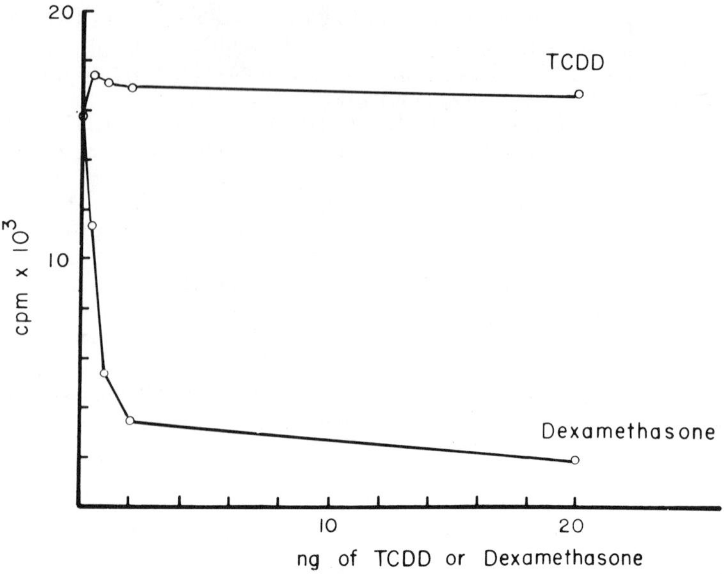

FIGURE 1. Effect of TCDD on dexamethasone binding to rat liver cytosol glucocorticoid receptor. The curves indicate the relative abilities of TCDD or unlabeled dexamethasone to displace ^{3}H-dexamethasone from the cytosol glucocorticoid receptor.

treated and pair-fed control rats not only exhibited increased basal levels of TAT activity but responded more strongly to dexamethasone challenge. In this instance TCDD-treated rats showed a somewhat greater response to dexamethasone than pair-fed controls. The reason for this effect is unexplained at present but similarly higher basal levels of mitochondrial α-glycerophosphate dehydrogenase in TCDD-

TABLE 5

EFFECT OF TCDD ON DEXAMETHASONE INDUCTION OF TYROSINE AMINOTRANSFERASE*

	nmoles *p*-hydroxyphenylpyruvate/10 min/mg protein		
	Days after TCDD		
Treatment	4	8	12
Pair-fed Control	190 ± 45	86 ± 25	91.7 ± 21
Pair-fed control† + Dexamethasone	1028 ± 206 (541)	252 ± 40 (293)	262 ± 55‡ (286)
TCDD	152 ± 24	152 ± 39	90.5 ± 20
TCDD +† Dexamethasone	1223 ± 296 (805)	353 ± 55 (232)	207 ± 45‡ (229)

*Adult male Sprague-Dawley rats received TCDD (50 μg/kg) in olive oil i.p. or olive oil alone. At the times indicated, the experimental and pair-fed control rats were injected i.p. with dexamethasone (50 μg/kg) or distilled water and were sacrificed 5 hours later. TAT activities are the mean ± S.D. of four individual rats per group. The numbers in parentheses indicate percent induction above the control value at that time point.

†All activities for dexamethasone-induced rats are significantly different from controls $p < .05$; Student's t-test.

‡Mean ± S.D. of three rats.

treated and control rats are seen at this same time point. This may be a transient metabolic response to decreased food intake which has begun to occur in TCDD treated rats (and pair-fed controls) at this time.

From an examination of the time course of plasma corticosterone levels and the effect of adrenalectomy on TCDD mortality it may be concluded that the toxicity of TCDD in rats is not mediated by a process involving adrenal hyperfunction. Administration of TCDD to rats at a near LD_{50} dose (50 $\mu g/kg$) causes no alteration in plasma corticosterone levels until between four and seven days after administration. Prior to this time the rats exhibit an anorexic response, the liver begins to show morphologic and histochemical alterations and the induction of enzymes responsive to TCDD has reached near maximal levels. If the increase in corticosteroid output were the causative agent in TCDD toxicity then one would expect the rise in plasma corticosterone to precede such effects as these. The adrenalectomized rats exhibited 100% mortality after administration of a normally sublethal dose of TCDD (10 $\mu g/kg$). These results suggest that TCDD is capable of imposing an acute stress upon rats relatively soon after its administration, and that rats lacking their adrenal glands are incapable of the normal stress response of increased glucocorticoid output.

The results of the competitive binding studies of TCDD and dexamethasone indicate that TCDD cannot mimic glucocorticoid activity in rat liver by binding to the hepatic cytosol glucocorticoid receptor. This is substantiated by the lack of the ability of TCDD administration to affect basal levels of a glucocorticoid responsive hepatic enzyme, tyrosine aminotransferase.

TCDD has a marked effect on lipid metabolism in all animal species examined. A change in the fatty acid composition of cellular membranes induced by TCDD could interfere with normal cellular processes. Therefore, we compared the fatty acid composition of the livers of normal and TCDD-treated guinea pigs to see if there were differences. The results are shown in TABLE 6. We were unable to detect any major differences. Also, no differences in the fatty acid composition of plasma and adipose tissues were seen.

It has been postulated by various investigators that an interference with the absorption of nutrients from the gut may be responsible for the "wasting syndrome" seen in animals intoxicated with TCDD. The fact that TCDD-treated animals fed orally lose weight whereas pair-fed controls generally do not strongly suggests that some essential nutrients may not be absorbed from the gut or may not cross cell membranes as readily in TCDD-treated animals. In some preliminary experiments we have examined CO_2 production in pair-fed control and TCDD-treated guinea pigs orally administered ^{14}C-glucose, ^{14}C-alanine and ^{14}C-oleate. The TCDD-treated guinea pigs had received a dose of 2 $\mu g/kg$ i.p. 7 days prior to the start of these experiments. At this time they had lost approximately 25% of their original body weight. No significant differences were noted in the ability of the TCDD-treated animals to metabolize these compounds to $^{14}CO_2$ as compared to pair-fed controls. Thus, it appears there is no generalized impairment of the absorption of carbohydrates, fatty acids, or amino acids from the intestine or an apparent decrease in the transfer of these materials into cells in TCDD treated guinea pigs. Also, the enzymes of glycolysis and the tricarboxylic acid cycle also appear to be functioning at a normal rate in the treated animals.

We have also examined for an increase in blood NH_3, evidence for increased superoxide anion levels, changes in cyclic nucleotide metabolism and an interference with the activity of enzymes containing riboflavin coenzymes. The latter were examined because of the structural similarity between the dioxin ring system and the isoalloxazine ring system of riboflavin. None of the studies revealed a significant effect of TCDD on these parameters with the exception that TCDD dramatically increased the activity of the flavin-enzyme DT-diaphorase.[18] However, the induction of this enzyme is probably not directly related to the mechanism by which TCDD exerts its toxicity.

The various experiments that have been carried out to date have not yet provided us with information on which to base a hypothesis as to the biochemical mechanism of the lethality of TCDD. An adverse effect on DNA synthesis, protein synthesis, ATP production, the oxidation-reduction state of cells, the action of thyroid and glucocorticoid hormones, the general absorption of carbohydrates, amino acids and fatty acids from the intestine and transfer through cell membranes, aerobic and anaerobic metabolism of carbohydrates, amino acids and fatty acids, and the fatty acid composition of tissues appear not to be the mechanism by which TCDD is exerting its toxicity.

There is good evidence to support the hypothesis that binding of TCDD to a receptor[15] followed by translocation into the nucleus is the initial event in TCDD-mediated toxicity. It is still not clear, however, which of the biochemical changes that

TABLE 6

FATTY ACID COMPOSITION OF THE LIVERS OF TCDD-TREATED GUINEA PIGS AND PAIR-FED CONTROLS

Fatty Acid	Pair-Fed Controls	+ TCDD*
14:0	0.5 ± 0.2	0.5 ± 0.1
15:0	0.3 ± 0.0	0.4 ± 0.1
16:0	14.2 ± 0.1	14.2 ± 1.7
16:1	0.6 ± 0.1	0.3 ± 0.1
17:0	0.8 ± 0.1	0.7 ± 0.2
18:0	18.1 ± 1.4	18.5 ± 0.7
18:1	9.2 ± 0.4	8.7 ± 0.8
18:2	40.0 ± 0.8	40.1 ± 1.8
18:3	3.8 ± 0.5	3.1 ± 0.3
20:3	0.6 ± 0.1	0.6 ± 0.2
20:4	8.4 ± 0.4	8.7 ± 0.7
20:5	0.2 ± 0.1	0.3 ± 0.2
22:3	0.7 ± 0.0	0.4 ± 0.3
22:4	1.0 ± 0.1	1.3 ± 0.5
22:5	0.8 ± 0.2	0.9 ± 0.3

*Animals were sacrificed 7 days after administration of TCDD.

occur as a result of translocation of TCDD into the nucleus are responsible for the toxic effect of TCDD. It is clear that a coordinate induction of a number of enzymes results from the transfer of TCDD into the nucleus (arylhydrocarbon hydroxylase and other monooxygenases, DT-diaphorase, glucuronyl transferase, aldehyde dehydrogenase, ALA synthetase, ornithine decarboxylase). However, the evidence to date does not indicate that the increased activity of any of these enzymes is responsible for toxic effects of TCDD. Thus, 3-methylcholanthrene binds to the same receptor[15] and induces arylhydrocarbon hydroxylase to the same degree as TCDD but does not produce the same symptoms of toxicity as TCDD even on daily administration of as much as 100 mg/kg for 20 days to rats. Also, TCDD does not consistently induce hepatic arylhydrocarbon hydroxylase or DT-diaphorase in the guinea pig, the most sensitive species to TCDD, in spite of the fact that most of the administered TCDD resides in the liver at 7 days post-administration.

It appears at this time the most reasonable hypothesis concerning the biochemical mechanism of the lethality of TCDD is that it binds to a receptor in the cytosol fraction of mammalian cells, is transferred into the nucleus and increases or decreases the synthesis of a critical protein(s) or enzyme(s). This increase or decrease in critical protein or enzyme levels leads eventually to a generalized dysfunction of a number of different cell types leading eventually to death. What biochemical function or functions are affected by this increase or decrease in protein or enzyme synthesis is unknown at this time.

References

1. Greig, J. B., D. M. Taylor & J. D. Jones. 1974. Effects of 2,3,7,8-tetrachlorodibenzo-*p*-dioxin on stimulated DNA synthesis in the liver and kidney of the rat. Chem. Biol. Interact. **8:** 31–39.
2. Lucier, G. W., O. S. McDaniel, G.E.R. Hask, B. A. Fowler, B. R. Sonawane & E. Faeder. 1973. TCDD induced changes in rat liver microsomal enzymes. Environ. Health Perspect. **5:** 199–209.
3. Cunningham, H. M. & D. T. Williams. 1972. Effect of tetrachlorodibenzo-*p*-dioxin on growth rate and the synthesis of lipids, and proteins in rats. Bull. Environ. Contam. Toxicol. **4:** 45–51.
4. Vos, G. J., J. A. Moore & J. G. Zinkl. 1973. Toxicity of 2,3,7,8-tetrachlorodibenzo-*p*-dioxin on the immune system of laboratory animals. Environ. Health Perspect. **5:** 149–162.
5. Vos, G. J., J. A. Moore & J. G. Zinkl. 1974. Toxicity of 2,3,7,8-tetrachlorodibenzo-*p*-dioxin (TCDD) in C57B1/6 mice. Toxicol. Appl. Pharmacol. **29:** 229–241.
6. Vos, G. J. & J. A. Moore. 1974. Suppression of cellular immunity in rats and mice by maternal treatment with 2,3,7,8-tetrachlorodibenzo-*p*-dioxin. Int. Arch. Allergy Appl. Immunol. **47:** 777–794.
7. Greig, J. B., G. Jones, W. H. Butler & J. M. Barnes. 1973. Toxic Effects of 2,3,7,8-tetrachlorodibenzo-*p*-dioxin. Food Cosmet. Toxicol. **11:** 575–595.
8. Jackson, W. T. 1972. Regulatory mitosis III cytological effect of 2,4,5-trichlorophenoxyacetic acid and of dioxin contaminants in 2,4,5-T formulations. J. Cell Sci. **10:** 15–25.
9. Beatty, P. W., K. J. Lembach, M. A. Holscher & R. A. Neal. 1975. Effect of 2,3,7,8-tetrachlorodibenzo-*p*-dioxin (TCDD) on mammalian cells in tissue cultures. Toxicol. Appl. Pharmacol. **31:** 309–312.
10. Niwa, A., K. Kumaki & D. W. Nebert. 1975. Induction of aryl hydrocarbon hydroxylase activity in various cell cultures by 2,3,7,8-tetrachlorodibenzo-*p*-dioxin. Mol. Pharmacol. **11:** 399–408.
11. Bastomsky, C. H. 1977. Enhanced thyroxine metabolism and high uptake goiters in rats after a single dose of 2,3,7,8-tetrachlorodibenzo-*p*-dioxin. Endocrinology **101:** 292–296.
12. Beatty, P. W., M. A. Holscher & R. A. Neal. 1976. Toxicology of 2,3,7,8-tetrachlorodibenzo-*p*-dioxin in larvae and adult forms of *Rana catesbeiana*. Bull. Environ. Contam. Toxicol. **5:** 578–581.
13. Ashmore, J. & D. Morgan. 1967. *In* The Adrenal Cortex. A. B. Eisenstein, Ed. 249–267. Little Brown, Boston.
14. Oliver, R. M. 1975. Toxic effects of 2,3,7,8-tetrachlorodibenzo-1,4-dioxin in laboratory workers. Br. J. Industr. Med. **32:** 49–53.
15. Poland, A., E. Glover & A. Kende. 1976. Stereospecific, high affinity binding of 2,3,7,8-tetrachlorodibenzo-*p*-dioxin by hepatic cytosol. J. Biol. Chem. **251:** 4936–4946.
16. Kensler, T. W., W. F. Busby, N. E. Davidson & G. N. Wogan. 1976. Effect of hepatocarcinogens on the binding of glucocorticoid-receptor complex in rat liver nuclei. Cancer Res. **36:** 4647–4651.
17. Lin, E. C. C. & W. E. Knox. 1957. Adoption of the rat liver tyrosine-α-ketoglutarate transaminase. Biochem. Biophys. Acta **26:** 85–88.
18. Beatty, P. W. & R. A. Neal. 1976. Induction of DT-diaphorase by 2,3,7,8-tetrachlorodibenzo-*p*-dioxin. Biochem. Biophys. Res. Commun. **69:** 197–204.

STUDIES ON THE MECHANISM OF ACTION OF THE CHLORINATED DIBENZO-*P*-DIOXINS AND RELATED COMPOUNDS*

Alan Poland†‡

McArdle Laboratory for Cancer Research
University of Wisconsin
Madison, Wisconsin 53706

William F. Greenlee

Department of Pharmacology and Toxicology
University of Rochester School of Medicine
Rochester, New York 14642

Andrew S. Kende

Department of Chemistry
University of Rochester
Rochester, New York 14627

Our intent in this report is (1) to review the investigations on one of the most sensitive biochemical responses evoked by 2,3,7,8-tetrachlorodibenzo-*p*-dioxin (TCDD)§ and structurally related chlorinated aromatic hydrocarbons, the induction of aryl hydrocarbon hydroxylase (AHH) activity; (2) to review the evidence that there is a stereospecific recognition site which mediates the induction of this enzyme activity; (3) to compare the structure activity relationship for binding to this receptor and the structure activity relationship for the toxicity of these chlorinated aromatic hydrocarbons; and (4) to suggest a model for the mechanism of the toxic action of these compounds.

The toxicology of the halogenated aromatic hydrocarbons (the halogenated dibenzo-*p*-dioxins, dibenzofurans, azo- and azoxybenzenes, and biphenyls) is reviewed elsewhere in this symposium; however, it is useful to summarize some of these effects first so that we share a common understanding when the term toxicity is used in this report. TCDD was selected as the prototype compound, because it is the most potent and the most thoroughly studied. TCDD produces a variety of toxic effects in laboratory animals and human beings. The histopathologic responses produced by TCDD differ quantitatively and qualitatively in different species and to some extent in acute versus chronic administration. (1) *Lethality*. The cause of death remains unknown. There is an order of species sensitivity with guinea pigs, and chickens being very sensitive, rats intermediate, and mice showing less sensitivity. Following the

*Supported in part by National Institute of Environmental Health Sciences Grant R01-ES0-01884, National Cancer Institute Program Project Grant P01-CA-22484.

†To whom reprint requests should be sent.

‡Recipient of Research Career Development Award. K04-ES-0017

§TCDD—2,3,7,8-tetrachlorodibenzo-*p*-dioxin; AHH—aryl hydrocarbon hydroxylase; MC—3-methylcholanthrene; ALAS—δ-aminolevulnic acid synthetase; ED_{50}—dose which produces one half the maximal response, in this case induction of hepatic AHH activity; K_D—equilibrium dissociation constant for binding; TCAB—3,4,3′,4′-tetrachloroazobenzene; TCAOB—3,4,3′,4′-tetrachlorazoxybenzene.

0077–8923/0320–02143 $01.75/0 1979, NYAS

administration of a toxic dose of TCDD, animals show a slow wasting syndrome, with a latent period before death that is species specific. (2) *Chloracne and hyperkeratosis.* This characteristic response is reported in humans, rabbits, monkeys, and hairless mice. (3) *Involution of the lymphoid organs.* This is another characteristic response of TCDD in most species. There is notably thymic atrophy and this is accompanied by suppression of the cellular immune response in young animals. (4) *Teratogenicity, embryotoxicity, and fetal wastage.* While TCDD is a teratogen (producing fetal malformations), it produces more embryotoxicity and fetal resorption in mice and rats. (5) *Chick edema syndrome.* Edema, especially hydropericardium, is a characteristic lesion in young chickens; edema is also seen in mice. (6) *Liver lesions.* Hepatocellular necrosis has been reported in rats and rabbits and focal lesions in mice, but little or no liver pathology in guinea pigs. Several species develop disturbances of hepatic porphyrin metabolism.

The distinct features of TCDD toxicity—species differences, delayed onset, chloracne formation, edema in young chickens, thymic involution, and tetratogenicity—have been reported for other halogenated dibenzo-*p*-dioxins. The early work of Schwetz *et al.*[1] and the more recent report of McConnell *et al.*,[2] suggest that a dibenzo-*p*-dioxin congener which is toxic, will elicit all of these responses if given at a sufficient dose. The data on specific isomers of halogenated dibenzofurans, azo- and azoxybenzenes, and biphenyls are more limited but also suggest that congeners

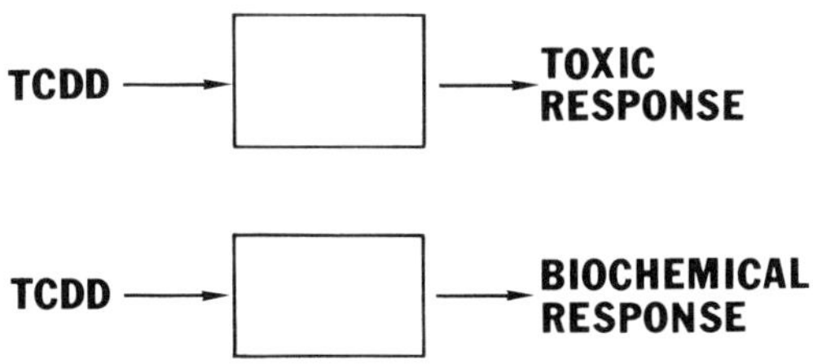

FIGURE 1. Schematic representation of studies on the toxic and biochemical responses produced by the administration of TCDD to animals.

isosteric with TCDD (see below for a fuller discussion of isosterism) will produce the same toxic syndrome.

Despite the characteristic spectrum of toxic effects produced by these compounds, investigations on their mechanism of toxicity have been fraught with difficulty. In FIGURE 1, is schematically diagrammed the "black box" approach to the problem, where the "black box" is the animal, the input signal is TCDD or some congener, and the output signal is the toxic response one chooses to study. The difficulties arise in *measuring* the toxic effect: (1) the ultimate "target organ," the cause of death, is unknown; (2) in studying a particular toxic effect (e.g. chloracne, chick edema) the selection of the animal species is most important, for species differ not only quantitatively as to their sensitivity to the drug, but also qualitatively, whether they express the response at all; and (3) perhaps most importantly there are few reports of toxic responses that are *dose-related, quantitative,* and *easily measured.* Obtaining this important information is difficult and costly.

We have chosen an alternative approach (FIGURE 1), to measure a biochemical response, the induction of AHH activity. This response occurs in most tissues of many species, is extremely sensitive to chlorinated aromatic hydrocarbons, and gives an easily measured dose-response relationship.

As reviewed below, there is a remarkably good correspondence between the potency of halogenated dibenzo-*p*-dioxins and other halogenated aromatic compounds to induce AHH activity and their toxic potency. The questions we wish to consider are: How do we explain this correlation? What is the induction of AHH telling us about the mechanism of toxicity of these compounds?

Induction of AHH Activity

The microsomal monooxygenase system is an enzyme complex embedded in the endoplasmic reticulum consisting of NADPH-cytochrome P-450 reductase, and cytochrome P-450, which metabolizes many lipophilic compounds to more polar metabolites. There are several (perhaps many) different types of cytochrome P-450, differing in the spectrum of substrates they catalyze and which are differentially induced by the administration of various compounds. Aryl hydrocarbon hydroxylase activity can be considered a measure of one of these cytochrome species, so-called cytochrome P_1-450 (or P-448).

When TCDD is administered to chicken embryos, and 24 hours later the liver is removed and AHH activity measured *in vitro,* there is a dose-related increase in hepatic AHH activity (FIGURE 2A). The dose of TCDD that produces one half the maximal response (ED_{50}) is approximately 3ng/egg (0.2 to 0.3 nmoles/kg). We initially examined a series of 15 halogenated dibenzo-*p*-dioxins for their capacity to induce hepatic AHH activity in the chicken embryo.[3] As seen in FIGURE 2B, the five isomers that induce AHH activity have two common properties: (1) halogen atoms occupy at least three, and for maximal potency, four of the lateral ring positions (positions 2, 3, 7 and 8); and (2) at least one ring position is nonhalogenated (Note the octachlorodibenzo-*p*-dioxin is inactive). The same series of 15 halogenated dibenzo-*p*-dioxins was tested for their capacity to induce hepatic δ-aminolevulinic acid synthetase (ALAS) activity, the initial and rate-limiting enzyme in heme synthesis, in the chicken embryo. An identical structure activity relationship was observed.[3] Subsequently, a much larger series of halogenated dibenzo-*p*-dioxins and dibenzofurans have been tested for their capacity to induce AHH activity in chick embryo liver,[4,5] and in cell culture in a rat hepatoma line,[6] and the results confirm the structure activity relationship.

The most important observation was that for chlorinated dibenzo-*p*-dioxin congeners there was an excellent correlation between their potency to induce AHH activity (and ALAS activity) and their toxic potency.

Genetic Expression of AHH Activity in Inbred Strains of Mice

The classical inducers of AHH activity and cytochrome P_1-450 are the polycyclic aromatic hydrocarbons, such as 3-methylcholanthrene (MC). In a comparison of MC and TCDD for their capacity to induce hepatic AHH activity in the rat, we found MC and TCDD produce parallel dose-response curves, both compounds produce the same maximum response, simultaneous administration of maximally inducing doses of MC and TCDD produces no greater response, but TCDD is 30,000 times as potent as MC, and a single administration produces sustained enzyme induction which lasts over 35 days.[7]

When randomly bred or certain inbred strains of mice are administered MC, they respond with the induction of hepatic AHH activity; certain other inbred strains when challenged with MC fail to respond.[8,9] The prototypical strain responsive to MC is C57BL/6J and the prototypical nonresponsive strain is DBA/2J. In crosses and backcrosses between C57BL/6J and DBA/2J mice, the trait of responsiveness to

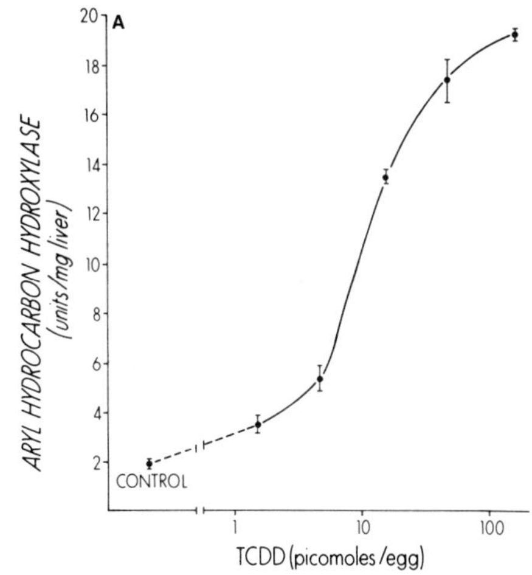

FIGURE 2A. Log dose-response curve for the induction of hepatic AHH activity in chicken embryos by TCDD. Eighteen-day-old embryos were injected with TCDD dissolved in *p*-dioxane and 24 hours later hepatic AHH activity was assayed. Each point is the mean ± S. E. of four groups of pooled livers.

aromatic hydrocarbons is inherited in a simple autosomal dominant mode and the genetic locus controlling this trait is called the Ah locus.

The extraordinary potency of TCDD relative to MC prompted us to examine its effects in inbred strains that are responsive and nonresponsive to MC. TCDD induced AHH activity in all strains tested regardless of their response to MC.[10] In C57BL/6J

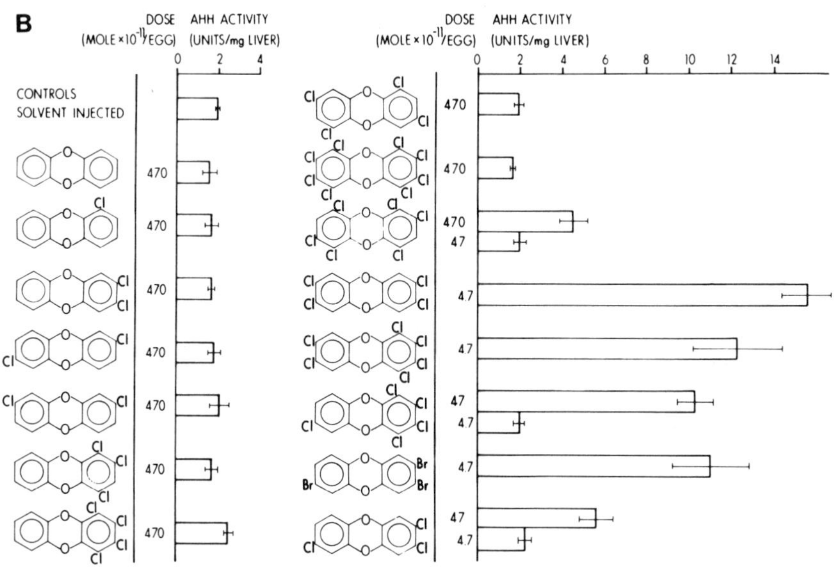

FIGURE 2B. Structure-activity relationship for induction of hepatic AHH activity by halogenated dibenzo-*p*-dioxins. Eighteen-day-old chicken embryos were injected with the test compounds dissolved in *p*-dioxane, and 24 hours later hepatic enzyme activity was assayed. Each bar represents the mean ± S. E. of four groups of pooled livers; all except the control and TCDD groups were n = 12.

mice the enzyme activity and the spectral properties of the cytochrome P_1-450 that were induced by MC or TCDD, or both compounds administered together were found to be similar to those induced in DBA/2J mice by TCDD. These results suggest that the same gene product(s) is induced in C57BL/6J mice by both compounds and in DBA/2J mice by TCDD.

Thus, mice that are nonresponsive to MC, do respond to a more potent stimulus, TCDD, and therefore these mice do have the structural and regulatory genes necessary for the expression of AHH activity. The failure of these mice to respond to MC suggests that they fail to recognize MC as a signal for induction. We postulated that the mutation in nonresponsive mice results in a defective recognition or receptor site for AHH induction, a receptor which has a diminished (or absent) affinity for MC. Consider the hypothesis that TCDD acts at the same receptor as MC, but because of its greater potency, and hence its presumed greater affinity, it is able to saturate this receptor even in nonresponsive mice, and initiate the induction of AHH activity. This hypothesis predicts that inbred mice which are nonresponsive to MC, should respond to TCDD, but should be uniformly less sensitive to TCDD, than strains that are responsive to MC.

We administered various doses of TCDD to six inbred strains of mice and 24 hours later measured their hepatic AHH activity.[11] The data are presented in FIGURE 3 as

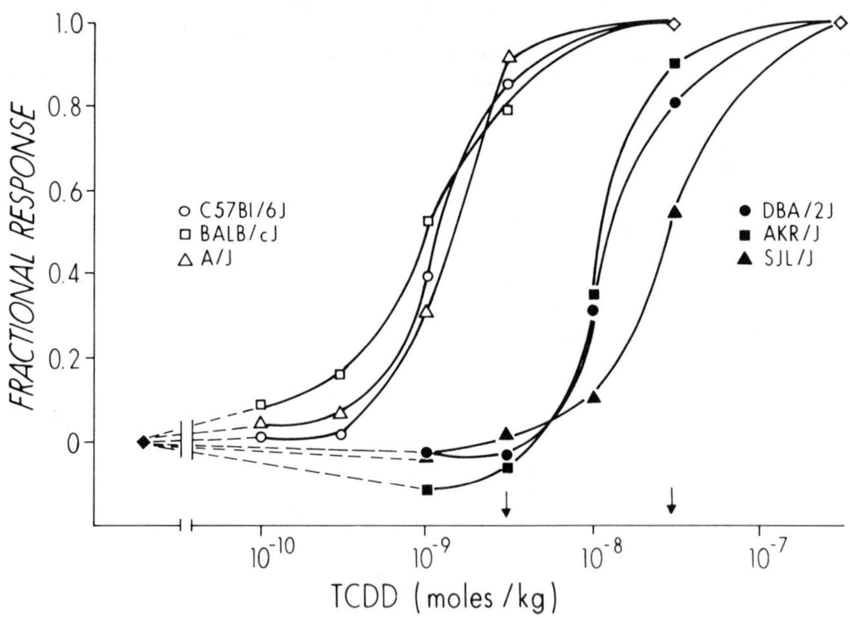

FIGURE 3. Log dose-response curves for the induction of hepatic AHH activity by TCDD in six inbred strains of mice. The mice were injected intraperitoneally with varying doses of TCDD dissolved in *p*-dioxane; 24 hours later the animals were killed and their livers were assayed for AHH activity. The data are expressed as fractional responses to eliminate strain differences in basal and maximally induced enzyme activities. The mean control activity was equated to zero, the mean maximally induced activity to 1.0; and the intermediate responses were calculated as a fraction of the maximal response. Each point represents the average of four or five animals.

the log of the dose versus the fractional response (control activity = 0, maximal activity = 1.0). The three inbred strains that are nonresponsive to MC (DBA/2J, AKR/J and SJL/J) are less sensitive to TCDD, than are the responsive strains (C57BL/6J, BALB/cJ and A/J); i.e., the log dose response curves for the nonresponsive strains are shifted to the right. The ED_{50} for AHH induction in responsive strains is approximately 1 nmole/kg, and for nonresponsive strains the ED_{50} is ≥ 10 nmoles/kg. The data support the hypothesis that TCDD and MC act on the same receptor, and that the mutation in nonresponsive strains¶ is due to a receptor with a diminished affinity for inducing compounds.

Cytosolic Binding Species

We next wished to search for a macromolecular species that had the *in vitro* binding properties one would predict for the hypothesized receptor based on the *in vivo* biology. Specifically, one would expect such a moiety to have the following *in vitro* binding properties: (1) reversibly bind TCDD with a high affinity and the binding affinity (K_D) should correspond to the *in vivo* potency (ED_{50}) for AHH induction; (2) the binding species from nonresponsive mice should have a lower affinity for TCDD, than the binding species from responsive mice; (3) the rank order of binding affinities of other halogenated dibenzo-*p*-dioxins, and other chlorinated aromatic hydrocarbons should correspond to their rank order of potencies to induce AHH activity; and (4) other compounds, such as the polycyclic aromatic hydrocarbons, which induce AHH activity should bind to this moiety.

An initial series of experiments, suggested that there was a moiety which bound ^{3}H-TCDD with a high affinity in the liver cytosol (105,000 × *g* supernatant fraction).

The binding of ^{3}H-TCDD (specific activity 52.5 Ci/mmole) to liver cytosol was measured by the charcoal-dextran binding assay (for details see reference 5). When hepatic cytosol from C57BL/6J mice was incubated with varying concentrations of ^{3}H-TCDD and ^{3}H-TCDD plus a 200-fold excess of unlabeled 2,3,7,8-tetrachlorodibenzofuran, we observed a small pool of high affinity sites (Figure 4A).[5] In Figure 4B the same data for specific binding (displacable, high affinity binding) is presented in a Scatchard plot. The binding affinity for TCDD, K_D, is 0.27 nM which compares quite favorably with the *in vivo* potency for AHH induction in C57BL/6J mice ($ED_{50} \sim 1$ nmole/kg).

Another expectation of this binding species, if it is the hypothesized receptor, is that one would expect the hepatic cytosol from responsive mice to have a greater affinity for ^{3}H-TCDD than liver cytosol from nonresponsive mice. As shown in Figure 5, the specific binding in C57BL/6J liver cytosol is greater at all concentrations of ^{3}H-TCDD, than is the specific binding in DBA/2J mice. Unfortunately, the limited aqueous solubility of ^{3}H-TCDD prevents us from achieving a sufficient concentration to saturate the cytosol binding species from nonresponsive mice and estimating the K_D.

¶In this report, we refer to the phenotypes of the Ah locus using the original terminology: responsive or nonresponsive to polycyclic aromatic hydrocarbons with the induction or failure of induction of hepatic AHH activity. Both phenotypes respond to TCDD with induction of hepatic AHH activity and differ only in their sensitivity (responsive mice, $ED_{50} \sim 1$ nmole TCDD/kg, "nonresponsive" mice $ED_{50} \geq 10$ nmoles TCDD/kg). Thus, when the nonresponsive phenotype is referred to, it is nonresponsive to MC and relatively less sensitive to TCDD.

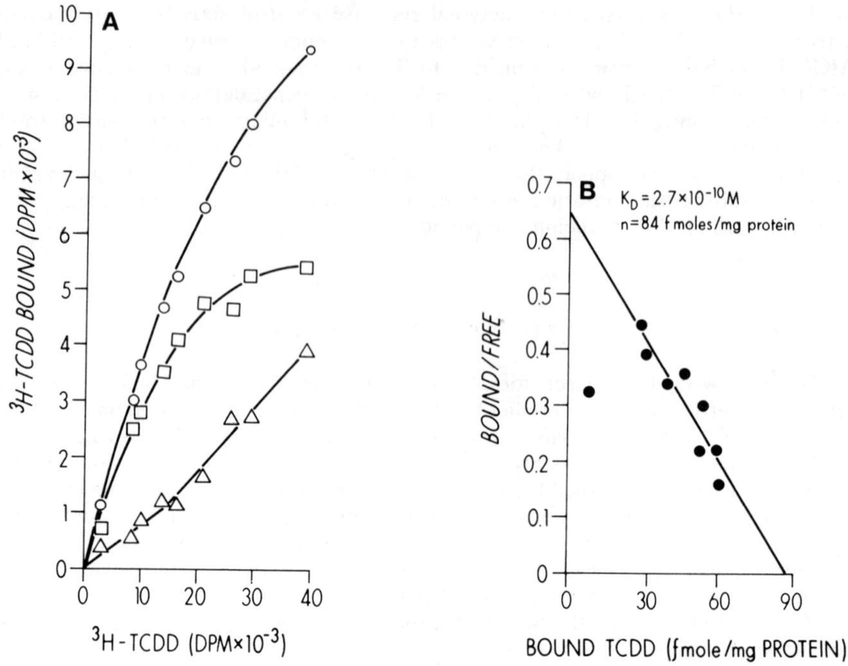

FIGURE 4. (A) Total nonspecific and specific binding of ^{3}H-TCDD to hepatic cytosol from C57BL/6J mice. The cytosolic fraction of liver from C57BL/6J mice at 2 mg protein/ml in buffer was incubated with varying concentrations of ^{3}H-TCDD and ^{3}H-TCDD plus a 200-fold excess of 2,3,7,8-tetrachlorodibenzofuran for 2 hours at 0°. Samples were treated with a suspension of charcoal-dextran to remove unbound ligand, and the radioactivity quantified. Total binding—^{3}H-TCDD alone (○); nonspecific or nondisplaceable binding—^{3}H-TCDD plus 2,3,7,8-tetrachlorodibenzofuran (△); specific or displaceable binding (□).

(B) Scatchard plot of the specific binding in (A).

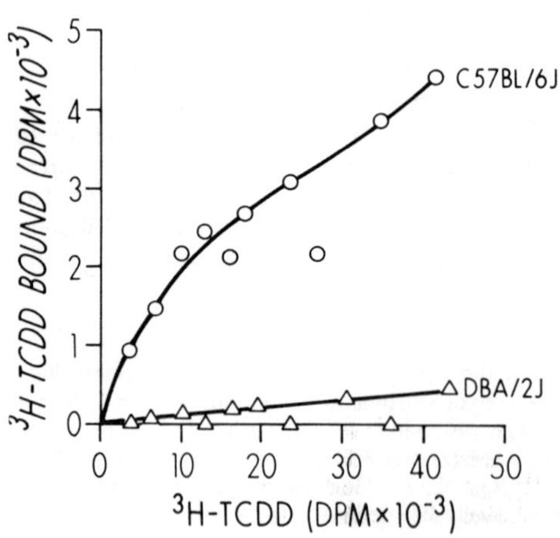

FIGURE 5. The specific binding of ^{3}H-TCDD to hepatic cytosol from C57BL/6J and DBA/2J mice. Specific or displaceable binding was measured as described in FIGURE 4. For details see reference 5.

If the hepatic cytosol binding species is the receptor, then we would expect the *in vitro* binding affinity of the halogenated dibenzo-*p*-dioxin congeners should correspond to their *in vivo* potency to induce AHH activity. The binding affinities of these unlabeled congeners are measured by their capacity to compete with ^{3}H-TCDD for specific binding sites, and the results are expressed relative to the binding affinity of TCDD (TCDD = 100) in FIGURE 6. The biologic potencies of these congeners to induce hepatic AHH activity in the chicken embryo (ED_{50}) are also expressed relative to TCDD. As seen in FIGURE 6, there is a very good correspondence between the rank order of binding affinity and biologic potency for these compounds.

We have observed that other compounds which induce AHH activity, such as MC, benzo[a]pyrene and β-naphthoflavone, compete for specific cytosol binding sites, but drugs that induce other types of microsomal monooxygenase activities (e.g. phenobarbital, p.p′-DDT, pregnenolone-16-carbonitrile), and the steroid hormones and their analogues, which have liver cytosol binding proteins, all fail to compete with ^{3}H-TCDD for specific cytosol binding sites.

Let us briefly summarize the studies on the mechanism of induction of AHH activity. TCDD or another inducing compound enters the liver cell and binds to a cytosolic protein. This binding species has the properties of the receptor, it provides stereospecific recognition of the inducing compound. Recent studies in our laboratory indicate the complex of the cytosol protein and ligand translocate to the nucleus** and presumably, in some unknown manner initiates the transcription and translation of the gene(s) which code for cytochrome P_1-450 (AHH activity). In inbred strains of mice, the trait of responsiveness or nonresponsiveness to polycyclic aromatic hydrocarbons (or greater or lesser sensitivity to TCDD) is determined by the Ah locus. The genetic and biochemical evidence all suggests, but does not prove, that the Ah locus is the structural gene for the cytosolic receptor protein. The mutation in nonresponsive mice appears to result in an altered receptor with a diminished affinity for the inducing compound.

COORDINATE GENE EXPRESSION

The administration of MC or TCDD to laboratory animals stimulates not only AHH activity, but a number of other hepatic enzyme activities. There is evidence for several of these enzymes, that this increase in activity represents true induction, i.e. *de novo* protein synthesis. MC and TCDD have been shown to induce the following hepatic enzymes: δ-aminolevulinic acid synthetase, UDP-glucuronosyl transferase, glutathione-S-transferase B, aldehyde dehydrogenase, DT-diaphorase, and ornithine decarboxylase. The induction of several of these enzymes activities have been shown to segregate with the Ah locus in inbred strains of mice, (for a more detailed discussion see reference 12). It would appear that the Ah locus controls not only the expression of AHH activity, but also the coordinate expression (and perhaps repression) of a number of other enzymes (FIGURE 7).

STRUCTURE-ACTIVITY RELATIONSHIP

We can extend our structure-activity relationship to other classes of halogenated aromatic compounds.

**Greenlee, W. F. & A. Poland, manuscript in preparation.

	RELATIVE BINDING AFFINITY	RELATIVE BIOLOGIC POTENCY		RELATIVE BINDING AFFINITY	RELATIVE BIOLOGIC POTENCY
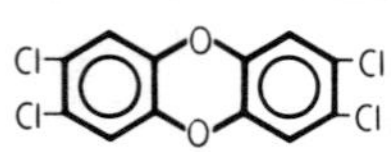	100†	100††	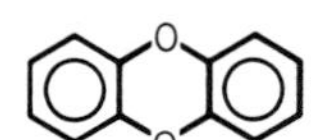	inactive (5.4×10^{-7})*	inactive (9.4×10^{-8})**
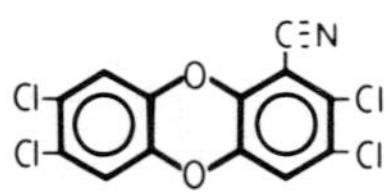	167	100	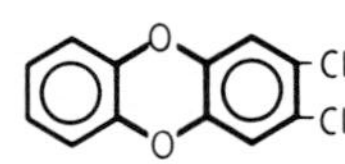	inactive (2.7×10^{-8})	inactive (9.4×10^{-8})
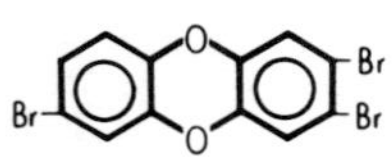	43	43	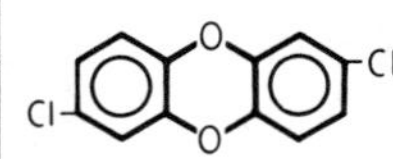	inactive (5.4×10^{-8})	inactive (9.4×10^{-8})
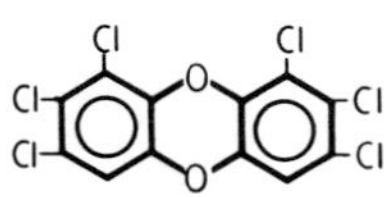	20	22	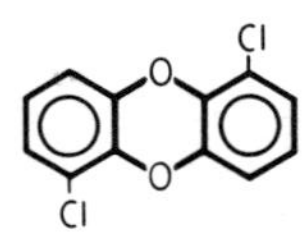	inactive (5.4×10^{-8})	inactive (4.7×10^{-7})

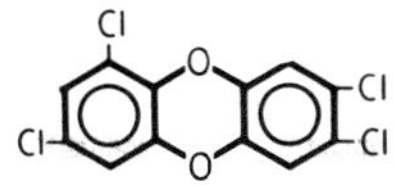	16	8	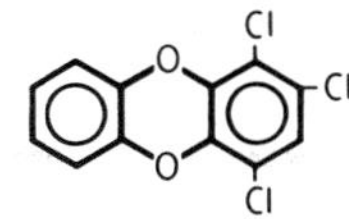	inactive (5.4×10^{-8})	inactive (9.4×10^{-8})
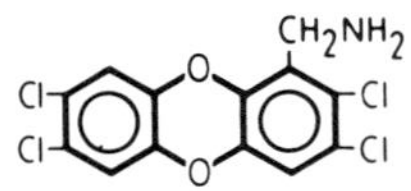	13	3	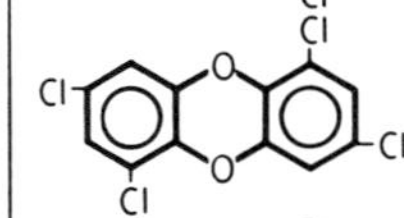	inactive (2.7×10^{-8})	inactive (9.4×10^{-8})
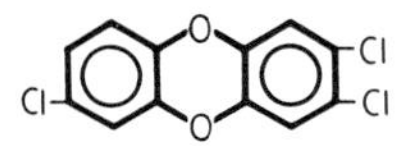	14	0.06	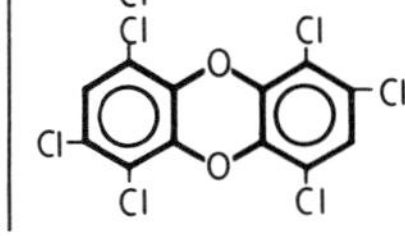	inactive (1.1×10^{-8})	inactive (9.4×10^{-8})

FIGURE 6. The cytosol binding affinities and biologic potencies of dibenzo-*p*-dioxin congeners relative to TCDD. The binding affinities of the dibenzo-*p*-dioxin congeners for hepatic cytosol were estimated by the capacity of these compounds to compete with ^{3}H-TCDD for specific binding sites in C57BL/6J liver cytosol. The binding affinities are expressed relative to TCDD, which is assigned a value of 100. For inactive analogues, the highest concentration tested that was judged to be soluble is given in parentheses.

The biologic potency (ED_{50}) of each congener was the dose that produced one-half the maximal induction of hepatic AHH activity in the chicken embryo, and the potency was expressed relative to TCDD (TCDD = 100). For inactive compounds, the highest dose tested (in moles/kg) is given in parentheses. This assumes the weight of an average chicken egg is approximately 50 g.

†The absolute value of the K_D for TCDD is 0.27nM.

††The absolute value of the ED_{50} for TCDD induction of hepatic AHH activity in the chicken embryo is 0.31 nmoles/kg.

*The highest concentration tested that was judged soluble in moles/liter.

**The highest concentration tested in moles/kg.

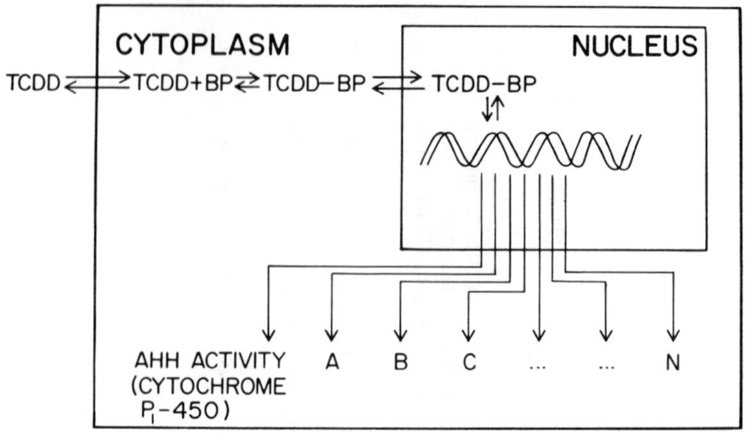

FIGURE 7. Schematic representation of the model proposed for the pleiotrophic response mediated by the TCDD-receptor complex. BP- is the TCDD-binding protein or receptor.

*Dibenzo-*p*-dioxins and Dibenzofurans*

To summarize the structure activity relationship for the halogenated dibenzo-*p*-dioxins, those congeners which induce AHH and ALAS activities and bind to the cytosol binding protein (1) have halogen atoms in at least three of the lateral ring positions (positions 2,3,7 and 8); (2) have an order of potency of substitution Br > Cl > F, NO_2; and (3) have at least one unsubstituted ring position. A similar structure activity is found for the halogenated dibenzofurans; again the 2,3,7,8-tetrachloro- and 2,3,7,8-tetrabromo- congeners are most active, and the octachloro-analogue is inactive. On the whole, the dibenzofurans follow the dibenzo-*p*-dioxins rather well.[5]

Azo and Azoxybenzene

3,4,3′,4′-Tetrachloroazobenzene (TCAB) and 3,4,3′,4′-tetrachloroazoxybenzene (TCAOB) have been reported as contaminants of 3,4-dichloroaniline or herbicides made from 3,4-dichloroaniline and are thought to be responsible for several episodes of chloracne in factory workers.[13,14] TCAOB, TCAB, and 3,4,3′,4′-tetrachlorohydrazobenzene are all approximate stereoisomers of TCDD, all bind with a high affinity to the cytosolic binding species, and are all potent inducers of hepatic AHH in the chicken embryo. Congeners such as 3,5,3′,5′-tetrachloroazo- 3,5,3′,5′-tetrachloroazoxy-, or 3,5,3′,5′-tetrachlorohydrazo- benzene, azobenzene or azoxybenzene were inactive as inducers or competitive ligands. TCAOB produced acne in hairless mice, but 3,5,3′,5′-tetrachloroazoxybenzene did not.

Halogenated Biphenyls

Of a series of 16 chlorinated and brominated biphenyl congeners, only three (3,4,3′,4′-tetrachloro-, 3,4,5,3′,4′,5′-hexachloro-, and 3,4,5,3′,4′,5′-hexabromo-) induced hepatic AHH activity and competed for the cytosolic binding species (FIGURE

8).[15] The structural requirements necessary for induction of AHH activity were (1) halogen atoms in at least two adjacent lateral positions on each benzene ring (3,4,3′,4′ or 3,4,5,3′,4′,5′); and (2) the lack of halogen substitutions adjacent to the biphenyl bridge (the 2,6,2′ or 6′ positions)—such substitutions lead to marked nonplanarity. A more extensive investigation of microsomal monooxygenase induction by chlorinated biphenyl congeners by Goldstein *et al.*,[16] supports these conclusions. There are a limited number of investigations on the toxicity of chlorinated biphenyl congeners, but one report by McKinney *et al.*,[17] is particularly interesting. The authors compared the toxicity of 5 hexachlorobiphenyl congeners in chickens: 2,3,6,2′,3′,6′-hexachloro-, 2,4,6,2′,4′,6′-hexachloro- 2,4,5,2′,4′,5′-hexachloro-, 2,3,4,2′,3′,4′-hexachloro- and

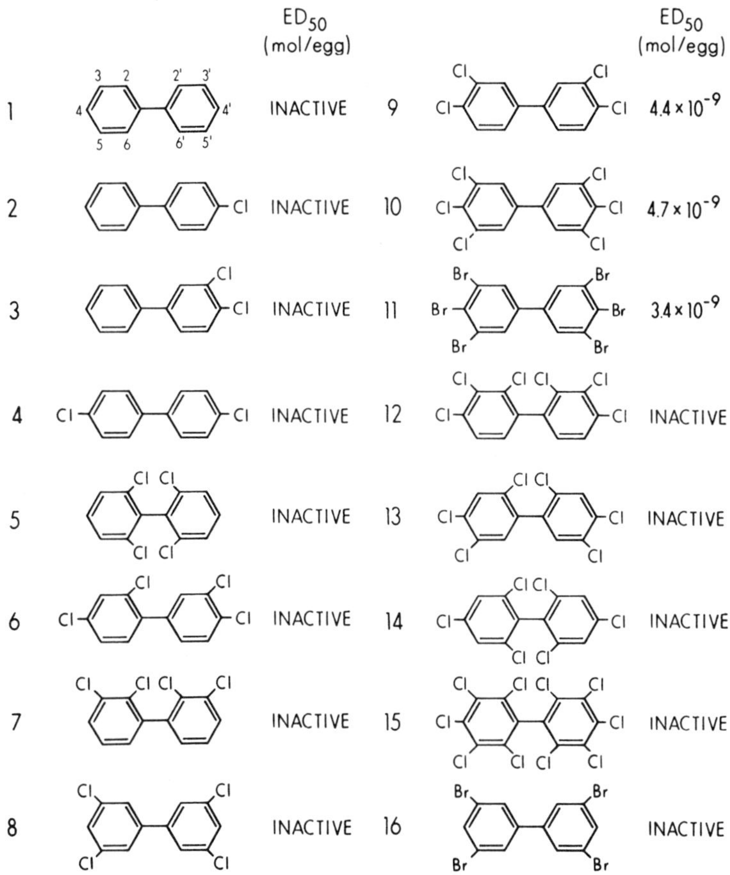

FIGURE 8. Structure activity relationship of halogenated biphenyl congeners for the induction of hepatic AHH activity in chicken embryos. Seventeen-day embryos were injected with halogenated biphenyl congeners dissolved in *p*-dioxane. Twenty-four hours later, the animals were killed and their livers assayed for AHH activity. Congeners were considered inactive if a dose of 4.66×10^{-7} moles/egg produced less than 10% of the maximal response. The biologically active congeners were tested at four doses, and the ED_{50} was estimated from the log dose-response curve.

3,4,5,3′,4′,5′-hexachloro-. The latter compound had the greatest toxic potency, and was the only one to produce thymic atrophy and edema in the chickens, toxic responses characteristic of the TCDD-syndrome.

We can generalize about the structure-activity relationship of each of these classes of halogenated aromatic hydrocarbon. Consider a prototype of each class—TCDD, 2,3,7,8-tetrachlorodibenzofuran 3,4,3′,4′-tetrachlorazoxybenzene and, 3,4,3′,4′-tetrachlorobiphenyl—depicted in FIGURE 9. These compounds are all approximate stereoisomers and their molecular structures can be thought of as roughly fitting into a rectangle 3 × 10Å with halogen atoms in the four corners. Planarity or near planarity seems to be essential. In FIGURE 9 the *in vitro* binding affinity for the cytosol binding species and the potency to induce AHH activity in the chick for each of these compounds, and for a number of other tetrahalogenated aromatic compound that are approximate stereoisomers, are compared. The toxicity of these latter compounds has not been tested.

While many points remain to be answered about the chemical features recognized by the receptor, the planar rectangular site depicted in FIGURE 9 seems a good first approximation.

A Model for the Mechansim of Toxicity

It appears that the halogenated dibenzo-*p*-dioxins, dibenzofurans azo- and azoxybenzenes, and biphenyls, which are stereoisomers of TCDD, produce a similar toxic syndrome.

We believe the correlation between the potency of these compounds to induce AHH activity and their toxic potency is more than fortuitous. It is difficult to envision how the induction of AHH activity, *per se,* might be involved in specific organ toxicities such as chloracne or thymic involution. Thus we wish to focus on the receptor—the correlation between the binding affinities of these compounds for the cytosol binding species and their toxic potencies.

We suggest that the toxicity of these chlorinated aromatic hydrocarbons is mediated through the receptor; that is, the initial event in their toxic action is the stereospecific recognition and binding of them by the cytosolic binding species.

One prediction from this model is immediately obvious. The genetic evidence suggests the Ah locus determines the receptor protein and its binding affinity. Thus, if the toxicity of TCDD (and related halogenated aromatic compounds) is mediated through the receptor, then inbred strains of mice with a high affinity receptor (and which are sensitive to induction of AHH activity by TCDD) should be sensitive to toxicity from TCDD, and mice with a lower affinity receptor should be less sensitive to the toxicity of TCDD. In short, the toxicity of TCDD should segregate with the Ah locus.

To examine this, we chose the involution of the thymus because (1) we feel this is a "toxic response"—not just a biochemical response—in that it represents tissue loss and in young animals it is associated with suppression of cellular immunity[18,19]; and (2) thymic atrophy is dose-related and quantifiable.

C57BL/6J, DBA/2J, and hybrid B6D2F_1/J male mice, all 5 weeks old, were given a single intraperitoneal injection of varying doses of TCDD.†† Six days later the animals were killed, their thymuses dissected out and weighed, and the data expressed as the thymic/body weight ratio (FIGURE 10A). TCDD produced a dose-related decrease in the ratio of thymus/body weight. The responsive C57BL/6J mice were

††Poland, A. & E. Glover, unpublished data.

	Induction of AHH Act. (chick) ED_{50} (nmoles/kg)	Binding Liver Affinity Mouse Liver Cytosol K_D (nm)		Induction of AHH Act. (chick) ED_{50} (nmoles/kg)	Binding Liver Affinity Mouse Liver Cytosol K_D (nm)
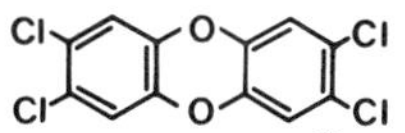	0.31	0.27	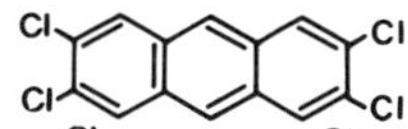	8.9	1.4
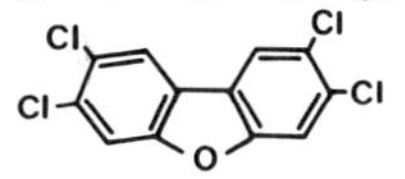	0.46	0.73		102	3.2
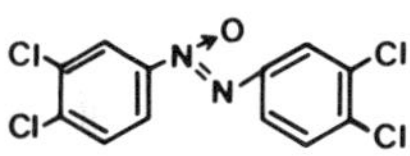	0.45	0.93	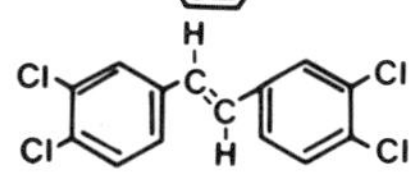	12.6	2.4
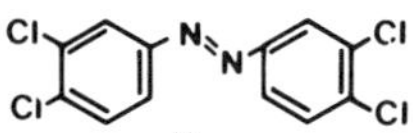	2.0	1.1	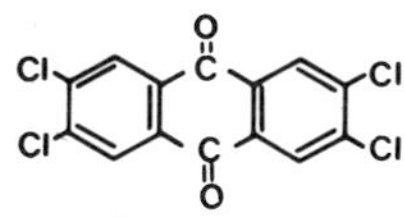	640	>225
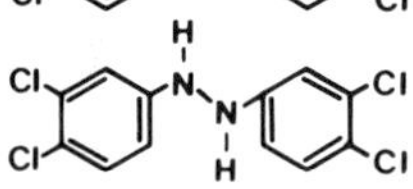	2.9	1.2	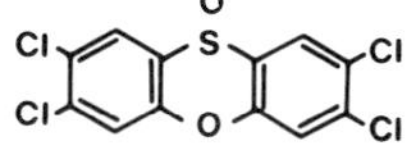	1.6	0.71
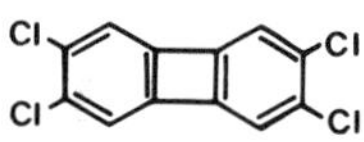	0.28	0.34	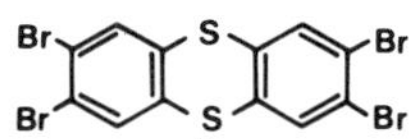	13.2	5.0
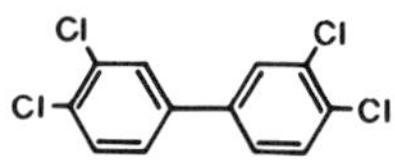	88	15	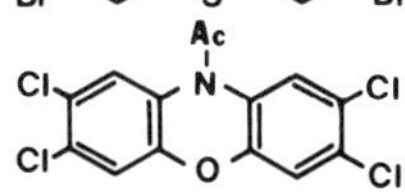	0.3	0.74

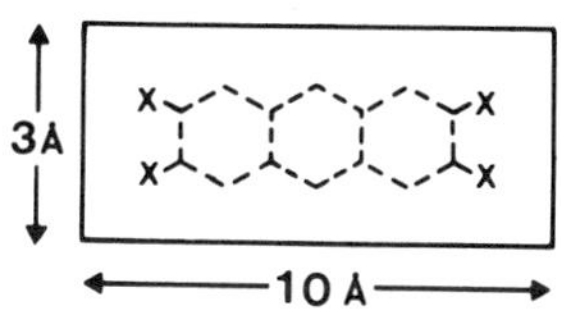

FIGURE 9. Isosteric tetrahalogenated aromatic hydrocarbons: comparison of their potencies to induce AHH activity and their cytosolic binding affinities. The biologic potency of each compound was determined from the log dose-response curve for induction of hepatic AHH activity in the chicken embryo. The binding affinity of each congener was determined by its capacity to compete with ^{3}H-TCDD for specific binding sites in liver cytosol from C57BL/6J mice.

more sensitive than the nonresponsive DBA/2J mice, and the hybrid mice were intermediate. We then bred B6D2F$_1$/J mice (heterozygous at the Ah locus, Aa) with DBA/2J mice (homozygous nonresponsive, aa) and the resultant offspring were phenotyped as heterozygous responsive (Aa) or homozygous nonresponsive (aa). At 6 weeks of age the animals were dosed with TCDD as above, killed 6 days later, and their thymus/body weight ratio determined (FIGURE 10B). Since sex influences the thymus/body weight ratio in control mice, data from males and females are presented separately. As seen in FIGURE 10B, for both males and females, heterozygous responsive mice (Aa) are more sensitive to TCDD, than are homozygous nonresponsive (aa) mice. The data indicate that thymic involution by TCDD segregates with the Ah locus.

These data support the model, but before generalizing, it is wise to see if other toxic responses produced by TCDD segregate with the locus.

We have reviewed two lines of evidence which suggest that toxicity of TCDD and related halogenated aromatic compounds is mediated by their stereospecific binding to the receptor for AHH induction: (1) the correlation between the binding affinities of these compounds for the cytosolic binding species and their toxic potencies; and (2) for TCDD, segregation of thymic involution with the Ah locus. Since the Ah locus controls not only the expression of AHH activity, but several (perhaps many) other genes, it is tempting to speculate that these compounds exert their toxicity through the expression (or repression) of one or more of the genes controlled by the receptor. However, the genetic evidence, only permits us to state that toxicity (thymic involution) segregates with the Ah locus.

In closing, we would like to consider some limitations of the proposed model and some predictions it permits us to make. The model is predicated on the correlation of the binding affinity of halogenated aromatic hydrocarbon congeners for receptor and the toxic potency of these compounds. The correlation is limited by the available

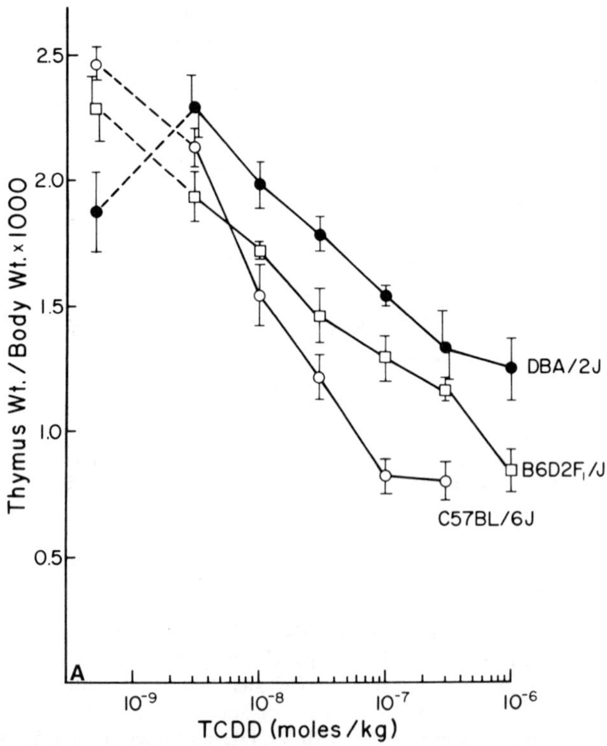

FIGURE 10A. Log dose-response curves for the effect of TCDD on the ratio of thymus-body weight in C57BL/6J, DBA/2J and B6D2F$_1$/J mice. Five-week-old male mice were given a single intraperitoneal injection of TCDD dissolved in *p*-dioxane or the solvent alone. Six days later the animals were killed, their thymuses weighed, and the data expressed as the ratio of the thymus/body weight. Each point is the mean ± S. E. of 5 to 8 animals. The solvent-injected (control) groups are connected by dashed lines to the rest of the dose-response curves.

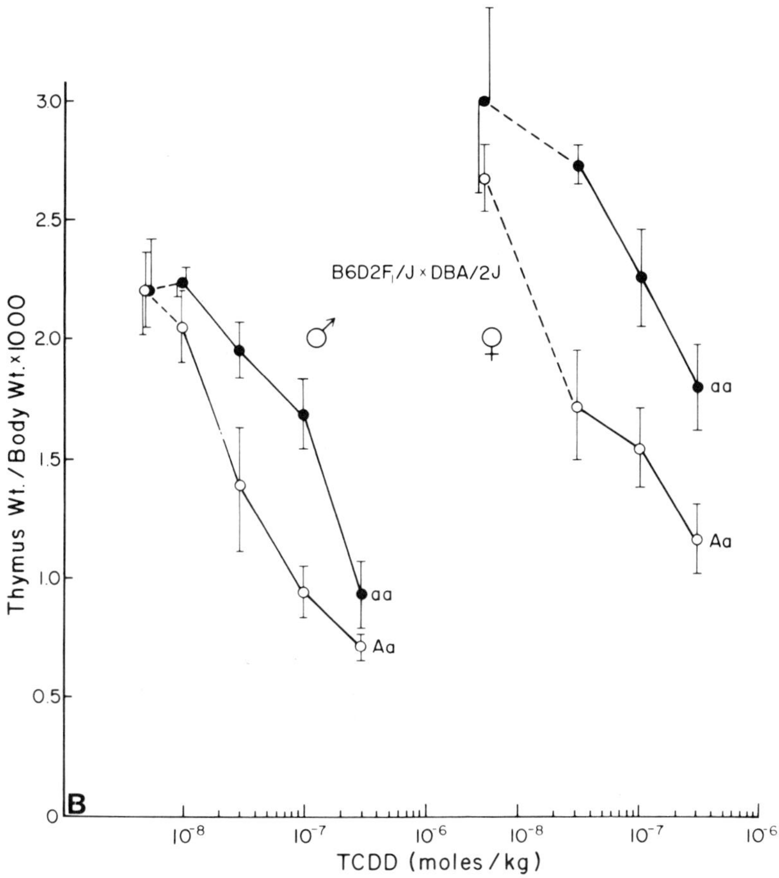

FIGURE 10B. Log dose-response curve of TCDD on thymus/body weight ratio in male and female offspring of the $B6D2F_1/J \times DBA/2J$ cross. The offspring of the $B6D2F_1/J \times DBA/2J$ cross were phenotyped as heterozygous responsive—Aa, or homozygous, nonresponsive—aa, at the Ah locus. At 5½ weeks of age the animals were dosed with TCDD as above and killed 6 days later. The data on males and females are presented separately, because the sex of the mice affected the ratio of thymus/body weight in the solvent-injected control groups. Each point is the mean ± S. E. of 4 to 8 animals. The solvent-injected (control) groups are connected by dashed lines to the rest of the dose-response curves.

toxicity data for individual congeners. For instance, we do not know if stereoisomers such as shown in FIGURE 9, column 2, produce the toxic syndrome that TCDD elicits. We do not understand why nonhalogenated compounds that bind to the induction receptor, such as polycyclic aromatic hydrocarbons and β-naphthoflavone, do not produce the spectrum of toxic responses characteristic to the chlorinated aromatic hydrocarbons. It is possible, that toxicity is related to the *sustained* occupation of the receptor, and the relatively long biologic half-lives of the halogenated aromatic compounds versus the rapidly metabolized polycyclic aromatic hydrocarbons accounts for the difference.

The model gives no clue to the ultimate biochemical lesion(s) responsible for the toxicity of these compounds, but merely states that toxicity is mediated through binding to the receptor. For a tissue or cell type to develop a toxic response to chlorinated aromatic hydrocarbons, the presence of the receptor, is essential, but may not be sufficient. Despite ignorance of the ultimate biochemical lesion, it may be possible to inhibit the toxicity of these compounds, by designing a competitive antagonist, an analogue that binds to the receptor, but doesn't initiate gene expression.

References

1. Schwetz, B. A., J. M. Norris, G. L. Sparschu, V. K. Rowe, P. J. Gehring, J. L. Emerson & C. G. Gerbig. 1973. Environ. Health Perspect. **5:** 87–99.
2. McConnell, E. E., J. A. Moore, J. K. Haseman & M. W. Harris. 1978. Toxicol. Appl. Pharmacol. In press.
3. Poland, A. & E. Glover. 1973. Mol. Pharmacol. **9:** 736–747.
4. Kende, A. S., J. Wade, D. Ridge & A. Poland. 1974. J. Org. Chem. **39:** 931–937.
5. Poland, A., E. Glover & A. S. Kende. 1976. J. Biol. Chem. **251:** 4936–4946.
6. Bradlaw, J. A., L. H. Garthoff, N. E. Hurley & D. Firestone. 1976. Toxicol. Appl. Pharmacol. **37:** 119.
7. Poland, A. & E. Glover 1974. Mol. Pharmacol. **10:** 349–359.
8. Nebert, D. W., F. M. Goujon & J. E. Gielen. 1972. Nature New Biol. **236:** 107–110.
9. Nebert, D. W. & J. E. Gielen. 1972. Fed. Proc **31:** 1315–1325.
10. Poland, A., E. Glover, J. R. Robinson & D. W. Nebert. 1974. J. Biol. Chem. **249:** 5599–5606.
11. Poland, A. & E. Glover. 1975. Mol. Pharmacol. **11:** 389–398.
12. Poland, A. & A. Kende. 1977. *In* Origins of Human Cancer. H. H. Hiatt, J. D. Watson, J. A. Winston, Eds. Cold Spring Harbor Laboratory, Vol. B:847–867.
13. Poland, A., E. Glover, A. S. Kende, M. DeCamp & C. M. Giandomenico. 1976. Science **194:** 627–630.
14. Taylor, J. S., R. C. Wuthrich, K. M. Lloyd & A. Poland. 1977. Arch. Dermatol. **113:** 616–619.
15. Poland, A. & E. Glover. 1977. Mol. Pharmacol. **13:** 924–938.
16. Goldstein, J. A., P. Hickman, H. Bergman, J. D. McKinney & M. P. Walker. 1977. Chem. Biol. Interact. **17:** 69–87.
17. McKinney, J. D., K. Chae, B. N. Gupta, J. A. Moore & J. A. Goldstein. 1976. Toxicol. Appl. Pharmacol. **36:** 65–80.
18. Vos, J. G., J. A. Moore & J. G. Zinkl. 1973. Environ. Health Perspect. **5:** 149–162.
19. Vos, J. G. & J. A. Moore. 1974. Int. Arch. Allergy Appl. Immunol. **47:** 777–794.

THE ROLE OF PLASMA PROTEINS IN THE TRANSPORT AND DISTRIBUTION OF CHLORDECONE (KEPONE®) AND OTHER POLYHALOGENATED HYDROCARBONS*

Harry L. Skalsky, Marc W. Fariss, Robert V. Blanke, and Philip S. Guzelian†

Departments of Pharmacology, Pathology, and Medicine
Medical College of Virginia
Richmond, Virginia 23298

INTRODUCTION

It has been amply documented that the blood of humans or animals exposed to organochlorine pesticides[1] or other polyhalogenated hydrocarbon chemicals[2,3] contains readily detectable quantities of these highly lipophilic substances. The presence of exceedingly hydrophobic chemicals in such an aqueous medium as blood might be explained by their dissolution in blood lipids. In support of this idea, Morgan and Roan[4] showed that the concentrations of DDT, DDE, and dieldrin in most tissues including blood are directly proportional to the tissue's concentration of lipid. An additional factor which contributes to the level of polyhalogenated hydrocarbons in the blood is binding of these chemicals to plasma proteins. Moss and Hathaway[5] demonstrated association of telodrin and dieldrin with albumin and α-globulin in the blood of rabbits and rats. Subsequent studies by others have confirmed that many organochlorine compounds are bound by plasma proteins, particularly plasma lipoproteins.[6-15] Skalsky and Guthrie[16-18] reported that binding of DDT and dieldrin by lipoproteins and albumin involved slowly reversible, hydrophobic interactions. These observations are important because evidence is accumulating that protein binding may be a key factor in establishing the distribution in the body of at least some of these organic chemicals. For example, Eliason and Posner[12] demonstrated that transplacental transport of dieldrin to the fetus is dependent not only upon the concentration of maternal plasma proteins, but also upon the ontogenic development of specific fetal plasma proteins. Morrison[19] showed that the concentration of DDT in cerebral spinal fluid is higher than would be expected from the total content of lipid, a finding which suggests that the pesticide may be bound to proteins in the spinal fluid.

We have reported another example of dissociation between the distribution of lipid and that of an organochlorine pesticide in chemical workers exposed to large quantities of chlordecone (Kepone®).[20] The ratio of the concentration of chlordecone in adipose tissue compared to blood was seven to one, a value strikingly lower when compared with the distribution of other organochlorine pesticides as, for example, DDT, which has a ratio of concentration in fat to blood of 280 to 1.[21] From the

*This project was supported by a grant (1R01-01519) and a contract (N01-ES-6-2117) from the National Institutes of Environmental Health Sciences, by a grant from the Allied Chemical Company, and by a grant from the Virginia Environmental Endowment. Dr. Skalsky is supported by a grant from Merck, Sharpe and Dohme. Dr. Guzelian is the recipient of a Clinical Investigator Award in Gastroenterology from the National Institutes of Health (AM-00128).

†Address Reprint Requests to: Philip S. Guzelian, M.D., Department of Medicine, Medical College of Virginia, Box 267, MCV Station, Richmond, Virginia 23298.

0077-8923/79/0320-0231 $01.75/0 © 1979, NYAS

observations that chlordecone was found almost exclusively in the plasma fraction of whole blood and that chlordecone was undetectable in urine, we inferred that chlordecone may be bound to plasma proteins.[20] The unusually high levels of chlordecone in blood suggested that this pesticide may be transported in a special manner by these proteins. To gain insight into this problem, the following investigation was undertaken to identify the chlordecone-binding proteins.

RESULTS

The proteins in serum taken from a patient 6 months after his last exposure to chlordecone were separated by gel filtration chromatography on Sephadex G-200 (FIGURE 1). As previously reported with this technique,[22] the serum proteins were resolved into three distinct peaks, with the first peak containing the lipoproteins, and the third peak containing albumin. The major portion of the eluted chlordecone was associated with the first protein peak. Twelve months later, the levels of chlordecone in adipose tissue and blood in this patient had both declined, although the ratio of their concentrations remained seven to one. Gel filtration of a second sample of serum obtained at this time revealed a similar profile, with chlordecone eluted largely with the proteins in the first protein peak (FIGURE 1). Since elution of chlordecone was

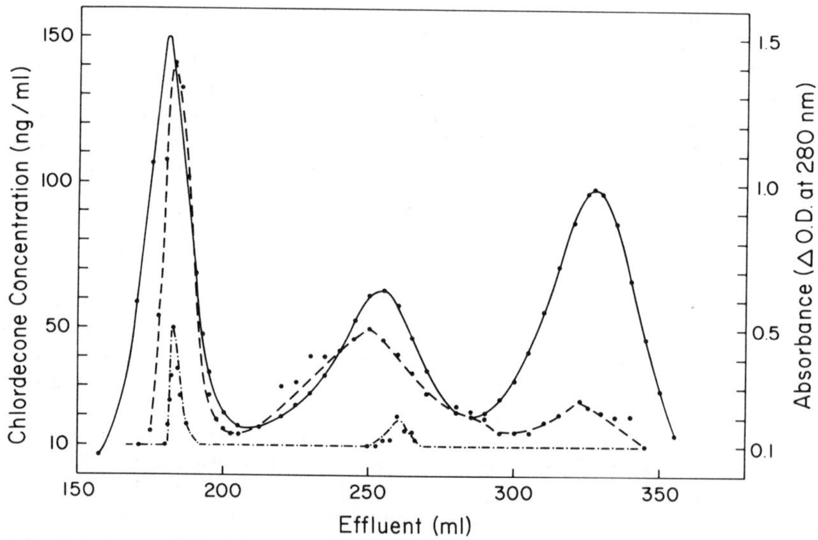

FIGURE 1. Analysis by gel filtration of serum proteins from a single patient exposed to chlordecone. One ml of serum was applied to a Sephadex G-200 column (2.5 × 91 cm.) previously equilibrated with 0.14M sodium chloride in 0.05M veronal buffer (pH 7.5) at 4°C, and was eluted with the same buffer at a rate of 7.5 ml/ hr. The fractions were analyzed for chlordecone by gas-liquid chromatography[32] and for protein by absorbance at 280 nm. Results are shown for absorbance at 280 nm (——) and chlordecone concentration (---) for a serum sample obtained under nonfasting conditions eight months after the patient's last exposure to chlordecone, and also for chlordecone concentration (–·–·) in a second sample obtained after an overnight fast from the same patient 16 months after his last exposure to chlordecone.

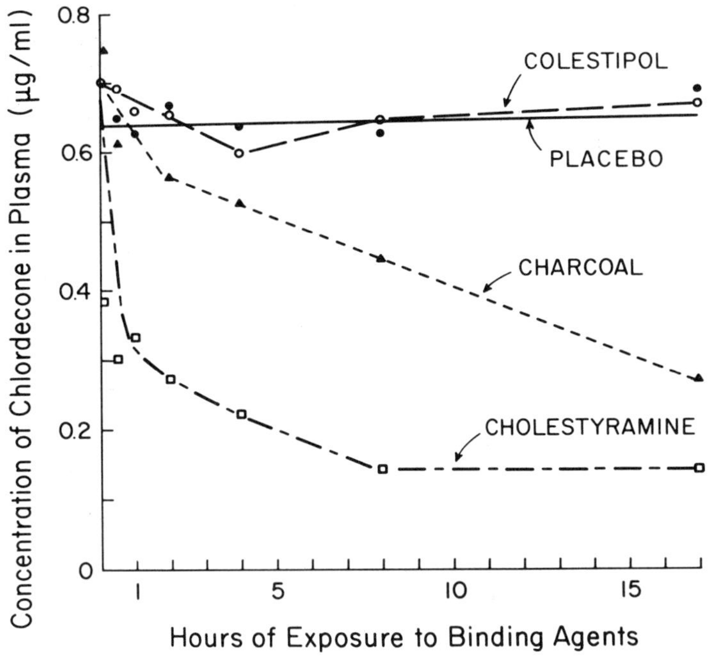

FIGURE 2. Rate of removal of chlordecone from human plasma by various binding agents. Plasma was obtained under nonfasting conditions from a patient exposed to chlordecone and was stored at −20°C until used for the experiments in FIGURES 2 and 3. For each experiment, 2 ml of plasma was combined with 0.1M Tris buffer (pH 8.0) at 4°C with stirring. To this mixture was added 100 mg of one of the binding agents previously equilibrated with the same buffer. At the indicated times, 0.5 ml was removed and centrifuged at 2000 × *g* for 5 minutes. The supernatant was removed and was analyzed for chlordecone.[32] Results are given for cholestyramine (—– □); charcoal (---- ▶); colestipol (—— ○); and Questran® placebo (—— •).

similar in samples obtained under fasted and fed conditions, we concluded that chylomicrons may be of lesser importance than other lipoproteins in binding chlordecone. It may be inferred also from the present elution profile (FIGURE 1), that binding of chlordecone to the proteins in the first peak involves slowly reversible interactions. This conclusion is based on the assumption that protein and the protein-ligand complex are of similar size and thus will migrate on the gel as one, whereas the "free" ligand is, comparatively, much smaller and will migrate in the gel as a separate band. Therefore, if the rate of dissociation of the protein-chlordecone complex were rapid, then chlordecone would be physically sieved away from proteins in the first peak and would be eluted in later fractions.[23]

Binding of chlordecone by human lipoproteins was also investigated by electrophoresis on agarose gels.[24] Plasma was obtained from a normal man after an overnight fast and, after addition of [^{14}C]-chlordecone, the lipoproteins were separated electrophoretically and analyzed for radioactivity. In several experiments, 80% of the added [^{14}C]-chlordecone was recovered with the lipoproteins. Of the lipoprotein associated radioactivity, 75% co-migrated with α-lipoproteins, 18% with β-lipoproteins, and the remainder with the pre-β-lipoproteins. The strikingly large fraction of chlordecone associated with the α-lipoproteins (which correspond to the high-density lipoproteins when separated by ultracentrifugation) stands in marked contrast to other organochlorine pesticides which bind largely to low-density lipoproteins (which correspond to β-lipoproteins), as well as to very-low-density lipoproteins (which correspond to pre-β-lipoproteins). In other experiments, separation of plasma lipoproteins by ultracentrifugation confirmed that the majority of lipoprotein-associated chlordecone was bound to high-density lipoproteins (data not shown).

Because of the difficulties of studying binding of hydrophobic compounds in aqueous systems, we examined the ability of various binding agents to remove chlordecone bound to plasma proteins, according to the procedure used by Schmid *et al.*,[25] for analysis of bilirubin binding to albumin. As illustrated in FIGURE 2, charcoal and an anion exchange resin, cholestyramine, each removed similar amounts of chlordecone from plasma, although the latter removed the pesticide at a markedly more rapid rate. In contrast to the effect of cholestyramine, little chlordecone was removed from plasma by colestipol, which is also an anion-exchange resin capable of binding bile salts. Among all agents tested, XAD-4 and XAD-2 exhibited the greatest capacity for binding chlordecone *in vitro* (FIGURE 3). These resins rely less on anion-exchange mechanisms than do cholestyramine or colestipol. The selectivity of these agents in binding chlordecone provides indirect evidence that specific forms of hydrophobic or ion exchange interactions (or both) may be involved in binding chlordecone to plasma proteins.

Binding of lipophilic halogenated hydrocarbons by plasma lipoproteins is not confined to the organochlorine pesticides. Benditt[26] demonstrated the association of

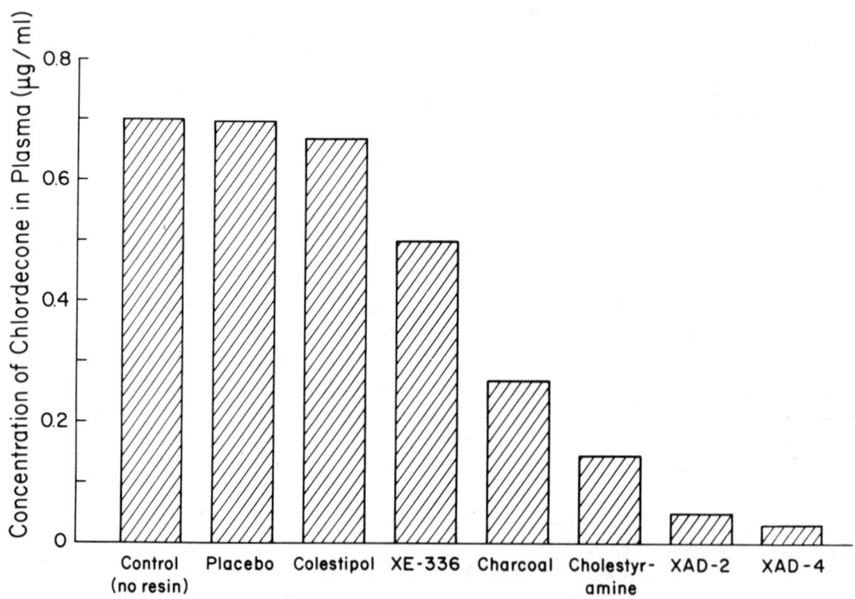

FIGURE 3. Concentration of chlordecone in human plasma exposed to binding agents for 17 hours. Experimental conditions were those described for FIGURE 2, except that 50 mg XE-336, XAD-2 and XAD-4 were added, and only a single analysis was made in each experiment after 17 hours of incubation of plasma with the binding agent.

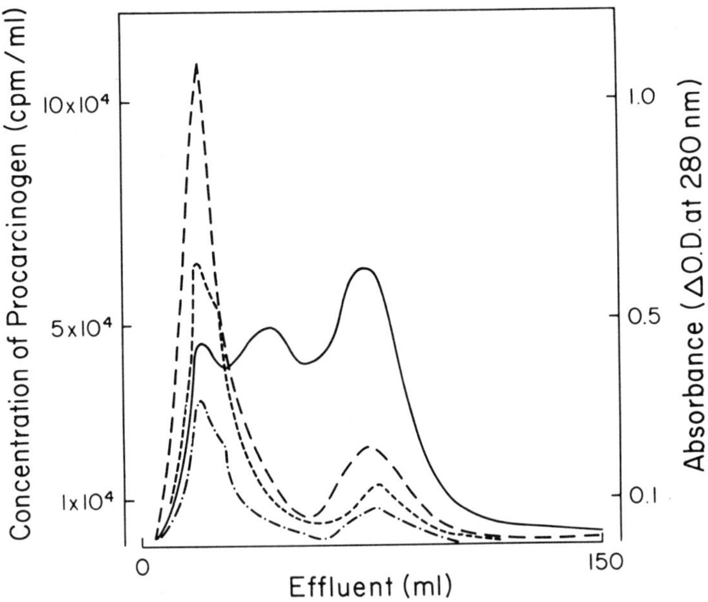

FIGURE 4. Sephadex G-200 gel filtration of rat serum containing added procarcinogens. Experimental conditions were similar to those described for FIGURE 1, except the buffer consisted of 0.5M sodium chloride in 0.1M Tris-HCl (pH 8.0). Results are shown for fasted rat serum to which a radiolabeled procarcinogen was added as follows: [^{3}H]-benzo[a]-pyrene (10 Ci/m mol) (——); [^{14}C]-dimethylbenz[a]anthracene (114.7 mCi/m mol) (----); [^{14}C]-3-methlcholanthrene (60.2 mCi/m mol) (—·—·); and for protein concentration as judged by absorbance at 280 nm (———).

benzo[a]pyrene with the serum lipoproteins by electrophoresis. Using gel filtration, we have confirmed that not only benzo[a]pyrene, but also dimethylbenz[a]anthracene and 3-methylcholanthrene bind to the serum proteins of rat. These procarcinogens appear to be preferentially associated with the first protein peak or lipoprotein fraction (FIGURE 4). Skalsky and Guthrie[17-18] and Benditt[26] have considered, at least speculatively, the toxicologic implications of the binding of xenobiotics of lipoproteins. Selective absorption of low-density lipoproteins by defined population of cells, and the subsequent metabolism of the lipoprotein by a specialized group of enzymes may provide a site of activation for attached procarcinogens in close proximity to the nucleus.

DISCUSSION

In some respects, binding of chlordecone to plasma proteins resembles the pattern of binding found with DDT or dieldrin. Each of these chemicals is bound by plasma lipoproteins and each appears to be attached by slowly reversible hydrophobic interactions. However, chlordecone appears to be bound by high-density lipoproteins, whereas DDT and dieldrin are reported to be associated with the β-lipoproteins[11,12,15-17] (corresponding largely to low-density lipoproteins). This difference may be an important factor in explaining the unusual partitioning of chlordecone between adipose tissue and blood in humans.[20] It is also possible that high-density lipoproteins play a special role in the transport of chlordecone, as has been suggested for cholesterol. Glomset[27] has proposed that high-density lipoproteins transport cholesterol from peripheral cells to the liver, whereas low-density lipoproteins transport cholesterol in the reverse direction. Schwartz *et al.*[28] found that biliary cholesterol in man is derived preferentially from high-density lipoproteins. Therefore, binding of chlordecone to high-density lipoproteins may help to explain the high levels of chlordecone in human bile and possibly the unusually high concentration of chlordecone in human liver relative to the concentration of chlordecone in blood.[20] In rats given a single dose of chlordecone, Egle *et al.*[29] have observed high rates of biliary excretion of chlordecone in bile and a progressive rise in the ratio of the concentration of chlordecone in liver compared to blood.

Studies with inert binding agents provide a means for indirectly investigating the attachment to plasma proteins of chlordecone and possibly other xenobiotics. These studies also may be important in designing methods for removal of halogenated hydrocarbons from humans poisoned with these chemicals. From the partition of chlordecone between blood and adipose tissue in humans, it may be calculated that blood contains 7% of the estimated total content of chlordecone in the body.[20] If some of the binding agents used in the present studies are capable of selectively removing chlordecone from plasma (FIGURES 2 and 3), then their use in hemoperfusion columns may provide a practical means of removal of chlordecone by continuous purification of the blood. Treatment via the hematogenous route may also be considered during the acute phase of poisioning with other organochloride pesticides when the partitioning of these chemicals resembles that of chlordecone.[30] An alternative approach is oral administration of the binding agents. Treatment with cholestyramine increases fecal excretion of chlordecone in man[20] and rats,[31] and reduces the half-life of the pesticide in the tissues. Matching an appropriate binding agent with the particular chemical toxin under consideration may be a widely applicable approach for purposes of detoxification. Further studies of protein binding of halogenated hydrocarbons are needed to validate these postulates. It is anticipated that, as illustrated by the present report, such studies will help to elucidate the role of plasma proteins in the toxicity and pharmacokinetics of these chemicals.

Acknowledgment

The authors are grateful to Dr. James Boylan for helpful suggestions.

References

1. HAYES, W. J., Jr. 1975. Toxicology of Pesticides. William & Wilkins Company. Baltimore, Md.
2. MATTHEWS, H. B. & M. W. ANDERSON. 1975. The distribution and excretion of 2,4,5,2′,5′-pentachlorobiphenyl in the rat. Drug Metab. Dispos. **3:** 211–219.
3. MATTHEWS, H. B., S. KATO, N. M. MORALES & D. B. TUEY. 1977. Distribution and excretion of 2,4,5,2′,4′,5′-hexabromobiphenyl, the major component of firemaster BP-6. J. Toxicol. Environ. Health **3:** 599–605.
4. MORGAN, D. P. & C. C. ROAN. 1970. Chlorinated hydrocarbon pesticide residues in human tissues. Arch. Environ. Health **20:** 452–457.
5. MOSS, J. A. & D. E. HATHAWAY. 1964. Transport of organic compounds in the mammal. Biochem. J. **91:** 384–392.
6. FERRY, D.G., D. OWEN, D. L. BALLARD & E. G. MCQUEEN. 1972. Pesticides and serum proteins. Proc. Univ. Otago Med. Sch. **50:** 10–11.

7. FERRY, D. G., D. OWEN & E. G. MCQUEEN. 1972. The effects of phenytoin on the binding of pesticides to serum proteins. Proc. Univ. Otago Med. Sch. **50:** 8–9.
8. DALE, W. E., A. CURLEY & C. CUETO. 1966. Hexane extractable chlorinated insecticides in human blood. Life Sci. **5:** 47–54.
9. OLSON, W. P. 1973. Dieldrin transport in the insect: an examination of Gerolt's hypotesis. Pest. Bochem. Physiol. **3:** 384–390.
10. MICK, D. L., K. R. LONG, J. S. DRETCHEN D. P. BONDERMAN. 1972. Aldrin and dieldrin in human blood components. Arch. Environ. Health **23:** 177–180.
11. MORGAN, D. P., C. C. ROAN & E. H. PASCHAL. 1972. Transport of DDT, DDE and dieldrin in human blood. Bull. Environ. Contam. Toxicol. **8:** 321–326.
12. ELIASON, B. C. & H. S. POSNER. 1971. Placental passage of ^{14}C-dieldrin altered by gestational age and plasma proteins. Am. J. Obstet. Gynecol. **111:** 925–929.
13. JAKUBOWSKI, T. & L. A. CROWDER. 1973. Binding of [^{36}Cl]-dieldrin to suspected target and non-target proteins *in vitro.* Bull. Enivron. Contam. Toxicol. **10:** 217–221.
14. SCHOOR, W. P. 1973. *In vivo* binding of p,p′-DDE to human proteins. Bull. Environ. Contam. Toxicol. **9:** 70–74.
15. WINTER, C. E., O. GIANNOTTI, & E. L. HOLZHACKER. 1975. DDT-lipoprotein complex in the American cockroach hemolymph: a possible way of insecticide transport. Pest Biochem. Physiol. **5:** 155–162.
16. SKALSKY, H. L. & F. E. GUTHRIE. 1975. Binding of insecticides to macromolecules in the blood of the rat and American cockroach. Pest. Biochem. Physiol. **5:** 27–34.
17. SKALSKY, H. L. & F. E. GUTHRIE. 1977. Affinities of parathion, DDT, dieldrin and carbaryl for macromolecules in the blood of the rat and American cockroach. Pest. Biochem. Physiol. **7:** 289–296.
18. SKALSKY H. L. & F. E. GUTHRIE. 1978. Binding of insecticides to human serum proteins. Toxicol. Appl. Pharmacol. **43:** 229–235.
19. MORRISON, G. 1971. penetrations of the blood-brain-cerebral spinal fluid. Bull. Environ. Contam. Toxicol. **6:** 48–51.
20. COHN, W. J., J. J. BOYLAN, R. V. BLANKE, M. W. FARISS, J. R. HOWELL & P. S. GUZELIAN. 1978. Treatment of chlordecone (Kepone®) toxicity with cholestyramine. N. Engl. J. Med. **298:** 243–248.
21. MORGAN, D. P. & C. C. ROAN. 1974. The metabolism of DDT in man. *In* Essays in Toxicology. W. J. Hayes, Jr., Ed. pp. 39–97. Academic Press. New York, N.Y.
22. GELOTTE, B., P. FLODIN & J. KILLANDER. 1962. Fractionation of human plasma proteins by gel filtration and zone electrophoresis or ion-exchange chromatography. Arch. Biochem. Biophys. Suppl. **1:** 319–329.
23. COOPER, D. F. & G. C. WOOD. 1968. Protein-binding of small molecules: a new gel filtration method. J. Pharm. Pharmacol. Suppl. **20:** 1505–1512.
24. ELEVITCH, F. R. & D. G. AUSTIN. 1969. Lipoprotein Electrophoresis in Thin Agarose Gel. Analytical Chemists, Inc. Palo Alto, Cal.
25. SCHMID, R., I. DIAMOND, L. HAMMAKER & C. B. GUNDERSEN. 1965. Interaction of bilirubin with albumin. Nature **206:** 1041–1043.
26. BENDITT, E. P. 1976. Implications of the monoclonal character of human artherosclerotic plagues. Beitr. Pathol. **158:** 405–416.
27. GLOMSET, J. A. 1968. The plasma lecithin: cholesterol acyltransferase reaction. J. Lipid Res. **9:** 155–158.
28. SCHWARTZ, C. C., L. G. HALLORAN, Z. R. VLAHCEVIC, D. H. GREGORY L. SWELL. 1978. Preferential utilization of free cholesterol from high-density lipoproteins for biliary cholesterol secretion in man. Science **200:** 62–64.
29. EGLE, J. L., S. B. FERNANDEZ, P. S. GUZELIAN & J. F. BORZELLECA. 1978. Distribution and excretion of chlordecone (Kepone®) in the rat. Drug. Metab. Dispos. **6:** 91–95.
30. GARRETTSON, L. K. & A. CURLEY. 1969. Dieldrin: studies in a poisoned child. Arch. Environ. Health **19:** 814–822.
31. BOYLAN, J. J., J. L. EGLE & P. S. GUZELIAN. 1978. Cholestyramine: use as a new therapeutic approach for chlodecone (Kepone®) poisoning. Science **199:** 893–895.
32. BLANKE, R. V., M. W. FARRIS, F. D. GRIFFITH & P. S. GUZELIAN. 1977. Analysis of chlordecone (Kepone®) in biological specimens. J. Anal. Toxicol. **1:** 57–62.

SELECTIVE ACCUMULATION AND DEPLETION OF POLYCHLORINATED BIPHENYL COMPONENTS: FOOD ANIMAL IMPLICATIONS

Larry G. Hansen

College of Veterinary Medicine
University of Illinois
Urbana, Illinois 61801

Having accumulated residue data from several species under variable conditions of polychlorinated biphenyl (PCB) dosing, we are forced to make one conclusion: the data in TABLE 1 are relatively meaningless if they are to be used to predict the potential hazards to consumers of products from these food animals. A lamb chop or egg from animals exposed to dietary Aroclor® 1242 does not translate to consumer exposure to Aroclor 1242.

Increasing information presented at this conference indicates that there are very important qualitative as well as quantitative differences in the biologic effects of different PCB analogs. Now that more toxicity data are available for individual analogs, we can see the critical importance of knowing the general composition of PCB residues in food products. The following discussion centers on a series of investigations concerned with the selective absorption, distribution, and elimination of individual components within Aroclor 1242 and 1254 mixtures.

METHODS OF PROCEDURE

Samples of specified lots of Aroclor 1242 (Electrical Grade, lot KB-05-415) and Aroclor 1254 (Electrical Grade, lot KB-04-612) were kindly supplied by Monsanto Chemical Company. We first determined the percentage composition of each lot by thermal conductivity gas chromatography and gas chromatography-mass spectometry (GC-MS).[1] Various extraction techniques were then evaluated for fidelity[2] and quantitation techniques were computerized to yield: (1) composition of the residue relative to the parent Aroclor, (2) absolute concentration of major components and (3) true total PCB.[3] A limited number of pure PCBs were synthesized or purchased.

RESULTS AND DISCUSSION

The limitations of generalized quantitation of PCB residues were recognized soon after results began to be reported.[4] Differences in accumulation between components has since been well documented.[5–13] FIGURE 1 illustrates the PCB profile resulting from feeding Aroclor 1242-contaminated feed (20 ppm) to growing swine for 91 days.[13] Except for peak 37 (primarily 3,4,4′-trichlorobiphenyl) and peak 48 (2,5,2′,5′-tetrachlorobiphenyl), the residue in fat more closely resembles the profile of Aroclor 1254 (FIGURE 2) than that of Aroclor 1242. Aroclor 1254 similarly administered, however, yields swine-fat residues quite distinct from the parent PCB mixture (FIGURE 2). Peak 176 accumulated to even greater relative amounts in swine liver, kidney, and muscle.[13] Sheep tended to accumulate peak 149 to a greater extent (FIGURE 3), broiler cockerels maintained the same general pattern for peaks 127

0077–8923/79/0320–0238 $01.75/0

TABLE 1

TOTAL PCB RESIDUES IN THE TISSUES OF GROWING ANIMALS FED STANDARD RATIONS CONTAMINATED TO 20 PPM AROCLORS 1242 AND/OR 1254

		PCB Concentration (ppm)							
	Time	Fat		Liver		Kidney		Blood	
	(days)	1242	1254	1242	1254	1242	1254	1242	1254
Catfish	252	118.4	—	2.5	—	11.7	—	1.2	—
Chickens	56	56.3	59.3	2.5	4.9	2.8	6.2	0.3	0.6
Swine	91	10.8	37.6	1.3	2.1	0.2	0.3	0.04	0.04
Sheep	115	15.6	30.6	1.6	1.6	0.1	0.2	0.04	0.06

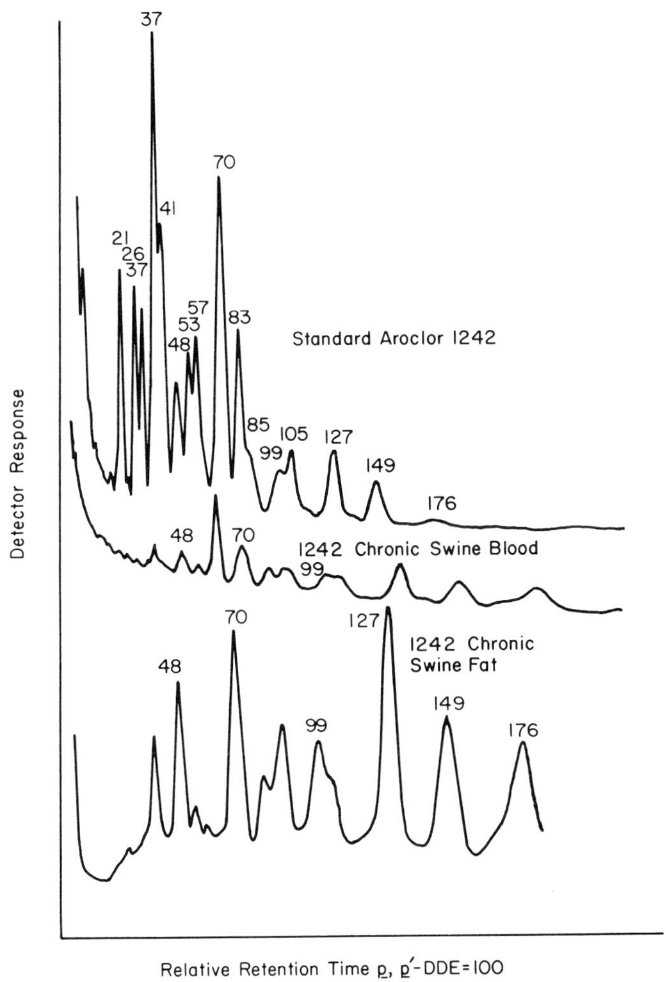

FIGURE 1. Comparison of residue profiles in swine fed 20 ppm Aroclor 1242 for 91 days. Blood sample was concentrated and fat sample diluted.

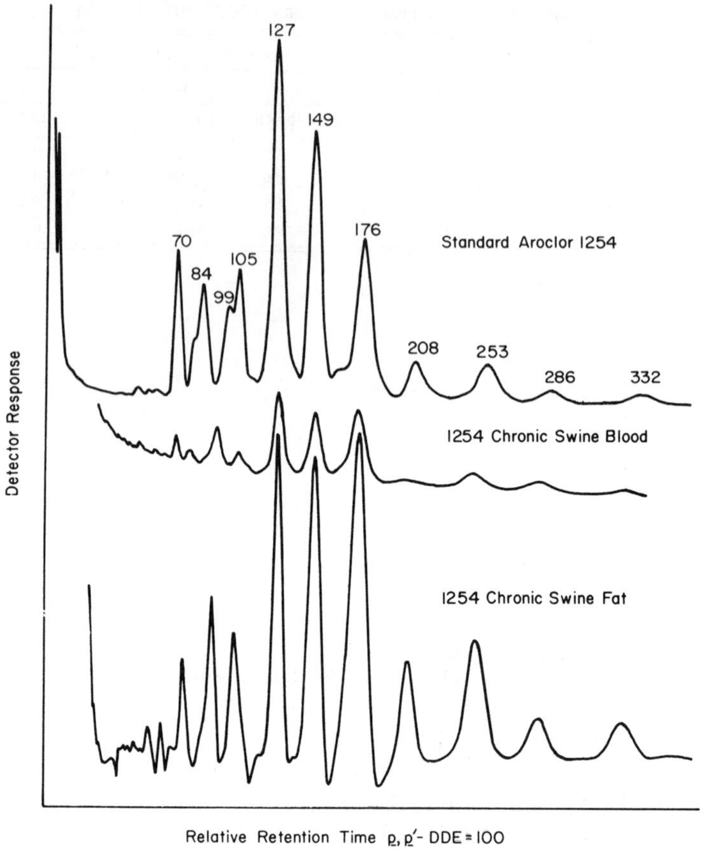

FIGURE 2. Comparison of residue profiles in swine fed 20 ppm Aroclor 1254 for 91 days.

through 332 but eliminated the earlier peaks,[12] and channel catfish fingerlings did relatively little to alter the profile of even Aroclor 1242.[11]

Several factors are involved in the selective accumulation of PCB components in adipose tissue. Certainly different PCBs are metabolized at different rates, primarily to hydroxylated products.[14–17] Within a given homologous series, the lower chlorinated components are metabolized more readily by hepatic microsomal oxidases;[13] however, subjecting the complete Aroclor 1242 mixture to the same microsomes results in selective metabolism leaving certain trichlorobiphenyls (peak 37) and tetrachlorobiphenyls (peak 48) relatively unscathed (FIGURE 4).

Absorption of different PCB components varies both with chemical structure and gastrointestinal physiology.[18] Distribution rates and patterns of PCB components within the animal differ both when the components are a part of an Aroclor mixture[18–20] or when they are administered as pure analogs.[20–22]

Rates of disappearance of PCB residues are dependent on absorption, metabolism, distribution, and excretion, all of which vary from analog to analog and from species to species.[13–22] These disappearance rates can be deceiving, however, because the long half-lives permit growth and changes in body composition during the period tissue concentrations are monitored. As pointed out by Dr. Lech in the previous paper and others,[11,20,24] concentrations must be converted to amounts to prevent significant underestimation of the real half-life.

Nutritional state is an important determiner of chlorinated hydrocarbon clearance. Residues of hexachlorobenzene (HCB) in adipose tissue are mobilized during periods of food restriction.[23] Mobilized PCBs are much more rapidly excreted[21] and, although this might be expected to provide an equalizing factor, it might also further perturb the component profile for Aroclor mixtures. Even a semistatic reduced adipose tissue compartment can have a profound influence on the body clearance of chlorinated hydrocarbons: the 7-day cumulative excretion of ^{14}C in a robust male beagle receiving i.v.^{14}C-HCB was 2% of the dose in urine and 8% of the dose in feces[24]; in a very lean male beagle with a gall bladder cannula, 7-day cumulative excretion was increased 11-fold in the urine (22.0% dose), even though 13.5% of the dose was collected with the bile and never available for reabsorption.[25] Only traces of parent HCB were present in the urine, so that this increased excretion was due to increased metabolism. Other factors may have been contributory, but one inescapable conclusion is that the metabolism of HCB was enhanced by eliminating a haven in which the HCB could be sequestered and not exposed to the oxidative enzymes; thus, even

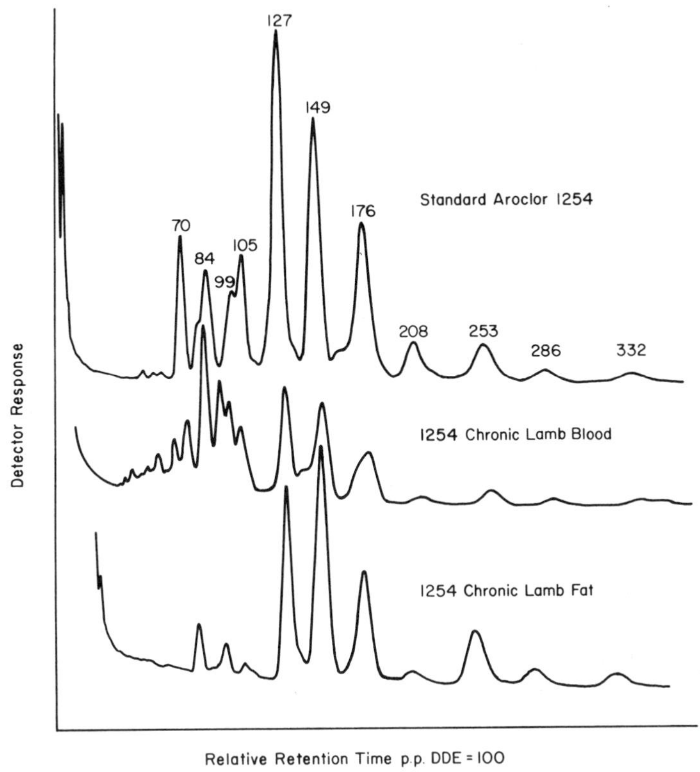

FIGURE 3. Comparison of residue profiles in lambs fed 20 ppm Aroclor 1254 for 105 days.

though metabolism rate constants may be incorporated into clearance models,[22] the amount of chemical metabolized is not a simple function of rate and concentration. Other factors such as the presence of microsomal enzyme(s) with multiple affinities and reaction velocities,[26,27] competition between analogs for enzyme sites, and the differential induction of drug metabolism enzymes by different PCBs[28,29] further complicate the picture and make availability, and thus lipophilicity, an important determinant for the selective metabolism and excretion of PCB components. Given the same intrinsic rate constant for metabolism, the less lipophilic of two PCB analogs will be cleared more rapidly from the body.

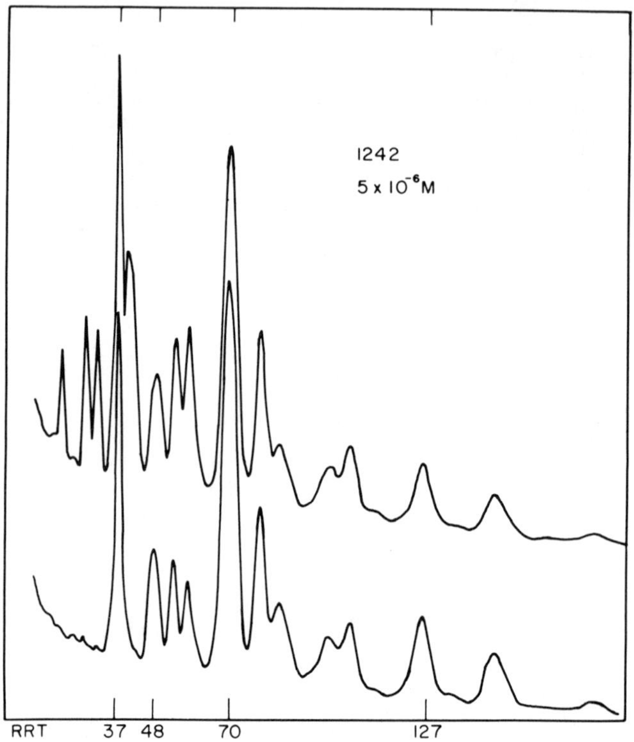

FIGURE 4. Extracts of sheep liver microsomal reaction mixtures containing Aroclor 1242 after 0 minutes (*top tracing*) and 30 minutes (*bottom tracing*) incubation time.

The effect of the size of the adipose compartment on chlorinated hydrocarbon elimination is emphasized by comparing swine to other domestic species with smaller proportions of body fat. Swine tend to decrease concentrations of Aroclor 1254 in the fat 2–3 times more rapidly than sheep,[18,20] but they also grow 4–7 times more rapidly than sheep.[13,18] Correcting for growth and expansion of the fat compartment increases the apparent $t_{1/2}$ for elimination from fat 2-fold for HCB[24] and 6- to 7-fold for Aroclor 1254.[20] Swine have hepatic mixed-function oxidase activity comparable to that of sheep[13,27] and somewhat greater than that of laying chickens.[30] However, results from our laboratory indicate that disappearance of HCB from swine fat lags far behind that

TABLE 2

DIFFERENTIAL TOXICITY AND PERSISTENCE OF PCB ANALOGS IN LAYING CHICKENS AND GROWING SWINE; TWO SEPARATE STUDIES FOR CONTROL AND 2,4,5,3′,4′-PCB IN CHICKENS

Parameter	Description of PCB and Mean Value for Parameter						
Chlorines	—	3	4	5	5	6	—
Pattern	Control	2,5,2′	2,5,2′,5′	2,3,6,2′,3′	2,4,5,3′,4′	2,4,5,2′,4′,5′	Aroclor 1254
Peak†	—	26	48	83	127	149	—
Embryo-toxicity‡							
Week 2	3;8	16	7	50*	24;73*	6	55**
Week 8	9;12	9	31*	100**	100;100**	6	67**
Swine§							
Growth	100	—	74	—	85	87	67*
Liver	100	—	146*	—	151**	109	144*
Thyroid	100	—	77	—	80	75	73
Adrenal	100	—	104	—	156*	110	100
Residues¶							
Liver	—	0.05	0.08	0.30	24;27	30	1.6
Fat	—	0.02	0.31	—	145	233	146
Half-life#							
Swine	—	—	15.4 wk	—	38.3 wk	>50 wk	40.5 wk
Chickens	—	—	1.0 wk	—	7.4 wk	4.4 wk	12.2 wk

* = Significantly different from control at 0.05; ** = significantly different at 0.01.

†GLC retention time relative to *p,p′*-DDE = 100; comparable peaks identified in Aroclors (see figures).
‡PCB in laying chicken diet at 20 ppm; expressed as % embryo mortality.[31,32]
§Growing swine received 2.14 mg/kg/day for 7 days and were terminated 16–18 weeks later; values are percent control values for percent 16-week weight gain or organ weight/body weight.
¶PPM fresh weight in laying chickens 2 weeks after administering 20 ppm in feed for 10 weeks.
#Elimination from fat based on total body burden.

of other domestic species; comparative half-lives of elimination of HCB from dissectable fat were: dogs = 10–20 days, laying chickens = 25–27 days, mature tomcats = 30–40 days, and swine = 66 days.

Laying chickens have a route of excretion (egg yolks) not available to growing swine. The blood, liver, and kidney of the laying chicken also contain more lipid than those of the pig. Thus, in spite of somewhat lower metabolism constants, more PCB is presented to hepatic enzymes and developing eggs so that PCBs are much less

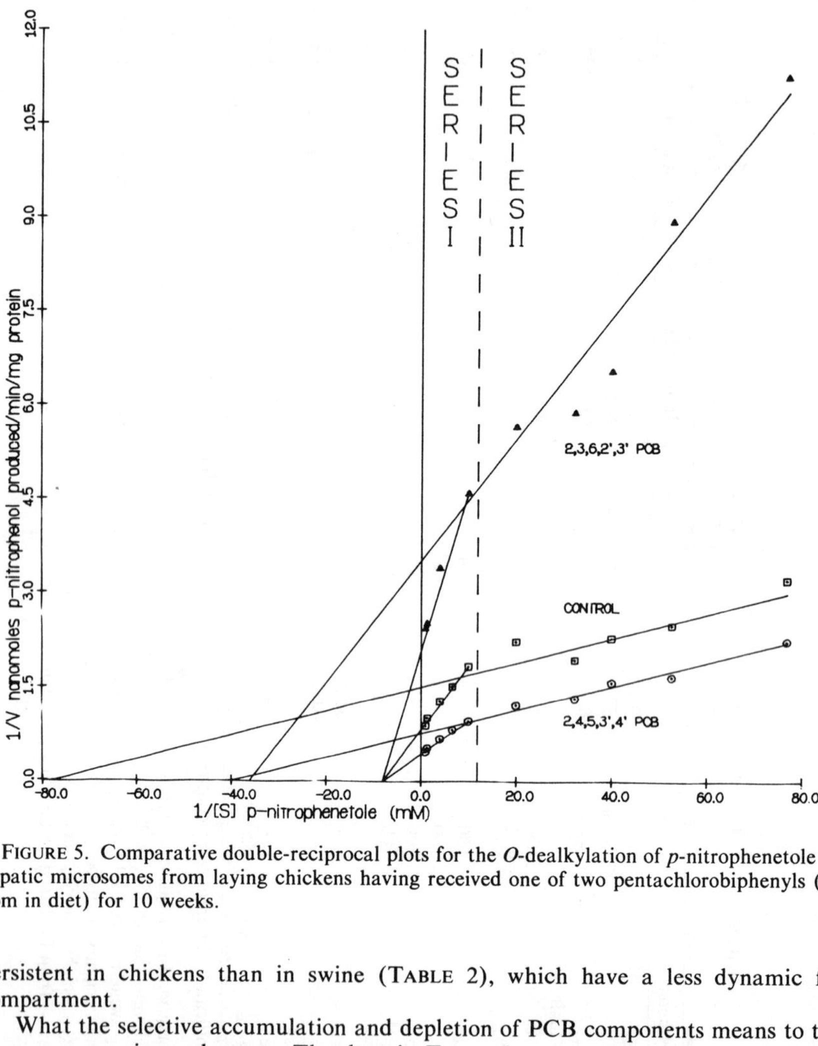

FIGURE 5. Comparative double-reciprocal plots for the *O*-dealkylation of *p*-nitrophenetole by hepatic microsomes from laying chickens having received one of two pentachlorobiphenyls (20 ppm in diet) for 10 weeks.

persistent in chickens than in swine (TABLE 2), which have a less dynamic fat compartment.

What the selective accumulation and depletion of PCB components means to the consumer remains to be seen. The data in TABLE 2 were collected from pure PCB preparations. In spite of earlier[31] nonspecific results, which indicated the possible presence of dibenzodioxins, the second set of data for 2,4,5,3′,4′-PCB was obtained with a very highly purified compound. In addition to these data, the two pentachlorobiphenyls behaved quite differently toward chicken liver microsomal oxidases: the labile 2,3,5,2′,3′-PCB suppressed activity while the persistent 2,4,5,3′,4′-PCB induced laying chicken microsomes (FIGURE 5).[30] Our data indicate that the pentachlorobiphenyls, be they labile microsomal suppressors or persistent inducers, are the toxic components of Aroclor mixtures; these data also illustrate that any number of conclusions can be reached depending on the species or parameter studied and, more importantly, the choice of PCB analogs.

REFERENCES

1. HIRWE, S. N., R. E. BORCHARD, L. G. HANSEN & R. L. METCALF. 1974. Gas-liquid chromatography-mass spectrometric characterization of Aroclor 1242 and 1254 components. Bull. Environ. Contam. Toxicol. **12:** 138–144.
2. WELBORN, M. E., R. E. BORCHARD, L. G. HANSEN & R. L. METCALF. 1974. Extraction and analysis of Aroclor 1254 components in blood. J. Assoc. Off. Anal. Chem. **57:** 1248–1252.
3. HANSEN, L. G. 1975. PCB's in Food Animals. *Final Report* to the U.S. Food and Drug Administration for Research Contract FDA 72–116 from the Bureau of Veterinary Medicine.
4. ROTE, J. W. & P. G. MURPHY. 1971. A method for the quantitation of PCB isomers. Bull. Environ. Contam. Toxicol. **6:** 377–384.
5. PLATONOW, N. & N. Y. CHEN. 1973. Transplacental transfer of polychlorinated biphenyls (Aroclor 1254) in a cow. Vet. Record, Jan. 20, 1973:69–70.
6. GREICHUS, Y, A., J. J. WORMAN, M. A. PEARSON & D. J. CALL. 1974. Analysis of PCB in bird tissues and Aroclor standards with gas chromatography and mass spectroscopy. Bull. Environ. Contam. Toxicol. **11:** 113–120.
7. CALL, D. J., Y. A. GREICHUS & J. J. WORMAN. 1974. A gas chromatographic-mass spectrometric comparison of polychlorinated biphenyl residues in the Japanese quail brain to an Aroclor standard. Bull. Environ. Contam. Toxicol. **11:** 333–338.
8. BUSH, B., F. D. BAKER, C. E. TUMASONIS, F.-C. LO & C. L. HOUCK. 1975. Modification of the homolog and isomer composition of a PCB mixture during passage through two biological systems. In Proceedings: Recent Advances in the Assessment of the Health Effects of Environmental Pollution. Commission of the European Communities, Publisher. Vol. 2, pp. 879–885. Luxembourg, 1975.
9. HANSEN, L. G., C. S. BYERLY, R. L. METCALF & R. F. BEVILL. 1975. Effect of a PCB mixture on swine reproduction and tissue residues. Am. J. Vet. Res. **36:** 23–26.
10. SUNDLOF, S. F. 1975. Aroclor 1254 in three laboratory model ecosystems. M. S. Thesis, University of Illinois, Urbana, IL.
11. HANSEN, L. G., W. B. WIEKHORST & J. SIMON. 1976. Effects of dietary Aroclor 1242 on channel catfish and the selective accumulation of PCB components. J. Fish. Res. Board Can. **33:** 1343–1352.
12. HANSEN, L. G., D. W. WILSON, R. L. METCALF & M. E. WELBORN. 1976. Residues of PCB components in broiler cockerels receiving two Aroclors in three dietary variations. J. Agric. Food Chem. **24:** 256–261.
13. HANSEN, L. G., M. E. WELBORN, R. E. BORCHARD, R. H. TESKE & R. L. METCALF. 1977. Tissue distribution of PCB components in swine and sheep fed three different rations containing Aroclors 1242 and 1254. Arch. Environ. Contam. Toxicol. **5:** 257–278.
14. HUTZINGER, O., *et al.*, 1972. Polychlorinated biphenyls: Metabolic behavior of pure isomers in pigeons, rats, and brook trout. Science **178:** 312–314.
15. GARDNER, A. M., J. T. CHEN, J. A. G. ROACH & E. P. RAGELIS. 1973. Polychlorinated biphenyls: Hydroxylated urinary metabolites of 2,5,2′,5′-tetrachlorobiphenyl identified in rabbits. Biochem. Biophys. Res. Commun. **55:** 1377–1383.
16. YOSHIMURA, H. & H. YAMAMOTO. 1973. Metabolic studies on polychlorinated biphenyls. I. Metabolic fate of 3,4,3′,4′-tetrachlorobiphenyl in rats. Chem. Pharm. Bull. (Japan) **21:** 1168–1169.

17. YOSHIMURA, H., H. YAMAMOTO & S. SAEKI. 1973. Metabolic studies on polychlorinated biphenyls. II. Metabolic fate of 2,4,3′,4′-tetrachlorobiphenyl in rats. Chem. Pharm. Bull. (Japan) **21:** 2231–2236.
18. BORCHARD, R. E., M. E. WELBORN, W. B. WIEKHORST, D. W. WILSON & L. G. HANSEN. 1975. Pharmacokinetics of Aroclor 1254 components in swine and sheep after a single oral dose. J. Pharmaceut. Sci. **64:** 1294–1302.
19. BORCHARD, R. E., L. G. HANSEN, W. G. HUBER, R. L. METCALF & M. E. WELBORN. 1974. Pharmacokinetics of Aroclor 1254 components after intravenous administration to swine and sheep. Arch. Environ. Contam. Toxicol. **2:** 179–192.
20. HANSEN, L. G. & M. E. WELBORN. 1977. Distribution, dilution and elimination of polychlorinated biphenyl analogs in growing swine. J. Pharm. Sci. **66:** 497–501.
21. MATTHEWS, H. B. & M. W. ANDERSON. 1975. Effect of chlorination on the distribution and excretion of polychlorinated biphenyls. Drug Metab. Dispos. **3:** 371–380.
22. LUTZ, R. J., R. L. DEDRICK, H. B. MATTHEWS, T. E. ELING & M. W. ANDERSON. 1977. A preliminary pharmacokinetic model for several chlorinated biphenyls in the rat. Drug Metab. Dispos. **5:** 386–395.
23. VILLENEUVE, D. C. 1975. The effect of food restriction on the redistribution of hexachlorobenzene in the rat. Toxicol. Appl. Pharmacol. **31:** 313–319.
24. WILSON, D. W. & L. G. HANSEN. 1976. Pharmacokinetics of HCB in growing swine. Pharmacologist **18:** 196.
25. SUNDLOF, S. F., L. G. HANSEN & M. E. WELBORN. 1976. Pharmacokinetics of HCB in male laboratory beagles. Pharmacologist **18:** 149.
26. HAYES, J. R., M. MGBODILE & T. C. CAMPBELL. 1973. Dependence of K_m and V_{max} on substrate concentration for rat hepatic microsomal ethylmorphine *N*-demethylase. Biochem. Pharmacol. **22:** 1517–1520.
27. WILSON, D. W. & L. G. HANSEN. 1976. Comparative kinetics and inhibition of swine and sheep liver microsomal enzymes. Gen. Pharmacol. **7:** 371–375.
28. CHEN, P. R., H. M. MEHENDALE & L. FISHBEIN. 1973. Effect of two isomeric tetrachlorobiphenyls on rats and their hepatic enzymes. Arch. Environ. Contam. Toxicol. **1:** 36–47.
29. GOLDSTEIN, J. A., J. D. MCKINNEY, G. W. LUCIER, P. HICKMAN, H. BERGMAN & J. A. MOORE. 1976. Toxicological assessment of hexachlorobiphenyl isomers and 2,3,7,8-tetrachlorodibenzofuran in chicks. II. Effects on drug metabolism and porphyrin accumulation. Toxicol. Appl. Pharmacol. **36:** 81–92.
30. BRUCE, R. J. 1978. The effects of two purified isomers of polychlorinated biphenyls in chickens. M. S. Thesis. University of Illinois, Urbana, IL.
31. AX, R. L. & L. G. HANSEN. 1975. Effects of purified PCB analogs on chicken reproduction. Poult. Sci. **54:** 895–900.
32. AX, R. L., W. L. MILLER, Jr. & L. G. HANSEN. 1976. Toxicity comparisons of two pentachlorobiphenyls. J. Anim. Sci. **42:** 1365.

MAMMALIAN BIOTRANSFORMATION OF CHLORDANE: *IN VIVO* AND PRIMARY HEPATIC COMPARISONS

Alan A. Brimfield and J. C. Street

Interdepartmental Program in Toxicology and Department of Animal, Dairy, and Veterinary Science Utah State University Logan, Utah 84322

Patented in Great Britain in 1949, chlordane was the first member of the cyclodiene family of chlorinated hydrocarbon pesticides to be used in agriculture.[1,2] The group also includes the familiar names dieldrin, heptachlor, aldrin, and endrin as well as several lesser known insecticides. Like other chlorinated hydrocarbon pesticides these compounds are characterized by long environmental half-life. This has contributed to both their pesticidal effectiveness and to the ubiquitous nature of the environmental contamination that has resulted from their use.

The cyclodiene family has been the subject of intensive research by Brooks in England and by Ludwig and Korte in Germany. Brooks' work has been concerned with the creation of more effective pesticides of reduced mammalian toxicity through increased understanding of metabolism and clearance by mammalian and insect enzyme systems. Work of the German group has dealt with mammalian metabolism and environmental chemistry. Work with the cyclodienes has been comprehensively reviewed by Brooks.[3,4]

The serious study of the mammalian metabolism of chlordane began in the United States in response to an environmental problem. Lawrence *et al.*[5] reported that an epoxide residue had been found in the milk and cheese from cows fed chlordane-treated alfalfa. Immediate research concentrated on the chemical characterization of this epoxide, oxychlordane, (V, FIGURE 1) and on its production and fate *in vivo*.[6–8] Since then, work has been directed at reaching an understanding of the overall metabolism of chlordane *in vivo* and *in vitro*.[9–14] A recent study of the chronic effects of low level feeding has placed chlordane on the list of suspected environmental carcinogens.[15]

OXYCHLORDANE PRODUCTION AND A HYPOTHETICAL PATTERN FOR OVERALL METABOLISM

The position of oxychlordane in the context of complete chlordane metabolism was clarified with the publication of the *in vitro* investigations of Street and Blau.[8,9] They exposed pure *cis*-chlordane (I, FIGURE 1) and *trans*-chlordane (II, FIGURE 1) isomers to metabolism by liver postmitochondrial supernatant from rats induced by pretreatment with DDT. By measuring the recovery of unaltered *cis*- or *trans*-chlordane and the production of the two known metabolites, dichlorochlordene (IV, FIGURE and oxychlordane, they found that a greater proportion of the *trans*-chlordane (37.7%) was converted to oxychlordane than of *cis*-chlordane (14.1%).[8] Total recoveries of unmetabolized *cis*- and *trans*-chlordane were 18% and 39% respectively. Oxychlor-

0077–8923/79/0320–0247 $01.75/0 © 1979, NYAS

dane was recovered virtually unmetabolized when used as a substrate for the *in vitro* liver system. Based on these facts they concluded that although both isomers are oxidatively metabolized to the inert epoxide, other paths must be responsible for the transformation of a large portion of both isomers; namely, 23.2% of the *trans*-chlordane and 67.9% of the *cis*-chlordane. Work by Brimfield and Street[12] confirmed this. The indication was that *cis*-chlordane was a more suitable substrate for these alternative paths than *trans*-chlordane.

Results of that work also led to the proposal of a metabolic route for the production of oxychlordane from either of the chlordane isomers through the common unsaturated intermediate dichlorochlordene. The desaturation step was proposed as being rate-limiting in oxychlordane production based on an experiment in which rabbit liver was shown to convert the chlordanes to dichlorochlordene with much less facility than rat liver *in vitro* but was equally capable of producing oxychlordane from dichlorochlordene when that desaturated compound was used as substrate.[9]

Evidence for the terminal nature of oxychlordane as well as for the existence of alternate routes of chlordane biotranformation also comes from the *in vivo* work of Barnett and Dorough.[11] No oxychlordane was found in the feces of rats fed either a 25:75 mixture of *trans*-chlordane and *cis*-chlordane or the individual isomers although other metabolites were present. There was some oxychlordane present in the urine of the animals fed the mixture. In a related study, animals given a single oral dose of oxychlordane excreted it intact in the feces. No other metabolites were found in the feces of those animals. These results support the conclusion of Street and Blau that oxychlordane represents a terminal residue in the rat. An explanation of the difference in relative amounts of epoxide production from *cis*- and *trans*-chlordane had to await the application of more sophisticated analytic techniques.

Direct Evidence for the Existence of Pathways Other Than Desaturation and Epoxide Production

The extent to which oxidative metabolism contributes to chlordane biotransformation by paths other than epoxide production became further apparent with the report of hydroxylated metabolites found in the urine of rabbits fed ^{14}C-labeled *trans*-chlordane.[10] Two compounds were found one of which was shown to be identical to one of the two chlorohydrins derived synthetically from chlordene (XIII, Figure 2). Based on the polarity it exhibited during thin-layer chromotography, this compound was assigned the structure 1-hydroxy-2-chlorodihydroheptachlor (VII, Figure 1). Brooks,[4] reviewing an earlier description of this work, concluded on steric grounds that this compound would have an exo-hydroxyl group. The second metabolite was thought to be 1,2-dihydroxydihydrochlordene (XVII, Figure 2), although confirmation of this was not pursued. It was assumed that this diol was the product of the hydrolytic removal of both chlorines.

Similar compounds have been identified in two additional studies involving metabolism of chlordane in different systems. Barnett and Dorough[11] identified a hydroxylated dihydroheptachlor with a 1-chloro-2-hydroxy configuration (VII, Figure 2) as a component in extracts of the feces of rats fed a mixture of 75% *cis*-chlordane and 25% *trans*-chlordane. Configuration, in this case, was based on comparison with the behavior of a reference sample with the same structure. Acetylation of the experimental material showed that it consisted of equal concentrations of two compounds. The second corresponded to a minor component of the acetylated reference sample. This second compound could be identical to the 1-hydroxy-2-chlorometabolite found in the work of Poonawalla and Korte. Brimfield and Street[12] identified the 1-chloro-2-hydroxydihydroheptachlor (VII, Figure 2) as a metabolite of *trans*-chlordane formed by rat liver post-mitochondrial supernatant.

Structural determination, in that case, was based on thin-layer chromatographic comparison with a synthetic standard characterized by nuclear magnetic resonance.

Actual metabolic events are probably most faithfully represented in the finding of Barnett and Dorough that both the 1-chloro-2-hydroxy and the 1-hydroxy-2-chloro compounds are produced. One configuration may tend to predominate over the other, however, depending on the metabolic system under investigation or the isolation and identification techniques employed.

Additional light has been shed on the metabolic fate of chlordane exclusive of the

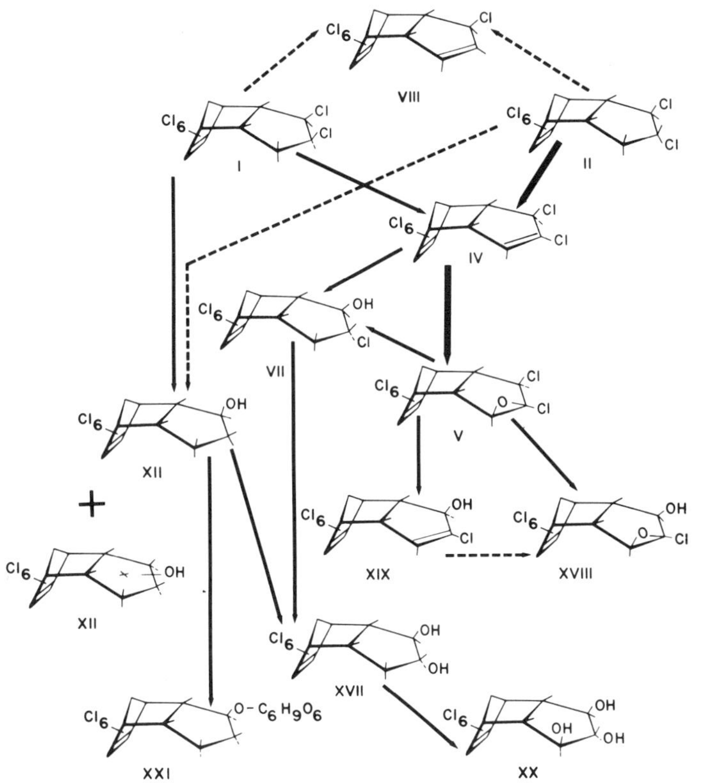

FIGURE 1. The proposed routes of metabolism for *cis*- and *trans*-chlordane by rats *in vivo*. The solid lines indicate the active routes and the broken lines represent weak metabolic routes. The thicknesses of solid lines indicate relative metabolic activity. (Adapted from Tashiro and Matsumura.[13] Reprinted by permission of the American Chemical Society.)

dichlorochlordene to oxychlordane conversion by the work of Barnett and Dorough.[11] Although they made no attempt to outline a pathway to account for their results, they were able to partially characterize four previously unreported metabolites of chlordane. These compounds were a hydroxylated chlordane (III, FIGURE 2), a dihydroxydihydroheptachlor having a molecular weight of 404 (X, FIGURE 2), the *cis*- and/or *trans*-diol of dihydrochlordene (XVII, FIGURE 2) and a triply hydroxylated dihydrochlordene (XX, FIGURE 1), Excreted conjugated of hydroxychlordane and the trihydroxy species were also observed, which constituted the first report of chlordane-generated conjugates.

The results of the work of Poonawalla and Korte,[10] Barnett and Dorough,[11] and

Brimfield and Street[12] lend further credibility to the conclusion reached by Street and Blau.[8] Considerable oxidative metabolism of both chlordane isomers occurs exclusive of epoxide formation.

The Overall Pathway of Chlordane Metabolism: Comparison of the Results from *in Vitro* and *in Vivo* Investigations

The in Vitro *Investigation*

The possibility of finding a mechanistic explanation for the production of different amounts of oxychlordane from *cis*- and *trans*-chlordane isomers led Brimfield *et al.* to re-study the primary hepatic metabolism of these isomers.[14] An *in vitro* system was employed consisting of washed microsomes, prepared from rats induced with the chlordane isomer of interest, and a system for the reduction of NADP. Washing microsomes were intended to remove soluble enzymes and glutathione which might have interferred with the identification of the resulting metabolites. It was felt that working *in vitro* would help to eliminate the uncertainty that often arises in establishing the origin and mode of production of metabolites because of possible effects of digestive tract physiology and microflora.

After incubation with one of the pure chlordane isomer, extracts of the *in vitro* metabolizing system were fractionated on neutral alumina columns and subjected to analysis by combined gas-chromatography—mass spectroscopy. The results of this work are summarized in TABLE 1. These results led to the proposed pathway shown in FIGURE 2. The pathway is an attempt to outline the primary hepatic metabolism of the chlordanes and explain the metabolic dichotomy with respect to oxychlordane production found by Street and Blau in a manner consistent with the present and previous work on cyclodiene metabolism.

Briefly, it was proposed that a molecule of chlordane can follow one of three courses during hepatic microsomal metabolism. It can be hydroxylated at the 3 position to give hydroxychlordane (III, FIGURE 2), presumably prior to conjugate formation per reference 11. Alternatively, desaturation to dichlorochlordene (IV, FIGURE 2) can occur followed by the formation of oxychlordane (V, FIGURE 2). Finally, it can proceed through reductive dehalogenation to give metabolites with increasingly fewer chlorines on the cyclopentane ring (VI, and XI, FIGURE 2). These dehalogenated molecules can then participate in the pathways of hydroxylation or desaturation and epoxidation that parallel the formation of dichlorochlordene and oxychlordane from chlordane. The major route taken by a given chlordane isomer is probably determined by whether the chlorine at position 2 is exo (*cis*) or endo (*trans*). The *trans*-isomer is predominately desaturated or hydroxylated whereas the *cis*-isomer is dehalogenated as its major mode of biotransformation.[14]

An additional metabolic transformation indicated by the scheme in FIGURE 1 is the hydrolytic cleavage of epoxide rings by the inducible enzyme, hepatic epoxide hydratase. This reaction has been shown to be sterospecific always yielding *trans*-dihydrodiols and is independent of molecular oxygen and $NADPH_2$.[4,16] No evidence was found in this work to indicate an interaction of epoxide hydratase with oxychlordane. However, induction of rats with a 25:75 mixture of *trans*- and *cis*-chlordanes caused an increase in microsomal epoxide hydratase activity.[17] Additionally, the cleavage of heptachlor epoxides, chlordene epoxide and the epoxides of various of the methanonaphthalene derivatives is well-documented.[4,16] Our finding of a dihydrodiol of heptachlor (X, FIGURE 2) from *cis*-chlordane[14] as well as the report of a six chlorine-containing diol from the work of Barnett and Dorough[11] provided evidence

TABLE 1A

COMPOUNDS IN ELUATES FROM COLUMN CHROMATOGRAPHY OF AN EXTRACT FROM THE *IN VITRO* METABOLISM OF *CIS*-CHLORDANE*

Compound	Empirical Formula	Molecular Weight	GLC Retention Time (min)	Column Fraction 25 ml
1. *Cis*-chlordane	$C_{10}H_6Cl_8$	406	2.7	All
2. Dichlorochlordene	$C_{10}H_4Cl_8$	404	1.8	I
3. Oxychlordane	$C_{10}H_4OCl_8$	420	2.1	I
4. Dihydroheptachlor	$C_{10}H_7Cl_7$	372	2.3	I
5. Hydroxychlordene	$C_{10}H_6OCl_6$	352	2.0	III
6. 1-Chloro-2-hydroxy-dihydrochlordene	$C_{10}H_7OCl_7$	388	3.5	III

*From Brimfield *et al.*[14]

TABLE 1B.

COMPOUNDS IN ELUATES FROM COLUMN CHROMATOGRAPHY OF AN EXTRACT FROM THE *IN VITRO* METABOLISM OF *TRANS*-CHLORDANE*

Compound	Empirical Formula	Molecular Weight	GLC Retention Time (min)	Column Fraction 25 ml
1. *Trans*-chlordane	$C_{10}H_6Cl_8$	406	2.5	All
2. Dichlorochlordene	$C_{10}H_4Cl_8$	404	1.8	I, II
3. Oxychlordane	$C_{10}H_4OCl_8$	420	2.1	I, II
4. Heptachlor	$C_{10}H_5Cl_7$	370	1.3	I
5. 1-Chloro-2-hydroxy-dihydrochlordene	$C_{10}H_7OCl_7$	388	3.5	III
6. Hydroxychlordene	$C_{10}H_6OCl_6$	352	2.0	III
7. Hydroxychlordane	$C_{10}H_6OCl_8$	422	4.8	III

*From Brimfield *et al.*[14]

that epoxide hydratase may be involved in chlordane metabolism at some level other than that of oxychlordane where it is inactive.

An Alternative Overall Pathway for Chlordane Biotransformation Based on the Outcome of an in Vivo *Investigation*

The most current investigation of the metabolism of pure *cis*- and *trans*-chlordane isomers to be found in the literature is that of Tashiro and Matsumura.[13] This was an *in vivo* study using rats in which a small amount of confirmatory *in vitro* work was carried out. The research included feeding pure chlordane isomers and collecting urine and feces for metabolite recovery, qualitative identification and quantitation, the determination of relative excretion rates using ^{14}C-labeled *cis*- and *trans*-chlordane and, finally, the determination of the relative insecticidal activity of isolated metabolites using a mosquito larvae bioassay.

The conclusions reached as a result of this investigation are in general agreement with the proposals of Street and Blau[8,9] and the results of the *in vitro* metabolism work of Brimfield and Street.[12] They found that while both *cis*- and *trans*-chlordane are metabolized to oxychlordane, via the common intermediate dichlorochlordene, more oxychlordane was generated in the metabolism of *trans*-chlordane. Tashiro and Matsumura also found that the major metabolic route for *cis*-chlordane involves forms of oxidative metabolism other than epoxide formation leading to more complete metabolism of *cis*-chlordane and a higher excretion rate for *cis*-chlordane derived material. Agreement between this investigation and the study of Brimfield *et al.*[14] lies in the identification of heptachlor as a chlordane metabolite. The results of this *in vivo* work were incorporated into a proposed pathway for the biotransformation of the chlordanes in rats. This pathway is reproduced in FIGURE 1.

Despite the general agreement between the *in vitro* work of Brimfield *et al.* and this *in vivo* investigation, when the details of the two pathways are compared correlation breaks down to some extent: specifically the sequence of events proposed to account for the various metabolites identified is considerably different.

The identification of a hydroxylated chlordane as a metabolite of *trans*-chlordane in the *in vitro* paper[14] marks a major departure from the Tashiro and Matsumura pathway. This compound was neither among the direct metabolic products identified from extracts of urine and feces or the aglycones generated as a result of the action of β-glucuronidase on isolated glucuronide conjugates from the *in vivo* work. On the basis of the results of Barnett and Dourough this might have been an expected source of the hydroxylated chlordane.[11] Brimfield *et al.* felt that the fact that this material arose as a product of the metabolism of *trans*-chlordane only made mechanistic sense from the point of view that the relatively facile desaturation of the *trans*-isomer indicates a tendency toward oxidative metabolism not shared by *cis*-chlordane. They are not able, unfortunately, to back up this point of view with the results of Barnett and Dorough because the latter did not study the metabolism of pure isomers but used a mixture. The absence of hydroxy chlordane in this investigation may be explainable on the basis of hepato-biliary circulation of this metabolite or its conjugation product leading to "remetabolism" and excretion as the dihydroxy or trihydroxy chlordenes. Both of these compounds were identified.[13]

Significantly, the production of heptachlor or a heptachlor-like substance is reported in both the *in vivo* and *in vitro* investigations (VIII, FIGURES 1 and 2) leading to the offering of overall pathways of metabolism. Tashiro and Matsumura found that heptachlor arises as the result of the metabolism of both *cis*-chlordane and *trans*-chlordane. Brimfield *et al.* detected it only in extracts of the microsomal metabolism of *trans*-chlordane although they indicate the probability that it arises as the result of the metabolism of both isomers in the pathway. The point of contention with respect to heptachlor production, in the two metabolic schemes, is where it arises in overall chlordane metabolism.

Tashiro and Matsumura have placed heptachlor in the position of a terminal residue undergoing no additional metabolism (FIGURE 1). Brimfield *et al.*, on the other hand, speculate that heptachlor belongs in the position of a metabolic

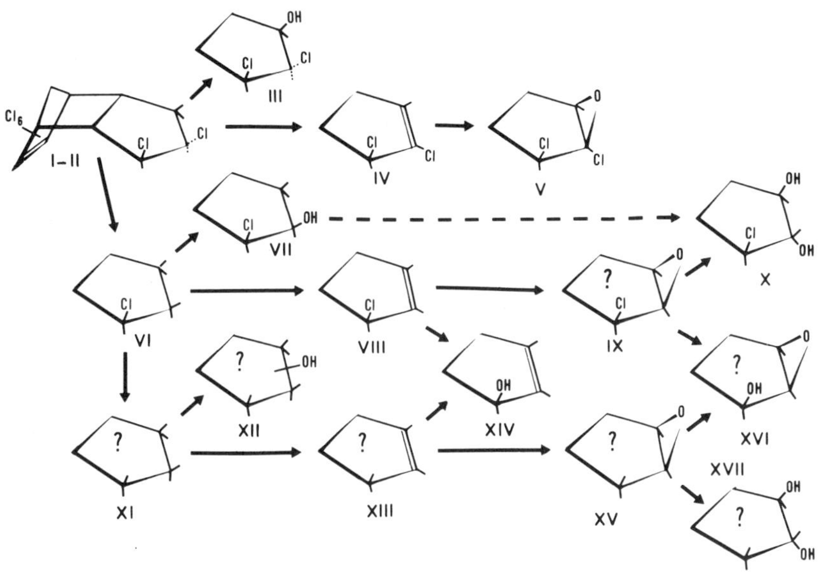

FIGURE 2. The proposed routes of metabolism for *cis*- and *trans*-chlordane in microsomal preparations from induced rat liver. Structure (I-II) represents the complete *cis*- and *trans*-chlordane molecules. Metabolites are shown only as the five-membered portion of the molecule with the hexachloronorbornene nucleus understood. Metabolites proposed, but not actually found in this or other work, are designated with a question mark (?). (From Brimfield *et al.*[14] Used by permission of Academic Press.)

intermediate (FIGURE 2). In the course of their study Tashiro and Matsumura subjected the heptachlor isolated from *in vivo* metabolism to metabolism by rat liver post-mitochondrial supernatant. Their results show only the production of a small quantity of 1,2-dihydroxydihydrochlordene although this is not reflected in the final pathway. It is undoubtedly this dearth of metabolic products that led them to place heptachlor in a terminal position. The literature shows that heptachlor is, indeed, a metabolically labile compound both *in vivo*[18] and *in vitro*[19] yielding heptachlor epoxide and a hydroxy epoxide. Neither Brimfield *et al.* nor Tashiro and Matsumura found that these compounds arise during the metabolism of chlordane. But it is possible that a heptachlor epoxide does arise as a result of chlordane metabolism and is rapidly hydrolysed to the diol or converted to the hydroxy epoxide and then bound to tissue macromolecules in the manner of the hydroxy epoxides formed from the carcinogenic polycyclic aromatic hydrocarbons.[20] The report by Brimfield *et al.*[14] of a dihydrodiol of heptachlor from the metabolic transformation of *cis*-chlordane and the production of a 6 chlorine-containing diol in the *in vivo* investigation by Barnett and Dorough[11] support the former point of view. The binding theory may have some merit though in the light of a report of the carcinogenic activity of heptachlor in mouse liver as the result of feeding heptachlor.[15] This is clearly an area where further work is called for.

The report of Brimfield *et al.*[14] of a *cis*-chlordane derivative with a fully saturated cyclopentane ring having only one chlorine (VI, FIGURE 2) marks another point disagreement between the results of the *in vivo* and *in vitro* investigations. There is no precedent for this type of metabolic product in any of the studies of chlordane metabolism under discussion here. This, in fact, may be the kind of easily overlooked detail on which a case can be built for studying metabolism both *in vivo* and *in vitro* before drawing conclusions about metabolic phenomena. The obvious conclusion is that this compound does not show up in urine and feces.

The assumption from the presence of this metabolite in the *in vitro* system was that reductive dehalogenation is involved in the hepatic metabolism of the chlordanes, primarily the *cis*-isomer.[14] Although this is at odds with the results of the *in vivo* studies, it was not an entirely novel suggestion. For example, it has been known for some time that DDT is converted to its reduced dehalogenated derivative DDD in rat liver homogenate.[21] Recently, Lay *et al.*[22] described the reductive dechlorination of the aldrin metabolite dihydroaldrin dicarboxylic acid in liver microsomes. Although Tashiro and Matsumura did not find this metabolite *per se*, a glance at their pathway indicates that dehalogenation is one of the dominant modes of *cis*-chlordane metabolism (XII, XVII, XXI, FIGURE 1) when compared to the largely mono and dichlorinated derivatives of *trans*-chlordane (IV, V, VII, FIGURE 1).

Confirmation of the results of Brimfield *et al.* will have to await additional *in vitro* work. For the time being, however, the obvious conclusion is that the dihydroheptachlor (VI, FIGURE 2) is rapidly and completely converted to other metabolites so that it does not appear in urine or feces.

One final troublesome spot in the Tashiro and Matusumura pathway (FIGURE 1) is the handling of oxychlordane. This represents a departure from all previously reported work with the chlordanes (see above). The several attempts to exhibit oxychlordane metabolism *in vivo*[11] and *in vitro*[9] have led to the conclusion that oxychlordane is a terminal residue. In looking at oxychlordane metabolism in induced male rat liver microsomes, Street and Blau recovered 95% of the added oxychlordane intact after 4 hours of incubation.[9] The addition of glutathione and ascorbic acid to their system, however, resulted in 100% recovery so the 5% loss may have been the result of some nonenzymatic transformation occurring over the course of a 4-hour incubation. The addition of glutathione and ascorbate to the same system metabolizing *trans*-chlordane had an enhancing effect on oxychlordane production so there was no inhibition of oxidative metabolism due to the presence of these two compounds.

Attempts by Brimfield and Street to implicate hepatic epoxide hydratase in oxychlordane metabolism showed that the presence of this epoxide in three concentrations had no effect or a slight enhancing effect on the conversion of the hydratase labile substrate ^{3}H-styrene oxide to ^{3}H-styrene dihydrodiol.[17] The implication is that oxychlordane is inert even to the activity of an epoxide-specific microsomal enzyme that has been shown to be active in the conversion of the epoxides of other less highly chlorinated cyclodiene molecules.[16]

Tashiro and Matsumura pathway depicts oxychlordane (V, FIGURE 1) as being transformed via minor routes of metabolism to 1-hydroxy-2-chlorodihydroheptachlor (VII, FIGURE 1), a 1-hydroxy-2-chloroheptachlor (XIX, FIGURE 1), and a 1-hydroxy-2-chloroheptachlor epoxide (XVIII, FIGURE 1). It is difficult to account for this sudden spate of metabolic products from what has, until now, been assumed to be a terminal product of chlordane metabolism. One possible explanation is the production of these new metabolites during preparative thin-layer chromatography on alumina G as was the practice during this work. Morley, in his discussion of the chromatographic behavior of the highly base labile chlorinated cyclodienes, says that gels of basic alumina give rise to dehydrohalogenation and dehydration products when these compounds are submitted to chromatographic separation on them.[23] This has been the experience in our laboratory as well.[17]

REFERENCES

1. WINDHOLZ, M. Ed. 1976. Merck Index 9th edition. Merck and Co. Rahway, N.J.
2. MELNIKOV, N. N. 1971. Chemistry of pesticides. Residues Reviews. Vol. 36. F. A. Gunther & J. D. Gunther, Eds. Springer Verlag. New York, Heidelberg, and Berlin.
3. BROOKS, G. T. 1974. Chlorinated Insecticides Vol II. CRC Press, Cleveland, Ohio.
4. BROOKS, G. T. 1966. Progress in metabolic studies of the cyclodiene insecticides and its relevance to structure-activity correlations. World Rev. Pest. Cont. **5**(2): 62–84.
5. LAWRENCE, J. H., R. P. BARRON, J. Y. T. CHEN, P. LOMBARDO & W. R. BENSON. 1970. Note on the identification of a chlordane metabolite found in milk and cheese. J. Assoc. Off. Anal. Chem. **53:** 261–262.
6. POLEN, P. B., M. HESTER & J. BENZIGER. 1971. Characterization of oxychlordane, animal metabolite of chlordane. Bull. Environ. Contam. Toxicol. **5:** 521–528.
7. SCHWEMMER, B., W. P. COCHRANE & P. B. POLEN. 1970. Oxychlordane, animal metabolite of chlordane. Science **169:** 1087.
8. STREET, J. C. & S. E. BLAU. 1972. Oxychlordane: accumulation in rat adipose tissue on feeding chlordane isomers and technical chlordane. J. Agric. Food Chem. **20**(2): 395–397.
9. STREET, J. C. & S. E. BLAU. 1971. Chlordane metabolism by rat liver *in vitro*. Paper presented at the 162nd national meeting of the American Chemical Society, Los Angeles, Ca.
10. POONAWALLA, N. H. & F. KORTE. 1971. Metabolism of trans-chlordane-^{14}C and isolation and identification of its metabolites from the urine of rabbits. J. Agric. Food Chem. **19:** 467–470.
11. BARNETT, J. R. & H. W. DOROUGH. 1974. Metabolism of chlordane in rats. J. Agric. Food Chem. **22**(4): 612–619.
12. BRIMFIELD, A. A. & J. C. STREET. Comparison of the metabolism of cis- and trans-chlordane *in vitro* in liver homogenates from the rat. Paper presented at the 167th national meeting of the American Chemical Society, Los Angeles, Ca. April, 1974.
13. TASHIRO, S. & F. MATSUMURA. 1977. Metabolic routes of cis- and trans-chlordane in rats. J. Agric. Food Chem. **25**(4): 872–880.
14. BRIMFIELD, A. A., J. C. STREET, J. FUTRELL & D. CHATFIELD. 1978. Identification of products arising from the metabolism of cis- and trans-chlordane in rat liver microsomes *in vitro:* outline of a possible metabolic pathway. Pestic. Biochem. Physiol. **9:** 84–95.
15. Drinking water and health. 1977. A report of the Safe Drinking Water Committee, Advisory Center on Toxicology, Assembly of Life Sciences of the National Research Council. National Academy of Sciences, Washington, D.C.
16. BROOKS, G. T., A. HARRISON & S. E. LEWIS. 1970. Cyclodiene epoxide ring hydration by microsomes from mammalian liver and houseflies. Biochem. Pharmacol. **19:** 255–273.
17. BRIMFIELD, A. A. & J. C. STREET. Unpublished observations.
18. KAUL, R. W., W. KLEIN & F. KORTE. 1970. Beitrage zu okologischen chemie XX

verteilung, ausscheidung und metabolismus von telodrin und heptachlor in ratten und mannlischen kaninschen, endprodukt des warmblutermetabolismus von heptachlor. Tetrahedron **26:** 331–337.
19. NAKATSUGAWA, T., M. ISHIDA & P. A. DAHM. 1965. Microsomal epoxidation of cyclodiene insecticides. Biochem. Pharmacol. **14:** 1853–1865.
20. SIMS, P., P. L. GROVER, A. SWAISLAND, K. PAL & A. HEWER. 1974. Metabolic activation of benzopyrene proceeds by a diol-epoxide. Nature **252:** 326–328.
21. PETERSON, J. E. & W. H. ROBINSON. 1964. Metabolic products of p, p′-DDT in the rat. Toxicol. Appl. Pharmacol. **6:** 321–327.
22. LAY, J. P., I. WEISGERBER & W. KLEIN. 1975. Conversion of the aldrin/dieldrin metabolite dihydroaldrin dicarboxylic acid-^{14}C in rats. Pestic. Biochem. Physiol. **5:** 226–232.
23. MORLEY, H. V. 1966. Adsorbants and their application to column cleanup of pesticide residues. Residue Reviews. Vol. 16. F. A. Gunther, Ed. Springer-Verlag, New York.

PHARMACOKINETICS OF HALOGENATED HYDROCARBONS

P. M. Bungay and R. L. Dedrick

Biomedical Engineering and Instrumentation Branch
National Institutes of Health,
Bethesda, Maryland 20014

H. B. Matthews

Laboratory of Pharmacokinetics
National Institutes of Environmental Health Sciences,
Research Triangle Park, North Carolina 27709

INTRODUCTION

Pharmacokinetics concerns the time course of acquisition, distribution, chemical conversion and elimination of a substance by the body, together with the concomitant variations of co-reactants and metabolic products. All of these processes occur at various specific sites (or regions) within (or on) the body. Generally, the rates of the processes are related to local concentration levels, as well as a multitude of properties of the milieu which also vary from location to location. The description of pharmacokinetic behavior at the whole-body level necessitates being able to follow the local variations. One means of achieving this is through the use of mathematical models whose structure corresponds to anatomic and physiological concepts, and whose parameters are characteristic of particular organs, fluids, or tissues. The models provide the capability to mathematically simulate the time course of local concentration throughout the body. These mathematical tools are referred to generically as physiological pharmacokinetic models. The modifier "physiological" is used to distinguish the modeling approach to be addressed in this paper from that of classical pharmacokinetics. The latter is typically an abstract curve-fitting process in which the parameters are not amenable to direct verification.

PHYSIOLOGICAL MODELING CONCEPTS

In formulating physiologically based models of the body, a mass conservation equation is written for each important, anatomically recognizable subdivision, or "compartment." Importance relates to whether the compartment is a target tissue (for a drug) or a site of toxicity, or significantly influences the kinetics. Each compartment is assumed to be sufficiently homogeneous that it can be considered well-mixed with respect to internal distribution of the chemical. For illustrative purposes, a typical organ is shown in FIGURE 1, arbitrarily subdivided into two compartments, intravascular, and extravascular. Other subdivisions are possible depending upon the way in which the particular substance distributes. Alternative compartment definitions might be chosen, such as plasma, interstitial fluid, extracellular or intracellular fluid. The essential point is that regardless of the choice of subdivision, each compartment has associated with it a real volume (denoted by V) which can in theory be directly measured.

The essence of the modeling approach is to compute, by means of simultaneous solution of the set of mass balance equations, the variation over time in the amount of

0077-8923/79/0320-0257 $01.75/0 © 1979, NYAS

the substance—or equivalently its concentration—in each compartment. Any number of substances can be followed. For example, to predict the kinetics of a parent compound *plus* one or more metabolites, a set of conservation equations would be formulated for each of the chemical constituents of interest. The equations are interrelated by terms representing mass exchange between compartments or interconversion between constituents by chemical reaction.

The most important process for exchange between organs is convection, usually by means of the circulatory system, but other fluids such as lymph, cerebrospinal fluid, or gastrointestinal contents can serve as significant vehicles for bulk transport. Associated with the convective movement is a flow rate, Q, of the carrier fluid in units of volume per time. The amount of the substance brought to the blood compartment within the organ in FIGURE 1 is the product of the flow rate and the average concentration, C_{B_i}, in the incoming blood. Likewise, the amount per unit time leaving the organ in the blood is QC_{B_o}. It is assumed that the blood within the organ is well mixed with uniform concentration C_{B_o}. The tissue is also considered to have a uniform concentration, C_T.

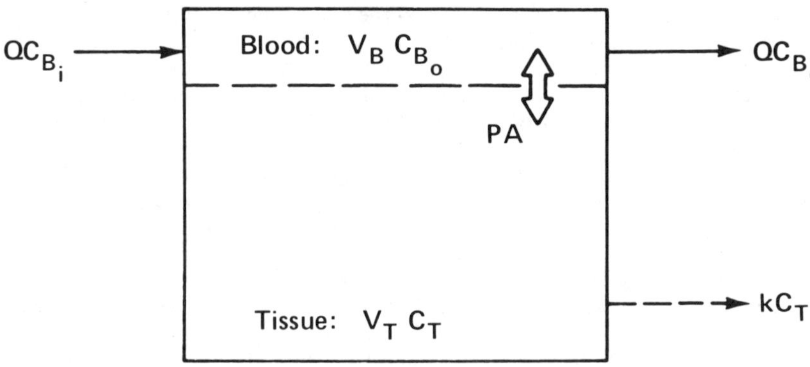

FIGURE 1. Schematic drawing of a typical organ subdivided into homogeneous compartments to permit generalized physiologically based pharmacokinetic analysis. Q, Blood flow rate; PA, permeability-area product; k, metabolism, or excretion rate constant; V_B, volume of organ blood compartment; C_{B_i}, concentration in incoming blood; C_{B_o}, concentration in outgoing blood.

In general, the substance exchanges between the capillary blood and surrounding tissue but experiences some resistance, however small, at interfaces such as the capillary endothelium or tissue cell membrane. The resistance may be associated with passive and/or active processes. The mechanisms may be complex and are frequently unknown. For modeling purposes it is usually satisfactory to describe the overall effect of such processes in simple mathematical terms. The example suggested in FIGURE 1 is a linear, symmetric diffusional exchange of the substance between blood and tissue across the capillary endothelium. The net amount of the substance entering the tissue per unit time would be given by $PA(C_{B_o} - C_T/R_T)$, in which A is surface area for exchange, P is the permeability, and R_T is the ratio of tissue-to-blood concentrations under equilibrium conditions. The product, PA, is treated as a single transport parameter, since it is usually not practicable to measure P and A separately. The rate of exchange may be either positive or negative, depending upon the direction of the exchange. No net exchange would be equivalent to an equilibrium distribution between the tissue and the blood leaving the organ. The exchange need not be linear or symmetric. Transport processes at cell membranes are often saturable giving rise to nonlinear behavior. For example, the nonlinear Michaelis-Menton type kinetic expressions have been used to describe transport in pharmacokinetic models of the drug methotrexate[1].

The mass balance for the organ blood compartment is obtained by summing the rates at which the substance is being delivered to the compartment by convection and diffusion and equating the sum to the time derivative of the amount, $V_B C_{B_o}$, in the compartment,

$$Q(C_{B_i} - C_{B_o}) + PA(C_T/R_T - C_{B_o}) = \mathrm{d}(V_B C_{B_o})/\mathrm{d}t \tag{1}$$

Usually average values of parameters such as Q, V, PA, and R for the animal species under study are used. For Q and V, particularly, a large body of data exists in the literature for a number of animals and man, so that a significant portion of any model can be specified *a priori*. The important feature is that these parameters and R are generally susceptible to independent measurement either on a species average normal basis or otherwise. Sometimes it is desirable to determine a parameter specific to the subject of a given experiment. For example, certain pathologic conditions might be associated with a PA altered from normal. Moreover, the parameters need not be constants. Body weight or individual organ weights can change sufficiently during the course of long-term experiments to require that this variation be incorporated in the compartment volumes.

Within compartments, the parent substance may be eliminated either through chemical transformation to another species or by excretion processes. While it is desirable to have knowledge of the mechanism of metabolism and excretion, simple empirical kinetic expressions may provide an adequate first approximation for simulating the overall behavior. The kinetic expressions are formulated in terms of the concentration variables which would be measurable in a pharmacokinetic experiment. For the hypothetical organ in FIGURE 1 this would be the average tissue concentration, C_T. For illustration, the rate of elimination is assumed to be linearly dependent upon C_T. The following mass balance for the parent substance thus contains the term, $-kC_T$, to describe its rate of disappearance with k being the first-order rate constant,

$$\mathrm{d}(V_T C_T)/\mathrm{d}t = PA(C_{B_o} - C_T/R_T) - kC_T \tag{2}$$

By assuming that compartments are well mixed at the whole-organ level, one of course loses the spatial variation in events that would be seen in observations at the cellular level. However, the advantage gained is a simplification in the description of the events which permits the construction of tractable whole animal kinetic models. A notable aspect of the physiologically based models is that they provide a mathematical framework into which more complex models can be imbedded as detailed mechanistic and kinetic information becomes available. Because the models are constructed with well-defined real physical parameters and variables, it is possible, for example, to directly incorporate the results of *in vitro* experiments on tissue slices, homogenates or cellular fractions.

The fact that the models can usually be greatly simplified and still retain the ability to simulate the essential pharmacokinetic behavior rests upon the existence of rate-limiting steps. In complex systems involving multiple processes that occur simultaneously in series and in parallel, certain processes control the overall kinetics because they proceed at relatively slow rates. This effect is illustrated by the competition between the convective and diffusive contributions to the exchange between blood and tissue. For simplicity it is assumed that metabolism is negligible ($k = 0$) and the tissue and incoming blood are far from equilibrium. If $PA \ll Q$, transport is said to be "diffusion-controlled." The actual magnitude of the organ blood flow rate would be unimportant, since the rate of exchange would be limited by how quickly the substance could cross the capillary endothelial barrier. If, on the other hand, $Q \ll PA$, then the rate of exchange would be determined by the rate at which the substance could be brought to the organ by the blood. Each unit of blood dwells long enough within the organ for the substance to essentially equilibrate with the tissue. The exit blood concentration can then be approximated by $C_{B_o} - C_T/R_T$. In this "flow-limited" extreme, the rate of exchange is insensitive to the magnitude of PA. The two equations, (**1**) and (**2**), collapse into one:

$$\frac{d}{dt}\left(\frac{V_B}{R_T} + V_T\right) C_T = Q\left(C_{B_i} - \frac{C_T}{R_T}\right) - kC_T \tag{3}$$

Thus, in this limit, both the number of equations and the numbers of parameters can be reduced. For most substances, particularly those that are lipid soluble and not appreciably bound to blood proteins, V_B/R_T is small compared to V_T, and hence V_B can also be eliminated as a parameter. Fortunately, the pharmacokinetics of a large fraction of the substances of interest can be described adequately by flow-limited models. In subsequent sections we will consider two representative physiological models. The flow-limited assumption is invoked throughout the first model. The second model illustrates a situation in which diffusional resistances must be considered.

FIGURE 2 shows a schematic flow diagram for a whole-body model that would be suitable for many halogenated hydrocarbons and lipophilic environmental agents. Fat, skin, and muscle compartments are included since these are among the most important storage organs for lipid-soluble compounds. The liver and kidney compartments appear because of their role in metabolism and elimination. The liver is also an important storage organ for some agents—for instance, Kepone®—as well as a site of toxicity. Many of the environmental agents are known to be absorbed into the body from the intestinal contents, and it has been shown that some pass into the gut lumen in the reverse direction. It is important to consider the gut in disposition models, since fecal excretion is a primary route of elimination for the agents with higher molecular weights. The gut in FIGURE 2 is subdivided into blood, tissue, and contents compartments. In particular applications, further subdivision may be desirable, such as into series segments of small and large intestines. This will be considered below in the section on Kepone.

Compartments can be added or removed depending upon the distribution of the particular substance under study and the type of experiment performed. Examples of the influence of the type of experiment are the addition of a stomach compartment in an oral feeding study, or a peritoneal cavity compartment for intraperitoneal administration. However, the formulation of the core equations and definition of parameters in each whole-animal model are independent of the route of entry of the substance into the body. The generalizability of the physiologically based models is one of the powerful features of the approach.

The various compartments in the whole-animal model are linked together by convective exchange channels, e.g. the circulatory and biliary excretion systems, and routes for exchange across interfaces between compartments which are in contact, such as the tissue-blood and gut tissue-contents surfaces. As we have indicated above, each process of exchange, metabolic conversion and excretion represented on the diagram has its mathematical counterpart in the model equations. The model is a simplified representation of the biologic system, but which insofar as practicable incorporates directly verifiable physical and biochemical information such as blood flow rates, organ sizes, constants representing chemical binding, etc.

Much of the difference in pharmacokinetic behavior between animal species is accounted for by differences in purely physical parameters, such as flow rates, organ volumes and exchange areas. Such parameters generally scale with body weight to a power ranging approximately from 2/3 to 1. These body-weight correlations support the use of physiological pharmacokinetic models to explain differences in disposition

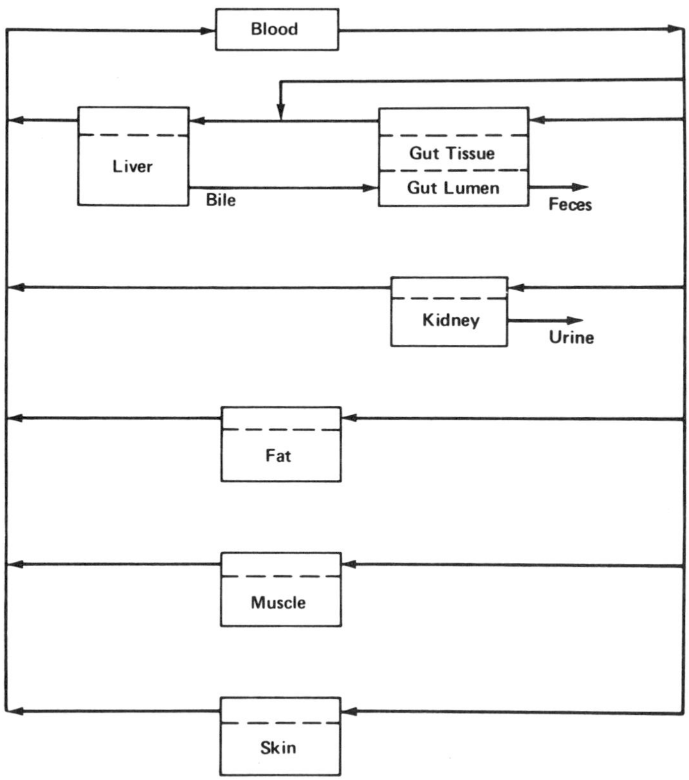

FIGURE 2. Schematic diagram for a typical whole-body physiological pharmacokinetic model.

behavior between animal species and man. Pharmacokinetic differences resulting from interspecies variation in metabolism are not so easily correlated, making this an active area of research. However, once the distribution, metabolism and excretion processes are satisfactorily described, the models provide a rational basis for pharmacokinetic predictions. Provided generalized relationships between the pharmacokinetics and pharmacodynamics can be described for toxic chemicals, the models could supply a predictive capability for toxicity as well.

Few physiological pharmacokinetic models have been developed for environmental contaminants. Aspects of models for two agents, a hexachlorobiphenyl isomer and Kepone, will be discussed in the next sections. Although both compounds are highly chlorinated hydrocarbons, they exhibit quite different distribution behavior.

HEXACHLOROBIPHENYL

Polychlorinated biphenyls (PCBs) are among the most prevalent and widely dispersed environmental contaminants. The reactivity, and hence the persistence, of the various species in this class of compounds is determined by the number and position of chlorine atoms on the molecule. A pharmacokinetic model for the rat has been developed[2,3] for each of four PCB compounds: 4-chloro, 4,4′-dichloro, 2,2′,4,5,5′-pentachloro, and 2,2′4,4′,5,5′-hexachlorobiphenyl. The models describe the disposition of the parent and the principal metabolite. The model differs among the four compounds solely in the magnitudes of the tissue-to-blood distribution ratios and the metabolism and excretion rate constants. For present purposes only the model for hexachlorobiphenyl (6-CB) will be considered.

In the rat the distribution kinetics of 6-CB and its metabolite, a glucuronide conjugate, are well approximated by a model that is simply flow-limited for most tissues. Since the metabolite is much more polar than the parent, its distribution is quite different, as can be seen by comparison of the equilibrium tissue-to-blood distribution ratios for the two chemical species given in TABLE 1. The very large distribution coefficient for the parent in adipose tissue is typical for highly lipophilic substances. The preferential distribution into fat not only dictates that a large proportion of 6-CB administered to the rat will tend to accumulate in this compartment, but this factor also exerts a predominant influence on the distribution kinetics. From Equation 3 the rate of approach to the quasisteady-state distribution following administration of 6-CB to the animal is governed by the time constant, $V_{FAT}R_{FAT}/Q_{FAT}$. The adipose compartment is poorly perfused compared to other compartments, i.e., the ratio of compartment volume to flow rate, V_{FAT}/Q_{FAT}, is relatively large, being of the order of 45 minutes. The poor perfusion combined with the high distribution coefficient results in very slow distribution transients. FIGURE 3 shows the time course of 6-CB concentration in the major tissues following a single intravenously administered dose. The liver and muscle concentrations peak within the first two hours and the skin concentration within a day. The redistribution into the fat occurs very slowly. The fat concentration is still climbing at the end of four days, and hence none of the tissues reach quasisteady state on this time scale. In FIGURE 3 the symbols represent experimental values. The solid lines are computer generated from the model equations showing how well the model simulates the distribution behavior.

The i.v. dosing experiment was carried out to 42 days. On this long a time scale

TABLE 1

6-CB TISSUE-TO-BLOOD DISTRIBUTION RATIOS[2,3]

Compartment	Parent	Metabolite
Adipose	400	2
Liver	12	4
Muscle	4	0.3
Skin	30	2

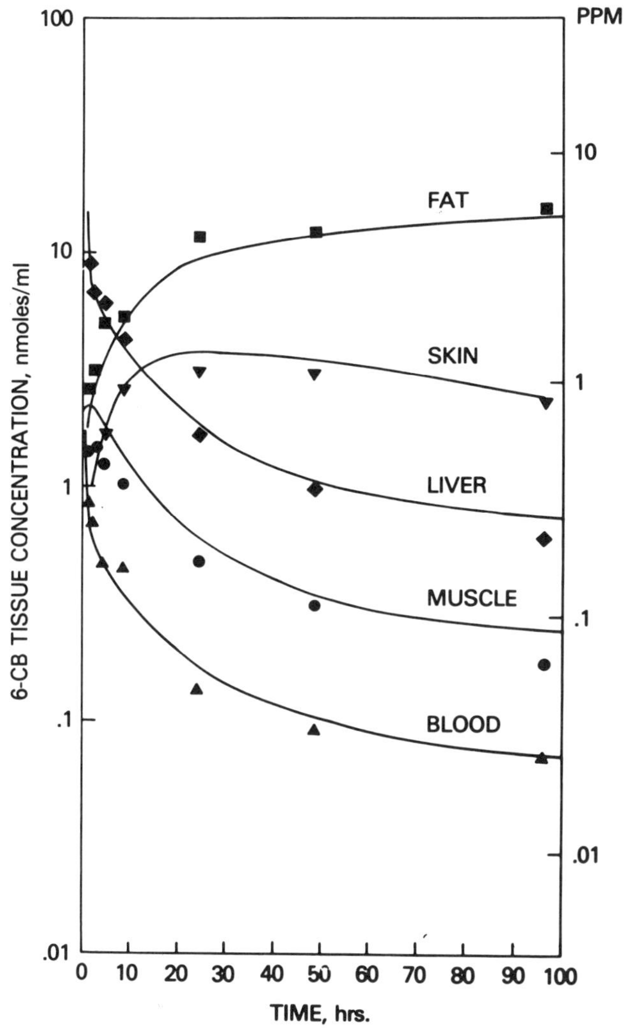

FIGURE 3. Variation in rat tissue concentrations of 6-CB during four days following intravenous administration of a single dose of 0.6 mg/kg body weight. Solid lines are generated by computer solution of pharmacokinetic model equations. Points represent experimental data. (Figure, which appeared originally in *Drug Metab. Disp.* **5**: 386 (1977), reprinted by courtesy of American Society for Pharmacology and Experimental Therapeutics.)

animal growth effects became significant and need to be taken into account in the analysis. Rats gain steadily in weight as they age. In addition, fat becomes an increasing proportion of the body. Both effects cause the 6-CB to be diluted, i.e., the concentrations would drop even in the absence of elimination. The influence of growth on fat and blood levels is shown in FIGURE 4. Both pairs of curves are solutions to the model equations. The dashed lines represent no growth, i.e., constant tissue volumes.

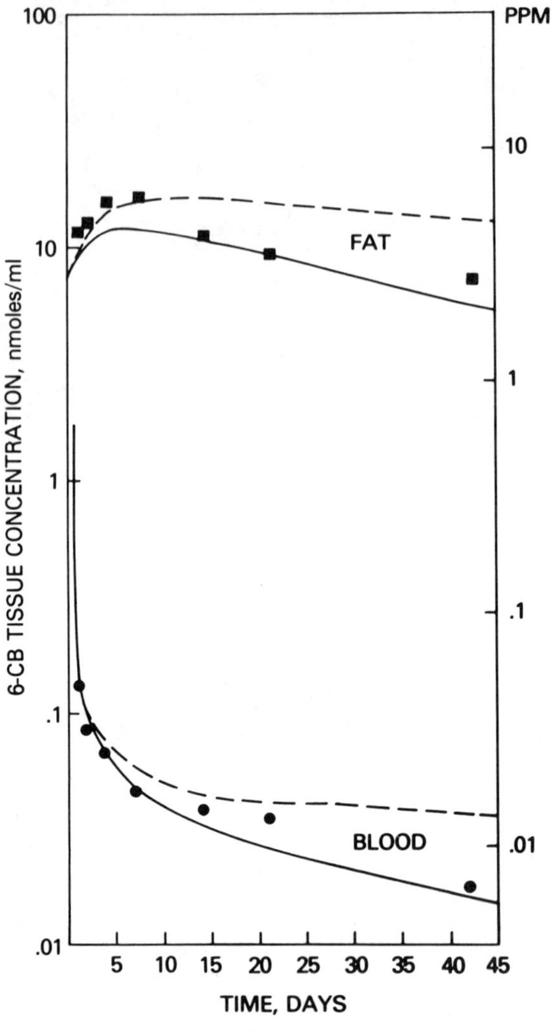

FIGURE 4. Concentration of 6-CB in the fat and blood during six weeks following a single intravenous administration of 0.6 mg/kg of body weight in the rat. The dashed lines are model simulations for no growth in any compartments. The solid lines are model simulations in which the fat compartment alone increased about 2.5 times its initial volume in six weeks according to an experimentally determined growth equation. Points represent experimental concentration data. (Figure, which appeared originally in *Drug Metab. Disp.* **5:** 386 (1977), reprinted by courtesy of American Society for Pharmacology and Experimental Therapeutics.)

Concentrations are overpredicted at long times. The improved long time predictions given by the solid lines are obtained by incorporating an experimentally determined relation for the increase in fat compartment volume as a function of age.

The 6-CB is eliminated in feces and urine, albeit very slowly. Over 80% of the i.v. dose was still retained by the rat at 6 weeks. That which is eliminated is mostly metabolite. Metabolism occurs in the liver. The flow-limited conservation equation for the parent in the liver is

$$V_L dC_L/dt = Q_L\,(C_B - C_L/R_L) - k_M C_L/R_L \tag{4}$$

in which the rate constant for metabolism, k_M, has been based on the concentration in venous blood from the liver. Under flow-limited conditions, the rate constant can be based on either the venous blood or the tissue concentration, since these are directly related through the distribution ratio. Basing the rate constant on venous blood concentration allows the direct comparison of the magnitude of k_M to that of the tissue flow rates. By use of Equation 4, together with the conservation equations for the other compartments, the rate constant was estimated to be k_M = .045 ml/min. The flow rate for the fat compartment for a 250-g rat is of the order of Q_F = .4 ml/min. Blood flow rates for the other compartments of FIGURE 2 are larger than that of the fat. Since k_M is much less than the flow rate of any of the tissues, it can be concluded that after the distribution phase is over, changes in parent 6-CB levels are controlled by the rate of metabolism.

Some of the metabolite diffuses into the blood perfusing the liver and thus is carried to the kidneys and other parts of the body. Metabolite enters the gut by excretion in the bile. Some metabolite appears to be reabsorbed through the gut wall and hence undergoes enterohepatic circulation. The remainder of the metabolite in the gut is excreted in the feces. Using the model equations, one can evaluate transport coefficients for the elimination in urine and feces. Comparison of the elimination rate coefficients to the metabolism rate constant leads to conclusions (borne out by the data) that the metabolite does not build up appreciably compared with parent, and that the rate of elimination of 6-CB is controlled by the rate of metabolism.

Data have recently been collected on PCB disposition in beagle dogs and cynomologous monkeys in order to determine which pharmacokinetic parameters are species dependent and which are species independent. The data indicate significant variation in metabolism among the three animal species tested. Attempts will be made to generalize the models thus far developed to determine if they will permit prediction of the pharmacokinetics in man.

CHLORDECONE (KEPONE®)

Like the hexachlorobiphenyl, Kepone (chemical formula $C_{10}Cl_{10}O$) is a highly chlorinated hydrocarbon. However, the disposition of Kepone in rats is in some respects markedly different from 6-CB. One striking difference is in the key distribution coefficients as can be seen by comparison of the values for these two substances in rats (TABLES 1 and 2). The data in TABLE 2 give the range of values

TABLE 2

KEPONE TISSUE-TO-BLOOD DISTRIBUTION RATIOS

Compartment	Rat[4,5]	Man[6]
Adipose	8–31	4–12(6.7)*
Liver	28–126	5–31(15.0)
Muscle	4–8	2–4.5(2.9)
Bile	1–5	1–4(2.5)

*Average values for man in parentheses.

reported in the literature[4,5] as well as the average value for man (in parentheses). For Kepone the highest distribution coefficient is associated with the liver, not the adipose, tissue. Furthermore, the coefficient for distribution into fat is quite low compared with that of 6-CB and many other highly halogenated hydrocarbons. Cohn *et al.*[6] have suggested that this is a consequence of the formation of chlordecone hydrate or strong binding of the Kepone to plasma proteins.

Values reported by Cohn *et al.*[6] for Kepone distribution ratios in man have also been included in TABLE 2 to show their similiarity to the values for rats. The liver values may reflect differences in metabolism between the two species. For the other tissues in which distribution is likely to be dominated by thermodynamic partitioning, the similarity is evident. Similarity in physical properties, as noted previously, improves the prospects for using pharmacokinetic models to extrapolate results from one species to another.

As yet, differences in Kepone metabolism between rat and man are not clear. Cohn *et al.*[6] identified and measured a significant amount of Kepone alcohol in the bile and feces of man. Indirect evidence of metabolite in rat bile was reported by Boylan *et al.*,[5] but no metabolite has been isolated from rat specimens.

Although Kepone is excreted by the kidneys, the percentage of body burden appearing in urine is small. The only route of elimination of consequence is in the feces. A fecal excretion rate of 0.16% of body burden per hour was observed in rats during a 7-day period following administration of 1 mg Kepone/kg body weight. Similarly treated rats whose bile ducts were cannulated excreted Kepone in the bile at a rate of 0.20%/hr. While the biliary excretion rate is sufficient to account for the appearance of Kepone in the feces, the bile is not the only route of access to the gut. Our initial experiments suggested that much more Kepone entered the gut lumen within the first hour after an intravenous administration than could possibly have reached there by way of the bile. The data suggested to us that Kepone was able to diffuse into the contents of the gut from the blood perfusing the gut wall, as well as being absorbed in the reverse direction. The existence of other pathways in parallel with biliary excretion is noteworthy in terms of intervention measures for treating victims of Kepone poisoning, as shown by Cohn *et al.*[6] It is plausible that the excretion mechanism is qualitatively similar for other halogenated hydrocarbons, and possibly for other lipid-soluble substances. Hence, we have been studying the transport of Kepone into and out of the gut by two types of experiments: intravenous administration to bile-duct cannulated rats and oral administration to intact rats. A pharmacokinetic model has been indispensable in interpreting the results of these experiments.

The organization of the gut portion of our compartmental model is indicated in FIGURE 5. The GI tract has been subdivided into six segments: stomach, three sections of small intestine of equal length, cecum, and large intestine. This subdivision corresponds to that employed in obtaining tissue and gut content samples for assay in the experiments. In the model, each segment is separated into three compartments: blood, tissue, and lumen. The six blood compartments are perfused in parallel. The blood coming into each blood compartment has the same arterial Kepone concentration, and the exiting blood is pooled to form the portal vein blood perfusing the liver. Because of lack of information, the contribution of the lymphatic system to transport of Kepone away from the gut has been neglected.

The lumina of the six GI segments are connected in series. The flow rate of the contents leaving each lumenal compartment is different, but each flow rate is assumed constant over the time course of the experiments. The difference between the contents' flow rates entering and leaving a particular compartment represents the net secretion or absorption of fluid in that compartment. Kepone is assumed to exchange between lumenal compartments only by convection of contents as indicated in FIGURE 5 by the single-bar arrows. These arrows are single ended to denote the net exchange is always directed down the GI tract.

Each lumenal compartment exchanges Kepone with the adjacent tissue compartment by a linear diffusional-type process characterized by a permeability-area rate parameter as described earlier. The diffusional exchange is denoted by a double-bar arrow crossing the mucosal interface. The arrow points in both directions because this exchange process is assumed to be symmetrical, i.e., Kepone crosses the interface either into or out of the lumen depending upon the sign of the driving force, $C_{GT_j} - C_{GL_j}/R_{LT_j}$. Since the contents change in composition along the GI tract, the gut lumen-to-tissue distribution coefficient, R_{LT_j}, may be different for each segment.

The entire gut tissue is assumed to be interposed between the gut blood and lumen compartments. All compartments are considered well mixed, i.e., each has a uniform

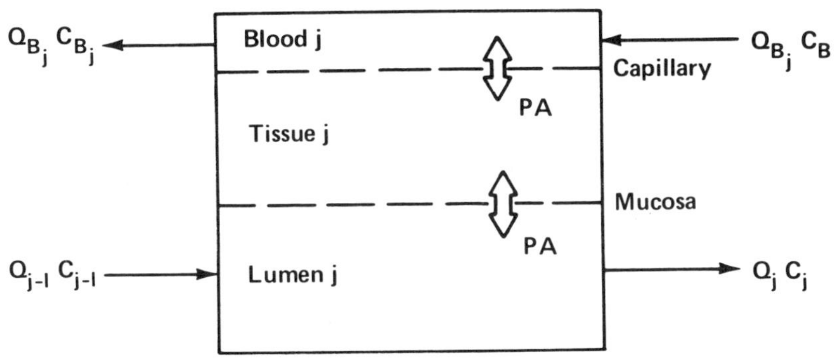

FIGURE 5. Schematic diagram of the *jth* segment of the gastrointestinal tract as used in a pharmacokinetic model for Kepone. The GI tract was divided into stomach, small intestine, cecum and large intestine segments. The small intestine was further subdivided into three segments of approximately equal length. Each segment was composed of three compartments: blood, tissue and lumenal contents. The lumen compartments were connected in series, the blood compartments in parallel. *Q*, flow rate; *PA*, permeability-area product; *C*, concentration.

Kepone concentration. A linear, symmetric exchange process is assumed for the capillary interface analogous to that for the mucosal interface. The permeability-area constants are in general different for each interface and each segment. Values for gut blood volumes were obtained as a fraction (3.4%) of tissue volume. A literature value for the overall blood flow rate to the gut was apportioned to the gut segments in relation to the average measured tissue weight of each segment.

Several alternative gut models have been proposed and their application to steady state absorption experiments has been summarized by Winne.[7] All of these other models assume only a fraction of the gut blood flow rate is effective in transporting absorbed substances. In the present study for simplicity no attempt has been made to split the blood perfusing the gut into portions which participate in the Kepone exchange and portions which do not. With the types of experiments employed the consequences of lower effective blood flow rates cannot be distinguished from higher capillary resistances, i.e., lower apparent capillary permeability-area constants.

The average weight of lumen contents was measured segment by segment. These values, together with estimates for the average residence time of the contents in each segment yield nominal values for the lumenal flow rates. The residence time estimates were obtained by fitting the model simulation of the time course of tracer concentration in each segment to the data of Sikov *et al.*[8] for the transport through the gut of an orally administered nonabsorbable tracer. The nonabsorbable tracer simulation is the solution of the full model for the special case in which the mucosa permeability is set equal to zero. With the volume and flow rate parameters determined, the full model was then used to fit the data from the Kepone experiments. By trial and error values were obtained for the permeability-area constants and the distribution ratios. The longest experiments used in this estimation procedure were four hours after administration. For such short-duration experiments, it is unlikely that much metabolite, if any, would be present. Hence, the parameters are probably characteristic of the transport and distribution of the parent Kepone.

The estimate obtained for the combined capillary-mucosa large intestine permeability-area coefficient was 0.8 ml/hr. By comparison the average rate of production of feces in the rat is 0.3 g/hr. Considering that the large intestine is the least permeable of the gut segments, these numbers suggest that in long-term quasisteady-state elimination, the parent compound may approach an equilibrium distribution between blood and feces. If so, the elimination of parent Kepone is a flow-limited process in which the rate of elimination r_E is given by,

$$r_E = Q_F C_F = Q_F R_F C_B \tag{5}$$

In the preceding equation, C_F is the Kepone concentration in the feces and R_F is the equilibrium feces-to-blood distribution coefficient. Under conditions described by Equation 5 the rate of elimination could be enhanced by increasing either Q_F or R_F. To some degree Q_F can be manipulated through diet. Dietary changes, however, are also likely to alter R_F. Alternatively, it is conceptually easier to imagine greatly increasing R_F without a concomitant change in Q_F. The latter objective could be achieved by continuous oral administration of a small amount of an inert, nonabsorbable, nondegradeable substance with a high capacity for selective absorption of Kepone.

Although R_F could in theory be very large, Equation 5 becomes invalid as R_F is increased. The elimination rate, instead of being limited by the rate of flow of the contents, would become limited by the rate of transport across the gut wall. The ideal adsorbent would keep the free Kepone concentration essentially zero throughout the gut lumen. The entire gut wall would then be passing Kepone at the maximum rate. At a quasisteady-state the maximum elimination can be described by

$$r_{E,\mathrm{MAX}} = (Cl_G + Cl_C)C_B \tag{6}$$

in which the Cl_G is the diffusion-limited "clearance" for transport across the gut wall, and Cl_C is the "clearance" obtainable by the biliary route. The steady-state solution of the pharmacokinetic equations for the gut wall yields an estimate of Cl_G = 25 ml/hr for the parent compound. Experimental measurement of the biliary excretion of Kepone in rats whose bile ducts were cannulated gave Cl_C = 5 ml/hr. Hence the maximum clearance of the parent Kepone predicted for rats on selective adsorbent therapy is 30 ml/hr. By comparison, our experiments with untreated rats have indicated a fecal clearance of 1–4 ml/hr so that it is theoretically possible to enhance the elimination rate substantially through administration of a selective adsorbent.

Boylan *et al.*[5] used the ion exchange resin, cholestyramine, in a test of the general approach of adsorbent treatment. Rats were provided with feed containing 4% resin by weight. Over a 2-week period the treatment doubled the rate of excretion of radioactivity associated with a previously administered dose of ^{14}C-Kepone, as compared with untreated rats. Cholestyramine is not selective for Kepone. In particular, it binds bile salts, among other constituents of gastrointestinal fluid, and consequently may alter normal processes in the gut and affect Kepone transport in secondary ways. Consequently the results at this point should not be considered representative of the limits of adsorbent therapy. Cholestyramine was employed in this study to support its use in the treatment of patients with Kepone poisoning. The resin was chosen in the latter instance because it had been approved for clinical use and gave better results than activated charcoal in preliminary clinical tests. In a clinical trial with these patients, Cohn *et al.*[6] reported finding that cholestyramine treatment led to a reduction in blood half-life of Kepone from an average of 165 days to 80 days, thus substantiating the feasibility of the therapy. Enhancing the rate of elimination was shown to be desirable in the Kepone poisoning cases because the symptoms of toxicity are apparently reversible and correlate qualitatively with the body burden of the agent. Also, they speculated that the reduced tissue exposure may be significant with respect to the potential carcinogenicity of Kepone.

From the data of Cohn *et al.*[6] one can calculate that in the absence of treatment the average whole body clearance of Kepone would be of the order of 40 ml/hr for a 70 kg man. As indicated earlier, body surface area varies among vertebrate species approximately in proportion to body weight raised to the 2/3 power (see, for example, Schmidt-Nielsen[9]). If Kepone permeability in the human gut is assumed to be similar to that in the rat, then the human permeability-area product scaled up from 270 g to 70 kg is of the order of 1000 ml/hr. This calculation suggests that the ideal adsorbent could achieve over a 25-fold increase in the rate of elimination of Kepone provided that transport out of the reservoir compartments does not become controlling. While this result is highly speculative, it does suggest that adsorbent therapy deserves closer examination. It would be desirable to obtain information on the gut permeabilities of other toxic substances, as well as searching for other clinically acceptable, but more selective, adsorbents.

Conclusion

We have looked at the development of physiological pharmacokinetic models and their application to the study of halogenated hydrocarbons. The models complement experimental investigation by: (1) giving precise meaning to distribution and disposition pharmacokinetic parameters and yielding insight into fundamental mechanisms of distribution and disposition; (2) assisting in the design of experiments which determine the values of unknown parameters; and (3) providing a mechanism for comparing the pharmacokinetic behavior in different animal species in order to extrapolate the results to man. We have also suggested through an examination of adsorbent therapy that the models can assist in the development of clinical measures for reducing the body burden in cases of overexposure to environmental agents and other toxic substances.

Summary

Pharmacokinetic modeling is being applied to chemicals posing health hazards because of their presence as contaminants in our environment. This paper presents a rationale for constructing mathematical models based on anatomic and physiological principles. Pharmacokinetic investigations incorporating such models are illustrated for two halogenated hydrocarbons that exhibit different patterns of disposition in the rat: 2,2′,4,4′,5,5′-hexachlorobiphenyl and Kepone®. The models permit interspecies comparison and prediction of normal pharmacokinetic behavior and alterations induced by intervention measures. As an example a theoretical limit is predicted in the case of a treatment for Kepone poisoning involving oral administration of adsorbents.

Acknowledgment

The authors wish to express their appreciation to Mrs. Teresa Troy for secretarial assistance.

References

1. Dedrick, R. L., D. S. Zaharko & R. J. Lutz. 1973. Transport and Binding of Methotrexate *in vivo*. J. Pharm. Sci. **62:** 882–890.
2. Lutz, R. J., R. L. Dedrick, H. B. Matthews, T. E. Eling & M. W. Anderson. 1977. A Preliminary Pharmacokinetic Model for Several Chlorinated Biphenyls in the Rat. Drug Metab. Disp. **5:** 386–396.
3. Anderson, M. W., T. E. Eling, R. J. Lutz, R. L. Dedrick & H. B. Matthews. 1977. The Construction of a Pharmacokinetic Model for the Disposition of Polychlorinated Biphenyls in the Rat. Clin. Pharm. Therap. **22:** 765–773.
4. Egle, J. L., S. B. Fernandez, P. S. Guzelian & J. F. Borzelleca. 1978. Distribution and Excretion of Chlordecone (Kepone) in the Rat. Drug Metab. Disp. **6:** 91–95.
5. Boylan, J. J., J. L. Egle & P. S. Guzelian. 1978. Cholestyramine: Use as a New Therapeutic Approach for Chlordecone (Kepone) Poisoning. Science **199:** 893–895.
6. Cohn, W. J., J. J. Boylan, R. V. Blanke, M. W. Fariss, J. R. Howell & P. S. Guzelian. 1978. Treatment of Chlordecone (Kepone) Toxicity with Cholestyramine. Results of a Controlled Clinical Trial. New Engl. J. Med. **293:** 243–248.
7. Winne, D. 1978. Blood Flow in Intestinal Absorption Models. J. Pharm. Biopharm. **6:** 55–78.
8. Sikov, M. R., J. M. Thomas & D. D. Mahlum. 1969. Comparison of Passage of a Tracer Through the Gastrointestinal Tract of Neonatal and Adult Rats. Growth **33:** 57–68.
9. Schmidt-Nielsen, K. 1970. Energy Metabolism, Body Size and Problems of Scaling. Fed. Proc. **29:** 1524–1532.

BODY CLEARANCE OF HALOGENATED HYDROCARBONS: WORKSHOP SUMMARY

Chairpersons:

Philip S. Guzelian,

Department of Medicine
Medical College of Virginia,
Richmond, Virginia 23298

Mary S. Wolff

Mount Sinai Medical Center,
New York, New York, 10029

Discussants: Alan Brimfield, Peter Bungay, Larry Hansen, John Lech, Jim Rose, Harry Skalsky, and David Villeneuve.

Many halogenated hydrocarbons are excreted slowly from humans and animals and thus may persist in the tissues for years. It has been assumed that these chemicals are inextricably deposited in adipose tissue; this assumption is based mainly on evidence of their limited biotransformation (with minimal urinary excretion) and high lipid solubility (relatively low blood-fat partition). Several exceptions to this general theory were identified at the Workshop. Photomirex, chlordecone (Kepone®), pentachlorophenol and tetrachlorodibenzodioxin are similar to other halogenated hydrocarbons in polarity and other physical properties, but are unusual in that they accumulate in the liver in higher concentrations than in adipose tissue despite the higher lipid content of the latter tissue. In fact, chlordecone, unlike most polychlorinated hydrocarbons, is found in blood at levels 20 times higher than would be expected based solely on the total lipid content of blood. A possible explanation offered for the latter finding is that chlordecone or its hydrate may bind preferentially to high density lipoproteins. From these observations, it was concluded that the distribution of halogenated hydrocarbons may depend on the specific types of lipid present in a given tissue (e.g., phospholipid vs. triglyceride) or possibly the presence of specific binding proteins, both membrane bound and intercellular.

Studies of the overall process of elimination of halogenated hydrocarbons from the tissues suggested that the rate of metabolism may be limiting for excretion of the halogenated biphenyls, chlordane or pentachloroanisole. An important methodologic consideration in studies of the disappearance of halogenated hydrocarbons from a given tissue (e.g., body fat) is that the concentration of the chemical may decline with time simply because of growth of the animal. Therefore, estimate of total content of the chemical in the tissue compartment is preferable to measurement of its concentration in the tissues alone.

Recent evidence suggests that the halogenated hydrocarbons are excreted not only by the biliary route, but also by the intestine itself. Oral administration of cholestyramine, a nonabsorbable binding agent, increased the net excretion by the intestine of chlordecone both in the presence and in the absence of an intact biliary circuit. Mathematical models of this phenomenon suggested that human gut is capable of excreting even larger amounts of chlordecone than have been measured experimentally. This nonbiliary pathway may also function for the excretion of chlordane from humans.

In summary, it was concluded that our knowledge of the factors controlling the binding, distribution, storage, metabolism, and excretion of polyhalogenated hydrocarbons is surprisingly limited considering the ubiquitous presence of these chemicals in the environment. Better understanding of the pharmacokinetics of halogenated hydrocarbons is of fundamental importance to designing rational means for accelerating the elimination of these substances from the body.

PRESENT STATE OF YUSHO PATIENTS

Harukuni Urabe, Hiromu Koda, and Masakazu Asahi

Department of Dermatology
Faculty of Medicine
Kyushu University
Fukuoka 812, Japan

INTRODUCTION

"Yusho" is the Japanese term for a notorious mass case of PCB poisoning that occurred in Japan in 1968 as the result of distributing cooking oil, Kanemi Rice Oil, which had been polluted with PCB. At the beginning of the outbreak the hazard was limited to the western part of Japan, particularly Kyushu, where the contaminated Kanemi Rice Oil was intensively marketed. Tracking down the culpable substance and the treatment of the patients were soon undertaken by large-scale collaborative teams from Kyushu University. The history and present status of the catastrophe was stated.

EPIDEMIOLOGIC STUDIES

At the end of 1977 the victims numbered 1665. They are now scattered throughout Japan because ten years have passed since the outbreak of the tragedy, but are still centered in Kyushu. Six hundred and eighty-four patients live in Fukuoka Prefecture, which has the largest population in Kyushu, and 563 in Nagasaki Prefecture.

Taking Fukuoka Prefecture as an example, in 1968, when the tragedy first occurred, 320 victims were identified (TABLE 1). This number increased year by year until the peak of 737 was reached at the end of last year. The sudden increase in the number of newly diagnosed patients from 1973 to 1975 is due to the revision of the criteria for diagnosis based on more reliable means of determining blood PCB levels.

CAUSES OF DEATH IN YUSHO PATIENTS

The death toll up to the end of 1977 is 51. Thirty-one victims whose cause of death has been confirmed were classified (TABLE 2). There have been 11 deaths from neoplasms, 35.4% of the total. This rate is substantially higher than the 21.1% that is the mortality rate from neoplasms in the same prefecture this year. However, it would be premature to conclude that this high mortality bears association with PCB poisoning.

TRANSITIONS OF CLINICAL MANIFESTATIONS

At the beginning of the outbreak, lesions that led to the diagnosis of Yusho were mainly limited to the skin and mucous membranes. They included black comedo and acneiform eruptions, pigmentation of the face, eyelids, gingivae, and nails, hypertro-

0077-8923/79/0320-0273 $01.75/0 © 1979, NYAS

TABLE 1

ANNUAL INCIDENCE OF YUSHO PATIENTS IN FUKUOKA PREFECTURE

Year	1968	1969	1970	1971	1972	1973	1974	1975	1976	1977
Number of patients										
New	320	88	28	15	1	35	69	123	45	13
Cumulative	320	408	436	451	452	487	556	679	724	737

phy of the meibomian glands and hypersecretion from the eyes. In addition to these mucocutaneous manifestations, systemic signs such as loss of appetite, nausea, vomiting, weakness, and numbness of the extremities were also reported, but at that time these were regarded as complications of minor importance.

During the six years from 1969 to 1975, the mucocutaneous lesions improved in 64% of the patients. These lesions gradually gave way to systemic disorders such as dullness, headache or heavy headedness, indefinite stomachache, numbness or pain of the extremities, swelling and pain of the joints, coughing, and bronchitis-like symptoms. Children had retarded growth, abnormal tooth development, and newborn babies showed systemic pigmentation and SFD, which stands for "small for date."

These transitions of the clinical manifestations therefore compelled us to alter the criteria for diagnosis in 1973. A new technique for determining blood PCB levels developed five years ago made it possible for us to establish more accurate criteria for diagnosis of Yusho, and to delve into various contributory findings including increased blood levels of triglycerides, anemia, lymphocytosis, hypoalbuminemia, adrenocortical hypofunction, and reduction of sensory nerve conduction velocity.

BLOOD PCB ANALYSIS IN YUSHO PATIENTS

Determination of blood PCB levels by gas chromatography was conducted by Masuda and co-workers (TABLE 3). They analyzed the blood specimens of patients

TABLE 2

CAUSES OF DEATH IN YUSHO PATIENTS

Malignant neoplasms		11
Stomach cancer	2	
Stomach cancer + Liver cancer	1	
Liver cancer + liver cirrhosis	2*	
Lung cancer	2*	
Lung tumor	1	
Breast cancer	1	
Malignant lymphoma	2	
Cerebrovascular lesions		7
Traffic accidents		3
Pneumonia + Pulmonary fibrosis		2
Myocardial degeneration + pericarditis		2*
Amyloidosis		1*
Osteodystrophia fibrosa		1*
Status thymicolymphaticus		1*
Liver cirrhosis		1
Suicide		1
Senility		1
Total		31

*Autopsied

TABLE 3

BLOOD LEVELS OF PCBs OF YUSHO PATIENTS AND HEALTHY PEOPLE

	PCB concentration (ppb)															Total number	Average conc.
	1	2	3	4	5	6	7	8	9	11	12	13	17	18	26	(%)	(ppb)
Yusho patients																	
Pattern A		1	7	8	2	5	6	5	2	1		2	1	2	1	43(59.7)	7.2 ± 4.9
B	1	5	9	5	1	1	1			1	1	1				26(36.1)	4.3 ± 3.1
C	1	2														3(4.2)	1.7 ± 0.2
A+B+C	2	8	16	13	3	6	7	5	2	2	1	3	1	2	1	72(100)	5.9 ± 4.5
Healthy people																	
Pattern	2	4	3													9	2.1 ± 0.8

and healthy persons, and showed that the patients' gas chromatograms fell into three different patterns: A, which is highly specific to Yusho patients; B, which is in the neighborhood of A; and C, which is close to the pattern for healthy people.

This table shows the blood PCB levels and the incidence of the three patterns in the 72 patients who visited Kyushu University Hospital from April 1973 to March 1974 and 9 contemporary healthy persons who served as controls. For the 72 patients, pattern A was recorded in 43 patients, or 59.7%, B in 26, or 36.1%, and pattern C was recorded only in 4.2% of the patients. The 9 healthy persons all showed C. These findings clearly show that A and B are specific to Yusho patients. In addition, A, B, and C, in decreasing order, were found to reflect the severity of mucocutaneous lesions.

The average blood levels of PCB in the patients and healthy people, as reported by Masuda, were 5.9 ppb and 2.1 ppb respectively. For the 72 patients classified by their gas chromatograms, the average blood level was 7.2 ppb for pattern A, still as high as 4.3 ppb for B, and 1.7 ppb for pattern C.

In 1976, we rechecked 52 patients, who had already been checked for blood PCB level three years earlier, for persisting PCB, and were surprised at how long the villain lurked in the patients' bodies. The gas chromatograms of the 52 patients clearly showed that the patterns remained unimproved in as many as 73% in spite of minute decreases in blood PCB in patients whose levels were 6 ppb or higher, and that it was only in the remaining 27% that the pattern shifted from A to B or C, or from B to C.

Conclusions

Yusho is acute or subacute poisoning due to ingestion of Kanemi Rice Oil contaminated with PCB. The mucocutaneous lesions observed in the early stages of the tragedy, such as acneiform eruptions, pigmentation, and eye disorders, have greatly improved over the 10 years since the outbreak. However, they have been gradually superseded by systemic disorders such as dullness, nervous and endocrinologic disturbances, respiratory tract disorders, retarded development in children, and fat metabolism disorders. Neoplasms account for the largest number of deaths; elucidation of whether the association between PCB poisoning and malignancy is casual must, however, await further studies. Gas chromatographic patterns of the blood specimens from Yusho patients differ from those of healthy people residing in the same areas. This characteristic pattern of the Yusho patients was found not to diminish, in spite of slight decreases in blood PCB levels.

References

1. Higuchi, K., Ed. 1976. PCB Poisoning and Pollution. Kodansha Ltd. Tokyo.
2. Masuda, Y., R. Kagawa, K. Shimamura, M. Takada & M. Kuratsune. 1974. Polychlorinated biphenyls in the blood of Yusho patients and ordinary persons. Fukuoka Acta Med. **65**(1): 25–27.
3. Koda, H. & Y. Masuda. 1975. Relation between PCB level in the blood and clinical symptoms of Yusho patients. Fukuoka Acta Med. **66**(10): 624–628.

DECREASE IN VITAL CAPACITY IN PCB-EXPOSED WORKERS IN A CAPACITOR MANUFACTURING FACILITY*

Raphael Warshaw,† Alf Fischbein,‡ John Thornton,§
Albert Miller,‡ and Irving J. Selikoff‡

*†Pulmonary Laboratory, Pulmonary Divison
Department of Medicine, Mount Sinai Hospital
New York, New York 10029*

*‡Environmental Sciences Laboratory
Mount Sinai School of Medicine
New York, New York 10029*

*§Department of Biostatistics
Mount Sinai School of Medicine
New York, New York 10029*

INTRODUCTION

Since their commercial introduction in the 1930s, polychlorinated biphenyls (PCBs) have been widely used in industry, especially as dielectric fluids in capacitors and transformers. As a result, they have become widely dispersed in the environment.

Although the effects of occupational exposure to PCB compounds had been previously described,[1-3] the toxicity and persistence of these substances became more fully appreciated following the "Yusho" incident in which more than 1,000 persons in southwestern Japan ingested PCB-contaminated rice oil. The toxic manifestations included acneiform eruptions, hepatic, endocrinologic, neurologic, and ophthalmic disorders.[4] Concerning respiratory involvement, 405 of the "Yusho" patients reported cough and sputum production suggestive of chronic bronchitis; pulmonary function studies were generally within normal limits.[4]

The present paper reports a study of workers occupationally exposed to PCBs, in whom restrictive impairment was found. This finding is discussed in the light of recent reports that some PCB isomers, as such or as metabolites, are heavily accumulated in the bronchial mucosa and lung parenchyma of mice[5,6]

MATERIALS AND METHODS

A survey of 326 capacitor workers was conducted in order to assess the health effects associated with occupational exposure to PCBs. Respiratory studies were part of the overall clinical field survey. Included were questionnaires concerning past medical and occupational histories and respiratory symptomatology. Physical exami-

*This investigation was supported by NIH Grant ES 00928-06 and The Jack Martin Fund.

†Requests for reprints should be directed to Raphael Warshaw, Pulmonary Laboratory, Annenberg 24–26, The Mount Sinai Hospital, 1 Gustave L. Levy Place, New York, New York 10029

0077–8923/79/0320–0277 $01.75/0 © 1979, NYAS

TABLE 1

PREVALENCE OF SYMPTOMS IN THE WORKPLACE AMONG 326 CAPACITOR MANUFACTURING WORKERS

Symptom	Number	Percent of Group Examined
Cough	45	13.8
Wheezing	11	3.4
Tightness in Chest	33	10.1
Upper Respiratory and Eye	157	48.2

nations were performed. Standard 14 × 17 inch chest films were read according to ILO/UC classification.

Measurements of forced vital capacity (FVC), forced expiratory volume 1 sec. (FEV_1) and forced expiratory flow over the mid-FVC (FEF 25–75) were made with a heated wire flow sensor (SRL mod M-10). The authors have found the results obtained with this system to be comparable to those obtained by conventional methods for FVC and FEV_1.[7] Maximal expiratory flow volume curves (MEFV) were plotted on an x-y recorder in order to assess the adequacy and reproducibility of each subject's effort.

Of the 309 workers for whom spirometric studies were obtained, 66 were removed from the analysis because of exposure to asbestos, talc, or textile dust, leaving 243 to be analyzed.

Calibration was checked after every eight subjects using a precision 2.808 L. syringe. Subjects were tested standing; nose clips were employed. At least two efforts were performed by each subject. If they did not agree within 5% for both FVC and FEV_1 or if the hard copy MEFV curve showed an artifact such as a second breath, early termination of expiration, or submaximal effort, additional efforts were obtained. Values for analysis were taken from the effort with the largest FVC and FEV_1. If a single curve did not provide both the largest FVC and FEV_1, a composite best effort was analyzed providing that the FVC of the curve from which the largest FEV_1 was taken was within 5% of the largest FVC obtained. A modification of the predicted values of Morris[8] was used. Calculation of predicted and percentage predicted were by central computer (IBM model 370).

TABLE 2

PREVALENCE OF CHRONIC BRONCHITIS*

in 315 CAPACITOR MANUFACTURING WORKERS

	Male (n = 159)			Female (n = 156)		
Chronic Bronchitis	Smokers & Ex-smokers	Non-smokers	All	Smokers & Ex-smokers	Non-smokers	All
yes	18 (11.3%)	5 (3.1%)	23 (14.5%)	13 (8.3%)	7 (4.5%)	20 (12.8%)
no	91 (57.2%)	45 (28.3%)	136 (85.5%)	83 (53.2%)	53 (34%)	136 (87.2%)

*M.R.C. definition

TABLE 3

PREVALENCE OF X-RAY ABNORMALITIES AMONG 326 CAPACITOR MANUFACTURING WORKERS*

Obliteration of Costophrenic Angle	5 (1.5%)
S ≥ 1/0*	6 (1.8%)
Pleural Thickening	4 (1.2%)
Other	14 (4.3%)

*Interstitial changes by ILO/UC classification.

RESULTS

Forty-eight percent of the examined group reported work-related eye or upper respiratory irritation, and 10% had experienced tightness in the chest. These symptoms are summarized in TABLE 1. Of the 315 workers for whom respiratory questionaires were completed, 205 (65%), 109 males (35%) and 96 (30%) females, were smokers or ex-smokers; 43 (13%), had chronic bronchitis (TABLE 2). Prevalence of radiographic abnormalities was low (TABLE 3); only 6 workers (1.8%) had interstitial involvement according to the ILO/UC classification.

Mean values for spirometric measurements, shown in TABLE 4, are below the most commonly used age-and sex-specific normals[8] for all measurements including FVC. The prevalence of spirometric abnormalities is presented in TABLE 5. Fourteen percent (14.9% of the males, 13.1% of the females) had an abnormal FVC, 11.1% (13.2% of the males, 9% of the females) had restrictive impairment (FVC < 80% predicted and $FEV_1/FVC \geq 0.70$) (TABLE 6).

TABLE 4

MEAN VALUES FOR AGE, HEIGHT, AND SPIROMETRIC MEASUREMENTS AMONG CAPACITOR MANUFACTURING WORKERS WITHOUT OTHER OCCUPATIONAL EXPOSURE

	Male			Female		
	Mean	S.D.	n	Mean	S.D.	n
Age (years)	41.1	± 12.6	124	47.3	± 12.5	131
Height (inches)	67.5	± 2.7	123	62.9	± 2.3	130
FVC (liters)	4.50	± 0.838	121	3.06	± 0.578	122
FVC % pred.	94.5	± 13.3		95.2	± 15.5	
FEV_1 (liters)	3.44	± 0.731	121	2.31	± 0.545	121
FEV_1 % pred.	94.8	± 15.2		94.8	± 19.3	
MMF (L/sec.)	3.45	± 1.165	111	2.46	± 1.039	115
MMF % pred.	88.4	± 28.5		81.8	± 31.4	

TABLE 5

PREVALENCE OF SPIROMETRIC ABNORMALITIES AMONG 243 CAPACITOR MANUFACTURING WORKERS

	Male			Female		
	Nonsmokers	Current & Ex-smokers	All	Nonsmokers	Current & Ex-smokers	All
FVC*	5/40 (12.5%)	13/81 (16%)	18/121 (14.9%)	2/47 (4.3%)	14/75 (18.7%)	16/122 (13.1%)
FEV_1*	2/40 (5%)	14/81 (17.3%)	16/121 (13.2%)	3/46 (6.5%)	17/75 (22.7%)	20/121 (16.5%)
MMF†	5/38 (13.2%)	30/73 (41.1%)	35/111 (31.5%)	11/46 (23.9%)	41/69 (59.4%)	52/115 (45.2%)

* <79.5000.
† <74.5000.

TABLE 6

PREVALENCE OF RESTRICTIVE IMPAIRMENT AMONG 243 CAPACITOR WORKERS WITHOUT OTHER OCCUPATIONAL EXPOSURE

Definition of Restrictive Impairment
Vital Capacity < 79.50% predicted; $FEV_1/FVC \geq 0.70$
Prevalence of Diminished Vital Capacity: 14% (n = 34)
Prevalence of Restrictive Impairment: 11.1% (n = 27)

Male			Female		
Smokers & Ex-smokers	Nonsmokers	All	Smokers & Ex-smokers	Nonsmokers	All
11/81 (13.6%)	5/40 (12.5%)	16/121 (13.2%)	9/75 (12%)	2/47 (4.3%)	11/122 (9.0%)

Mean values for total time of employment in capacitor manufacturing was 14.9 years (S.D. ± 8.11 years) for men and 16.1 years (S.D. ± 8.3 years) for women. Duration of employment was longer for those with restrictive impairment (males 17.5 years, S.D. ± 8.8 years; females 19 years ± 7.5 years) than for workers without such impairment, although the differences were not significant (p = 0.2).

DISCUSSION

Restrictive impairment in the absence of radiographic change is unusual in occupational exposures. The prevalences of abnormal FVC in the study group, in the three other occupational cohorts studied by the same investigators, and in Morris' normal population are shown in TABLE 7. Only the present group and the asbestos insulation workers had an increased prevalence of this abnormality. Restrictive impairment is well known in asbestos workers who show a high prevalence of radiographic abnormalities as well.

Morris' nonsmoking normal subjects have a similar prevalence of decreased FVC (5.6%) to the polyvinyl chloride (4.67%) and styrene (5.9%) groups, which are made up of both smokers and nonsmokers. The capacitor manufacturing workers (14%

TABLE 7

PREVALENCE OF ABNORMAL FVC IN FOUR OCCUPATIONAL GROUPS AND MORRIS'S NORMAL POPULATION

Group	n	Abnormal FVC Number	Abnormal FVC Percent
Morris's normal (8) population	957	54	5.6
Asbestos (11) insulation workers	182	22	12.1
Polyvinyl chloride (10) polymerization workers	348	16	4.6
Styrene synthesis and (12) polymerization workers	488	29	5.9
Capacitor manufacturing workers (PCB)	243	34	14.0

prevalence) are comparable to the asbestos workers (12.1%) despite the exclusion from the former of the subjects with talc, asbestos, and textile exposure. A reduction in FVC is not a characteristic change in cigarette smokers. Smoking is associated with a decrease in FVC when generalized airway obstruction and air trapping are present. The 34 workers who demonstrated a reduced FVC were analyzed for significant airway obstruction (FEV_1/FVC .69) which was present only in 7. The remaining 27 were accordingly classified as "restrictive." By comparison, the PVC workers[10] had a high prevalence of airway obstruction (57% of male workers with a smoking history demonstrated a reduction in MMF and 39% of nonsmokers).

Autoradiographic studies following injection have shown a selective concentration and retention of certain PCBs in the lung parenchyma and bronchial mucosa of mice.[5] These studies suggest that the compounds found in the respiratory tract are, at least in part, metabolic products as well as unaltered compounds. The accumulation of chlorinated biphenyls in lung tissue varies according to their structure and is saturable.[6] The mechanism of the toxic action of PCBs is not known and may differ for various chlorobiphenyls; the metabolites (chlorobiphenyls) may be more toxic than the parent compounds.[9]

The finding of an increased prevalence of restrictive impairment in workers exposed to PCBs, taken in conjunction with reports of selective concentration in the lung parenchyma of mice following injection, points the need for further clinical, epidemiologic, and experimental studies.

Summary

Pulmonary function was evaluated in 243 workers exposed to PCB in the manufacture of capacitors. Mean employment was > 15 years.

Thirty-four of the workers (14%) were found to have a reduced Forced Vital Capacity (FVC <80% of Morris' predicted). Of the 34 with reduced FVC, 27 (80%) demonstrated a restrictive pattern of impairment ($FEV_1/FVC > 0.7$). Only one of these 27 workers had an abnormal chest roentgenogram ($\geq 1/0$ by ILO UC Classification of Radiographs of Pneumoconioses).

These findings are of interest in view of recent experimental data indicating the accumulation of PCBs and PCB metabolites in lung tissue (Brandt[5] and Jansson[6]). Restrictive spirometric impairment with no radiographic change is unusual in occupational exposure.

References

1. Schwartz, L. 1936. Dermatitis from Synthetic Resins and Waxes. Am. J. Publ. Health **26:** 586–592.
2. Meigs, J. W., J. J. Albom & B. L. Kartin. 1954. Chlornacne from an Unusual exposure to aroclor. JAMA **554:** 1417–1418.
3. Ouw, H. K., G. R. Simpson & D. S. Siyali. 1976. Use and health effects of aroclor 1242, a polychlorinated biphenyl, in and electrical industry. Arch. Environ. Health, **31:** 189–194.
4. Higuchi, K., Ed. PCB Poisoning and Pollution. 1976. Kodansha Ltd. (Tokyo) and Academic Press (New York, London).
5. Brandt, I. Tissue localization of polychlorinated biphenyls chemical structure related to pattern of distribution. 1977. Acta Pharmacol. Toxicol. **40** (Suppl. II): 62–78.
6. Jansson, B. 1978. Transformation of some organochlorine compounds in the environment studied by gas chromatography and mass spectrometry. Report from the National

Swedish Environment Protection Board, Special Analytical Laboratory, Wallenberg Laboratory, University of Stockholm, @-106 91 Stockholm, Sweden.

7. MILLER, A., M. CHUANG, R. WARSHAW, H. SMITH & I. J. SELIKOFF. Validation of automated spirometry used in field survey of large occupational groups: comparison with conventional water spirometers. Arch. Environ. Health. In press.
8. MORRIS, J. F., A. KOLSKI & L. C. JOHNSON. 1971. Spirometric standards for healthy nonsmoking adults. Am. Rev. Respirat. Dis. **103:** 57–67
9. HUTZINGER, O., S. SAFE & V. ZITKO. 1974. The chemistry of PCB's, CRC Press, p. 2.
10. MILLER, A., A. TEIRSTEIN, M. CHUANG, I. J. SELIKOFF & R. WARSHAW. 1975. Changes in pulmonary function in workers exposed to vinyl chloride and polyvinyl chloride. Ann. N.Y. Acad. Sci. **246:** 42–52.
11. MILLER, A. & J. ELIOTT. Comparison of spirometry performed in the field by two teams: an investigation of inter-observer variability. Unpublished.
12. LORIMER W. & R. LILIS, 1976. Clinical studies of styrene workers: initial findings. Environ. Health Prospect. **17:** 171–181.

COHORT STUDY OF MICHIGAN RESIDENTS EXPOSED TO POLYBROMINATED BIPHENYLS: EPIDEMIOLOGIC AND IMMUNOLOGIC FINDINGS

Philip J. Landrigan,* Kenneth R. Wilcox, Jr.,† Joseph Silva, Jr.,‡
Harold E. B. Humphrey,† Carol Kauffman,‡
and Clark W. Heath, Jr.*

**Center for Disease Control*
Atlanta, Georgia 30333

†Michigan Department of Public Health
Lansing, Michigan 48933

‡University of Michigan Medical Center
Ann Arbor, Michigan 48104

Introduction

The polybrominated biphenyls (PBB) are a group of halogenated aromatic hydrocarbon compounds employed until recently as flame retardants in the plastics industry.[1] PBB have low acute toxicity,[2] but are fat-soluble, biologically persistent,[3] and capable of causing chronic injury to the liver,[4] kidneys,[5] skin,[5] and lymphatic tissue.[6] In addition, PBB have been found in chronic rat feeding studies to induce neoplastic liver nodules.[4]

In 1973, several hundred pounds of PBB were introduced into cattle feed as the result of a shipping accident in Michigan.[7] Thousands of dairy animals[8] died or were destroyed, and widespread human exposure resulted from contact with contaminated feed and from consumption of PBB in dairy foods. While human exposure was most intense on quarantined farms,[9] and in chemical workers,[10] low-level exposure was widespread throughout the state, and in October, 1976, a survey of randomly selected samples of human breast milk showed that 51 (96%) of 53 samples from the lower peninsula of Michigan contained detectable levels of PBB.[11]

To determine the effects of PBB exposure on human health, the Michigan Department of Public Health (MDPH), the Center for Disease Control (CDC), the National Institutes of Health (NCI and NIEHS), the Food and Drug Administration (FDA), and the Environmental Protection Agency (EPA) have established a cohort of persons with varying levels of PBB exposure. The members of this cohort have been examined systematically to determine whether or not they have experienced an increased incidence of acute or subacute illness, of biochemical aberrations, or of alterations in the outcome of pregnancy in relation to their PBB exposure. In addition, the health status of the cohort is to be followed prospectively over the next several decades to determine whether they will experience an increased incidence of chronic disease, particularly of cancer. In this report, we shall summarize the epidemiologic and clinical data collected on the Michigan cohort in the first 4 years since the onset of PBB exposure. In addition, we shall present the results of an immunologic study conducted in 1977 to evaluate peripheral lymphocyte function in the most heavily exposed subset of the cohort.

0077–8923/79/0320–0284 $01.75/0 © 1979, NYAS

METHODS

Selection of Participants

Four categories of Michigan residents were invited by mail and by interview to participate in this prospective cohort study:

Farm Residents on Quarantined Farms

Invited to join this group were the 2,248 persons who were members of families residing on quarantined farms at the time of those farms' quarantine for PBB contamination; included here were 110 residents of quarantined farms who had participated in a 1974 MDPH pilot study of PBB exposure.[9] Quarantine was established when any single specimen of meat, eggs, or milk from a farm exceeded quarantine action limits; the number of specimens submitted for PBB testing varied from farm to farm. In May, 1974, the maximum allowable limit for PBB in milk or meat fat was set at 1.0 parts per million (ppm), and in eggs at 0.1 ppm, and 183 farms were quarantined. In November, 1974, these quarantine action limits were decreased

TABLE 1
SURVEY PARTICIPATION RATES BY ENROLLMENT GROUP
(MICHIGAN PBB STUDY, 1976–77)

Group	Invited (No.)	Enrolled (No.)	(%)
1. Quarantined farm residents	2248	2148	95.6
2. Farm product recipients	1495	1421	95.1
3. Chemical workers (and families)	322	251	78.0
4. Pilot study control participants	60	57	95.0

to 0.3 ppm for milk and meat fat, and to 0.05 ppm for eggs, and another 229 farms were quarantined. Additional farms have subsequently been placed under quarantine. The survey participation rate in Group 1 was 95.6% (TABLE 1).

Farm Product Recipients

Invited into this group were 1,495 persons who had not themselves resided on quarantined farms, but who had received meat, eggs, or dairy products directly from quarantined premises in 1973 or 1974. The participation rate in Group 2 was 95.1% (TABLE 1).

Chemical Workers

This group includes workers, who had been exposed occupationally to PBB in a chemical manufacturing plant in Michigan, as well as the members of these workers' families. The total number invited was 322; the participation rate was 78.0% (TABLE 1).

Pilot Study Participants

Invited to join this group were 60 persons who had been identified in the MDPH pilot study[9] as being resident on farms with low-level PBB contamination; they had served as a control group in the pilot study. The participation rate in this group was 95.0% (TABLE 1).

Finally, a small number of additional persons were enrolled into the cohort who were not originally invited to participate:

(1) 331 self-referred persons who had either resided on farms identified as being contaminated by PBB in levels below quarantine limits or who had eaten food produced on such farms (Group 5); and

(2) 337 self-referred volunteers who had no direct connection with contaminated farm premises (Group 6).

Field Studies

All persons enrolled in the study were visited in their homes over a 12-month period between September, 1976 and August, 1977, by trained interviewers from MDPH. Repeated visits (up to 4 per household) and repeated mailings were made in an attempt to reach families who could not initially be contacted, or who refused in the first instance to participate in the cohort study. After informed consent had been obtained for each person, a questionnaire was administered which sought data on potential sources of PBB exposure, as well as on the occurrence since 1973 of 17 symptoms and conditions considered possibly related to PBB contact. This symptom list was developed from anecdotal data on the possible human toxicity of PBB[8] as well as from published information[12] on the human toxicity of the polychlorinated biphenyls (PCB). Symptoms and conditions sought were weight loss, fatigue, headaches, skin rash, changes in skin pigmentation, changes in the nails, heart disease, gastrointestinal complaints (nausea, vomiting, diarrhea, or abdominal pain), liver disease, Reye's syndrome, diabetes mellitus, thyroid disease, peripheral neuropathy, convulsions, joint disease, benign tumors, and cancer. Data were obtained also on tobacco and alcohol consumption. Detailed information, including birth weight data and data on any congenital malformations, was sought on the outcome of all pregnancies since 1963.

Toxicologic Laboratory Studies

A venous blood sample for PBB analysis was requested of each study participant, and 3,639 were obtained. Samples were analyzed for PBB concentration at MDPH using gas chromatography with electron capture detection; the limit of detection for PBB was 1 part per billion (ppb).

Quality control on PBB analysis was conducted jointly by MDPH and the Toxicology Branch, Clinical Chemistry Division, Bureau of Laboratories, CDC. In addition to internal quality control at MDPH, which consisted of repeated analyses of quality control specimen pools, external surveillance was provided by (a) repeat analysis at CDC of 13% of samples, and (b) blind insertion by CDC of quality evaluation specimens into the regular system of specimen collection. For 466 serum samples analyzed in duplicate at MDPH and CDC, mean PBB values were 16.9 and 16.2 ppb respectively; the coefficient of correlation between paired results was 0.9982.

Immunologic Studies

To evaluate peripheral lymphocyte function in the members of the cohort identified previously as having highest serum PBB values, an immunologic investigation was undertaken by MDPH and CDC in collaboration with the University of Michigan (UM) in October, 1977.

Invited to participate were all 41 cohort members shown previously to have serum PBB values $\geq$ 300 ppb; 34 of these 41 (83%) agreed to participate. Invited as a comparison group were all 7 persons shown to have a serum PBB concentration less than 1 ppb, as well as a random sample of 59 of the 915 enrollees with serum values of 1–9 ppb; 56 (85%) of these 66 persons agreed to participate.

Fasting early morning venous blood specimens, taken from 8–10 persons per day, were collected in heparinized vacuum tubes over a 3-week period in participants' homes and at nearby field stations. Members of both exposure groups were included each day. Samples were coded in the field and then transported at ambient temperature by automobile to the UM laboratory where test procedures were begun 3–5 hours after venipuncture. To examine for any possible decrement in lymphocyte function during transport, blood samples were obtained in the field from 1–3 MDPH staff members each day (total, 9 persons, 33 samples) from whom blood samples had previously been taken in the UM laboratories and tested immediately.

Mononuclear cells were obtained by centrifugation through a ficoll-isopaque gradient.[13] The mononuclear cell layer was removed and washed twice in Hanks' balanced salt solution (HBSS). Cells to be used for T and B lymphocyte enumeration were resuspended in HBSS, counted, and diluted to a final concentration of 4×10^6 mononuclear cells/ml. Cells to be used for lymphocyte transformation were resuspended in RPMI-1640 (GIBCO, Grand Island, N.Y.), and counted and diluted to a concentration of 10×10^6 cells/ml. These procedures gave greater than 99% cell viability as determined by exclusion of 1% trypan blue dye.

B cells were enumerated by detecting complement receptors by means of complement-coated zymosan particles according to thc methods of Mendes *et al.*[14] Two-hundred lymphocytes were counted in duplicate specimens using a phase microscope; cells that had three or more zymosan particles attached were recorded as B cell rosettes.

T lymphocytes were enumerated by determining the number of cells that formed rosettes spontaneously with sheep erythrocytes according to the method of Jondal *et al.*[15] Two-hundred lymphocytes were counted in duplicate specimens using a phase microscope. Cells that had three or more erythrocytes attached were recorded as T cell rosettes.

Peripheral blood lymphocytes were studied for transformation to three mitogens, phytohemagglutinin (PHA), concanavalin A (Con A), and pokeweed mitogen (PWM) in wells of flat-bottomed microtiter plates. Each mitogen was run in quadruplicate wells that contained 10^5 lymphocytes/well in a total volume of 0.2 ml RPMI-1640 with 20% autologous plasma.

Microtiter plates were placed in a 5% CO_2 humidified incubator at 37°C. After incubation for 96 hours cells were pulsed with 1 microcurie (μCi) of ^{3}H-thymidine and

TABLE 2
MEAN AGE AND PERCENTAGE MALE AND FEMALE BY ENROLLMENT GROUP
(MICHIGAN PBB STUDY, 1976–77)

Group	Mean Age (yrs)	Percentage Male	Percentage Female
1. Quarantined farm residents	28.0	52.6	47.4
2. Farm product recipients	28.3	48.4	51.6
3. Chemical workers (and families)	22.6	51.6	48.4
4. Pilot study control participants	33.7	50.0	50.0
5. Low-Level PBB farm residents	27.9	53.1	46.9
6. Volunteers	29.4	50.8	49.2

harvested 6 hours later with an Otto-Hillier cell harvester. Cells were then washed with saline, dried, resuspended in scintillation fluid and counted. Data were expressed as mean counts per minute (cpm) per 10^5 lymphocytes.

EPIDEMIOLOGIC RESULTS

Members of the Michigan cohort ranged in age from 1 to 89 years (median, 24 years) at the time of enrollment; 2,317 were male and 2,220 female (8 of unspecified sex). Age and sex distributions were approximately comparable in the 6 groups (TABLE 2).

Analysis of data on the prevalence of symptoms showed that fatigue, headaches, paresthesias, and joint problems were the conditions most frequently reported. For nearly all conditions the volunteer group (Group 6) had the highest prevalence, followed closely by the group who had resided on farms with low-level PBB contamination (Group 5) (TABLE 3).

Data on serum PBB levels showed a range of values from 0–1900 parts per billion (ppb); 106 values were greater than 100 ppb. Examination of results by age-group showed that children aged 10 years or less tended to have the highest serum PBB concentrations. Above age 10, there was no significant gradient by age ($F = 1.81$, $p = 0.11$). Males had significantly higher values than females ($\chi^2 = 256.5$; $p < 10^{-10}$).

TABLE 3
PREVALENCE RATES (%) FOR SELECTED SYMPTOMS BY ENROLLMENT GROUP
(MICHIGAN PBB STUDY, 1976–77)

Symptoms and Conditions	Group 1	2	3	4	5	6
Fatigue	36.4	32.4	22.0	15.8	41.4	54.4
Rashes	8.1	7.5	6.4	1.8	14.5	3.1
Joint pains	25.1	26.4	18.9	11.1	32.0	39.9
Hepatitis	1.5	1.8	1.6	0.0	3.3	3.6
Diabetes mellitus	1.9	2.2	2.4	1.8	2.7	2.4
Benign tumors	4.2	5.4	5.8	1.8	5.5	9.8
Cancer—all sites	0.4	0.5	0.4	0.0	0.6	0.6
Number of subjects	2428	1421	251	57	331	337

Serum PBB levels showed a strong tendency to cluster by families (intra-class coefficient of correlation = 0.85).

Examination of serum PBB levels by enrollment group showed that highest values were found in the chemical workers and in the members of their families (TABLE 4) followed by the residents on quarantined farms (Group 1). In the chemical worker group, the workers themselves had a serum PBB level of 108.7 ppb (range, 0–1,240 ppb), significantly higher than that in members of their families. Volunteers (Group 6) had the lowest serum PBB levels.

Paired serum samples, one collected in 1974 and the other in 1977, were available for 148 members of the cohort. These data indicate that serum levels were generally stable over the 3-year period. The mean change in serum ppb value was −16 ppb, and the coefficient of correlation (r) between the 2 sets of values was 0.96.

Simultaneous paired specimens of blood and adipose tissue were obtained on 221 Michigan residents, 19 of whom were members of the cohort. A high coefficient of correlation (r) was seen between these 2 sets of values (0.951), and in the 132 instances where both values were above detection limits the ratio of adipose to serum PBB concentrations was 362.8:1.0 (FIGURE 1).

TABLE 4

SERUM PBB LEVELS* BY ENROLLMENT GROUP (MICHIGAN PBB STUDY, 1976–77)

Group	Number	Serum PBB Range	Mean	Median
1. Quarantined farm residents	1750	0–1900	26.9	4.0
2. Farm product recipients	1114	0–659	17.1	3.0
3. Chemical workers (and families)	216	0–1240	43.0	4.5
4. Pilot study control participants	44	1–13	3.5	2.0
5. Low-level PBB farm residents	242	0–24	3.5	2.0
6. Volunteers	273	0–111	3.2	1.0
Total	3639	0–1900	21.2	3.0

*PBB levels in parts per billion (ppb).

An evaluation of dose-response relationships was undertaken by dividing the cohort into seven segments on the basis of serum PBB levels. No positive associations were found between serum concentrations of PBB and reported symptom frequencies (TABLE 5). Symptom-prevalence rates (excluding volunteers) were slightly higher in persons with no detectable PBB in serum than in those with measurable quantities. Relationships between symptom prevalence rates and serum PBB level were also examined within each enrollment group, and no positive trends were found; in all groups, including chemical workers and quarantined farm residents, highest prevalence rates occurred in persons with lowest serum PBB levels.

Since 1973, 65 children have been born to women in Michigan with potential exposure to PBB (TABLE 6). Serum PBB concentrations in the 52 women examined at the time of delivery ranged from not detectable (<1 ppb) to 1150 ppb (mean, 26.2 ppb). Cord serum PBB levels at birth in 58 infants ranged from not detectable to 104 ppb (mean 3.2 ppb). The mean ratio of maternal/cord serum PBB values for the 13 maternal-infant pairs where both had detectable values was 7.04:1.0 (range, 1.5–

10.3). Breast milk samples were obtained from 32 women, and PBB values (fat basis) ranged from 32 to 93,000 ppb (mean 3614 ppb). The mean ratio of serum to breast milk PBB values in the 21 women where both concentrations were detectable was 122.0:1.0 (range, 62.6–256.7).

Immunologic Results

Total leucocyte counts did not differ significantly among the high PBB, low PBB, and MDPH staff groups. The mean percentage of circulating lymphocytes was significantly elevated in the high PBB exposure group as compared to the low ($p < 0.0001$). However, this increase was found only in children less than 12 years of age who made up a greater proportion of the high than of the low exposure group.

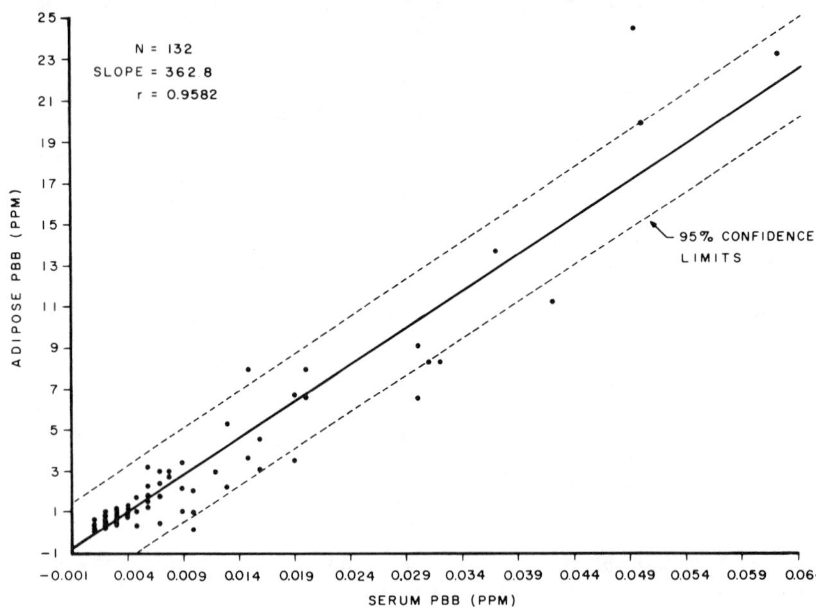

FIGURE 1. Correlation between simultaneous paired serum and adipose PBB levels. Michigan PBB Study, 1976–1977.

The high exposure group had no depressions in absolute numbers of T or B lymphocytes, or in responses to any of the 3 mitogens as compared to the group with low PBB exposure (TABLE 7). However, 15% of persons from the 2 exposure groups were found to have 2 or more abnormalities in measures of *in vitro* lymphocyte function.

Significant differences were found (TABLE 7) between results obtained in the field on MDPH staff and results obtained on these same individuals in the nontransported samples taken at the UM laboratory; this discrepancy suggests the possible existence of a "transportation effect."

TABLE 5

PERCENTAGE INCIDENCE OF REPORTED SYMPTOMS AND CONDITIONS* SINCE 1973 BY SERUM PBB† LEVEL (MICHIGAN, PBB STUDY, 1976–77

Symptoms and Conditions	0	1	2–3	4–9	10–19	20–99	100+	Mean
Fatigue	46.5	40.1	40.7	38.1	37.8	28.7	27.4	38.3
Rashes	9.0	8.5	8.4	7.2	8.3	6.3	4.1	7.8
Joint pains	42.9	32.7	29.8	26.4	28.4	21.6	17.1	28.6
Hepatitis	4.5	1.4	1.8	1.4	1.3	0.0	0.0	1.4
Diabetes mellitus	4.7	2.9	2.4	1.8	1.9	3.3	0.0	2.4
Benign tumors	2.7	7.5	5.6	4.3	5.3	3.8	2.6	5.3
Cancer-all sites	0.0	0.9	0.5	0.5	1.0	0.4	1.8	0.6
Number of subjects*	89	716	941	892	316	276	126	3,356

*Excludes volunteer subjects.
†PBB concentrations in parts per billion (ppb).

DISCUSSION

The finding in this study of elevated serum PBB levels in families from quarantined farms and in Michigan chemical workers confirms earlier reports of increased PBB absorption in these groups.[9–11] Although a number of exposed individuals reported symptoms, as has been the case in previous studies,[16,17] it was noteworthy that symptom prevalence rates bore no relationship to serum PBB levels.

TABLE 6

PBB CONCENTRATIONS* IN MATERNAL SERUM, CORD SERUM, AND BREAST MILK† (MICHIGAN PBB STUDY, 1976–77)

		PBB Concentrations			
Specimen	Number	Range	Mean	Median	Average Ratio to Maternal Serum (and Range)
Maternal serum	52	0–1,150	26.2	2.5	—
Cord serum	58	0–104	3.2	1.0	7.04 (1.5–10.3)
Breast milk	32	32–93,000	3614	225	122.0 (62.2–256.7)

*PBB concentrations in parts per billion (ppb).
†Breast milk PBB values expressed on fat basis.

TABLE 7

LYMPHOCYTE FUNCTION TESTS BY EXPOSURE GROUP (MICHIGAN PBB STUDY 1977)

	Low Exposure Group		High Exposure Group		p-Value (Student's t-Test)	MDPH Staff Bloods Drawn in Lab		MDPH Staff Bloods Drawn in Field		p-Value (Paired t-Test)
	n		n			n	Mean ± 1 S.D.	n	Mean ± 1 S.D.	
Mean serum PBB level, ppb (range)	(51)	2.8 <1–11	(32)	787 188–2560		(8)*	1.88 ± 1.55 (<1–5.0)	(8)*	1.88 ± 1.55 (<1–5.0)	
T and B Cell Quantitation										
% T cells (mean ± 1 S.D.)	(56)	61.4 ± 10.2	(34)	67.2 ± 11.8	.015	(9)*	72.9 ± 8.1	(33)*	65.0 ± 13.0	.15
% B cells (mean ± 1 S.D.)	(56)	5.01 ± 2.36	(34)	5.11 ± 2.51	.857	(9)*	6.4 ± 2.5	(33)*	5.6 ± 3.5	.40
Lymphocyte Function Tests										
Maximal blastogenic response of peripheral blood lymphocytes (in scintillation counts per minute) to:										
PHA (mean ± 1 S.D.)	(54)	38224 ± 16803	(31)	45797 ± 19349	.056	(9)*	63170 ± 16308	(31)*	38101 ± 20742	.001
PWM (mean ± 1 S.D.)	(54)	41936 ± 21871	(31)	43505 ± 16523	.731	(9)*	86763 ± 25814	(31)*	36766 ± 15636	.0001
Con A (mean ± 1 S.D.)	(54)	44721 ± 21731	(31)	54359 ± 26750	.074	(9)*	65858 ± 37265	(31)*	44737 ± 23217	.025

*Same individuals constitute both groups.

Symptom frequency was much more closely related to mode of entry into the cohort, with highest frequencies occurring in volunteers (Group 6) and in persons from farms that were contaminated by PBB at levels insufficient to result in quarantine (Group 5). These observations suggest that factors other than PBB absorption may have been responsible for the production of symptoms and that selection factors may have played an important role. Also, however, it may be necessary to examine samples of the commercial PBB mixture dispersed in Michigan to determine whether it contained other toxins in addition to PBB, possibly in concentrations not correlated directly with PBB levels.

The failure of this investigation and of a previous immunologic study[18] to demonstrate dose-related depression of lymphocyte function in persons exposed to PBB suggests that there may be no causal relation between the two. If, however, an association does exist, our failure to detect it could have resulted from any of several possibilities. One is that the methods used in this study were insufficiently sensitive. Another is that the 3–5 hours' transportation time required here may have depressed cell function in both of our groups to a point where intergroup differences were no longer evident. A third is that persons in both study groups (as well as MDPH field staff) may have had exposure to PBB at concentrations above a threshold level for lymphocyte depression; hence all may have been equally depressed. A final possibility is that recovery of immunologic function may have occurred in the exposed population in Michigan in the 8 months which elapsed between this and the previous, more strongly positive study.[18] Further investigation will be required to distinguish among these possibilities, and to explore the clinical significance of the immunologic abnormalities observed in members of the two exposure groups.

PBB is a highly lipophilic compound and extremely persistent in the human body. Persons exposed to PBB may be at potentially increased risk of developing delayed reactions, particularly cancer, since PBB fed to rats, has produced neoplastic liver nodules.[4] Because of the potential for delayed disease, MDPH and the federal agencies plan to continue to follow the Michigan cohort for at least another 10–15 years through repeated contact with the exposed persons and their family physicians. Also, within the next year, a cohort of approximately 2000 Iowa farm residents without known exposure to PBB is to be recruited and interviewed using procedures identical to those used in Michigan; this group will be followed prospectively as a comparison cohort.

Summary

Polybrominated biphenyls (PBB) were dispersed widely in Michigan by a 1973 shipping accident in which PBB was introduced into cattle feed. Human exposure resulted principally from ingestion of contaminated dairy food products. To determine whether PBB exposure has or will cause acute or chronic illness, a prospective cohort study of 4545 persons has been undertaken. Three exposure groups were sought: all persons living on PBB-quarantined farms; persons who had received food directly from such farms; workers (and their families) engaged in PBB manufacture. Enrollment rates were 95.6, 95.1 and 78.0%. Also enrolled were 725 persons with low-level PBB exposure. All were queried concerning 17 symptoms and conditions related possibly to PBB. Venous blood was drawn on 3639 and analyzed for PBB by gas chromatography. Mean serum PBB levels were 26.9 ppb in quarantined farm families, 17.1 in recipients, 43.0 ppb in workers, and 3.4 ppb in the low exposure groups. No associations were found between serum PBB levels and symptom prevalence rates. To evaluate peripheral lymphocyte function, T and B cell quantitation and *in vitro* responses to 3 nonspecific mitogens were studied in 34 persons with highest PBB levels (mean, 787 ppb), and in 56 with low values (mean, 2.8 ppb). No statistically significant differences in lymphocyte number or function were noted.

References

1. Sanders, H. J. 1978. Flame retardants—special report. Chem. Eng. News, April 24, 1978: 22–36.
2. Aftosmis, J. G., R. Culik, K. P. Lee, H. Sherman & R. S. Waritz. 1972. Toxicology of brominated biphenyls. I. Oral toxicity and embryotoxicity (Abstr.) Toxicol. Appl. Pharmacacol. **22:** 316.
3. Matthews, H. B., S. Kato, N. M. Morales & D. B. Tuey. 1977. Distribution and excretion of 2, 4, 5, 2, 4, 5,-hexabromobiphenyl, the major component of Firemaster BP-6. J. Toxicol Environ. Health **3:** 599–605.
4. Kimbrough, R. D., V. W. Burse & J. A. Liddle. 1977. Toxicity of polybrominated biphenyl. Letter. Lancet **2:** 602–603.
5. Moorhead, P. D., L. B. Willett & F. L. Schanbacher. 1978. Effects of PBB on cattle. II. Gross pathology and histopathology. Environ. Health Perspect. **23:** 111–118.
6. Luster, M., R. E. Faith & J. A. Moore. 1978. Effects of polybrominated biphenyls (PBB) on immune response. Environ. Health Perspect. **23:** 227–232.
7. Carter, L. J. 1976. Michigan's PBB incident: chemical mix-up leads to disaster. Science **192:** 240–243.
8. Jackson, T. F. & F. L. Halbert. 1974. A toxic syndrome associated with the feeding of polybrominated biphenyl- contaminated protein concentrate to dairy cattle. J. Am. Vet. Med. Assoc. **165:** 437–439.
9. Humphrey, H. E. B. & N. S. Hayner. 1975. Polybrominated biphenyls: an agricultural incident and its consequences, an epidemiological investigation of human exposure. Presented at the Ninth Annual Conference on Trace Substances in Environmental Health, Columbia, Missouri, June, 1975.
10. Anderson, H. A., M. S. Wolff, A. Fischbein & I. J. Selikoff. 1978. Investigation of the health status of Michigan Chemical Corporation employees. Environ. Health Perspect. **23:** 187–192.
11. Brilliant, L. B., G. Van Amburg, J. Isbister, H. Humphrey, K. Wilcox, J. Eyster, A. W. Bloomer & H. Price. 1978. Breast milk monitoring to measure Michigan's contamination with polybrominated biphenyls. Lancet **2:** 643–646.
12. Kuratsune, M., T. Yoshimura, J. Matsuzaka & A. Yamaguchi. 1972. Epidemiologic study on Yusho, a poisoning caused by ingestion of rice oil contaminated with a commercial brand of polychlorinated biphenyls. Environ. Health Perspect. April, 1972: 119–128.
13. Boyum, A. 1968. Isolation of mononuclear cells and granulocytes from human blood. Scand. J. Clin. Lab. Inves. **20:** 77–89.
14. Mendes, N. F., S. S. Miki & Z. F. Peixinho. 1974. Combined detection of human T and B lymphocytes by rosette formation with sheep erythrocytes and zymosan-C3 complexes. J. Immunol **113:** 531–536.
15. Jondal, M., G. Holm & H. Wigzell. 1972. Surface markers on human T and B lymphocytes. I. A large population of lymphocytes forming non-immune rosettes with sheep red blood cells. J. Exp. Med. **136:** 207–215.
16. Anderson, H. A., R. Lilis, I. J. Selikoff, K. D. Rosenman, J. A. Valciukas & S. Freedman. 1978. Unanticipated prevalence of symptoms among dairy earners in Michigan and Wisconsin. Environ. Health Perspect. **23:** 217–226.
17. Meester, W. D. & D. J. McCoy. 1977. Human toxicology of polybrominated biphenyls. Clin. Toxicol. **10:** 474
18. Bekesi, J. G., J. F. Holland, H. A. Anderson, A. S. Fischbein, W. Rom, M. S. Wolff & I. J. Selikoff. 1978. Lymphocyte function of Michigan dairy farmers exposed to polybromianted biphenyls. Science **199:** 1207–1209.

ENVIRONMENTAL CHLORACNE: UPDATE AND OVERVIEW

James S. Taylor

Section of Industrial Dermatology
Department of Dermatology,
The Cleveland Clinic Foundation
Cleveland, Ohio, 44106

Since 1974, when this author reviewed chloracne (*Chloracne—A Continuing Problem*)[1] a number of new outbreaks of the disorder have occurred, both in the United States and abroad. Additionally, the scope of the problem has expanded to include involvement of other nonindustrial populations and precipitation by brominated aromatic compounds. Thus, a more appropriate term might be environmental halogen acne or haloarylacne after the terminology of Weirich.

Since chloracne has been extensively reviewed by Crow[2] in 1970 and Weirich[3] in 1978 this report will present findings from the more recent outbreaks as well as an overview of the subject.

Chloracne is a refractory follicular dermatosis produced by environmental exposure to certain halogenated aromatic compounds, and which may be accompanied by systemic toxicity.[1]

Between 1897, when Von Bettman first observed chloracne, and World War II most cases of chloracne were caused by the chloronaphthalenes and polychlorinated diphenyls, the latter now referred to as PCBs. Since then most attention has focused on the role of trace contaminants of PCBs, herbicides, herbicide intermediates, and flame retardants, among others.[1–3] The chemicals which have been associated with chloracne are listed in TABLE 1.

POLYHALOGENATED NAPHTHALENES

The polychloronaphthalenes have been used as boat hull coatings, because of their antimagnetic properties, in the electronics, radio, and television industries as insulating waxes, sealing compounds, and dielectrics for condensers, and also as high-pressure additives for lubricants, and as wood preservatives. Although these functions have now largely been replaced by plastics and silicones, chloracne from chlorinated naphthalenes was reported in New York State in 1972 by Kleinfeld *et al.*[4] Fifty-six cases were found in workers exposed to tetra- and pentachloronaphthalene used in the application of wax insulation to wire coils at an electric plant. Exposure by contact and inhalation resulted from inadequate exhaust ventilation and poor personal hygiene.

Crow has emphasized that of the chloronaphthalene compounds, fumes are the most potent chloracnegens; solutions are less so, and solids rarely cause chloracne.[2] Shelly and Kligman,[5] in experiments on human subjects, showed that penta- and hexachloronaphthalenes were potent acnegens, whereas tri-, tetra-, hepta- and octachloronaphthalenes were nonacnegenic. Chlorinated naphthalene oils and wax-like solids are manufactured in the United States under the trade name Halowax®.[6]

Hexachloronaphthalenes may occur as contaminants of PCBs and the polybromonaphthalenes may occur as contaminants of PBBs and may be chloracnegens in those circumstances.[7]

0077–8923/79/0320–0295 $01.75/0 © 1979, NYAS

Polyhalogenated Biphenyls

The PCBs have been used industrially since 1929, but it was not until 1966 that PCB was identified as an environmental contaminant. Since then PCBs have been found almost everywhere and are some of the most persistent chemicals known.[8] They were recently found in a sample of microscope immersion oil.[9] The chemistry and uses of PCBs has been extensively studied and reviewed. PCBs are a complex heterogeneous series of compounds whose chemical composition has not been clearly defined.[10] They consist of various polychlorinated phenyl compounds and in addition to

Table 1

Chloracne-Producing Chemicals

1. Polyhalogenated naphthalenes*
   a. polychloronaphthalenes
   b. polybromonaphthalenes†,‡
2. Polyhalogenated biphenyls*
   a. polychlorobiphenyls (PCBs)
   b. polybromobiphenyls (PBBs)
3. Polyhalogenated dibenzofurans*
   a. polychlorodibenzofurans especially tri-, tetra-, penta- and hexachloro- dibenzofuran
   b. polybromodibenzofurans-especially tetra-bromodibenzofuran
4. Contaminants of polychlorophenol compounds-especially herbicides (2,4,5-T and penta-chlorophenol) and herbicide intermediates (2,4,5-trichlorophenol)
   a. 2,3,7,8-tetrachlorodibenzo-*p*-dioxin (TCDD)
   b. hexachlorodibenzo-*p*-dioxin
   c. tetrachlorodibenzofuran
5. Contaminants of 3,4-dichloroaniline and related herbicides (Propanil, Methazole, etc.)
   a. 3,4,3′4′-tetrachloroazoxybenzene
   b. 3,4,3′4′-tetrachloroazobenzene
6. Other
   a. 1,2,3,4-Tetrachlorobenzene (experimental)
   b. Dichlobenil (Casoran)—a herbicide (clinical only)
   c. DDT (crude trichlorobenzene‡)

*The polychlorodibenzofurans and hexachloronapthalenes may occur as contaminants in some PCBs.

†The polybromonaphthalenes may occur as contaminants in some PBBs.

‡Not confirmed as chloracnegens.

hexachloronaphthalenes[11] may also be contaminated with the hepatotoxic and acnegenic polychlorodibenzofurans.[12] In the United States, PCBs have been withdrawn from sale other than for closed-system applications such as electrical transformers and capacitors.[2,11,13] Monsanto sells PCBs under the trade name Aroclors®[14]; European PCBs are sold under the trade names Phenochlor® and Clophen®.[11]

Certain PCBs are acnegenic when applied to the rabbit's external ear,[15] and some are porphyrinogenic.[12] The presence of PCBs in human adipose tissue and human milk raises questions of potential effects of these chemicals on human health.[9]

Undoubtedly the largest single epidemic of chloracne occurred in Japan in 1968 from the ingestion of rice oil contaminated with Kanechlor 400®, a tetrachlorobiphenyl.[16] By 1971 over 1000 patients were affected with Yusho, or oil disease. The cutaneous and systemic findings of this epidemic have been reviewed elsewhere.[16,17] Clinically, the findings of cystic swelling and hypersecretion of the meibomian glands have been emphasized with the PCBs more than with the other chloracnegenic chemicals. Further studies of the Yusho epidemic have revealed that dibenzofurans contaminated the biphenyl in the oil. Urabe *et al.* also recently reported that the mucocutaneous lesions have gradually given way over the past ten years to systemic symptoms: nervousness, dullness, dysmenorrhea, and retarded development.[18]

Hara[19] in 1969 and Hasegawa[20] in 1973 reported acneform eruptions among Japanese capacitor plant workers. In the plant descibed by Hasegawa, persons working in environments that contained 0.2–0.3 mg PCBs/m^3 (0.02–0.03 ppm) also had "brown chromodermatosis" of the dorsal joints of the hands, fingers, and nail beds in addition to the acne. In a factory where Kanechlor 300® was employed as a heat exchange medium, the environmental concentration of PCBs was reported to be 0.02 mg/m^3 and no dermatologic manifestations were observed among its employees.[21]

Fischbein *et al,*[22] have recently studied PCB-exposed workers in a capacitor manufacturing facility and found a relatively high prevalence of dermatologic findings (39.3%), including pigmentary disturbances and an acneform eruption. There was a correlation between the prevalance of cutaneous findings and PCB exposure. Ophthalmologic findings were also noted and included increased eye discharge and palprebral edema.

Polybrominated biphenyls (PBBs) came to light as an environmental hazard in Michigan in 1973. Firemaster BP-6®, a hexabrominated biphenyl flame retardant, was inadvertently mixed with food for dairy cattle. The PBB manufacturer, Michigan Chemical Company, used similar tradenames and packaging for the toxic flame retardant (Firemaster) and the feed constituent magnesium oxide (Nutrimaster). It took several months from the appearance of the first toxic symptoms in the farm animals to identify the cause of the poisoning. About 30,000 cattle and a large number of other farm animals died or had to be killed and food products from them had to be destroyed.[8] The Environmental Sciences Laboratory of Mt. Sinai School of Medicine has studied health effects among the exposed farm populations as well as the Michigan Chemical Company Workers.[23] In the latter group, Chanda found a 13% incidence of halogen acne and among the farm families the overall incidence of halogen acne was 3%. Additional cutaneous findings included abnormal sweating, increased fingernail growth, and an unexplained nonscarring alopecia.[24]

Kimbrough confirmed the acne-producing capabilities of PBB. Sixty micrograms applied to the rabbit ear produced follicular hyperkeratosis.[25]

Polyhalogenated Dibenzofurans

The polychlorodibenzofurans (PCFs) occur as contaminants in technical formulations of chlorinated phenols at levels of about 100 to 1000 ppm,[11] in several PCB formulations and in fly ash.[26] Although it has been stated that PCFs are no longer manufactured by industry because of their extreme toxicity, Norström mentions the remarkable claim of the use of polyhalogenated dibenzofurans as flame retardants in a German patent from 1974.[28]

The tri- and tetrachlorodibenzofurans are potent chloracnegens and hepatotoxins. The penta- and hexachloro-derivatives are also potent acnegens.[3]

Moore has reported finding a methyl polybrominated furan in the polar fraction of a Firemaster® PBB product of unstated designation. Kimbrough has shown tetrabromodibenzofuran to be acnegenic when 20 μgm were applied to the rabbit ear.[25]

Dioxins

Chlorinated phenols are produced in large amounts and have widely varying applications as insecticides, fungicides, herbicides, mold inhibitors, antiseptics, and dyes and pigments. Their effectiveness generally increases with the degree of chlorination. Annual world production has been estimated at 200,000 tons.[24]

The earliest observations of toxicity from the production of chlorophenols occurred during the late 1940s and 1950s in the chemical industry. In an accident at a 2,4,5-T plant in the United States in 1949, 228 workers were affected. All but one developed chloracne. Suskind indentified the sodium trichlorophenate fraction as the chloracnegen in these cases.[31] In 1957 Kimmig and Schulz identified the contaminant, 2,3,7,8-tetrachlorodibenzo-*p*-dioxin (TCDD) as the acnegenic compound formed during the synthesis of 2,4,5-trichlorophenol, an intermediate in the synthesis of the herbicide 2,4,5-T.[1-3] TCDD is one of the most toxic chemicals known to man. Its single oral dose LD^{50} in guinea pigs is slightly less than one part per billion body weight, and in rats toxic effects are observed at about 1/100th of the LD_{50}.[32] In animal experiments TCDD has been shown to be 10,000 times more toxic than hexachloronaphthalene.[33]

At least 20 episodes of TCDD exposure have occurred worldwide, affecting industrial, and more recently, nonindustrial populations. Many of these outbreaks have recently been included in reviews by Weirich[3] and Firestone.[30] Several others have gone unreported.[34] Poisoning with TCDD can manifest itself with various degrees of intensity depending upon the type and extent of exposure. Acute exposure symptoms are nausea and vomiting, headache, and signs of respiratory tract, eye and skin irritation. This initial stage has usually occurred from production accidents during the synthesis and processing of 2,4,5-trichlorophenol with explosive liberation of sodium hydroxide, triclorophenol, TCDD, and perhaps other unknown substances. It should be emphasized that this initial cutaneous reaction is comparable to a chemical burn and is not chlorance.[1-3] Following the intital acute stage there maybe a chronic form of TCDD poisoning. Chloracne is most often the first and most constant finding in the chronic type. In Suskind's experience those with the shortest induction period for acne after exposure are those who have also suffered from the severe skin irritiation and burns as a result of the trichlorophenol exposure.[35] Systemic disturbances which may also appear include (1) porphyria cutanea tarda; (2) hyperpigmentation and hypertrichosis without porphyria; (3) systemic involvement of the neuromuscular system, peripheral and central nervous system, liver, kidneys, and pancreas; (4) depression and other psychiatric disturbances; (5) altered clinical chemistry parameters—bilirubin, SGOT, SGPT: additionally aberrations in lipid metabolism with hyperlipidemia and/or hypercholesterolemia have been reported.[1-3]

The accident at the ICMESA chemical plant near Seveso, Italy on July 10, 1976 again focused attention on TCDD and its potential long-term effects. The factory was manufacturing trichlorophenol when an explosion that discharged a toxic cloud of TCDD (more than 2kg), sodium trichlorophenate, caustic soda, sodium chloride, etc., in the form of micelles of ethylene glycol occurred in a reactor. In Seveso, the reactor was vented to the atmosphere causing the cloud to be deposited over 2000 acres of land surrounding the plant, whereas in similar accidents that have occurred in Europe and the United States, the exposure contents were confined within the factories.[36–37]

A Medical Epidemiologic Commission of the Lombardi Regional Government was established. Dermatologic investigation at screening clinics initially established the presence of 447 patients with skin lesions specifically attributable to the accident. These acute lesions were diagnosed between July 10 and early August, 1976 and were characterized by severe erythema and edema of exposed area of the body and caustic burns with epidermal necrosis in the most severly exposed children. Within twenty days of the accident, sheets of follicular hyperkeratosis covered these areas and granuloma annular like lesions were also noted on the dorsum of the hands.[38,39] The hyperkeratosis was so severe that in some cases deep curettage was performed under general anesthesia.[40]

The following complaints were also recorded in some of the 477 subjects with acute skin lesions: anorexia, asthma, nausea, vomiting, headache, diarrhea, hyperhidrosis, and conjunctival irritation.[38]

Chloracne appeared in late July and early August in some of the children who had been more seriously affected by the acute burns.[41] However, most chloracne cases were not diagnosed until September, and by October, 1976 at least 37 cases had been confirmed; more than one-half of these cases occurred in chilren under 12 years of age. From February through April, 1977 more than 30,000 school children were screened and a number of new cases of chloracne appeared. Some of these occurred in children who had been absent from the Seveso area at the time of the accident.[38]

Several points warrant special mention regarding the Seveso chloracne cases:

1. Only a small percentage of those affected with acute burns subsequently developed chloracne.[38] This is in contrast to the industrial cases of TCDD chloracne, where according to Crow,[42] persistent chlorance lesions almost invariably followed such acute lesions.

2. The appearance of new chloracne cases more than 8 months after the accident is epidemiologically significant and may point to the continued contact with,[41] ingestion,[38] or inhalation of dioxin.

3. The clinical and histopathologic findings in some of the children in Seveso bear some resemblance to the cases of epidemic follicular keratosis reported by Schuppli from Basel in 1947. In both instances granuloma annulare-like lesions were observed in conjunction with follicular hyperkeratosis.[39]

4. Gianotti has also reported sweat gland metaplasia with horny cysts in the acrosyringium in skin biopsy specimens from some of the acute skin lesions in children.[39]

5. The early, acute erythematous skin lesions and subsequent hyperkeratosis in some of the children was probably more severe than those reported from industrial TCDD accidents.[41]

Another recent episode of environmental dioxin contamination also came from an industrial source.[43] The outbreak involved three horse arenas, a farm road, and other sites with excavated arena soil and killed 57 horses, 5 colts and numerous other animals. Several children developed chloracne and another child who used the arena soil for a sand box developed epistaxis, diarrhea, lethargy, proteinuria, and hemorrhagic cystitis without evidence of chloracne. A 5.3 year follow-up report of this child was published in 1977 which showed complete recovery with no sequelae, except for a calyceal diverticulum, which had not changed from the time of the original examination. The patient's mother and sister, whose original illness was limited to abdominal pain, diarrhea, and headaches, also recovered completely.[44] The outbreak was initially traced to the spraying of the arenas with salvage oil contaminated with PCDD; trichlorophenol and PCBs were later identified. The contamination of the salvage oil resulted from the improper disposal of toxic industrial waste.[43]

The environmental ramifications of continued use of chlorinated phenoxy acids and their dioxins was the subject of a recent conference arranged by the Royal Swedish Academy of Sciences. In addition to human environmental exposure epidemics of systemic toxicity and death in chickens (chick-edema) have been traced to commerical sources of dioxin. Herbicide use in forests may have effects on both vertebrate and invertebrate populations.[45]

Chloracne has also been associated with technical pentachlorophenol (PCP), either during its production[46] or subsequent use.[47] PCP is registered with the U.S. Environmental Protection Agency as an antimicrobial agent for agricultural seed treatment, on cordage and leather, masonry, in paints and varnishes, on cable covering, nets, for mold control, in drilling and production of petroleum, and as a wood preservative.[48] Technical PCP may be contaminated with hexa-, hepta-, and octa-chlorodibenzo-*p*-dioxin and with tetra-, penta-, hexa-, hepta, and octa-chlorodibenzofurans.[49] Pure PCP may inhibit oxidative phosphorylation resulting in acute toxicity with fever, diaphoresis, respiratory distress, and progressive coma. However, toxic effects such as chloracne and hepatocellular necrosis may result from chronic exposure to the toxic trace contaminants. PCP is readily absorbed through intact skin in amounts that can cause death. Low concentrations of the dust produce painful irritation to the mucous membranes of the nose, throat, and eyes; these warning symptoms are usually sufficient to prevent serious systemic effects. PCP has caused marked irritation and burns to rabbit skin as well as chloracne when applied to rabbit ears.[47] The sodium salt of PCP may also be a skin sensitizer.[50]

Chloracne associated with neurologic pain of the lower extremities and bronchitis was reported from Germany in 1951 during the technical production of PCP.[46]

In 1974 the U.S. National Institute for Occupational Safety and Health found seven cases of chloracne among 15 workers examined in a lumber preservative operation. The hazardous exposure resulted from the sporadic manual opening and dumping of bagged powdered PCP. One worker had generalized chloracne. Most affected workmen showed no evidence of the systemic abnormalities of the type mentioned in previous reports.[47]

Azo- and Azoxybenzenes

This author reported the first documented cases of chloracne from a chlorobenzene compound 3,4,3′,4′-tetrachloroazoxybenzene (TACOB) in 1974.[1] Further details of the clinical aspects of this exposure[51] as well as the microsomal oxygenase enzyme inducing capacity of TCAOB and 3,4,3′,4′-tetrachloroazobenzene (TCAB) were subsequently reported.[52] These two compounds are formed either as trace contaminants during the synthesis of 3,4-dichloroaniline or its further conversion to herbicides, when the conditions of heat and mild oxidation are employed. 3,4-Dichloroaniline is the starting material for the synthesis of a number of commercially important herbicides (acylphenylamides, phenylcarbamates and phenylureas).[52]

A number of outbreaks of chloracne have occurred among workers in chemical plants manufacturing 3,4-dichloroaniline or its derivatives (TABLE 2). The initial reports by this author described chloracne from TCAOB, an extraneous intermediate formed during the manufacture of Methazole® ((2-(3,4-dichlorophenyl)-4-methyl-1,2,4-oxadiazolidine-3,5-dione)). More than 90% (41 workers) of the workers in the plant had chloracne. Exposure to the TCAOB most probably occurred by direct skin contact, but aerosolization of the chemical during centrifugation, with resultant inhalation or ingestion was also possible. The acne involved exposed as well as covered areas of the body in 75% of the cases. The extreme potency of the TCAOB is evident since family members of four workers developed chloracne even though they had never been in or near the plant. Direct contact occurred when contaminated work clothes and tools were brought home. There was no evidence of systemic toxicity including porphyria, although the evaluation was incomplete.[51] I am aware of at least three other episodes of chloracne involving TCAOB (TABLE 2).

The only report of chloracne associated with TCAB exposure is contained in a health hazard evaluation conducted by the U.S. Center for Disease Control and the National Institute for Occupational Safety and Health in 1977.[53] A sudy of 102 workers at a pesticide and herbicide production plant in Arkansas revealed that 11% had been hospitalized over a two-year period for acute illnesses related to chemical

TABLE 2

CHLORACNE OUTBREAKS FROM 3,4-DICHLOROANILINE AND RELATED HERBICIDES

Location	Year	Source of Exposure	Probable Chloracnegenic Chemical
1. Illinois	1960s and/or 1970s	Chemical plant	TCAOB
2. England	1970s	Chemical plant	TCAOB
3. Ohio	1972–73	Chemical plant	TCAOB
4. Michigan	1970s	Chemical plant	TCAOB
5. New Jersey	1960s and/or 1970s	Chemical plant	TCAB
6. Southern U.S.A.	1960s and/or 1970s	Railroad cars delivering herbicide	TCAB
7. Arkansas	1974–77	Chemical plant	TCAB
8. Europe	1970s	Chemical plant	TCAB

TCAOB—tetrachloroazoxybenzene.
TCAB—tetrachloroazobenzene.

exposures at the plant. Of these hospitalizations 41% were for treatment of acute methemoglobinemia following exposure to dichloroaniline. The plant was manufacturing Propanil® from 3,4-dichloroaniline and 38% of the workers were found to have chloracne. The highest incidence of chloracne (71%) was reported by the production workers exposed to TCAB. Urinary uroporphyrins were not measured in the study.[53]

Previous studies have shown that several 3,4-dichloroacylanilide herbicides can be degraded in soil with the release of 3,4-dichloroaniline, which can then be further transformed by soil fungi to TCAB and TCAOB.[52] Any herbicide that is metabolized to an aromatic amine might form azobenzenes.[54]

Despite awareness of the formation of TCAB and TCAOB both in chemical production and in the degradation of herbicides in soil, no studies of toxicity of these compounds were published until 1976.[52] The chemical companies involved refused to release any such data. Accordingly, the author sent some TCAOB to Dr. Alan Poland for study.[52] He and his coworkers demonstrated that TCAOB and TCAB are approximately isoteric to TCDD and TCDF. All four compounds are acnegens and potent inducers of hepatic aryl hydrocarbon hydroxylase (AHH) activity, and compete for stereospecific binding sites in the hepatic cytosol, which are thought to be the receptor sites for the induction of this enzyme. The ability of TCAB and TCAOB and their congeners to induce AHH activity correlates with their binding affinity for the hepatic-cytosol-specific binding sites and their capacity to induce acne; this structure activity relation parallels that demonstrated for the chlorinated dibenzo-*p*-dioxins and dibenzofurans.[52]

Poland and Glover have more recently studied similar structure activity releationships with various biphenyls. Only three congeners, 3,4,3′,4′-tetrachlorobiphenyl, 3,4,5,3′,4′,5′-hexachlorobiphenyl, and 3,4,5,3′,4′,5′-hexabromobiphenyl were active.[45]

Bickers *et al.*[56] and Alvarez *et al.*[57] have demonstrated similar enzyme induction by PCBs in rat skin and the polycyclic hydrocabon, benz(a)-anthracene in human skin. They postulate that skin may accordingly play a more significant role in overall detoxification mechanisms than originally thought.

Hsia *et al.*[58] have recently reported that TCAB is toxic to and transforms mouse embryo fibroblasts and is weakly mutagenic by the salmonella/mammalian microsome test. Further studies of the possible oncogenic nature of TCAB are needed.

The acnegenicity of TCAOB and TCAB was confirmed in animals by Taylor *et al.*[51] and Poland *et al.,*[52] respectively. The herbicides themselves are not acnegens.

There is an old report by Bowen *et al.,*[59] of two chlorobenzene compounds producing chloracne during dichlorodiphenyltrichlorethane (DDT) production. This has never been confirmed and is of doubtful validity. More recently chloracne has been reported in American workers producing the herbicide 2,6-dichlorobenzonitrile (Dichlobenil®).[60] However, there were no studies on the toxicity of intermediate compounds[60] and the chloracnegenicity of Dichlobenil® has been denied by Dr. G. DeMaar, industrial physician for Philips-Duphar B.V. of the Netherlands, European producers of the herbicide.[61]

Experimentally, Rimington and Ziegler produced keratotic plugging of follicles with sebaceous gland atrophy after feeding rats 1,2,3,4-tetrachlorobenzene.[62]

During chronic toxicity studies of a new Japanese fungicide, 3-(3,5-dichlorophenyl)-5,5-dimethyl-oxazolidine-dione-2,4 or DDOD (chemically resembling Methazole®) test animals developed "possible chloracne." Accordingly, production and sale of the chemical was halted.[63]

Clinical Findings

While details of the clinical picture of chloracne have been reviewed elsewhere, several salient points deserve emphasis:

1. In general it is not possible to distinguish differences in the cutaneous pattern of chloracne itself from among the various causative chemicals. Typically, straw colored cysts are said to be almost pathognomonic.[1,2] However, comedones often abound involving every follicular orifice in affected areas. Inflammatory pustules and abscesses also occur as in acne vulgaris. Although exposed body areas are preferential sites of involvement, lesions often occur in covered areas.[38]

2. Chloracne lesions may continue to appear after all exposure to the inciting chemical has ceased. This may result from release of dioxin or PCBs from fat or hepatic body stores.

3. Accompanying or preceding cutaneous signs may point to certain chemical

exposures. Hyperpigmentation may follow exposure to TCDD and may or may not be a manifestation of porphyria cutanea tarda[15] but has also been prominent in PCB exposure, especially in the Yusho episode where no porphyria occurred.[16–18] The acute burns with marked facial edema and erythema with subsequent development of hyperkeratosis and chloracne seems peculiar to accidents during 2,4,5-trichorophenol manufacture. However, earlier reports of "cable rash" resembling photosensitivity, preceded chloracne from exposure to cables insulated with various mixtures of chlorinated naphthalenes and biphenyls.[2]

4. Although chloracne typically occurs from external cutaneous contact, ingestion or inhalation may be important in some cases. This is best illustrated in the Yusho epidemic where all manifestations occurred from ingestion alone of contamined rice oil.[16–18] Puchinelli feels that ingestion of TCDD may have also played a role in some of the Seveso cases.[38]

5. Although certain chloracnegenic chemicals readily induce porphyria in laboratory animals, they have infrequently been associated with porphyria cutanea tarda in man. In some cases porphyria may have been overlooked. However, with the addition of thin layer chromatography of urinary porphyrins, abnormal porphyrin patterns may be detected in the absence of significantly elevated total urinary porphyrins.[64] In laboratory animals experimental porphyria characterized by delayed onset after exposure and increased excretion and hepatic accumulation of uroporphyrins have been produced by technical pentachlorophenol, hexachlorobenzene, polychlorinated biphenyls and TCDD.[49] The most recent clinical outbreak of porphyria cutanea tarda was reported in a 2,4,5-T plant by Jirasek *et al.*[65] Eleven of 55 workers with chloracne also had porphyria cutanea tarda manifested by hyperpigmentation, hypertrichosis, and uroporphyrinuria. The accumulation of uroporphyrin may be the result of inhibition of the enzyme uroporphyrinogen decarboxylase as has recently been shown in familial porphyria cutanea tarda by Benedetto *et al.*[66]

6. Chloracne is one of the most sensitive indicators of chronic exposure to certain halogenated aromatic compounds. This certainly has been true in most of the industrial accidents associated with dioxin exposure. Exceptions to this rule have been reported and included the one case each of possible systemic toxicity without chloracne in the Missouri TCDD accident[43,44] and the laboratory accident reported by Oliver.[67] In Seveso polyneuropathy has been reported but the presence or absence of chloracne is not commented upon.[37] In a report of an investigation of PCBs exposure in a capacitor plant elevated triglycerides and altered hepatocellular enzymes occurred at exposure levels insufficient to produce clinical symptomatology. This may be because chloracne has rarely been encountered at serum PCB levels of less than 200 ppb.[68]

7. The diagnosis of chloracne not only requires a review of the clinical cutaneous and systemic findings but also study of the epidemiology of various environmental factors. In mild isolated cases differentiation from acne vulgaris or acne venenata of nonindustrial cause may be difficult or impossible. Systemic toxicity may be either acute or chronic depending upon degree of exposure and should be monitored on a long-term basis following exposure to these chemicals irrespective of the presence or absence of chloracne. More recently immune suppression (especially the cell-mediated type) has been emphasized as well as the potential for mutagenicity and carcinogenicity.[45]

Pathogenesis

For years the rabbit's external ear has been used as the animal model for assay of compounds causing chloracne as well as an excellent model for studying acne vulgaris.[69] More recently the hairless mouse has been used as an animal model.[17] Chloracne-like lesions also have been produced in monkeys.[45]

In systematic, sequential histopathologic studies of large numbers of experimentally induced chloracne lesions in man certain differences from acne vulgaris have been demonstrated.[5] However, there are a number of similarities and histologic differentiation of the two entities may be difficult. Initially the epithelial lining of chloracne comedones is thicker (acanthotic) than in acne vulgaris and is tortuous with out-pouchings. Complete disappearance of the sebaceous glands is more likely to be seen in chloracne whereas in acne vulgaris sebaceous glands usually can be seen following serial sections. Furthermore, in open and closed comedones of acne vulgaris sebaceous glands may remain intact or hypertrophied.[70] Shelly and Kligman have shown that in experimental chlornaphthalene chloracne in man the inflammatory lesions develop from a clinically prominent, well-developed comedone, which is not true of acne vulgaris.[5] Sweat-gland metaplasia was reported by Gianotti in Seveso children.[39] An electron microscopic study of pilosebaceous ducts in TCAB chloracne was similar to the findings previously reported for acne vulgaris. Biochemical analysis of chloracne lesions showed statistically different amounts of skin surface lipids than in acne vulgaris.[71]

Prevention

The prevention and control of chloracne requires a totally enclosed manufacturing process with no opportunity for direct skin contact or inhalation of the toxic chemicals—a difficult environmental engineering task. The only other alternative is to attempt to alter the chemical synthetic process to eliminate or minimize exposure to the contaminant chloracnegens.

Treatment

Chloracne has been reported to persist for as long as 15 years after all chemical exposure has ceased.[1] In many of the more severe cases, the disorder is refractory to therapy with the usual methods employed in acne vulgaris. Of these, antibiotics for inflammatory lesions, acne surgery (manual expression of comedones and cysts with a comedone extractor) and dermabrasion, if indicated, for severe scarring are probably most beneficial. Some success in treating chloracne has been reported by Plewig[72] and by the author with use of varying concentrations (0.05% to 0.3%) of vitamin A acid (retinoic acid). Excellent results in various keratinizing disorders and in acne vulgaris (severe conglobate acne) have been reported utilizing the oral synthetic retinoid, 13-cis retinoic acid.[72] A trial of this drug in selected cases of chloracne is certainly indicated.[41]

Conclusion

In conclusion, chloracne is a cutaneous disease representing one of the most sensitive indicators of biologic response to certain halogenated aromatic compounds. In many cases chloracne serves as a marker of the medical and environmental impact of contamination of technical grade chemicals with potentially highly toxic intermediates.

Environmental questions will continue to be raised about these contaminants and metabolites of various herbicides and related chemicals, and long-term medical and environmental evaluation is imperative following exposure to them. Correlation of these data with the chemical classification system of these compounds, animal toxicity studies, and bioassay of various biochemical parameters (microsomal oxygenase enzyme induction) offer valuable tools for selecting related chemicals for study.

REFERENCES

1. TAYLOR, J. S. 1974. Chloracne—a continuing problem. Cutis **13:** 585–591.
2. CROW, K. D. 1970. Chloracne, a critical review, including a comparison of two series of acne from chloronaphthalene and pitch fumes. Trans St. Johns Hosp. Dermatol. Soc. **56:** 79–99.
3. WEIRICH, VON E. G. 1978. Die kontaktakne: beispiel ener zwilisationsdermatose. Dermatosen Beruf Umwelt **26:** 7–21, 45–52.
4. KLEINFELD, M., J. MESSITE & R. SWENCICKI. 1972. Clinical effects of chlorinated naphthalene exposure. J. Occup. Med. **14:** 377–379.
5. SHELLY, W. B. & A. M. KLIGMAN. 1957. Experimental production of acne by penta and hexachloronaphthalenes. Arch. Dermatol. **75:** 689–695.
6. HALOWAX® chlorinated naphthalene Oils and Wax-like Solids. Koppers Company, Inc. Pittsburgh, PA, 15219.
7. KIMBROUGH, R. Personal Communication, June 1978.
8. U.S. DEPARTMENT of HEALTH, EDUCATION & WELFARE. Final Report of the Subcommittee on the Health Effects of Polychlorinated Biphenyls and Polybrominated Biphenyls. Washington, D.C., July 1976.
9. BICKERS, D. R., J. EISEMAN, A. KAPPAS & A. P. ALVARES. 1975. Microscope immersion oils: Effects of skin application on cutaneous and hepatic drug metabolizing enzymes. Biochem. Pharmacol. **24:** 779–783.
10. COOK, J. W. 1972. Some chemical aspects of polychlorinated biphenyls (PCBs). Environ Health Perspect. Exp. Issue **1:** 3–4.
11. KIMBROUGH, R. D. 1972. Toxicity of chlorinated hydrocarbons and related compounds. Arch. Environ. Health **25:** 125–131.
12. VOS, J. G. 1972. Toxicology of PCB's for mammals and for birds. Environ. Health Perspect. Exp. Issue **1:** 105–118.
13. BROADHURST, M. G. 1972. Use and replaceability of polychlorinated biphenyls. Environ. Health Perspect Exp. Issue **2:** 81–.
14. Arochlor Plasticizers, Tech Bulletin O/PL–306A, Monsanto Industrial Chemicals Co., St. Louis, Mo.
15. KOMATSU, F. & M. KIKUCHI. 1972. Skin lesions by 3,4,3′,4′-tetrachlorobiphenyl in rabbits. Rukuoki-Igaku-Zasshi **63:** 384–395.
16. KURATSUNE, M., T. YOSHIMURA, J. MATSUKAKA & A. YAMAGUCHI. 1972. Epidemiologic study on Yusho, a poisoning caused by ingestion of rice oil contaminated with a commercial brand of polychlorinated biphenyls. Environ. Health Perspect. Exp. Issue **1:** 119–128.
17. KIMBROUGH, R. D. 1974. The toxicity of polychlorinated polycyclic compounds and related chemicals. Crit. Rev. Toxicol. **2:** 445–498.
18. URABE, H., H. KODA & M. ASAHI. 1979. Present state of Yusho patients. Proc. N.Y. Acad. Sci. This volume.
19. HARA, I. 1969. (Health supervision of workers exposed to chlorobiphenyl in an electrical condenser factory.) Proc. Osaka Prefect Inst. Pub. Ed., Industry Health **7:** 26–31, as mentioned in reference 8 above.
20. HASEGAWA, J., M. SATO & H. TSURUTA. 1973. (Report on survey of work area environment where PCB is handled and of the health of workers handling PCB. Ministry of Labor, Bureau of Labor Standards and Institute of Industrial Health, Republic of Japan), as mentioned in reference 8 above.
21. HASEGAWA, H., M. SATO & H. TSURUTA. 1973. (An investigation on the toxicity of some new substances used as PCB replacement, concentrations in air of SAS, KME oil and

PCB's in carbonless paper producing plants and health examination of workers. Special Research Report 141–211, Research Coordination Bureau, Science and Technology Agency, Republic of Japan), as mentioned in reference 8 above.

22. FISCHBEIN, A., M. S. WOLFF, J. THORNTON, R. LILIS & I. J. SELIKOFF. 1979. Clinical findings among PCB exposed capacitor manufacturing workers. Proc. N.Y. Acad. Sci. This volume.
23. Health Effects of Exposure to Polybrominated Biphenyls. Results of a Clinical Field Survey, Nov. 4–10, 1976 Interim Summary Report to the National Institute of Environmental Health Sciences. ES00928. Environmental Science Laboratory. Mount Sinai School of Medicine, New York.
24. CHANDA, J. J. Personal Communication, June 1978.
25. KIMBROUGH, R. D., V. W. BURSE & J. A. LIDDLE. 1977. Toxicity of polybrominated biphenyl. Lancet **2:** 602–603.
26. RAPPE, C., H. R. BUSER & H. P. BOSSHARDT. 1979. Dioxins, dibenzofurans, and other polyhalogenated aromatics-production, use formation, and destruction. Proc. N.Y. Acad. Sci. This volume.
27. NORSTRÖM, A. 1977. Studies of potential environmental hazardous halogenated aromatics. Dissertation. Umeå University, Umeå, Sweden.
28. MOORE, J. & KAY, K. Polybrominated biphenyls (PBB) environmental contamination in Michigan, 1973–1976. 1977. Environ. Res. **13:** 74–93.
29. NILSSON, C.-A. 1977. Studies of polychlorinated 2-phenoxyphenols precursors of chlorianted dibenzo-*p*-dioxins. Dissertation. Umeå University, Umeå, Sweden.
30. FIRESTONE, D. 1978. The 2,3,7,8-tetrachlorodibenzo-para-dioxin problem: a review. Ecol. Bull. (Stockholm) **27:** 39–52.
31. SUSKIND, R. R. 1973. Personal communication.
32. BAUGHMAN, R. & M. MESELSON. 1973. An analytical method for detecting TCDD (Dioxin): Levels of TCDD in samples from Viet Nam. Environ. Health Perspect. Exp. Issue **5:** 27–36.
33. GOLDMAN, P. J. 1973. Severe acute chloracne—a mass intoxication by 2,3,6,7,-tetrachlorobenzodioxin. Hautarzt **24:** 149–152.
34. MAY, G. as quoted in New Scientist, August 19, 1976, p. 371.
35. SUSKIND R. R. 1977. Written communication.
36. POCCHIARI, F. 1978. 2,3,7,8,-Tetrachlorodibenzo-para-dioxin decontamination. Ecol. Bull. (Stockholm) **27:** 67–70.
37. POCCHIARI, F, V. SILANO & A. AZMPIERI. 1979. Human health effects from accidental release of tetrachlorodibenzo-*p*-dioxin (TCDD) at Seveso, Italy. Proc N.Y. Acad. Sci. This volume.
38. PUCCINELLI, V. 1978. Skin disorders in the Seveso area during the first year after the accident at the ICMESA plant. Commission for the European Communities, Health and Safety Directorate, Document 4383, Kirchburg, Luxemborg.
39. GIANOTTI, F. 1977. Chloracne on tetrachloro-2,3,7,8-dibenzo-*p*-dioxine chez les enfants. An. Dermatol. Venereol. **104:** 825–829.
40. GIANOTTI, F. 1977. Personal comments to the author at a meeting of the Chloracne Panel of the Commission for the European Communities, Milan, Italy, July 11, 1977.
41. Personal conclusions of the author based on visits to Milan in November 1976 and July 1977.
42. CROW K. D. 1977. Personal communication.
43. KIMBROUGH, R, C. D. CARTER, J. A. LIDDLE & R. E. CLINE. 1977. Epidemiology and pathology of a tetrachlorodibenzodioxin poisoning episode. Arch. Environ. Health **25:** 125–131.
44. BEALE, M. G., W. T. SHEARER, M. M. KARL & A. M. ROBSON. 1977. Long-term effects of dioxin exposure. Lancet **1:** 748.
45. RAMEL, C., Ed. 1978. Chlorinated phenoxy acids and their dioxins. Ecol. Bull. (Stockholm) No. 27.
46. BAADER, H. C. E. W. & H. J. BAUER. 1951. Industrial intoxication due to pentachlorophenol. Ind. Med. Surg. **20:** 286–290.
47. LUCAS, J. B. Health Hazard Evaluation, No. 74–117, 1975. National Institute for Occupational Safety and Health, U. S. Public Health Service, Cincinnati, Ohio.

48. KIMBROUGH, R. D. Personal communication, 1977.
49. GOLDSTEIN, J, M. FRIESSEN, R. E. LINDER, P. HICKMAN, J. R. HASS & H. BERGMAN. 1977. Biochem. Pharmacol. **26:**1549–1557.
50. Technical Information Bulletin, Dow Chemical Co., Midland, Michigan.
51. TAYLOR, J. S., R. C. WUTHRICH, K. M. LLOYD & A. POLAND. 1977. Chloracne from manufacture of a new herbicide. Arch. Dermatol. **113:** 616–619.
52. POLAND, A., E. GLOVER, M. DECAMP & E. M. GIANDOMENCO. 1977. 3,4,3′,4′-tetrachloroazoxybenzene and azobenzene: Potent inducers of aryl hydrocarbon hydroxylase. Science **194:** 627–630.
53. MORSE, D. L. & E. L. BAKER. 1977. Health Hazard Evaluation and technical assistance report No. TA 76-59. U.S. Center for Disease Control, National Institute for Occupational Safety and Health, Cininnati, Ohio.
54. ALBRIGHT, B. Personal communication, 1978.
55. POLAND, A. & E. GLOVER. 1977. Chlorinated biphenyl induction of aryl hydrocarbon hydroxylase activity: a study of the structure activity relationship. Mol. Pharmacol. **13:** 924–938.
56. BICKERS, D. R., A. KAPPAS & A. ALVARES. 1974. Differences in inducibility of cutaneous and hepatic drug metabolizing enzymes and cytochrome P-450 by polychlorinated biphenyls and 1,1,1,-trichloro-2,2-bis (p-chlorophenyl) ethane (DDT). J. Pharm. Exp. Therap. **188:** 300–309.
57. ALVARES, A., D. R. BICKERS & A. KAPPAS. 1973. Polychlorinated biphenyls: a new type of inducer of cytochrome P-448 in the liver. Proc. Nat. Acad. Sci. **70:** 1321–1325.
58. HSIA, M.-T. S., F. V. Z. BAIRSTOW, L. C. T. SHH, J. G. POUNDS & J. R. ALLEN. 1977. 3,4,3‘,4‘-tetrachloroazobenzene: a potential environmental toxicant. Res. Commun. Chem. Pathol. Pharmacol. **17:** 225–236.
59. BOWEN, S. S. & M. P. MOURSUND. 1957. Chloracne in the manufacture of DDT. Arch. Dermatol. **75:** 743–746.
60. DEEKEN, J. H. 1974. Chloracne induced by 2,6-dichlorobenzonitrile. Arch. Dermatol. **109:** 245–246.
61. DEMAAR, G. as quoted in a written communication to the author dated November 20, 1978 from L. L. DeAtley, Thompson-Hayward Chemical Co.,, Kansas City, Missouri.
62. RIMINGTON, D. & G. ZIEGLER. 1963. Experimental porphyrins in rats induced by chlorinated benzenes. Biochem. Pharmacol. **12:** 1387–1397, 1963.
63. SATO, H. Written communication to the author. Hokko Chemical Industry Co. Ltd., Tokyo, Japan, Sept. 25, 1975.
64. STRIK, J. J. T. W. A. 1979. Porphyrins in urine as indication for exposure to chlorinated hydrocarbons. Ann. N.Y. Acad. Sci. **320.** This volume.
65. JIRASEK, L., J. KALENSKY, K. KUBEC, J. PAZDEROVA & E. LUKAS. 1974. (Acne chlorina, porphyria cutanea tarda and other manifestations of general intoxication during the manufacture of herbicides II). Ceska Dermatol. **49:** 145–157.
66. BENEDETTO, A. V., J. P. KUSHNER & J. S. TAYLOR. 1978. Porphyria cutanea tarda in three generations of a single family. N. Engl. J. Med. **298:** 358–362.
67. OLIVER, R. M. 1975. Toxic effects of 2,3,7,8-tetrachlorodibenzo 1,4-dioxin in laboratory workers. Brit. J. Ind. Med. **32:** 49–53.
68. LANDRIGAN, P. J., C. W. HEATH, J. A. LIDDLE & D. D. BAYSE. 1978. Exposure to polychlorinated biphenyls. U.S. Publich Health Service EPI-77-35-2, Center for Disease Control, Atlanta, Ga.
69. ADAMS, E. M., D. D. IRISH & H. C. SPENCER. 1941. The response of rabbit skin to compounds reported to have caused acneform dermatitis. Ind. Med. Surg. **2:** 1–4.
70. PLEWIG, G. 1970. Kinetics of the formation of comedos in chloracne (Halowax acne). Arch. Klin. Exp. Dermatol. **238:** 228–241.
71. CUNLIFFE, W. J., M. WILLIAMS, J. C. EDWARDS, S. WILLIAMS, K. T. HOLLAND, C. D. ROBERTS, R. L. HOLMES, D. WILLIAMSON & W. C. PALMER. 1975. An explanation for chloracne—an industrial hazard. Acta Derm Venereol. (Stockholm) **55:** 211–214.
72. PLEWIG, G. 1971. Vitamin A acid treatment of chloracne. Hautarzt **22:** 341.
73. WINDHORST, D. B. Personal communication, 1977.

PORPHYRINS IN URINE AS AN INDICATION OF EXPOSURE TO CHLORINATED HYDROCARBONS

J. J. T. W. A. Strik

Department of Toxicology
Agricultural University
Wageningen, the Netherlands

Chronic Hepatic Porphyria

Chronic hepatic porphyrias, always accompanied by chronic liver damage, exist in clinically latent and manifest stages, which can be recognized and differentiated on the base of the characteristic urinary porphyrin excretion, porphyrin fluorescence of liver biopsy material, the accumulation of uro- and heptacarboxylic porphyrin in the liver.[1]

Porphyrinogenic Compounds as Environmental Contaminants

Some of the polyhalogenated aromatic compounds that are able to produce porphyria in experimental animals are presently known as environmental contaminants. Some polyhalogenated aromatics and -aliphatics have been involved in human poisoning episodes: hexachlorobenzene (Turkey, 1955), polychlorinated biphenyls (Japan, 1968), polybrominated biphenyls (USA, 1973/1974), chlorodibenzo-*p*-dioxins (Italy, 1976), in industrial hygiene: methylchloride (England, 1940), vinylchloride (Germany, 1976), hexachlorobenzene (Australia, 1973), chlorodibenzo-*p*-dioxins (different countries, 1892, 1949, 1953, 1954/55, 1963, 1964, 1965–1969, 1968, 1970) and in wildlife (reviewed in reference 2).

Diagnosis of Chronic Hepatic Porphyria

In the most severe form of porphyria, porphyria cutanea tarda (PCT), a diagnostic indicator is the simultaneous increase of both uro- and heptacarboxylic porphyrin in urine. PCT is present when 45 to 80% of the total porphyrins is uroporphyrin and 15 to 35% is heptacarboxylic porphyrin. In studies of the porphyrins in liver biopsies and urine, it has been found that chronic hepatic porphyrias without clinical symptoms begins with accumulation of uro- and heptacarboxylic porphyrin in the liver followed by a gradually increasing, pathologically elevated excretion of the porphyrins in the urine.

According to the type of porphyrin accumulation in the liver, the extent of porphyrinuria, and the relative distribution of uroporphyrin, coproporphyrin, and porphyrins with five to seven carboxylic groups in the urine, the chronic hepatic porphyrias (CHP) are divided in types A, B and C. Type A, the mildest form, marks the beginning of the progression.[3]

Porphyrins in Urine of Factory Workers Exposed to Chlorinated Hydrocarbons

Porphyrins in urine were determined from workers producing hexachlorocyclopentadiene, allylchloride, epichlorohydrin, and endrin. In the group of workers exposed to one of the chemicals, no increase in urinary total porphyrin level was found as compared to a control group of office employees. The porphyrin patterns of the exposed groups were all normal. The chemicals mentioned were tested for porphyrinogenic capacity in Japanese quail. Only allylchloride seemed to be weak porphyrinogenic.[4]

Coproporphyrinuria and Chronic Hepatic Porphyria Type A in Farm Families from Michigan (USA) Exposed to Polybrominated Biphenyls (PBBs)

Morning urine (sampled in June 1977) from 142 members of farm families from Michigan, who were exposed to polybrominated biphenyls (PBBs), has been analyzed for total porphyrins and porphyrin pattern. The same analysis has been done with a similar group, unexposed to PBBs, from Wisconsin (sampled in November 1977).

There was no significant difference between the urinary total porphyrin values of the Michigan and Wisconsin groups; both values are, however, significantly elevated compared to a Dutch control group. At least 47% of the Michigan group has a porphyrin pattern that is different from normal (indicative for coproporphyrinuria and chronic hepatic porphyria type A). In the Wisconsin group only 6% showed an abnormal porphyrin pattern.[5,6]

Porphyrins in the Urine of Yusho Patients

An annual survey of 16 Yusho patients exposed to rice-oil contaminated with polychlorinated biphenyls (Kanechlor®) and some showing chloracne symptoms was extended with urinary total porphyrin and porphyrin pattern analysis. A significant difference of the liver function tests and urinary total porphyrins and porphyrin pattern between the Japanese controls and cases was not found.[3]

Coproporphyrinuria and Chronic Hepatic Porphyria Type A in TCDD-Exposed People from Seveso, Italy

Responding to a disastrous explosion on the 10th of July, 1976 at the Icmesa plant in Seveso, Italy first morning urine of people living close to the most strongly polluted area has been analyzed for total porphyrins and porphyrin pattern. Sampling took place in March 1978. The toxic cloud released had been contaminated with TCDD (2,3,7,8-tetrachloro-dibenzo-*p*-dioxin).

Urinary total porphyrin values from Seveso samples did not differ significantly from data from PCB-exposed persons (Yusho patients) or controls from Japan or Europe. TLC of urinary porphyrins did reveal an abnormal porphyrin pattern indicative of chronic hepatic porphyria type A.[8]

Conclusion

Just looking at the urinary porphyrin excretion in the range up to 200 μg/liter can not give an indication of exposure to porphyrinogenic chlorinated hydrocarbons. However, the urinary porphyrin pattern can be used to indicate such exposure.

References

1. DOSS, M. In press. Chronic hepatic porphyrias in humans. *In* The Diagnosis and Occurrence of Chronic Hepatic Porphyria in Man Caused by Chemicals. J. J. T. W. A. Strik & J. H. Koeman, Eds. North-Holland, Amsterdam.
2. STRIK, J. J. T. W. A. In press. Chronic hepatic porphyria in humans (exogenic factors, chemicals). *In* The Diagnosis and Occurrence of Chronic Hepatic Porphyria in Man Caused by Chemicals. J. J. T. W. A. Strik & J. H. Koeman, Eds. North-Holland, Amsterdam.
3. DOSS, M. 1975. Clinical Biochemistry and Regulation of Porphyrin Metabolism. M. Doss, Eds. Proc. German-Brazilian Seminar on Medicine and Biomedicine, Deutscher Akademischer Auschtauschdienst (DAAD), Bonn—Bad Godesberg, p. 13.
4. NAGELSMIT, A., P. W. VAN VLIET, W. A. M. VAN DER WIEL-WETZELS, M. J. WIELARD & J. J. T. W. A. STRIK. In press. Porphyrins as Possible Parameter for Exposure to Hexachlorocyclopentadiene, Allylchloride, Epichlorohydrin and Endrin. *In* The Diagnosis and Occurrence of Chronic Hepatic Porphyria in Man Caused by Chemicals. J. J. T. W. A. Strik & J. H. Koeman, Eds. North-Holland, Amsterdam.
5. STRIK, J. J. T. W. A., M. DOSS, L. W. ROBERTSON, R. VON TIEPERMANN, & E. G. M. HARMSEN. In press. Coproporphyrinuria and Chronic Hepatic Porphyria Type A Found in Farm Families in Michigan, U.S.A., Exposed to Polybrominated Biphenyls (PBB). *In* the Diagnosis and Occurrence of Chronic Hepatic Porphyria in Man Caused by Chemicals. J. J. T. W. A. Strik & J. H. Koeman, Eds. North-Holland, Amsterdam.
6. SCHRAA, G., J. J. T. W. A. STRIK & L. W. ROBERTSON. In press. Coproporphyrinuria and Chronic Hepatic Porphyria Type A in Farm Families from Michigan (U.S.A.) Exposed to Polybrominated Biphenyls (PBB). II Control Study. *In* the Diagnosis and Occurrence of Chronic Hepatic Porphyria in Man Caused by Chemicals. J. J. T. W. A. Strik & J. H.
7. STRIK, J. J. T. W. A., H. KIP, T. YOSHIMURA, Y. MASUDA & E. G. M. HARMSEN, In press. Porphyrins in urine of Yusho patients. In The Diagnosis and Occurrence of Chronic Hepatic Porphyria in Man Caused by Chemicals. J. J. T. W. A. Strik & J. H. Koeman, Ed. North-Holland, Amsterdam.
8. CENTEN, A. H. J., J. J. T. W. A. STRIK & A. COLOMBI, In press. Coproporphyrinuria and Chronic Hepatic Porphyria Type A in People from Seveso (Italy) Exposed to 2,3,7,8,-tetrachlorodibenzo-*p*-dioxin (TCDD). *In* The Diagnosis and Occurrence of Chronic Hepatic Porphyria in Man Caused by Chemicals. J. J. T. W. A. Strik & J. H. Koeman, Eds. North-Holland, Amsterdam.

HUMAN HEALTH EFFECTS FROM ACCIDENTAL RELEASE OF TETRACHLORODIBENZO-*p*-DIOXIN (TCDD) AT SEVESO, ITALY

Francesco Pocchiari, Vittorio Silano, and Alfredo Zampieri

Istituto Superiore di Sanità
Rome, Italy

Extent of TCDD Contamination

On July 10, 1976, early in the afternoon, there was an explosion at the ICMESA chemical plant in Meda (about 25 miles north of Milan, Italy), which released a cloud of toxic material that contaminated the immediate environment. A run-away reaction in a reactor for the synthesis of trichlorophenol led to the discharge of hot vapors into the atmosphere. One of the safety valves had blown, and there had been no additional provisions made for trapping the exhaust. The strongly alkaline cloud, which contained TCDD among its other components, fell south-southeast of the plant, covering an area of about 2,000 ha.[1-4] The size of the contaminated area was established mainly by assaying TCDD in the soil and by using as additional criteria the occurrence of dead animals and of dermal lesions among human beings. The area was divided into three zones depending on the concentration of TCDD in the soil. Zone A, the most heavily contaminated section, contained about 2 kg of TCDD unevenly distributed over approximately 110 ha located within a range of 1,500 m from the ICMESA plant (Figure 1). Within Zone A, the highest TCDD soil contamination was found in Zone A_1 (adjacent the ICMESA plant) and the lowest in zone A_7 (Table 1). Zone B, a long narrow strip of about 270 ha located south of Zone A, contained about 20 g of TCDD and was, therefore, much less contaminated than Zone A; however, there were some sectors in this zone with TCDD levels up to 43 $\mu g/m^2$. A further 20 g of TCDD were found over an area of 1,430 ha surrounding Zone B and coded as Zone R. As evaluated on September 9, 1976 (when all animals were sacrificed), animal mortality following the accident was very high in zone A ($\sim$ 25%), whereas in Zones B and R it was in the range 0.3–0.5%.

The figures given for the TCDD contents of Zones A, B, and R must be considered only approximate, owing to the uncertainties involved in TCDD analysis in the soil and the statistical processing of the figures obtained. The amount of TCDD released into the environment should be more precisely evaluated from analysis of the reactor itself. However, up to now no adequate sampling of the reactor has been possible. The analysis of one sample from the head of the reactor showed a TCDD content in the trichlorophenol of 0.35%. As the estimated amount of trichlorophenol released was 500 kg, a theoretical amount of TCDD of 1.7 kg could be inferred. This value is not substantially different from the one obtained by soil analysis.

Population Affected and Epidemiologic Screening

Segments of two municipalities (Seveso and Meda) contribute to Zone A and of three (Seveso, Cesano Maderno, and Desio) to zone B; six municipalities (Seveso, Cesano Maderno, Desio, Seregno, and Barlassina) contribute to varying extents to Zone R. The total population of these six municipalities is about 120,000 residents, of

0077-8923/79/0320-0311 $01.75/0 © 1979, NYAS

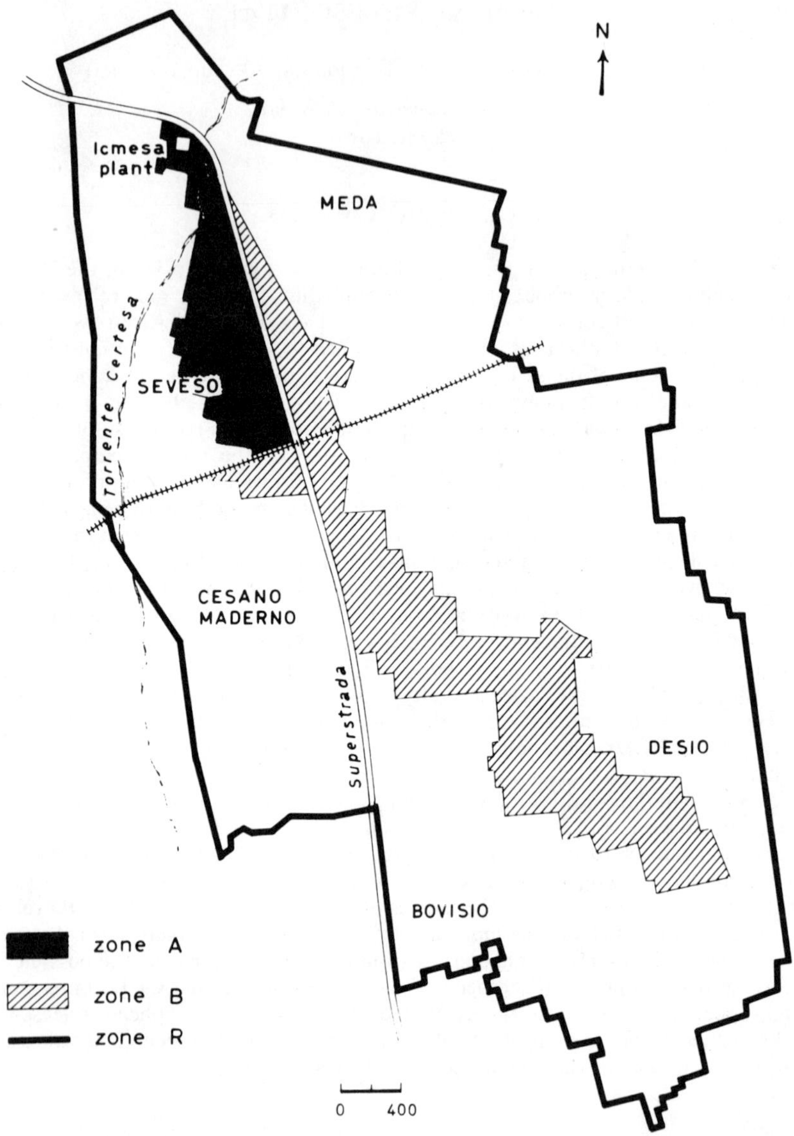

FIGURE 1. TCDD-contamination at Seveso. About 2 kg of TCDD were found in Zone A, whereas Zones B and R contained about 20 g of TCDD.

whom only 733 were in Zone A, 4,800 in Zone B and about 22,000 in Zone R. In any case, eleven municipalities (including the six mentioned above) with a total population of about 220,000 were taken into account for the epidemiologic screening. Zone A was evacuated 14 days after the accident, whereas Zones B and R were not evacuated. Since the ICMESA accident people living in Zones B and R had to observe a number of hygienic regulations; in particular, no cultivation and consumption of local agricultural products and no raising and keeping of poultry and other animals were allowed. Inhabitants of Zone A represent a population group exposed to higher levels of TCDD for only a limited period of time. As decontamination operations in Zone B are still in progress, people from this zone represent a population group exposed to lower levels of TCDD for a much longer period of time.

The epidemiologic program undertaken by the Lombardy Regional Authority can be summed up in three parallel lines of action:

- clinical follow up of the total population living in the accident area;
- longitudinal and systematic health control of the different groups at risk (a few thousand persons from Zones A, B, and R);
- long term morbidity cohort study.

TABLE 1

TCCD CONTAMINATION ($\mu g/m^2$) OF THE SEVESO AREA AS ESTIMATED FROM SOIL SAMPLE ANALYSIS

Zone	Average Value	Maximum Value	Minimum Value
A_1	580.4	5,447.0	not detectable
A_2	521.1	1,700.0	6.1
A_3 (North)	453.0	2,015.0	1.7
A_3 (South)	93.0	441.0	not detectable
A_4	134.9	902.0	not detectable
A_5	62.8	427.0	not detectable
A_6	29.9	270.0	not detectable
A_7	15.5	91.7	not detectable
B	3.0	43.8	not detectable

All of the epidemiologic data have not yet been processed adequately. There are several reasons for the delay in the processing of the mass of existing data. In the first place, the nature of the accident made it very difficult to estimate in time the extent of the population at risk. For safety purposes the Administrators were forced to place 200,000 people under surveillance for the first year after the accident. Moreover, the toxicity characteristics of TCDD were the subject of heated debates among experts both at a national and at international levels and no clear indications for the epidemiologic screening were readily available. Finally, there were various problems of coordination of national, regional and municipal provisions that required some time to be solved. Therefore, this paper is to be considered as a progress report.

DERMATOLOGY

At the moment, 187 cases of skin lesions have been clearly established in a sample of 32,200 people screened.[5] Of these, 50 were in Zone A; most of these cases were observed after October 1976 (only a few months after the accident). In Zone A the percentage of skin lesions was about 7% of the total population residing or present at the moment of the accident and as much as 13% of the screened population from this zone. In the chloracne group ⅔ of the people were under 12 years and ⅓ over 12 years, with the majority being under 21 years. There were more females than males. The dermatologic pathology differed in severity and extent. In 31 cases chloracne was mild, in 11 cases severe, and only in 8 cases was it very severe.

In early February 1977, as there were various persons with chloracne in the contaminated zones who did not come to be checked at the Seveso Dermatological Out-patient's Department, a survey was carried out for children in the nursery, primary and high schools. In a rapid screening of school children, 32,200 children from the schools of Seveso, Meda, Cesano Maderno, Desio, and neighboring municipalities were checked. This investigation made it possible to detect another 85 cases of mild-to-serious chloracne and 153 cases which could not be classified (that is, subjects in which, because of either anamnestic factors or irregularity of symptoms, a definite specific pathology could not be clearly identified). Further investigations established 52 more cases of mild chloracne. Thus, 0.6% of all the people screened were affected by chloracne; however, the percentage of chloracne cases among residents of Seveso, Meda, Cesano Maderno, and Desio (zone B) was equal to about 1.2%.

A third screening has been undertaken since January 1978 and up to now 16,868 people from Seveso, Meda, Cesano Maderno, and Desio have been screened. Only 6 mild chloracne cases were diagnosed, whereas 95 cases could not be classified and deserve further investigations. Even if all these cases were to be classified as chloracne, maximal percentage of chloracne in these four TCDD-contaminated municipalities would be 0.5%, which is very close to the value 0.3% found in other cities of northern Italy such as Como and Lecco not involved in the ICMESA accident.

It appears that the incidence of chloracne is decreasing and that affected people are improving (TABLE 2).

The possibility that, minor skin pathology of the acneform type, appeared in later periods, following contact with the contaminated environment has been suggested by Puccinelli.[5] Some support to such a hypothesis seems to derive from certain cases where very mild symptoms of chloracne, not present in 1977, appeared in the 1978 screening.

NEUROLOGY

The results of the first neurologic screening, carried out in 1977, show (TABLE 3) that the percentage of cases with idiopathic clinical or subclinical neurologic damage,

TABLE 2

EVOLUTION OF FREQUENCY OF CHLORACNE SYMPTOMS OF 75 AFFECTED INDIVIDUALS WITH TIME FROM THE DAY OF TCDD RELEASE

Chloracne symptoms	Time from the ICMESA accident (months) 6–12	18–24
Absent	—	19
Very mild	29	34
Mild	31	21
Severe	6	1
Very severe	9	—

TABLE 3

NEUROLOGIC SCREENING

	Number of Screened Individuals	Number of Individuals with Idiopathic Clinical Neurologic Damage	Number of Individuals with Idiopathic Subclinical Neurologic Damage
1st Screening (1977)			
Zone A	446	30 (6.7%)	14 (3.1%)
Zone B & R	255	3 (1.2%)	3 (1.2%)
2nd Screening (1978)			
Zone A	205	24 (11.7%)	10 (4.9%)
Zone B & R	—	—	—

that might have been caused by TCDD, is higher in Zone A than in Zones B and R for both adults and children.[6] The most frequently encountered pathologic signs were detected in the peripheral nervous system; specifically, there were signs of subclinical neurologic damage such as reduced nerve conduction velocity. Cases of polyneuropathy detectable on clinical examination were confirmed by instrumental examination. At present the pathologic signs detected in the central nervous system, include reduced vision, anisocoria, and nystagmus; these are judged to be of dubious significance.

In 1978 a second neurologic screening has been carried out[6] on about one-half of the individuals from Zone A screened in 1977. The more recent screening showed an increased percentage of cases with idiopathic clinical and subclinical damage. Among the new cases detected in the second screening, polyneuropathy and nystagmus were common findings. No evidence supporting a higher incidence of neurologic symptoms among individuals with chloracne is available.

About two hundred workers from the ICMESA plant and another factory located in Zone A were submitted to neurologic examination and electromyoneurographic study of motor conduction velocity (MCV) and slow fiber conduction velocity (SFCV) of the common peroneal and ulnar nerves; the m. abductor minimi digiti of the hand was studied with needle electromyography.[7] The observed clinical signs showed the presence of damage to the peripheral nerve fibers with a polyneuropathic distribution in 8 subjects (4.0%). In 3 subjects subsequently hospitalized in the Institute of Occupational Health of the University of Milan (Clinica del Lavoro), a polyneuropathy of the lower extremities was diagnosed with clinical and electromyographic criteria. The most common causes of polyneuropathy i.e. alcoholism, diabetes, kidney failure, treatment with drugs with potential neurotoxic action, and previous infectious diseases were ruled out. The mean value of MCV and SFCV of the ulnar and common peroneal nerves were not significantly different from known reference values. However, a trend to an increase of dispersion of MCV values of the ulnar nerve in the workers of ICMESA was observed.

IMMUNOLOGY

Forty-five children aged 3 to 7 years exposed to TCDD in Zone A and 45 unexposed children of the same age (controls) were periodically subjected to a battery of laboratory tests (TABLE 4) to evaluate immune response. These tests started in September 1976 and were repeated every 4 months. Of the exposed children, 21 had chloracne. Up to now 4 screenings have been completed. Statistical evaluation of the results obtained at any time tested did not show any significant immunologic difference between exposed and control children.[8]

GYNECOLOGY, OBSTETRICS, AND PEDIATRICS

No adequate evaluation of the effect of TCDD exposure on frequency of spontaneous abortions is possible at the moment. Some increase in the abortion incidence following the ICMESA accident has been reported by some authors.[4,9] The significance of such an increase is doubtful because of the limited reliability of

TABLE 4

IMMUNOLOGIC AND HEMATOLOGIC SCREENING

Immunologic Screening
- —serum levels of the main classes of immunoglobulins
- —levels of circulating complements
- —lymphocyte response *in vitro* to aspecific stimulation with mitogen (phytohemagglutinin and Pokewood)
- —percent evaluation of the lymphocyte sub-populations T and B (by evaluation of the rosettes E and $Z_y^m C$)
- —lymphocyte response *in vitro* to alloantigens in the mixed lymphocyte culture test

Hematologic Screening
- —GOT
- —GPT
- —alkaline phosphatases
- —γ-GT
- —total bilirubin
- —total serum proteins
- —protein electrophoresis
- —creatinine
- —total cholesterol
- —triglycerides
- —full blood count
- —full urine test
- —differential count
- —urine ALA
- —Au antigen
- —total porphyrin (on special request)
- —pregnancy test
- —glycemia, HCG, HPL, α-fetoprotein in pregnant women

abortion frequency estimates before TCDD release. Moreover, the decrease in birth rate observed in the first 6 months of 1977 suggests that a number of pregnant women from the contaminated area resorted to therapeutic abortion. The possibility also exists that many women resorted to abortion outside the official organizations. Tuchmann-Duplessis[4] has pointed out that the percentage of spontaneous abortions in the contaminated area is below the values generally reported for European countries and for the USA.

Morphologic analysis of fetuses from 30 therapeutic and 4 spontaneous abortions were carried out in Lubeck.[10] No fetal damage indicating a mutagenic, teratogenic, or embryotoxic effect of TCDD exposure was detected. Even morphologic alterations of fetuses from the 4 spontaneous abortions were considered to be independent from TCDD exposure. As it has been pointed out by Hay,[2] the examinations had certain inherent limitations. Minor malformations and particularly organ damage are hardly detectable in the very young fetus, and in all the cases it was difficult to establish the extent to which women had been exposed to TCDD. Therefore, the fetal examinations cannot be considered as a true epidemiologic survey and do not allow any conclusion concerning the embryotoxic activity of TCDD in man.

A systematic obstetrical and pediatric control showed 38 malformations in the twelve months of 1977 among 2,774 newborns in the 11 municipalities of the ICMESA area. These malformations, mainly found in less contaminated municipalities, did not show repetitive characteristics and, in any case, fall within the rate (1–3%) generally accepted for the Western countries.

Postnatal development of a large part of the babies born in Zones A, B, and R (about 2,000 up to now) after the ICMESA accident is being followed through an extensive clinical follow up including a thorough pediatric examination at birth and at 1, 3, 6, 9, 12, and 18 months.[12] Urine and blood analyses have been carried out on about 10% of the babies at 6 to 12 months of age. No statistical and computerized evaluation of the data collected has been possible yet. However, no major pathology that might be attributed to TCDD exposure has been reported by physicians.

Internal, Hematologic, and Cytogenetic Screening

The available data refer to a group of 1,654 subjects from Zones A, B, and R subjected to internal check-up by Ideo's team. Eight percent of the total showed hepatomegaly with idiopathic etiology, while about 24% showed hepatomegaly with an etiology depending on alcohol or viral hepatitis. However, no information is available on the criteria by which the hepatomegaly was evaluated. None of the subjects examined presented a symptomatology that could be attributed with any certainty to TCDD contamination, if the nausea, vomiting, itch or headache sometimes experienced in the days following the accident are excluded. Only time will provide the answer to the question of the significance of this idiopathic hepatomegaly, if present. As far as the preschool and school children are concerned, in about 10% there was moderate hepatomegaly; in 8% vision problems were encountered (in about half a visual deficiency ignored by the patient and in the rest irritative pathology or lowered visual efficiency).[11]

After the ICMESA accident a number of hematologic parameters including transaminases (SGOT and SGPT), γ-glutamyl transferases (γ-GT) and cholesterol were determined at three- or six-months intervals on a number of residents in Zones A and B as well as on ICMESA workers and on workers involved in decontamination of the contaminated area.[13] Statistical evaluation of the data collected is still in progress and only preliminary results are available. With regard to the comparison between the first (July–October 1976) and second (October 1976–January 1977) screenings, an increase was noted in Zone A in the percentage of abnormal values of transaminases and γ-GT. As far as the third (January–August 1977) and fourth (August 1977–March 1978) screenings are concerned, the percentage of abnormal values decreased to values lower than those observed in the first screening. Parallel to these variations, some changes were also observed in the average values of SGOT, SGPT, and γ-GT; however, these average values fell within the normal range. A comparison between children from Zone A and Zone B indicated, as a common trend, a decrease of frequency of abnormal values of these three hematologic parameters, from the first to the last screening. These findings suggest that, particularly in Zone A, TCDD exposure caused an increase of some significant enzymatic activities of the liver in some individuals. This increase was more evident during the months following the ICMESA accident and reached a peak at the end of 1976. An increased frequency of abnormal SGPT and γ-GT values was also observed at the second screening among ICMESA workers. The significance of these findings is not clear at the moment; in fact, inconsistent alterations of these three hematologic parameters of comparable order of magnitude have also been observed among groups of workers involved in the decontamination of the contaminated zone which had not been previously exposed to

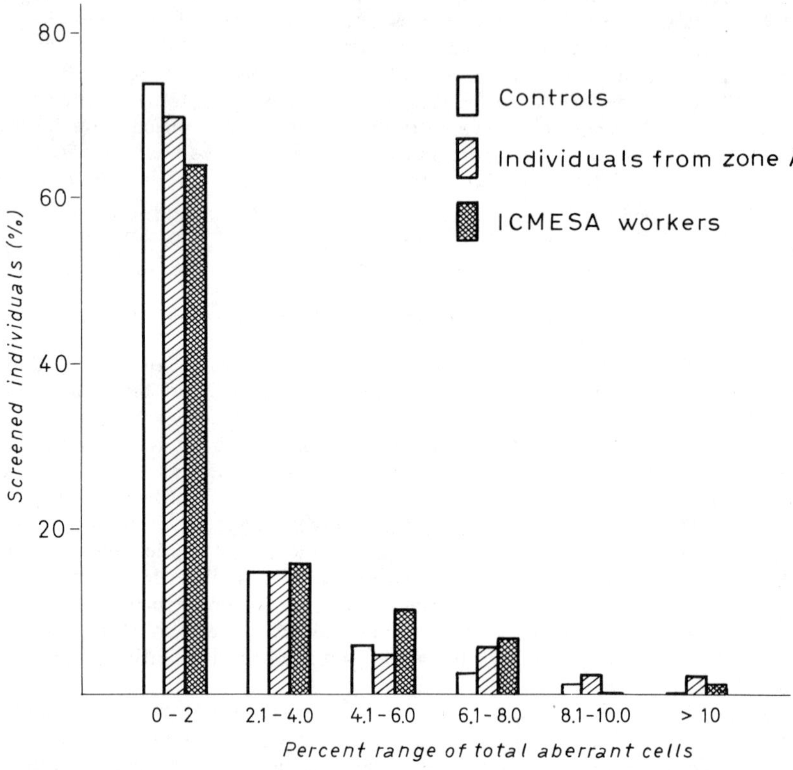

FIGURE 2. Comparative cytogenetic screening of a control group and of groups exposed to TCDD in acute (individuals from Zone A) or chronic (ICMESA workers) forms. Reported cell aberrations include gaps, breaks and structural rearrangements; they were established in lymphocytes from peripheral blood stimulated *in vitro* in the presence of phytohemagglutinin. (Based on De Carli's data).

TCDD and were all chosen as normal healthy individuals on the basis of preliminary clinical examinations.

Cytogenetic research has been carried out by De Carli's team. The data available concern 69 controls, 125 persons from Zone A, and 59 ICMESA workers; the frequency of gaps, breaks and structural rearrangements was studied in lymphocytes from peripheral blood stimulated *in vitro* in the presence of phytohemagglutinin. As shown in FIGURE 2, there was a slight increase of dispersion of total cell aberration frequency in individuals from Zone A and ICMESA workers, as compared with the controls. The increase was also evident when only breaks and rearrangements were taken into account. A comparison of cell aberrations between two very small groups of 15 individuals with and without chloracne has also been carried out. A trend to an increase of dispersion of cell aberration frequency in the chloracne group was observed. Given the cell aberration frequency range usually reported for healthy individuals and the variability commonly observed among analyses from different Laboratories, the differences observed between controls and ICMESA workers or individuals from Zone A do not appear to be significant. However, no final conclusion can be drawn until the statistical significance of these findings has been evaluated.

Infectious Diseases

A notable increase has been observed in the reports of infectious childhood diseases with no apparent relationship to environmental contamination. This can probably be attributed to the greater attention paid by doctors to compulsory reports. This interpretation is also supported by the previously mentioned data, which indicate no alteration in immune response in a consistent group of children from Zone A.

Conclusion

The whole picture of health effects from TCDD accidental contamination at Seveso is far from being complete. A tremendous amount of epidemiologic data concerning short-term effects of TCDD exposure have been collected. However, part of the relevant information pertaining to the health of residents in the contaminated area has not yet been adequately evaluated. Moreover, the possible correlation among population groups characterized by peculiar diseases deserves further investigations. This work is now in progress and will be accomplished as soon as possible. In the meantime, epidemiologic plans for long-term surveillance of TCDD exposed people are in progress. The importance of the Seveso cohorts to study TCDD carcinogenicity in man has been recently stressed by a joint NIEHS/IARC working group,[14] and the setting of mortality register as well as population files suitable for adequate descriptions of morbidity data have been strongly recommended.[14] The clinical follow-up is to be continued for at least 5 years.

Summary

This paper is a progress report of the epidemiologic work carried out under the supervision of the Lombardy Regional Authority during the two years elapsed from the accident in a TCP-producing factory (ICMESA) in Meda (Italy), which resulted in the contamination of several towns of a large, densely populated area called the Brianza di Seveso with a total population of 220,000 inhabitants. A wide follow-up program is in progress in the Seveso area; it includes a clinical screening of the population living in the contaminated area and longitudinal and systematic health control of different groups at risk; a long-term morbidity cohort study has been also undertaken. TCDD exposure following the ICMESA accident resulted in an increased chloracne frequency. Neurologic examinations showed both signs of idiopathic subclinical neurologic damage and cases of clinically detectable idiopathic polyneuropathy in adults. A limited percentage of idiopathic hepatomegaly was reported to be present on clinical investigation; no information, however, is given on the criteria by which the hepatomegaly was investigated. Some alterations were observed in some exposed people in one or more liver tests (mainly transaminases and γ-GT). So far, immunologic investigations, cytogenetic examination and embryomorphology analysis on cases of therapeutical or spontaneous abortions have not given abnormal results.

References

1. Pocchiari, F. 1976. Relazione introduttiva sugli aspetti analitici. *In* Proceedings of the Expert Meeting on the Problems Raised by TCDD Pollution. A. Berlin, A. Buratta & M. Th. Van der Venne, Eds. pp. 31–37.
2. Hay, A. W. M. 1977. Tetrachlorodibenzo-*p*-dioxin release at Seveso. Disasters **1:** 289–308.
3. Pocchiari, F. 1977. Tetrachlorodibenzo-*p*-dioxin decontamination. *In* Chlorinated Phenoxy Acids and their Dioxins. Mode of Action, Health Risks and Environmental Effects. Ecol. Bull. (Stockholm). C. Ramel, Ed. Vol. 27.
4. Tuchmann-Duplessis, H. 1978. Pollution de l'environment et descendance. A propos de l'accident de Seveso. Méd. et Hyg. **36:** 1758–1766.
5. Puccinelli, V. 1977. Report to Lombardy Regional Authority.
6. Boeri, E. 1978. Report to Lombardy Regional Authority.
7. Chiappino, G., R. Gilioli & V. Foa'. 1978. Report to Lombardy Regional Authority.
8. Sirchia, G. 1978. Report to Lombardy Regional Authority.
9. Tognoni, G. 1977. Health survey at Séveso. NATO Ecotoxicology Workshop, University of Surrey, Guilford, Surrey, U.K.
10. Rehder, H., L. Sanchioni, F. Cefis & A. Gropp. 1977. Report to Lombardy Regional Authority.
11. Remotti, V. 1977. Report to Lombardy Regional Authority.
12. Marni, E. 1977. Report to Lombardy Regional Authority.
13. 1978 Report from Lombardy Regional Authority.
14. Coordination of epidemiological studies on the long-term hazards of chlorinated dibenzodioxins/chlorinated dibenzofurans. IARC Internal Technical Report No. 78/001. March 1978.

ACCIDENTS IN TRICHLOROPHENOL PLANTS: A NEED FOR REALISTIC SURVEYS TO ASCERTAIN RISKS TO HEALTH

Alastair Hay

Department of Animal Physiology and Nutrition
University of Leeds
Leeds, LS5 3HL, England

I think that the sensitivity of this particular subject has been indicated by the discussion. I have been following the accident at Séveso as an observer and I was asked to make some comments on that particular situation, and also to refer to some of the other industrial accidents where workers were exposed to a variety of dibenzodioxins and dibenzofurans during the manufacture of trichlorophenol. However, before I do that there are some points I would just like to bring out about the previous study and the one done by Dr. Taylor, which raised one of the more pertinent questions about the sensitivity of indicators to dioxin. (I shall refer to dibenzodioxins simply as dioxin, and for the tetrachlorinated isomer, I shall use TCDD.) And, according to Dr. Taylor, chloracne is one of the more sensitive indicators of TCDD exposure. It is also applicable to exposure to certain aromatic hydrocarbons.

Now, at Séveso the children are the ones that have been most seriously affected by TCDD and they are the ones who have chloracne. It is—and I do not know whether it is premature to say this—reassuring to see that the chloracne is disappearing in these children. But what is still confusing, is the fact that the neurologic damage that has been reported in the population of Séveso is confined mainly to the adult population, with little observed amongst the children. And I must say that I find this very confusing. If chloracne is a sensitive indicator of dioxin exposure (or aromatic hydrocarbon exposure) I would have thought that the children would have shown some signs of neurologic damage, and that this could have been related to exposure. I would very much welcome some comments on that.

The immunologic studies that have been done do not show any signs of real damage which can be linked to TCDD exposure. This is the one group at Séveso where there is a proper control population. The studies, however, will take some time because, as most of you know, lymphocytes and leukocytes have a fairly long half-life. If, therefore, one is looking for signs of thymus atrophy, or some other measure of damage to the immune system, it is going to be some time before this information will be available.

I do have one serious criticism to make concerning the studies that have been carried out so far. I understand the reason for looking at a large population at Séveso and there are very sound political reasons for doing so. In my opinion, however, something is missing, and that is a more limited study restricted solely to the people who are (or were) exposed to what might be classed as highish concentrations of TCDD. This group should have a matched control. I do feel that there is a real danger of being swamped by too much data from an open-ended study, where a more limited study of certain populations could be more revealing. I would urge the Italian authorities to consider this point.

As for birth abnormalities, in the 34 fetuses from the women who either had therapeutic abortions, or spontaneous abortions at Séveso there are none. All this material was analyzed by Dr. Gropp at the Lübeck Medical School in Germany and in

0077–8923/79/0320–0315 $01.75/0 © 1979, NYAS

his final report he suggested that there might be one possible case of fetal Down's syndrome. However, he did not link this with TCDD exposure.

The one serious limitation in the study of the fetuses is the fact that there is no way of predicting just how much TCDD the mothers were exposed to. I think there were six women from Zone A, and various others from Zone B; the rest said that they had passed through the contaminated area. The therapeutic abortions were permitted on the grounds that the women were in some kind of danger, or I should say, that it was their mental condition which was considered to be at risk. With these constraints it is obvious that the study done on abnormalities was a study done on data to hand. It was not a controlled study in any way, and for that reason we are still not in a position to say whether TCDD is embryotoxic in man, or whether this dioxin would cause birth abnormalities in the human population. That, unfortunately, is something I feel we have lost. Many of the pregnant women from Séveso went to other countries for abortions. Many came to the U.K., and there were many more who went to other countries to have an abortion performed simply because they could not have the operation done in Italy. Only one of the fetuses was examined in the U.K., and I do not know the results of that study.

These, then, were just a few comments I had to make on the previous contributions. I shall sum up at the end.

TABLE 1 simply catalogues the number of trichlorophenol accidents that have taken place over the years. They are the ones that I have managed to track down, and you will see that they start fairly early on. The first occurred in 1949 and the others happened fairly regularly thereafter, until 1976. Some of the information is not complete and some needs revising. In the Monsanto accident in 1949 there were 228 workers exposed to dibenzodioxins and dibenzofurans, and not 117 as shown in the table. Thirty-six of these are being followed up. Dr. Suskind is, as you all know, involved in that particular study. He said that no significant long-term observations are available at the moment, but that he is hoping to do a mortality cohort study. This is most important as this is the largest group of industrial workers ever exposed to these chemicals.

One of the German companies had 75 workers exposed to trichlorophenol. The most significant thing about this particular study is that the accident happened fairly early on, in 1953 to be exact. Seventeen of the 75 people have died. In that group, however, 6 died from malignant neoplasms. Four of these were from cancers of the gastrointestinal tract. It has been pointed out—and the International Agency for Research in Cancer are very much aware of the problem—that this is a very high incidence of gastrointestinal cancer. It is one study that will certainly be followed up and the IARC should be collecting the data.

In Czechoslovakia some 80 people were involved. Fifty-five have been monitored, five of whom have died from a variety of causes; many of the remaining subjects are still showing signs of hypertension, hyperlipidemia and hypercholesterolemia. But, the follow-up study is in progress and the 10-year report is about to be completed by Dr. Jirasek.

In England, we have also had our problems—at a chemical company in Derbyshire. Eighty-five to 90 people developed chloracne at this chemical plant. In the actual accident one worker was killed by the explosion. There has only been one death since then as far as can be ascertained. This man had chloracne. He died from a coronary thrombosis. However, the whole chloracne population is now being monitored. Those people who have, or who have had chloracne, are being watched. The workers who were exposed to TCDD, but did not develop chloracne, are being monitored too, and a control group is also being studied.

In one of the other European incidents, the accident in Holland, there were some

TABLE 1

ACCIDENTS IN CHEMICAL PLANTS MANUFACTURING TRICHLOROPHENOL

Date	Manufacturer	Country	Personnel Injured	Cause of accident
1949	Monsanto	U.S.A.	117	Overheating leading to explosion
1952/3	Boehringer	W. Germany	37	Exposure during manufacturing process
1953	Badische Anilin und Soda Fabrik AG (BASF)	W. Germany	55	Overheating leading to explosion
1953–1971	Rhone Poulenc	France	97 (Including 17 injured on a 1956 explosion and 21 in another in 1966)	Exposure during manufacturing process
1956	Hooker	U.S.A.	Staff employees	Overheating
1960	Diamond Sham-Roch	U.S.A.	Figure unknown	Overheating
1963	Philips Duphar	Holland	30	Overheating leading to explosion
1964	Spolana	Czechoslovakia	72	Exposure during manufacturing process
1964	Dow Chemical	U.S.A.	30	Exposure during manufacturing process
1968	Coalite and Chemical Products	U.K.	79	Overheating leading to explosion
1970(?)	Bayer	W. Germany	5	Exposure during manufacturing process
1972/73	Chemie/Linz	Austria	50	Exposure during manufacturing process
1976	Icmesa (Givaudan)	Italy	106 Children	Overheating
Before 1976	Thompson Hayward	U.S.A.		Overheating leading to explosion

See Properties of TCDD for chemical reaction. Above the critical temperature – about 180°C – if the solvent is distilled off a significant exothermic reaction will occur. This leads to rapid overheating and may result in an explosion.

106 people involved, and of the 44 who worked in the factory, 26 developed chloracne. Thirteen years later, some 20 still have the skin disease.

There were various other populations involved in the Dutch study, including the people who were involved in the clean-up process, some of whom have chloracne. Some of this group have since died. And, then there was a third group of factory employees, only a fraction of whom had chloracne. The one significant point about the deaths of workers in these particular groups is that they seem to present a higher-than-normal incidence of myocardial infarction. This is another trend that has emerged. We now have two: an increase of gastrointestinal cancer and a possible increase in myocardial infarction in people exposed to dibenzodioxins/dibenzofurans. The Dutch workers are also being monitored and a variety of tests will be performed on them. The information received will subsequently be sent to the International Agency for Research in Cancer. The IARC should be able to collect much of this information in a reasonably short time.

There is an additional report involving workers in the Soviet Union. Of these workers, 128 had what was called occupational skin disease. This was a report that appeared in 1970. It referred to workers involved in the manufacture of the herbicide 2,4,5-T. This group exhibited a variety of clinical symptoms similar to those reported by Dr. Taylor. As Dr. Taylor said, industrial hygiene is one of the important ways of getting rid of industrial exposure to halogenated hydrocarbons and that, in fact, is what this report suggests. The solution arrived at was to stop production of the herbicide altogether.

Summing up I would just like to say that at Séveso we have a situation where a large population was exposed to chlorinated dibenzodioxins, and primarily TCDD. But it is probably low level exposure. In fact it is almost certainly low level exposure. However, my own view is that it is probably a little premature to say that there will be no long-term effects to the Séveso population. I think that must remain speculation at the present time.

And, the one other point I would make is the following. The workers who have been exposed to dioxins in the manufacture of trichlorophenol do represent an important group to be monitored. The International Agency for Research in Cancer is to be congratulated on agreeing to collate information relating to this group. Confidentiality is obviously an important question, and so many of the companies involved in this exercise want to remain anonymous. It is vital that this information be collected and I suppose confidentiality is one thing we must accept. We hope that the IARC will be able to provide a great deal of information on TCDD exposure, and the one thing that we do really need to get out of all this is some measure of the body burden. We do need to have some kind of objective assessment of TCDD exposure, and we need to be able to relate clinical symptoms and biochemical findings to some kind of chemical measurement of dioxins/TCDD in the body.

I appreciate that the chemical analysis of biologic samples containing TCDD is fraught with difficulty, but I would urge the chemists to persevere because we do need to know some of these answers.

STUDIES ON THE NEUROBEHAVIORAL EFFECTS OF POLYBROMINATED BIPHENYLS IN RATS

Hugh A. Tilson and Patrick A. Cabe

Laboratory of Behavioral and Neurological Toxicology
National Institute of Environmental Health Sciences
Research Triangle Park, North Carolina 27709

INTRODUCTION

In 1973, Firemaster FF-1®, a mixture of polybrominated biphenyls (PBBs) used as a flame retardant in plastics and synthetic fibers, was mixed accidentally into animal feeds. It has been estimated that several thousand people were exposed to PBBs through the ingestion of contaminated meat and animal products. In many people having elevated serum and adipose levels of PBBs, there have been reports of alterations in circulating blood lymphocytes,[1] changes in liver enzyme function,[2] neurobehavioral deficits,[3] and skeletomuscular abnormalities.[4]

Over the last year, our laboratory has been attempting to develop animal models capable of assessing neurobehavioral deficits following exposure to relatively low amounts of PBBs. In a series of experiments, rats exposed chronically or subchronically to FF-1 or 2,4,5,2′,4′,5′-hexabromobiphenyl (HBB) have been examined for the presence or absence of neurobehavioral deficits frequently reported by humans exposed to PBBs. These variables, which include muscular impairments, irritability, deficits in cognitive abilities, and decrements in the ability to perform daily, on-going activities, and the corresponding tests employed in rats are summarized in TABLE 1.

MATERIALS AND METHODS

Male and female rats of the Fisher 344/N strain (Charles Rivers Breeding Co., Wilmington, Mass.) were used as subjects. All subjects were approximately 5–6 weeks of age at the start of each study and were housed in rooms having a constant light-dark cycle (light, 7 a.m. to 7 p.m.), temperature (21 ± 2°C) and relative humidity (50 ± 10%). Rat chow (NIH diet #31) and water were freely available throughout the experiment.

Three separate experiments were conducted on the behavioral and neurologic toxicity of the PBBs as summarized in TABLE 2. In Experiment I, Firemaster FF-1 (Michigan Chemical Co., lot no 1312FT) and 2,4,5,2′,4′,5′-hexabromobiphenyl (HBB) were suspended in corn-oil vehicle and given by gavage, five days per week for a total of 22 doses. Rats were given 30 mg/kg of FF-1, 16.8 mg/kg of HBB or 1 ml/kg of corn-oil vehicle. After 30 days of dosing, the rats were tested for neurobehavioral deficits.

Animals in the second experiment received 3 or 10 mg/kg of FF-1 or corn-oil vehicle, 5 days per week, for 26 weeks. After cessation of dosing, the neurobehavioral functioning of the animals was evaluated in a battery of tests. The rats in the third study were trained to postpone electric footshock on an unsignaled continuous avoidance schedule.[5] After a baseline of response was established, the animals were randomly assigned to treatment groups and received 10 mg/kg of FF-1, 10 mg/kg of HBB or 1 ml/kg of corn-oil vehicle 3 days per week, for 8 weeks. All dosing followed

0077–8923/79/0320–0325 $01.75/0 © 1979, NYAS

TABLE 1

SYMPTOMS OF PBB TOXICOSIS IN HUMANS AND THE ANIMAL MODELS EMPLOYED

Symptoms in Humans	Tests Employed in Rats
Muscular weakness, apathy, lethargy	Spontaneously occurring motor activity Forelimb grip strength Hindlimb extensor response Inclined screen performance
Irritability	Startle response to air puff
Cognitive deficit	Acquisition of one-way avoidance
Impaired daily (on-going) activities	Operant behavior (Unsignaled avoidance responding)

daily behavioral sessions and any changes in baseline as a function of exposure to PBBs were measured.

A mixed-model analysis of variance (ANOVA) for repeated measures[6] was used where appropriate to evaluate overall effects of treatment, sex, and interactions between the fixed variables. If a significant treatment effect was observed, differences between means of treated subjects and controls were tested for statistical significance by means of Fisher's least significant difference test.[6] In cases where responses could be divided into two categories, differences between means were assessed using Fisher's exact probability tests, while chi-square (χ^2) and Mann-Whitney U-tests were used for rating scales and response latencies, respectively.[7] The accepted level of significance was $p < 0.05$, using two-tailed tests.

SPONTANEOUS MOTOR ACTIVITY

Methods

Rats receiving 30 mg/kg of FF-1, 16.8 mg/kg of HBB or corn-oil for 30 days (Experiment I) were tested for spontaneous motor activity in an openfield maze. The maze (3 ft x 3 ft x 1 ft deep) was constructed of plexiglas and was marked off into 36

TABLE 2

SUMMARY OF EXPERIMENTAL PROTOCOLS USED TO TEST NEUROBEHAVIORAL EFFECTS OF PBBs IN RATS

Experiment	Dosing Regimen	Total Dose	Time of Testing
I	FF-1 (30 mg/kg) and HBB (16.8 mg/kg) given 5 days per week for 22 doses	FF-1 = 660 mg/kg HBB = 370 mg/kg	After 30 days of dosing
II	FF-1 (3 or 10 mg/kg) given 5 days per week for 130 doses	FF-1 = 390 or 1300 mg/kg	After 6 months of dosing
III	FF-1 (10 mg/kg) or HBB (10 mg/kg) given 3 days per week for 40 doses	FF-1 = 400 mg/kg HBB = 400 mg/kg	Measure operant behavior daily throughout dosing

equal segments with black paint. At the start of the test, the subjects were placed in one of the four corners facing the wall. The number of square crossings made in a 3 minute period was counted.

Animals dosed for a period of 6 months with 3 or 10 mg/kg of FF-1 or corn-oil vehicle were tested in commercially available activity monitors (Automex; Columbus Instruments Co., Columbus, Ohio) placed inside sound-attenuating outer chambers (Coulbourn Instruments, Lehigh Valley, Pa.). Rats were placed individually into darkened chambers and motor activity was measured for a period of 9 minutes.

Results

The number of crossings made by rats in an open field during a 3 minute observation period was decreased by PBB treatment ($F = 32.72$; $df = 2,47$; $p < 0.0010$). Female rats were more active than male rats ($F = 8.18$; $df = 1,47$; $p < 0.0010$), and the sex by treatment interaction was statistically significant ($F = 6.98$; $df = 2,47$; $p < 0.0024$).

Pairwise comparison of the activity scores indicated that female rats given FF-1

TABLE 3

EFFECTS OF FF-1 AND HBB GIVEN OVER A PERIOD OF 30 DAYS ON ACTIVITY OF RATS IN THE OPEN FIELD

Treatment	N	Average Square Crossings in 3 min ± S.E.M. Males	Females
Controls	12	44 ± 6	98 ± 7
FF-1 (30 mg/kg)	6	45 ± 6	42 ± 12*
HBB (16.8 mg/kg)	6	26 ± 3*	67 ± 12*

*Statistically different from controls (Fisher's least significant difference test, $p < 0.05$, two-tailed).

(30 mg/kg) or HBB (16.8 mg/kg) had significantly lower activity scores after 30 days of dosing (TABLE 3). The activity of male rats was affected significantly at 30 days in subjects receiving 16.8 mg/kg of HBB.

The motor activity of rats treated with FF-1 over a period of 6 months was also significantly decreased ($F = 7.76$; $df = 2,66$; $p < 0.0010$). Females were more active than males ($F = 31.44$; $df = 1, 66$; $p < 0.0001$) and the sex by treatment interaction was significant ($F = 24.88$; $df = 2, 66$; $p < 0.0001$). Pairwise comparisons indicated that female rats receiving 3 and 10 mg/kg of FF-1 for 6 months had significantly lower activity counts than controls (TABLE 4). Male rats were not affected significantly by FF-1.

Chronic and subchronic exposure to PBBs reduced locomotor activity as measured by two different procedures. In a behavioral sense, this effect may be viewed as analogous to the apathy or lethargy reported by humans exposed to PBBs. The decrement in activity observed in this experiment may also be interpreted as an indication of muscular weakness.

It was noteworthy that the motor activity of the female rats was affected by PBBs to a greater extent than that of the males. This may have been due to the fact that the male rats were much less active under control conditions. Thus, the observed sex

TABLE 4
EFFECTS OF FF-1 GIVEN FOR 6 MONTHS ON THE SPONTANEOUS MOTOR ACTIVITY OF RATS

Group	Average Activity Counts in 9 min ± S.E.M. Males	Females
Controls	158 ± 18(15)*	425 ± 34(9)
FF-1 (3 mg/kg)	188 ± 20(14)	159 ± 26(7)†
FF-1 (10 mg/kg)	177 ± 16(14)	213 ± 16(8)†

*Number of subjects per group.
†Statistically different from control (Fisher's least significant difference test, $p < 0.05$, two-tailed).

difference in response to PBBs was most likely due to a "basement" effect resulting from the low activity of the males. In subsequent experiments, various aspects of fore- and hindlimb functioning were assessed separately in order to characterize the skeletomuscular effects of the PBBs more precisely.

FORELIMB GRIP STRENGTH

Methods

Muscular strength in the forelimbs was measured using a recording grip meter procedure.[8] Animals were held by the tail in one hand and supported in the other about 10–15 cm above a wire ring 45 mm in diameter connected to a push-pull strain gauge (J. Chatillon and Sons, Kew Gardens, N.Y.). The rat was lowered by the tail at a constant rate until it reached out and grasped the ring with both paws. The rat was then pulled in a horizontal attitude away from the ring until it released it. An average of 3 readings was recorded as the grip strength score.

Results

Exposure to FF-1 (30 mg/kg) or HBB (16.8 mg/kg) in Experiment I significantly decreased forelimb grip scores ($F = 14.53$; $df = 2, 47$; $p < 0.0001$). The sex effect ($F = 12.93$; $df = 1, 47$; $p < 0.0008$) was statistically significant, but the treatment by sex interaction was not significant ($F = 1.01$). Pairwise comparisons indicated that 30 mg/kg of FF-1 decreased the grip strength of both male and female rats after 30 days of dosing (TABLE 5). HBB (16.8 mg/kg) produced a significant decrease in the grip scores of male rats after 30 days of dosing while the slight decrease observed in the female rats was not statistically significant.

PBBs also reduced the forelimb grip scores of rats in Experiment II ($F = 24.74$; $df = 2, 69$; $p < 0.0001$). Although there was a sex difference ($F = 96.47$; $df = 1, 69$; $p < 0.0001$), the sex by treatment interaction was not statistically reliable ($F = 1.93$; $df = 2, 69$). Pairwise comparisons showed that male rats were affected significantly by both 3 and 10 mg/kg of FF-1, while only females receiving 10 mg/kg were affected significantly (TABLE 5).

The results of this study indicate that the forelimb grip strength of male and, to a lesser extent, female rats was decreased by subchronic and chronic exposure to PBBs.

In the next experiment, the hindlimb extensor response of rats treated with FF-1 for 6 months was assessed.

Hindlimb Extensor Response

Methods

The hindlimb extensor response (HLER) of rats was studied using a method described in detail elsewhere.[9] Briefly, a T-bar constructed of 5 inch × ⅛ inch brass rod through a hole in one end of 3 inch hexagonal threaded aluminum standoff is attached to a push-pull strain gauge (Chatillon and Sons, Kew Gardens, N.Y.). The strain gauge is mounted at a 45° angle by means of pretapped holes in the instrument such that the T-bar is parallel to the edge of a 9 inch × 12 inch plexiglas platform covered with coarse sandpaper. The platform is 13 inches, above the bench top and is 1½ inches higher and 3 inches laterally displaced from the T-bar.

The rat is held in the experimeter's right hand around the thorax under the shoulder girdle, while the rat's tail is taken in the left hand and the plantar surfaces of the rat's hind feet placed symmetrically on the bar. The forelimbs are placed on the platform and then released. After sitting in this position long enough for the experimenter to zero the meter, a sharp air puff is delivered to the area of the rat's rump. The deflection observed after the animal jumps to the platform is recorded. The largest 3 recordings out of 5 trials is taken as the measure of HLER.

Results

The HLER of rats treated with FF-1 was significantly decreased ($F = 3.99$; $df = 2, 66$; $p < 0.0236$). There was no difference between sexes ($F = 2.65$; $df = 1, 66$) and the sex by treatment interaction was not significant ($F = 0.78$). Pairwise comparisons indicated that rats receiving 10 mg/kg of FF-1 over a period of 6 months had reduced hindlimb responses (FIGURE 1). Thus, the results of this study indicate that PBBs depress hindlimb, as well as forelimb, muscular function. In the next experiment,

TABLE 5

THE EFFECTS OF PBBs ON FORELIMB GRIP STRENGTH OF RATS

Study and Group	Average Grip Strength (g) ± S.E.M. Females	Males
1. After 30 Days of Dosing (I)		
Controls	258 ± 8(12)*	318 ± 11(12)
FF-1 (30 mg/kg)	195 ± 17(6)†	229 ± 20(6)†
HBB (16.8 mg/kg)	227 ± 19(6)	248 ± 25(6)†
2. After 6 Months of Dosing (II)		
Controls	515 ± 21(12)	708 ± 17(15)
FF-1 (3 mg/kg)	461 ± 24(7)	633 ± 20(14)†
FF-1 (10 mg/kg)	429 ± 16(8)†	548 ± 9(14)†

*Number of subjects per group.

†Statistically different from controls (Fisher's least significant difference test, $p < 0.05$, two-tailed).

PBB-exposed rats were tested on an inclined screen procedure.[10] This technique measures general motor function and motor coordination.

INCLINED SCREEN

Method

Rats treated with 3 or 10 mg/kg of FF-1 over a period of 6 months were tested on an inclined screen procedure as described by Tilson and Cabe.[10] Briefly, each rat was picked up by the tail and gently flung with an upward motion toward the middle of a rectangular (10 in × 14 in) wire screen (¼ in mesh) located at a distance of approximately 3 feet. The subjects were rated on a 0–5 point scale as summarized in TABLE 6.

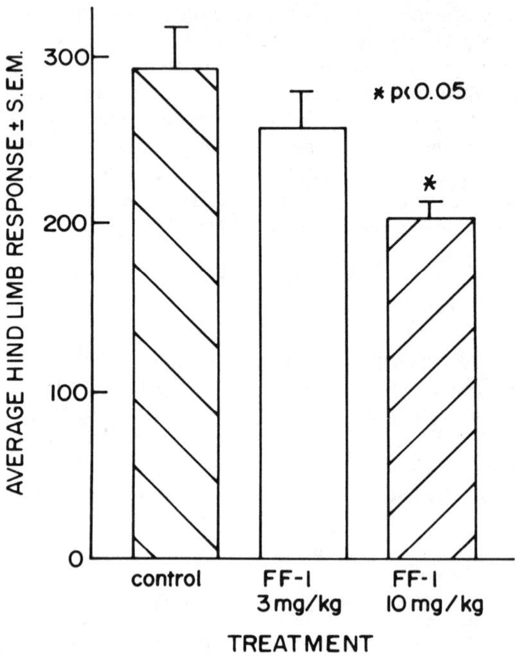

FIGURE 1. The effects of FF-1 on the hindlimb extensor response. Rats were given either 3 mg/kg of FF-1 (14 males and 7 females), 10 mg/kg of FF-1 (14 males and 8 females), or corn-oil vehicle (15 males and 12 females) by gavage, 5 days per week, for 26 weeks. The asterisk indicates a statistically significant difference between the treated and control groups (Fisher's least significant difference, $p < 0.05$, two-tailed).

Results

The frequency with which rats were rated as minimally (0–1), moderately (2–3), or severely (4–5) impaired was not different between groups as tested by Chi-square analysis (TABLE 7). These data indicate that, although rats showed decreased motor activity and impaired fore- and hindlimb strength, there was no evidence of gross muscular weakness or impaired coordination.

TABLE 6

RATING SCALE FOR INCLINED-SCREEN PROCEDURE

Score	Description
0	The subject displayed no obvious motor dysfunction upon landing on the screen.
1	Upon landing, the subject showed no signs of motor dysfunction, but appeared behaviorally depressed (not alert, listless) or displayed a hunched posture.
2	Upon landing, the animal did not appear to grasp securely and a slight backward motion or bounce was observed. No other signs of motor dysfunction were noted.
3	The subject usually bounced upon landing and in subsequent movements, the animal was noticeably unsteady in its movements, but did not slip from the screen.
4	At some point during the test, the subject lost its grip on the screen and slipped, but did not fall off of the screen.
5	The subject fell off and slid to the bottom of the screen at some point during the test.

STARTLE RESPONSIVENESS

Methods

Rats in Experiment II treated with 3 or 10 mg/kg of FF-1 for 6 months were tested for startle responsiveness to an air-puff stimulus. A strain gauge was mounted in a 90° angle by pre-tapped holes to a plexiglas frame. A plexiglas platform ($5\frac{1}{4}$ in × 5 in × $\frac{3}{16}$ in) was mounted on top of the strain gauge. The distance from the platform to the bench top was 18 inches. Rats are placed on the platform for approximately 30–45 seconds, while the strain gauge was zeroed and set in the push-record mode. An air puff from a standard source was delivered from a distance of 5 inches at an angle of 45° to the face of the subject. The average deflection recorded on the strain gauge as a consequence of 5 trials, each separated by at least 10 seconds, was taken as a measure of startle magnitude.

Results

ANOVA indicated that FF-1 decreased startle responsiveness significantly ($F = 6.94$; $df = 2, 69$; $p < 0.0019$). The sex effect ($F = 26.26$; $df = 1, 69$; $p < 0.0001$) and

TABLE 7

BEHAVIORAL RATING OF RATS RECEIVING FF-1 OVER A PERIOD OF 6 MONTHS AND TESTED ON THE INCLINED SCREEN

	Number of Rats Receiving Score Rating Score			
Treatment	0–1	2–3	4–5	Total
Controls	27	0	0	27
FF-1-3 mg/kg	20	1	0	21
FF-1-10 mg/kg	22	0	0	22

sex by treatment interaction ($F = 3.48$; $df = 2, 69$; $p < 0.0367$) were also statistically significant.

Pairwise comparisons of the groups indicated that male rats receiving 10 mg/kg of FF-1 had significantly decreased startle responsiveness (FIGURE 2). Three mg/kg of FF-1 decreased the startle of female rats ($p < 0.05$) while a similar effect at the 10 mg/kg dose was suggestive ($p < 0.06$). These data indicate that rats exposed to FF-1 over a period of 6 months do not demonstrate signs of irritability or hyperreflexia.

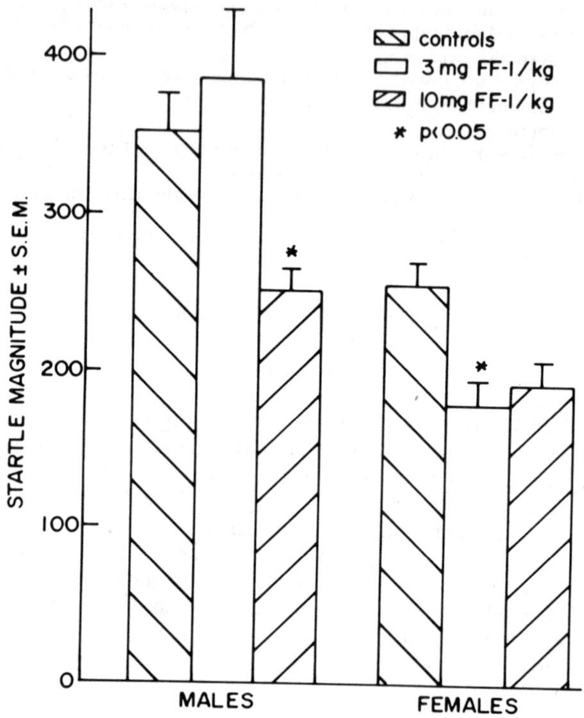

FIGURE 2. The effects of FF-1 on the startle responsiveness of rats to an air puff. Rats were given either 3 mg/kg of FF-1 (14 males and 7 females), 10 mg/kg of FF-1 (14 males and 8 females), or corn-oil vehicle (15 males and 12 females) by gavage, 5 days per week, for 26 weeks. The asterisk indicates a statistically significant difference between the treated and control groups (Fisher's least significant difference, $p < 0.05$, two-tailed).

ACQUISITION OF AVOIDANCE

Methods

After 6 months of chronic dosing with 10 mg/kg of FF-1, 14 male rats from Experiment II were tested for the ability to learn a one-way avoidance response using a procedure described by Clark.[11] Each subject was first placed into a small plexiglass cubicle (7 in × 5 in × 4 in) having grid floors to which a scrambled electric shock (0.8 mA) from a standard source could be applied. A speaker and two white lamps were located at the rear of the cubicle opposite a 3 inch × 3 inch guillotine door leading to a larger, darkened chamber having a solid floor. Each was given 4 acquisition or escape trials, which consisted of presentation of light, tone, and footshock concurrent with the raising of the guillotine door. The trial ended after 40 seconds of shock or when the subject left the small cubicle and entered the larger chamber. Avoidance trials consisted of an initial 20 second presentation of light and tone in the absence of shock. If an avoidance response was not made in 20 seconds, the subject received electric shock for 20 seconds or until a response was observed. The latencies of avoidance and escape responses were recorded using electronic timing circuitry.

Results

Administration of 10 mg/kg of FF-1 for a period of approximately 6 months had no effect on the average escape latency during learning (Mann-Whitney U-test). The average latency of FF-1 treated rats was 1.03 ± 0.07 seconds as compared to 1.24 ± 0.12 seconds (mean ± S.E.M.) for the control rats. These data indicated that FF-1 exposed rats were responsive to aversive electric footshock.

The average response latencies during the avoidance trials were 1.86 ± 0.33 sec and 1.54 ± 0.21 sec for FF-1 and control rats, respectively. The groups did not differ statistically. This observation indicates that FF-1-exposed rats were not impaired in the ability to associate the pairing of two previously neutral stimuli (light and tone) with the onset of an aversive stimulus (shock).

Continuous Avoidance Responding

Methods

The effects of PBBs on baseline performance of rats was measured using an unsignaled continuous avoidance procedure.[5] Subjects were trained to press a lever in standard operant chambers (Coulbourn Instruments, Lehigh Valley, Pa.) to avoid a 0.8 mA footshock applied to the grids of the chamber. Each response postponed electric shock for a period of 20 seconds. If the time between responses lapsed to 20 seconds, the subject was shocked every 5 seconds until a lever press was made. The shock duration was 0.5 seconds and each response terminated all shock and initiated another 20-second interval.

Twenty-one rats were run on the operant schedule over a period of 2 months to establish a baseline of responding. They were then divided into three equal groups and received 10 mg/kg of FF-1, 10 mg/kg of HBB or 1 ml/kg of corn-oil vehicle 3 times per week for a period of 8 weeks. At the end of the dosing period, rats were given 7.5, 15, or 30 mg/kg of phenobarbital or 0.25, 0.50, 1.0, or 2.0 mg/kg of *d*-amphetamine, intraperitoneally, immediately prior to each behavioral session. Drug sessions were separated by at least two days and were preceded by a session in which distilled water was administered.

Results

The average rate of avoidance responding was not affected significantly during the 8 weeks of dosing with PBBs. ANOVA indicated no treatment (F = 0.84) or

TABLE 8
AVOIDANCE RATES OF RATS DURING REPEATED DOSING WITH FF-1 AND HBB OVER AN 8-WEEK PERIOD*

Group	Average Rate (Responses/min) ± S.E.M. Week of Dosing 2	4	6	8
Vehicle	4.9 ± 0.5	4.6 ± 0.7	4.7 ± 0.6	4.9 ± 0.5
FF-1 (10 mg/kg)	4.6 ± 0.5	4.4 ± 0.4	4.2 ± 0.4	4.1 ± 0.3
HBB (10 mg/kg)	4.7 ± 0.6	4.7 ± 0.5	4.7 ± 0.5	5.0 ± 0.5

period (F = 0.05) effect or treatment by period interaction (F = 0.12). Avoidance rates for each group averaged over two week periods are summarized in TABLE 8.

In spite of the absence of behavioral effects following repeated dosing with PBBs, alterations in the responsiveness of the rats to phenobarbital and *d*-amphetamine were evident. In the case of phenobarbital, ANOVA of the dose response curves of the 3 groups indicated a significant treatment effect ($F = 5.97$; $df = 2, 50$; $p < 0.0052$), but the dose ($F = 2.99$; $df = 2, 50$) and treatment by dose interaction ($F = 0.91$) were not significant. Pairwise comparisons of the FF-1 and HBB data indicated that those rats dosed chronically with PBBs were not as suppressed as control rats by 30 mg/kg of phenobarbital (FIGURE 3).

ANOVA of data from the *d*-amphetamine experiment indicated a significant dose

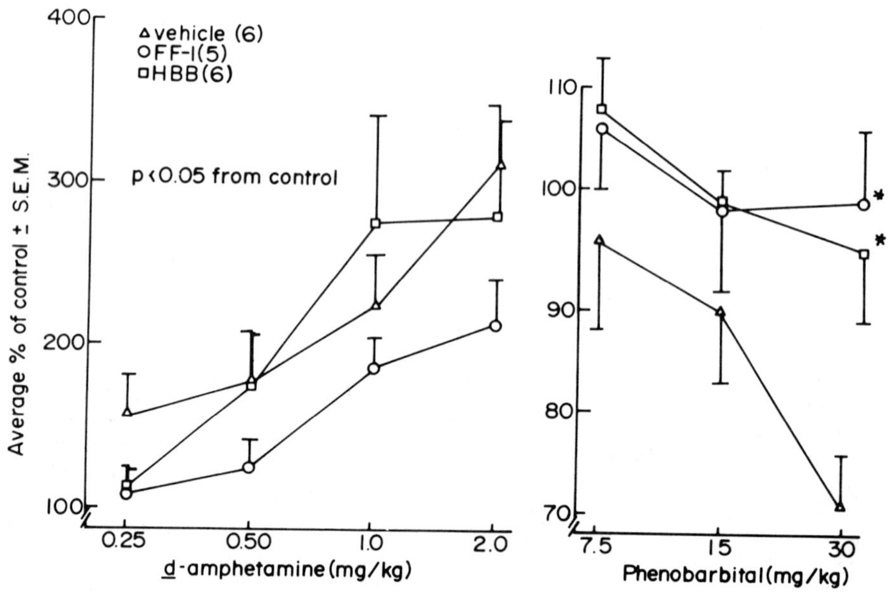

FIGURE 3. The effects of repeated exposure to PBBs on the responsiveness of rats to the behavioral effects of *d*-amphetamine and phenobarbital. Data are the average ± S.E.M. percentage of response rates observed on the day prior to treatment with phenobarbital or *d*-amphetamine. Asterisks indicate a significant difference from vehicle-treated rats (Fisher's least significant difference test, $p < 0.05$, two-tailed).

effect ($F = 8.88$; $df = 3, 74$; $p < 0.0001$). The treatment effect ($F = 2.57$; $df = 2, 74$) and treatment by dose interaction ($F = 0.50$) were not statistically significant.

DISCUSSION

The repeated administration of the Firemaster (FF-1) mixture of polybrominated biphenyls (PBBs) produced a consistent pattern of neurobehavioral deficits. Locomotor activity in the open field and the automex, as well as forelimb grip strength was decreased by FF-1 in total cumulative doses of 390–1300 mg/kg in 6 months and 660 mg/kg over a 30-day period. The hindlimb extensor response was not affected at 390 mg/kg total dose in 6 months, but was affected significantly at the higher cumulative dose of 1300 mg/kg. Performance on the inclined screen, which is sensitive to the peripheral neuropathy and corresponding muscular incoordination produced by acrylamide,[10] was not affected by doses of FF-1 up to 1300 mg/kg. These data indicate that repeated exposure to PBBs leads to behavioral suppression and signs of muscular weakness in laboratory animals.

The mechanism by which PBBs affect neuromuscular functioning is not known. Studies are currently underway in our laboratory to identify the neurophysiologic effects of PBB exposure on various parameters of peripheral nerve and muscle function. It is noteworthy that in a previous study,[10] acrylamide was found to depress the hindlimb extensor response and inclined screen ratings, but did not affect forelimb grip strength. Thus, the rather nonspecific neurobehavioral effects of the PBBs observed in the present study suggests that the PBBs act on organ systems other than the central nervous system. PBBs have been shown to induce hepatic microsomal enzymes and produce hepatohistopathology,[12] suppress immune responses,[13] and affect the function of the thyroid and testes.[14] Marked effects on one or more of these systems might contribute to the behavioral toxicity we observed.

The animal models chosen for study appear to substantiate claims of muscular weakness and impaired skeletomuscular function in humans exposed to PBBs.[4] However, increased irritability or hyperreflexia was not suggested by our results. The startle response, which is taken as a measure of the general responsiveness of the animal to the environment, was suppressed rather than augmented by exposure to PBBs. Likewise, there was no indication of impaired ability of PBB-treated rats to acquire a simple one-way avoidance test. The possibility that PBBs may affect acquisition of more complex tasks or influence higher levels of cognitive processing, such as that occurring in primates, deserves investigation. Finally, chronic exposure to PBBs did not disrupt the performance of a negatively reinforced operant task requiring the animal to make an internally based timing response. The unsignaled continuous avoidance procedure has been shown to be relatively sensitive to the effects of various psychoactive drugs.[5]

The results of the present study indicate that 2,4,5,2′,4′,5′-hexabromobiphenyl (HBB), the major component of the FF-1 mixture, is generally less toxic than the FF-1 mixture. Administration of HBB in concentrations equal by weight to the HBB in the FF-1 mixture had relatively few consistent effects on behavior. When HBB and FF-1 were compared directly (equal doses, 10 mg/kg for 8 weeks), no behavioral effects were observed in the operant study. In the drug interaction experiment, rats treated with either FF-1 or HBB appeared to be less sensitive to the behavioral effects of phenobarbital. The attenuation of the effects of phenobarbital by PBBs may be explained by the fact that PBBs are potent inducers of hepatic metabolic enzymes.[15] Although the operant behavioral procedure used in this study was not able to detect subtle changes in neurobehavioral functioning following repeated dosing with FF-1 and HBB, the use of a pharmacologic challenge unmasked functional changes in the responsiveness of PBB-exposed rats to phenobarbital.

There were two occasions in these experiments in which one sex of rats appeared more sensitive to PBBs than the other. Female rats, for example, were generally affected more than males when the behavioral variable was locomotor activity. Male rats, on the other hand, were apparently more sensitive to PBBs than females when forelimb grip strength was measured. Both instances appear to represent cases in which the behavioral effect was dependent upon the existing frequency or magnitude of the baseline response. That the observed sex differences were, in fact, dependent upon differing baselines was indicated by ANOVA in both studies, i.e., the sex effect and sex by treatment interaction was significant.

In summary, the oral administration of FF-1 in a cumulative dose of 390 or 1300 mg/kg over 6 months or 660 mg/kg over 30 days produces a decrement in skeletomuscular functioning including a depression in locomotor activity, impaired forelimb grip strength, and decreased hindlimb extensor responses.

Acknowledgments

The technical assistance of Mr. Richard Rhoderick, Mr. Tom Burne, Ms. Lyn Reed, and Mr. W. Gordon is gratefully acknowledged.

Special thanks go to Ms. Sue Bolton for her skillful preparation of this manuscript.

The 2,4,5,2′,4′,5′-hexabromobiphenyl was isolated from FF-1 by Ms. Carol E. Parker of the Chemistry Section, Environmental Biology and Chemistry Branch, NIEHS.

References

1. Bekesi, J. G., J. F. Holland, H. A. Anderson, A. S. Fischbein, W. Rom, M. S. Wolff & I. J. Selikoff. 1978. Science **199:** 1207–1209.
2. Anderson, H. A., E. C. Holstein, S. Daum & I. J. Selikoff. 1978. Environ. Health Perspect. **23:** 333–340.
3. Valciukas, J. A., R. Lilis, H. A. Anderson & M. S. Wolff. 1978. Environ. Health Perspect. **23:** 199–210.
4. Meester, W. D. & D. J. Mccoy. 1976. Proceedings of a Symposium on Environmental Toxicology, Seattle, Washington.
5. Tilson, H. A., J. H. Chamberlain & J. A. Gylys. 1977. Pharmacol. Biochem. Behav. **6:** 627–630.
6. Winer, B. J. 1962. Statistical Principles in Experimental Design. McGraw-Hill. New York.
7. Siegel, S. 1956. Non-Parametric Statistics. McGraw-Hill. New York.
8. Cabe, P. A., H. A. Tilson, C. L. Mitchell & R. Dennis. 1978. Pharmacol. Biochem. Behav. **8:** 101–102.
9. Cabe, P. A. & H. A. Tilson. 1978. Pharmacol. Biochem. Behav. **9:** 133–136.
10. Tilson, H. A. & P. A. Cabe. 1978. Toxicol. Appl. Pharmacol. In press.
11. Clark, R. 1966. Psychonomic Sci. **6:** 11–12.
12. Sleight, S. D. & V. L. Sanger. 1976. J. Am. Vet. Med. Assoc. **169:** 1231–1235.
13. Luster, M. I., R. E. Faith, & J. A. Moore. 1978. Environ. Health Perspect. **23:** 227–232.
14. Ringer, R. K. & D. Polin. 1977. Fed. Proc. **36:** 1894–1898.
15. Goldstein, J. A. & P. Hickman. 1978. Proceedings of the 17th Annual Meeting of the Society of Toxicologists, San Francisco, California, p. 155 (Abstract).

THE NEUROTOXICITY OF POLYBROMINATED BIPHENYLS: RESULTS OF A MEDICAL FIELD SURVEY*

José A. Valciukas, Ruth Lilis, Henry A. Anderson, Mary S. Wolff, and Michael Petrocci

Environmental Sciences Laboratory
Mount Sinai School of Medicine
The City University of New York
New York, New York 10029

Introduction

Purpose of the Present Study

The purpose of the present study was to investigate the neurotoxic effects of PBB using epidemiologic and toxicologic methods. These included the study of differences in prevalence of neurologic symptoms in exposed and control populations with comparable demographic characteristics, and the correlation between indices of environmental exposure (in this case, food ingestion), biologic indicators of exposure (PBB in blood serum) and effects on central nervous system function as determined by psychometric procedures.

Specifically, the major topics addressed in this report are: (1) the prevalence and incidence of neurologic symptoms in males and in females in the Michigan population exposed to PBB (with comparison to similar data in the Wisconsin, non-PBB-exposed group); (2) the correlation between presence of the most prevalent neurologic symptoms in males and females and the results of objective performance tests used for the assessment of neuropsychologic dysfunction; (3) the relationship between consumption of home-produced and store-bought products and performance test scores; and (4) the correlation of serum PBB levels with neurologic symptoms and performance.

A review of human health effects of polybrominated and polychlorinated biphenyls with particular emphasis on neurotoxicity in animal studies can be found elsewhere.[1]

Material and Methods

Subjects

The prevalence and incidence of neurologic symptoms were analyzed in 626 adults from Michigan and 153 from Wisconsin. The characteristics of these two populations have been reported elsewhere.[2] The subsample examined by means of performance tests consisted of 95 males and 67 females from Michigan and 50 males and 43 females from Wisconsin.

This subsample, was examined during a comprehensive clinical survey of these two

*This work was supported by a grant from the National Institute of Environmental Health Sciences (ES 00928).

0077-8923/79/0320-0337 $01.75/0 © 1979, NYAS

populations. All examinees were whites. Their educational levels ranged from a few years of schooling to college degrees. The distribution of educational levels in the two populations was almost identical, with a modal rating unit between 3–3.9, corresponding to a high school degree or a high school degree plus at least one year in college. The mode of the age distribution for both the Michigan and the Wisconsin adult populations was located in the 35–44 year range. The weight attributed to educational level and age in these tests has been discussed elsewhere.[3] The average performance scores of the Wisconsin sample population were expected to be slightly lower than the Michigan group because of differences (not statistically significant) in age. In a few cases and for a variety of reasons (such as interruption of the testing for compelling reasons or physical handicap), data for individual tests among some examinees are missing.

TABLE 1

LIST OF ALL SYMPTOMS ON THE SPECIAL QUESTIONNAIRE USED TO STUDY THE COMPARATIVE PREVALENCE OF SYMPTOMS IN MICHIGAN AND WISCONSIN FOR THE PERIOD 1972–1976

Eye irritation	Red urine
Nasal irritation	Brown urine
Sore throat	Headaches
Cough	Blurred vision
Wheezing	Dizziness
Tightness in chest	Depression
Skin rash	Tiredness
Acne	Perceptual changes
Sensitivity to sun	Nervousness
Burning sensation on skin	Sleeplessness
Darkening of skin	Sleepiness
Thickening of skin	Muscle weakness
Discoloration or deformation of fingernails	Difficulty in walking
Slower or poor healing of cuts	Paresthesia
Lost more than ten pounds without dieting	Loss of balance
Appetite lost	Swelling of joints
Nausea	Pain in lower back and legs
Abdominal pain	Dry skin
Pain related to meals	Thyroid
Crampy abdominal pain	Change of voice
Constipation	Increased perspiration
Diarrhea	

Symptoms

TABLE 1 lists all symptoms recorded on a special questionnaire used in the examination of the Michigan and Wisconsin sample populations. This questionnaire was administered by physicians; the presence or absence—and, if present, the intensity—of each symptom was recorded for the period preceding the year 1973, and for each consecutive year thereafter including 1976.

The neurologic symptoms considered in this study are: tiredness, sleepiness, nervousness, depression, dizziness, headaches, paresthesia, difficulties in balance, blurred vision and muscle weakness. This list of neurologic symptoms includes the symptoms for which a statistically significant difference in prevalence was found when comparing the Michigan PBB-exposed population and the Wisconsin control group.

Performance Tests

Performance tests were used in order to obtain an objective assessment of brain functions. Block Design (B.D.), Digit-Symbol (D.S.) and Embedded Figures (E.F.) tests were used to measure performance of the populations studied. The Block Design test is widely used in the evaluation of the effects of neurotoxic substances. The test provides a quick and objective measure of the subject's ability to visualize spatial relations. The examinee is presented with a set of colored cubes. Each cube has two red sides, two white sides and two sides half red and half white. The test requires arranging the cubes so that the top surface reproduces a displayed design. In the first six trials, only four of the cubes are used and the designs to be reproduced are relatively simple. In the last four trials, nine cubes are used and the designs are more complex. The score is based on the number of correctly completed trials, as well as the speed with which the test is accomplished.[4]

The Digit Symbol test assesses the subject's ability to associate symbols with a code. The subject is given a list in which symbols are associated with digits from 1 to 9 and is asked to enter the symbols into the blank spaces next to a list of random digits on a form. The score is the number of symbols correctly entered in 90 seconds.[4]

The Embedded Figures test has been extensively used in neurologic research, particularly during World War II. The test, consisting of four sets of superimposed outlines of common objects, assesses visual perception in masked conditions. The score is the total number of objects successfully identified in the four sets.[5]

A composite index of test scores (SMX for males and SFX for females) was also used. This was calculated as the average Z-score for each of the three tests. Further details justifying the use of these indices are found below. Hereafter, the word "performance" is used instead of "indices of performance."

In comparing performance test scores with neurologic symptoms and blood chemistry, it must be borne in mind that the higher the score, the "healthier" the person and vice versa. The same concept applies to the interpretation of the composite index of test scores.

Reliability and Validity of Tests

Test-retest reliability data for the Block Design test and for the Digit Symbol test have been provided elsewhere.[4] The test-retest correlation coefficient is about 0.82 for ages 18–54 and 0.80 respectively for ages 16–24. The Embedded Figures test has been developed in our laboratory in studies of over 600 subjects classified by age, sex, race, education, occupation, geographic area, rural or urban living and native language (English and Spanish). This battery of tests has been shown to be sensitive in detecting subclinical central nervous system dysfunction as a result of long-term exposure to lead.[3]

All test scores correlate positively and significantly among themselves within both the control and exposed populations. Age correlated significantly and negatively with all behavioral test scores in both groups.

Food Intake

The weekly consumption of each of the following foods was recorded according to whether the product was *home produced* or *store bought;* consumption during 1972–1974 and during 1975–1976 was separately recorded. The distinction between home-produced and store-purchased food was considered important, since consumption of home produce from a quarantined farm could have produced a larger PBB burden in humans. Milk (number of 8-ounce glasses per week), eggs (number per week), butter and cheese (ounces per week), beef, chicken, pork and fish (servings per week) were considered in this analysis.

The selection of foods was made on the basis of possible or known sources of PBB contamination. Dietary data were obtained through self-administered questionnaires. The information recorded in this questionnaire was checked by a trained interviewer at the time the subject presented himself for the medical examination. Home products consumed by males during 1972–1974 were taken as a reference and all the values in that period transformed to 100. By this procedure, male and female differences in consumption for the two periods could be analyzed independently of the units of measurements for each product.

Biologic Indicators of Exposure

PBB in blood serum was analyzed according to a procedure described elsewhere.[6,7]

The metabolic pathway followed by PBB is poorly understood. Early experimental data results suggest that decay functions for serum and fat PBB, after a single exposure to PBB, initially follow quite different time courses and then tend to equilibrate. However, these laboratory findings are difficult to extrapolate to human populations, since in each individual case the period of actual PBB exposure and its pattern and variations cannot be reconstructed. It is also not clear whether and to what extent Michigan residents are still exposed to PBB. Because of the lipophilic property of PBB, PBB levels as detected in subcutaneous fat deposits may be more relevant for the assessment of PBB neurotoxicity. However, fat samples are more difficult to obtain, and not all subjects are willing to cooperate. These limitations must be borne in mind while trying to interpret the quantitative data where PBB levels are used as biologic indicators of exposure.

Moreover, individual contaminants in a PBB mixture might be as toxic or more toxic than the total mixture. In the PCB incident (Yusho) in Japan, the contaminant chlorinated dibenzofuran seemed to have an impact on the clinical effects of PCB exposure.

Blood Chemistry

Urea, creatinine, sodium, potassium, chloride, uric acid, calcium, inorganic phosphorus, total protein, albumin, cholesterol, total bilirubin, alkaline phosphatase, SGOT, SGPT, LDH, and CEA were included in the study protocol. A higher proportion of SGPT and SGOT abnormalities was found in the Michigan PBB-exposed population as compared to the Wisconsin control group.[2]

Since liver toxicity might be one of the target organ effects of PBB, consideration was given to the association between neurotoxic effects and hepatotoxic effects. The evaluation of neurotoxic effects was attempted within as large a frame of reference as possible, particularly in relation to possible interference with homeostatic mechanisms and possible confounding factors such as intercorrelation among blood chemistries.

Other Variables Considered in This Study

We considered the possible confounding effects of anthropometric characteristics, educational status and category (to be explained below). The following characteristics were taken into account: sex, race, age, height and weight. However, most of the analysis of measures of body mass are still incomplete. All subjects in the present analysis are 18 years old or older. Educational level was classified using the following scale: (1) no formal education; (2) finished grammar school; (3) high school equivalent; (4) finished formal high school; (5) finished a junior college with an associate degree; (6) Bachelor's degree; (7) a Master's degree; (8) a doctorate degree and *not* currently working in his/her field (retired, married and attend the house full time, etc.); and (9) doctorate working in his/her field.

The selection of the Michigan sample was made on the basis of the following categories: (1) residents of quarantined farms selected at random; (2) residents of nonquarantined farms selected at random; (3) a special (nonrandom) group living on quarantined farms; (4) a special (nonrandom) group living on nonquarantined farms; (5) residents of quarantined farms where the highest PBB levels (in farm animals) were detected; (6) consumers of products of quarantined farms; (7) consumers of products of nonquarantined farms; (8) a group of individuals drawn from the farmers advisory committee (nonquarantined farms); and a group of Michigan Chemical Company workers. Further details on the characteristics and number in each group can be found in Anderson.[2]

Throughout this report, "Michigan" and "Wisconsin" are the two study groups, Wisconsin being the reference (control) population.

Statistical Analysis of Data

All data processing was performed through an IBM 370/168 computer system (University Computer Center, The City University of New York). Data files were created through WYLBUR (an editor language), and statistical tests were run using SPSS, Statistical Package for the Social Sciences.[8] Wherever appropriate, rules for treatment of missing values were observed, resulting in different sample sizes in different tables. The nature of the data (e.g., categorical, interval, continuous, etc.) determined the selection of suitable statistical methods of analysis.[9]

Test scores were combined in males (SMX) by adding the standardized scores of each individual test and dividing the number by the number of tests as follows:

$$SMX = (Z_{BD} + Z_{DS} + Z_{EF})/3$$

SFX (index of performance for females) was calculated in a similar manner.

Weekly ingestion of combinations of foodstuffs (FOX indices) were calculated using similar principles. Thus, when the presence or absence of neurologic symptoms is related to ingestion of different combinations of foodstuffs, the indices are average departures from the group mean.

FIGURE 6 illustrates the vector representation of correlation coefficients relating ingestion of food produced at home or bought in stores during two different periods. The horizontal axis (positive side) always represents the home-produced food consumption of individual products in 1972–1974. Correlation coefficients (r) can be represented by vectors because of the relation $r = \cos \alpha$, where α is the angle between the two vectors. Further details on the use of these procedures can be found in reference 10. Matrices of Pearson's correlation coefficients were used as the bases from which these vectors were calculated, and although available on request, have not been included below.

A logarithmic transformation (log base 10) was used for serum PBB levels. This transformation condenses the extremely large range of serum PBB levels (up to seven logarithmic units) and compensates for the "pull" toward higher values exerted by high values of individual subjects. Thus, the mean value of serum PBB levels was approximately ten times higher than the antilog of the mean value of the serum log PBB values (the geometric mean).

Throughout this paper, "correlation coefficient" means "Pearson's correlation coefficient." In the calculation of degrees of freedom for partial correlation coefficients the following formula was used:

$$DF = \min\,[n_1, n_2, n_3] - \text{number of variables involved in the partial coefficient}$$

where n_i is the number of nonmissing observations of the ith variable ($i = 1,2,3$).

Certain statistical tests could not be performed in the Wisconsin group, whenever "neurologic symptoms" were considered in the analysis. In the Wisconsin group neurologic symptoms were either absent or occurred so infrequently that the significance of differences between mean values for blood chemistry and test scores of subjects who did and subjects who did not exhibit neurologic symptoms (as examined by t tests) became statistically meaningless.

Considerations Regarding the Analysis of Data in Light of the Experimental Design

FIGURE 1, the equivalent of an experimental design, outlines the parameters that have been analyzed. In choosing the dimensions of analysis, practical considerations have forced us to be selective. As stated above, we had *two* populations (Michigan and Wisconsin), *eight* subgroups in Michigan, *nine* of the most prevalent symptoms, *five* different years, *four* assessments of performance and educational level, *two* sexes, *three* age groups, *seventeen* blood chemistries, *eight* foodstuffs and *two* biologic indicators of exposure (serum and fat PBB). A $2\times8\times9\times5\times4\times2\times3\times17\times8\times2$ factorial design amounts to approximately five million comparisons, if all the cells were complete. Therefore, in selecting those dimensions which we thought would be most enlightening in describing the neurotoxicity of PBB and some possible mechanisms of assessing toxicity, we have relied on our experience in similar studies.

RESULTS

Prevalence of Neurologic Symptoms among Michigan and Wisconsin Residents

TABLE 2 shows the percentage of cases in the Michigan and Wisconsin groups exhibiting neurologic symptoms. The total number of people so examined in Michigan and Wisconsin was 162 and 93, respectively. While the number of symptoms during the period 1972–1976 varied little in Wisconsin, in Michigan there was a significant increase in the prevalence of these symptoms during the same period. It is interesting to note that prior to 1973 certain symptoms such as tiredness, headaches, sleepiness, nervousness, dizziness, blurred vision, muscle weakness and loss of balance were reported *less* frequently by Michigan residents than by the Wisconsin group.

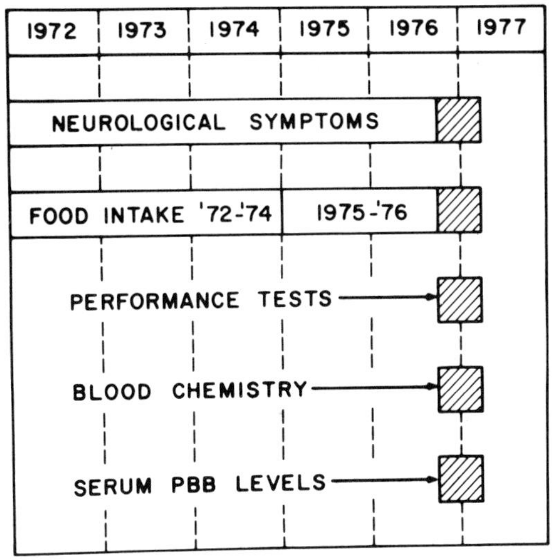

FIGURE 1. Time Frame for Data Collection. The diagram is an attempt to relate the dimensions of analysis in the present study to their appropriate time references. The medical field surveys were conducted in November of 1976 (Michigan) and in March 1977 (Wisconsin). Neurologic symptoms and food intake were reported for the period 1972–1976. Performance tests, blood chemistry and serum PBB were those of November 1976 (Michigan) and March 1977 (Wisconsin).

TABLE 2

NUMBER AND PERCENTAGE OF NEUROLOGIC SYMPTOMS AMONG MICHIGAN (N = 162) AND WISCONSIN (N = 92) RESIDENTS DURING 1972–1976 AS OBTAINED BY HISTORY

		1972		1973		1974		1975		1976	
		M	W	M	W	M	W	M	W	M	W
Tiredness	N	2	9	16	9	62	9	88	9	90	11
	%	1.2	9.7	9.6	9.7	37	9.7	53	9.7	54	11.9
Headaches	N	10	7	17	7	41	7	53	8	64	8
	%	6.0	7.6	10.2	7.6	25	7.6	32	8.7	39	8.7
Somnolence	N	1	3	13	4	45	5	63	5	62	5
	%	0.6	3.3	7.8	4.3	27	5.4	38	5.4	37	5.4
Nervousness	N	3	3	12	3	30	3	40	1	44	1
	%	1.8	3.3	7.2	3.3	18	3.3	24	1.1	27	1.1
Depression	N	3	1	11	1	29	1	38	1	42	2
	%	1.8	1.1	6.6	1.1	17	1.1	23	1.1	25	2.2
Dizziness	N	1	3	9	3	22	4	22	4	43	5
	%	0.6	3.3	4.5	3.3	13	4.3	13	4.3	26	5.4
Paresthesia	N	3	2	7	2	22	2	26	4	36	3
	%	1.8	2.2	4.2	2.2	13	2.2	16	4.3	22	3.3
Blurred vision	N	2	3	9	3	16	4	27	3	36	4
	%	1.2	3.3	5.4	3.3	10	4.3	16	3.3	22	4.3
Muscle Weakness	N	1	5	4	4	15	3	24	3	31	3
	%	0.6	5.4	2.4	4.3	9	3.3	14	3.3	19	3.3
Loss of Balance	N	2	4	7	4	20	2	29	1	30	1
	%	1.2	4.3	4.2	4.3	12	2.2	17	1.1	18	1.1
Insomnia	N	1	2	4	2	15	2	25	1	30	2
	%	0.6	2.2	2.4	2.2	9.0	2.2	15	1.1	18	2.2
Perceptual change	N	1	0	2	0	5	1	10	1	10	2
	%	0.6	0.0	1.2	0.0	3.0	1.1	6.0	1.1	6.0	2.2
Difficulty walking	N	0	2	2	3	6	1	8	0	10	0
	%	0	2.2	1.2	3.3	3.6	1.1	4.8	0.0	6.0	0.0

FIGURE 2 illustrates the method used to calculate the approximate time of onset of each symptom. The proportions of persons with symptoms and the proportions of persons without symptoms in the Michigan and Wisconsin populations were compared for each year (1972–1976) using the chi-square test. The chi-square value was plotted against time, and the statistical significance at the 0.05 and 0.01 level was indicated by the two parallel lines. The points of intersection of these lines with the curve resulting from successive chi-square values indicate the time at which a symptom became significantly more prevalent in the Michigan population. By this method it was possible to determine the approximate time of onset of the various symptoms reported by the examined persons (i.e., the time when they became significantly more prevalent in the Michigan population) at the 5% and 1% level of significance.

FIGURE 3 illustrates the time course of each symptom as determined by the method described above. A temporal sequence was reconstructed according to a relative time

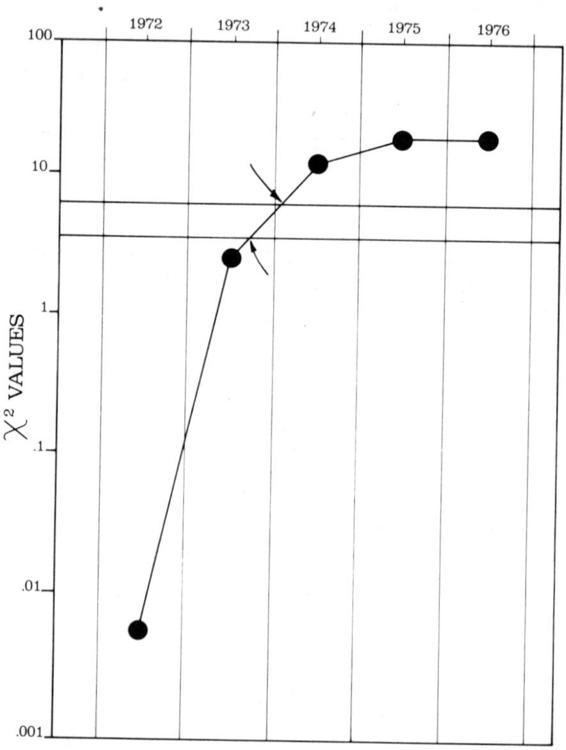

FIGURE 2. Method Used to Calculate the Time of Onset of Symptoms in the Michigan Population. The significant differences in the number of persons with symptoms and in the number of persons without symptoms in the Michigan and Wisconsin population were assessed for each year (1972–1976) using the chi-square test. The value of the chi-square was plotted against time and the statistical significance at the 0.05 and 0.01 levels is indicated by the two parallel horizontal lines. The points of intersection of these lines with the curve drawn between successive chi-square values indicate the times at which a symptom became significantly more prevalent in the Michigan population. By this method it was possible to determine the approximate time of onset of the various symptoms reported by each group at the 5% and 1% levels of significance.

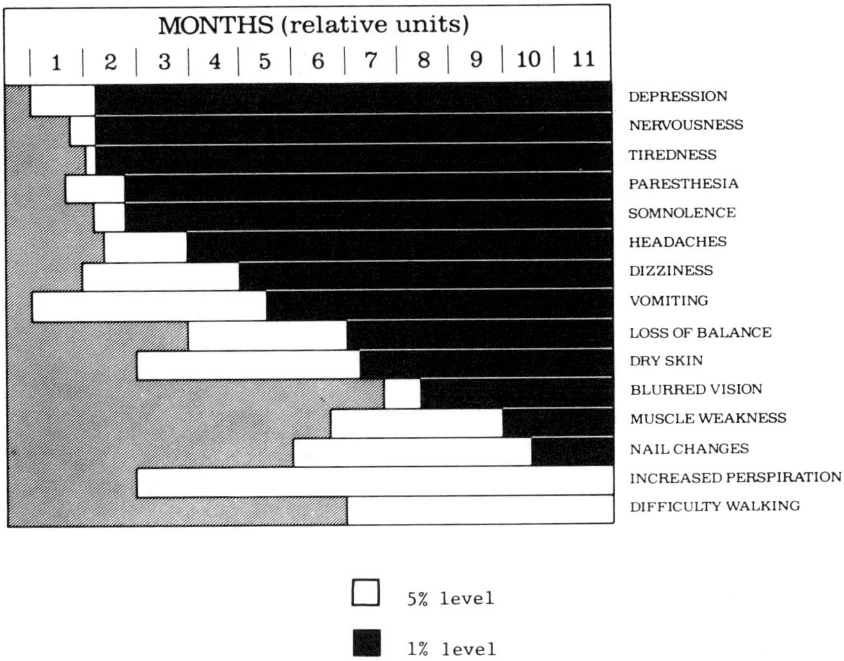

FIGURE 3. Time of Onset of All Significant Symptoms in the Michigan Population. The time of onset was determined as described in FIGURE 2. The upper scale indicates months as a relative unit. Although no attempt was made to determine the precise time of onset of the symptoms, their order of onset could be ascertained. Significance at the 5% and 1% level are indicated by white and dashed bars respectively and the arrangement is made according to the 1% level of significance.

of onset (the number in the scale indicates months as relative units). The dashed and solid lines indicate areas where the criterion for significance of 5% and 1%, respectively, was met. All the symptoms listed in TABLE 1 were considered for this analysis, and only those showing significant differences when compared to the Wisconsin population are illustrated.

Onset of neurologic symptoms preceded that of all other symptoms in the Michigan population. The frequency of some of the symptoms, such as nausea and dizziness, increased slowly over the months whereas other symptoms, such as depression, nervousness, paresthesia, tiredness, sleepiness and headaches showed a more rapid increase in prevalence. "Motor involvement" such as loss of balance, muscle weakness, difficulty in walking, and blurred vision occurred later in time or showed a slower rate of increase. Finally, symptoms such as increased perspiration, slower healing of cuts, and nail discoloration progressed even more slowly (3 to 7 months). Some of these symptoms, such as increased perspiration, swelling of joints, and difficulty in walking never exceeded the 5% level of significance when their prevalence was compared to that in the Wisconsin population.

TABLE 3 lists significant symptoms as observed in November of 1976. The left column indicates symptoms reported in the Michigan group listed in decreasing order of prevalence. The column on the right is a list of significant symptoms in the

TABLE 3

SYMPTOMS OBSERVED AMONG MICHIGAN DAIRY FARMERS IN 1976

Proportion of Cases (1976, Michigan)	Significance of Difference between Groups*
Tiredness	Tiredness
Headaches	Nervousness
Sleepiness	Sleepiness
Drier skin	Headaches
Nervousness	Paresthesia
Paresthesia	Dizziness
Perspiration	Depression
Dizziness	Vomiting
Depression	Nail discoloration
Blurred vision	Loss of balance
Muscle weakness	Blurred vision
Loss of balance	Drier skin
Healing cuts	Muscle weakness
Vomiting	Healing cuts
Nail discoloration	Difficulty walking
Difficulty walking	Perspiration
Swelling of joints	Swelling of joints

*Chi-square tests for significance of difference in frequency of symptomatic individuals in the sample populations.

Michigan sample as compared to the Wisconsin group, as detected by the chi-square test ($p \leq .01$) in decreasing order of significance. Both methods indicated that neurologic symptoms were the earliest manifestation of abnormalities in the Michigan population.

FIGURE 4 shows the changes in the magnitude of chi-square values for population differences in the frequency of neurologic symptoms for males and females among residents of Michigan and Wisconsin. The symptoms were ordered according to the value of the chi-square for the last year (1976), and the 0.05 and 0.01 level of significance are indicated by two parallel lines.

There were marked differences in the evolution of neurologic symptoms in males and females. The magnitude of the chi-square value—when compared to the Wisconsin group—was higher in males than females for all the symptoms. In fact, the chi-square values in females for certain symptoms such as depression, dizziness, difficulties in balance, blurred vision and muscle weakness were never significantly different from the controls.

Population Differences in Objective Measures of Performance

FIGURE 5 gives the mean scores for three performance tests (B.D., D.S. and E.F.) in males (black bars) and females (white bars) in Michigan and Wisconsin for three age groups (18–29; 30–49; and 50-over). Serum PBB mean values are also indicated. Statistical differences between males and females at the 5% and 1% level of significance are indicated by one and two asterisks respectively.

TABLE 4 contains numerical data corresponding to this illustration. It was found that females scored higher than males in the D.S. test both in Michigan and Wisconsin in all age groups. However, among Michigan subjects aged 50 and over, males scored lower than females in two additional tests (B.D. and E.F.). Serum PBB level differences between males and females were highly significant in all three age groups in Michigan. Male-male and female-female comparisons in both populations did not indicate significant differences (TABLE 5).

TABLE 6 shows the correlation between age, serum PBB, performance test scores for Michigan males and females divided among three age groups. Calculation of Pearson's correlation coefficients for the Wisconsin population was inappropriate, since 95% of Wisconsin residents studied had undetectable levels of PBB. Group size and critical values for the evaluation of the significance of the correlation coefficients are shown at the bottom of each column.

Correlation coefficients between performance test scores and serum PBB levels were negative, with one exception (the correlation coefficients of the B.D. test with serum PBB in females), suggesting that a higher serum PBB level was associated with a lower performance test score. The only statistically significant correlation with PBB was that for E.F. test scores among males in the 50-years old and over group. The complete correlation matrices for Michigan and Wisconsin for three age groups from which these data were derived have been published elsewhere.[11]

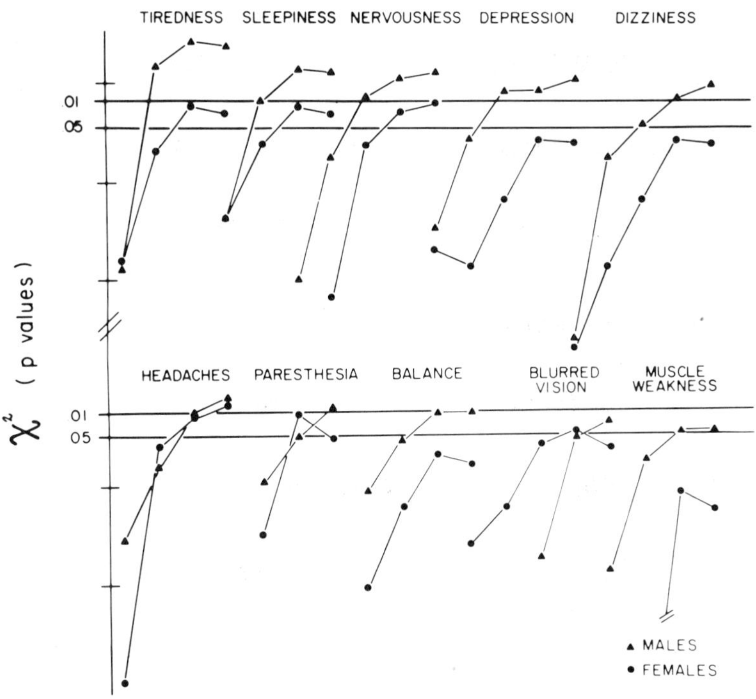

FIGURE 4. Time Course of Magnitude of Chi-square Changes Indicating Significant Population Differences (Males and Females) in Neurologic Symptoms. The curves shown represent the changes in chi-square values over time for individual symptoms. All symptoms in which there was a significant difference between the Michigan and Wisconsin populations in 1976 are represented. As can be seen, neurologic symptoms predominated particularly in males. Among males all chi-square values seemed to reach a plateau but none decreased over the years studied.

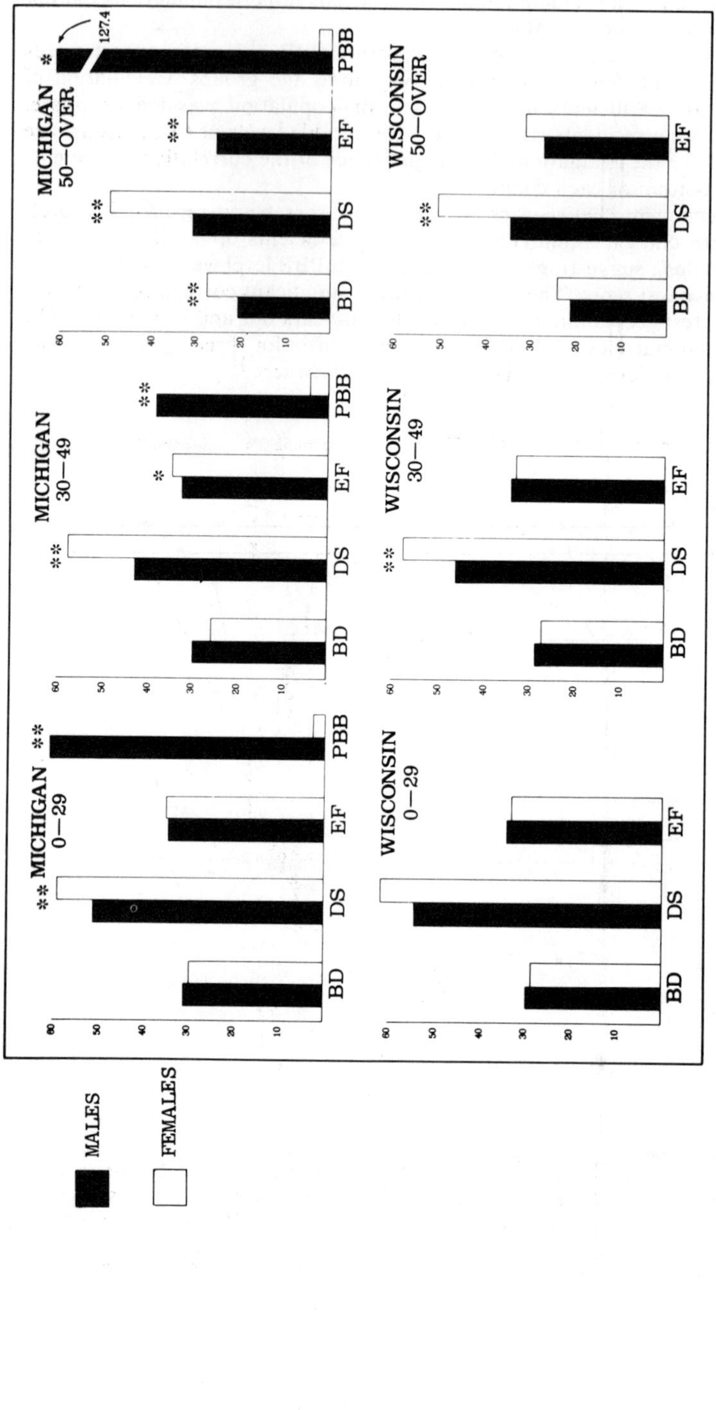

FIGURE 5. Results of three performance tests, Block Design (B.D.), Digit Symbol (D.S.) and Embedded Figures (E.F.) in males and females in Michigan and Wisconsin for three age groups as shown. Asterisks indicate differences which are significant at the 1% and 5% levels respectively. Note that while Wisconsin females scored better than males in one test (D.S.), among Michigan subjects and (particularly those aged 50 and over), males scored below females in all tests. Serum PBB levels were significantly higher in males than females in all age groups. Ninety-five percent of the Wisconsin sample population has undetectable serum PBB levels.

TABLE 4

MALE-FEMALE COMPARISONS IN EDUCATION, PERFORMANCE TEST SCORES, AND PBB LEVELS IN MICHIGAN AND WISCONSIN IN THREE AGE GROUPS

	0–29						30–49						50–+					
MICHIGAN	Males (N = 16)		Females (N = 19)				Males (N = 46)		Females (N = 27)				Males (N = 15)		Females (N = 11)			
	Mean	S.D.	Mean	S.D.	t	p	Mean	S.D.	Mean	S.D.	t	p	Mean	S.D.	Mean	S.D.	t	p
Educ.	3.33	.66	3.26	.93	+.25	n.s.	3.01	.99	2.94	.81	+.31	n.s.	2.86	.91	3.13	.78	−.79	n.s.
B.D.*	30.3	9.7	29.3	7.50	+.34	n.s.	28.8	8.6	24.8	9.68	+1.83	<.05	19.7	6.56	26.9	7.48	−2.61	<.005
D.S.	50.6	10.4	58.9	6.89	−2.83	<.005	42.3	14.2	56.6	12.46	−4.37	<.005	30.5	7.81	48.18	10.51	4.93	<.005
E.F.	34.1	4.4	34.3	4.58	−1.29	n.s.	31.4	5.2	33.9	3.08	−2.27	<.025	24.9	8.15	31.5	5.13	2.35	<.025
PBB	60.2	195.6	2.5	2.54	+1.28	n.s.	37.8	195.0	3.26	6.78	+.92	n.s.	127.4	456.7	2.93	1.99	+7.31	<.001
WISCONSIN	Males (N = 16)		Females (N = 17)				Males (N = 14)		Females (N = 18)				Males (N = 22)		Females (N = 16)			
	Mean	S.D.	Mean	S.D.	t	p	Mean	S.D.	Mean	S.D.	t	p	Mean	S.D.	Mean	S.D.	t	p
Educ.	2.87	.93	3.3	.85	−.19	n.s.	2.94	1.26	3.2	.61	−.74	n.s.	2.38	.83	2.97	1.06	+1.70	<.05
B.D.	29.7	7.5	28.8	5.84	+.28	n.s.	27.9	9.2	26.9	9.22	+.30	n.s.	20.9	10.3	23.9	9.19	−.78	n.s.
D.S.	54.4	14.9	61.6	7.91	+1.19	n.s.	45.3	10.9	57.0	11.26	−2.95	<.005	34.0	10.5	50.0	11.02	−3.93	<.005
E.F.	34.2	4.3	33.6	4.35	+.30	n.s.	33.6	3.7	32.8	4.55	+.53	n.s.	27.7	7.7	31.7	4.34	−1.52	≃.05

*B.D. = Block Design; D.S. = Digit Symbol; and E.F. = Embedded Figures.

Table 5

Michigan-Wisconsin Comparisons in Education, and Performance Test Scores in Males and Females in Three Age Groups

	Male																								
	0–29						30–39						50–+					
	Wis. (N = 16)		Mich. (N = 23)				Wis. (N = 14)		Mich. (N = 46)				Wis. (N = 22)		Mich. (N = 15)			
	Mean	S.D.	Mean	S.D.	t	p	Mean	S.D.	Mean	S.D.	t	p	Mean	S.D.	Mean	S.D.	t	p
EDUC	2.87	.93	3.33	.66	−1.81	n.s.	2.94	1.26	3.01	0.99	−0.22	n.s.	2.38	.83	2.86	.91	−1.66	<.05
B.D.*	29.7	7.5	30.3	9.7	−0.21	n.s.	27.9	9.2	28.8	8.6	−0.33	n.s.	20.9	10.3	19.7	6.56	+0.40	n.s.
D.S.	54.4	14.9	50.6	10.4	−0.94	n.s.	45.3	10.9	42.2	14.2	+0.75	n.s.	34.0	10.5	30.5	7.81	+1.09	n.s.
E.F.	34.2	4.3	34.1	4.4	−0.07	n.s.	33.6	3.7	31.4	5.2	+1.46	n.s.	27.7	7.7	24.9	8.15	+1.06	n.s.
PBB	—	—	60.2	195.6			—	—	37.8	195.0			—	—	127.4	456.7		
	Female																	
	Wis. (N = 7)		Mich. (N = 19)				Wis. (N = 18)		Mich. (N = 27)				Wis. (N = 16)		Mich. (N = 11)			
	Mean	S.D.	Mean	S.D.	t	p	Mean	S.D.	Mean	S.D.	t	p	Mean	S.D.	Mean	S.D.	t	p
EDUC	3.23	0.85	3.26	.93	+0.20	n.s.	3.19	.61	2.94	.81	+1.14	n.s.	2.97	1.06	3.13	.78	−0.43	n.s.
B.D.	28.8	5.84	29.3	7.50	−1.58	n.s.	26.9	9.22	24.8	9.68	+0.72	n.s.	23.9	9.19	26.9	7.48	−0.90	n.s.
D.S.	61.6	7.91	58.9	6.89	+0.85	n.s.	57.0	11.26	56.6	12.46	+0.11	n.s.	50.0	11.02	48.18	10.51	+0.43	n.s.
E.F.	33.6	4.35	34.3	4.68	−0.34	n.s.	32.8	4.54	33.9	3.08	−0.97	n.s.	31.7	4.34	31.5	5.13	+0.11	n.s.
PBB	—	—	2.52	2.54			—	—	3.26	6.78			—	—	2.93	1.99		

*B.D. = Block Design; D.S. = Digit Symbol; and E.F. = Embedded Figures.

TABLE 6

MICHIGAN—CORRELATION BETWEEN PBB AND PERFORMANCE TEST SCORES IN MALES AND FEMALES IN THREE AGE GROUPS AS SHOWN*

	0–29				30–49				50–Over			
	Males		Females		Males		Females		Males		Females	
	DDE	PBB	DDE	PBB	DDE	PBB	DDE	PBB	DDE	PBB	DDE	PBB
B.D.†	+.02	+.18	+.16	+.05	+.02	−.06	+.02	+.03	−.30	−.32	−.14	+.41
D.S.	−.35	+.23	−.20	+.05	−.05	−.007	−.06	−.23	−.41	−.41	−.04	−.09
E.F.	−.07	+.13	−.21	+.14	+.08	−.06	−.15	−.16	−.59	−.57	+.29	−.15
N	23		19		46		27		15		11	
$r_{.01}$	.50		.43		.29		.36		.48		.55	
$r_{.05}$	.39		.55		.37		.47		.60		.68	

*Wisconsin residents have no detectable serum PBB levels.

†B.D. = Block Design; D.S. = Digit Symbol; and E.F. = Embedded Figures.

As previously mentioned, the mean serum PBB level for males was significantly higher than that of the females.

The behavioral findings, analyzed by means of correlational techniques, were convergent with those obtained from assessment of population differences: severity of central nervous system dysfunction, as evaluated by means of performance tests, increased with age in the male group.

Correlation between Neurologic Symptoms and Performance Test Scores

TABLES 7 and 8 show the means and standard deviations of test scores (B.D., D.S. and E.F.) in subjects who exhibited and in subjects who did not exhibit the most prominent neurologic symptoms (tiredness, sleepiness, nervousness, depression, dizziness, headaches, loss of balance, blurred vision and muscle weakness) in November 1976 and t-values for the significance of differences between the means.

When mean ages for individuals with neurologic symptoms were compared to those individuals without neurologic symptoms, no significant differences were found except for males who experienced headaches, loss of balance, and blurred vision at the time of the examination.

The mean educational level of males who experienced neurologic symptoms was significantly lower than those without neurologic symptoms. No similar trend was observed in females.

Among males, the index of performance (SMX) was found to be significantly lower in those who experienced any of the most prominent neurologic symptoms, except in the comparison involving sleepiness. TABLE 8 also shows the means, standard deviations and t-values of individual test scores. In females, no similar trend relating the index of performance (SFX) and presence of neurologic symptoms was found.

The mean serum PBB levels in males and females who experienced neurologic symptoms in 1976 were not significantly different than those of the asymptomatic group. However, when a logarithmic transformation was used to describe the serum PBB levels, it was found that in males and for some neurologic symptoms such as tiredness, headaches, loss of balance, blurred vision, and muscle weakness, those who had neurologic symptoms had a significantly *lower* serum PBB level. No such trend was observed in females. Interpretation of these findings is difficult; more insight is expected from the analysis of similar correlations with PBB levels in subcutaneous fat.

FIGURE 6 is the graphic representation of correlation coefficients as explained above. FIGURE 7 is a summary of data representing shifts in the consumption of home-produced and store-bought products in 1972–1974 and in 1975–1976. Details on how to read this figure were given above. Consumption of home-produced foodstuffs was slightly negatively correlated with consumption of store-bought foodstuffs. This indicates that subjects who produced their own products at home were unlikely to buy the same products at the store. This pattern was found both in Michigan and Wisconsin.

Michigan residents showed significant change in consumption of milk, butter, cheese, fish, and eggs as compared to the Wisconsin sample population. Michigan residents also altered their consumption of beef, pork, and chicken, but these changes were not significantly different from Wisconsin residents.

As a rule, Wisconsin residents who purchased their products in stores in 1972–1974 continued to do so in 1975–1976. This is shown by the two overlapping arrows in the left half of the semicircle. In Michigan, however, the consistent rightward shifting of the 1974–1976 "store-bought products" arrows and the

TABLE 7

PERFORMANCE TEST SCORES IN MICHIGAN MALES WITH AND WITHOUT NEUROLOGIC SYMPTOMS

		Tiredness		Sleepiness		Nervousness		Depression		Dizziness		Headache		Balance		Blr Vision		Muscle Wk	
		−	+	−	+	−	+	−	+	−	+	−	+	−	+	−	+	−	+
Age	MN	36.5	38.2	38.3	36.5	37.0	39.1	37.2	38.3	36.8	40.0	35.3	40.1	36.2	42.0	35.9	42.1	36.5	40.8
	SD	15.1	11.9	14.2	11.8	13.3	12.6	13.1	13.1	13.4	12.0	14.3	11.2	13.4	11.0	13.6	10.3	13.5	11.2
	T	−.58		.62		−.65		−.35		−.99		−1.66		−1.74		−1.92		−1.31	
Educ.	MN	3.55	2.81	3.36	2.78	3.24	2.65	3.25	2.75	3.24	2.61	3.27	2.87	3.20	2.67	3.23	2.65	3.24	2.60
	SD	1.13	.71	1.08	.68	1.01	.57	.91	.93	.98	.65	1.03	.80	.98	.75	.97	.73	.98	.60
	T	3.60		2.87		2.66		2.31		2.70		1.91		2.26		2.53		2.85	
B.D.*	MN	30.5	25.3	28.2	26.1	28.2	24.4	28.0	25.5	28.2	23.9	28.5	25.6	28.0	24.2	28.7	23.0	28.1	24.4
	SD	9.3	8.6	9.35	9.08	9.0	9.2	8.6	10.1	8.5	10.3	9.4	8.7	8.9	9.6	8.7	9.3	8.8	9.8
	T	2.52		.99		1.67		1.15		1.89		1.45		1.62		2.55		1.57	
D.S.	MN	47.1	41.0	43.0	43.7	44.4	40.4	44.7	40.4	44.3	40.2	46.7	39.4	45.0	37.7	45.7	36.2	45.0	38.3
	SD	14.8	12.2	15.3	11.4	14.8	8.6	13.4	13.1	13.0	14.2	13.6	12.2	13.1	13.2	13.2	11.6	13.6	11.9
	T	1.96		−.22		1.20		1.38		1.21		2.50		2.11		2.91		2.00	
E.F.	MN	31.8	30.0	30.4	30.8	31.4	28.8	31.0	30.0	31.6	27.9	31.3	30.0	31.2	28.9	31.6	28.1	31.2	29.0
	SD	7.54	5.55	6.76	6.00	6.6	5.3	6.6	5.9	5.8	6.9	7.0	5.5	6.4	5.8	6.4	5.4	6.7	5.0
	T	1.22		−.31		1.65		.71		2.42		.93		1.44		2.28		1.40	
SMX	MN	.29	−.14	.01	.02	.13	−.29	.11	−.18	.14	−.37	.17	−.17	.11	−.28	.17	−.42	.13	−.34
	SD	1.03	.83	.99	.88	.98	.70	.92	.91	.85	1.03	.96	.87	.88	1.01	.87	.94	.94	.80
	T	2.03		−.03		1.86		1.39		2.23		1.68		1.64		2.64		2.04	
PBB	MN	78.0	9.88	51.4	16.4	42.8	11.4	48.2	6.9	43.8	5.45	58.6	7.6	44.1	2.72	44.3	5.8	44.4	3.8
	SD	334.9	25.7	280.1	40.2	235.7	30.5	247.1	8.8	232.0	7.4	277.5	19.0	230.0	1.87	234.0	8.2	231.9	5.24
	T	1.47		.77		.63		.88		.75		1.15		.80		.77		.80	
Log PBB	MN	.86	.48	.72	.51	.68	.46	.68	.50	.68	.44	.73	.49	.72	.30	.69	.42	.73	.30
	SD	.70	.63	.57	.77	.67	.69	.71	.60	.71	.55	.78	.54	.72	.41	.70	.57	.69	.55
	T	2.42		1.38		1.28		1.09		1.37		1.59		2.46		1.60		2.61	

*B.D. = Block Design; D.S. = Digit Symbol; and E.F. = Embedded Figures.

TABLE 8

PERFORMANCE TEST SCORES IN MICHIGAN FEMALES WITH AND WITHOUT NEUROLOGIC SYMPTOMS

		Tiredness		Sleepiness		Nervousness		Depression		Dizziness		Headache		Balance		Blr Vision		Muscle Wk	
		−	+	−	+	−	+	−	+	−	+	−	+	−	+	−	+	−	+
	MN	37.5	35.7	36.6	36.5	36.2	37.2	37.1	34.7	36.3	37.0	37.6	35.2	36.4	37.1	35.5	42.7	37.3	32.2
Age	SD	13.6	11.4	12.1	13.3	12.9	11.3	12.5	12.1	12.9	11.6	13.8	10.2	12.7	11.3	12.1	12.8	12.6	10.3
	T	.52		.04		−.27		.60		−.21		.73		−.15		−1.55		1.07	
	MN	3.29	3.08	3.28	2.93	3.22	3.08	3.13	3.36	3.31	2.94	3.29	3.04	3.2	3.1	3.22	2.94	3.23	2.87
Educ.	SD	.77	.68	.82	.25	.77	.62	.67	.92	.71	.70	.80	.61	.76	.57	.74	.56	.76	.35
	T	1.08		1.63		.68		−.93		1.83		1.28		.39		1.03		1.31	
	MN	27.4	27.1	27.4	27.0	27.0	27.8	26.5	30.0	28.5	24.8	28.0	26.3	27.3	27.1	28.0	23.1	27.4	26.7
B.D.*	SD	7.78	9.87	8.51	9.99	9.31	8.22	8.89	8.78	8.28	9.74	7.31	10.7	9.03	8.79	8.4	11.2	9.19	7.50
	T	0.13		.15		−.32		−1.20		1.49		.72		.07		1.44		.18	
	MN	57.0	55.2	57.7	52.3	56.2	55.7	56.4	54.9	57.5	53.3	55.6	56.7	56.3	54.8	57.0	50.5	55.8	57.2
D.S.	SD	11.9	10.9	11.3	10.6	11.4	11.5	11.8	9.32	9.59	13.9	9.89	13.1	11.7	9.61	11.2	10.9	11.51	10.5
	T	.59		1.68		.18		.39		1.34		−.35		.38		1.52		−.32	
	MN	33.8	33.6	33.5	34.05	33.4	33.4	33.5	34.4	33.8	33.5	33.0	34.6	33.5	34.7	33.9	32.2	33.9	34.2
E.F.	SD	4.71	3.97	4.44	4.01	4.61	3.51	4.45	3.70	4.37	4.22	4.3	4.1	4.33	4.11	4.19	4.80	4.19	5.04
	T	.17		−.42		−.71		−.66		.21		−1.44		−.82		1.03		−.40	
	MN	.08	.00	.08	−.05	.01	.10	.00	.18	.14	−.15	−.00	0.09	.03	.08	.12	−.40	.03	.10
SFX	SD	.72	.82	.78	.78	.83	.65	.81	.60	.70	.88	.56	.99	.81	.60	.71	1.03	.79	.68
	T	.40		.60		−.40		−.70		1.40		−.46		−.17		1.87		−.24	
	MN	2.22	3.11	2.96	2.13	3.23	1.60	2.24	4.37	3.01	2.11	2.42	3.06	2.02	5.73	2.22	5.54	2.82	2.02
PBB	SD	2.15	6.26	5.67	1.97	5.78	1.11	2.12	9.66	5.86	1.65	2.26	6.91	1.86	10.5	2.06	11.9	5.17	1.84
	T	−.68		.59		1.18		−1.36		.66		−.48		−2.28		−1.84		.43	
Log	MN	.14	.15	.15	.13	.19	.04	.13	.18	.13	.17	.17	.11	.10	.34	.13	.22	.16	.05
PBB	SD	.47	.53	.52	.48	.53	.43	.48	.59	.55	.41	.48	.53	.48	.60	.48	.63	.49	.58
	T	−.06		.18		1.02		−.30		−.31		.46		−1.34		−.47		.59	

*B.D. = Block Design; D.S. = Digit Symbol; and E.F. = Embedded Figures.

consistent leftward shifting of the 1974–1976 "home-produced products" arrows suggest that the widely divergent home-store consumption pattern observed in 1972–1974 changed subsequently to a pattern characterized by a substantial positive correlation between the consumption of home-produced and store-bought foodstuffs in 1974–1976. Given the decrease in average consumption among Michigan residents of all home products during that period, these data indicate a trend among the exposed population to extensively substitute store-bought food for home-produced items after 1974.

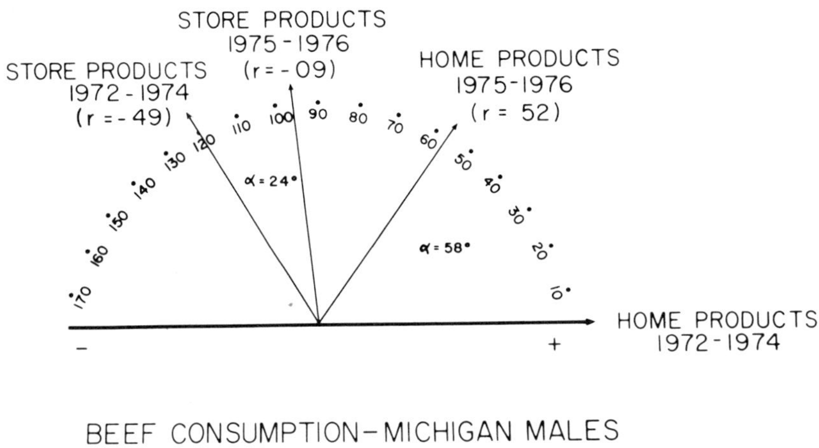

FIGURE 6. Vector Representation of Correlation Coefficients. Vector representation of correlation coefficients relating ingestion of home-produced and store-bought products in 1972–1974 and 1975–1976. Correlation coefficients can be represented by means of vectors because of the relation r = cos α, where α is the angle between the two vectors. For example, given sample size N = 100, if α is equal to 81 degrees, then r is equal .16 and the probability level is .05.

Correlation between Consumption of Home-Produced and Store-Bought Products and Performance Test Scores

In Michigan males, lower performance index was correlated with higher ingestion of home-produced foodstuffs in 1972–1974 (TABLES 9 and 10). What follows are the significant correlation coefficients between indices of performance and specific foodstuffs arranged according to a decreasing order of significance: eggs (r = −0.29; p = 0.005), butter (r = −0.27; p = 0.008), beef (r = −0.22; p = 0.02), cheese (r = −0.20; p = 0.03), pork (r = −0.19; p = 0.05), fish being a borderline case (r = −0.18; p = 0.06).

Correlations among home-produced foodstuffs and performance index for 1975–1976 and store-bought products and performance index for 1972–1974 and for 1975–1976 were not significant, except for a borderline correlation between store-bought butter in 1975–1976 and performance index r = −.18; p = −.06. In Michigan females and in Wisconsin residents (both males and females) no similar significant trends were detected.

To control for possible confounding effects due to age and education, partial

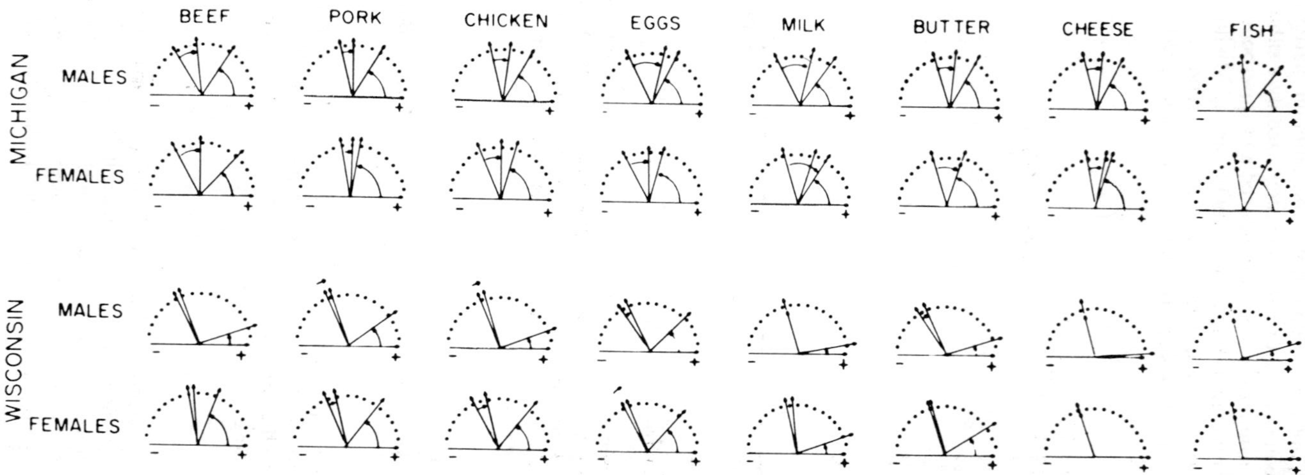

FIGURE 7. Vectorial Display of Changes in Food Consumption Among Michigan and Wisconsin Residents. Details on how to read the figure are given in FIGURE 6. The arrow pointing to the right is the reference point and indicates the consumption of a given home-produced foodstuff in 1972–1974. Further details in text.

TABLE 9

CORRELATIONS BETWEEN INTAKE OF FOODSTUFFS AND AGE, EDUCATION, PBB, LOG PBB AND PERFORMANCE (SMX) IN MALES

	Michigan (N = 79)										Wisconsin (N = 46)*					
Home Produced	1972–1974					1975–1976					1972–1974			1975–1976		
	Age	Educ.	PBB	LPBB	SMX	Age	Educ.	PBB	LPBB	SMX	Age	Educ.	SMX	Age	Educ.	SMX
Milk	−.18	−.13	.08	.15	−.04	−.12	−.09	.13	.04	.11	−.24	−.10	.28	−.11	.15	.28
Eggs	.12	−.32	−.10	−.09	−.29	.04	−.18	−.05	−.01	−.01	−.06	−.10	.11	.03	.04	−.15
Beef	.16	−.16	−.05	−.07	−.22	.09	−.07	.07	.07	−.03	−.25	−.19	.17	−.02	.17	−.05
Chicken	.06	−.10	−.04	.04	−.10	−.03	−.08	−.04	−.06	.05	−.15	−.04	.24	.22	−.02	−.24
Pork	.01	−.30	−.09	−.05	−.18	−.02	−.15	−.05	.01	.05	−.12	−.01	.27	−.02	−.24	−.20
Fish	.12	−.05	−.05	−.10	−.18	.08	−.14	−.05	−.07	−.08	−.14	−.11	−.01	−.22	−.10	.19
Butter	.17	−.03	.10	.04	−.26	.17	.05	.26	.18	−.10	.14	−.06	−.06	−.25	−.36	−.07
Cheese	.01	−.13	−.02	.04	−.20	−.15	−.01	−.02	.03	.10	.09	.03	-09	−.02	−.05	−.04
Store Bought	1972–1974					1975–1976					1972–1974			1975–1976		
Milk	.16	.11	−.07	−.20	−.02	−.02	−.02	−.08	.12	−.05	−.02	−.11	.27	−.08	.15	.24
Eggs	−.07	.15	−.07	−.03	.15	.02	−.05	−.11	−.01	−.09	−.10	.12	.03	.12	−.12	−.22
Beef	−.21	.12	−.00	.12	.17	−.07	−.01	−.12	−.07	−.05	−.26	−.12	.27	−.04	−.07	−.21
Chicken	−.19	.07	−.08	−.01	.16	−.15	−.03	−.12	−.12	.05	−.05	−.01	.17	.15	−.15	−.18
Pork	−.08	.12	−.06	.00	.14	−.04	.06	−.08	.07	.00	−.20	−.03	.29	−.02	−.25	−.31
Fish	−.23	−.13	−.07	−.12	.14	−.19	−.10	−.07	−.12	.13	−.14	−.11	−.01	−.21	−.14	.17
Butter	.03	.04	−.09	.06	−.06	.08	−.01	−.10	.05	−.18	.14	−.06	−.06	−.28	−.36	−.04
Cheese	.01	.07	−.14	−.16	.11	.01	.10	−.14	−.06	.01	.09	.03	−.08	−.04	−.01	−.02

*PBB levels in Wisconsin are nondetectable.

Michigan: $|r| = .18; p \leq .05$; $|r| = .25; p \leq .01$

Wisconsin: $|r| = .24; p \leq .05$; $|r| = .34; p \leq .01$

TABLE 10

CORRELATIONS BETWEEN INTAKE OF FOODSTUFFS AND AGE, EDUCATION, PBB, LOG PBB AND PERFORMANCE (SFX) IN FEMALES

	Michigan (N = 55)										Wisconsin (N = 44)*					
	1972–1974					1975–1976					1972–1974			1975–1976		
	Age	Educ.	PBB	LPBB	SFX	Age	Educ.	PBB	LPBB	SFX	Age	Educ.	SFX	Age	Educ.	SFX
HOME PRODUCED																
Milk	−.31	.13	−.16	−.14	.36	−.11	−.04	.02	.08	.00	−.20	.18	.14	.06	−.07	.18
Eggs	.19	−.12	.11	.14	−.15	−.10	.04	−.07	.05	.20	.05	−.23	−.11	.14	−.03	−.19
Beef	−.04	−.03	−.15	−.28	.17	−.23	−.08	.04	.11	−.06	−.16	.00	.19	−.10	.11	−.05
Chicken	−.02	.13	.19	.16	.10	−.06	−.10	−.09	−.02	.09	−.22	.16	.07	.16	.13	−.11
Pork	.21	.08	.16	.21	−.10	−.20	−.09	−.02	.00	−.03	−.10	.03	−.20	.18	−.27	−.03
Fish	.01	−.06	.18	.20	.19	−.08	−.08	−.12	−.19	−.17	−.02	.14	.13	−.09	.01	.00
Butter	−.03	−.04	.01	.13	.02	−.01	−.02	−.05	−.05	.06	.27	−.30	−.26	−.17	.16	.10
Cheese	.07	−.03	.06	.19	.01	−.11	−.07	−.07	−.10	.18	.28	−.27	−.22	.02	−.03	−.20
STORE BOUGHT																
Milk	−.13	.33	−.09	−.10	.19	−.34	−.07	−.10	−.14	.14	−.14	.13	.09	−.02	−.01	.19
Eggs	−.04	−.06	.08	.20	.08	.02	.04	.02	−.02	.25	−.07	−.22	−.23	.21	.10	.07
Beef	.08	.03	−.13	−.15	−.02	−.28	−.13	−.04	−.21	.37	−.18	−.08	.13	−.13	.26	.08
Chicken	−.02	.34	.02	.13	−.15	−.08	−.16	.01	−.02	.12	−.23	.10	.02	.16	.16	−.04
Pork	−.03	.20	.19	.10	−.04	−.06	−.10	.08	.16	.00	−.07	−.07	−.24	.18	−.23	−.01
Fish	.18	−.06	.28	.21	.07	−.16	−.06	−.10	−.24	−.07	−.02	.14	.13	−.07	.00	−.02
Butter	.21	−.04	−.05	−.05	−.13	−.07	−.02	.02	.14	.12	.27	−.30	−.26	−.15	−.01	.11
Cheese	—	—	—	—	—	−.18	−.05	−.01	.01	.25	.28	−.27	−.21	−.03	−.03	−.17

*PBB levels in Wisconsin are nondetectable.

Michigan: $|r| = .23; p \leq .05$; $|r| = .32; p \leq .01$

Wisconsin: $|r| = .24; p \leq .05$; $|r| = .34; p \leq .01$

correlations were computed between ingestion of home-produced and store-bought foodstuffs in 1972–1974 and in 1975–1976 controlling for age and education alone and for the possible joint effects of both age and education. TABLES 11 and 12 show the results of this analysis. Even after applying the partial correlations, the consumption of home-produced eggs, butter and cheese were negatively correlated with performance index. No similar trends were detected in Michigan females and among Wisconsin residents (both males and females).

Correlation between Consumption of Home-Produced and Store-Bought Products and Blood Chemistries

Intercorrelations between ingestion of home-produced and store-bought products in 1972–1974 and 1975–1976 and blood chemistry have been reported elsewhere.[1] No distinctive patterns were observed. Some individual correlations did in fact reach significant levels in males, but the absence of any distinct pattern prevents the assignment of biologic significance to these statistics.

Correlations between Blood Chemistries and Performance Tests Scores and Neurologic Symptoms

In TABLE 13 significant correlations between blood chemistries and performance tests scores are grouped: tests of hepatic function (SGPT, SGOT, bilirubin, total protein), renal function (creatinine, uric acid), electrolytes possibly reflecting endocrine and renal function (calcium and sodium), cholesterol and carcinogenic embryonic antigen (CEA).

In Michigan males significant correlation coefficients were found with three of the hepatic function tests, one of the renal function tests, both electrolytes, cholesterol, and CEA. None of these correlations was significant for Wisconsin males. In Michigan females fewer significant correlations were found; only the creatinine, uric acid and CEA coefficients were significantly different from those of the Wisconsin female population.

The mean and standard denotion of 17 blood chemistries of Michigan subjects who exhibited the most prevalent neurologic symptoms in 1976, those who did not exhibit these symptoms, as well as the t-values associated with the differences between these means have been reported elsewhere.[1] In both cases there were no consistent "column" or "row" effects; that is, the differences in means of a given blood test result were not singularly associated with a distinctive pattern of neurologic symptoms, and the presence of a given neurologic symptom was not associated with significant differences in a distinctive pattern of blood chemistries.

This type of analysis could not be performed in the non-PBB exposed Wisconsin group because neurologic symptoms were either absent or exhibited a very low frequency, rendering the statistical analysis meaningless.

DISCUSSION

Neurologic symptoms were the earliest and most prominent symptoms reported by Michigan residents exposed to PBB, compared with a non-PBB exposed population in Wisconsin. Among Michigan residents (particularly males), those with the most marked symptoms tended to show diminished performance as assessed by special tests.

TABLE 11

MICHIGAN MALES: PARTIAL CORRELATIONS BETWEEN INGESTION OF FOOD PRODUCTS AND SERUM PBB, LOG SERUM PBB, AND PERFORMANCE, CONTROLLING FOR AGE, EDUCATION AND AGE PLUS EDUCATION (N = 76)*

	Control for Age			Control for Educ.			Control for Age & Educ.		
	PBB	LPBB	SMX	PBB	LPBB	SMX	PBB	LPBB	SMX
HOME PRODUCTS 1972–1974									
Milk	.14	.17	−.18	.12	.19	.01	.21	.23	−.13
Eggs	−.15	−.10	−.27	−.02	.00	−.19	−.05	−.01	−.18
Beef	−.10	−.09	−.16	−.01	−.03	−.18	−.06	−.05	−.12
Chicken	−.07	.04	−.08	−.02	.07	−.07	−.03	.07	−.05
Pork	−.10	−.05	−.23	−.02	.03	−.08	.00	.04	−.14
Fish	−.10	−.12	−.13	−.05	−.10	−.17	−.10	−.12	−.13
Butter	.05	.03	−.21	.11	.06	−.28	.05	.03	−.22
Cheese	−.03	.04	−.25	.01	.08	−.17	.01	.08	−.22
STORE PRODUCTS 1972–1974									
Milk	−.12	−.22	.09	−.10	−.24	−.07	−.18	−.28	.04
Eggs	−.05	−.02	.15	−.11	−.07	.10	−.10	−.07	.10
Beef	.06	.15	.05	−.03	.10	.13	.04	.13	.03
Chicken	−.02	.01	.06	−.10	−.03	.14	−.03	.00	.05
Pork	−.04	.01	.12	−.10	−.03	.11	−.08	−.02	.09
Fish	−.01	−.11	.01	−.04	−.10	.21	.06	−.06	.08
Butter	−.10	.06	−.06	−.10	.05	−.09	−.12	.05	−.08
Cheese	−.16	−.16	.15	−.17	−.18	.09	−.19	−.19	.13

	Control for Age			Control for Educ.			Control for Age & Educ.		
	PBB	LPBB	SMX	PBB	LPBB	SMX	PBB	LPBB	SMX
HOME PRODUCTS 1975–1976									
Milk	.17	.05	.04	.16	.06	.15	.22	.09	.09
Eggs	−.07	−.01	.02	−.01	.04	.07	−.01	.04	.09
Beef	.04	.06	.03	.09	.09	−.01	.06	.08	.05
Chicken	−.03	−.05	.04	−.02	−.04	.09	−.01	−.03	.07
Pork	−.04	.01	.04	−.01	.05	.12	.00	.06	.10
Fish	−.08	−.08	−.05	−.02	−.04	−.03	−.04	−.05	.00
Butter	.22	.16	.01	.26	.17	−.12	.21	.14	−.03
Cheese	.03	.04	.02	−.01	.03	.12	.05	.06	.03
STORE PRODUCTS 1975–1976									
Milk	−.08	.13	−.08	−.08	.14	−.05	−.07	.14	−.08
Eggs	−.12	−.01	−.10	−.10	.00	−.08	−.11	.00	−.09
Beef	−.11	−.07	−.11	−.12	−.07	−.05	−.10	−.06	−.11
Chicken	−.08	−.11	−.04	−.12	−.12	.07	−.07	−.10	−.02
Pork	−.08	.08	−.02	−.10	.06	−.02	−.10	.06	−.04
Fish	−.02	−.11	.03	−.05	−.10	.19	.03	−.07	.08
Butter	−.12	.04	−.17	−.10	.05	−.19	−.13	.04	−.18
Cheese	−.16	−.06	.01	−.18	−.09	−.04	−.20	−.10	−.02

*N is different from that of TABLE 9 because of missing age and education values.

$|r| = .18; p \leq 0.05$

$|r| = .25; p \leq 0.01$

TABLE 12

MICHIGAN FEMALES: PARTIAL CORRELATIONS BETWEEN INGESTION OF FOOD PRODUCTS AND SERUM PBB AND SERUM LOG PBB LEVELS AND PERFORMANCE, CONTROLLING FOR AGE, EDUCATION, AND AGE PLUS EDUCATION (N = 51)

	Home Products 1972–1974									Home Products 1975–1976								
	Control for Age			Control for Educ.			Control for Age & Educ.			Control for Age			Control for Educ.			Control for Age & Educ.		
	PBB	LPBB	SFX	PBB	LPBB	SFX	PBB	LPBB	SFX	PBB	LPBB	SFX	PBB	LPBB	SFX	PBB	LPBB	SFX
Milk	−.12	−.06	.28	−.14	−.09	.29	−.10	−.02	.21	−.07	−.08	.16	−.07	−.09	.16	−.06	−.07	.13
Eggs	.09	.10	−.09	.10	.14	−.13	.08	.10	−.08	.09	.22	.07	.08	.20	.10	.09	.22	.09
Beef	−.15	−.30	.17	−.15	−.29	.19	−.15	−.29	.19	−.15	−.18	.01	−.14	−.17	.02	−.16	−.20	.04
Chicken	.20	.17	.09	.18	.14	.16	.19	.15	.17	.04	.14	−.18	.03	.13	−.17	.04	.13	−.18
Pork	.14	.16	−.03	.15	.19	−.04	.12	.14	.03	.20	.10	−.06	.19	.07	.00	.19	.08	−.01
Fish	.18	.21	.21	.18	.21	.21	.18	.21	.23	.26	.18	.15	.28	.22	.07	.26	.18	.15
Butter	.01	.14	.01	.01	.12	.04	.01	.13	.03	−.08	−.11	−.05	−.06	−.07	−.10	−.09	−.12	−.04
Cheese	.06	.18	.04	.05	.17	.06	.05	.16	.08	—	—	—	—	—	—	—	—	—
	Store Products 1972–1974									Store Products 1975–1976								
Milk	.04	.12	−.04	.06	.15	−.13	.07	.18	−.17	−.05	−.05	.02	−.08	−.10	.06	−.04	−.01	−.07
Eggs	−.06	.08	.18	−.07	.06	.20	−.06	.09	.18	.02	−.03	.27	.02	−.01	.24	.02	−.02	.27
Beef	.07	.19	−.15	.02	.09	−.01	.06	.16	−.12	−.01	−.14	.30	−.03	−.18	.35	.01	−.12	.28
Chicken	−.09	.01	.07	−.08	.01	.03	−.07	.02	.01	.02	.00	.10	.00	−.04	.17	.01	−.01	.15
Pork	.00	.06	−.11	−.04	−.02	.03	−.01	.03	−.05	.09	.19	−.03	.06	.13	.09	.07	.16	.06
Fish	−.11	−.18	−.22	−.13	−.21	−.15	−.12	−.19	−.20	−.08	−.20	−.14	−.11	−.25	−.05	−.09	−.22	−.12
Butter	−.05	−.04	.06	−.02	.00	−.03	−.03	−.01	−.03	.04	.16	.10	.06	.20	.02	.06	.22	.01
Cheese	−.06	−.07	.16	−.07	−.10	.19	−.06	−.07	.16	.01	.06	.20	.01	.04	.22	.03	.08	.17

TABLE 13
CORRELATIONS BETWEEN INDEX OF PERFORMANCE (SMX AND SFX) AND BLOOD CHEMISTRIES IN MALES AND FEMALES

		Males				Females			
		Michigan (N = 76)		Wisconsin (N = 44)		Michigan (N = 53)		Wisconsin (N = 35)	
	Blood test	r	p	r	p	r	p	r	p
Indicators of Hepatic function	SGPT	−.19	.05	.07	n.s.	.06	n.s.	−.20	n.s.
	SGOT	−.08	n.s.	−.10	n.s.	−.05	n.s.	−.30	n.s.
	LDH	.14	n.s.	−.10	n.s	−.28	.05	−.24	n.s.
	Bilirubin	.23	.025	−.09	n.s.	−.05	n.s.	−.04	n.s.
	Total Protein	.20	.05	−.10	n.s.	−.02	n.s.	.19	n.s.
Indicators of Renal Functions	Creatinine	−.17	n.s.	−.02	n.s.	−.35	.01	−.24	n.s.
	Uric Acid	−.01	n.s.	−.10	n.s.	−.33	.01	−.23	n.s.
	CEA	.15	n.s.	−.19	n.s.	−.27	.05	−.24	n.s.
	Cholesterol	.31	.005	−.03	n.s.	−.31	.05	−.29	n.s.
	Calcium	.20	.05	−.11	n.s.	−.17	n.s.	.05	n.s.
	Sodium	−.25	.025	.16	n.s.	−.05	n.s.	−.12	n.s.

They also had significantly higher intakes of home-produced foodstuffs—particularly beef, pork and butter—in 1972–1974 and of store-purchased foodstuffs in 1975–1976, as compared to the asymptomatic Michigan males. In recent years, Michigan residents have made significant changes in the habitual patterns of intake of food products suspected to be contaminated with PBB. Serum PBB levels were negatively correlated with performance tests scores among Michigan males. Serum PBB levels were not found to be significantly different in males showing the most prominent neurologic symptoms, when compared with asymptomatic males.

Neurologic symptoms of the type observed among Michigan residents (tiredness, sleepiness, depression, dizziness, headaches, difficulties in balance, blurred vision, and muscle weakness) could be termed subjective. The extensive questionnaire that was used was specifically designed to be comprehensive, and it was meant to be used in the evaluation of the total health status of each individual participant. Exploration of neurologic symptoms was only a part of the evaluation. Moreover, this evaluation attempted to assess a situation with no previous record in the world literature. Wisconsin residents examined by identical procedures and by the same professional team did not report the symptoms with a prevalence comparable to that found in Michigan residents. The fact that objective performance tests scores were consistent with the clinical findings argues against the view that the reported neurologic symptoms were entirely subjective and that the respondents were influenced by motivational factors. Subjects were not familiar with the performance tests, and they did not know the scoring procedures.

The curves showing the time course of differences in prevalence of neurologic symptoms between Michigan residents and the Wisconsin reference group (as measured by the chi-square tests) are based on the subjects' recollection of the presence or absence of several individual symptoms. FIGURE 8 illustrates the change in chi-square test value for differences in prevalence of "tiredness" between the two populations. In addition to the empirical curve, several alternative hypothetical curves could be constructed describing the time course of the chi-square value for that symptom. It could be legitimately argued that the increase in difference between the two populations reflected by these curves is affected by memory, and views in favor or against the concave downward function as the true description of the time course of neurologic symptoms could be presented. However, the presence of two empirical points—the nonsignificant difference in 1972 and the highly significant difference in 1976—supplies strong evidence that a marked increase in the prevalence of neurologic symptoms in the Michigan population occurred between 1972 and 1976, compared with the Wisconsin population.

Critical readers of this report will undoubtedly notice the heavy reliance upon mnemonic data, particularly with reference to symptomatology and individual food consumption. While we make no claim about the inherent "softness" or "hardness" of these data, they remain essential to a coherent, logical investigation—given that human PBB contamination proceeded via ingestion of contaminated foodstuffs. Our three-tiered analysis illustrated by the model

FOOD ⟶ PBB BODY BURDEN ⟶ NEUROLOGIC SYMPTOMS / PERFORMANCE TESTS SCORES / ABNORMAL BLOOD CHEMISTRIES

1 2 3

attempted to relate measurements at the primary level to secondary and ternary consequences. Individual metabolic differences and exposure circumstances with

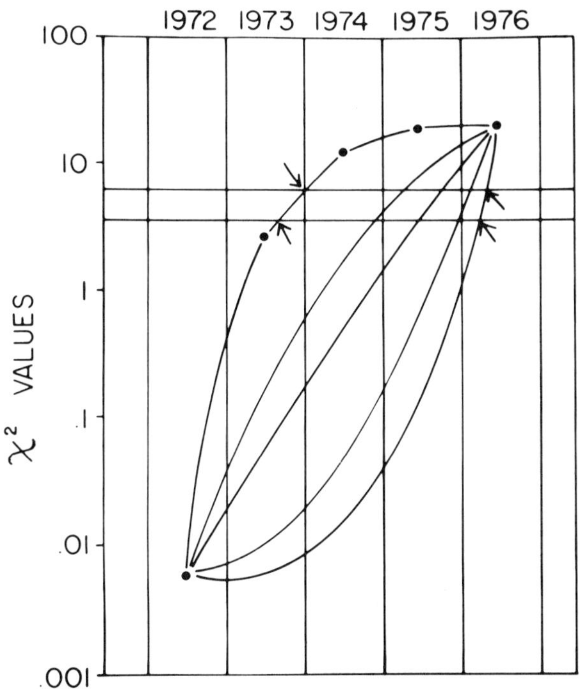

FIGURE 8. Family of Curves Illustrating Time Course of a Neurologic Symptom. Concave downward function (top two), linear function (middle) and concave upward function (bottom two). The uppermost curve has been empirically determined; the others are hypothetical.

regard to PBB may introduce random variation between the first and second levels; therefore, the correlations between level **1** parameters with clinical results of level **3** were also explored. Intercorrelations among the latter afforded further insight into the consistency of objective findings and subjective complaints.

We should not ignore the possible bias in our sampling procedures. As explained above, the original selection of the subjects was made on the basis of different categories, including farmers living in quarantined and nonquarantined farms and consumers of products purchased from these farms. It is hard to estimate the number of healthy or extremely sick people who did not appear at the study site. In fact, the probability for healthy persons to be present or absent and for sick persons to be present or absent is similar to that expected in an outbreak of food poisoning. Subjects will appear with symptoms which are similar, but not due, to food poisoning and subjects will be very sick without linking their condition to food poisoning.

Having sampled by invitation from a stratum of relatively heavily exposed individuals, we cannot extrapolate our findings to the general population of Michigan. Concerns about effects of PBB contamination forced us to examine subjects whose past history (particularly in those whose farms had been quarantined) suggested a very high level of exposure; it is worth considering that within the sampled group, demographic characteristics such as age, did not appear skewed. A study under way based on a randomized cross-sectional sample of the Michigan state population will provide adequate data to examine the extent of PBB contamination in the general population of that state.

Our analysis of data centered on the study of main effects of prominent factors, such as food ingestion and serum PBB levels, on neurologic symptoms and performance tests scores. It is important to consider the fact that the starting point of the contamination episode occurred several years before the examination, and that in all probability the human health effects are best described by a complex model of interactions of factors. This report does not deal with higher-order interactions. Although the critical reader may have noticed the extensive use of "elementary statistics," one must point out that this unprecedented exposure to PBB forced us to rely much more heavily upon exploratory data analysis. Indeed, although under laboratory conditions it is possible to randomize populations and deliver cause-effect statements based on inferential statistics, the broad spectrum of PBB symptomatology and the diverse human population under study were best investigated with a descriptive methodology sufficiently flexible to allow integration of several variables into a more simple, though less rigid, analytic design.

More specifically, within the context of this report, we question the advisability of interpreting higher-order interaction effects involving nonmetric factors (data not measured in scales), such as the presence or absence of neurologic symptoms. We believe, however, that these latter data will prove ideal for discriminant analysis; multivariate exploratory techniques (e.g., factor analysis) will undoubtedly afford future insight into the relative importance of the variables discussed herein to the prevalence of neurologic dysfunction among populations exposed to PBB.

It has been asked whether neurologic symptoms were associated with stress, particularly in the case of farmers whose farms were quarantined. As a result of the PBB contamination, many of them had to sacrifice their livestock and at the time of examination had incurred heavy financial losses. Data reported elsewhere in this volume[12] do not support this view: there were no significant differences in symptomatology between farmers whose farms had been quarantined and those whose farms had not been quarantined, and—more importantly—between farmers and consumers.

A comparison of neurologic symptoms and their relative prevalence in the

Michigan population versus those reported by individuals who had experienced very severe stress (combat veterans) indicates important differences. The most striking one was related to the high prevalence of hypersomnia in the Michigan group, while sleeplessness or other sleep disturbances prominently characterized the combat veterans group.[13]

The question of stress as a factor appears again when considering the significant shift in sources of food (home versus store). Shifting from home to store as a source of food involves major financial consequences. We are not aware of current studies that have documented these changes. However, our data seem to point out that the total average weekly food consumption remained fairly constant for the period 1972–1976. The vector representation of the shifting in patterns of food consumption was found to be a convenient method to illustrate these changes.

The list of most prevalent neurologic symptoms is consistent with other reports on symptomatology in Michigan farmers,[14] although the order is not the same. Differences in sample size and the probability level chosen for statistical significance may have influenced the ordering of the symptoms. FIGURE 4 shows that different time courses of symptoms could have been derived, had we used a 5% or a 1% level of significance in the chi-square test.

The results indicate that some of the neurologic symptoms—such as fatigue and depression—showed a relatively early increase in prevalence, while some other symptoms—such as blurred vision and muscle weakness—were characterized by a much slower and protracted increase in prevalence.

Thus far, animal studies on PBB toxicity have not explored neurotoxic effects in depth. There are inherent limitations in any animal model of neurologic symptomatology, particularly since many such symptoms are obtained through the subject's use of introspection, a uniquely human capacity. Furthermore, there exists an apparent lag between animal research and clinical studies on the neurotoxic effects of PBB. Memory changes, balance problems, and even eye movement dysfunctions that have been observed in humans could be documented on animal models by currently available behavioral or electrophysiologic techniques.

The correlations between performance tests scores and blood chemistries, although to be interpreted with caution, seem to indicate the following: the negative correlation with SGPT and the positive correlation with total protein seem to indicate an association between neurotoxicity and liver toxicity; the positive correlation with bilirubin may be related to the effect of PBB on the biliary clearing mechanism; the negative correlation with creatinine and uric acid may reflect an association between neurotoxicity and renal toxicity. In animals, renal toxicity due to PBB has been amply demonstrated.[15] The negative correlation of cholesterol is consistent with similar effects reported in PCB or halogenated-dibenzodioxin-exposed individuals;[16,17] interference with thyroid function may also have to be considered. In the case of correlations between performance and cholesterol, creatinine, and uric acid, interpretations of these results have to be made with restraint since these biochemistries are age-dependent. Correlations with CEA cannot be interpreted without further follow-up.

Finally, we have to address ourselves to the question of the correlations between PBB and neurologic or behavioral changes. As expected from our experience with other human populations and with other biologic indicators of exposure, dose-response relationships cannot be easily established. For example, when studying neurologic symptoms and performance tests scores in lead exposed workers, statistically significant negative correlations with a biologic indicator of exposure—zinc protoporphyrin—were found, while correlations with blood lead levels were less significant.[18] The complexities of the metabolic model for lead give valid reasons for these findings.

In the case of PBB exposure such relationships must take into account complex time-dependent metabolic pathways. As new data become available, we would have to re-address this question using PBB in fat as a biologic indicator of exposure. Even with these shortcomings, it seems apparent that the contaminant present in key home-produced foodstuffs between 1972 and 1974 was associated with neurologic and behavioral changes. Individuals with high PBB serum levels (particularly if male) seemed to have a decreased ability to perform basic mental tasks. It also appears that the highly symptomatic persons were neither exclusively farmers nor were they living only on quarantined farms.[12]

SUMMARY

Neurologic symptoms were the earliest and most prominent symptoms recorded in Michigan farm residents exposed to PBB as compared to a non-PBB exposed control farm population in Wisconsin. In Michigan (particularly among males) those who exhibited the most marked symptoms tended to show diminished performance as assessed by special tests, although population differences in performance were not as marked. Low indices of performance were also significantly correlated with intake of home-produced foodstuffs, particularly during the years 1972–1974 and store-bought products during the years 1975–1976. Between 1972 and 1976 the Michigan farm residents studied made significant changes in their consumption patterns of products suspected to be contaminated with PBB, as compared to those of Wisconsin farm residents. Serum PBB levels were not found to be significantly higher in Michigan males and females exhibiting the most prominent neurologic symptoms. Serum PBB levels were negatively correlated with performance test scores, particularly in males in older age groups.

ACKNOWLEDGMENTS

We extend our appreciation to Ms. Nancy Gilbert for her valuable assistance in obtaining the behavioral test data. We also thank Dr. Bertrand Aubrey and Ms. Francie Camper for assistance in the PBB analysis. We would like to express our thanks to Mrs. Julia Roberts for her skillful assistance in the preparation and editing of this manuscript.

REFERENCES

1. VALCIUKAS, J. A., R. LILIS, H. A. ANDERSON, M. S. WOLFF & M. PETROCCI. 1978. Polybrominated biphenyls: neurological and behavioral effects. Report to National Institute of Environmental Health Sciences (May 26, 1978).
2. ANDERSON, H. A. 1976. Prevalence of unanticipated symptoms among dairy farmers in Michigan and Wisconsin. *In* Health Effects of Exposure to Polybrominated Biphenyls, Results of a Clinical Field Survey, November 4–10, 1976. Interim summary report to the National Institute of Environmental Health Sciences, ES 00928.
3. VALCIUKAS, J. A., R. LILIS, J. EISINGER, W. E. BLUMBERG, A. FISCHBEIN & I. J. SELIKOFF. 1978. Behavioral Indicators of Lead Neurotoxicity. Int. Arch. Environ. Occup. Health. **41:** 217–236.
4. WECHSLER, D. 1955. *In* Adult Intelligence Scale. The Psychological Corporation, New York.

5. TEUBER, H.-L. 1960. Perception *In* Handbook of Physiology, Sec. 1, Vol. III, Neurophysiology. J. Field, H. W. Magoun and V. E. Hall, Eds., American Physiological Society, D.C. pp. 1595–1668.
6. LIDDLE, J. A., H. A. PRICE, & D. D. BAYSE. 1976. Protocol for Specimen Analysis of Quality Assurance Program, Center for Disease Control, Atlanta, Ga. (We thank Dr. Liddle for a copy of this procedure.).
7. WOLFF, M. S. & B. AUBREY. 1977. PBB homologues in sera of Michigan dairy farmers. Workshop on Scientific Aspects of Polybrominated Biphenyls. Michigan State University, East Lansing, Michigan, Oct. 23–24.
8. NIE, N. H., C. H. HULL, J. G. JENKINS, K. STEINBRENNER & D. H. BENT. 1975. Statistical Package for the Social Sciences, Second Edition. New York. McGraw-Hill Book Co.
9. STEVENS, S. S. 1946. On the theory of scales of measurement. Science **103:** 677–680.
10. BAGGALEY, A. R. 1964. Intermediate Correlational Methods. New York. John Wiley and Sons, Inc.
11. VALCIUKAS, J. A., R. LILIS, M. S. WOLFF & H. A. ANDERSON. 1978. Comparative neurobehavioral study of a polybrominated biphenyl-exposed population in Michigan and a non-exposed group in Wisconsin. Environ. Health Perspect. **23:** 199–210.
12. ANDERSON, H. A., R. LILIS, I. J. SELIKOFF & M. PETROCCI. 1979. Clinical symptoms and abnormalities from ingestion of PBB-contaminated food products. *In* Ann. N.Y. Acad. Sci. This Volume.
13. ARCHIBALD, H. C. & R. D. TUDDENHAM. 1965. Persistent stress reaction after combat. Arch Gen. Psychiat. **12**(5): 475–481.
14. MEESTER, W. D. Summary Statement of Human Toxicology of Polybrominated Biphenyls. Before the Committee of the U.S. Senate Committee Investigating the PBB Incident in Michigan. (Held in Kent County Commission Chambers in Grand Rapids, Michigan on March 29, 1977).
15. KAPLINGER, M. L. 1976. Meat and milk residue study with Firemaster FF-1 in dairy cattle. Industrial Bio-Test Laboratories, Inc., Northbrook, Ill.
16. OLIVER, R. M. 1975. Toxic effects of 2,3,7,8-tetrachlorodibenzo 1,4 dioxin in laboratory workers. Brit. J. Ind. Med. **32:** 49–53.
17. POLAND, A. P., D. SMITH, C. METTER & P. POSSICK. A health survey of workers in a 2,4-D and a 2,4,5-D plant with special attention to chloracne, porphyria tarda and psychological parameters. Arch. Environ. Health. **22:** 316–327.
18. LILIS, R., A. FISCHBEIN, J. EISINGER, W. E. BLUMBERG, S. DIAMOND, H. A. ANDERSON, W. ROM, C. RICE, L. SARKOZI, S. KON & I. J. SELIKOFF. 1977. Prevalence of lead disease among secondary lead smelter workers and biological indicators of lead exposure. Environ. Res. **14:** 255–285.

NEUROPSYCHIATRIC FINDINGS IN PATIENTS EXPOSED TO POLYBROMINATED BIPHENYLS

Jeoffrey K. Stross

University of Michigan Medical School
Ann Arbor, Michigan 48109

Robert K. Nixon and Marvin D. Anderson

Henry Ford Hospital
Detroit, Michigan 48201

INTRODUCTION

The inadvertent mixing of polybrominated biphenyls (PBBs) into animal feed in 1973 resulted in the contamination of thousands of Michigan farms. Farm animals who ingested this feed suffered various consequences including weight loss, decreased productivity, frequent infections, skin changes, and reproductive abnormalities.[1] This error was not initially known, and widespread contamination of humans occurred through the consumption of animals and animal products. Families from quarantined farms were thought to have the heaviest exposure, with consumers of products from quarantined farms and chemical workers directly associated with PBBs also receiving heavy exposure.

At the time of this incident, little was known about the possible health effects of PBBs in humans. The only related experience was an accidental ingestion of polychlorinated biphenyls (PCBs) that occurred in Japan and caused many health problems.[2–3]

In Michigan, practicing physicians reported adverse health effects in patients, and the Department of Public Health initiated a study in 1974 to try to document these effects. This study failed to identify any acute effects on health, and neither a definite dose-response relationship nor a pattern of illness consistent with the extent of PBB exposure was identified.[4] Because of a concern for delayed or late consequences of PBB ingestion, the Michigan Department of Public Health, in collaboration with several agencies of the Department of Health, Education, and Welfare initiated a multi-year comprehensive study of the possible human health consequences of exposure to PBBs. This study began in 1976 and historical data and blood specimens were collected and analyzed. Although data analysis is not complete, there appears to be no relationship between PBB serum levels and the presence or absence of symptoms. In fact, those people with no measurable PBBs in their serum reported the highest prevalence of health care complaints.[5] Many of these complaints were neurobehavioral in nature and included poor memory, difficulty in concentration, personality changes, irritability, and hypersomnolence. Since laboratory studies in cattle revealed low PBB concentrations in the central nervous system,[6] the role of PBB ingestion in the etiology of these complaints was in question. A recent study of farmers exposed to PBBs has shown a statistically significant increase in neurobehavioral complaints, as well as poorer performance in various tests of neurologic function as compared to controls. These results were also correlated with serum PBB levels.[7] Since these findings were at odds with each other, this study was designed to evaluate these neurobehavioral complaints and this paper describes the findings.

0077–8923/79/0320–0368 $01.75/0 © 1979, NYAS

METHODS AND MATERIALS

Sixty-four persons with known exposure to PBBs and incapacitating health complaints were identified from previous studies. These people complained of a serious deterioration in their health status and were unable to engage in their previous occupations. They were contacted by letter and asked to participate in a research project aimed at determining the health effects of PBBs. Fifty persons agreed to participate, and 46 actually were evaluated in the study, 37 men and 9 women. Patients were evaluated in the Clinical Research Center of University Hospital in Ann Arbor or at Henry Ford Hospital in Detroit.

The 46 patients were studied with a protocol that included a history, physical examination, X-ray, film of the chest, and electrocardiogram. Extensive biochemical and hematologic studies were carried out, as were endocrine and immunologic testing, including lymphocyte-stimulation studies. Serum, bile, feces and fat specimens were obtained for PBB analysis. X-rays of the skull, pituitary function studies, muscle enzymes, rheumatoid factor, antinuclear antibodies, LE preparations, electromyograms and nerve conduction velocities were routinely obtained.

Patients also underwent extensive psychologic testing and psychiatric interviews. The psychologic tests included the Minnesota Multiphase Personality Inventory, Reitan-Indiana Aphasia Screening Test, Beck Depression Inventory, Wechsler Adult Intelligence Scale, H Words, Color-Naming Tests, Trail-Making Tests, Hand Dynamometer, Grooved Pegboard, Finger Tapping Test, Finger Agnosia and Dysgraphesthesia Test, Benton Facial Recognition Test, Wechsler Memory Scale and a Short-Term Memory Distractor Test. The results of these tests were compared with normal values established for age and sex differences.

RESULTS

Historical Complaints

The patients' complaints were grouped into five categories: general, musculoskeletal, skin, gastrointestinal, and neurologic. TABLE 1 lists the frequency of these findings.

Physical Findings

Complete examinations were done on all participants, and the only positive findings were the presence of hepatomegaly (72%), skin abnormalities (28%), objective joint abnormalities (13%), and neurologic abnormalities (15%). The degree of liver enlargement was mild in most cases and was not associated with tenderness or stigmata of chronic liver disease. The skin abnormalities consisted of conditions that are commonly found in the general population and there were no cases of chloracne noted. The findings in the six patients with joint abnormalities consisted of active synovitis in one patient with rheumatoid arthritis, traumatic changes in another and degenerative changes in the rest. The neurologic abnormalities consisted of sensory changes in the lower extremities. There were no abnormalities in the cranial nerves, reflexes or cerebellar testing studies.

Laboratory Findings

Although extensive laboratory testing was undertaken, few abnormalities were found. Elevation of liver enzyme studies was the most common abnormality noted. These changes were usually minimal and only exceeded twice normal values in those patients with extensive alcohol intake. The alkaline phosphatase was normal in nine patients (20%), lactic dehydrogenase in 10 (22%), and serum transaminases (SGOT and SGPT) in three (7%). Serum triglycerides were elevated in 14 patients (30%), and serum iron was elevated in 12 patients (26%). No other studies were abnormal in more than 10% of the patients. Serum PBB levels ranged from 1 to 180 parts per billion (ppb) with an average of 14, while fat levels ranged from 30 to 34,400 ppb with an average of 3,260. When compared with other published studies, these patients had higher average levels.[5,7]

TABLE 1
HISTORICAL FINDINGS IN 46 PBB-EXPOSED PATIENTS

Historical Finding	Number of Patients*
Fatigue	40 (87)
Weakness	30 (65)
Myalgias	25 (54)
Arthralgias	31 (67)
Skin rashes	17 (37)
Frequent infections	11 (24)
Abdominal pain	14 (30)
Irregular bowel	10 (22)
Memory change	30 (65)
Decreased concentration	28 (61)
Personality change	20 (44)
Increased irritability	32 (70)
Hypersomnolence	23 (50)
Headache	21 (46)
Dizziness	9 (20)
Decreased coordination	18 (39)
Depression	24 (52)
Paresthesias	24 (52)

*Figures in parentheses indicates percentage.

Neurologic Studies

Electromyograms were abnormal in six patients (13%) with no consistent or diagnostic findings. Nerve conduction studies were abnormal in 19 patients (41%), with slowing in sensory nerve latencies the predominant finding. The abnormal values averaged 4.7 milliseconds compared to our normal value of $<$ 3.9 milliseconds. There was an excellent correlation between those patients with objective findings on neurologic examination and abnormal nerve conduction studies. There was no relationship between the presence of these abnormalities and serum or fat PBB levels.

Psychologic Evaluation

Despite the fact that an extensive battery of tests was administered, few objective abnormalities were documented. The most common findings were those of somatic preoccupation, irritability and mild depression. The tests of motor function were normal, while the tests of sensory modalities showed minor differences that were not outside normal limits. Most had I.Q.s between 100 and 140 with no differences between estimated and observed levels. Although most patients complained of memory difficulties, no objective deterioration in memory could be elicited.

Psychiatric interviews confirmed the aforementioned findings that 31 patients (67%) were depressed. No evidence of endogenous depression was noted, and it was the opinion of the psychiatrists involved that the findings were characteristic of reactive depression. This could have resulted from the patients' material losses, the legal proceedings, their inability to obtain help, or from their multiple medical complaints which could be related to toxic effects of PBB on other organ systems.

DISCUSSION

Many studies have been, or are now being, done to determine the health effects of PBB ingestion. Initial studies failed to document any acute changes in the health status of the subjects, and there was no relationship between the presence of symptoms and blood or fat levels of PBBs. Because of continued complaints about health, a series of follow-up studies was undertaken in an attempt to more accurately define the clinical problems. Unfortunately, most of these studies were initiated two or three years after the first contamination occurred, so that no satisfactory baseline studies were available.

A number of difficulties have befallen the farmers as a result of the contamination, and these must enter into any analysis of the health data. Among these are public charges of an initial cover-up of the entire issue, difficulty in obtaining medical care, financial ruin with the loss of livestock, and little financial aid to assist in reestablishing homesteads. Since these have the potential for increasing adverse health effects, it becomes imperative to attempt to sort out the direct effects of a toxic chemical from the behavioral problems induced by this incident and its management.

This study addresses itself to the neurobehavioral effects of PBBs and attempts to correlate them with other parameters. The study population was selected to include those patients who were previously identified as having the greatest change in health status that corresponded chronologically to the PBB contamination. It was hoped that if abnormalities truly existed, they would be most evident in this group of people who appeared to have suffered the most incapacitating effects. We did not have an appropriate control group that was matched for age, sex, and socioeconomic background, so it was necessary for us to compare our findings with normal values previously established for the tests involved.

The high prevalence of symptoms should not be surprising in view of how our patient population was selected. The similarity of these symptoms in published studies is noted. The only striking finding in the physical examination was the presence of hepatomegaly in almost three-fourths of the group. However, minimal enlargement was the rule and few abnormalities of liver function were found. In fact, the paucity of laboratory findings was remarkable when considered in light of the historical complaints.

The electromyographic findings were not clinically useful and did not correlate with physical findings. The most common abnormality was the presence of isolated fasciculations in three patients. The nerve conduction studies revealed normal motor latency and velocity but abnormal delays in the sensory latencies. These sensory findings are often the most sensitive indications of early neuropathy, and it is not uncommon to have selective involvement of only sensory fibers similar to what is seen in diabetic neuropathy. The etiology of this is unclear, but is is possible that PBBs, with their high affinity for adipose tissue, could select our peripheral nerve tissue with its high concentration of phosphatides and sphingomyelin.

The extensive psychologic work-ups failed to identify any consistent findings aside from the presence of reactive depression. It is important to stress that despite the multiple complaints, objective evidence of organic changes in the central nervous system could not be found. Since no major changes could be identified, it is unlikely that they would be found in less exposed groups such as the public at large in the state of Michigan. The depression noted was usually better now than it was two years ago and psychopharmacologic agents have not been used in the management of these people.

We concluded that in this group of patients with known exposure to PBBs and incapacitating health care complaints, there was a high prevalence of hepatomegaly (72%), sensory neuropathies (41%), and reactive depression (67%). We found no evidence of changes suggestive of organic brain syndromes and based upon historical information presented, the depressive reactions have improved over time.

These conclusions suggest that the people of the state of Michigan with little exposure to PBBs are not at great risk of suffering any adverse health effects from this exposure.

References

1. Jackson, T. F. & F. L. Halbert. 1974. J. Am. Vet. Med. Assoc. **165:** 437.
2. Atsuko, Y. 1971. Investigation concerning babies born from women who consumed oil contaminated with chlorobiphenyl. Fukuoka Igakkai Shi Fukuoka Acta Medica **62:** 117–122.
3. Masuada, Y. 1974. Comparison of polychlorinated biphenyls in Yusho patients and ordinary persons. B. Environmental Contamination and Toxicology. Vol. II: 213–216.
4. Michigan Department of Public Health. 1975. The short term effects of PBB on health.
5. Michigan Department of Public Health. 1978. An epidemiological study of human health effects associated with the consumption of milk, meat or other foods contaminated with polybrominated biphenyl. CDC Contract # 200-76-0625, 1978.
6. Willett, L. B. and H. A. Irving. 1975. Distribution and clearance of polybrominated biphenyls by cows. J Dairy Sci. **58:** 764.
7. Valciukas, J. A., R. Lileis, M. S. Wolff, *et al.* 1978. Comparative neurobehavioral study of a PBB-exposed population in Michigan and a non-exposed group in Wisconsin. Interim Summary Report to the National Institute of Environmental Health Sciences ES 00928.

NEUROPATHOLOGY OF "SPINNING SYNDROME" INDUCED BY PRENATAL INTOXICATION WITH A PCB IN MICE*

S. M. Chou, T. Miike,† W. M. Payne, and G. J. Davis‡

Neuropathology Laboratory
West Virginia University Medical Center
Morgantown, West Virginia 26506

‡Raltech Scientific Services
Madison, Wisconsin 53707

Introduction

Polychlorinated biphenyls (PCBs) are industrial pollutants widely distributed in the environment. In 1968, a human PCB intoxication of epidemic proportion occurred in western Japan.[1] The intoxication and the resultant malady, Yusho (oil disease), were found to be caused by ingestion of rice oil that had been accidentally contaminated with a commercial brand of polychlorinated biphenyl (Kanechlor 400) composed primarily of tetrachlorobiphenyl. The latter was used as a heat exchanger in a heating pipe that leaked during the process of removing the odorous matter from the rice oil. The principal organs and tissues affected by this PCB intoxication include: skin; nervous system; eyes; liver; and endocrine organs.[2] Neurologic symptoms comprised headaches, numbness, hypoesthesia, and neuralgia; a peripheral neuropathy selective for sensory nerves was detected by studies on nerve conduction velocity in Yusho patients.[3] Specific pathologic processes causing these neurologic symptoms have not yet been elucidated. Transplacental PCB intoxication was described in humans with several infants being born with Yusho syndrome and some of these to unaffected mothers;[1] this has been supported by animal experiments.[4,5] If transplacental PCB intoxication is possible, damage to the developing CNS must be suspected. In the present study our attention was directed towards the effects of PCB intoxication on the developing CNS. A permanent motor disturbance, "spinning syndrome," developed with a high frequency in weaning mice whose dams received oral 3,4,3′,4′-tetrachlorobiphenyl (4-CB) during gestation (days 10–16).[6] The purpose of this report is to describe neuropathologic findings in "spinners" and to correlate them with the syndrome. The study advocates that prenatal PCB intoxication may potentially induce analogous damage in the developing human CNS and proposes that the resultant dysfunction may be comparable to "minimal cerebral dysfunction" in children, especially the hyperactivity syndrome.

Materials and Methods

From over 100 spinners developed from 60 CD-1 (The Charles River Breeding Lab Inc., Wilmington, Mass) litters, 24 spinners of varied ages (4 to 34 weeks) provide the material for this study. Pregnant CD-1 mice were dosed orally on days 10 through

*Supported in part by Contract NIEHS, No. 1-ES6-2111.

†Holder of Post-doctoral Fellowship from the Muscular Dystrophy Association of America, Inc.

0077–8923/79/0320–0373 $01.75/0 © 1979, NYAS

16 of gestation with 32 mg/kg/day of 3,4,3′,4′-tetrachlorobiphenyl (4-CB). The 4-CB compound (tested by GCmass-spectrometry and supplied by the National Institute of Environmental Health Sciences, Research Triangle Park, N.C.) is 97.7% pure, containing trace amounts of other PCBs but none of the toxic impurities such as dibenzofurans, dibenzodioxin, or other compounds which are common contaminants in commercial PCB products. The chemical structure of 4-CB is as follows:

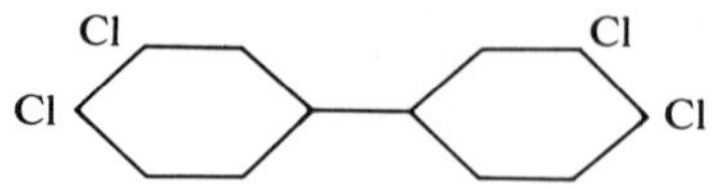

3,4,3′,4′-tetrachlorobiphenyl (4-CB)

The 4-CB-dosed pregnant mice were caged individually with food (Purina Lab Chow) and tap water *ad libitum*. Twenty-four spinners from 60 litters (aged 4, 5, 6, 8, 9, 10, 11, 15, 28, 30, 32, 34 weeks) along with 12 age-matched controls and 12 litter mates of the spinners who did not develop the spinning syndrome (nonspinners) were subjected to pathologic studies, including light and electron microscopy and histofluorescent study for catecholamines. For light and electron microscopy anesthetized animals were killed by intracardiac perfusion of a combined formaldehyde-glutaraldehyde fixative in 2 different concentrations according to the two step perfusion method described by Peters.[7] Cryostat sections of unperfused brain, spinal cord, and muscle from 4 spinners and 2 control mice were used for studying the CNS distribution of biogenic amines according to the recently modified glyoxlic acid method,[8] and for histochemistry of skeletal muscle.

After performing light microscopy on serial sections from representative levels of the CNS, morphologic studies were concentrated in the following areas: spinal cord, spinal nerve roots, substantia nigra, and nucleus accumbens. Selected tissue blocks from these areas were osmicated after rinsing with cacodylate buffer, dehydrated in a graded series of ethanol, and embedded in Epon-Araldite resin. One-micron sections, cut with an LKB Ultratome with glass knives and stained with Paragon, were screened with the light microscope. Ultra-thin sections were double stained with uranyl acetate and lead citrate and examined in an AEI (Corinth 275) microscope.

For morphometric analysis of nerve fibers, spinal root specimens (L1, L2, & L3) from controls, 4-CB nonspinners, and 4-CB spinners were post-fixed in buffered osmium tetroxide for 2 hours and embedded in Epon-Araldite mixture. Transverse and longitudinal sections of the proximal segments of the roots were stained with Paragon and photographed at a final magnification of ×1000. The number of myelinated fibers, both CNS- and PNS-type, in the spinal roots were counted and tabulated according to the diameters for plotting histograms. The ratios (d/D) of the axon diameter (d) to fiber diameter (D) of the CNS-type and PNS-type fibers were obtained from approximately 200 myelinated fibers in the spinal roots. The periodicity of the myelin lamellae was measured and compared for both the CNS-type and PNS-type fibers from the same section at an identical magnification in at least 50 measurements from each given spinal root.

The growth rates of the spinners, the nonspinning litter mates, and age-matched control mice were compared by weighing each mouse twice weekly. The rate and direction of the spinning behavior were recorded by a 4-channel rotation counter with 3 interrupters to prevent oscillation of the pulse and to monitor the direction of rotation. The pulse generated from the interruptors was recorded with a Grass Model 5 Polygraph recorder. The mouse was fastened with gauze connected to a steel wire which in turn was held securely in a roller bearing as previously described.[9] The mouse was placed in a cylindrical Tupperware® container, 22 cm in diameter and 15 cm in depth. After the baseline recording was taken from each of 10 spinners, 2 nonspinning litter mates, and 2 age-matched control mice, the mice were tested to see if the spinning behavior would be modified by administration of dopaminergic agonists and antagonists. Two dopaminergic agonists were used: apomorphine HCl (Merck) and *d*-amphetamine sulfate (Smith-Kline-French) the one antagonist used was haloperidol (Haldol, McNeil). The dosage range for apomorphine was 1–5 mg/kg and that for d-amphetamine was 2–14 mg/kg. All drugs were mixed with saline (0.9%) and then injected i.p. using a tuberculin syringe with a 27 g needle. The rather unstable apomorphine solution was kept refrigerated and in the dark, and a fresh solution was prepared every 4 hours. Haloperidol was dissolved in a small amount of propylene glycol to which HCl was added; this solution was then dissolved in saline (0.9%). pH was adjusted to 7.1 by adding NaOH, and the solution was then injected i.p. concentrations ranging from 0.1 mg/kg to 2 mg/kg.

RESULTS

Growth Rate

As previously described,[6,10] the average size of litters from 4-CB intoxicated dams was smaller (about 60% of controls), and the neonatal mortality higher (40% at day 20 of life). The average body weight of the spinners was significantly lower (FIGURE 1) (71% to 81%, average 75%) of age-matched control mice; however, the litter mates of the spinners were often slightly heavier than the controls. This latter finding was thought attributable to the smaller litter size. The most common macroscopic anomaly associated with 4-CB spinners was bilateral hydronephrosis,[10] which was also observed in the litter mates free of motor dysfunction.

Spinning Syndrome

The spinning syndrome, characterized by repetitive jerking of the head, hyperkinesia, and swift stereotyped circular movement, developed in about 55% of members of the 60 4-CB litters. No evidence of maternal neurotoxicity developed with the 7 doses of 4-CB given during gestation. The spinning syndrome became manifest approximately at weaning (at 3 weeks of age), and the spinners survived well into adult life (over 9 months). The "spinning" was arbitrarily defined as spontaneous, sustained turning in one direction at a minimum rate of 40 turns/minute for a period of at least 3 minutes with only brief interruptions. The rate and direction of spinning varied considerably among the spinners, however, the majority displayed unidirectional spinning, occasionally up to 150 turns per minute. With the rotation counter used in this study, about 1 to 2 rotations to either direction per minute was randomly registered by control mice during the 20 minute test period; no significant difference in this behavior was observed in the nonspinning litter mates. The spinning movement could be best described as "chasing one's own tail" since spinning was in a small circle. It often occurred spontaneously or was elicited by slight agitation or noise. The average rate of spinning was 40 to 80 turns per minute and there was no tendency toward change in the speed of spinning or attenuation of the syndrome as the spinners grew older. When the spinners were not spinning, they appeared restless and moved constantly and aimlessly in the cage, with shaking of the head.

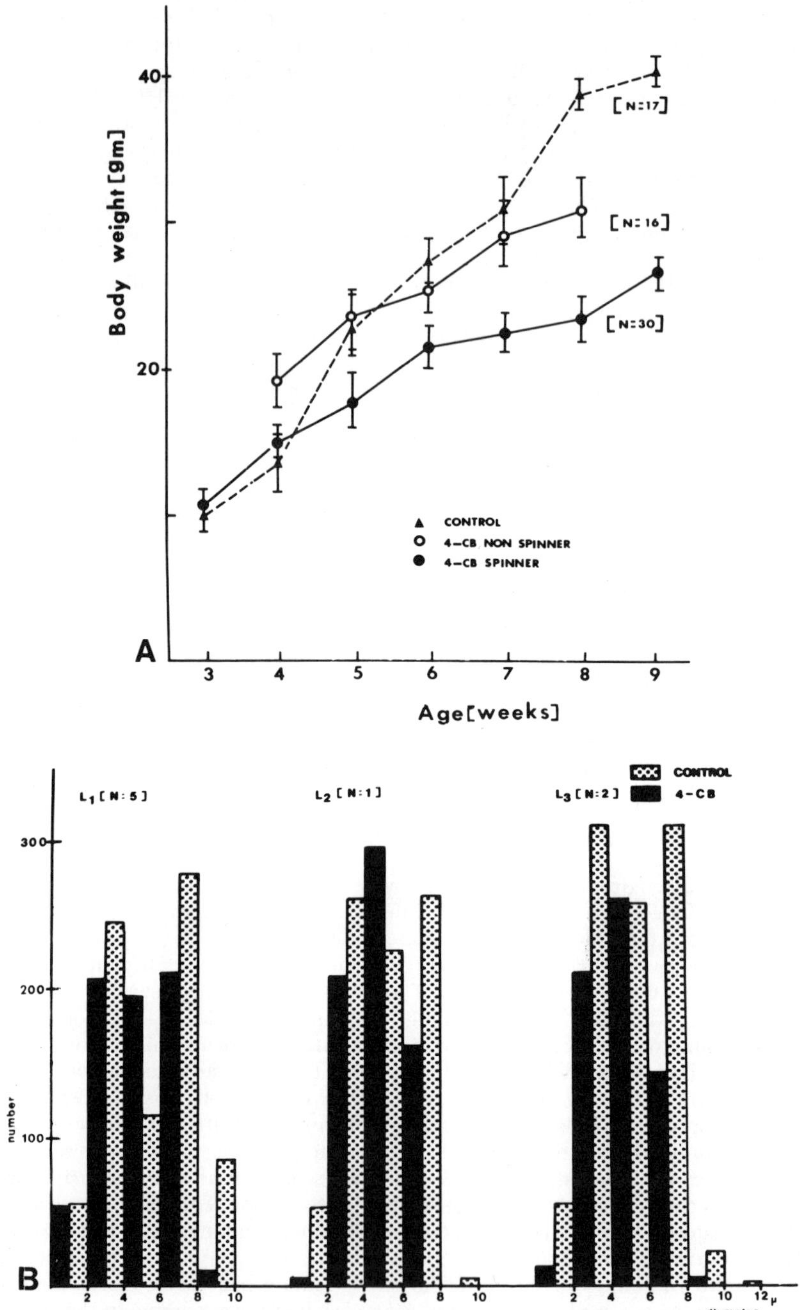

FIGURE 1. (A) Growth rate curve of 4-CB spinners, 4-CB nonspinners, and age-matched controls. (B) Histograms of myelinated fibers in the ventral spinal roots (L1, L2, & L3) of 4-CB spinners and age-matched controls.

Effects of Dopamine Agonists and an Antagonist on Spinning

The controls and nonspinners responded to dopamine agonists and the antagonist as an homogenous group and in contrast to spinners. Of note is that spinning, as defined previously, was never observed in either controls or nonspinners after any dose of amphetamine, apomorphine, or haloperidol, while these drugs had profound effects on the spinners, as summarized in TABLE 1.

1. *d-Amphetamine* induced a biphasic dose-dependent effect on the rate of spinning in the spinners: doses of 2–6 mg/kg either had no effect on the rate or increased it, whereas doses of 6–14 mg/kg decreased the rate of spinning. In controls and 4-CB nonspinners, activity, in general, in the absence of spinning, increased steadily with increasing doses in the 2–10 mg/kg range. Amphetamine also reversed

TABLE 1

EFFECTS OF DOPAMINE AGONISTS AND ANTAGONISTS ON SPINNING

	4-CB Spinners		4-CB Nonspinners	Controls
Dose of	Rate of Spinning	Direction of Spinning	Activity	Activity
Amphetamine	(action at 30–120 min)			
2–6 mg/kg	sl ↑	6 reversed	↑	↑
6–14 mg/kg*	↓	5 (-)	↑	↑
Apomorphine	(action at 20–40 min)			
0.025 mg/kg	↑	(-)	↓	↓
0.05 mg/kg	↑	(-)	↓	↓
0.1 mg/kg	↑	(-)	(-)	(-)
1.0 mg/kg	↑	(-)	(-)	(-)
5.0 mg/kg†	↓	(-)	sl↑	sl↑
Haloperidol	(action at 30–150 min)			
0.1 mg/kg	(-)	(-)	(-)	(-)
0.5 mg/kg	↑ (30–60 min)	(-)	very sl↓	very sl↓
	stopped (90–150 min)			
1.0 mg/kg	stopped (90–150 min)	(-)	sl↓	sl↓

(−): unaltered; ↑: increased; ↓: decreased.

*Moderate stereotypy at Amphetamine 14 mg/kg.

†Very slight stereotypy at Apomorphine 5 mg/kg.

the direction of spinning in 6 of 11 unidirectional spinners. The effects of *d*-amphetamine were maximal from 30 to 120 minutes after administration.

2. *Apomorphine* showed a similar biphasic action on spinners: all spinners showed no change or an increase in the baseline rate of spinning at doses of 0.025–1 mg/kg, whereas a dose of 5 mg/kg resulted in a decrease in the rate in all spinners tested. In nonspinners and controls doses of 0.025 and 0.050 mg/kg decreased (nonspinning) activity, in general, while doses of 0.1–5 mg/kg were associated with a mild and dose-dependent increase in such activity. Apomorphine effects were most prominent 15 to 40 minutes after injection.

3. *Haloperidol* in doses of 0.1 mg/kg appeared to have little effect on any of the 3 groups, while doses of 0.5 mg/kg and 1.0 mg/kg resulted in very slight and moderate depression of overall activity, respectively. In spinners, however, spinning was virtually halted during the period from 90 to 150 minutes after either the 0.5 or 1.0 mg/kg doses. Interestingly, haloperidol at these higher doses caused bursts of rapid (2–4/sec)

spinning 20 to 60 minutes after dosage. The details and interpretation of these drug actions on the spinners will be the subject of a separate publication.

Light Microscopy

With special stains routinely applied for neuropathology, no detectable neuronal changes, gliotic foci, or demyelinative lesions were detected in the CNS of either 4-CB spinners or 4-CB nonspinners. Occasionally, however, misplaced or heterotopic motor neurons and astroglia were seen within the ventral white matter (FIGURE 2B) of the spinal cord along the emerging intramedullary ventral spinal fibers. Most unusual was the occurrence of the cylindrical CNS peninsulas (CCPs) (FIGURE 2C) which projected into the cranial and spinal nerve roots of all the 4-CB spinners and some of the 4-CB nonspinners. None of the control mice showed the CCPs. The CCPs were characterized by a cylindrical column of nerve fibers, myelinated or nonmyelinated, and/or glial bundles (FIGURE 2D) in varying proportions, which were well-circumscribed and delineated from the neighboring PNS-type nerve fibers, which were ensheathed by thicker myelin. CCPs composed purely of either myelinated fibers or of glial bundles were commonly encountered. The diameter and length of CCPs varied considerably: many measured up to 100 μ in diameter and reached up to 5 mm in length in ventral lumbar roots and then tapered off. They were seen predominantly in the ventral roots at all levels, and in a few cranial nerve roots, especially of the hypoglossal nerve. They were also seen, to a lesser extent, in the dorsal spinal roots (FIGURE 3B). The CCPs were not necessarily seen in each and every spinal root, but were most abundantly noted at lumbar levels. Large, apparently displaced, motor neurons were often detected within CCPs (FIGURE 2C), and also at the exit of the emerging spinal nerve roots. Vesicular, elongated astrocytic nuclei with their longitudinal axes parallel to the direction of axons could be found in the glial CCPs (FIGURE 3C). Signs of degeneration were increasingly noted with age in the myelinated fibers of CCPs, and dense, small myelin ovoids could be detected, especially in longitudinal sections of CCPs at higher magnification (FIGURE 4 A & B). As the age of the animal increased, the CCPs with glial bundles increased in number while the number of CNS-type myelinated fibers within the CCPs tended to decrease.

Comparison of the number of myelinated fibers (including both CNS and PNS-type fibers in the CCPs) in the ventral spinal roots at L1, L2, and L3 levels between the 4-CB spinners and controls revealed a decrease of up to 33% (average 20%) in the spinners. The total number of myelinated fibers varied from 686 to 1179 in control ventral spinal roots at L1, for example, whereas in 4-CB spinners it varied from 633 to 847. Their histograms (FIGURE 1B) displayed an overall tendency to lose the normal bimodal distribution of myelinated fibers and a shift of frequency distribution to small fibers. There was no apparent difference in frequency distribution of small myelinated fibers (below 6 μ in diameter), but there was a marked reduction in the number of thick fibers (above 6 μ in diameter) of up to 57% in 4-CB spinners. No appreciable difference was noticed on monoamine histofluorescence in nucleus accumbens or substantia nigra between 4-CB spinners and age-matched controls.

Comparison of 4 muscles (soleus, extensor digitorum longus, gracilis, and sartorius) from 4-CB spinners and controls consistently revealed smaller bulk of each muscle in the spinners. Histochemistry of muscles showed the white muscles (the extensor digitorum longus and sartorius) to be more affected than the red muscles (the soleus and gracilis). The myofibers in the white muscles from the spinners were generally rounded and demonstrated marked variation in fiber size. No apparent type

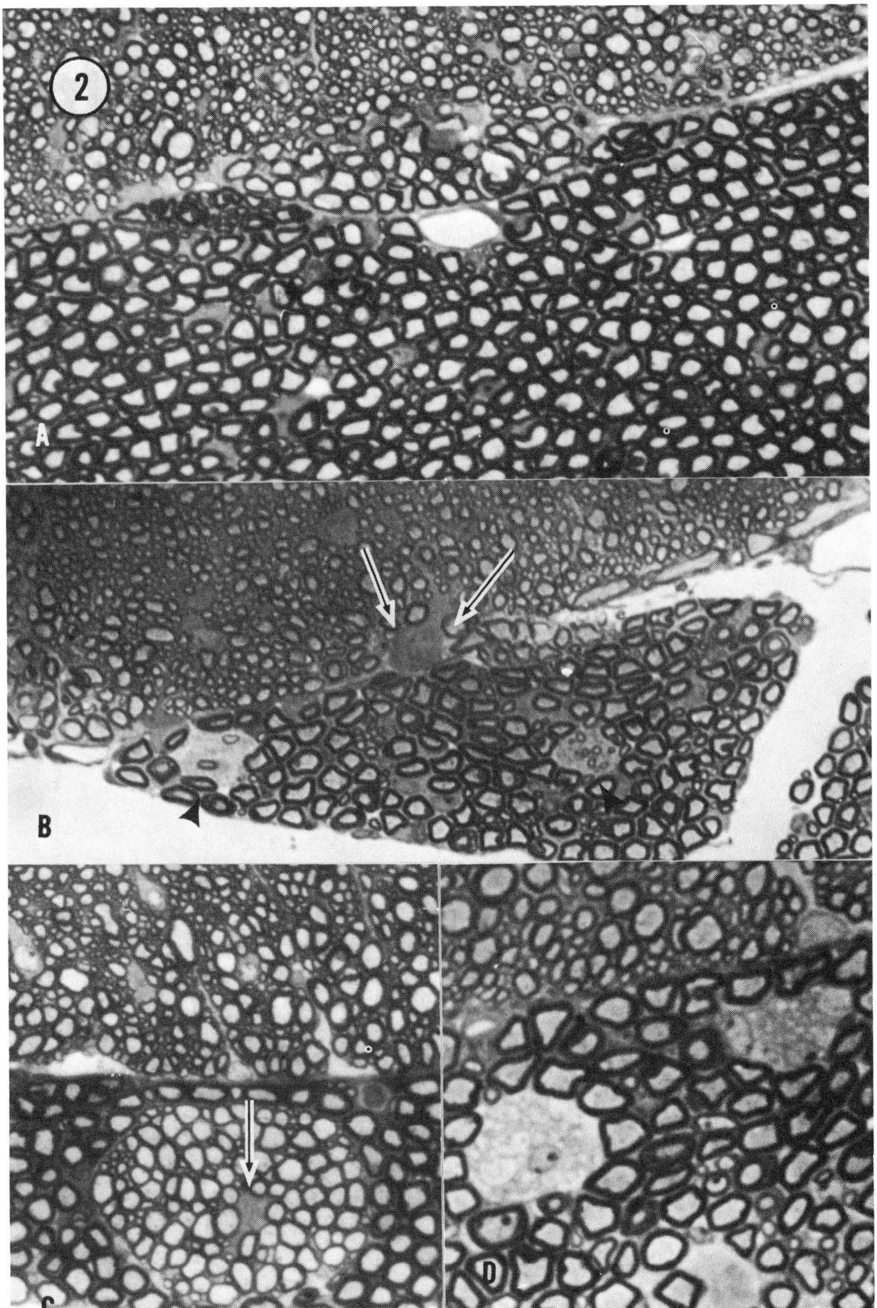

FIGURE 2. Ventral spinal root (L1) of: (A), a 10-week-old control mouse, mag. × 510; (B), a 10-week-old spinner, with 2 CCPs (arrow heads) and a heterotopic motoneuron at the exit zone (arrows), mag. × 510; (C), a 5-week-old spinner with a CCP with CNS-type myelinated axons and a heterotopic neuron (arrow), mag. × 500; and (D), a 24-week-old spinner with 2 glial CCPs, mag. × 790.

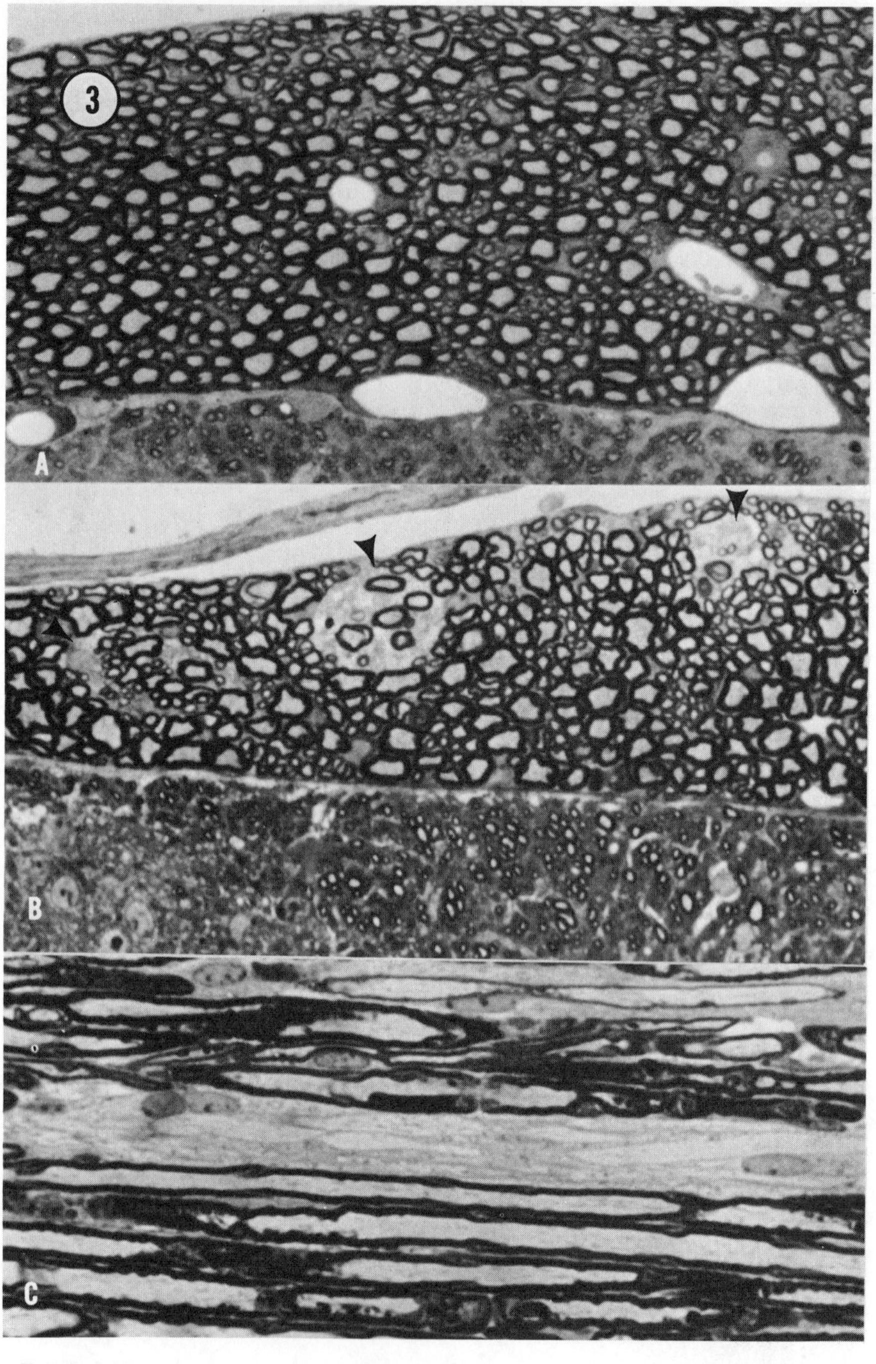

FIGURE 3. Dorsal spinal root (L1) of (A) an 8-week-old control, mag. × 560 and (B) of an 10-week-old spinner, with 3 CCPs (arrow heads), mag. × 560. (C) Longitudinally cut glial CCP with 3 large elongated astrocytic nuclei in the ventral root in a 30-week-old spinner, mag. × 640.

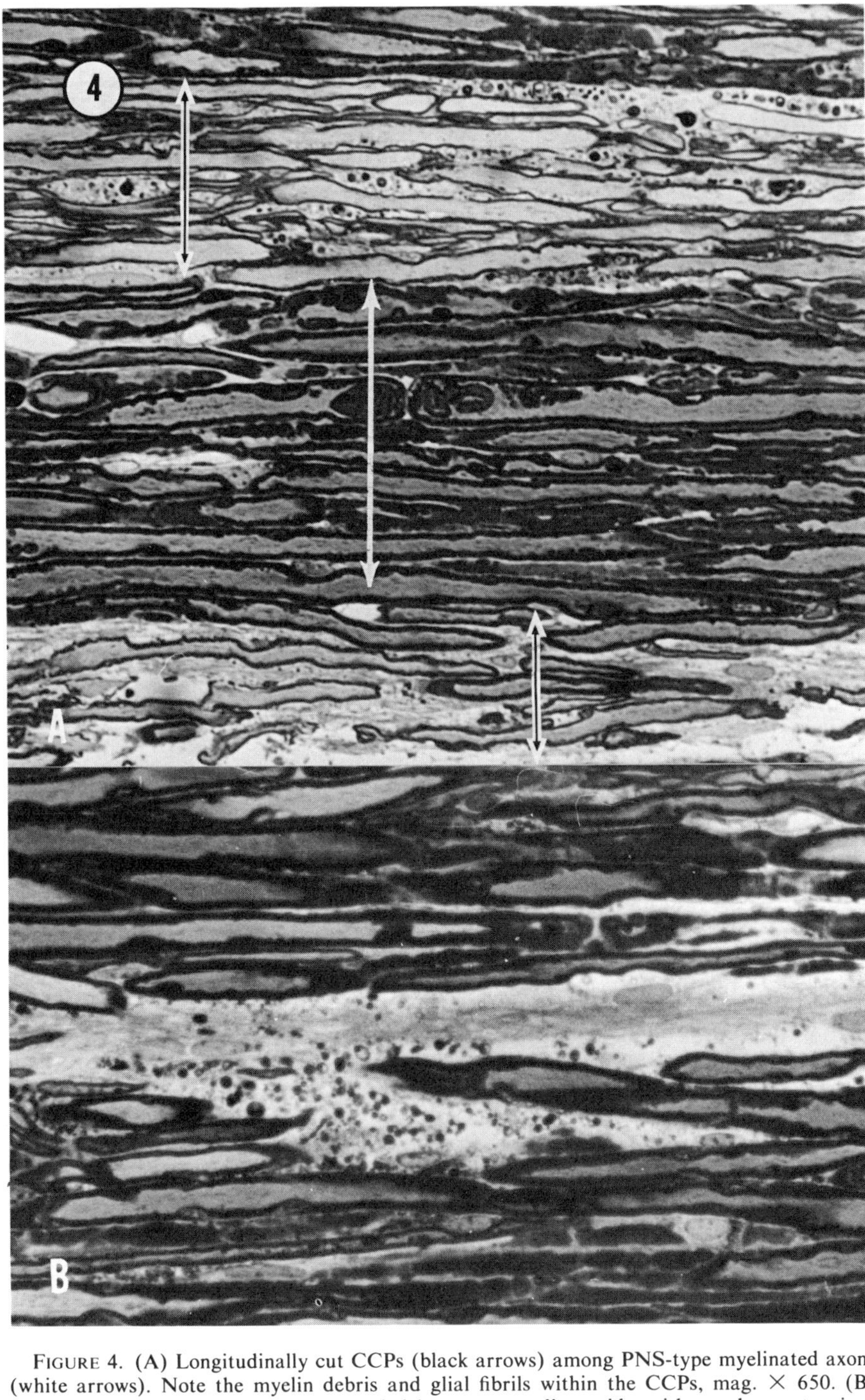

FIGURE 4. (A) Longitudinally cut CCPs (black arrows) among PNS-type myelinated axons (white arrows). Note the myelin debris and glial fibrils within the CCPs, mag. × 650. (B) Longitudinally cut glial CCP accompanied by many myelin ovoids, with no phagocytes, in a 34-week old spinner, mag. × 720.

fiber disproportion was noticed; however, there was an increased occurrence of the normally rare IIC fibers, marked atrophy with an increased oxidative enzyme activity of type IIB fibers (FIGURE 5B), and moderate hypertrophy of type IIA fibers (FIGURE 5D). Quantitative analysis of myofiber changes is now in progress.

Electron Microscopy

Myelinated Fibers in CCPs

The CCPs in both dorsal or ventral spinal roots were invariably enclosed by a glycocalyx layer indistinguishable from the basement membrane (FIGURE 6) and invested externally by scattered collagen fibrils; collagen fibrils were never seen within the CCPs. The myelinated fibers within the CCPs were exclusively of the CNS-type, readily distinguishable from the PNS-type fibers by: (1) the lack of an outer basement membrane (FIGURE 7); (2) the presence of a small cytoplasmic tongue of oligodendroglia instead of the encircling Schwann cytotoplasm characteristic of the PNS-type fibers (FIGURE 7); and (3) the shorter periodicity of the myelin lamellae (10.4 nm ± 0.16 nm S.D.) in comparison with that of the PNS-type myelin lamellae (11.4 nm ± 0.15 nm S.D.). Occasionally, a large unmyelinated axon was surrounded by lamellae of astroglial processes, which were admixed with myelinated CNS-type fibers (FIGURE 7). Although signs of myelin-sheath degeneration were very common in the CCPs, in contradistinction to the preserved PNS-type myelinated fibers around the CCPs, no macrophage infiltration was seen. Myelin debris or ovoids were scattered freely within the CCPs. Myelinated CCP axons which became partly demyelinated and invested by glial fibrils (FIGURE 8A) were often associated with myelin ovoids in close proximity.

Glial Bundles and Unmyelinated Axons in CCPs

The CCPs composed solely of glial bundles were seen more frequently in aged 4-CB spinners (FIGURE 9). These glial CCPs were generally smaller in diameter than the CCPs with myelinated fibers. Such a glial CCP consisted of many slender, cylindrical astroglial processes, which were often tightly packed with numerous glial filaments (75 nm in diameter), and occasionally included microtubules and mitochondria (FIGURE 8B). These glial processes, or glial fibrils, were delineated by a plasmolemma gathered together with or without punctate adhesions (FIGURE 9, inset). Astrocytic nuclei with perikaryal cytoplasm (FIGURE 9) containing abundant glial filaments were frequently observed along with many surrounding glial fibrils. Such glial fibrils were differentiated with ease from unmyelinated axons which contained neurofilaments (they are thicker than glial filaments, being 100 nm in diameter and less electron-dense than glial filaments), more mitochondria, and membrane-bound vesicles. At the distal portion of the CCPs, abundant interwoven glial fibrils, containing fewer glial filaments and more glycogen granules, formed multiple layers around unmyelinated axons resembling "onion bulb" formation described in the process of repair in demyelinated PNS fibers (FIGURE 10A). Transition from glial ensheathing of axons to Schwann cell ensheathing was not encountered.

Cellular Elements in CCPs

All the cellular elements of the CNS, i.e., neurons, astroglia, and oligodendroglia, were seen within the CCPs. Some of the heterotopic neurons appeared normal

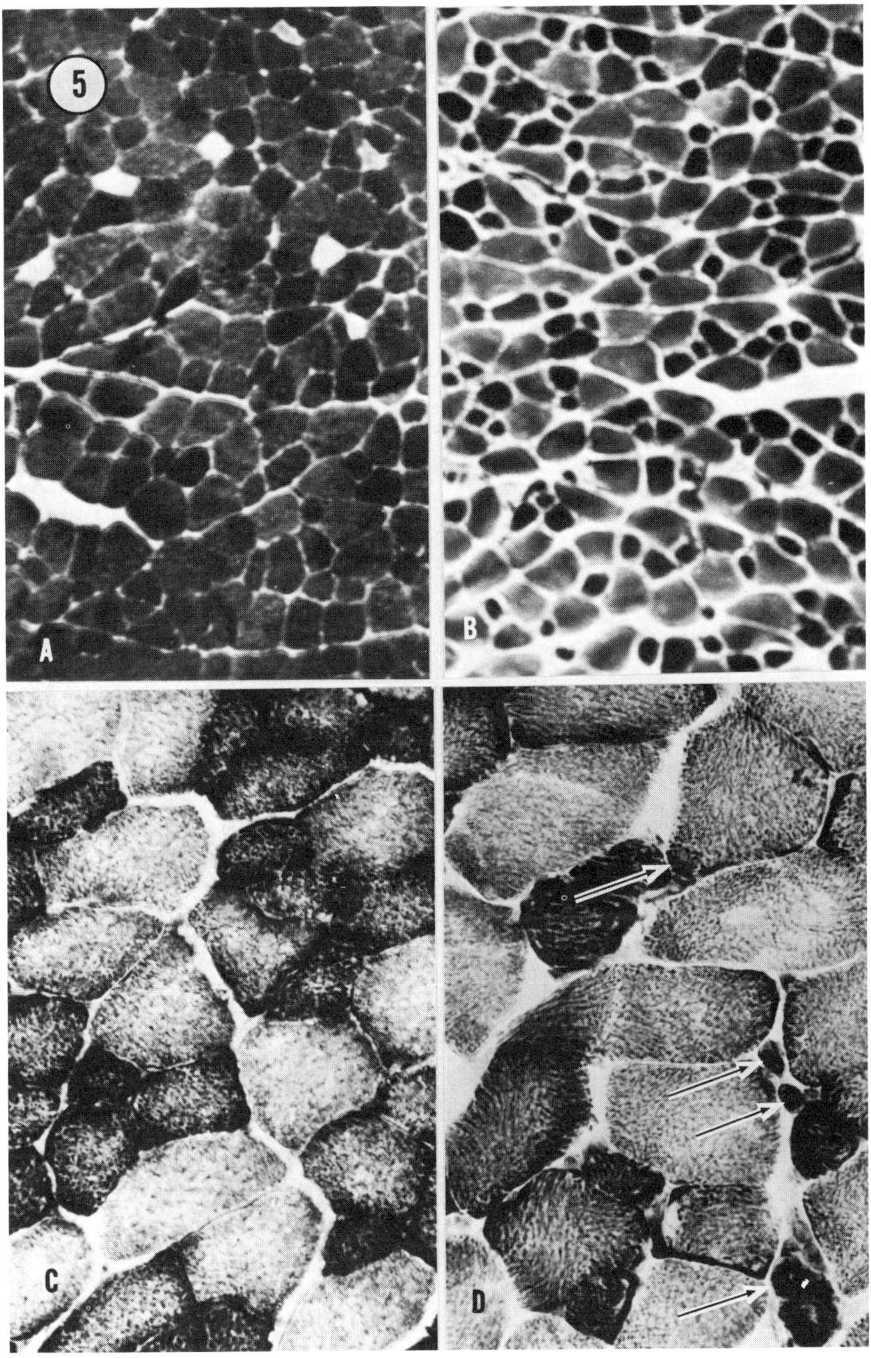

FIGURE 5. Extensor digitorum longus muscle stained for myofibrillar ATPase (pH 9.6) of control (A), and a 12-week-old spinner with atrophic IIB and IIC fibers (B), mag. × 140. The same muscles stained for NADH of control (C), and of the spinner showing atrophic fibers with increase enzyme activity (arrows) (D), mag. × 370.

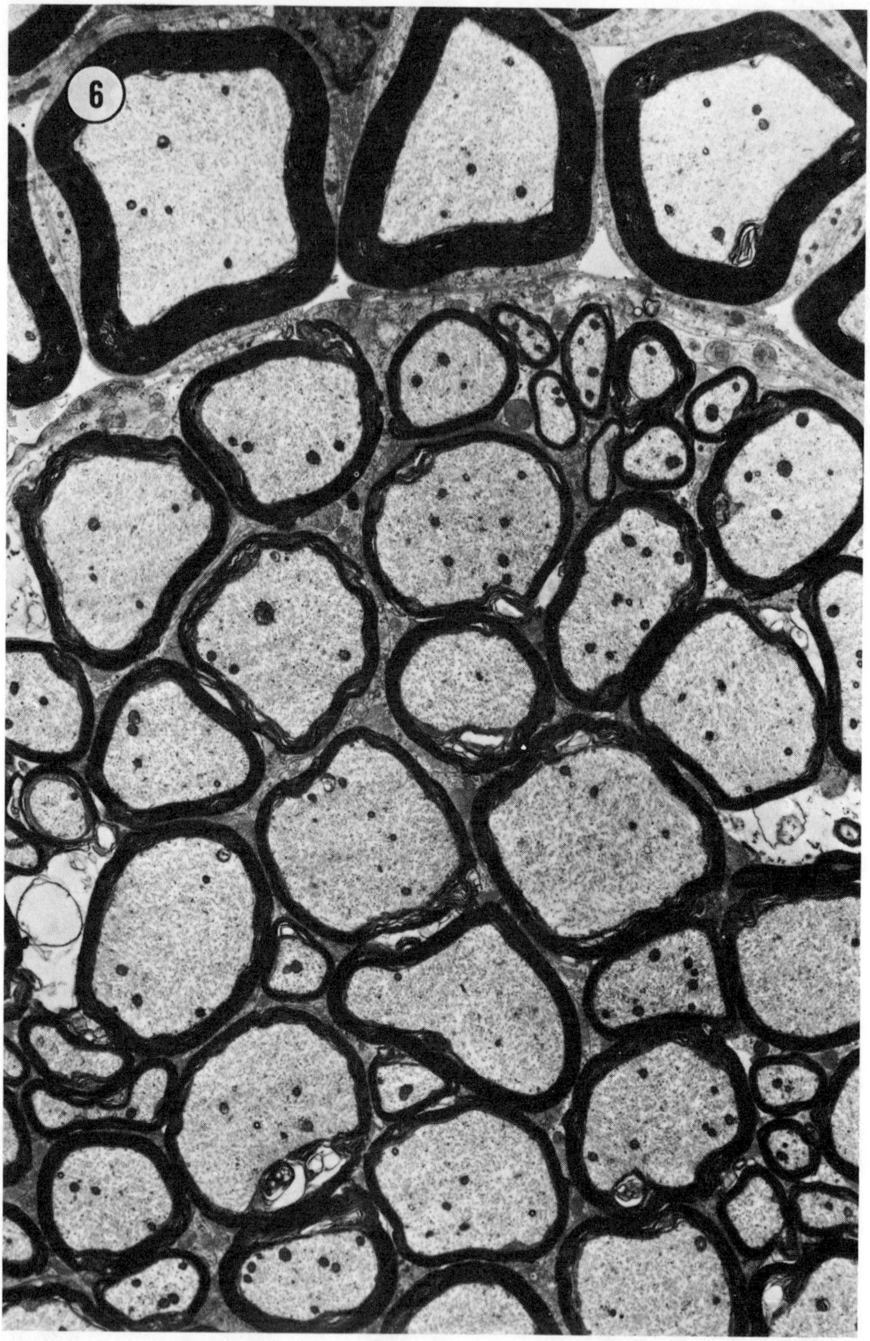

FIGURE 6. A well-circumscribed CCP with thinly myelinated CNS-type fibers enclosed with a basement membrane and surrounded by collagen fibrils and PNS-type myelinated fibers, mag. × 8250.

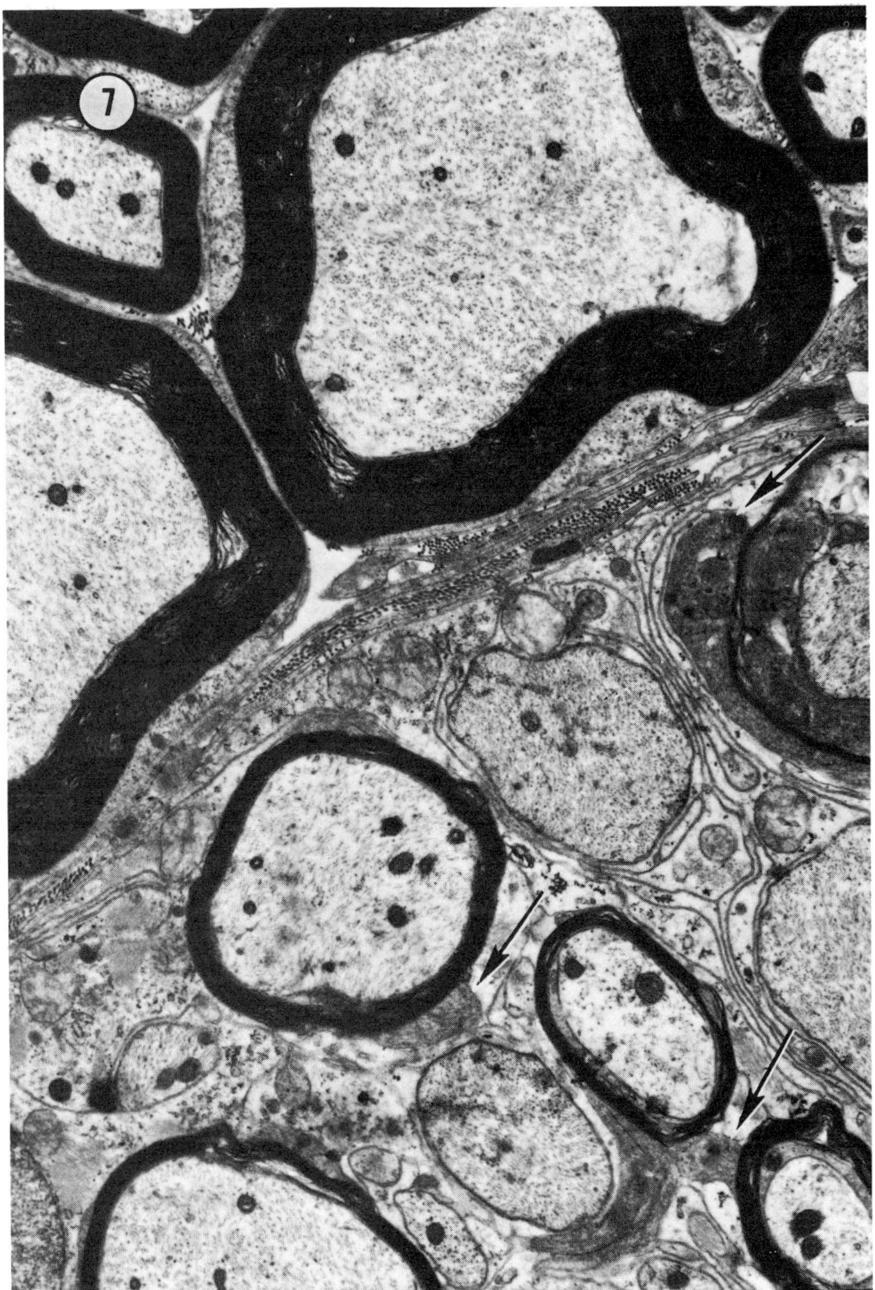

FIGURE 7. A higher magnification than in Figure 6, at the borderline between a CCP and PNS-type myelinated fibers encircled by Schwann cytoplasm, whereas the CNS-type fibers with oligodendroglial tongues (arrows) attached with the myelin sheath. Note unmyelinated axons surrounded by astroglial processes. Note also absence of collagen fibrils within the CCP, mag. × 9800.

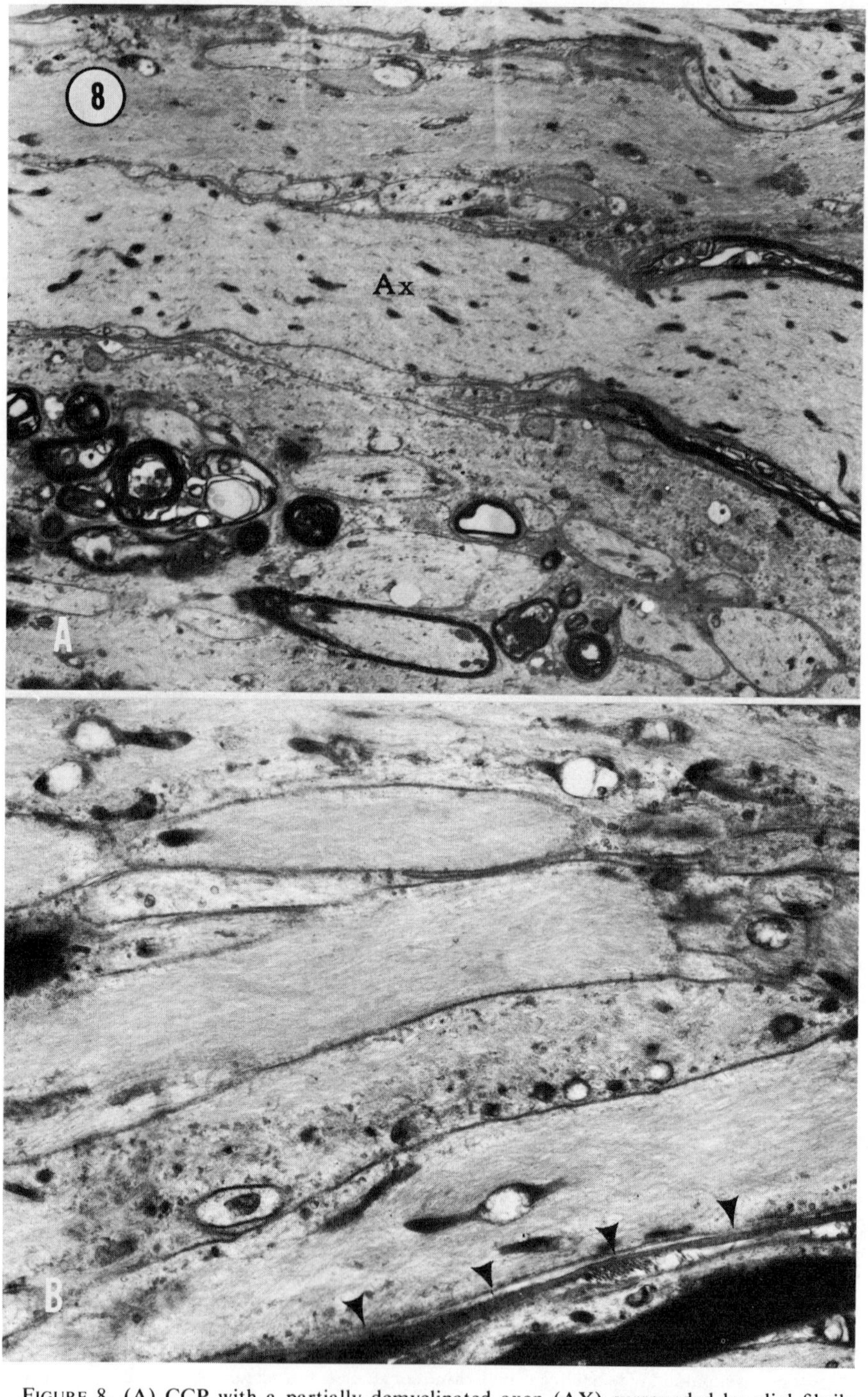

FIGURE 8. (A) CCP with a partially demyelinated axon (AX) surrounded by glial fibrils, accompanied by freely scattered myelin ovoids, mag. × 5250 (B) Longitudinally cut glial CCP, borderlined with a PNS-type, myelinated axon and with collagen fibrils (arrow heads) mag. × 10,400.

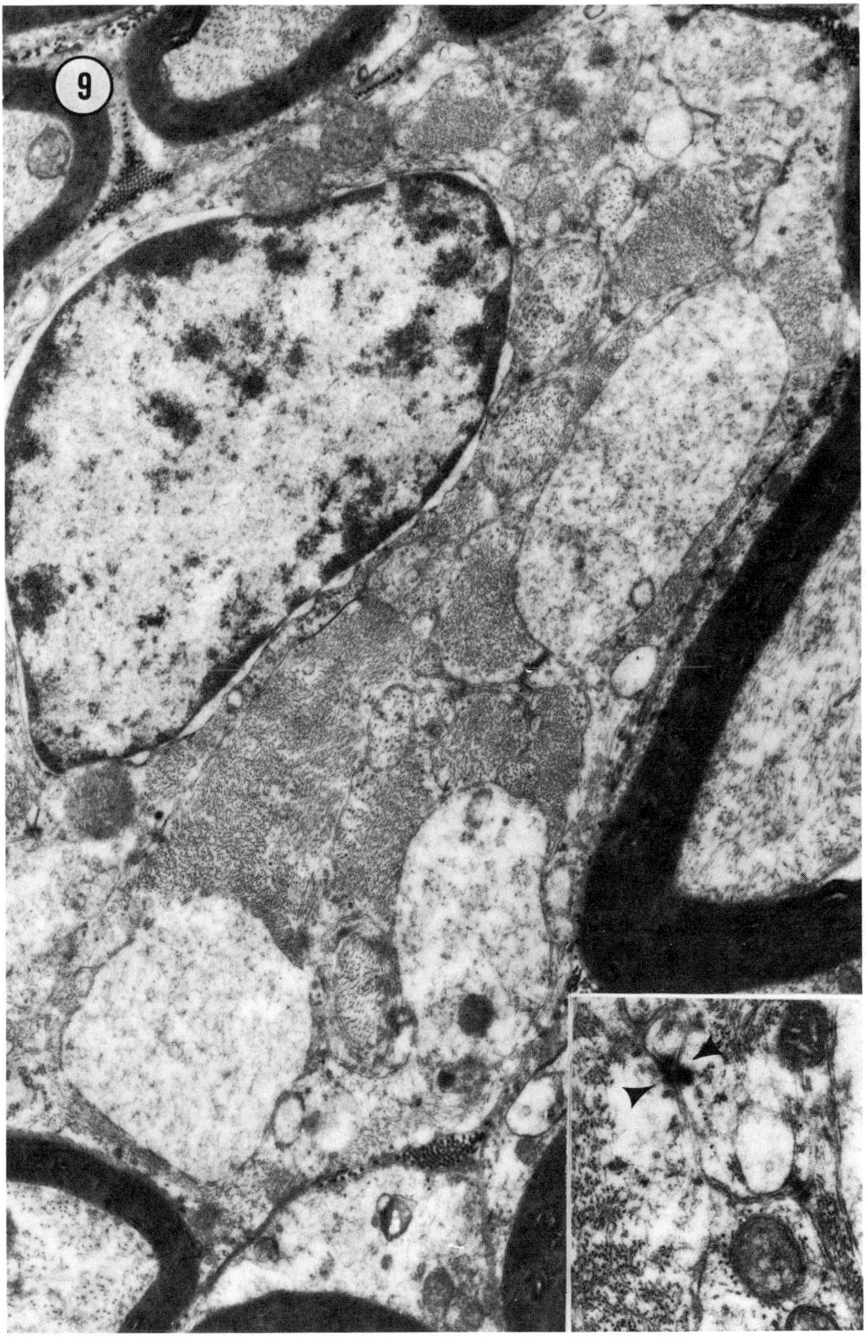

FIGURE 9. Glial CCP, enclosed by a basement membrane and surrounded by collagen fibrils, with a large astrocytic nucleus, mag. × 12,400. *Inset:* Rare punctate adhesions (arrow heads) between neighboring glial fibrils, mag. × 28,000.

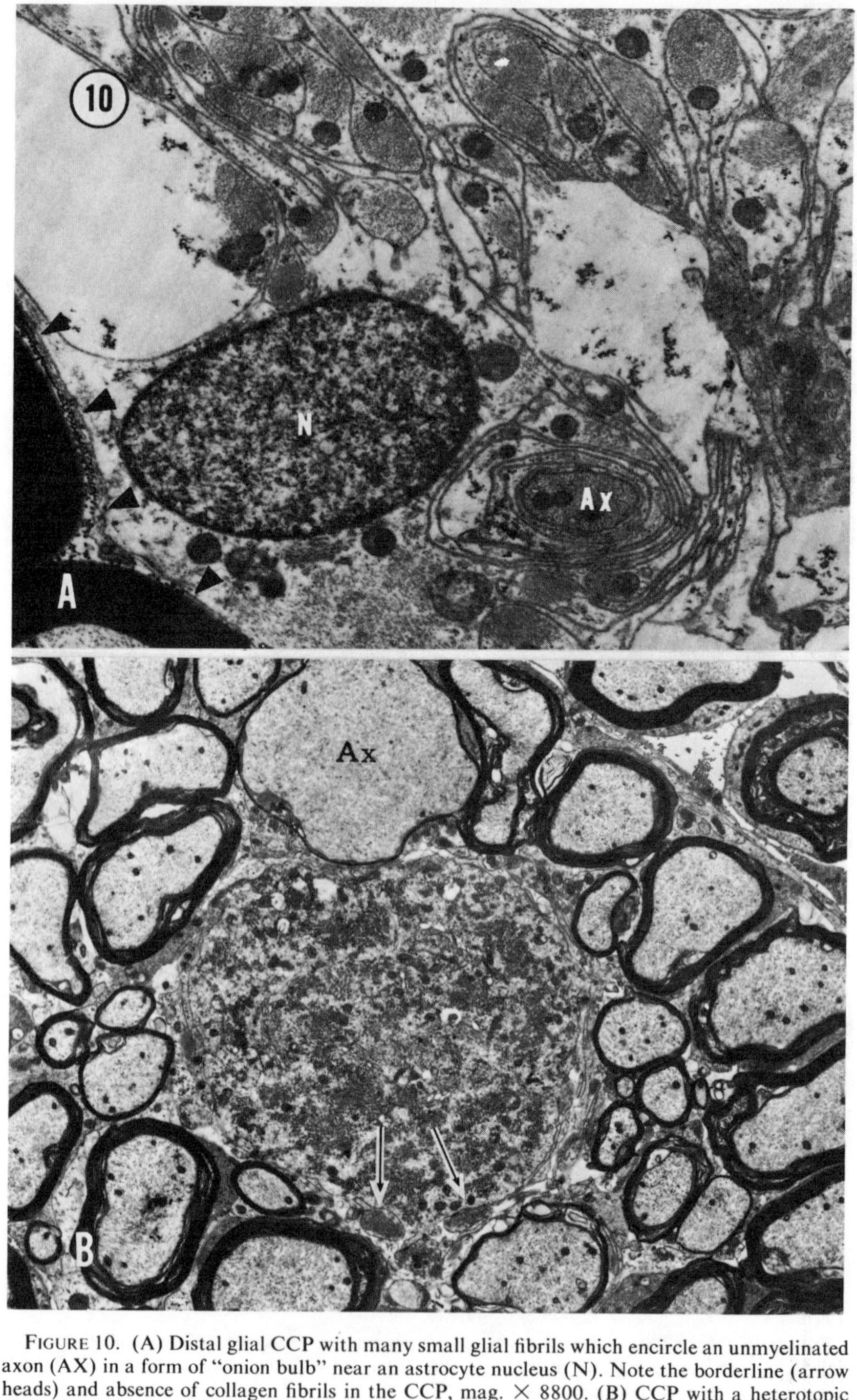

FIGURE 10. (A) Distal glial CCP with many small glial fibrils which encircle an unmyelinated axon (AX) in a form of "onion bulb" near an astrocyte nucleus (N). Note the borderline (arrow heads) and absence of collagen fibrils in the CCP, mag. × 8800. (B) CCP with a heterotopic neuron accompanied by a swollen, thinly-myelinated axon (AX). Note the terminal boutons (arrows) still attached to the neuron, mag. × 3750.

(FIGURE 10B) and well-preserved terminal boutons could be detected along the cell surfaces. Some of the CNS-type axons near such an heterotopic neuron were markedly swollen, stuffed with neurofilaments, and thinly-myelinated relative to other CNS-type myelinated fibers in the vicinity. Intraaxonal collections of dense granular and membranous bodies suggestive of axonal degeneration were occasionally seen in both intraspinal axons (FIGURE 11A) and in the CCP axons. Signs of neuronal degeneration in the CCPs were characterized by an unique intranuclear vacuolation, which was also observed in the CCP oligodendroglia (FIGURE 11B), but not in the CCP astrocytes. No cells resembling either fibrocytes of Schwann cells were encountered in the CCPs. Variation in the CCP appearance obviously resulted from the disproportion in the contained cellular elements and it was seemingly affected by the age of the spinners. Both neurons and oligodendroglia seemed to decrease in number, while the number of astroglia appeared unaltered by the age of spinners.

Neurons in Anterior Horns, Nucleus Accumbens, and Substantia Nigra

Except for frequent distal displacement in the ventral white matter, ventral horn motor neurons showed no ultrastructural changes suggestive of degenerative process in the perikaryon itself. However, along the cell surfaces of motor neurons were scattered dense, amorphous, shrunken nerve terminals. Less frequently, the perikaryal surface was surrounded by glial fibrils (FIGURE 12A). Neurons in the substantia nigra displayed little change. Neurons in the nucleus accumbens demonstrated peculiar deposits or formation of electron-dense membranous material at the synaptic junctions (FIGURE 12, B, C & D), and focal vacuolation of boutons suggestive of previous focal injuries.

DISCUSSION

The motor dysfunction induced by prenatal intoxication with 4-CB is characterized by spontaneous swift rotation, spinning and generalized hyperactivity. Since many authors have implicated the dopaminergic system in both rotatory behavior and the level of locomotor activity,[11,12] we decided to test dopaminergic agonists and antagonists for their effects on spinning. All three drugs tested (see TABLE 1) were able to modify spinning, and the ability of haloperidol to stop spinning is especially noteworthy. Indeed, with an optimal dose of haloperidol, the behavior of spinners became normal and indistinguishable from controls. These findings suggest a certain biochemical lesion in the dopaminergic system. To elucidate the pathomechanism of 4-CB-induced lesions in the dopaminergic system, biochemical and correlative ultrastructural studies on dopamine receptors and terminal axons are now underway. It must be emphasized that the 4-CB induced spinning syndrome is apparently different from the waltzing syndrome induced by IDPN ($\beta\beta'$-iminodipropionitrile) in adult mice. Prenatal neurotoxicity of IDPN in mice has not been reported, and preliminary attempts at our laboratory have been unsuccessful. Also, attempts to induce spinning in adult CD-1 mice have failed consistently (unpublished data). Unlike the IDPN-induced waltzing syndrome,[13,14] retropulsion and ataxia are absent and hyperkinesia in 4-CB spinners appears more extensive and coordinated than that of IDPN-waltzers. The induced motor dysfunction persists at a more constant level and amplitude for 4-CB spinners than IDPN waltzers. The proximal axonal swellings in spinal and brainstem motor neurons, the characteristic pathoanatomic concommitant of IDPN-induced waltzing, are not seen in the 4-CB spinners. Instead, the spinal motoneurons and their axons appear intact and there is no sign of a defect in

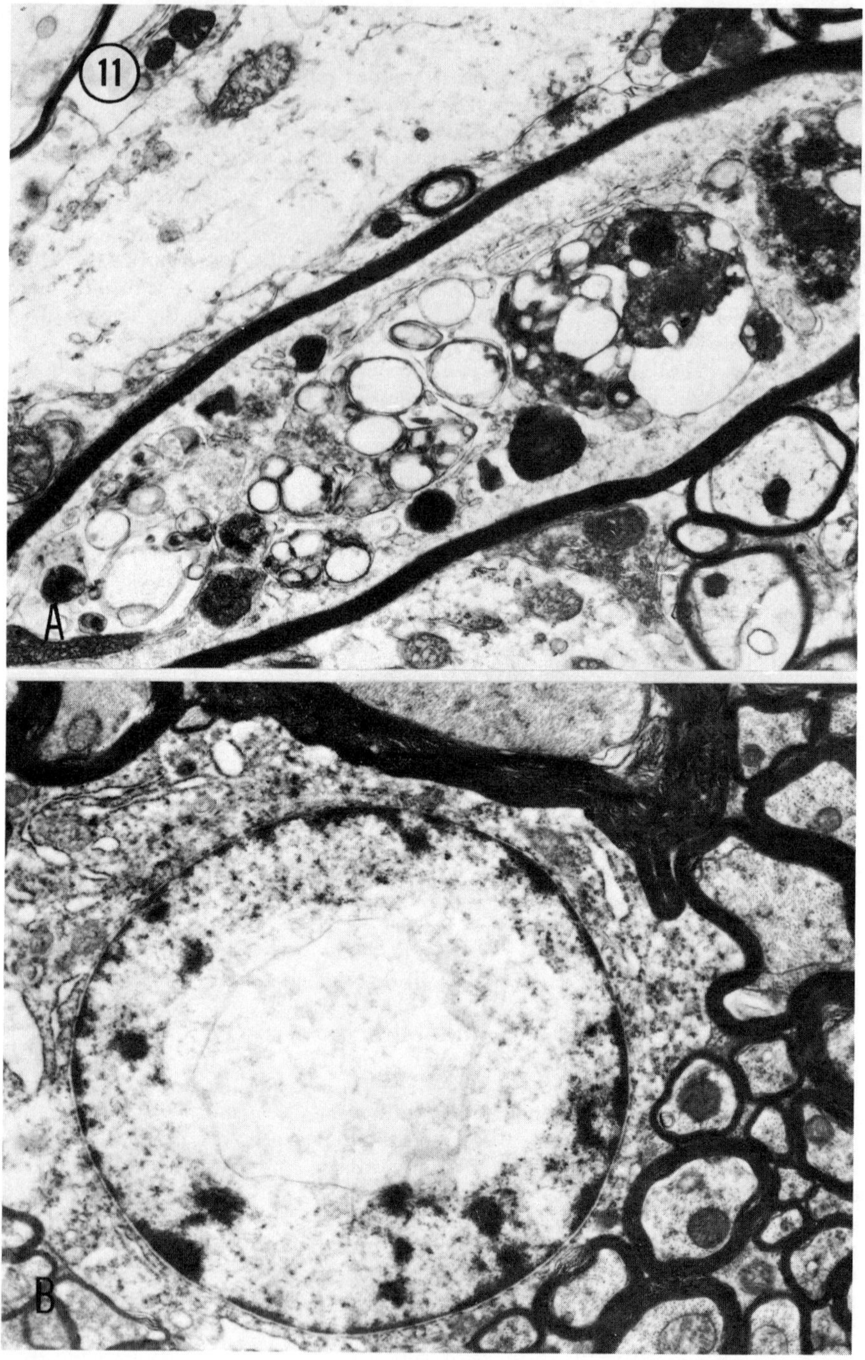

FIGURE 11. (A) Intraaxonal degenerative bodies, both dense amorphous and granule-vesicular bodies, suggestive of neuro-axonal degeneration in the anterior horn of an 8-week-old spinner, mag. × 13,200. (B) Vacuolated oligodendroglial nucleus within a CCP, an early sign of cell degeneration, mag. × 8000.

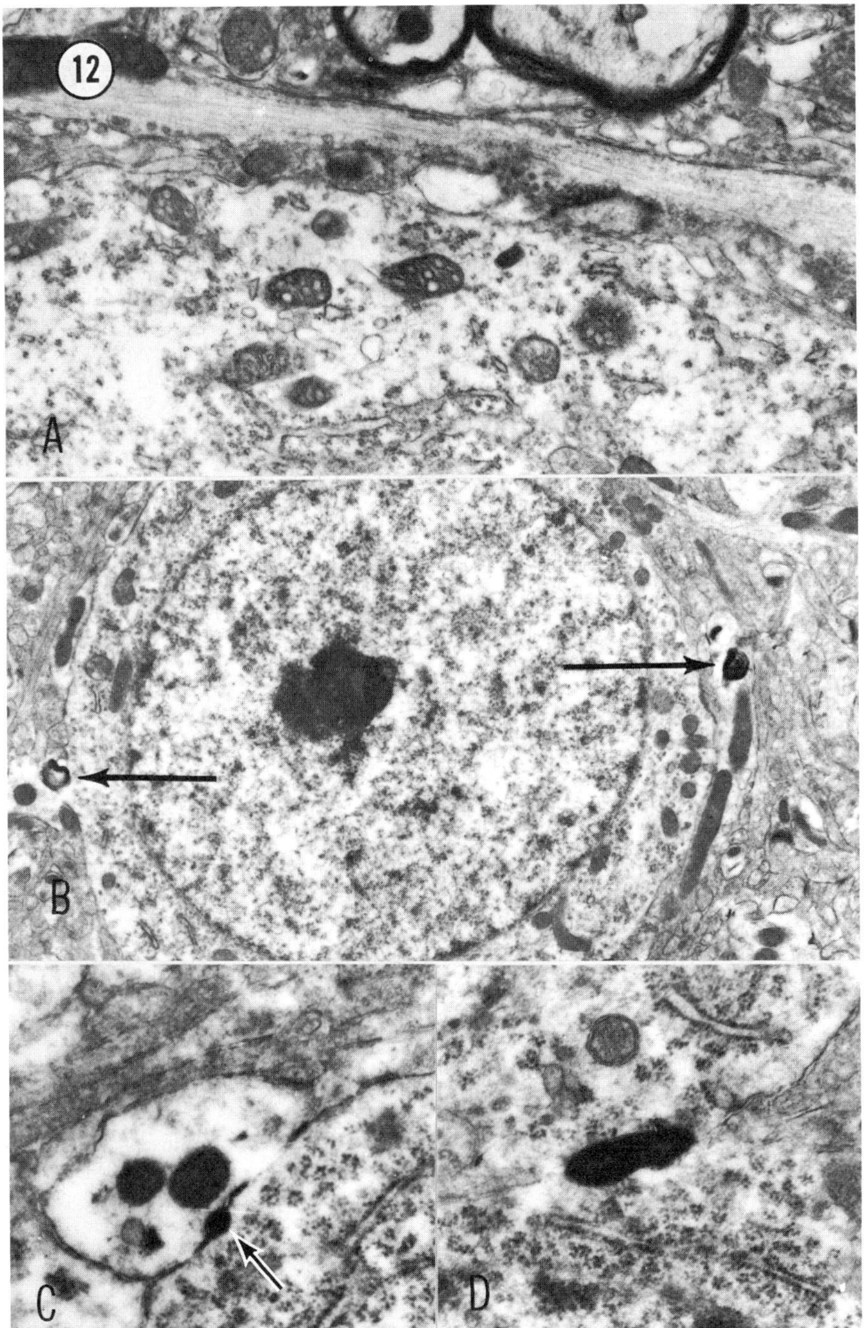

FIGURE 12. (A) Motor neuron in a spinner, partly surrounded by a glial fibril. The perikaryon itself appears intact, mag. × 15,600. (B) Neuron in the nucleus accumbens of a spinner, showing electron-dense degenerated material at the junctional zones (arrows), mag. × 8000. (C & D) Similar electron dense material at synaptic junctions of neurons in the nucleus accumbens from a spinner, mag. × 26,000.

myelination, although axons within the CCPs were occasionally ballooned (FIGURE 10B). The basic cytotoxic effects of IDPN and 4-CB may be different.

As in polybrominated biphenyls (PBBs),[15] transplacental transportation of PCBs has been demonstrated in rats,[4] rabbits,[16] cows,[17] monkeys,[5] and humans.[1] Since the mammalian fetal brain lacks a blood-brain-barrier, damage to the developing CNS by polyhalogenated biphenyls must be potentially suspected, if transplacental intoxication is possible. In fact, induction of a permanent spinning syndrome in mice strongly indicates that 4-CB is not only fetotoxic and teratogenic but also neurotoxic for the developing CNS. Neurotoxicity of polyhalogenated biphenyls to the adult CNS has not been reported. Growth retardation is one of the most constant manifestations in infants and children with PCB poisoning,[2] and this is the case also in 4-CB spinners. Only a certain class of PCBs appears to possess the unique fetal neurotoxicity.[10] Studies on tissue distribution for C^{14}-4-CB in mice revealed that only very low levels of C^{14}-4-CB were detected in newborns following transplacental exposure and radioactivity was no longer detected in 8-week-old pups. It was concluded that the spinning was induced by 4-CB action during the critical developmental period on the fetal CNS and was excreted or metabolized long before the syndrome developed. A permanent structural and biochemical change may, therefore, be induced by 4-CB during the fetal period, since the spinning syndrome persists for the life-time of the spinners long after 4-CB is excreted. The present study disclosed the unique histopathologic marker for a 4-CB-induced CNS injury during the fetal life. This pathognomonic lesion is characterized by the longitudinal projections of cylindrical CNS peninsulas (CCPs) in both ventral and dorsal roots and to a lesser extent in cranial nerve roots. The CCPs have been observed in 4 of 12 4-CB nonspinners studied, but never seen in any of 12 control mice. This finding suggests that the CCPs may serve as a histopathologic marker for prenatal 4-CB injury but may not necessarily be the histopathologic substrate for the spinning syndrome. The CCPs must have been formed as a sequela of primary CNS lesion afflicted during the fetal life.

Abnormally thin myelinated CNS-type nerve fibers in the ventral spinal roots similar to those in the CCPs have been recently described in the mutant mouse, Trembler.[18] The abnormality was considered to be secondary to defects in the Schwann cell of the PNS. No sign of defective Schwann cell or myelination in the PNS is noted in 4-CB spinners. Sensory neuropathy was documented in 10 out of 12 patients with known PCB (probably tetrachlorobiphenyl) poisoning[3]; however, the possibility that the peripheral neuropathy in those cases was caused by the known contaminants such as dibenzofurans or dibenzodioxin was not ruled out.

Glial outgrowth with CNS-type myelinated fibers strikingly similar to the CCPs was induced in pig spinal roots following transection (1.5 cm away from the cord) and anastomosis.[19] The glial outgrowth is thought to be induced by degeneration of axons passing the junctional zone between the CNS and PNS, which stimulates a peripheral proliferation of astroglial and oligodendroglia along the inner empty basal-lamina tubes of degenerated axons. While such a mechanical injury by 4-CB during the prenatal period is unlikely, of great interest is that such a pathomechanism has been proposed in Werdnig-Hoffmann disease (WH).[20–22]

Loss of large myelinated fibers with the resultant change in the frequency distribution of fiber sizes from bimodal to unimodal is characteristic feature in both WH and 4-CB spinal nerve roots.[21,22] Cylindrical peninsulas of glial bundles in both spinal and cranial nerve roots have been considered pathognomonic for WH.[21–22] Like the CCPs, glial peninsulas are found most commonly in the anterior spinal roots, but posterior spinal roots are not spared in WH. There were glial bundles in each of 46 selected cases of WH (6 from WVU, 2 from the Children's Hospital of Philadelphia, and 38 from the AFIP) in the recent survey by one of us (S.M.C.). Motor neurons entrapped within the glial bundles and heterotopic neurons in the anterior columns near the root exit region were commonly seen in WH. CNS-type myelinated fibers, however, have never been observed in WH glial peninsulas. Myelin ovoids similar to those observed in the CCPs were also described in WH spinal root nerve fibers[21–25] and were thought to be secondary to axonal degeneration. The possibility that CNS-type myelinated fibers might have existed prior to the formation of glial bundles can not be eliminated since both neuronal and astroglial elements are consistently observed in WH glial bundles; there is no reason for oligodendroglia to have been excluded. Oligodendroglia may be less viable than astroglia in peripheral tissue, and this may be the cardinal reason why the role of myelin production in the PNS is transferred from oligodendroglia to Schwann cells. This contention is well demonstrated in the different types of CCPs: in CCPs, oligodendroglia, and therefore myelin sheaths, appear to degenerate first; the neuronal elements degenerate next; and astroglia remain and proliferate within the confined cylindrical tract enclosed by a basement membrane. The latter appears to preclude invasion by collagen fibrils. The significance of this peculiar phenomenon is unknown, yet it seems to reflect the sovereign compartmentalization of the nervous system.

Understanding the neurotoxic mechanisms by which 4-CB induces similar neuroglial peninsulas should cast light on the pathogenesis of WH. Alternately, insight into the pathogenesis involved in the formation of glial peninsulas may elucidate the fundamental mechanism inducing the spinning syndrome. It is conceivable that 4-CB affects organization rather than differentiation of neuroglial elements, since no histopathologic abnormalities can be detected in the 4-CB-CNS. The early organization of interneuronal junctions, or synaptogenesis, may be affected by 4-CB, since PCBs are known to induce a membranous dysfunction of electrolyte transportation[23,24] which is essential for both the physiologic interneuronal transmission and the structural formation of synaptic junction. Selective interference with synaptogenesis of inhibitory terminals by 4-CB must be considered. In this regard it is pertinent to recognize the implication that among two distinct types of dopamine receptors,[25] the inhibition-mediating receptors are formed earlier than the excitation-mediating receptors. The inhibition-mediating receptors, therefore, may be susceptible to 4-CB intoxication. This may, in part, explain the hyperkinesia and rotatory movement of the spinners. Future biochemical study is needed to correlate dysfunction of the dopaminergic system with the spinning syndrome. The present electron microscopy on the nucleus accumbens, in particular, and the responses to dopamine agonists and antagonists in the spinners suggest such a relationship. It may be implied that 4-CB-induced defects in selected interneuronal junctions, such as those in the nigrostriatal system and ventral motor neurons, play a causal role in the spinning syndrome and formation of the CCPs in spinal roots, respectively. The motor neurons with "loose" junctions may be vulnerable to centrifugal displacement along with glial elements toward the periphery as the mice grow older and the discrepancy in the growth rate between the skeletal and CNS systems becomes greater.

Thus, the prenatal 4-CB induction of such a unique clinico-pathologic anomaly provides us a singular model for investigating two human neurologic diseases, Werdnig-Hoffmann disease and minimal brain dysfunction (MBD) or hyperactive syndrome. Of great interest is the recent documentation of a high incidence in anomalies of motor development in hyperactive boys.[26,27] The fact that the stimulant, amphetamine, is the most efficacious pharmacologic agent in both the hyperactive syndrome in children and the spinning syndrome in 4-CB mice may not be sheer coincidence.

Summary

A striking motor dysfunction, "spinning syndrome," developed with a high frequency in weaning mice whose dams received oral 3,4,3′,4′-tetrachlorobiphenyl (4-CB) during gestation (day 10 through day 16). The syndrome is permanent and is characterized by swift circling movements sustained in one direction at a minimal rate of 40 turns/min (usually 50 to 150 turns/min), restlessness, and hyperkinesia. Twenty-four spinners and 4-CB nonspinners and age-matched controls were subjected to histopathologic, histofluorescent, histochemical, and electron microscopic studies. The most reliable histopathologic marker for prenatal 4-CB injury to the CNS is the presence of cylindrical CNS peninsulas (CCPs) in the spinal and cranial nerve roots. The CCPs consist of either CNS-type myelinated fibers, unmyelinated fibers, or astroglial bundles in varying proportions, and are enclosed by a basement membrane. The CCPs are also observed in 4-CB nonspinners but in none of 12 controls studied. A selective defect in synaptogenesis induced prenatally by 4-CB is proposed as the primary event pursuant to the development of the CCPs, while interference with synaptogenesis may have occurred selectively in the striatonigral dopaminergic system. This is suggested by electron microscopy on the nucleus accumbens and also by the responses to administration of dopaminergic agonists and antagonist. The 4-CB induced clinico-pathologic anomaly may serve as a singular model for understanding human neurologic disorders, in particular, Werdnig-Hoffmann disease and minimal brain dysfunction syndrome.

Acknowledgments

The authors wish to gratefully acknowledge Dr. T.W. Crosby for reviewing the manuscript and for his comments, Mr. Gerald Voice for his technical assistance and Ms. Linda Kent for her excellent clerical assistance.

References

1. Kuratsune, M., T. Yoshimura, J. Matsuzaka & A. Yamaguchi. 1972 Epidemiologic study on Yusho, a poisoning caused by ingestion of rice oil contaminated with a commercial brand of polychlorinated biphenyl. Environ. Health Perspect. **1:** 119–128.
2. Hirayama, C. 1976. Clinical aspects of PCB poisoning. *In* PCB Poisoning and Pollution. K. Higuchi, Ed. pp. 87–104., Kodansha Ltd. Tokyo & Academic Press. New York, San Francisco.
3. Murai, Y. & Y. Kuroiwa. 1971. Peripheral neuropathy in chlorobiphenyl poisoning. Neurol. **21:** 1173–1176.
4. Takagi, Y., D. Otake, M. Kataska, Y. Murate, S. Aburada, S. Akasaka, K. Hasimoto, H. Uda & T. Kitaura. 1976. Studies on the transfer and distribution of ^{14}C-polychlororinated biphenyls from maternal and fetal and suckling rats. Toxicol. Appl. Pharmacol. **38:** 549–558.
5. Allen, J. R. & D. A. Barsotti. 1976. The effects of transplacental and mammary movement of PCB's on infant rhesus monkeys. Toxicol. **6:** 331–340.
6. Davis, G. J., S. M. Chou, T. Miike, G. W. Lucier & J. A. McLachlan. 1978. The "spinning syndrome" in mice following prenatal exposure with a PCB. Science. In press.
7. Peter, A. 1976. The fixation of central nervous tissue and the analysis of electron microscographs of the neuropil with special reference to the cerebral cortex. *In* Contemporary Research Methods in Neuroanatomy. W.T.H. Nauta, Ed. pp. 56–76. Springer-Verlag.

8. De La Torre, J. C. & J. W. Surgeon. 1976. A methodological approach to rapid and sensitive monoamine histofluorescence using a modified glyoxylic acid technique: The SPG Method. Histochemie **49:** 81–93.
9. Cheng, H. C., J. P. Long, L. S. Van Orden III, J. G. Cannon & J. P. O'Donnell. 1976. Dopaminergic activity of some apomorphine analogs. Res. Commun. Chem. Pathol. Pharmacol. **15**(1): 89–106.
10. Lucier, G., G. J. Davis & J. A. McLachlan. 1978. Transplacental toxicology of the polychlorinated and polybrominated biphenyls: Methodological approach. *In* Hartford Symposium on Developmental Toxicology on Energy related Pollutants. NITS, Springfield, Va., D. D. Mohlum. et al., Eds. In press.
11. Ungerstedt, U., S. G. Butcher, N. E. Anden & K. Fuxe. 1969. Direct chemical stimulation of dopaminergic mechanisms in the neostriatum of the rat. Brain Res. **14:** 461–471.
12. Glick, S. D., T. P. Jerussi & L. N. Fleisher. 1978. Turning in Circles: The neuropharmacology of rotation. Life Sci. **18:** 889–896.
13. Chou, S. M. & H. A. Hartman. 1964. Axonal lesions and waltzing syndrome after IDPN administration in rats, with a concept "Axostasis". Acta Neuropathol. **3:** 428–450.
14. Chou, S. M. & H. A. Hartman. 1965. Electron microscopy of focal axonal lesions produced by $\beta\beta'$-iminodipropionitrile (IDPN) in rats. I. The advanced lesions. Acta Neuropathol **4:** 590–603.
15. Beaudoin, A. R. 1977. Teratogenicity of polybrominated biphenyl in rats. Environ. Res. **14:** 81–85.
16. Villeneuve, D. C., D. L. Grant, K. Khera, D. J. Clegg, H. Baer & W. E. J. Phillips. 1971. The fetotoxicity of a polychlorinated biphenyl mixture (Aroclor 1254) in the rabbit and in the rat. Environ. Physiol. **1:** 67–69.
17. Platonow, N. & N. Y. Chen. 1973. Transplacental transfer of polychlorinated biphenyls (Aroclor 1254) in a cow. Vet. Rec. (Jan): 69–70.
18. Harrison, B. M. 1977. The spinal cord-ventral root junction in the trembler mouse. Acta Neuropathol. **38:** 33–38.
19. Meier, C. & H. Sollmann. 1978. Glial outgrowth and central-type myelination of regenerating axons in spinal nerve roots following transection and suture: Light and electron microscopic study in the pig. Neuropathol. Appl Neurobiol. **4:** 21–35.
20. Chou, S. M., & A. V. Fakadej. 1971. Ultrastructure of chromatolytic motoneurons and anterior spinal roots in a case of Werdnig-Hoffmann disease. J. Neuropathol. Exp. Neurol. **30:** 368–379.
21. Chou, S. M. 1973. Infantile muscular atrophy: Correlation between alteration in anterior spinal roots and muscle fiber atrophy. *In* Clinical Studies in Myology. pp. 206–217 Excerpta Medica. Amsterdam.
22. Chou, S. M. & I. Nonaka. 1978. Werdnig-Hoffmann disease: Proposal of a pathogenetic mechanism. Acta Neuropathol. **41:** 45–54.
23. Davis, P. W., J. M. Friedoff & G. A. Wodemeyer. 1972. Organochlorine insecticide, herbicide and polychlorinated biphenyl (PCB) inhibition of Na, K-ATPase in rainbow trout. Bull. Environ. Contam. Toxcol. **8:** 69–72.
24. Desaiah, D., L. K. Cutkomp, H. H. Yap & R. B. Koch. 1972. Inhibition of oligomyeim-sensitive and -insensitive magnesium adenosine triphosphatase. Biochem. Pharmacol. **21:** 857–865.
25. Cools, A. R. & J. M. Vanrossum. 1976. Excitation-Mediating and inhibition-mediating dopamine-receptors: A new concept towards a better understanding of electrophysiological, biochemical, pharmacological, functional and clinical data. Psychopharmacology **45:** 243–254.
26. Denckla, M. D. & R. G. Rudel. 1978. Anomalies of motor development in hyperactive boys. Ann. Neurol. **3:** 231–233.
27. Kinsbourne, M. 1973. Minimal brain dysfunction as a neurodevelopmental lag. Ann. N.Y. Acad. Sci. **205:** 268–273.

LONG-TERM TOXICOLOGIC STUDIES OF 2,3,7,8-TETRACHLORODIBENZO-*p*-DIOXIN (TCDD) IN LABORATORY ANIMALS

R. J. Kociba, D. G. Keyes, J. E. Beyer, R. M. Carreon, and P. J. Gehring

Toxicology Research Laboratory
Health and Environmental Research
Dow Chemical, U.S.A.
Midland, Michigan 48640

INTRODUCTION

The compound 2,3,7,8-tetrachlorodibenzo-*p*-dioxin (TCDD) is a highly toxic impurity that may be formed as an unwanted contaminant under certain conditions during the production of 2,4,5-trichlorophenol. TCDD has been considered one of the causes of chloracne, which has been associated historically with the industrial production of 2,4,5-trichlorophenol and other products made from 2,4,5-trichlorophenol.

Over the past decade, a number of toxicologic studies with TCDD were conducted to assess the potential for acute toxicity and teratogenesis. Results of these earlier studies have been summarized in a previous publication by Kociba *et al.*[1] In that publication, we also reported the results of a subchronic study in which rats were given 1.0, 0.1, 0.01, 0.001 or 0 μg TCDD/kg 5 days/week for 13 weeks. Doses of 1.0 μg TCDD/kg/day caused multiple toxicologic effects, with significant morphologic changes in the liver, thymus and reproductive organs. A dose level of 0.1 μg TCDD/kg/day caused lesser degrees of toxicity, and rats given 0.01 or 0.001 μg TCDD/kg/5 days/week for 13 weeks had no alterations considered of any toxicologic significance.

Since that time, studies have been in progress to assess the potential for chronic toxicity associated with long-term exposure to TCDD. Allen *et al.*[2] reported on a subchronic study in which monkeys consumed a diet containing 500 ppt of TCDD for up to 9 months. It was calculated that these monkeys ingested 2–3 μg TCDD/kg over the course of the 9-month study. Clinically, these monkeys had changes similar to those described by McConnell *et al.*[3] as well as some hematologic depression and hemorrhages in various tissues. Hypertrophy, hyperplasia and/or metaplasia were noted in the epithelium of the bile ducts, salivary glands, bronchi, pancreatic ducts, sebaceous glands, skin, gastric lining and urinary tract of these monkeys given diets containing 500 ppt of TCDD.

In 1969, Innes *et al.*[4] reported on a long-term carcinogenic study conducted in mice given 2,4,5-trichlorophenoxyacetic acid contaminated with a level of TCDD sufficient to supply approximately 0.27 μg TCDD/kg/day. Neither of the two strains of mice had an increase in tumors as a result of the treatment. DiGiovanni *et al.*[5] reported on a study in which TCDD was reported to be only a weak tumor initiator in the two-stage system of mouse skin carcinogenesis with 7,12-dimethylbenz(a)anthracene (DMBA).

Van Miller and Allen[6] issued a preliminary report on a study of small groups of male rats fed diets containing TCDD for 65 weeks. All 10 rats of each group receiving 1.0, 0.5 or 0.05 ppm TCDD in the diet died within 4 weeks, with acute toxic effects.

0077–8923/79/0320–0397 $01.75/0 © 1979, NYAS

Groups of male rats on diets containing 5000 or 1000 ppt of TCDD had increased mortality, decreased weight gain and liver toxicity. In a more recent publication by this same group of researchers, Van Miller *et al.*[7] reported lung and/or liver tumors in those rats given 1000 or 5000 ppt of TCDD. Neoplasms of various organs, but not of lung or liver, were reported in rats given diets containing 500, 50 or 5 ppt of TCDD. Van Miller *et al.* reported a zero incidence of tumors in the group of rats given 1 ppt of TCDD and in a group of 50 control rats.

In view of the need for an evaluation of the chronic toxicity and potential for carcinogenicity of TCDD, the study summarized herein was conducted. For the sake of brevity, only the more salient data are included in this presentation. The full report can be found in a recent paper by Kociba *et al.*[8]

Materials and Methods

The TCDD sample was prepared by the Dow Chemical Company, Midland, Michigan, USA, and had a purity exceeding 99%. On the basis of results from a previous 13-week long toxicity study[1] in rats dosage levels of 0.1, 0.01 and 0.001 μg of TCDD/kg/day were given in the diets. The test diets were adjusted at intervals to provide these dose levels of TCDD as required by body weight and food comsumption determinations. Groups of 100 rats (50 male, 50 female) of the Sprague-Dawley strain were maintained on diets supplying these dose levels for two years. The control group consisted of 85 males and 86 females.

Samples of the prepared diets were analyzed periodically to ascertain if the dietary levels of TCDD were being maintained as scheduled.

Parameters monitored during the course of the study included body weights, food consumption, hematologic parameters including total erythrocyte count (RBC), total and differential leukocyte counts (WBC), thrombocyte and reticulocyte counts, packed cell volumes (PCV) and hemoglobin concentration, routine urinalyses, and urinary excretion of creatinine, coproporphyrin, uroporphyrin and delta-amino levulinic acid (delta-ALA). Serum samples were collected preterminally or terminally for determination of urea nitrogen (BUN), glutamic pyruvic transaminase (SGPT), bilirubin (total, direct and indirect), cholesterol, triglycerides, alkaline phosphatase (AP), gamma-glutamyl transferase (GGT), total protein, albumin and globulin.

All rats killed after two years, as well as those dying or culled during the study, were subjected to a complete necropsy examination by a veterinary pathologist. Representative tissues from all organ systems of the body of each rat were preserved in formalin fixative, with subsequent preparation of paraffin-embedded sections for light microscopy examination. Additional sections of livers collected at terminal necropsy from three female rats per group were processed for transmission electron microscopy. The weights of the liver, kidney, brain, heart, thymus, spleen and testes or ovaries/uterus were recorded at the time of terminal necropsy.

Liver and fat samples collected from three females per group at terminal necropsy were subsequently analyzed for TCDD content, using gas chromatography low-resolution spectrometry.

The various data were statistically analyzed according to the procedures listed in the previously cited publication by Kociba *et al.*,[8] which includes the full details of this research.

Results

Analyses of the test diets indicated that rats given 0.1, 0.01 and 0.001 μg TCDD/kg/day were ingesting an average of 2,193, 208 and 22 ppt of TCDD, respectively, in the diet for two years.

The rate of mortality was increased in female rats given 0.1 μg TCDD/kg/day. The body weights of male and female rats given 0.1 μg TCDD/kg/day and also the females given 0.01 μg TCDD/kg/day were decreased. However, these parameters were not affected at the lower dose levels.

Hematologic data indicated some effects at the high dose level of 0.1 μg TCDD/kg/day: this included slight decreases in PCV and Hgb of males and decreases in PCV, total RBC and WBC counts and Hgb values of females; reticulocyte counts appeared to be slightly increased. No effect was noted on thrombocyte counts and WBC differentials at all the dose levels studied. No hematologic changes were noted in rats given 0.01 or 0.001 μg TCDD/kg/day. Changes in urinary parameters were limited only to females given 0.1 or 0.01 μg TCDD/kg/day; these groups had increased urinary excretion of coproporphyrin and uroporphyrin. Urinary excretion of delta-ALA was increased only in females given 0.1 μg TCDD/kg/day.

TABLE 1

MAJOR GROSS NECROPSY FINDINGS IN RATS ATTRIBUTED TO INGESTION OF DIETS CONTAINING TCDD FOR TWO YEARS

Dose Level of TCDD (μg/kg/day)		
0.1	0.01	0.001
Emaciation (M, Fe)	—*	—
Icterus (Fe)	—	—
Liver toxicity (M, Fe)	Liver toxicity (M, Fe)	—
Liver nodules (Fe)	Liver nodules (Fe)	—
Thymic atrophy (Fe)	—	—
Splenic atrophy (M, Fe)	—	—
Anemic paleness of tissue and focal hemorrhages (Fe)	—	—
Increased incidence of mesenteric periarteritis (M, Fe)	Mesenteric periarteritis (Fe)	—
Pulmonary congestion, edema, effusion and mass formation (Fe)	—	—
Decrease in incidence of pituitary enlargement, endometrial hyperplasia, uterine polyps, and subcutaneous masses (Fe)	—	—
Decrease in severity of chronic renal disease (M)	—	—
Increased incidence of keratinized proliferative lesions near pharynx (M, Fe)	—	—

*(—) Indicates observation was not noted at that dose level, or incidence was comparable to control data.

Tabulation excludes voluminous data considered unrelated to treatment with TCDD.

Of all the clinical chemical parameters evaluated, the only treatment-related changes were noted in females given 0.1 μg TCDD/kg/day, which had increases in SGPT, AP, and GGT activities.

For the sake of brevity, treatment-related gross and microscopic observations on tissues have been summarized in TABLES 1 and 2. Gross necropsy examination indicated that the target organs of chronic TCDD toxicity in rats were primarily the liver, lymphoid organs, vascular and respiratory systems (TABLE 1). Light microscopy examination revealed the liver to be the organ most consistently affected, and rats

TABLE 2

MAJOR HISTOPATHOLOGIC FINDINGS IN RATS ATTRIBUTED TO INGESTION OF DIETS CONTAINING TCDD FOR TWO YEARS

Dose Level of TCDD (μg/kg/day) 0.1	0.01	0.001
Liver		
Hepatic degenerative inflamatory, necrotic and proliferative changes. Cytomegaly, distortion of lobular pattern and atropy of cords, cytoplasmic vacuolation, fatty change, altered tinctorial properties, hepatic necrosis and inflammation, multinucleated hepatocytes, foci or areas of hepatocellular alteration (swollen hepatocytes), pigment aggregates, bile duct hyperplasia, periportal inflammation, fibrosis	Livers had much lesser degree of change.	No changes. Livers of females had statistical increase in foci or area of hepatocellular alteration (swollen hepatocytes), and livers of males had statistical decrease in area of hepatocellular alteration (swollen hepatocytes).
Hepatocellular carcinoma (Fe)		
Hepatocellular hyperplastic (neoplastic) nodules (Fe)	Hepatocellular nodules (Fe)	—
Reproductive		
Decreased incidence of uterine hyperplasia, cyst formation and polyp formation (Fe)	Questionable decreased incidence of uterine polyps (Fe)	—
Mammary Tissue		
Decreased incidence of mammary tumors	—	—
Endocrine		
Pituitary—decreased incidence of pituitary hemangiectasis and adenoma formation (Fe)	—	—
Adrenal—decreased incidence of medullary hyperplasia (M, Fe) and pheochromocytoma (M)	—	—
Increased incidence of cortical necrosis and hemorrhage (Fe), cortical adenoma and hematocyst (M)	—	—
Pancreas—decreased incidence of acinar adenoma (M)	—	—
Thyroid—occurrence of various isolated follicular changes of questionable significance (M)	—	—
Parathyroid—decreased incidence of parathyroid hyperplasia, secondary to chronic renal disease (M)	—	—
Lymphoid		
Isolated cases of thymic/splenic atrophy (Fe)	—	—
Respiratory		
Focal alveolar hyperplasia, pigment aggregates, accumulation of alveolar macrophages and cholesterol clefts, pulmonary edema, focal interstitial inflammation and fibrosis	Focal alveolar hyperplasia (Fe)	—

TABLE 2 (*Continued*)

Dose Level of TCDD (μg/kg/day) 0.1	0.01	0.001
Keratinizing squamous metaplasia (Fe)	—	—
Keratinizing squamous cell carcinoma (Fe)	—	—
Cardiovascular		
Isolated cases of hemorrhage in CNS (Fe)	—	—
Increased incidence of mesenteric and thoracic periarteritis, with cases of thrombosis or hematoma formation (M, Fe)	Increased incidence of periarteritis (M)	—
Increased incidence of myocardial degenerative changes (Fe)	—	—

*(—) Indicates observation was not noted at that dose level, or incidence was comparable to control data.

Tabulation excludes voluminous data considered unrelated to treatment with TCDD.

given 0.1 or 0.01 μg TCDD/kg/day had a dose-dependent incidence of hepatic degenerative, inflammatory and necrotic changes listed in TABLE 2. Female rats given 0.1 μg TCDD/kg/day also had a statistical increase in lesions described morphologically as hepatocellular carcinomas (no metastasis) and hepatocellular nodules; females given 0.01 μg TCDD/kg/day had an increase in hepatocellular nodules.

Histopathologic examination of livers from rats given 0.001 μg TCDD/kg/day revealed none of the treatment-related effects that were noted at the higher dose levels. On a statistical basis, the incidence rate of a slight hepatocellular alteration (focal swollen hepatocytes) was increased in females and decreased in males of the group given 0.001 μg TCDD/kg/day; these statistical variations were not considered indicative of any toxicologic significance.

Ultrastructural examination of hepatocytes from females of the 0.1 μg TCDD/kg/day revealed the most consistent change to be in the rough endoplasmic reticulum (RER) which showed proliferation, distortion and fragmentation. Smooth endoplasmic reticulum (SER) and mitochondrial structures were within control limits. Other changes in hepatocytes from this group given 0.1 μg TCDD/kg/day included focal cytoplasmic vacuolization, increased lysosomal activity with residual body formation and an occasional multinucleated hepatocyte. In hepatocytes from females given 0.01 μg TCDD/kg/day, the most notable changes included a lesser degree of proliferation of the RER and some proliferation of the SER. This was accompanied by a slight increase in the number of hepatocytes with lipid droplet accumulation. The hepatocytes from females given 0.001 μg TCDD/kg/day were within the limits of variation seen in the hepatocytes of the control group. There was no general increase in lipid droplet content of the hepatocytes, although an occasional hepatocyte had some lipid droplet content.

As predicted from previous studies, gross and histopathologic examination of some rats given 0.1 μg TCDD/kg/day had isolated cases of thymic and/or splenic atrophy. Lower dose levels did not produce this effect.

The respiratory system of both male and female rats was affected by ingestion of 0.1 μg TCDD/kg/day, with focal alveolar hyperplasia, pigment aggregates, alveolar macrophages, cholesterol clefts, edema, focal interstitial inflammation, fibrosis, keratinizing squamous metaplasia, and keratinizing squamous cell carcinomas. At 0.01 μg TCDD/kg/day, these effects were not noted, except for focal alveolar hyperplasia in female rats. At the dose level of 0.001 μg TCDD/kg/day, there were no treatment-related changes in the respiratory tract.

Examination of the reproductive organs of females given 0.1 (and possibly those given 0.01 μg TCDD/kg/day) revealed a decreased incidence of the uterine changes typically encountered as spontaneous lesions in this strain of rat. The high dose group of females also had decreased incidences of both mammary and pituitary tumors. Adrenal changes noted only at the high dose level of 0.1 μg TCDD/kg/day included a

TABLE 3

INCREASED INCIDENCE OF TUMOR AND TUMOR-LIKE LESIONS ATTRIBUTED TO INGESTION BY RATS OF DIETS CONTAINING TCDD FOR TWO YEARS

	Dose Level of TCDD (μg/kg/day)			
	0	0.1	0.01	0.001
	Number of Males with Lesion/Number in Group			
Hard Palate/Nasal Turbinates				
Stratified squamous cell carcinoma	0/85	4/50*	0/50	0/50
Tongue				
Stratified squamous cell carcinoma	0/85	3/50*	1/50	1/50
Lung				
Squamous cell carcinoma	0/85	1/50	0/50	0/50
	Number of Females with Lesion/Number in Group			
Hard Palate/Nasal Turbinates				
Stratified squamous cell carcinoma	0/86	4/49*	1/50	0/50
Tongue				
Stratified squamous cell carcinoma	1/86	2/49	0/50	0/50
Lung				
Squamous cell carcinoma	0/86	7/49*	0/50	0/50
Liver				
Hepatocellular carcinoma	1/86	11/49*	2/50	0/50
Hepatocellular nodule(s)	8/86	23/49*	18/50*	3/50

*Statistically increased above control data, $p < 0.05$.

Table excludes additional data on incidence of tumors considered unrelated to treatment with TCDD.

decreased incidence of medullary hyperplasia and pheochromocytoma formation, and an increased incidence of cortical necrosis, hemorrhage, hematocyst and adenoma formation.

The incidence of pancreatic acinar adenoma was decreased in males given 0.1 μg TCDD/kg/day. This group of males also had a decrease in the incidence rate of secondary parathyroid hyperplasia; this was due to a decrease in the severity of chronic renal disease noted in males given this high dose level of 0.1 μg TCDD/kg/day.

Examination of the cardiovascular system revealed several effects assumed to be due to ingestion of 0.1 μg TCDD/kg/day. These effects included isolated cases of hemorrhage in different organs of the body, and an increase in the incidence of periarteritis and myocardial degenerative changes, both of which occur spontaneously

TABLE 4

DECREASED INCIDENCE OF TUMORS ATTRIBUTED TO INGESTION BY RATS OF DIETS CONTAINING TCDD FOR TWO YEARS

	Dose Level of TCDD (μg/kg/day) 0	0.1
	Number of Males with Lesions/Total Number in Group	
Pancreas		
Acinar adenoma	14/85	2/50*
Pheochromocytoma	28/85	4/50*
	Number of Females with Lesion/Total Number in Group	
Uterus		
Benign tumor	30/86	8/49*
Mammary Gland		
Benign tumor	73/86	24/49*
Pituitary		
Adenoma	43/86	12/49*

*Statistically decreased from control data, $p < 0.05$.

Data at lower dose levels comparable to controls, except for possible decrease in incidence of benign uterine tumors in females given 0.01 μg TCDD/kg/day.

in this strain of rat. None of these observations were noted at the two lower dose levels, except for an increase in incidence of periarteritis in male rats given 0.01 μg TCDD/kg/day.

Other treatment-related effects included an increase above the background incidence of squamous cell carcinomas of the tongue and hard palate/nasal turbinate region of rats given 0.1 μg TCDD/kg/day. Historically, squamous cell carcinomas have occurred in these organs of this strain of rat at a spontaneous incidence rate of 1–3%. Only the high dose level of 0.1 μg TCDD/kg/day increased the incidence of this type of neoplasm.

All the data on incidence rates of neoplasms that were increased or decreased by treatment with the high dose level of 0.1 μg TCDD/kg/day are summarized in TABLES 3 and 4.

Review of organ weights recorded at the terminal necropsy indicated the following effects considered related to treatment (a) an increase in absolute and/or relative liver weight of rats given 0.1 or 0.01 μg TCDD/kg/day and (b) a decrease in weight of the thymus of females given 0.1 μg TCDD/kg/day. There were no organ weight changes at the dose level of 0.001 μg TCDD/kg/day.

TABLE 5 lists the TCDD content of liver and fat samples collected from female rats at terminal necropsy after two years of ingesting diets containing TCDD.

TABLE 5

TCDD CONTENT IN RAT LIVERS AND FAT SAMPLES COLLECTED AFTER TWO YEARS INGESTION OF DIETS CONTAINING TCDD

Dose Level of TCDD (μg/kg/day)	TCDD Content (ppt) Fat	Liver
0.1	8100	24000
0.01	1700	5100
0.001	540	540

Discussion

The results of this study of rats ingesting TCDD for a lifetime serve as a basis for assessing the long-term chronic toxicity of TCDD. Continuous ingestion of a high dose level of 0.1 μg TCDD/kg/day (approximately 2200 ppt in diet) predictably caused multiple toxicologic effects. Liver toxicity was the most consistent observation, and this was accompanied by morphologic changes of the lymphoid, respiratory, and vascular tissues of the body. The incidence of hepatocellular carcinomas of the liver and squamous cell carcinomas of the lung, hard palate/nasal turbinates or tongue was increased at this dose level. Conversely, the incidence of tumors of the pituitary, uterus, mammary gland, pancreas and adrenal medulla was decreased at this high dose level of treatment. Similarly, the incidence of other spontaneous lesions such as chronic renal disease was also decreased at this high dose level.

Lifetime ingestion by rats of 0.01 μg TCDD/kg/day (approximately 210 ppt in diet) caused a lesser degree of toxicity, primarily of the liver. However, there was no increase in the incidence of neoplasia at this dose level.

Rats ingesting 0.001 μg TCDD/kg/day for two years had no adverse effects in spite of the fact the liver and fat each contained 540 ppt of TCDD at termination of the study.

Thus, these data indicate a good dose response for the long-term toxicity of TCDD. Whereas higher doses were predictably toxic and did alter the incidence of tumors, an intermediate dose level caused lesser toxicity but no neoplasia. A lower dose was tolerated for a lifetime with no adverse effects noted in any of the parameters evaluated.

References

1. Kociba, R. J., P. A. Keeler, C. N. Park & P. J. Gehring. 1976. 2,3,7,8-Tetrachlorodibenzo-*p*-dioxin (TCDD): Results of a 13-week oral toxicity study in rats. Toxicol. Appl. Pharmacol. **35:** 553–574.
2. Allen, J. R., D. A. Barsotti, J. P. Van Miller, L. J. Abrahamson, & J. J. Lalich. 1977. Morphological change in monkeys consuming a diet containing five hundred parts per trillion of 2,3,7,8-tetrachloro-dibenzo-*p*-dioxin. Food Cosmet Toxicol. **15:** 401–410.
3. McConnell, E. E., J. A. Moore & D. W. Dalgard. 1978. Toxicity of 2,3,7,8-tetrachlorodibenzo-*p*-dioxin (TCDD) in rhesus monkeys (Macaca mulatta) following a single oral dose. Toxicol. Appl. Pharmacol. **43:** 175–187.
4. Innes, J. R. M., B. M. Ulland, M. G. Valerio, L. Petrucelli, L. Fishbein, E. R. Hart, A. J. Pallotta, R. R. Bates, H. L. Falk, J. J. Gart, M. Klein, I. Mitchell & J. Peters. 1969. Bioassay of pesticides and industrial chemicals for tumorigenicity in mice: A preliminary note. J. Nat. Cancer Inst. **42:** 1101–1114.
5. DiGiovanni, A. Viaje, D. L. Berry, T. J. Slaga & M. R. Juchau. 1977. Tumor initiating ability of 2,3,7,8-tetrachlorodibenzo-*p*-dioxin (TCDD) and Arochlor 1254 in the two-stage system of mouse skin carcinogenesis. Bull. Environ. Contam. Toxicol. **18:** 552–557.
6. Van Miller, J. P. & J. R. Allen. 1977. Chronic toxicity of 2,3,7,8-tetrachlorodibenzo-*p*-dioxin in rats. Fed. Proc. **36:** 396 (Abstract 673).
7. Van Miller, J. P., J. J. Lalich & J. R. Allen. 1977. Increased incidence of neoplasms in rats exposed to low levels of 2,3,7,8-tetrachlorodibenzo-*p*-dioxin. Chemosphere **9:** 537–544.
8. Kociba, R. J., D. G. Keyes, J. E. Beyer, R. M. Carreon, C. E. Wade, D. A. Dittenber, R. P. Kalnins, L. E. Frauson, C. N. Park, S. D. Barnard, R. A. Hummel & C. G. Humiston. Results of a two-year chronic toxicity and oncogenicity study of 2,3,7,8-tetrachlorodibenzo-*p*-dioxin (TCDD) in rats. Toxicol. Appl. Pharmacol. In press, 1978.

STUDIES WITH CHLORINATED DIBENZO-*p*-DIOXINS, POLYBROMINATED BIPHENYLS, AND POLYCHLORINATED BIPHENYLS IN A TWO-STAGE SYSTEM OF MOUSE SKIN TUMORIGENESIS: POTENT ANTICARCINOGENIC EFFECTS*

David L. Berry, Thomas J. Slaga, John DiGiovanni,† and Mont R. Juchau†

Cancer and Toxicology Program
Biology Division, Oak Ridge National Laboratory
Oak Ridge, Tennessee 37830

†Department of Pharmacology, School of Medicine
University of Washington
Seattle, Washington 98195

The chlorinated dibenzo-*p*-dioxins and chlorinated dibenzo-furans are among the most potent teratogens and toxins that are found as environmental contaminants.[1,2] The polychlorinated biphenyls (PCBs) and polybrominated biphenyls (PBBs) are not nearly as toxic as the dioxins but they are environmentally more prevalent than the dioxins.[3–5] Both classes of compounds are inducers of the cytochrome P_{450}-mediated microsomal monooxygenases[5–10] and of δ-aminolevulinic acid synthetase.[1,10] The dioxins induce aryl hydrocarbon hydroxylase (AHH, E.C. 1.14.14.2), the enzyme system responsible for converting polycyclic aromatic hydrocarbons (PAH) to active and inactive metabolites, much the same as 3-methylcholanthrene (3-MC) but for periods up to 35 days after a single i.p. injection and at much lower doses[1,10] PCBs and PBBs are mixtures of various isomers and induce the monooxygenase system with properties similar to both phenobarbital and 3-MC but the duration of induction is much shorter than with 2,3,7,8-tetrachlorodibenzo-*p*-dioxin (TCDD).[2,10,11] The toxicologic response is similar for both classes of compounds in that they are hepatotoxic,[4,12–14] cause a decreased growth rate[3,15] and produce chloracne in test animals and man.[1,13,16]

The effects of PCBs on carcinogenicity of various chemicals have been investigated by numerous groups. Kanechlor-500 in combination with several hepatocarcinogens (3′methyl-4-dimethylaminoazobenzene, *N*-2-fluorenylacetamide and diethylnitrosamine) in the diets of rats markedly decreased the formation of hepatocarcinomas.[17] Kimura *et al.*[18] demonstrated that pretreatment with Kanechlor-400 in diets of rats 4 months prior to and 2 months during treatment with 3′methyl-4-dimethylaminoazobenzene protected the rats against the formation of hepatocarcinomas induced by this carcinogen. Treatment of animals with the organochlorine insecticide 1,1,1-trichloro-2,2-bis-(*p*-chlorophenyl)-ethane (DDT) has been shown to inhibit 7,12-dimethylbenz(a)anthracene (DMBA)-induced mammary tumors.[19] These findings as well as other reports[20–23] suggest that in terms of effects on carcinogenesis, timing is important in the application of PCB to the diets of laboratory animals. Treatment with the PCBs prior to the administration of carcinogen results in

*Research sponsored by the Division of Biomedical and Environmental Research, U.S. Department of Energy under contract W-7405-eng-26 with the Union Carbide Corporation.

0077–8923/79/0320–0405 $01.75/0 © 1979, NYAS

reduction of the tumor response whereas treatment subsequent to application of the carcinogen results in enhancement of the tumorigenic response. The inhibitory effects of the PCBs in these studies were attributed to the induction of hepatic microsomal enzymes while the enhancement was ascribed to the promoting properties of the PCBs.

The two-stage system of mouse skin tumorigenesis allows one to evaluate critically the initiation and promotion phases of carcinogenesis individually.[24-26] The initiation phase requires a single subthreshold dose of a polycyclic aromatic hydrocarbon and is irreversible, while the promotion stage requires multiple applications of a promoter and is reversible up to a given time point.[27,28] This system allows one to study the effects of modifiers on initiation and promotion separately in a skin carcinogenesis assay. The results presented here demonstrate that certain environmental chemicals (PCBs, PBBs, and TCDD) possess the capacity to modify tumor initiation in a mouse skin assay. At the doses utilized, these chemicals exhibited little or no initiating or promoting properties.

Materials and Methods

Animals

Female CD-1 mice were purchased from the Charles River Mouse Farms (North Wilmington, MA). Mice 6 to 8 weeks old were carefully shaved with surgical clippers 2 days prior to treatment, and only those mice in the resting phase of the hair cycle were used for the biochemical and tumor experiments. Animals were housed in plastic cages, received water and food *ad libitum,* and were weighed weekly.

Chemicals

DMBA was purchased from Sigma Chemical Company (St. Louis, MO). Benzo(a)pyrene (BaP) was supplied by the NCI Carcinogenesis Research Program (Bethesda, MD). The phorbol diester promoter 12-0-tetradecanoylphorbol-13-acetate (TPA) was purchased from Dr. P. Borchert, University of Minnesota, Minneapolis, MN. PCB (Aroclor 1254) was purchased from Analabs, Inc. (North Haven, CT). PBB was supplied by A. C. Kolbye, Jr. of the FDA. TCDD (Lot #851-144-2) was generously supplied by the Dow Chemical Company (Midland, MI). A. Poland generously supplied samples of 2,7-dichlorodibenzo-*p*-dioxin (DCDD) and 3,4′,3′,4′-tetrachloroazobenzene (TCAB).

Tumor-Induction Experiments

Each experimental group consisted of 30 pre-shaved mice. DMBA (200 nmol) was dissolved in acetone immediately before use, and 0.2 ml of the acetone solution was applied topically to the shaved area of the back. One week after DMBA initiation, a control group received application twice weekly of TPA (10 μg) in 0.2 ml acetone. The following test compounds: PBB (100 μg); PCB (100 μg) and TCDD (0.1 μg) were applied twice weekly topically in acetone to the shaved dorsal skin area of the DMBA-initiated mice. Promotion was continued for 30 weeks, when the experiments were terminated. In another set of animal experiments, animals were pretreated with: PCB (100 μg and 625 μg) 18 hours prior to initiation; TCDD (1.0 μg) at 1, 3 and 5 days prior to initiation; DCDD (100 μg) 3 days prior to initiation; and TCAB (10 μg) 3 days prior to initiation. The animals were initiated with 10 nmol DMBA and promoted with TPA. Incidences of both papillomas and carcinomas were observed weekly and papillomas and carcinomas were removed at random for histopathologic verification. Initial histologic experiments were performed with PCBs, PBBs, and TCDD to assess toxicity and changes in intrafollicular epidermis prior to dose selection for the experiments described.

Enzyme Induction by PCBs and TCDD

In order to induce epidermal monooxygenase systems functional in the oxidative biotransformation of DMBA, mice were pretreated with single doses of the following inducing agents: 100 μg of PCB 18 hours prior to sacrifice, 200 nmol of DMBA, dibenz(a,c)anthracene or 3-methylcholanthrene 18 hours prior to sacrifice and 1 μg of TCDD 72 hours prior to sacrifice. High-pressure liquid chromatography was utilized to assess the profiles of metabolites of ^{14}C-DMBA. Details of the methodology have been published by DiGiovanni *et al.*[29] To further assess the bioactivation process, binding to epidermal protein, RNA and DNA by ^{3}H-DMBA was measured using the method of Huberman and Sachs.[30]

RESULTS

Visual and histopathologic examination of mouse skin treated with TPA at 2 μg per mouse indicated marked inflammation and an increase in the number of layers in the intrafollicular epidermis from 1 to between 4 and 5 layers. PCBs and PBBs at either 100 μg or 625 μg per mouse did not affect the intrafollicular epidermis. TCDD at 1 μg per mouse increased the numbers of cells in the intrafollicular epidermis but the effect was not pronounced. DCDD or TCAB did not affect the thickening of the epidermis as compared to the strong hyperplastic response of TPA.

The chlorinated dioxins and biphenyls were tested for promoting activity in the mouse skin with a high (200 nmol) initiating dose of DMBA. In a 30-week treatment period, the normal DMBA-initiated, TPA-promoted controls yielded approximately 8 papillomas per mouse (TABLE 1). The two biphenyls (at doses of 100 μg/mouse given twice weekly) did not promote the development of skin tumors nor did TCDD at a dose of 0.1 μg twice weekly. When tested without DMBA initiation, neither the dioxin nor the biphenyls demonstrated any carcinogenic activity. By the end of 30 weeks of treatment, the dioxin-treated mice showed a reddening of the skin as well as ear lesions resembling acne. The biphenyls, however, did not produce any observable skin lesions.

A series of biphenyl- and dioxin-related inducers of aryl hydrocarbon hydroxylase were tested for their ability to act as tumor initiators or for their capacity to modify the initiating activity of DMBA (see TABLE 2). Aroclor 1254 and TCDD were very weak initiators as was shown previously by DiGiovanni *et al.*[31] Aroclor 1254 exhibited no effect on tumorigenesis when given 5 minutes prior to an initiating dose of DMBA. Aroclor 1254 given at 18 hours and 72 hours prior to administration of DMBA inhibited initiation by as much as 45% in a time-dependent fashion. The much stronger monooxygenase inducer, TCDD (at 1 μg) decreased DMBA initiation by 93% when applied 3 days prior to initiation. The less active DCDD inhibited DMBA initiation only slightly. A close structural analogue of the dioxins, 3,4,3′,4′-tetrachloroazobenzene,[32] which is much less potent than TCDD in terms of AHH

TABLE 1*

LACK OF PROMOTING ABILITY OF PCB, PBB, AND TCDD ON SKIN-TUMOR INITIATION WITH DMBA†

Initiator‡	Promoter	Doses§	Papillomas per Mouse	Percent with Papillomas
DMBA	None	—	0	0
None	TPA	10 μg¶	0.03	3
DMBA	TPA	2 μg	8.1	92
DMBA	PCB	100 μg	0	0
None	PCB	100 μg	0	0
DMBA	PBB	100 μg	0	0
None	PBB	100 μg	0	0
DMBA	TCDD	0.1 μg	0	0
None	TCDD	0.1 μg	0	0

*Data taken from Res. Commun. Chem. Pathol. Pharmacol. **20:** 101–107 (1978).
†30 female mice per group; promoted for 30 weeks.
‡DMBA was applied at a dose of 200 nmol per mouse.
§Promoter applied twice weekly in acetone.
¶TPA applied weekly in acetone.

induction in the mouse, slightly enhanced DMBA initiation at a dose of 10 μg given 3 days prior to initiation.

To further study the inhibitory effects of the dioxins on PAH initiation, a series of tumor experiments were conducted using DBMA or BaP as initiators. As shown in TABLE 3, pretreatment times of 1, 3 and 5 days (at a dose of 1 μg per mouse) resulted in a 94% inhibition of the tumor response to DMBA. To test the generality of the inhibition, the same experimental design was used with BaP as the initiator. A similar inhibitory action of TCDD on BaP tumor-initiation was also observed (unpublished observations). With BaP as the initiator, the dependence on the time of pretreatment with TCDD was similar to that observed with DMBA as the initiator. The maximum

TABLE 2

EFFECTS OF VARIOUS ORGANOHALIDES ON SKIN-TUMOR INITIATION BY DMBA (Pretreated mice were initiated with 2.56 μg of DMBA and promoted with twice weekly applications of 10 μg TPA*)

Experiment	Pretreatment (μg/mouse)	Pretreatment Time	Papillomas per Mouse†	Percent of Control‡
1	Acetone	5 min	3.80	100
2	Aroclor 1254 (100)	5 min	3.60	95
3	Aroclor 1254 (100)	18 hr	2.10	55
4	Aroclor 1254 (100)	3 days	3.30	80
5	Aroclor 1254 (625)	18 hr	2.70	71
6	DCDD (100)	3 days	3.44	91
7	TCAB (10)	3 days	4.73	125
8	TCDD (1)	3 days	0.34	7

*30 mice were used per experimental group. Promotion was continued for 20 weeks.
†Average number of papillomas per mouse after 20 weeks of promotion.
‡The average number of papillomas per mouse expressed as a percentage of the DMBA-initiated, TPA-promoted group (experiment 1).

inhibition of tumor formation occurred with pretreatment at 5 days prior to initiation, but the inhibitory response of TCDD pretreatment on BaP initiation was less than that observed with DMBA as the initiator. In all cases where the animals received 1 μg of TCDD, the weight gain, morphologic and histologic characteristics of the skin were normal. At topical doses of 2 μg, toxicity was observed and approximately one-third of the mice died at the end of the 20-week period.

To further show that the inhibition of PAH initiation by TCDD was related to the induction of the epidermal monooxygenase system, a dose-response study was designed. As illustrated above, maximum inhibition was observed at 3 to 5 days prior to DMBA initiation. As shown in TABLE 3, TCDD, in a dose-dependent fashion, inhibited skin-tumor initiation by DMBA. At doses of 0.01 μg, the inhibition of DMBA-initiation was approximately 80%, whereas 0.1 μg of TCDD was almost as effective as 1.0 μg. At this writing, these experiments are only in the 15th week and the

TABLE 3

EFFECTS OF 2,3,7,8-TETRACHLORODIBENZO-*p*-DIOXIN (TCDD) ON SKIN-TUMOR INITIATION BY DMBA

(Pretreated mice were initiated with 2.56 μg of DMBA and promoted with twice weekly applications of 10 μg of TPA.*)

Experiment	Pretreatment (μg/mouse)	Pretreatment Time	Papillomas per Mouse†	Percent of Control‡
1	Acetone	5 min	3.80	100
2	TCDD (2)	5 min	3.70	97
3	TCDD (1)	1 day	0.53	14
4	TCDD (1)	3 days	0.34	9
5	TCDD (1)	5 days	0.23	6
6	Acetone	5 min	2.43	100
7	TCDD (0.01)	3 days	0.27	11
8	TCDD (0.10)	3 days	0.10	4
9	TCDD (2.0)	3 days	0.07	3

*30 mice were used per experimental group. Promotion was continued for 20 weeks for experiments 1–5 and 15 weeks for experiments 6–9. Promotion is still in progress for both groups of experiments.

†Average number of papillomas per mouse after 15 or 20 weeks of promotion.

‡The average number of papillomas per mouse expressed as a percentage of the DMBA-initiated, TPA-promoted groups for each period of promotion (i.e., experiments 1 and 6).

animals have not yet reached plateau tumor levels. The trend is, however, readily apparent.

FIGURE 1 shows high-pressure liquid chromatographic profiles of metabolites formed upon incubation of ^{14}C-DMBA with epidermal homogenates from inducer-pretreated CD-1 mice. The doses of TCDD and Aroclor 1254 utilized and the pretreatment times were identical to the conditions employed in the tumor experiments in which maximal inhibition of tumor initiation was observed. These high-pressure liquid chromatographic profiles demonstrate that chlorinated hydrocarbons such as Aroclor 1254 and TCDD are capable of inducing epidermal monooxygenases that are responsible for converting DMBA to a variety of hydroxylated products. Furthermore, a good correlation between the increased rates of oxidative metabolism induced by Aroclor 1254 and TCDD and the inhibition of DMBA tumor-initiation was observed.

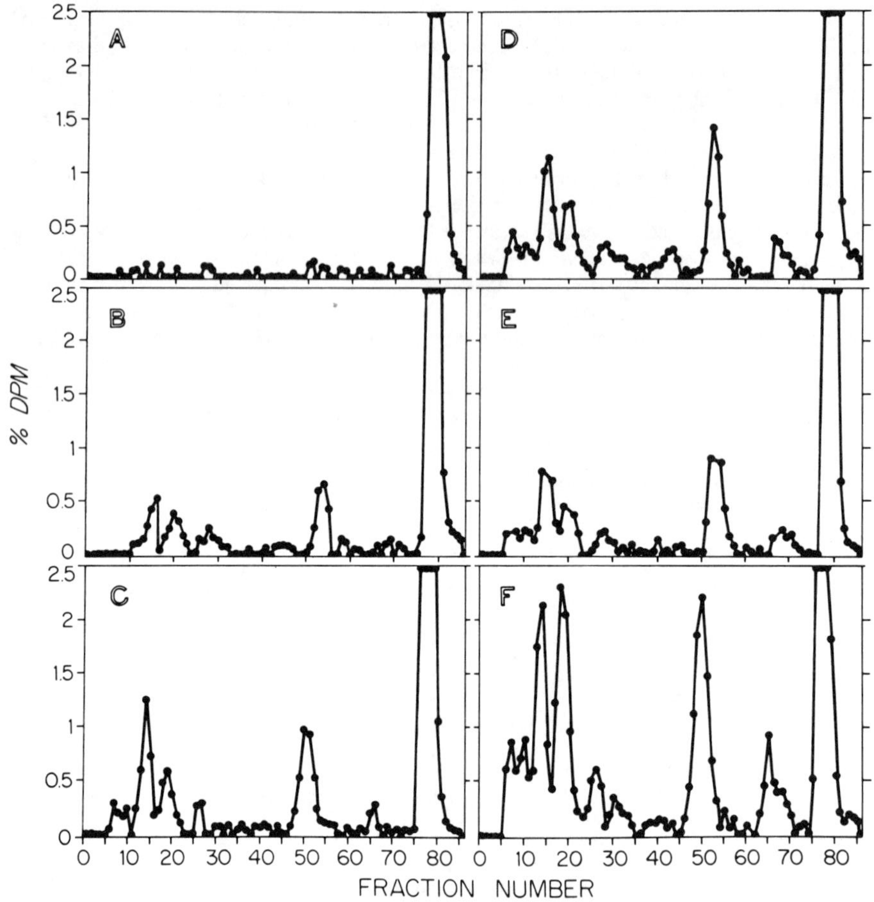

FIGURE 1. High-pressure liquid chromatographic profiles of [^{14}C]DMBA metabolites generated *in vitro* using epidermal homogenates from variously pretreated, female CD-1 mice. *Panel A,* heat inactivated control; *panel B,* DMBA-pretreatment (200 nmol 18 hr prior to sacrifice); *panel C,* 3-methylcholanthrene-pretreatment (200 nmol 18 hr prior to sacrifice); *panel D,* dibenz(a,c)anthracene-pretreatment (200 nmol 18 hr prior to sacrifice); *panel E,* Aroclor 1254-pretreatment (100 μg 18 hr prior to sacrifice); and *panel F,* TCDD-pretreatment (1 μg, 72 hr prior to sacrifice). All profiles have been normalized with respect to protein concentration. The peak fractions in which various synthetic standard compounds elute in this system are as follows: DMBA, 78; 5-hydroxy-DMBA, 72; 4-hydroxy-DMBA, 70; 2-hydroxy- and 3-hydroxy-DMBA, 61; 12-hydroxy-methyl-7-methylbenz(a)anthracene, 57; 7-hydroxymethyl-12-methylbenz(a)anthracene, 52; 5,6-dihydro-5,6-dihydroxy-DMBA, 22; 7,12-dihydroxymethylbenz(a)anthracene, 19; 8,9-dihydro-8,9-dihydroxy-DMBA, 13. Further details of the methodology can be found in reference 27.

To further test the possibility that the effects of TCDD on DMBA-mediated tumor initiation were due to changes in metabolism of the procarcinogen, a series of experiments were designed to monitor covalent binding of ^{3}H-DMBA to epidermal DNA, RNA and protein (TABLE 4). TCDD was applied topically to the shaved backs of CD-1 mice (1 μg per mouse) 3 days prior to application of 2.56 μg ^{3}H-DMBA (10 μCi). Control animals received acetone alone. At 3 and 24 hours after treatment, epidermal DNA, RNA and protein were extracted using the method of Huberman and Sachs.[30] TCDD pretreatment reduced the binding of ^{3}H-DMBA to DNA at 3 and 24 hours respectively by 63% and 72% (TABLE 4). Similar results were observed with respect to covalent binding of ^{3}H-DMBA to RNA although the magnitude was less. However, the binding of ^{3}H-DMBA to epidermal protein was not affected by TCDD pretreatment.

DISCUSSION

TCDD and Aroclor 1254 previously were shown to possess little or no tumor-initiating properties[31] or tumor-promoting properties[33] in mouse skin using the two-stage, initiation-promotion system of carcinogenesis. Previous studies by Kociba[3] and Van Miller[34] indicated that TCDD was carcinogenic in the rat; however, this carcinogenic effect appeared to be related mainly to the promotion of spontaneously occurring tumors. Furthermore, with a hepatic system utilizing an initiation-promotion regimen,[20–23] PCBs appeared to act as promoters.

This report demonstrates that TCDD possesses remarkable inhibitory actions on skin tumor-initiation to polycyclic aromatic hydrocarbons (PAH). Almost complete inhibition of DMBA tumor initiation is achieved with a single nontoxic topical dose of 0.1 μg and with a dose of 0.01 μg approximately 80% inhibition was achieved. This potent anticarcinogenic effect may be related to the ability of TCDD to induce epidermal enzyme pathways responsible for detoxifying PAH carcinogens in the skin. Induction of microsomal enzymes in several tissues has been postulated as a mechanism for the anticarcinogenic effects of a wide variety of compounds such as PAH, flavones, coumarins, phenothiazines, and phenobarbital as reviewed by Wattenberg.[35] Induction of microsomal enzyme systems also has been implicated in the anticarcinogenic activity of DDT[19] and various polychlorinated biphenyl mixtures.[17] However, investigations such as these have not demonstrated conclusively that the inhibitory actions are a result of induction within the target tissue for a particular carcinogen or that the effect is on tumor-initiation *per se*.

The unusually potent inhibitory action of TCDD on skin tumor-initiation by PAH described in this report stands in contrast to a report by Kouri[36] who showed an enhancing effect of TCDD on the formation of 3-MC-induced sarcomas at the site of subcutaneous injection. Both compounds were administered simultaneously and the

TABLE 4

COVALENT BINDING OF ^{3}H-DMBA TO MOUSE EPIDERMAL MACROMOLECULES AS MODIFIED BY PRETREATMENT WITH TCDD

		Hydrocarbon bound to macromolecule (fmoles/mg)†		
Treatment	Time (Hr)*	DNA	RNA	Protein
DMBA	3	48	29	695
DMBA + TCDD†	3	18	16	638
DMBA	24	126	72	388
DMBA + TCDD†	24	35	32	394

*Time of sacrifice after application of ^{3}H-DMBA; 40 mice were used for each experimental group.

†Mice in this group received 1 μg of TCDD, applied topically, 72 hours prior to application of ^{3}H-DMBA.

increased tumor formation was attributed to the ability of TCDD to enhance the metabolism of 3-MC *in vivo.* In view of the time of administration, this explanation seems somewhat unlikely. Furthermore, as shown in TABLE 3, when TCDD was applied 5 minutes prior to initiation with DMBA, little or no effect on the tumor response was observed.

The anticarcinogenic effects of TCDD and Aroclor 1254 appear to correlate well with their ability to induce monooxygenase enzymes of the skin. Furthermore, the quantity of ^{3}H-DMBA bound to DNA and RNA (but not protein) in the presence and absence of TCDD-pretreatment correlated well with the tumor response under similar conditions. When taken together, these experiments suggest that pretreatment with TCDD gives rise to an increased rate of inactivation of the DMBA molecule relative to the rate of activation in mouse skin. Although these data point to induction of oxidative biotransformation as a possible mechanism for the inhibitory effects exhibited by TCDD, other mechanisms could be operating. Possibilities include induction in epidermal tissues of nonoxidative metabolic pathways such as epoxide hydratase, glutathione-S-transferase, UDP-glucuronosyltransferase and others. Other possibilities include effects on DNA-repair systems and on the distribution of the carcinogen to the critical target site(s).

It is deemed possible that future research will enable the development of compounds that evoke minimal toxicity but retain the capacity to inhibit tumor initiation by virtue of their enzyme-inducing properties. Of particular advantage would be the production of chemicals capable of selectively inducing various inactivating enzymes in particular target tissues. Such chemicals could feasibly serve as important chemopreventive agents in cases in which contact with tumor initiators is necessary or unavoidable.

REFERENCES

1. POLAND, A. & A. KENDE. 1976. 2,3,7,8-Tetrachlorodibenzo-*p*-dioxin: environmental contaminant and molecular probe. Fed. Proc. **35:** 2404–2411.
2. POLAND, A. & E. GLOVER. 1977. Chlorinated biphenyl induction of aryl hydrocarbon hydroxylase activity: a study of the structure-activity relationship. Mol. Pharmacol. **13:** 924–938.
3. KOCIBA, R. J., D. G. KEYES, J. E. BEYER, R. M. CARREON, C. E. WADE, D. A. DITTENBER, R. P. KALNINS, L. E. FRAUSON, C. N. PARKE, S. D. BARNARD, R. A. HUMMEL & C. G. HUMISTON. Results of a two-year chronic toxicity and oncogenicity study of 2,3,7,8-tetrachlorodibenzo-*p*-dioxin (TCDD) in rats. Toxicol. Appl. Pharmacol. In press.
4. MCKINNEY, J. D., K. CHAE, B. N. GUPTA, J. A. MOORE & J. A. GOLDSTEIN. 1976. Toxicological assessment of hexachlorobiphenyl isomers and 2,3,7,8-tetrachlorodibenzofuran in chicks. I. Relationship of chemical parameters. Toxicol. Appl. Pharmacol. **36:** 65–80.
5. GOLDSTEIN, J. A., J. D. MCKINNEY, G. W. LUCIER, P. HICKMAN, H. BERGMAN & J. A. MOORE. 1976. Toxicological assessment of hexachlorobiphenyl isomers and 2,3,7,8-tetrachlorodibenzofuran in chicks. II. Effects on drug metabolism and porphyrin accumulation. Toxicol. Appl. Pharmacol. **36:** 81–92.
6. BICKERS, D. R., L. C. HARBER, A. KAPPAS & A. R. ALVARES. 1972. Polychlorinated biphenyls: Comparative effects of high and low chlorine containing Aroclors® on hepatic mixed function oxidase. Res. Commun. Chem. Pathol. Pharmacol. **3:** 505–512.
7. BICKERS, D. R., A. KAPPAS & A. P. ALVARES. 1974. Differences in inducibility of cutaneous and hepatic drug metabolizing enzymes and cytochrome P-450 by polychlorinated biphenyl and 1,1,1-trichloro-2,2-bis-(*p*-chlorophenyl)-ethane (DDT). J. Pharmacol. Exp. Therap. **188:** 300–309.
8. MARNIEMI, J., M. NOKKALA, H. VAINIO & K. J. W. HARTIALA. 1977. Stimulation of

hepatic drug hydroxylation and conjugation by a cutaneously applied PCB mixture (Clophen A50). Chem. Biol. Interactions **18:** 247–251.

9. Vaino, H. 1974. Enhancement of microsomal drug oxidation and glucuronidation in rat liver by an environmental chemical, polychlorinated biphenyl. Chem. Biol. Interact. **9:** 379–387.
10. Poland, A. & E. Glover. 1974. Comparison of 2,3,7,8-tetrachlorodibenzo-*p*-dioxin, a potent inducer of aryl hydrocarbon hydroxylase, with 3-methylcholanthrene. Mol. Pharmacol. **10:** 349–359.
11. Goldstein, J. A., P. Hickman, H. Bergman, J. D. McKinney & M. P. Walker. 1977. Separation of pure polychlorinated biphenyl isomers into two types of inducers on the basis of induction of cytochrome P-450 or P-448. Chem. Biol. Interact. **17:** 69–87.
12. Gupta, B. N., J. G. Vos, J. A. Moore, J. G. Zinkl & B. C. Ballock. 1973. Pathologic effects of 2,3,7,8-tetrachlorodibenzo-*p*-dioxin in laboratory animals. Environ. Health Perspect. **5:** 125–140.
13. Vos, J. G., J. A. Moore & J. G. Zinkl. 1974. Toxicity of 2,3,7,8-tetrachlorodibenzo-*p*-dioxin (TCDD) in C57BL/6 mice. Toxicol. Appl. Pharmacol. **29:** 229–241.
14. Lee, K. P., R. R. Hebert, H. Sherman, J. G. Aftomis & R. S. Waritz. 1975. Octabromodiphenyl-induced ultrastructural changes in rat liver. Arch. Environ. Health **36:** 465–471.
15. Kociba, R. J., P. A. Keller, C. N. Park & P. J. Gehring. 1975. 2,3,7,8-Tetrachlorodibenzo-*p*-dioxin (TCDD): Results of a 13-week oral toxicity study in rats. Toxicol. Appl. Pharmacol. **35:** 553–574.
16. Kay, K. 1977. Polybrominated biphenyls (PBB) environmental contamination in Michigan, 1973–1976. Environ. Res. **13:** 74–93.
17. Makiura, S., H. Aoe, S. Sugihara, K. Hirao, M. Arai & N. Ito. 1974. Inhibitory effect of polychlorinated biphenyl on liver tumorigenesis in rats treated with 3′-methyl-4-dimethylaminoazobenzene, *N*-2-fluorenyl-acetamide and diethylnitrosamine. J. Nat. Cancer Inst. **53:** 1253–1257.
18. Kimura, N. T., T. Kanematsu & T. Baba. 1976. Polychlorinated biphenyl(s) as a promoter in experimental hepatocarcinogenesis in rats. Z. Krebsforsch. **87:** 257–266.
19. Okey, A. B. 1972. Dimethylbenzanthracene-induced mammary tumors in rats. I. Inhibition by DDT. Life Sci. **11:** 833–843.
20. Kimbrough, R. D., R. A. Squire, R. E. Linder, J. D. Strandberg, R. J. Montali & M. W. Burse. 1975. Induction of liver tumors in Sherman strain female rats by polychlorinated biphenyl Aroclor 1260. J. Nat. Cancer Inst. **55:** 1453–1459.
21. Shulte-Herman, R. 1977. Two-stage control of cell proliferation induced in rat liver by α-hexachlorocyclohexane. Cancer Res. **37:** 166–171.
22. Peraino, C., R. J. M. Fry, E. Staffeldt & J. P. Christopher. 1975. Comparative enhancing effects of phenobarbital, amobarbital, diphenylhydantoin and dichlorodiphenyltrichloroethane on 2-acetylaminofluorene induced hepatic tumorigenesis in the rat. Cancer Res. **35:** 2884–2892.
23. Uchiyama, M., T. Chiba & K. Noda. 1974. Co-carcinogenic effect of DDT and PCB feedings on methylcholanthrene-induced chemical carcinogenesis. Bull. Environ. Contam. Toxicol. **12:** 687–693.
24. Berenblum, I. & P. Shubik. 1974. A new quantitative approach to the study of the stages of chemical carcinogenesis in mouse's skin. Br. J. Cancer **1:** 383–391.
25. Boutwell, R. K. 1964. Some biological aspects of skin carcinogenesis. Prog. Exp. Tumor Res. **4:** 207–250.
26. Scribner, J. D. 1973. Tumor initiation by apparently noncarcinogenic polycyclic aromatic hydrocarbons. J. Nat. Cancer Inst. **50:** 1717–1719.
27. VanDuuren, B. L., A. Sivak & B. M. Goldschmidt. 1970. Initiating activity of aromatic hydrocarbons in two-stage carcinogenesis. J. Nat. Cancer Inst. **44:** 1167–1173.
28. Baird, W. M. & R. K. Boutwell. 1971. Tumor-promoting activity of phorbol and four diesters of phorbol in mouse skin. Cancer Res. **31:**1074–1079.
29. DiGiovanni, J., T. J. Slaga, D. L. Berry & M. R. Juchau. 1977. Metabolism of 7,12-dimenthylbenz(a)anthracene in mouse skin homogenates analyzed by high pressure liquid chromatography. Drug Metab. Dispos. **5:** 295–301.

30. Huberman, E. & L. Sachs. 1977. DNA binding and its relationship to carcinogenesis by different polycyclic hydrocarbons. Int. J. Cancer **19:** 122–127.

31. DiGiovanni, J., A. Viaje, D. L. Berry, T. J. Slaga & M. R. Juchau. 1977. Tumor initiating ability of 2,3,7,8-tetrachlorodibenzo-*p*-dioxin (TCDD) and Aroclor 1254 in a two-stage system of mouse skin carcinogenesis. Bull. Environ. Contam. Toxicol. **18:** 552–557.

32. Poland, A., E. Glover, A. S. Kende, M. DeCamp & C. M. Giandomenico. 1976. 3,4,3′4′-Tetrachloroazoxybenzene and azobenzene: potent inducers of aryl hydroxylase. Science **194:** 627–630.

33. Berry, D. L., J. DiGiovanni, M. R. Juchau, W. M. Bracken, G. L. Gleason & T. J. Slaga. 1978. Lack of tumor-promoting ability of certain environmental chemicals in a two-stage mouse tumorigenesis assay. Res. Commun. Chem. Pathol. Pharmacol. **20:** 101–108.

34. VanMiller, J. P., J. J. Lalich & J. R. Allen. 1977. Increased incidence of neoplasms in rats exposed to low levels of 2,3,7,8-tetrachlorodibenzo-*p*-dioxin. Chemosphere **10:** 625–632.

35. Wattenberg, L. W. 1978. Inhibitors of chemical carcinogenesis. *In* Advances in Cancer Research. G. Klein and S. Weinhouse, Eds. Vol. 26: 197–226. Academic Press. New York.

36. Kouri, R. E. 1976. Relationship between levels of aryl hydrocarbon hydroxylase activity and susceptibility to 3-methylcholanthrene and benzo(a)pyrene-induced cancers in inbred strains of mice. *In* Polynuclear Aromatic Hydrocarbons. R. Freudenthal and P. Jones, Eds. pp. 139–151. Raven Press. New York.

THE CARCINOGENIC AND OTHER CHRONIC EFFECTS OF PERSISTENT HALOGENATED ORGANIC COMPOUNDS

Renate D. Kimbrough

Toxicology Branch, Center for Disease Control
United States Public Health Service
Department of Health, Education, and Welfare
Atlanta, Georgia 30333

A number of persistent halogenated organic compounds have been shown to produce live tumors in rodents, such as DDT, dieldrin, mirex, Kepone®, hexachlorobenzene, polychlorinated and polybrominated biphenyls (PCBs and PBBs) and now also 2,3,7,8 tetrachlorodibenzo-*p*-dioxin (TCDD) as well as hexachlorodibenzodioxin. TCDD also induced tumors of the lungs, hard palate or nasal turbinates in rats. It is possible that other isomers of the halogenated dibenzodioxins are also carcinogenic. Representatives of other groups of compounds in this category, the halogenated dibenzofurans, naphthalenes and terphenyls have not been tested for a carcinogenic effect. Most of the chemicals mentioned are persistent in mammals once they have been absorbed. They are also poorly degraded in the environment. For this class of compounds it is generally assumed that not the chemicals themselves but their metabolites are the proximate carcinogens. TCDD may be an exception. While most of the chemical compounds mentioned are commercial products, the halogenated dibenzodioxins and dibenzofurans are contaminants of a number of technical products. Some of these products that may theoretically be contaminated are listed in TABLE 1.

The persistent halogenated organic compounds under discussion induce mixed-function oxidases and cause hepatomegaly in rodents and other mammals. The hepatomegaly is the result of the hypertrophy of individual hepatocytes. Some hyperplasia also occurs and increased mitotic activity can be noted. The hepatocytes may accumulate lipid in their cytoplasm. At the ultrastructural level the enlarged hepatocytes show an increase in smooth endoplasmic reticulum and inclusions within the cytoplasm, which appear like concentric whorls, surrounding lipid vacuoles. Morphologic changes in the mitochondria have also been described (Gaines *et al.*[1]; Kimbrough *et al.*[2,3]). In addition to these alterations, the PCBs, PBBs, some isomers of the halogenated dibenzodioxins as well as hexachlorobenzene induce experimental hepatic porphyria. In the rat experimental hepatic porphyria only occurs in the female, but in other species it can be induced in both sexes. On microscopic examination an increase in macrophages and prominent Kupffer cells containing brown ceroid pigment and necrobiosis of liver cells is prominent. Lipid accumulates in the cytoplasm of hepatocytes resulting at times in hepatocytes with foamy cytoplasm. Similarly, lipid accumulation, lipofuchsin (ceroid pigment) have also been reported in livers of humans suffering from porphyria cutanea tarda (Biempica *et al.*[4]).

Adenofibrosis (cholangiofibrosis) (Kimbrough *et al.*[3]) has been observed in rats fed PCBs and technical but not pure pentachlorophenol (Kimbrough and Linder[5]). The PCBs were contaminated with trace amounts of chlorinated dibenzofurans and the pentachlorophenol with chlorinated dibenzodioxins and dibenzofurans. The significance of adenofibrosis and its relationship to the development of hepatocellular carcinomas is presently not understood. It usually occurs concomitantly with

0077–8923/79/0320–0415 $01.75/0 © 1979, NYAS

hepatocellular carcinoma in rodents (Kimbrough and Linder[6]; Kimbrough *et al.*[7]) and has also been described in rats exposed to other carcinogens such as *p*-dimethylaminoazobenzene (Edwards and White[8]). Although hepatocellular tumors have also been reported in rodents after exposure to DDT, dieldrin, mirex, and Kepone, these compounds do not produce the same type of liver pathology as outlined above. It is possible that PCBs, PBBs and related compounds cause lipid peroxidation while mirex, Kepone, DDT, and dieldrin do not. Whether the experimental hepatic porphyria produced by these compounds is in any way related to the induction of hepatocellular carcinoma is presently not known. A study reported by Jones and Sweeney[9] is interesting in this regard. These authors treated genetically responsive (DBA/2J) or nonresponsive (DBA/2J) mice to the induction of arylhydrocarbon hydroxylase with TCDD. The nonresponsive mice had normal hepatic uroporphyrinogen decarboxylase activity while it was decreased in the responsive mice, and uroporphyrin excretion was increased in this group. In the study of Kociba *et al.* presented in this volume, only the female rats showed hepatic porphyria, and they also developed hepatocellular carcinomas while the male rats had neither. It should be explored whether an effect on certain enzymes related to porphyrin metabolism is coupled with an effect on enzymes somehow related to the induction of hepatocellular carcinomas, or whether hepatic porphyria and hepatocellular carcinoma only coexist.

TABLE 1

EXAMPLES OF COMMERCIAL PRODUCTS THAT MAY BE CONTAMINATED WITH HALOGENATED DIBENZODIOXINS OR DIBENZOFURANS

2,4,5-Trichlorophenol	Fenchlorphos (ronnel)
Pentachlorophenol	Erbon
Hexachlorobenzene	Trichloronate
2,4-Dichlorophenyl-*p*-nitrophenyl ether (TOK)	Hexachlorophene
2,4,5-Trichlorophenoxyacetic acid	2,4,4′-Trichloro-2′ hydroxydiphenyl ether (Irgasan DP300)
Fenoprop (Silvex)	Polychlorinated biphenyls

Hepatic porphyria (porphyria cutanea tarda) has also been reported in humans following occupational exposure to some of these chemicals and porphyrin excretion should be screened for carefully in such cases in the future. The sex difference for the induction of hepatic porphyria found in rats does not exist in humans. The PCBs, TCDD and perhaps also the PBBs affect reproduction. The reduction of hormone-dependent tumors in the study by Kociba *et al.* may be indicative of an influence of these compounds on steroids. Other halogenated persistent organic compounds also affect endocrine glands. Kepone has been shown to cause hyperplasia of the adrenal cortex in rats (Cannon and Kimbrough[10]) and in quail (Eroschenko and Wilson[11]). High doses of technical hexachlorobenzene also caused hyperplasia of the adrenal cortex in rats (Kimbrough and Linder[12]).

All of the halogenated organic compounds that I have mentioned are persistent in biologic systems. Certain isomers of PBBs and PCBs seem to be retained indefinitely (Kimbrough *et al.*[13]). If a single high dose of a mixture of PBBs is given to weanling rats, hepatocellular tumors can be induced. It is therefore not necessary to conduct 2-year feeding studies with these compounds to show that they are carcinogenic. A single high dose at least of the PBBs will have the same effect. When rats were given

1 g PBB/kg body weight by stomach tube as weanlings, neoplastic nodules began to appear at about 6 months after dosing (Kimbrough *et al.*[13]). It was also found that PBB levels in the liver did not regress. It is presently not known whether the isomers that persist are the ones that are carcinogenic. If the persistent isomers prove to be carcinogenic, then the tumor induction would be due to the continuous internal exposure of the target organ to these isomers.

Neoplastic nodules are considered to represent part of the spectrum of response elicited by hepatocarcinogens in rodents. First, areas of alteration will develop, some of these will become neoplastic nodules and some of the neoplastic nodules will in time transform into hepatocellular carcinomas (Squire and Lewitt[14]). A number of carcinogens have produced this same spectrum of response and there is no difference in the response, whether the chemical producing it is a carcinogen of long standing or whether it is one of the halogenated compounds discussed here. Although it has been shown many times that the neoplastic nodules are part of the carcinogenic response they are not always included in the statistical evaluation of bioassays, which may lead to erroneously interpreted results, particularly when they are classified as hyperplastic nodules or "nodular hyperplasia" and when the number of animals studied was small (Carcinogenesis Testing Program, 1977[15]).

Since all of these compounds are also inducers of mixed-function oxidases and would enhance the metabolism of other xenobiotics, they may, depending on the circumstances, prevent or enhance the tumor induction by other chemicals. Active metabolites would have to be formed for many of the halogenated compounds, tetrachlorodibenzo-*p*-dioxin perhaps being an exception. The dermal application of these compounds that have to be activated would therefore not necessarily result in tumor formation. Further studies will have to be done to elucidate these problems. The question of skin carcinogenesis is particularly important in workers that have had occupational dermal exposure to TCDD, PCBs or chlorinated naphthalenes as evidenced by chloracne.

References

1. Gaines, T. B. & R. D. Kimbrough. 1970. Oral toxicity of mirex in adult and suckling rats. Arch. Environ. Health **21:** 7–14.
2. Kimbrough, R. D., T. B. Gaines & R. E. Linder. 1971. The ultrastructure of livers of rats fed DDT and dieldrin. Arch. Environ. Health **22:** 460–467.
3. Kimbrough, R. D., R. E. Linder & T. B. Gaines. 1972. Morphological changes in liver of rats fed polychlorinated biphenyls. Arch. Environ. Health **25:** 354–364.
4. Biempica, L., N. Kosower, M. H. Ma & S. Goldfischer. 1974. Hepatic porphyrias, cytochemical and ultrastructural studies of liver in acute intermittent porphyria and porphyria cutanea tarda. Arch. Pathol. **98:** 336–343.
5. Kimbrough, R. D. & R. E. Linder. 1978. The effect of technical and purified pentachlorophenol on the rat liver. Toxicol. Appl. Pharmacol. **46:** 151–162.
6. Kimbrough, R. D. & R. E. Linder. 1974. Induction of adenofibrosis and hepatomas of the liver in Balb/cd mice by polychlorinated biphenyl (Aroclor 1254). J Nat. Cancer Inst. **53:** 547–552.
7. Kimbrough, R. D., R. A. Squire, R. E. Linder, J. D. Strandberg, R. J. Montali & V. W. Burse. 1975. Induction of liver tumors in Sherman Strain female rats by polychlorinated biphenyl Aroclor 1260. J. Nat. Cancer Inst. **55:** 1453–1459.
8. Edwards, J. E. & J. White. 1941. Pathologic changes with special reference to pigmentation and classification of hepatic tumors in rats fed *p*-dimethylaminoazobenzene (butter yellow). J. Nat. Cancer Inst. **2:** 157–183.
9. Jones, K. G. & G. D. Sweeney. 1977. Association between induction of aryl hydrocarbon

hydroxylase and depression of uroporphyrinogen decarboxylase activity. Res. Commun. Chem. Pathol. Pharmacol. **17:** 631–637.

10. CANNON, S. B. & R. D. KIMBROUGH. Short-term Kepone toxicity in rats: including reproduction, organ pathology and reversibility. Toxicol. Appl. Pharmacol. In press.
11. EROSCHENKO, V. P. & W. O. WILSON. 1975. Cellular changes in the gonads, livers and adrenal glands of Japanese quail as affected by the insecticide Kepone. Toxicol. Appl. Pharmacol. **31:** 491–504.
12. KIMBROUGH, R. D. & R. E. LINDER. 1974. The toxicity of technical hexachlorobenzene in the Sherman strain rat. A preliminary study. Res. Commun. Chem. Pathol. Pharmacol. **8:** 653–664.
13. KIMBROUGH, R. D., V. W. BURSE & J. A. LIDDLE. 1978. Persistent liver lesions in rats after a single oral dose of polybrominated biphenyls (Fire Master FF-1) and concomitant PBB tissue levels. Environ. Health Perspect. **23:** 265–273.
14. SQUIRE, R. A. & M. H. LEVITT. 1975. Report of a workshop on classification of specific hepatocellular lesions in rats. Cancer Res. **35:** 3314–3223.
15. Carcinogenesis Testing Program, Division of Cancer Cause and Prevention, National Cancer Institute, National Institute of Health, Bethesda, Md. 20014. October 20, 1977. Bioassay of Aroclor 1254 for possible carcinogenicity.

REPRODUCTIVE EFFECTS OF HALOGENATED AROMATIC HYDROCARBONS ON NONHUMAN PRIMATES*

J. R. Allen, D. A. Barsotti, L. K. Lambrecht, and J. P. Van Miller

Department of Pathology and Regional Primate Research Center
University of Wisconsin
Madison, Wisconsin 53706

During the past several years experiments have been conducted in this laboratory to determine the toxicologic effects in animals exposed to three halogenated hydrocarbons—2,3,7,8-tetrachlorodibenzo-*p*-dioxin (TCDD), polychlorinated biphenyls (PCBs) and polybrominated biphenyls (PBBs). TCDD and PCBs are considered to be worldwide environmental contaminants. PBBs have caused major environmental problems in the state of Michigan. The health implications of these compounds are further complicated by their extremely slow biodegradation and their ability to accumulate in body tissues and remain there for an indefinite period.

As a result of their environmental and public health significance, a large amount of literature has been published on TCDD and PCB intoxication in humans and lower animals. There is also considerable research in progress to further clarify the toxicologic effects of the PBBs. However, there is a dearth of information on the effects of these halogenated hydrocarbons on reproduction in humans and closely related species, particularly when the exposure to these compounds is at low levels and occurs over an extended period.

The rhesus monkey is a suitable animal model with which to study the effects of exposure to halogenated hydrocarbons. Many of the toxicologic signs produced by these halogenated hydrocarbons are similar to those observed in humans. Thus, the data obtained from research conducted on the nonhuman primate are helpful in postulating the injurious effects that may arise in the human population following exposure to these compounds.

In the presently reported research, the three compounds were evaluated for their toxicologic effects on nonhuman primates. Attempts have been made to simulate the levels to which humans may be exposed. Data are presented which indicate that low-level, long-term exposure to these compounds gives rise to reproductive abnormalities in the adult primate as well as fetal and neonatal toxicity.

POLYCHLORINATED BIPHENYLS

Adult female rhesus monkeys were fed a diet containing 2.5 and 5.0 ppm of a PCB mixture (Aroclor 1248). Within 4 months and after having consumed from 60 to 120 mg of the PCBs, their menstrual cycles were decidedly altered. Menses was 5 to 7 days

*This investigation was supported in part by United States Public Health Service grants ES00958, ES00472, ES01339, ES01839 and RR00167 from the National Institutes of Health, and the University of Wisconsin Sea Grant Program. A portion of this research was conducted at the University of Wisconsin-Madison Biotron, a controlled environmental facility funded by the National Science Foundation and the University of Wisconsin-Madison. Primate Center Publication No. 18-004.

0077–8923/79/0320–0419 $01.75/0 © 1979, NYAS

longer in the experimental animals than in the controls. There was also an increase in menstrual bleeding. In addition, flattening and prolongation of the serum progesterone peak during the menstrual cycle was observed.

After being on the experimental diets for 7 months and having consumed from 105 to 210 mg of PCB,[2,3] attempts were made to breed the females to control males. All the animals fed 2.5 ppm PCB and the control animals conceived (TABLE 1). Six of the 8 animals on the 5.0 ppm diet conceived. Three of the 8 females on the 2.5 ppm diet and 4 of the 8 females on the 5.0 ppm diet experienced early abortions or resorptions. One of the females on the 5.0 ppm diet delivered a stillborn infant. Five females from the 2.5 ppm group and 1 female from the 5.0 ppm group were able to carry their infants to term.[2]

All of the infants born to the experimental animals were small (1 to 2 standard deviations from normal) but appeared normal in other respects. At birth there were detectable levels of PCB in the skin of these infants (1.0 to 4.8 μg/g tissue). In addition, analysis of the tissues of the stillborn infant of a female from the 5.0 ppm group showed large amounts of PCBs in the adrenals (25 μg/g) and pancreas (64 μg/g).[4]

TABLE 1

REPRODUCTIVE PERFORMANCE OF MONKEYS EXPOSED TO POLYCHLORINATED BIPHENYLS IN THEIR DIETS FOR 18 MONTHS

	Breeding Performance			
	During PCB Exposure		One Year After PCB Exposure	
	2.5 ppm	5.0 ppm	2.5 ppm	5.0 ppm
Total impregnated	8/8	6/8	8/8	7/7
Absorptions/resorptions	3/8	4/8	1/8	1/7
Stillborn	0/8	1/8	0/8	1/7
Normal births	5/8	1/8	7/8	5/7

The surviving infants were permitted to remain with their mothers, and their sole source of nourishment was through nursing. During this period the infants showed a consistent weight gain although it was less than that of the controls (TABLE 2). Within 2 months the infants began to show signs of PCB intoxication that included the development of acne, swelling of the eyelids, loss of eyelashes and increased pigmentation of certain portions of the skin.[4]

Following the development of PCB intoxication in the infants, milk samples were obtained from the females to determine PCB content. Levels in the whole milk ranged from 0.154 to 0.397 ppm PCB.[5]

The infant from the mother on the 5.0 ppm diet and two infants from mothers fed the 2.5 ppm diet died from PCB intoxication. At necropsy, the animals were small for their age and exhibited focal hyperpigmented areas of the skin. The skin of the face was dry and crusty with isolated acneform lesions. The upper eyelids were swollen, and there was an absence of eyelashes. There were only rudimentary thymuses, and the spleen and lymph nodes were small. There was also a decided thickening of the gastric mucosa. Microscopically, thymocytes were sparse, the lymph nodules of the spleen were small with inapparent germinal centers, and there was a decided hypocellularity of the bone marrow. The Meibomian glands of the eyelids were hypertrophic, and the hair follicles of the eyelashes were keratinized. In addition, the stomach showed

TABLE 2

BODY WEIGHTS OF INFANT MONKEYS BORN TO MOTHERS RECEIVING 2.5 or 5.0 ppm PCB (AROCLOR 1248) IN THEIR DIETS PRE- AND POSTPARTUM

Age (weeks)	Body Weight (grams)* Control	Experimental
0	507 ± 59	399 ± 22
2	601 ± 47	463 ± 40
4	711 ± 58	517 ± 12
6	796 ± 81	601 ± 38
12	1003 ± 71	765 ± 57

*Mean ± one standard deviation.

marked hyperplasia of the mucous-secreting cells of the gastric mucosa with glandular elements penetrating the underlying muscularis mucosa.[4]

The three surviving infants were weaned at 4 months of age and placed on a milk replacer. These infants showed improvement within 4 months after being removed from their mothers. Their eyelashes became more plentiful, the swelling of the eyelids disappeared, and the acne was less apparent. However, the hyperpigmentation particularly about the hair line of the face persisted.[6]

After being weaned the three infants were evaluated for their learning and behavioral development. It was found that these infants were hyperactive and did not learn reversal tasks as readily as control infants. These changes have persisted in the animals for over 2 years.[7]

After separation of the infants, all of the adult females were removed from the experimental diets. During the subsequent year there was marked improvement in their physical status. Their menstrual cycles became normal, and the level of serum estradiol and progesterone were similar to pretreatment levels. One year after their removal from the experimental diets the females were bred again to control males. All of the animals conceived (TABLE 1). Seven of the 8 animals from the 2.5 ppm group, and 5 of the 7 animals from the 5.0 ppm group gave birth to live infants. However, the infants were much smaller than the control infants. These infants have not grown as rapidly as the controls and have experienced a greater frequency of illnesses. Focal areas of hyperpigmentation have developed in these infants. Two infants from mothers in the 5.0 ppm group have died. The thymuses were small, and the lymph nodes and spleens were rudimentary in these animals.

In subsequent experiments female rhesus monkeys were fed diets containing 0.5 and 1.0 ppm of Aroclor 1248 three times per week. These animals showed no

TABLE 3

REPRODUCTIVE PERFORMANCE OF MONKEYS FED 0.5 AND 1.0 PPM POLYCHLORINATED BIPHENYLS (AROCLOR 1248) THREE TIMES PER WEEK DURING THE PRE- AND POSTPARTUM PERIODS

No. Animals	Level in Diet (ppm)	No. Conceived	Live Births	Birth Weights	PCB Intake at Conception (mg)	PCB Intake After 4 Months Lactation (mg)
8	0.5	8/8	6/8	463	12	16
8	1.0	8/8	7/8	466	18	46

irregularities in their menstrual cycles or alterations in serum estradiol or progesterone. After being on the experimental diets for 7 months and having consumed 8 to 16 mg of PCBs, the females were bred to control males. All of the animals conceived (TABLE 3), and 6 of the 8 animals on the 0.5 ppm diet and 7 of the 8 animals on the 1.0 ppm diet gave birth to live infants. These infants were somewhat smaller than their controls and have gained weight less rapidly. During the nursing period they developed focal areas of skin hyperpigmentation. However, they were devoid of other signs of PCB exposure. The initial psychologic evaluations on these infants indicate behavioral patterns similar to infants exposed to higher levels of PCB.

2,3,7,8-TETRACHLORODIBENZO-*P*-DIOXIN (TCDD)

2,3,7,8-Tetrachlorodibenzo-*p*-dioxin (TCDD) has been fed to adult female rhesus monkeys at a level of 500 parts per trillion (ppt).[8] After 7 months of exposure, the length of the menstrual cycle as well as the intensity and duration of menstruation were not altered appreciably. However, decreases in serum estradiol and progesterone were observed in 5 of the animals.[9] The females were bred to control males starting in the seventh month of treatment. The two animals with normal serum estradiol and progesterone at the time of breeding conceived. However, only one of these animals was able to carry her infant to term. A third animal conceived and aborted early in gestation. The remaining two animals failed to conceive (TABLE 4).[10] All of the controls conceived and gave birth to normal infants.

In a second experiment 8 female rhesus monkeys were fed diets containing 50 ppt TCDD. During the initial 6 months of the experiment the animals experienced normal menstrual cycles, and the levels of serum estradiol and progesterone were unaltered. Following 7 months of TCDD exposure when the animals had consumed approximately 1.8 μg of TCDD, attempts were made to breed the animals. Four of the animals conceived but aborted their fetuses (TABLE 3). Two of the animals have been bred on repeated occasions and have not conceived. Two of the 8 animals conceived and were able to carry their infants to term. All of the 8 control animals conceived and gave birth to normal infants.

POLYBROMINATED BIPHENYLS

Seven rhesus monkeys were fed a diet containing 0.3 ppm of a mixture of polybrominated biphenyls (Firemaster® FF-1). After being on the experimental diet for 6

TABLE 4

REPRODUCTIVE PERFORMANCE OF RHESUS MONKEYS FED TETRACHLORODIBENZO-*p*-DIOXIN (TCDD) AND POLYBROMINATED BIPHENYLS (PBBS) IN THE DIET FOR SEVEN MONTHS PRIOR TO CONCEPTION AND DURING PREGNANCY

	TCDD		PBB
	50 ppt*	500 ppt	0.3 ppm
Total impregnated	6/8	3/8	7/7
Absorptions/resorptions	4/8	2/8	1/7
Stillborn	0/8	0/8	1/7
Normal births	2/8	1/8	5/7

*ppt = parts per trillion.

TABLE 5

GROWTH RATE OF INFANT MONKEYS BORN TO FEMALE MONKEYS THAT CONSUMED 0.3 ppm POLYBROMINATED BIPHENYL (PBB) FOR SEVEN MONTHS PRIOR TO CONCEPTION AND DURING PREGNANCY

Age	Control (grams)	PBB (grams)
Birth	519 ± 53*(7)†	455 ± 34 (5)
4 weeks	666 ± 89 (7)	518 ± 67 (5)
8 weeks	750 ± 65 (7)	605 ± 78 (5)
12 weeks	901 ± 48 (7)	706 ± 106 (5)
16 weeks	982 ± 62 (7)	807 ± 116 (5)

*Mean ± one standard deviation.
†Number in parentheses = number of animals.

months, 4 of the animals showed a flattening of the serum progesterone peak and a lengthening of the menstrual cycle.[11] After 7 months on the diet (total consumption per animal, 12 mg) the animals were bred to control males. All of the experimental animals conceived after 1 to 4 breedings (TABLE 4). Two of the animals showed a prolonged implantation bleeding. One animal aborted a mummified fetus at 146 days gestation and a second animal had a stillborn infant on the 154th day. The remaining 5 animals delivered small, normal-appearing infants. These infants did not gain weight as rapidly as the control infants (TABLE 5).

EFFECTS OF THE THREE HALOGENATED HYDROCARBONS ON MALE RHESUS MONKEYS

Polychlorinated Biphenyls

Four adult male rhesus monkeys were given a diet containing 5.0 ppm PCB (Aroclor 1248) for 18 months. During this time the animals consumed from 530 to 692 mg of PCB. During the initial year the animals showed only minor gross signs of PCB exposure.[2] The sperm counts and breeding performance of these four males were similar to those of the control males. However, between the 12th and 18th month one of the four animals developed clinical signs of PCB intoxication. During this period there was a marked decrease in the sperm counts, and he failed to impregnate females after repeated breedings. A testicular biopsy revealed a hypocellularity of the seminiferous tubules and absence of spermatozoa. A second biopsy taken one year after the animals were removed from the experimental diet showed a complete recovery of spermatogenesis.

TCDD

Male rhesus monkeys fed diets containing a mixture of dioxins including TCDD for periods in excess of 1 year experienced a decrease in spermatogenesis. Tissue samples of the testes showed an absence of spermatids or mature spermatozoa.[12]

Polybrominated Biphenyls

There are no data available on the effects of the PBBs on the reproductive capability of the male nonhuman primate.

Discussion

All three of the halogenated hydrocarbons when administered at levels ranging as low as parts per billion in the case of the PBBs and PCBs and parts per trillion of TCDD in the diet are capable of causing reproductive abnormalities. The menstrual cycles become irregular and in most instances prolonged. There is prolonged and excessive hemorrhage during menstruation. These changes are associated with a reduced conception rate as well as a high incidence of early abortions usually following extremely heavy postconceptional bleeding.

All of the above alterations are thought to be related to the changes in the sex hormones. The serum progesterone peak is flattened and lengthened in most cases. Those animals showing no alterations in serum estradiol and progesterone experienced none of the reproductive problems. However, in the animals with the most dramatic changes in sex hormones the reproductive alterations were most severe and in some instances progressed to an anovulatory state. The specific reasons for the modifications in the sex hormones are not presently known. Alterations in levels of enzymes involved in the breakdown of these hormones may, at least in part, be involved. Such enzyme alterations are characteristic in chlorinated hydrocarbon toxicity.

Female humans exposed to PCB-contaminated rice oil also showed symptoms suggesting abnormal ovarian function.[13–15] Over 60% of the patients showed abnormalities in the menstrual cycle, with the interval being prolonged, shortened or irregular. In approximately 55% of the patients, abnormal durations of menstrual flow were observed. In addition, the quantity of menstrual flow was either increased or decreased. The basal body temperature showed an abnormal pattern in 85% of the patients as characterized by a poor high phase and a prolonged low phase, suggesting the existence of corpus luteum insufficiency, retarded follicular maturation, or no ovular cycle. There are also preliminary reports that suggest menstrual irregularities in PBB-exposed persons.[16]

Reduced birth weight of infant nonhuman primates born to mothers treated with halogenated hydrocarbons is a consistent observation. In addition, toxic effects typical of these compounds have been observed in neonates. This latter finding is probably due to ingestion of milk containing the compounds since the infants recover rapidly after weaning. Similar findings have been reported in human infants born to mothers exposed to PCBs through consumption of contaminated rice oil.[15] It is of interest that fetotoxicity is observed in PCB-treated monkeys even when the mothers are removed from the experimental diets and all overt signs of reproductive dysfunction have disappeared.

Further complications of exposure to PCBs *in utero* and from mothers' milk have been suggested in the studies on nonhuman primates. In addition to the development of acne, swollen eyelids, loss of eyelashes, and reduced growth rate, these infants have inadequately developed lymphoid tissues which seemingly were responsible for their increased susceptibility to infection. More subtle effects of PCB exposure on the fetus and neonate occurred in the central nervous system. There are data which indicate that permanent learning and behavioral deficits are complications on PCB exposure during fetal and neonatal development in nonhuman primates.

There is only a limited amount of information on the effects of the halogenated hydrocarbons on reproductive capabilities of the male primate. In nonhuman primate studies one of four males that consumed approximately 600 mg of PCBs over 18 months became sexually inactive and aspermatic. The limited amount of data available on dioxin toxicity in nonhuman primates suggest that acute as well as chronic exposure results in a marked reduction in spermatogenesis. The Japanese have reported decreased libido in human males that were exposed to the PCB-contaminated rice oil.[15] In addition, there are reports of decreased libido in some Michigan farmers that were exposed to the PBBs.

References

1. Allen, J. R. 1975. Response of primates to polychlorinated biphenyl exposure. Fed. Proc. **34:** 1675–1679.
2. Barsotti, D. A., R. J. Marlar & J. R. Allen. 1976. Reproductive dysfunction in rhesus monkeys exposed to low levels of polychlorinated biphenyls. Food Cosmet. Toxicol. **14:** 99–103.
3. Barsotti, D. A. & J. R. Allen. 1976. The association of transplacental and mammary movement of PCBs with infant morbidity and mortality. Am. J. Pathol. **82:** 21a–22a.
4. Allen, J. R. & D. A. Barsotti. 1976. The effects of transplacental and mammary movement of PCBs on infant rhesus monkeys. Toxicology **6:** 331–340.
5. Allen, J. R. & D. H. Norback. 1976. Pathobiological responses of primates to polychlorinated biphenyl exposure. *In* Proceedings of the National Conference on Polychlorinated Biphenyls, EPA-560/6-75-004, p. 43–49. U.S. Environmental Protection Agency Office of Toxic Substances, Washington, D.C.
6. Carstens, L. A., D. A. Barsotti & J. R. Allen. 1978. Exposure to infant rhesus macaques (*Macaca mulatta*) to polychlorinated biphenyls. International Symposium on Pathobiology of Environmental Pollutants. In press.
7. Bowman, R. E., M. P. Heironimus & J. R. Allen. 1978. Correlation of PCB body burden with behavioral toxicology in monkeys, Pharmacol. Biochem. Behav. **9:** 49–56.
8. Allen, J. R., D. A. Barsotti, J. P. Van Miller, L. J. Abrahamson & J. J. Lalich. 1977. Morphological changes in monkeys consuming a diet containing low levels of 2,3,7,8-tetrachlorodibenzo-*p*-dioxin. Food Cosmet. Toxicol. **15:** 401–410.
9. Barsotti, D. A., L. J. Abrahamson & J. R. Allen. 1978. Hormonal alterations in female rhesus monkeys fed diets containing 2,3,7,8-tetrachlorodibenzo-*p*-dioxin. Bull. Environ. Contam. Toxicol. In press.
10. Allen, J. R. & J. P. Van Miller. 1978. Health implications of 2,3,7,8-tetrachlorodibenzo-*p*-dioxin exposure in primates. *In* Pentachlorophenol. K. R. Rao, Ed., pp. 371–379. Plenum Publishing Corp. New York.
11. Lambrecht, L. K., D. A. Barsotti & J. R. Allen. 1978. Responses of nonhuman primates to a polybrominated biphenyl mixture. Environ. Health Perspect. **23:** 139–147.
12. Allen, J. R. & L. A. Carstens. 1967. Light and electron microscopic observations in *Macaca mulatta* monkeys fed toxic fat. Amer. J. Vet. Res. **28:** 1613–1629.
13. Hirayama, C. 1976. Clinical aspects of PCB poisoning. In PCB Poisoning and Pollution. K. Higuchi, Ed., pp. 87–104. Academic Press. New York.
14. Kuratsune, M., T. Yoshimura, J. Matsuzaka & A. Yamaguchi. 1972. Epidemiologic study on Yusho, a poisoning caused by ingestion of rice oil contaminated with a commercial brand of polychlorinated biphenyls. Environ. Health Perspect. **1:** 119–128.
15. Kuratsune, M., Y. Masuda & J. Nagayama. 1976. Some of the recent findings concerning Yusho. *In* Proceedings of the National Conference on Polychlorinated Biphenyls, EPA-560/6-75-004, p. 14–29. U.S. Environmental Protection Agency Office of Toxic Substances, Washington, D.C.
16. L. Robertson. Personal communication.

SUSPECTED LINK BETWEEN EXPOSURE TO HEXACHLOROPHENE AND MALFORMED INFANTS

Hildegard Halling

Department of Chronic Somatic Diseases
Södertälje Hospital
Södertälje, Sweden

INTRODUCTION

The thalidomide catastrophe taught us that drugs which appear to be reasonably safe in adults can, even in small doses, severely injure a fetus. This may be wholly divorced from pharmacologic properties for which the drug is used therapeutically. The thalidomide catastrophe offered proof, for the first time, that the teratogenic dose may be lower in man than in any other mammal, possible because of faster metabolism in rats and other laboratory animals.[1]

The clinical impression that employees in the chronic unit of our hospital where hexachlorophene (HCP) was introduced for handwashing seemed to have an increased incidence of babies with congenital malformations inspired this study.

METHODS AND RESULTS

Malformations included in this study are so grave that they seriously affect the individual's capacity to survive, call for medical intervention, constitute a considerable physical handicap, or are disfiguring. On this basis there are malformations diagnosed at birth in 3% of all neonates in Sweden.

In hospitals A and B in Sweden, at the Department of Chronic Diseases, it was noted between 1970–1976 that of a population of 65 neonates born to mothers who had used the soap Sanitvål, containing 0.5% hexachlorophene, and pHisoHex with 3% hexachlorophene, there were six severely malformed infants (two infants with analatresia; one infant each with atresia of the esophagus and bilateral renal polycystic disease accompanied by pulmonary hypoplasia, one bilateral cleft lip and palate, and one case of unilateral microophthalmia with bilateral uveal coloboma.) In addition, there were several infants born with less severe malformations. (TABLE 1)

During this time, one nurse gave birth to an infant, dead *in utero*. At *post-mortem* examination, no malformations were found. A control maternal population was studied, consisting of staff who worked in the same hospital under similar conditions except that the soap used did not contain hexachlorophene. In this latter group, there were no severely malformed infants in 68 births; there was one minor anomaly with good prognosis. Examination of records of both hexachlorophene-using mothers and controls revealed no apparent differences in age, intercurrent infections or drug usage.[2]

These findings prompted examination of similar populations in four other hospitals, C, F, H, and V. These hospitals are similar to the chronic units of Hospitals A and B.

As a result of information from the heads of the departments, concerning staff on

0077–8923/79/0320–0426 $01.75/0 © 1979, NYAS

maternity leave and estimated partus, the case records were scrutinized at the respective obstetrical departments.

Because of special circumstances—re-arrangement of case records or departure of staff to home town in connection with childbirth—there is a shortfall in case reports from the obstetrical departments of about 15% for groups F, H, and V, whose case sheets were accordingly not examined.

TABLE 1

SUMMARY OF INFANTS WITH DIAGNOSED MALFORMATIONS
(Medical Personnel at Hospitals, A and B*)

	No.	Mother's Year of Birth	Child's Date of Birth (Year, Month)	Diagnosis of Malformation	Remarks
Major malformations	1	1944	1971, 04	Cleft lip and palate bilateral	Took 10–15 tablets of promethazinchlorid
	2	1944	1971, 08	Anal atresia + hypospadia	Chromosome examination 46 XY
	3	1948	1972, 11	Hypoplasia lung + cystic kidney bilateral	Died after 13 hours
	4	1946	1972, 05	Microphtalmia unilateral + coloboma bilateral	Actual visus: 0.0; 0.5 with corr. (1977)
	5	1952	1973, 08	Anal atresia	Contraceptive pillows when 1 month pregnant
	6	1935	1973, 03	Esophageal atresia + double ureter sin. + right kidney absence	Died 10 days later after surgical operation; mother immune to Rubella.
Minor deformities	7	1951	1973, 03	Talipes equino varus + slight ventricular septal defect	
	8	1941	1974, 03	Talipes calcaneo valgus	
	9	1952	1975, 12	Talipes plano valgus	
	10	1953	1974, 03	Slight ventricular septal defect	
	11	1953	1974, 11	Persistent ductus arteriosus	
	12	1944	1970, 09	Slight to moderate ventricular septal defect	

Total number of births with exposure to hexachlorophene: 65

Total number of births without exposure to hexachlorophene: 68; one dislocation of the hip.

*Exposure to hexachlorophene during the first trimester: 3% pHisoHex and 0.5% liquid soap Sanitvål during 1970 and to the end of November 1972. From December 1972 only 0.5% Sanitvål.

The requisition of hexachlorophene was checked, partly by means of the store-keepers' order forms at the respective hospitals, partly at the manufacturer, with regard to quantity and delivery date. A check was also made on preparations in which the manufacturer altered the content from e.g. hexachlorophene to irgasan without changing the name of the detergent or skin cream.

From 1969 until the beginning of 1975 the nursing staff at hospital C washed their hands 14–67 times per day with a liquid soap containing 1% hexachlorophene. During 1969, a handcream containing 0.5% hexachlorophene was also used. The mothers examined were exposed to hexachlorophene, at least, during the first trimester of pregnancy.

A total of 82 case reports from the Department of Obstetrics of Hospital C were examined from 1970 to the end of 1975. Among these, 14 cases of malformation were identified, four of them being severe: one hand absence; two cardiac malformations; one total transposition with malformed ventricular septum, who died within three hours; one defective atrial septum, who died at home 2.5 months later; and one diaphragmatic hernia with Dextrocardia, who died after four hours. (TABLE 2)

The "control group" contained 46 women who worked at Departments of Chronic Diseases under the same local conditions but without exposure to hexachlorophene. No malformed children of these mothers were found.

The nursing staff at the Hospitals F, H and V have reported 20–60 handwashings per day. Summaries of at birth diagnosed malformations are to be found in TABLES 3, 4, and 5.

There are three infants with chromosome deviations, Down's syndrome, whose mothers, 24, 28 and 30 years old, were exposed also one year before pregnancy. They are not included in the material.

A summary of neonates with minor deformities born to nonexposed mothers is given in TABLE 6.

TABLE 2

SUMMARY OF INFANTS WITH DIAGNOSED MALFORMATIONS

(Medical personnel at the Department of Chronic Somatic Diseases of Hospital C*)

	No.	Mother's Year of Birth	Child's Date of Birth (year, month)	Diagnosis of Malformation	Remarks
Major malformations	1	1946	1970, 08	Left hand absence.	Took Norfenefrini chlorid in early pregnancy
	2	1942	1970, 04	Cardiac malformation: total transposition + ventricular septal defect	Died after three hours
	3	1939	1970, 03	Diaphragmatic hernia + dextrocardia	Died after four hours
	4	1942	1973, 09	Cardiac malformation: atrial septal defect	Died at home aged 2.5 months; mother immune to Rubella
Minor deformities	5	1948	1971, 02	Undescended testis	
	6	1949	1972, 07	Bilateral hydrocele testes	
	7	1946	1974, 08	Preauricular skin tag	
	8	1946	1972, 04	Preauricular skin tag + fistula colli	
	9	1942	1972, 02	Bilateral hydrocele testes	
	10	1951	1975, 02	Talipe calcaneo-valgus	
	11–14		1971–1974	Four dislocations of the hip	

Total number of births with exposure to hexachlorophene: 82.

Total number of births without exposure to hexachlorophene: 46. No malformation.

*Exposure to hexachlorophene: 1% liquid soap during at least the first trimester from 1969 until the beginning of 1975. 0.5% hand-cream added during 1969.

TABLE 3

SUMMARY OF INFANTS WITH DIAGNOSED MALFORMATIONS

(Medical personnel at the Department of Chronic Somatic Diseases of Hospital F*)

	No.	Mother's Year of Birth	Child's Date of Birth (year, month)	Diagnosis of Malformation	Remarks
Major malformations	1	1948 } same parents	1972, 09	Hydrocephaly, probably Arnold Chiari's syndrome	Died a few days before birth Both parents chromosomally normal; rubella negative; no medicine; no cigarettes; eczematous hands Two spontaneous abortions in 1973
	2	1948	1974, 11	Anencephaly	Died after one hour
	3	1947	1973, 08	Meningocele	Missed abortion
	4	1946	1973, 03	Pulmonary stenosis	Surgical operation
	5	1954	1974, 09	Bilateral cleft lip and palate (CLP)	Father's grandmother had CLP

Total number of births with exposure to hexachlorophene: 44.

Total number of births without exposure to hexachlorophene: 15. There is one slight Metatarsus varus.

*Exposure to hexachlorophene during at least the first trimester: 3% pHisoHex and 0.5% liquid soap in 1970 and to the end of February 1975.

The statistical analyses of the data are based on the so called Mantel-extension of the Mantel-Haenszel procedure for the calculation of p values (TABLE 7).

EVALUATION

The control groups are not always from the same period, usually *after* the use of hexachlorophene cleansing agents was stopped, sometimes before they were introduced, occasionally simultaneously, although on another ward.

Only 22 mothers were interviewed with regard to heredity, use of medical preparations, smoking and drinking habits, virus diseases with fever, and on place of sojourn at the critical time for the respective malformation. No common teratogenic factor emerged.

All three women who had given birth to infants with cleft lip and palate (CLP) have CLP in the family. The incidence of CLP in families of children with CLP need not exclude the possibility that an exogenous factor is significant. An exogenous factor may be added to the genetic factors, giving rise to a manifest CLP. Analogous conditions have been described in connection with anticonvulsants.

Blood tests for hexachlorophene could not be carried out on hospital personnel until March 1976, and measurable quantities of hexachlorophene were confirmed in serum.

Exposure to other hazardous chemicals at the hospitals could not be detected.

Two neonates with dislocation of the hip were born at a time when it was not clear whether or not the mothers, one of whom was a diabetic, had been exposed at the relevant stage of pregnancy; these two were therefore excluded from the material.

Discussion

The history of hexachlorophene goes back to 1941 when the substance was patented. Its use gradually increased through the years; it was included in more and more products, such as soaps, hair tonics, face lotions, lip salves, creams, powder, anti-perspirants, gargles, skin creams, toothpastes and hygienic sprays. Hexachlorophene is also often used in gardening as a fungicide on citrus fruits and vegetables.

Table 4

Summary of Infants with Diagnosed Malformations

(Medical personnel at the Department of Chronic Somatic Diseases of Hospital H*)

	No.	Mother's Year of Birth	Child's Date of Birth (year, month)	Diagnosis of Malformation	Remarks
Major malformations	1	1951	1971, 12	Microcephaly + meningocele, polydactyly (6 fingers and 6 toes), microphtalmia bilat. + CLP bilat. + cystic kidneys bilat. Probably Trisomy E	Died one hour after birth
	2	1945	1974, 03	Analatresia + choanalatresia + abn. pupil + cardiac malformation?	
	3	1954	1974, 06	Esophageal atresia	
	4	1949	1974, 09	Omfalocele	
Minor deformities	5	1952	1974, 01	Digitus superadductus dig V pedis bilateral	
	6	1945	1972, 06	Hydrocele testes bilateral	
	7	1933	1972, 08	Hydrocele testes bilateral	
	8	1939	1973, 11	Hydrocele testes bilateral	
	9	1952	1973, 12	Probably slight ventricular septal defect	
	10–14		1970–1974	5 dislocations of the hip	

Total number of births with exposure to hexachlorophene: 97.

Total number of births without exposure to hexachlorophene: 26. There are two dislocations of the hip and one hydrocele testis.

*Exposure to hexachlorophene during at least the first trimester: 3% and 0.5% liquid soap during 1969 and from January 1970 to the end of December 1974 only 0.5% liquid soap. 0.3% Marlén handcream added from the beginning of 1972 to the end of 1974.

Isolated cases of fatal poisoning were reported, but they attracted little interest. A few cases of fatal poisoning after local applications to burn patients were also published.[3,7] In November, 1959, pHisoHex, with 3% hexachlorophene, was introduced in Sweden.

Crystalline hexachlorophene emulsion contains a substance to reduce surface tension which is absorbed by the keratin layer of the skin and penetrates the pores, hair follicles, and sebaceous glands. Hexachlorophene can penetrate unbroken skin

TABLE 5

SUMMARY OF INFANTS WITH DIAGNOSED MALFORMATIONS
(Medical personnel at the Department of Chronic Somatic Diseases of Hospital V*)

No.	Mother's Year of Birth	Child's Date of Birth (year, month)	Diagnosis of Major Malformation	Remarks
1	1951	1974, 01	Oesophageal atresia	
2	1946	1974, 01	Cleft lip and palate (CLP)	Among the relatives of the parents there are 3 cases of CLP.
3	1953	1974, 06	Bilateral talipe equino-varus	
4	1942	1974, 11	Cardiac malformation (single ventricle + truncus) + asplenia + malrotation	Died after 30 hours. Multiple sclerosis diagnosed 1961. There is one meningocele and one anal atresia among relatives.
5	1955	1975, 06	Artrogryposis multiplex	
6	1950	1975, 06	Hypoplastic left ventricular syndrome with mitral atresia + ventricular septal defect (VSD)	Died after 24 hours. Immune to Rubella.

No.	Mother's Year of Birth	Child's Date of Birth (year, month)	Diagnosis of Minor Deformities	Remarks
7	1943	1971, 01	Severe external ear dysplasia	
8	1949	1969, 10	Probably ventricular septal defect	
9	1943	1969, 12	Hypospadia laevis	
10	1946	1970, 03	Slight cardiac defect	
11	1947	1972, 03	Hydrocele testis	
12	1949	1973, 01	Polydactyly manus + polydactyly pedis + syndactyly pedis	
13	1939	1973, 11	Hydrocele testis	
14	1951	1973, 10		
14	1951	1973, 10 1973, 10	Slight cardiac defect, probably VSD	Twins
15	1946	1973, 12	Slight cardiac defect, probably persistent ductus arteriosus	
16	1951	1975, 02	Undescended testis	
17	1941	1975, 08	Slight cardiac deformity, probably VSD	
18–27		1969–1974	10 dislocations of the hip	

Total number of births with exposure to hexachlorophene: 172.

Total number of births *without* exposure to hexachlorophene: 78, 1 minor cardiac defect, 1 hydrocele testis, 1 dislocation of the hip.

*Exposure to hexachlorophene during at least the first trimester: 3% and 0.5% liquid soap from February 1969 to the end of February 1975. 0.3% handcream added from the beginning of 1972 to the end of 1974.

TABLE 6

SUMMARY OF INFANTS WITH MINOR DEFORMITIES
(Nonexposed Medical Personnel at Departments of Chronic Somatic Diseases of 6 Hospitals)

No.	Child's Date of Birth (year, month)	Diagnosis of Malformation	Remarks
1	68, 08	Bilateral hydrocele testis	Hospital H
2	68, 06	Dislocation of the hip	Hospital H
3	76, 01	Dislocation of the hip	Hospital H
4	77, 01	Metatarsus varus	Hospital F
5	76, 08	Dislocation of the hip	Hospital V
6	69, 02	Hydrocele testis	Hospital V
7	77, 04	Slight VOC	Hospital V
8	76, 02	Dislocation of the hip	Hospitals A and B

Total number of births without exposure to hexachlorophene: 233.

and small quantities are found in the blood, adipose tissue, brain, liver, and other organs.[4,5]

The half-life of hexachlorophene is 19 hours; it is rapidly excreted within 24 hours after absorption, and completely eliminated after 4–7 days. Most of it (84%) is excreted unchanged in the feces and urine within 96 hours after topical application.[6]

In an investigation in the USA of 75 volunteers who washed their hands with 5 ml of 3% hexachlorophene 4 times a day, the mean blood level after 21 days was 0.21 μg hexachlorophene per ml blood.[7] On the 10th day of a similar investigation with 16 volunteers, 8 men and 8 women, the mean value was in all cases 0.37 μg hexachlorophene per ml blood. But a number of the women consistently had high values, 0.78 μg/ml, 0.84 μg/ml and 0.90 μg/ml.[7]

Hexachlorophene can penetrate the placental barrier, and significant quantities have been discovered in the blood from the umbilicus of neonates.[4]

Hexachlorophene has an antibacterial effect on gram-positive organisms, especially staphylococci and streptococci. Thus, the resident bacteria present in the skin are inhibited. Use of hexachlorophene has reduced the number of staphylococcal infections at many hospitals throughout the world.[8] Its effectiveness can be diminished

TABLE 7

FREQUENCY OF MALFORMED INFANTS, WHOSE MOTHERS WERE EXPOSED AND NONEXPOSED TO HEXACHLOROPHENE DURING PREGNANCY
(Medical Personnel at Departments of Chronic Somatic Diseases of 6 Hospitals)

Degree of Malformation (and Score)	Exposed	Nonexposed
Major (2)	25	0
Minor (1)	46	8
None (0)	389	225
Total births	460	233

Scoring the severeness of malformation as 0, 1, 2 the Mantel extension of the Mantel-Haenzsel test gives $\chi^2(1) = 23.5$ $p < 0{,}001$ (two-tailed).

by low pH, and inhibited by serum proteins and Tween 80.[8] Hexachlorophene has little or no effect on coliform bacteria and no effect on tubercle bacilli.

Pathologic Effect

It has been demonstrated that local application of hexachlorophene to skin can cause vacuolation and edema, status spongiosus in the heavily myelinated long tracts of the brain-stem (medial longitudinal fasciculus, spinotectal tract, inferior medial lemniscus and superior cerebellar peduncles) and intramyelinic vacuolization of the interperiod line[9]; 20–80 mg/kg causes irreversible vacuolation of myelin in the optic nerve in sheep, with loss of myelin and axons and eventual gliosis, atrophy, and blindness.[10]

Hexachlorophene is a toxic and photosensitizing substance with an oral LD_{50} for rats of 56 mg/kg for females and 66 mg/kg for males.[11] When administered intravenously to rats, the LD_{50} is 7.5 mg/kg.[12] The oral LD_{50} for dogs is 40–50 mg/kg,[13] and for sheep 30–60 mg/kg.[14]

When suspensions of 45%, 12%, and 3% hexachlorophene were introduced into the vagina of pregnant rats on the seventh day of gestation, the highest concentration gave a significant number of malformed fetuses, namely 44%. The 12% concentration was less teratogenic and produced 19% malformations. The 3% solution had no teratogenic effect. The control group yielded 2% malformations. Plasma concentration of hexachlorophene was 6–10 times higher following vaginal administration when compared to dermal application. The hexachlorophene concentration was highest in the vagina and liver after vaginal administration but extremely low in the maternal brain and the embryos. Two female rats from the 45% hexachlorophene group died two days after the administration. In this group, the most common malformations were hydrocephalus, anophthalmia, microphthalmia, wavy ribs, urogenital deformities, and cleft palate.[15]

Winthrop Laboratories requested an investigation of the possible teratogenic effect of hexachlorophene on rats; the results were reported on August 7, 1972 by Industrial Bio-Test Laboratories, Inc., Northbrook, Illinois, USA; 15 mg/kg body weight and 30 mg/kg respectively were administered *per os*. Three of 45 rats in the 30 mg/kg group died. The incidence of abnormal fetuses in the 15 mg/kg group was 27.2% and 27.9% in the 30 mg/kg group in contrast to 7.6% in controls. In addition, abnormalities of the eye (4/219) and angulated ribs (10/219) occurred only at the higher dosage.

Fertilized female rabbits were exposed to 6 mg/kg hexachlorophene, administered *per os*, and low frequency of rib defects (3/175) appeared in their offspring.[16] The teratological response to hexachlorophene after oral administration was adjudged minimal when the dose approached the amount which caused maternal toxicity.

Oakley and Shephard[17] observed 7% frequency of cleft palate in offspring of rats who had received 500 ppm hexachlorophene *per os* from day 7–8 of gestation. Rats receiving 100 ppm produced normal offspring. A 500 ppm administered to rats *per os* was regarded as probably teratogenic.

Human peripheral blood lymphocytes treated with hexachlorophene *in vitro* did not give evidence for increased chromosome aberrations, but hexachlorophene suppressed mitosis at concentrations as low as 40 μg/ml applied for 6 hours.[19]

The toxicity and justification of hexachlorophene in various products was not challenged until autumn 1971. The controversy was triggered by a pamphlet from the Food and Drug Administration (FDA) in the USA, which took exception to the washing of neonates with 3% hexachlorophene.[7] The blood levels for hexachlorophene in neonates washed with hexachlorophene were far too near those causing microscopic changes in the white brain matter of monkeys and rats. FDA recommended that hexachlorophene should not be included in cosmetics except as a preservative in a content of 0.1% at most, that a safety test be compulsory on products containing 0.1–0.7%, and that products with over 0.75% be placed on prescription. It may be used for washing neonates only if infection is present that cannot be treated in any other way.

The most important pharmacologic principle is the dramatic demonstration that hexachlorophene applied topically is a drug with systemic toxicity and that it can pentrate the placental barrier. This study will suggest hexachlorophene to be teratogenic.

Summary

Retrospective studies, performed in 6 hospitals in Sweden, demonstrated that severe congenital malformations occurred more frequently in neonates born to mothers who used hexachlorophene soaps, 10–60 times a day by handwashing and hand-creams during at least the first trimester of pregnancy (25 severe malformations in 460 neonates), when compared to neonates born to similarly employed mothers who did not use hexachlorophene soaps (0 severe malformations in 233 neonates). Minor congenital anomalies were also more commonly noted in the hexachlorophene-exposed group when contrasted with controls. Other known adverse influences, such as age, infections during pregnancy, other drug usage, or smoking habits did not appear different in the two groups.

References

1. Kalter, H. Supplement to Teratology workshop manual: A collection of lectures and demonstrations from the second workshop in teratology. Berkeley, California, January 25–30, 1965.
2. Halling, H. 1977. Suspected Link Between Exposure to Hexachlorophene and Birth of Malformed Infants. Läkartidningen **74:** 542–546. Sweden
3. Chilcote, R., et al. 1977. Hexachlorophene storage in a burn patient associated with encephalopathy. Pediatrics **59**(3): 457–459, Mar.
4. Curley, A., R. E. Hawk, R. D. Kimbrough, G. Nathenson & L. Finberg. 1971. Dermal absorption of hexachlorophene in infants. Lancet **2:** 296.
5. Ulsamer, A. G., P. D. Yoder, R. D. Kimbrough & F. N. Marzulli. 1975. Effects of hexachlorophene on developing rats: toxicity tissue concentrations and biochemistry. Food Cosmet. Toxicol. **13:** 69–80.
6. Guck, L. 1975. A perspective on hexachlorophene. Pediatrics **51:** 400–412.
7. Lockhart, J. D. 1972. Hexachlorophene and the food and drug administration. J. Clin. Pharmacol. **9:** 445–450.
8. Walter, G. & W. S. Gump. 1963. Effect of pH on hexachlorophene. Soap Chemical Specialties **39:** 55.
9. Pleasure, D., J. Towfight, D. Silberberg & J. Parris. 1974. The pathogenesis of hexachlorophene neuropathy, *in vivo* and *in vitro* studies. Neurology **24:** 1068–1075.
10. Udall, V. 1972. Drug-induced blindness in some experimental animals and its relevance to toxicology. Proc. R. Soc. Med. **65:** 197–200.
11. Gaines, T. B., R. D. Kimbrough & R. E. Linder. 1973. The oral and dermal toxicity of hexachlorophene in rats. Toxicol. Appl. Pharamacol. **25:** 332–343.
12. Kimbrough, R. D. 1973. Review of the toxicity of hexachlorophene, including its neurotoxicity. J. Clin. Pharmacol. **13:** 430–444.

13. GUMP, W. S. 1969. Toxicological properties of hexachlorophene. J. Soc. Cosmetic Chemists **20:** 173.
14. UDALL, V. & J. C. MALONE. 1970. Optic nerve atrophy after drug treatment. *In* The problems of Species Difference and Statistics in Toxicology. Vol XI Excerpta Medica Foundation, Amsterdam, p. 224.
15. KIMMER, C. A., W. MOORE, JR. & J. F. STARA. 1972. Hexachlorophene Teratogenicity in Rats. Lancet, Oct. **7:** 765.
16. KENNEDY, G. L., S. H. SMITH, M. L. KEPLINGER & J. C. CALANDRA. 1975. Evaluation of teratological potential of hexachlorophene in rabbits and rats. Teratology **12:** 83–87.
17. OAKLEY, G. P. & T. H. SHEPHARD. 1972. Possible teratogenicity of hexachlorophene in rats. Teratology **5:** 264.
18. VIG, B. K. 1972. No chromosome aberrations induced in cultured human leucocytes by hexachlorophene. EMS Newsletter 6, 5.

NUMBER AND IDENTITY OF ANTHROPOGENIC SUBSTANCES KNOWN TO BE PRESENT IN BALTIC SEALS AND THEIR POSSIBLE EFFECTS ON REPRODUCTION

Sören Jensen and Bo Jansson

National Swedish Environment Protection Board
Special Analytical Laboratory
Wallenberg Laboratory
University of Stockholm
Stockholm, Sweden

Mats Olsson

Swedish Museum of Natural History
Section for Vertebrate Zoology
Stockholm, Sweden

Since the middle of the 1960s, we have been working continously on the identification of anthropogenic substances in Swedish wildlife. Most often, species from the top of the food chain have been selected. Thus, for the identification of the polychlorinated biphenyls (PCBs), muscle tissue from a white-tailed eagle was used.[1] Before long, however, seals from the Baltic became the selected species. The reasons were that the Baltic was found to be highly polluted, it was easier to get samples, and the fat content was large enough to ensure that even very low levels of pollutants could be identified.

In the following, we will try to summarize the work performed on seal blubber up to the present.

DDT AND PCBs

In a baseline study using herring as the test organism, it was found that the PCB levels were of the same magnitude all over the Baltic, while the levels of 1,1-bis(4-chlorophenyl)-2,2,2,-trichloroethane (DDT) and its main metabolites 1,1-bis(4-chlorophenyl)-2,2-dichloroethane (DDD) and 1,1-bis(4-chlorophenyl)-2,2-dichloroethene (DDE) reached a maximum in the southern part and decreased steadily when going toward the north. On the west coast of Sweden, the levels were one order of magnitude lower, as can be seen in FIGURE 1.[2]

The corresponding figures for DDT substances and PCBs in seals, which live mainly on herring, gave the same pattern, as shown in TABLE 1.[3]

The number of PCB substances in herring, was somewhat higher than in the predator, the seal, where it was around 50. The pattern also differed so that many of the low-chlorinated homologues present in the herring were almost absent in the seal, as shown in the chromatograms in FIGURE 2.[4] As the composition of the polychlorinated biphenyls was known through the work by Sissons and Welti[5] and Jensen and Sundström[6], it was possible to identify the biphenyls present in both seal and herring.

0077-8923/79/0320-0436 $01.75/0

TABLE 1

MEAN VALUES OF DDT AND PCB COMPOUNDS (PPM) AND STANDARD ERROR $s/\sqrt{n}$ IN EXTRACTABLE FAT OF BLUBBER*

Area	Grey seal n	Grey seal s DDT	Grey seal s PCBs	Ringed seal n	Ringed seal s DDT	Ringed seal s PCBs	Harbor seal n	Harbor seal s DDT	Harbor seal s PCBs
1. West coast							6	46 ± 11 (12–82)	60 ± 12 (28–110)
2. Northern Baltic Proper	18	420 ± 53 (170–970)	140 ± 17 (47–290)						
3. Åland Sea	27	270 ± 31 (68–850)	100 ± 11 (20–320)						
4. The Gulf of Bothnia	15	210 ± 28 (110–560)	100 ± 18 (49–330)	33	200 ± 28 (31–770)	110 ± 15 (27–390)			

$p < 0.05$ at correlation of the content of DDT-compounds between areas 2 and 3
$p < 0.01$ ↕ ↕ ↕ ↕ ↕ ↕ ↕ ↕ ↕ 2 ↕ 4
$p < 0.001$ 2 1
$p < 0.01$ at correlation of the content of PCB-compounds between areas 2 and 1

*The material has been divided into different collection areas and species; n = number of specimens analyzed. The extremes are shown within brackets (Olsson *et al.*[3]).

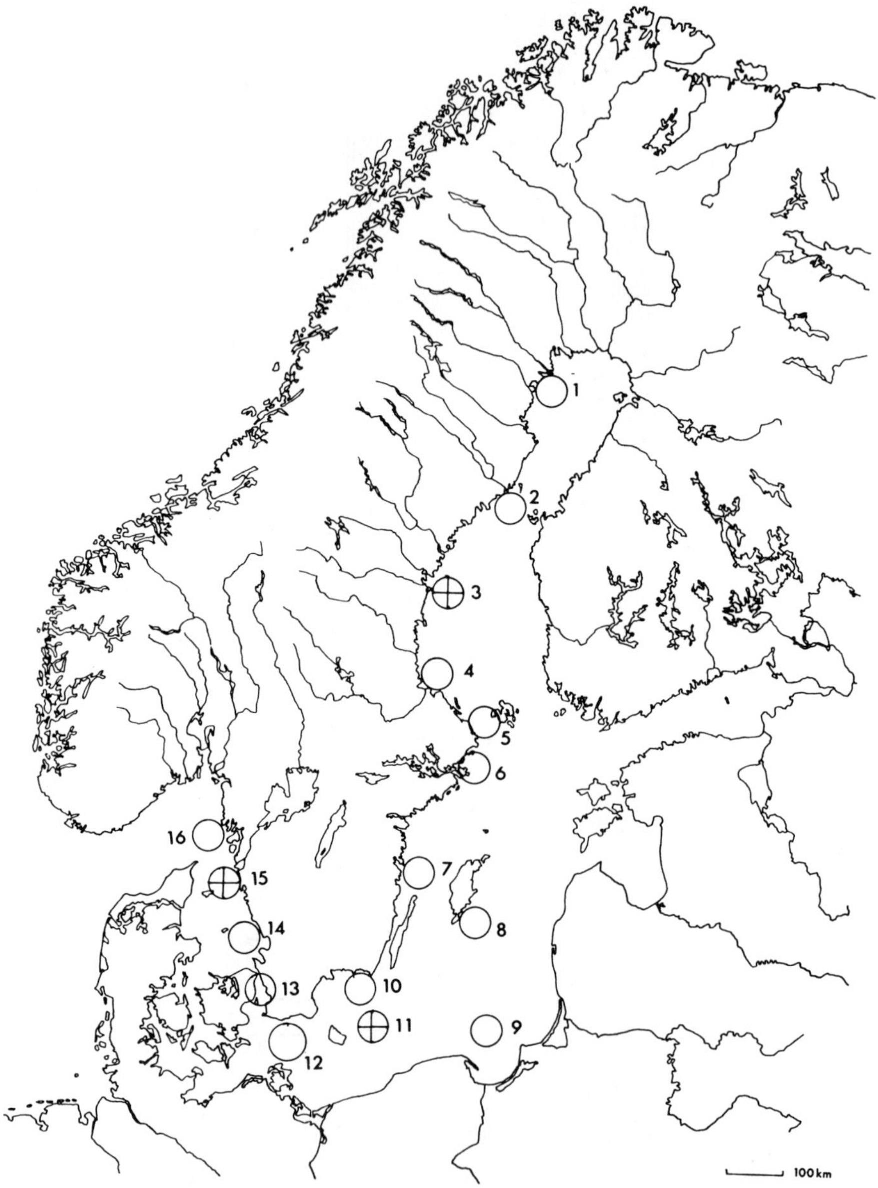

FIGURE 1. Sampling areas for herring in the Baltic and off the Swedish west coast. The bar diagram shows the obtained levels of DDT-compounds (black bars) and PCBs (white bars) (Jensen *et al.*,[2] reproduced with permission).

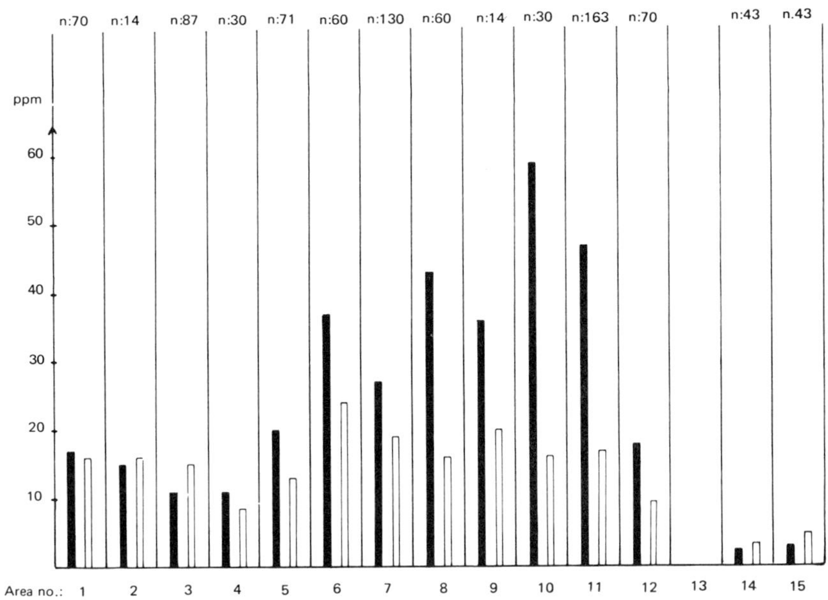

FIGURE 1. (*continued*)

HYDROXYLATED METABOLITES OF PCBS AND DDT

As many of the PCBs present in herring decrease on a relative basis in the seal, it was reasonable to believe that a metabolism of PCBs occurs in the seal. From laboratory experiments with some polychlorobiphenyls, it was known that PCBs could be converted to chlorinated phenols in the animal (for a review, see Sundström *et al.*[7]). The metabolites did not accumulate to any great extent, but were mainly excreted in feces. To study the metabolism of PCBs in the seal, feces samples were collected. These were extracted in the normal way with hexane and acetone. After isolation of the acidic fraction and treatment with diazomethane to convert any phenols to their methylethers, the extract was analyzed in a combined gas chromatograph-mass spectrometer, and the metabolites characterized by complete mass spectra. The number of biphenyls of different chlorination degrees were determined by mass chromatograms, and the result is given in TABLE 2.

In the total ion chromatogram, one major and two minor peaks were recognized, which did not correspond to any hydroxylated PCB. The molecular weights were 346 (^{35}Cl) and the number of chlorine atoms was 4, which indicated hydroxylated DDE, representing an unrecognized route for the metabolism of DDT. Through experiments with rats, it was found that the major component had the hydroxy group in a 3-position. In one of the minor components, however, the hydroxy group was situated in a 4-position, and the chlorine in a 3-position (FIGURE 3). Consequently, a so-called NIH shift had taken place. Also a metabolite with the hydroxy group in a 2-position could be traced.

Reproduction Problems in Seals

Parallel to the analytic attempt to identify and estimate the levels of pollutants in seals from the Baltic, the ecologists investigated if any changes in the population densities of seal had taken place in modern times. There are three species of seal in the Baltic, namely, the grey seal, the harbor seal, and the ringed seal. It turned out that the grey seal, whose hauling grounds are in the Swedish territorial waters at the

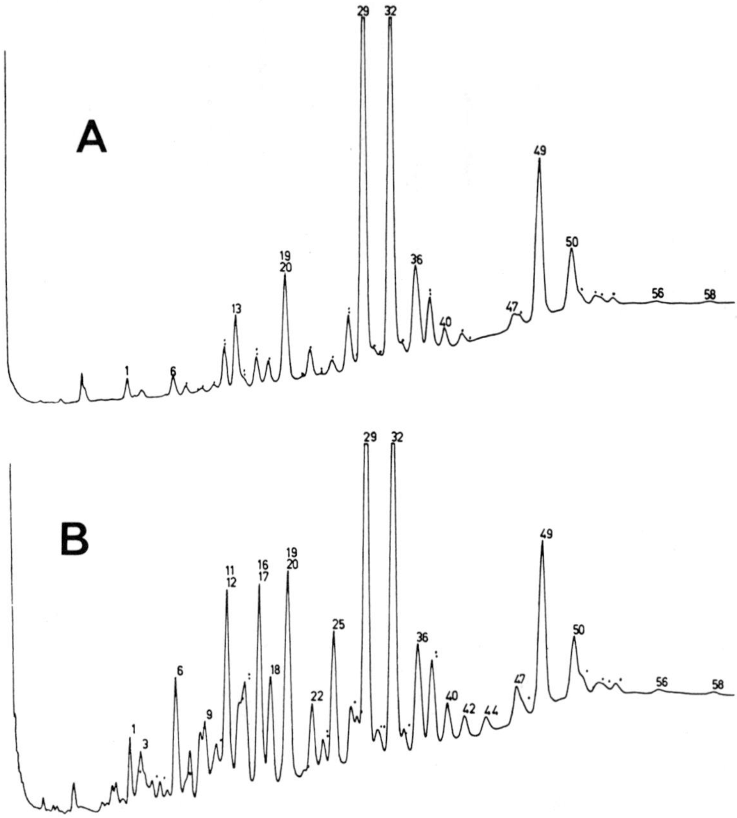

FIGURE 2. Gas chromatograms of extract from seal (A) and herring (B) tissues. The numbering refers to identified PCB compounds (see Jensen & Sundström[6]). Many of the substances with short retention times are, on a relative basis, less prevalent in the seal than in the herring (Jansson *et al.*[4]).

skerries of the southeastern coast, had decreased from >20,000 individuals in the 1940s, to a few thousands today. The harbor seal population does not exceed 200 individuals.[8]

Seal hunting might explain part of the decrease, but after World War II, very few seals have been shot, and seal hunting is today totally prohibited in Sweden. Therefore, this cannot be the explanation of the decrease during the late 1950s and 1960s.[9] Other possible causes are disturbances on the skerries from holiday boats,

TABLE 2
ACIDIC METABOLITES OF PCBs FOUND IN SEAL FECES
(Number of isomers containing n Cl and m OH substituents)*

nCl	m OH	Isomers
3	1	2
4	1	5
5	1	9
6	1	7
7	1	2
6	2	1

*(Jansson *et al.*[4])

aircraft, and commercial fishing and shipping, but these activities alone are not suspected to cause this effect on the seals.

The most probable explanation for the still decreasing number of seals, is that their reproduction is affected. Thus Helle[10] found that only 27% of the mature female ringed seals were pregnant. The corresponding figures for the same species in the Chotska inlet, and the Ochotskan Sea were 62.5 and 85–95%, respectively.[11,12]

The high levels of DDT and PCBs led to the suspicion that either one of these substance classes or both, possibly in combination, might effect reproduction. A report

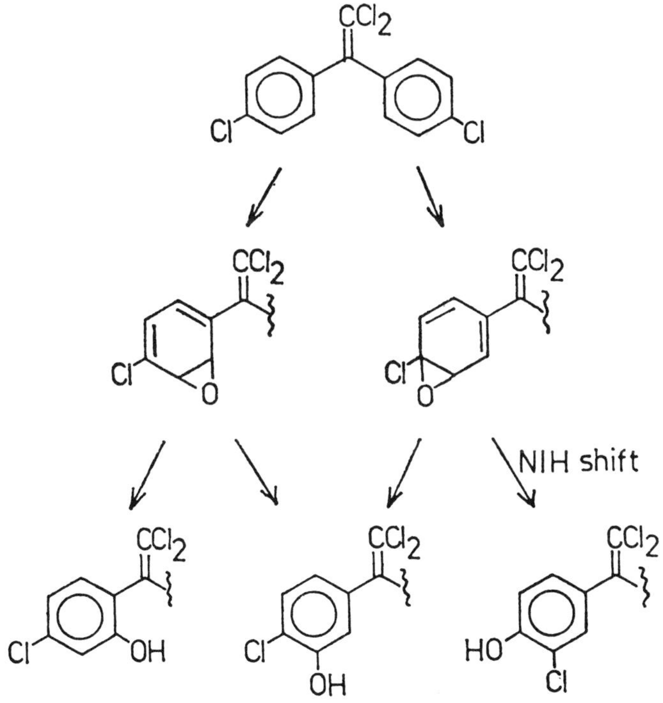

FIGURE 3. Metabolic routes for p,p′-DDE in rats (Sundström *et al.*[26]). These three phenolic metabolites have also been found in seal feces.

TABLE 3

LEVELS OF DDT AND PCB SUBSTANCES IN EXTRACTABLE FAT FROM BLUBBER OF RINGED SEALS FROM SIMO IN THE BOTHNIAN BAY†

Group	Sample	n	sDDT (mg/kg)‡	PCB (mg/kg)‡
I	Pregnant females	24	88 ± 9.7	73 ± 6.6
II	Nonpregnant females with stenosis and occlusions	29	130 ± 10	110 ± 7.8
III	Nonpregnant females with normal uteri	8	100 ± 15	89 ± 11
IV	Fetuses	24	62 ± 4.3	49 ± 3.0
V	Males	24	130 ± 18	100 ± 13
The probability of similarity between test groups (Student's t-test).				
I–II			$p < 0.01$**	$P < 0.01$**
I–III			$p > 0.05$	$p > 0.05$
I–IV			$p < 0.05$*	$p < 0.01$**

†(Helle *et al.*[14])
‡Mean level and S.E. $s/\sqrt{n}$.
**Groups significantly different at the 99% level.
*Groups significantly different at the 95% level.

on the increased number of aborted sea lion pups among the California sea lion populations supported this hypothesis.[3] Therefore, a joint Finnish-Swedish research program was started in 1975, in which the reproductive success of ringed seals, was related to their levels of DDT and PCBs (TABLE 3). It was found that both DDT and PCB levels were significantly higher in the nonreproducing group than in the reproducing one[13,14] and of the same magnitude as those reported to cause reproductive disturbances in California sea lions.[15] During the gestation period it was found that 35% of the seals were pregnant, 47% were nonpregnant with stenosis and occlusions in the uteri, and 27% were nonpregnant with normal uteri.

EXPERIMENTS WITH MINK

In 1971, Auerlich *et al.*[16] found decreasing reproduction and high mortality among mink fed Great Lake fishes. Ringer *et al.*[17] and Platonow and Karstad[18] stated that these effects were caused by PCBs. We therefore spiked mink food with a PCB mixture, similar to the composition in Baltic herring, and fed two groups of mink with 3.3 mg (corresponding to a normal level in Baltic fishes) and 11 mg per day, respectively, in 66 doses during the pregnancy.

It was found, as shown in TABLE 4, that the number of whelps in the low-dosage group decreased from 5.1 to 2.9 per female, and no whelps were born in the high-dosage group. The residues in mink from the high-dosage group were 280 ppm PCBs (fat weight basis), a quite normal value for seals from the Baltic. It was also shown that the number of implantations giving scars in the uterine wall was not changed in the PCB group. This means that resorbtion and maceration obviously cause the low reproduction rate.[19] In the minks fed DDT, no effects on reproduction could be observed.

Further Analytic Work

The relationship between the levels of PCBs and nonpregnancy in combination with the evidence for its effect on mink reproduction strongly indicate that PCBs are also the noxious agent in the seal. But a statistical relationship may give a false result if the effect is caused by an unknown substance covariating with PCBs. Furthermore, it is known that, with respect to the teratogenicity of a substance, it is difficult to compare the effect between species.

The analytic chemists were therefore urged to continue their unbiased search for unknown compounds in seals, to give a more complete list of accumulating substances. From the knowledge of the harmful effects of very low levels of 2,3,7,8-tetrachlorodibenzo-*p*-dioxin, it is necessary to start up with at least 100 grams of seal fat, to be able to identify small quantities. This can easily be accomplished by a partition procedure where the pollutants are partitioned between the fat itself and acetonitrile. Three liters of acetonitrile is enough to extract 99% of DDE, leaving behind 99% of the fat.[20] By means of adsorbtion chromatography, the presence of 1,1-bis(4-chlorophenyl)-2-chloroethene (DDMU), 4,4′-dichlorobenzophenone (DBP), dieldrin, heptachlorepoxide, hexachlorobenzene (HCB) and several isomers of hexachlorocyclohexane (BHC) could be verified.

Methyl Sulfones of PCBs and DDE

By means of a concentrated extract of seal and a gas chromatograph that was not switched off during the night, it was learned that a great number of unknown compounds were present. The retention times were very long at the temperature program normally used in PCB analysis. The substances disappeared from a hexane extract when treated with concentrated sulfuric acid, but could be recovered after dilution of the acid with ice. This property made their isolation rather simple.[20]

The result of a GC-MS investigation gave two series of substances:

Series 1:	$M(^{35}Cl)$	334	368	402	436	470
	Cl	3	4	5	6	7
Series 2:	$M(^{35}Cl)$		394			
	Cl		4			

The members of the first series all had molecular weights 78 mass units above the

Table 4

Reproduction of Mink Given PCBs for 66 Consecutive Days*

	Group A	Group B	Group C
No. of bitches	25	22	22
ppm PCBs in feed	0.05	3.3	11
ppm PCBs in extractable fat, mean value	14	86	280
Liver somatic index, mean values	0.344	0.408	0.554
Delivering bitches, percent of mated	92	73	0
No. of implantation sites/pregnant bitch	6.6	6.1	4.5
No. of whelps born/pregnant female	5.1	2.9	0.0

*(Jensen et al.[19])

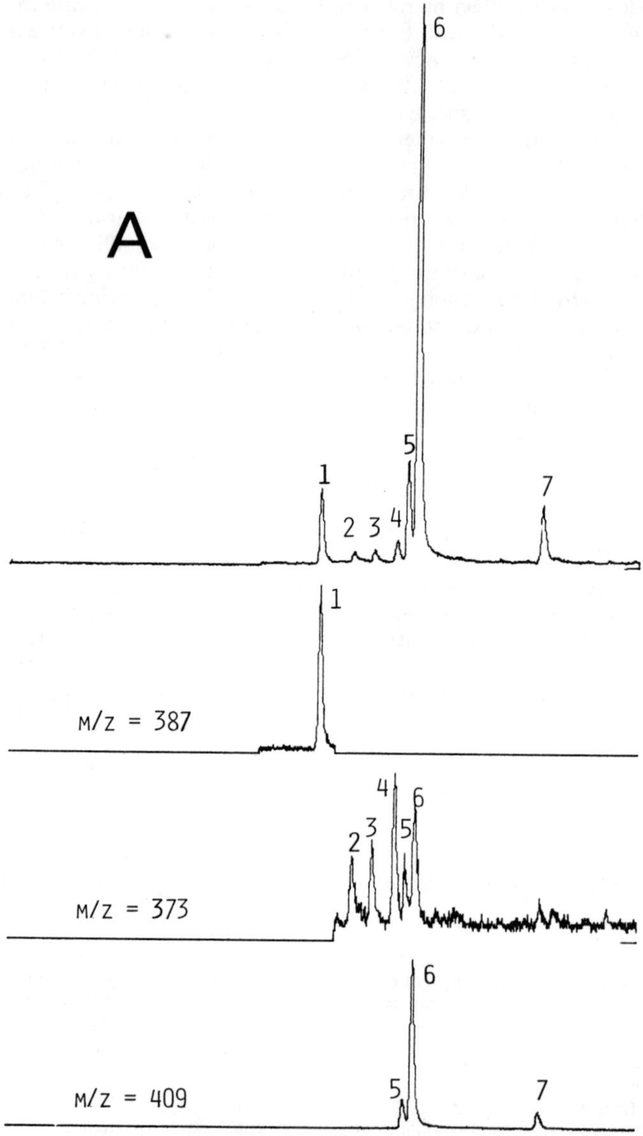

FIGURE 4. Compounds derived from chlordane in seal from the Baltic (A) and in a rat given technical chlordane (B). Compound 1 is oxychlordane, 2–4 are chlordanes and 5–7 nonachlors. The upper chromatograms represent the sum of the intensities of the measured ions.

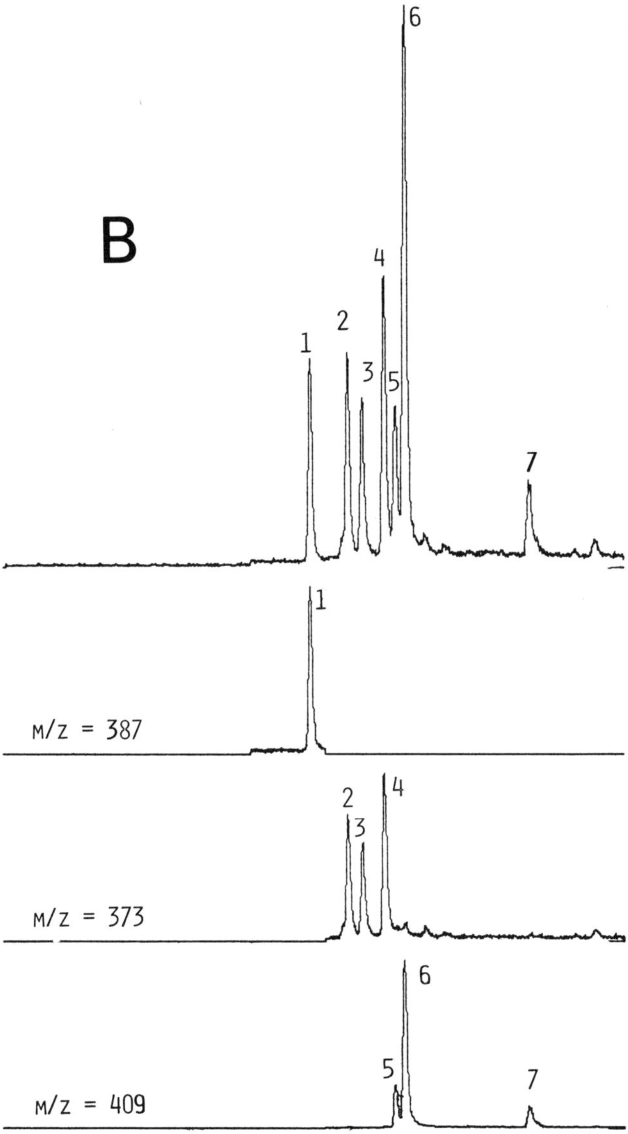

FIGURE 4. (*continued*)

corresponding chlorinated biphenyl, and the second 78 mass units above that of DDE.

This could be explained if the PCB and DDE compounds were substituted with a methyl sulfone group.

This hypothesis was confirmed by comparison with synthetic substances.[21] From independant animal experiments[22] using 2,2′,5,5′-tetrachlorobiphenyl it has been proved that the introduction of the methyl sulfone group takes place via a methylthioether, and that methionine is involved in the process.

In contrast to the behavior of the phenolic metabolites of PCB and DDE, which are rapidly excreted from the animal body, these methyl sulfones are accumulated. Thus, in the seal they are present in amount of 10–20 ppm, which is about one magnitude of ten below the concentration of PCBs.

Recently, both the methylthioethers and the methyl sulfones of PCBs were given to mink during the reproduction season to study whether the sulfones have effects similar to those of PCBs, and thus may be the active agents also when the mink is fed PCBs, as methyl sulfones could be shown to be present also in these minks. At least 30 methyl sulfones are present in the seal.

Another group of chlorinated hydrocarbons have recently been found in seals from the Baltic.[23] This group consists of substances derived from chlordane, a complex mixture of chlorinated cyclodienes used as a pesticide. Some of the components in this mixture are known to be more persistent than others, and these plus some metabolites are found accumulated.

In FIGURE 4, a mass fragmentographic analysis of an extract from seal blubber is compared with corresponding extract from a rat given technical chlordane. Three isomers of each chlordane and nonachlor can be observed as well as oxychlordane, a metabolite of chlordane.[24]

NUMBER OF CHARACTERIZED SUBSTANCES IN THE SEAL

This means that today, about 130 different chlorinated compounds are known to be pollutants in seals from the Baltic (TABLE 5).

TABLE 5

TYPE AND NUMBER OF KNOWN CHLORINATED POLLUTANTS IN SEAL FROM THE BALTIC

Type	Number
DDT,* DDE,* DDD,* DDMU, DBP	8
Phenols of DDE	3
Methyl sulfones of DDE	2
PCB	~ 50
Phenols of PCB	~ 26
Methyl sulfones of PCB	~ 30
HCB	1
BHC	3
Dieldrin	1
Chlordane	3
Nonachlor	3
Heptachlorepoxide	2
Oxychlordane	1
Total	~ 133

*Both *p,p′*- and *o,p′*-isomers.

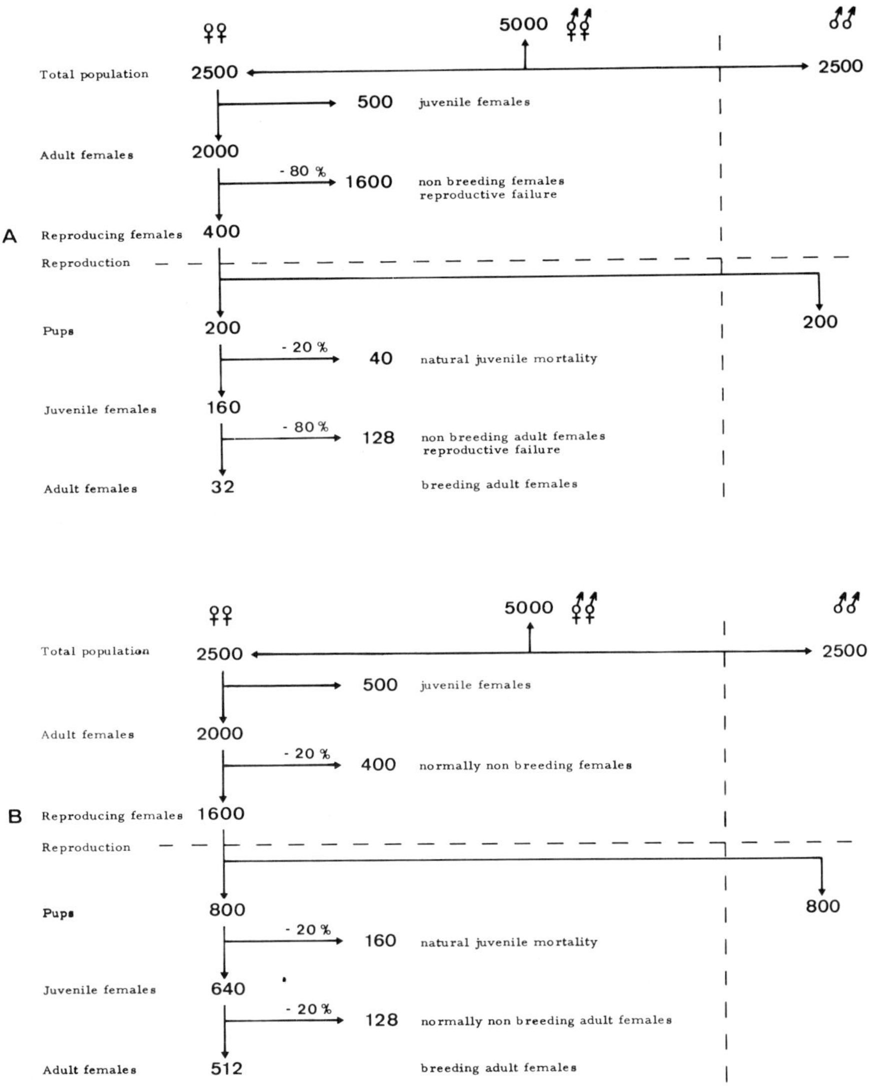

FIGURE 5. Two simple models of reproduction capacity: (A) in a seal population of 5000 animals where 80% of the adult females are nonpregnant. Of an annual reproductive rate of about 400 pups only about 30 reproducing females will enter the population after the juvenile period. (B) in a normal seal population of the same size. In the normal population the corresponding figure is 510 (Olsson[25]).

FATE OF THE BALTIC SEAL POPULATION

A model showing how a biomagnifying substance, affecting the reproduction system without killing any individual, will threaten the population is presented in FIGURE 5.[25] There are about 2000 adult females in the population, but only about 400 are normally reproducing animals. This means that simple population surveys are not always relevant.

CONCLUSIONS

It can be said that it is not fully proved, but highly probable, that PCBs interfere with the reproductive system of the seal, and if the Baltic does not recover quite soon, at least the seal population is bound to die out.

As most of the substances mentioned here are not used in Sweden today, the only thing that society can do in the future is to be still more restrictive with any further release of anthropogenic substances into this susceptible sea.

REFERENCES

1. JENSEN, S. 1966. New Sci. **32:** 612.
2. JENSEN, S., A. G. JOHNELS, M. OLSSON & G. OTTERLIND. 1972. Ambio Spec. Rep. **1:** 71.
3. OLSSON, M., A. G. JOHNELS & R. VAZ. 1975. National Swedish Environment Protection Board, PM **591:** 43.
4. JANSSON, B., S. JENSEN, M. OLSSON, L. RENBERG, G. SUNDSTRÖM & R. VAZ. 1975. Ambio **4:** 93.
5. SISSONS, D. & D. WELTI. 1971. J. Chromatogr. **60:** 15.
6. JENSEN, S. & G. SUNDSTRÖM. 1974. Ambio **3:** 70.
7. SUNDSTRÖM, G., O. HUTZINGER & S. SAFE. 1976. Chemosphere **5:** 267.
8. ALMKVIST, L. 1977. Personal communication.
9. SÖDERBERG, S. 1975. National Swedish Environmental Protection Board, PM **591:** 104.
10. HELLE, E. 1975. National Swedish Environment Protection Board, PM **591:** 38.
11. NAZARENKO, Y. 1965. Morskie Mlekopitayuskchie, 171; Fisheries Research Board of Canada, Translation Series No. 1461 (1970).
12. FEDOSEEV, G. A. 1964. Zoologicheskii Zhurnal **XLIII**(7): 1230.
13. HELLE, E., M. OLSSON & S. JENSEN. 1976. Ambio **4:** 188.
14. HELLE, E., M. OLSSON & S. JENSEN. 1976. Ambio **4:** 261.
15. DELONG, R. L., W. G. GILMARTIN & J. G. SIMPSON. 1973. Science **181:** 1168.
16. AUERLICH, R. J., R. K. RINGER, H. L. SEAGRAN & W. G. YOUATT.1971. Can. J. Zoology **49:** 611.
17. RINGER, R. K., R. J. AUERLICH & M. ZABIK. 1972. 164th National Meeting Am. Chem. Soc. **12:** 149.
18. PLATONOW, N. S. & L. H. KARSTAD. 1973. Can. J. Comp. Med. **37:** 391.
19. JENSEN, S., J. E. KIHLSTRÖM, M. OLSSON, C. LUNDBERG & J. ÖRBERG. 1977. Ambio **6:** 239.
20. JENSEN, S., & B. JANSSON. 1976. Ambio **5:** 257.
21. BERGMAN, Å., & C.-A. WACHTMEISTER. 1977. Acta Chem. Scand. **B31:** 90.
22. MIO, T., K. SUMINO & T. MIZUTANI. 1976. Chem. Pharmacol. Bull. **24:** 1958.
23. JANSSON, B., S. JENSEN, R. VAZ, G. BLOMKVIST & M. OLSSON. In preparation.
24. LAWRENCE, J. H., R. P. BARRON, J.-Y. T. CHEN, P. LOMBARDO & W. R. BENSON. 1970. JAOAC **53:** 261.
25. OLSSON, M. 1977. National Swedish Environment Protection Board, PM 900.
26. SUNDSTRÖM, G., B. JANSSON & S. JENSEN. 1975. Nature **255:** 627.

DEVELOPMENTAL TOXICOLOGY OF THE HALOGENATED AROMATICS: EFFECTS ON ENZYME DEVELOPMENT

George W. Lucier and Otelia S. McDaniel

National Institute of Environmental Health Sciences
Research Triangle Park, North Carolina 27709

INTRODUCTION

2,3,7,8-Tetrachlorodibenzo-*p*-dioxin (TCDD)* has been detected as a contaminant of some organochlorines, most notably the herbicide 2,4,5-T.[1] This chemical is extremely toxic[2,3] and is also a potent inducer of specific microsomal mixed—function oxidases and UDP glucuronyltransferase in several tissues of adult rats.[4–8] Teratogenic and fetotoxic properties of TCDD have been characterized; the most common anomalies are cleft palate and kidney malformations.[9–11]

The polychlorinated biphenyls (PCBs) are produced commercially as a mixture of many congeners and are generally considered to be biologically—and ecologically—persistent industrial pollutants. Most of the toxicology studies have been derived using these mixtures. Such studies provide important and necessary toxicologic data but reveal little about structure-activity relationships of individual congeners differing in the number and position of chlorine substitutions. PCBs have teratogenic potential as evidenced by skin, liver, and growth disorders in offspring of pregnant women exposed to PCBs in Yusho, Japan.[12] Moreover, PCB mixtures do cross the placenta, are secreted in significant concentrations in breast milk, and cause reproductive tract dysfunction in offspring exposed to PCBs during the prenatal period.[13–18] PCBs, like TCDD, induce activities of some hepatic mixed—function oxidase enzymes in adult animals,[19–21] and the PCBs also potentiate the hepatotoxic actions of carbon tetrachloride.[22]

Our studies are designed to investigate changes in enzyme ontogeny following prenatal and postnatal (milk) exposures of pure PCB congeners and TCDD. Since biochemical parameters are often more sensitive indicators of toxicity than morphologic criteria, these studies might provide insight regarding the developmental toxicity potential of specific organohalogens. Accordingly, the effects of the halogenated aromatics, particularly the PCBs, on the ontogeny of the drug-metabolizing enzymes are assessed by comparing enzyme changes with teratogenic actions of these chemicals.

EFFECTS OF TCDD ON ENZYME DEVELOPMENT

The effects of administration of 3 μg TCDD/kg to pregnant rats (5th, 10th, or 16th day of gestation) on the ontogeny of some hepatic drug-metabolizing enzymes was investigated.[23] Significant fetotoxicity or teratogenicity was not evident following doses less than 6 μg TCDD/kg at any of the gestational ages used in the treatment

*ABBREVIATIONS: TCDD, 2,3,7,8-tetrachlorodibenzo-*p*-dioxin; PCB, polychlorinated biphenyl; 1-CB, 4-chlorobiphenyl; 4-CB, 3,4,-3′,4′-tetrachlorobiphenyl; 6-CB, 2,4,5-2′,4′,5′-hexachlorobiphenyl; UDPGT, UDP glucuronyltransferase; AHH, aryl hydrocarbon hydroxylase; DES, diethylstilbestrol; PBB, polybrominated biphenyl.

0077–8923/79/0320-0449 $01.75 © 1979, NYAS

schedule. The maternal LD_{50} was approximately 75μg/kg. In our studies we concentrated on TCDD-mediated alterations of the ontogeny of AHH and UDPGT. AHH activity is dependent on the P450 monoxygenase system whereas UDPGT is not cytochrome-P450-dependent. Since UDPGT exhibits enzyme multiplicity,[24] we measured enzyme activity using steroid and hormonally-active substrates (B-estradiol, DES, testosterone, estrone, and phenolphthalein) or nonhormonally-active substrates (*p*-nitrophenol, 4-methylumbelliferone and 1-naphthol). Summarized and condensed results (TABLE 1) reveal that AHH was markedly induced in the fetal rat liver. Moreover, the fetal onset of enzyme activity was detected in livers 4 days prior to birth in the treated group compared to 2 days prior to birth in the control group.[23]

Later studies have investigated the transplacental induction of carcinogen hydroxylating systems by TCDD.[25] AHH was induced nearly 100-fold in fetal rats, but only minimal increases were evident in placentas. Rates of ring and *N*-hydroxylation of *N*-2 fluorenylacetamide were also markedly induced. Increased formation of benzypyrene diol in both the fetal and maternal compartments was associated with the increased AHH activity in fetal rat livers. Therefore, it appears that the lack of response of fetal enzymes to many inducing agents, such as phenobarbital[26] and 3-methylcholanthrene[27] is not due to the lack of an appropriate genetic mechanism in fetal liver but is likely a function of pharmacokinetic factors such as maternal metabolism, placental transfer, and/or TCDD-receptor interactions.

In contrast to AHH, fetal hepatic steroid and nonsteroid UDPGT are unaffected by TCDD treatment of the pregnant animal (TABLE 1). Increased activity of AHH and nonsteroid UDPGT is evident postnatally, whereas steroid UDPGT was unchanged in the same animals. Differences in susceptibility to TCDD inductive actions, tissue distribution, and enzyme ontogeny have been used as criteria for the contention that steroid and nonsteroid UDPGT are at least two separate enzyme systems.[24] However, solubilization and partial purification of hepatic UDPGT from control and TCDD-treated rats did not resolve physically distinct enzyme systems in early studies.[7] More recently, Tukey *et al.*, using rabbit liver microsomes, have separated estrone and *p*-nitrophenol UDPGT on DE-52 columns.[28] Elevated nonste-

TABLE 1

EFFECTS OF TCDD TREATMENT (3 μg/kg) OF PREGNANT RATES ON THE ONTOGENY OF HEPATIC FETAL AND NEWBORN AHH AND UDPGT†

	Developmental Stage					
	Gestation Day 21		8 Days Postnatal		21 Days Postnatal	
Enzyme ‡	Control	TCDD	Control	TCDD	Control	TCDD
AHH	0.003	0.032*	0.214	0.719*	0.224	0.863*
Non-steroid UDPGT	13.4	14.4	28.4	99.5*	23.1	164.2*
Steroid UDPGT	ND§	ND§	2.89	1.85	2.46	2.73

†Pregnant rats treated on the 16th day of gestation. Enzyme activity in nmole/min/mg microsomal protein. Data summarized from reference 23.

‡AHH measured by 3-hydroxy benzpyrene formation, nonsteroid UDPGT measured by *p*-nitrophenyl B-D glucuronide formation (similar results were obtained using 1-naphthol or 4-methylumbelliferone as substrate, steroid UDPGT measured by testosterone glucoronide formation (similar results were obtained using DES, β-estradiol or phenophthalein as substrate). An asterisk indicates that activities from treated animals are significantly different from controls at least at $p < 0.01$.

§Enzyme activity not detectable.

TABLE 2

POSTNATAL INDUCTION OF HEPATIC AND EXTRAHEPATIC NONSTEROID UDPGT BY TCDD TREATMENT (3 μg TCDD/kg) OF PREGNANT RATS‡

Tissue	nmoles/Min/Mg Microsomal Protein‡ Control	TCDD	Induction Factor
Liver	27.3	159.4*	5.8
Lung	0.3	2.9*	9.7
Kidney	8.4	87.7*	10.4
Intestine	1.2	9.6*	8.0
Testes	6.6	7.0	1.1

†Pregnant animals treated on the 16th day of gestation and enzyme activity measured in offspring 8 days after birth.

‡Activity measured using *p*-nitrophenol as the substrate. An asterisk indicates that treated values are significantly different from controls ($p < 0.01$).

roid UDPGT and AHH activities were still evident in 55-day old offspring, which appears to reflect the long biologic half-life of TCDD[29] as well as the unique sensitivity of hepatic AHH and UDPGT to TCDD tissue burdens.[4,5] For example, a single oral dose of 20 ng TCDD/kg to adult rats was sufficient to induce hepatic AHH activity and a bioassay measuring AHH activity in H-4-II-E cells has been suggested for the detection of 10^{-14} mole levels of TCDD.[30]

Studies in adult rats demonstrate that AHH activities in extrahepatic tissues are also induced by TCDD, with the exception of reproductive tract tissue, which is nonresponsive.[8] Data presented in TABLE 2 illustrate the effects of TCDD on extrahepatic enzyme activity in offspring of dams which had been given TCDD. The neonates have elevated levels of UDPGT in liver, kidney, lung, and intestine, but as in adult rats, testicular tissue is nonresponsive. Increased levels of enzyme activity are most pronounced in the kidney. Further studies are needed to determine if tissue-specific, as well as species-specific (guinea pig and rabbit are nonresponsive)[8] responses to TCDD are related to pharmacokinetic or genetic factors. Of particular interest would be the presence of an effective blood-testis barrier for TCDD or the lack of a high affinity cytosolic receptor for TCDD in testicular tissue.

Since nonsteroid UDPGT is induced postnatally but not prenatally following treatment of pregnant rats with TCDD, pups were cross-fostered immediately after birth to determine if the postnatal stimulation of UDPGT is related to newborn exposure to TCDD in milk or to the postnatal activation of an inducing mechanism. Summarized results[23] (TABLE 3) reveal that the major factor is newborn exposure to TCDD in milk although UDPGT in the postnatal liver does appear to be more responsive to TCDD than the prenatal liver. Tissue distribution experiments using ^{14}C radiolabel demonstrate that TCDD does not readily cross the placenta but is apparently secreted in high concentrations in milk.[31] For example, following administration of 5 μg ^{14}C-TCDD/kg to pregnant rats on the 11th day of gestation, each fetus contains approximately one ng TCDD body burden compared to 40 ng TCDD by 14 days after birth. Furthermore, each newborn received a dose of 1.8 μg TCDD/kg via suckling following a dose of 5.0 μg TCDD/kg to the dam (litter size was standardized to eight). Since high concentrations of TCDD are secreted in milk, it might be expected that enzyme activities in the dam would decrease as a function of decreased maternal TCDD body burden. Results demonstrate that treatment of pregnant rats with 3 μg TCDD/kg (5th day of gestation) induces maternal hepatic UDPGT approximately 15-fold from the 8th day of gestation until 3 days after birth. However,

TABLE 3

POSTNATAL INDUCTION OF HEPATIC MICROSOMAL ENZYMES BY TCDD TREATMENT (3 μg kg) OF PREGNANT RATS FOLLOWING CROSS-FOSTERING OF CONTROL AND TCDD-TREATED GROUPS†

Animals‡	PNP Glucuronidation §	Benzpyrene Hydroxylation §	Cytochrome P450§
Group 1	18.6 ± 1.5	0.227 ± 0.016	0.58 ± 0.07
Group 2	99.5 ± 6.44*	1.115 ± 0.0834*	1.60 ± 0.294*
Group 3	100.4 ± 8.94*	1.143 ± 0.0912*	1.52 ± 0.144*
Group 4	33.3 ± 2.244*	^ 0.604 ± 0.0734*	0.77 ± 0.055**
Group 5	95.6 ± 12.34*		1.16 ± 0.204*

†Pregnant rats were treated on the 5th day of gestation and the newborn were sacrificed 8 days after birth. Data summarized from reference 23.

‡Group 1, control mothers nursing control pups; Group 2, TCDD-treated mothers nursing TCDD-treated pups; Group 3, TCDD-treated mothers nursing control pups; Group 4, control mothers nursing prenatally TCDD-treated pups; Group 5, mother rats administered TCDD immediately after giving birth.

§PNP glucuronidation expressed in nmoles/min/mg microsomal protein. Benzypyrene hydroxylation expressed in nmoles 3-hydroxybenzpyrene formed/min/mg microsomal protein. Cytochrome P450 expressed in nmoles/mg microsomal protein.

*Significantly different from Group 1, $p < 0.01$.

**Significantly different from Group 1, $p < 0.05$.

maternal UDPGT activities decline rapidly in the treated group during the suckling period so that by 14 days after birth enzyme activity is increased threefold in the TCDD-treated group.

Postnatal induction of microsomal enzymes by newborn exposure to organohalogens in milk is not a unique property of TCDD. Neonatal mice that had received kepone and mirex exhibit dramatic elevations in oxidative *O*- and *N*-demethylation pathways.[32] The PBBs are also effective transplacental inducers (induction characteristics are mixed, with some phenobarbital-type induction and some 3-methylcholanthrene-type induction) of the hepatic monoxygenase system[33] but are also secreted in high concentrations in milk resulting in increased enzyme activity in pups receiving PBBs only via lactation.[33,34] Later in this report the role of placental transfer and milk secretion of pure PCB congeners in developmental alterations of hepatic metabolism will be examined.

A number of experiments have been performed to determine if enhanced UDPGT and AHH activities represent a "true" enzyme induction. Solubilization and partial purification of hepatic UDPGT from control and TCDD-treated rats reveal that UDPGT activities are increased sevenfold whether enzyme assays are conducted using endoplasmic reticulum-bound or solubilized preparations.[7] These results indicate that TCDD-mediated increases in UDPGT activity are not related to alterations in membrane environment of microsomal enzymes. Furthermore, actinomycin D effectively blocks elevation of AHH and UDPGT activity and kinetic evaluations on UDPGT are similar for preparations from control and treated rats; K_m values for UDPGT (co-factor) or *p*-nitrophenol (substrate) are unchanged, whereas V max is significantly increased.[7]

Since TCDD is such a potent and specific inducer of hepatic monooxygenase components, it was theorized that TCDD might interact with a specific high affinity cytosol receptor which could be translocated to the nucleus, bind to specific sites on the chromatin and thus stimulate gene expression and the synthesis of AHH or UDPGT. Recent studies[35] using ^{3}H-TCDD have characterized such a cytosolic receptor for polycyclic hydrocarbons and have shown a positive correlation between dissociation constants for TCDD-receptor binding and AHH inducibility in different mouse strains. These studies demonstrate that chemicals which induce cytochrome P448 compete with TCDD for the receptor but those that induce P450 do not compete with TCDD. To date, no information is available on the ontogeny of this receptor system in hepatic or extrahepatic tissues. It would be especially interesting to correlate the quantity of this receptor for polycyclics with age-dependent differences in AHH and UDPGT inducibility.

EFFECTS PCBS ON ENZYME DEVELOPMENT

Pharmacokinetic studies reveal that the presence of two unsubstituted carbon atoms on the biphenyl ring markedly enhance the metabolic rate constant of PCBs in adult rats.[36] For example, the metabolic rate for 1-CB is 200 times greater than 6-CB. Recent studies in our laboratory have characterized some of the pharmacologic properties of pure PCB congeners in developing rats.[37] We investigated several congeners; 4-chlorobiphenyl (1-CB), 4,4′-dichlorobiphenyl (2-CB), 3,4-3′,4′-tetrachlorobiphenyl (4-CB), 2,3,6-2′,3′6′-hexachlorobiphenyl (6-CB′) and 2,4,5-2′,4′,5′,-hexachlorobiphenyl (6-CB). It appears that the perinatal clearance of 1-CB, 2-CB, 4-CB, and 6-CB′, like adult systems is more rapid than that of 6-CB. However, we found some striking differences between fetal and adult animals with regards to the rapidly hydroxylated analogues. Radioactivity (^{14}C-labeled congeners were used in these studies) from those analogues that do not accumulate in adipose tissue (1-CB,2-CB,4-CB, and 6-CB′) is found in high concentrations in fetal intestine, although the PCBs are cleared rapidly from the intestine following parturition.[37] Metabolism studies demonstrate that the major fetal intestinal metabolites are glucuronides of hydroxylated PCBs; hydroxylation apparently takes place in the maternal compartment and glucuronidation in the fetal liver. Blood concentrations of 4-CB are consistently higher than those for other congeners in both the maternal and fetal compartment (TABLE 4).[37] The mechanism responsible for this finding is not clear, but 4-CB also has a different biochemical spectrum of effects in adult rats[21,38] including competition for the hepatic receptor that binds TCDD and 3-methylcholanthrene.[35]

Analysis of structure-activity relationships governing the perinatal pharmacology of the PCBs and PBBs provided guidance for selecting representative congeners for evaluating the effects of PCBs on ontogeny of hepatic enzymes. Accordingly, we treated pregnant rats with 1-CB, 4-CB, and 6-CB and then investigated developmental patterns of several hepatic enzymes; ethoxyresorufin deethylase (cytochrome P448-dependent), aminopyrine demethylase (cytochrome P450-dependent), 16α-hydroxylase (cytochrome P450-dependent), 5α-reductase, and UDPGT. Data

TABLE 4

FETAL AND MATERNAL BLOOD LEVELS OF PCBS IN RATS†

	DPM/Ml × 10^{-2}				
	1-CB	2-CB	4-CB	6-CB′	6-CB
Fetal	5	3	51	6	7
Maternal	3	2	9	2	3

†Pregnant rats administered 2 μCi ^{14}C-labeled PCB on the 18th day of gestation and sacrificed on the 21st day of gestation. Data summarized from reference 37.

summarized (manuscripts in preparation) in TABLE 5 reveal several interesting findings. 4-CB induces cytochrome P448-dependent enzymes and 6-CB induces P450-dependent pathways in livers of offspring exposed to these congeners transplacentally and during lactation whereas 1-CB has no detectable effect on enzyme ontogeny. Thus, the perinatal and adult liver[21,38] respond to PCBs in much the same way. Following treatment with Aroclor 1254 (mixture of many PCBs) Alvares and Kappas[39] demonstrated induction of both P448 and P450 systems in the fetal-placental unit and in neonatal livers during lactation. PBB mixtures also cause a mixed type of induction and both forms of induction are seen in neonates following transplacental or milk exposures.[33,34] Our studies demonstrate that in developing systems, like adults, individual congeners are specific P450 or P448 inducers and that neonates may be more sensitive to PCB-mediated increases in enzyme activities than adults. However, conclusions regarding age-dependent sensitivities to induction of hepatic enzymes must await quantitation of parent compound and/or toxic metabolites of individual congeners in various tissues during perinatal development.

TABLE 5

EFFECTS OF 1-CB, 4-CB, AND 6-CB ON THE ONTOGENY OF HEPATIC MICROSOMAL ENZYMES*

Enzyme	Offspring Age	Effects on Enzyme Activity 1-CB	4-CB	6-CB
Ethoxyresorufin-deethylase	6	No effect	Induction	No effect
	21	No effect	Induction	No effect
Aminopyrine demethylase	6	No effect	No effect	Induction
	21	No effect	No effect	Induction
p-Nitrophenol-Glucuronyltransferase	6	No effect	Induction	Induction
	21	No effect	Induction	Induction
	55-Male	No effect	Repression	Repression
	55-Female	No effect	No effect	No effect
16α-Hydroxylase	6	No effect	No effect	Induction
	21	No effect	No effect	Induction
5α-Reductase	6	No effect	Repression	No effect
	21	No effect	Repression	No effect

*PCBs administered orally on days 8–18 of gestation to pregnant rats at the following doses: 1-CB, 30 mg/kg/day; 4-CB, 3 mg/kg/day; 6-CB, 30 mg/kg/day. No fetotoxicity or alterations in the weight gain of offspring were observed by these doses.

UDPGT, which is not P450- or P448-dependent is induced by both 4-CB and 6-CB in neonatal rat livers. It has been suggested that the single genetic trait of aromatic hydrocarbon responsiveness determines the inducibility of both AHH and UDPGT.[35] While our data do not conflict with this statement, it appears that UDPGT inducibility may be regulated by a multiplicity of factors rather than by a single locus.[40] In contrast to other hepatic enzymes, activities of the hepatic steroid-metabolizing enzyme, 5α-reductase, are markedly repressed by 4-CB but not 6-CB in neonatal rats.[40] The mechanism of repression and the physiologic significance of this finding have not been characterized.

Enzyme induction, as discussed in an earlier part of this report, probably results from direct action of the chemical or chemical-receptor with genetic material resulting in an immediate modulation of enzyme levels as a function of increased enzyme synthesis. A second form of enzyme modulation that is initiated neonatally has been termed "imprinting" or programming of enzyme activity. Imprinting is manifested through neonatal exposure to hormones or hormonally-active agents and results in an irreversible determination of behavioral characteristics or enzyme levels in the subsequent adult animal.[41,42] These imprinted endpoints are, however, not evident in the prepubertal animal, suggesting a latency of gene expression resulting from hormone action during early development. The developmental patterns for many enzymes exhibit no sex differences in prepubertal animals. However, a postpubertal sexual differentiation of metabolic competence often occurs with the activities of one sex being markedly different than the other. Sexual differentiation of enzyme activity appears to be a function of neonatal estrogen or androgen exposure. UDPGT and many oxidative pathways exhibit higher activities in adult males than females.[43,44] However, neonatal exposure to either 4-CB or 6-CB transplacentally and through lactation results in decreased (feminized) levels of hepatic UDPGT in the resultant adult male rat. In contrast, no differences were evident in hepatic UDPGT from adult females. Thus, it appears that neonatal PCB exposure might be capable of altering the normal hepatic sexual differentiation of hepatic metabolic competence. It might be theorized that PCBs exert their programming action through their estrogenic properties since PCB mixtures administered *in vivo* are capable of increasing uterine wet weight.[45]

Further studies in mice have assessed the developmental toxicity of pure PCB congeners and defined some structure-activity relationships. 4-CB is a potent teratogen as evidenced by kidney malformations, reproductive tract dysfunction, and behavioral abnormalities in offspring exposed to 4-CB prenatally.[46,47] The behavioral alterations persist for the lifetime of the mice and are characterized by a specific neuropathologic lesion: cylindrical CNS outgrowths into the peripheral nervous system.[47]

In summary, it appears that chlorine position and number on the biphenyl ring play an integral role in the developmental pharmacology and toxicology of the PCBs.

Summary

Our studies were designed to define structure-activity relationships for the effects of pure PCB congeners and 2,3,7,8-tetrachlorodibenzo-*p*-dioxin (TCDD) on hepatic enzyme development. The organohalogens were usually administered to pregnant rats and enzyme activities measured in fetuses and offspring (6, 20, and 55 days after birth). Offspring of pregnant rats exposed to TCDD had elevated levels of benzpyrene hydroxylase and nonsteroid UDP glucuronyltransferase (UDPGT). Postnatal inductive actions of TCDD were caused primarily by newborn exposure to TCDD in milk. 3,4-3′,4′-Tetrachlorobiphenyl (4-CB) induced P448-dependent enzymes in 6- and 20 day-old rats whereas 2,4,5-2′,4′,5′-hexachlorobiphenyl (6-CB) induced P450-dependent enzymes. 4-CB was a potent teratogen (behavioral and kidney abnormalities) but 6-CB produced no such developmental defects. 4-CB might also alter the normal imprinting of sexual differentiation of hepatic metabolism as evidenced by a feminization (decrease) of UDPGT in adult male rats which were exposed neonatally to 4-CB. Therefore, the organohalogens might alter the ontogeny of hepatic metabolism in two ways: induction and imprinting.

References

1. Baughman, R. & M. Meselson. 1973. Environ. Health Perspect. **5:** 27–36.
2. Harris, M. W., J. A. Moore, J. G. Vos & B. N. Gupta. 1973. Environ. Health Perspect. **5:** 101–110.
3. Schwetz, B. A., J. M. Norris, G. L. Sparschu, V. K. Rowe, P. J. Gehring, J. L. Emerson & C. G. Gerbig. 1973. Environ. Health Perspect. **5:** 87–100.
4. Lucier, G. W., O. S. McDaniel, G. E. R. Hook, B. Fowler, B. R. Sonawane & E. Faeder. 1973. Environ. Health Perspect. **5:** 199–209.
5. Poland, A. & E. Glover. 1974. Mol. Pharmacol. **10:** 349–359.
6. Poland, A. P., E. Glover, J. R. Robinson & D. W. Nebert. 1974. J. Biol. Chem. **249:** 5599–5606.
7. Lucier, G. W., O. S. McDaniel & G. E. R. Hook. 1974. Biochem. Pharmacol. **24:** 325–335.
8. Hook, G. E. R., J. K. Haseman & G. W. Lucier. 1975. Chem. Biol. Interact. **10:** 199–214.
9. Courtney, K. D. & J. A. Moore. 1971. Toxicol. Appl. Pharmacol. **20:** 396–403.
10. Neubert, D., P. Zens, A. Rothenwallner & H. J. Merker. 1973. Environ. Health Perspect. **5:** 67–80.
11. Smith, F. A., B. A. Schwetz & K. D. Nitschke. 1976. Toxicol. Appl. Pharmacol. **38:** 517–523.
12. Kuratsune, M., T. Yoshimura, J. Matsuzaka & A. Yamaguchi. 1973. Environ. Health Perspect. **5:** 119–128.
13. Grant, D. L., D. C. Villeneuve, K. A. McCully & W. E. Phillips. 1971. Environ. Physiol. **1:** 61–66.
14. Linder, R. E., T. B. Gaines & R. D. Kimbrough. 1974. Food Cosmet. Toxicol. **12:** 63–77.
15. Torok, P. & H. Kriegel. 1975. Arch. Toxicol. **23:** 199–203.
16. Allen, J. R. & D. A. Barsotti. 1976. Toxicol. **6:** 331–340.
17. Barsotti, D. A., R. J. Marlar & J. R. Allen. 1976. Food Cosmet. Toxicol. **14:** 99–103.
18. Takagi, Y., T. Otake, M. Kataoka, Y. Murata, S. Aburada, S. Akasaka, K. Hashimoto, H. Uda & T. Kitaura. 1976. Toxicol. Appl. Pharmacol. **38:** 549–558.
19. Argus, M. F., G. M. Bryant, K. M. Pastor & J. C. Arcos. 1975. Cancer Res. **35:** 1574–1579.
20. Stonard, M. D. & J. B. Greig. 1976. Chem. Biol. Interact. **15:** 365–379.
21. Goldstein, J. A., P. Hickman, H. Bergman, J. D. McKinney & M. P. Walker. 1977. Chem. Biol. Interact. **17:** 69–84.
22. Carlson, G. P. 1975. Toxicol. **5:** 69–77.
23. Lucier, G. W., B. R. Sonawane, O. S. McDaniel & G. E. R. Hook. 1975. Chem. Biol. Interact. **11:** 15–26.
24. Lucier, G. W. & O. S. McDaniel. 1977. J. Ster. Biochem. **8:** 867–872.
25. Berry, D. L., P. K. Zachariah, M. J. Namkung & M. R. Juchau. 1976. Toxicol. Appl. Pharmacol. **36:** 569–584.
26. Welch, R. M., B. Gomni, A. P. Alvares & A. H. Conney. 1972. Cancer Res. **22:** 1120–1125.
27. Bresnick, E. & J. G. Stevenson. 1968. Biochem. Pharmacol. **17:** 1815–1822.
28. Tukey, R., R. Billings & T. Tephly. 1978. Fed. Proc. **37(6):** 1542.
29. Piper, W. N., J. Q. Rose & P. J. Gehring. 1973. Environ. Health Perspect. **5:** 241–244.
30. Niwa, A., K. Kumaki & D. W. Nebert. 1975. Mol. Pharmacol. **11:** 399–408.
31. Moore, J. A. Personal communication.
32. Faracher, D. L. & E. Hodgson. 1976. Toxicol. Appl. Pharmacol. **38:** 71–77.
33. Dent, J. G. 1978. Environ. Health Perspect. **23:** 301–308.
34. Moore, R. W., G. A. Dannan & S. D. Aust. 1978. Environ. Health Perspect. **23:** 159–165.

35. POLAND, A. & E. GLOVER. 1976. J. Biol. Chem. **251:** 4936–4946.
36. LUTZ, R. J., R. L. DEDRICK, H. B. MATTHEWS, T. E. ELING & M. W. ANDERSON. 1977. Drug Metab. Dispos. **5:** 386–396.
37. LUCIER, G. W., O. S. MCDANIEL, C. M. SCHILLER & H. B. MATTHEWS. 1978. Drug Metab. Dispos. **6:** 584–590.
38. POLAND, A. & E. GLOVER. 1977. Mol. Pharmacol. **13:** 924–934.
39. ALVARES, A. P. & A. KAPPAS. 1975. FEBS Lett. **50:** 172–174.
40. DIERINGER, C. S., C. M. SCHILLER, C. A. LAMARTINIERE & G. W. LUCIER. 1977. Pharmacologist **19:** 201.
41. EINARSSON, K., J. A. GUSTAFSSON & A. STENBERG. 1973. J. Biol. Chem. **248:** 4987–4997.
42. MCCEWEN, B. S. 1976. Sci. American **235:** 48–58.
43. CHHABRA, R. S. & J. R. FOUTS. 1974. Drug Metab. Dispos. **2:** 375–379.
44. KATO, R. 1974. Drug Metab. Rev. **3:** 1–32.
45. BITMAN, J. & H. C. CECIL. 1970. J Agric. Food Chem. **18:** 1108–1112.
46. LUCIER, G. W., G. J. DAVIS & J. A. MCLACHLAN. 1978. Proceedings of Hanford Symposium on Developmental Toxicology of Energy-Related Pollutants, Richland, Washington, Oct. 1977. In press.
47. DAVIS, G. J., S. M. CHOU, T. MIIKE, J. A. MCLACHLAN & G. W. LUCIER. 1978. In preparation.

SUMMARY OF THE WORKSHOP ON PERINATAL AND POSTNATAL DEFECTS AND NEUROLOGIC ABNORMALITIES FROM CHEMICAL EXPOSURES

Chairpersons:

Mason Barr, Jr.

University of Michigan School of Medicine
Ann Arbor, Michigan 48104

Carl A. Keller

National Institute of Child Health and Human Development
National Institutes of Health
Bethesda, Maryland 20014

Walter J. Rogan

National Institute of Environmental Health Sciences
National Institutes of Health
Research Triangle Park, North Carolina 27709

Invited Participant:

Jennie Kline

Columbia University
School of Public Health
New York, New York 10027

INTRODUCTION

Both chemical pollution and human reproduction are complex topics. Inquiry into chemical-related damage to human reproduction is hampered by this complexity. In general, the whole area of environmental chemicals versus human reproduction is one of great concern but, at this time, there are few hard data. The following provides an analytic framework for discussion of these topics, and also provides an outline for this presentation.

1. Specific chemical substances that occur in the environment constitute exposures. These may be widespread or occur only in special circumstances.
2. The events of the reproductive process constitute outcomes. These will be ordered temporarily, together with some comment on feasibility of ascertainment.
3. Clinical examinations and tests, chemical analyses, and historical information constitute measures. These may be gathered routinely and easily available, or require specific special efforts.

0077–8923/79/0320–0458 $01.75/0 © 1979, NYAS

Opportunities for study occur when ascertainment of the above occur jointly, together with the investigator's suspicion of an etiologic relationship between exposure and damage. This presentation will explore each of the areas, with comments on the plausibility of etiologic relationships and feasibility of their study.

Exposures

Exposures to environmental chemicals occur in a number of ways. However, we will concentrate here on exposures that are chronic, such as in the workplace, or persistent, such as to chemicals that are not readily metabolized or excreted. Except as examples of mechanisms, we will not discuss short-term, acute exposures nor exposures to drugs. Exposures to environmental chemicals take place in one of the three following ways; there is, of course, some overlap for specific chemicals.

1. *Occupational Exposure.* This usually involves a rather well-defined group of workers, and sometimes their families and associates by home contamination.
2. *Accidental Exposure.* This occurs through a specific contamination incident, such as the PBB incident in Michigan[1] or the dioxin incident in Seveso, Italy.[2] The population may be easily identifiable, but is generally less well defined and more difficult to follow than an occupational group.
3. *Ambient Exposure.* This occurs through media, such as air, water, and food, variously contaminated by environmental chemicals. Specific populations of exposed and unexposed individuals are often not available for study, and the clear establishment of etiologic relationships thus is limited to severe disease (i.e., lead encephalopathy).

Occupational Exposures

Occupational exposures to toxic chemicals include both exposures to unusually toxic chemicals and exceptionally heavy exposures to less toxic substances. Both of these possibilities provide opportunities for observing the effects of chemical exposure in a unique situation, i.e., one that makes it relatively easy to identify a population at risk. The major problem, however, is to identify those chemicals for which appropriate toxic effects are to be measured or tested. For those chemicals that we know to be teratogenic, the intention is usually to decrease the exposure to the reproductively active worker presumptively. Similar reasoning applies to those exposures such as lead[3] that are known to produce fetal wastage. Substances that are suspect for producing adverse pregnancy outcomes also include carcinogens and mutagens. Since this group of chemicals is so large, it is difficult to know where to begin.

One approach is to measure outcomes such as congenital malformations and pregnancy wastage in certain employment situations where a significant proportion of workers are female, and where a number of chemicals are used. Examples would include nurses, dry cleaning workers, garment workers, and textile workers. Another approach is to select certain chemicals as being suspicious on the basis of animal studies, widespread usage, or the observations of an alert practitioner. Although not so systematic, the latter method has most often been responsible for discovering occupation-related exposures that induce serious adverse health effects. Examples of this approach would be the association of adverse pregnancy outcomes and exposure to lead,[3] hexachlorophene,[4] and polychlorinated biphenyls.[5,9] With the current emphasis on reproductive outcome as a major health issue, and with the increasing participation of women in the workplace, additional chemical toxins in the workplace will likely be identified.

Ambient Exposure

Ambient exposure to toxic chemicals takes place in two ways; one is basically purposeful, although not necessarily to a known toxic substance, and the other is accidental.

Purposeful exposures include smoking, beverage alcohol consumption and the use of other consumer producers such as saccharin, cyclamates, or red dye number 2. The major component of purposeful exposure is a behavioral one in which the risks of toxic exposure are either thought to be offset by benefits or are not known. In other words, the exposure, via ingestion or inhalation or absorption through the skin, is done with knowledge of the exposure itself but without an adequate assessment of the consequences to reproductive outcome (or other health effects). Assessment of the extent of exposure is generally accomplished by measurement of intake, such as how many cigarettes are smoked, how many soft drinks consumed, etc.

Accidental exposures include those in which the individual is not necessarily aware that he or she is being exposed to the toxic substance. These constitute a very long list and include pesticides, industrial chemicals, contaminants and food additives. Assessment of exposure often requires measurements of the chemical in tissues or fluids, such as fat, blood, hair, milk, etc.

1. Pesticides

Among the more common pesticide residues found in human fat tissue are DDT,[6] heptachlor, dieldrin, and hexachlorobenzene.[7] These chemicals tend to persist in the environment and can reach human beings via the ingestion of animal products. Some pesticides are ingested directly as residues on vegetable material or are absorbed through the skin by careless use of commercial preparations used as consumer products.

2. Industrial Chemicals

Probably the most common and best known industrial chemicals that constitute a significant ambient exposure to humans are the polyhalogenated biphenyls, particularly PCBs.[8] These chemicals were used in the manufacture of a number of products, especially electrical transformers and electronic parts. Electrical equipment is eventually discarded, and the PCBs can become air or water pollutants. Paper products may be recycled, and the contained PCBs may become part of food packaging. Exposures have been detected throughout the U.S. through the analysis of fat and milk samples. In Japan, rice oil was contaminated by PCBs and other chemicals. Consumers of the contaminated oil developed chloracne, the characteristic skin lesion of chlorinated hydrocarbon exposure. Children born to exposed mothers were small for dates and had pigmented skin.[9] Administration of PCBs to pregnant monkeys have produced similar problems with infertility, fetal loss, and low birth weight.[10] The polyhalogenated biphenyls as a group have varying degrees of toxicity depending on the number and position of the halogen substitutions.[11] Since most of the industrial preparations are mixtures, this presents some problems in determining the potential toxicity of various compounds and for interpreting experimental and laboratory data.

Another source of exposure to toxic chemicals is through contamination during the manufacturing of other industrial chemicals. Examples of contaminated products that might have adverse effects on reproductive outcome include the furans in PCBs, dioxins in herbicides and other chemicals, chlorinated naphthalenes, chloroform and carbon tetrachloride. The latter two have been found in water supplies and pose an additional potential problem through reuse via recharging ground water with treated waste water.[12]

Other sources of air and water contamination by the disposal of waste effluents from industrial installations can involve almost any chemical. Humans may be exposed through direct use of contaminated water and air as well as indirectly through the consumption of plant and animal products which have been, in turn, contaminated through air or water. This latter route is of particular concern for those chemicals, particularly the halogenated hydrocarbons, which are capable of biomagnification in the food chain.

Another ambient source of human exposure to chemicals capable of inducing adverse reproductive outcomes is the addition of various substances to products used for human consumption. The indirect "food additives" include diethylstilbesterol used to increase weight gains in feedlot cattle, antibiotic residues (particularly the tetracyclines) used as animal food additives to provide vitamins and reduce diarrheal disease, and lead used in pipes for transmitting water supplies.

In addition to the variety of substances and the various routes of human exposure, there are some chemicals that can and do enter the human body through multiple routes. As an example, and by way of summarizing the various sources of exposure, the case of lead contamination will be considered.

Lead is known to cross the placenta[13] and is thought to induce fetal neurologic damage at blood levels that are not associated with encephalopathy.[14,15] It is stable in both the body and the environment and is a widely used material. It has had a number of different applications over a long period of time.

Lead can be absorbed by ingestion and/or inhalation in certain manufacturing situations, including battery making and smelting. It can be carried home on clothes[16] and this causes secondary as well as direct occupational exposure. It is found as an ambient air pollutant near motorways and refineries from its use in gasoline. It has constituted an inadvertant source of poisoning from ingestion of paint on old buildings, and to a much lesser extent, by its use in modern copper tubing joined with solder lead containing. It has been found in the past in canned milk and baby formula. Human exposure to lead can, thus, be either occupational or ambient. It can be a contaminant and a "food additive." Exposure can be either purposeful (albeit by young children, primarily) or accidental.

It should be noted that most of the current knowledge about the adverse reproductive effects of chemicals in the environment has come through the study of occupational groups or as a result of large scale accidental exposures, particularly those in which the effects were, or are, rather dramatic. There remains much to be learned about the effect of low-level exposure and subtle effects.

OUTCOMES

Classically, a teratogenic agent is one that induces structural malformations during the gestational stage of development. Recently, the definition of teratogenesis has been expanded to include aberrations of structure, growth and function, whether caused by genetically determined factors, by environmental factors, or by an interaction of genetic and environmental factors. Clearly, a death during the embryonic, fetal, or neonatal periods might be the result of aberrant prenatal development produced by a teratogenic agent, and thus, all pregnancy losses should be considered in the investigation of an environmental exposure. In like manner, alterations in fertility due to chemical agents might be classed as teratogenic effects. And finally, effects of exposures during development that lead to behavioral problems or physical disease such as cancer should be included. Thus, the complete spectrum of adverse reproductive outcomes which might be related to exposure to toxic agents can range from decreased fertility through fetal loss, growth retardation, and congenital malformations to childhood or even adult morbidity.

Fertility

Although fertility is measured as the efficiency with which a generation of people reproduces itself, it is more meaningful to divide this into components reflecting different stages in which adverse outcomes can occur as a result of exposure to toxic chemicals.

The earliest events which reduce reproductive efficiency are those that lead to a failure of gametocytes to develop into viable gametes. This can occur in either the male or the female, but, since gametogenesis proceeds differently in male and female mammals, toxic insults can have different consequences. In the sexually mature male, primary gametocytes are continuously converted through meiotic division into gametes. A disruption in this process can result in a reduced number of viable gametes. Such an effect has been observed with X-irradiation and with some industrial chemicals, a recent example being Kepone®.

The female gametogenic process starts during fetal development and oogonia reach a maximum in number (approximately 5×10^6) during the fifth month of gestation. A million or so oogonia differentiate into primary oocytes of which only a few hundred will subsequently form mature ova capable of being fertilized. Since only a relative few of the potentially available primary oocytes develop further, generally one at a time, it is unlikely that total infertility induced by the destruction of all available oocytes will occur from sublethal exposures to toxic substances. Recent evidence does suggest that an earlier menopause may occur as a result of more rapid depletion of primary oocytes, and cigarette-smoking women have been observed to experience menopause at younger ages than nonsmoking women in at least one investigation. Mouse studies also indicate that in some circumstances oocytes may be destroyed by chemical toxins and result in permanent total infertility.

The biologic processes involved suggest that temporary primary infertility or subfertility is most likely to occur in males exposed to toxic chemicals. Furthermore, detection of nonviable or morphologically abnormal gametes is considerably easier in the male by direct examination. Reduction in fertility can certainly occur in either sex, although it is more difficult to detect as primary in the female unless there is disruption or termination of the menstrual cycle. The unaided death of the conceptus is classified by the time of occurrence into spontaneous abortion ($<$20 weeks) and stillbirth ($>$20 weeks).

Death

Deaths occurring during the first trimester are spontaneous abortions and include any loss from failure of implantation of the blastocyst to the death and expulsion of a 12-week fetus. Unfortunately these are often unrecorded events, and if early enough, are almost undetectable. Special studies can and have been undertaken to measure the occurrence and condition of spontaneous abortions. Current studies have reported that approximately 30% of spontaneous abortions are chromosomally abnormal and anatomic abnormalities may be present in an even greater proportion of such abortions. It is known that maternal age and radiation are both associated with aneuploidy (an abnormal number of chromosomes), particularly trisomy (one extra chromosome). Since most aneuploid embryoes die rather early during fetal development, they are considerably more common in spontaneous abortions than in later fetal deaths or live births. Although this affords an opportunity for the early recognition of adverse reproductive outcome due to toxic exposures, tissue cultures are required, the cost is high, and considerable expertise is needed. In one current study, a single technician is able to handle about three cultures per week. The sensitivity of such a system is potentially very good, however, if a good collection and pregnancy monitoring system is available and if the type of abnormality, either chromosomal or anatomic, can be specified for a given exposure. Unfortunately, an additional problem with the examination of first trimester spontaneous abortions is that they are often degenerated by the time examination is possible, which hampers both the detection of morphologic anomalies and the successful culture of tissues.

Losses during the second trimester are much easier to examine since they are larger and their existence is more likely to be detected. From 20 weeks onward, there is a requirement that a fetal death certificate be filed in all vital statistic jurisdictions in the U.S. There is evidence, however, that many fetal losses go unreported in spite of these requirements, and few are examined in detail for the presence of abnormality. Third trimester fetal losses, or stillbirths, are probably fairly well reported in most sections of the country, but as with second trimester fetuses, are often not studied in any detail.

After a live birth, death may occur during the neonatal period, infancy or childhood. The vast majority of these deaths ordinarily occur within the month following birth, particularly in the first few days. While reporting is good for these events, there are many factors other than exposure to toxic substances which are known to be associated with infant and childhood mortality. Furthermore, these events are relatively rare and very large populations must be followed to detect differences in either stillbirth or infant and childhood mortality rates in differentially exposed groups. This difficulty may be in part offset by the fact that data are readily available.

Malformation

With some limitations on the collection of data, measurement of birth defects is a direct and useful means of assessing the potential toxic effects of teratogenic agents. Special examinations and a more reliable reporting system have been initiated in some parts of the U.S. and studies in this and other countries continue to verify that some agents are teratogenic.

Major structural malformations occur in approximately 3% of all live births in the U.S. and in Europe, and if added to those among abortions and stillbirths, they make up a significant proportion of all human conceptions. Furthermore, some of them have clearly been found to be the result of exposure to chemical toxins during fetal development, although the vast majority of birth defects at present are of unknown cause. One of the problems with measuring the rate of congenital malformations in different populations, however, is that, even though they are often obvious on examinations, they are frequently not recorded in routine record systems, i.e., birth certificates and fetal death certificates, and thus, require special reporting systems. It has also been well demonstrated that the detection of malformations is enhanced by repeated examination at intervals during infancy and early childhood.

Growth

One of the significant effects which can be due to exposure to toxic chemicals *in utero* is a decrease in growth rate of the developing fetus, and this effect may extend to postnatal development as well. Examples of two agents that have been implicated in fetal growth retardation are PCBs and cigarette smoking. PCB exposure in Japanese women consuming contaminated rice oil resulted in babies of lower than usual birthweight, and experiments with monkeys have led to similar results.[17] In studies all over the world, cigarette smoking and exposure to chemical compounds found in cigarette smoke have been shown to reduce birth weight in human babies by about 200 grams not attributable to premature delivery. Furthermore, there is some evidence that growth rates of children born to smoking mothers are subnormal as late as 8 years later.

One of the attractive features of an effect on growth rate is that it is relatively easy to measure, since birthweight is almost universally recorded on birth certificates. This should be a good measurement for comparison of variously exposed groups, but, as with perinatal deaths, there are other factors such as altitude, race, and maternal nutrition which can influence birthweight. The value of birthweight data would be considerably enhanced by the simultaneous recording of the length of gestation, the crown-heel length and the head circumference.

Functional Defects

The largest class of effects that can be due to exposure to toxic agents *in utero* are those which adversly affect normal functioning. These may also constitute the most numerous problems associated with such exposure since they comprise biochemical, behavioral, and other defects, many of which are uncataloged.

1. Biochemical defects include abnormal functioning in both endocrine and enzymatic systems. Many of these are the result of genetic abnormalities and often require special test systems to detect the nature of the functional deficit. The systems themselves are interdependent and a number of subtle changes can occur that escape detection.
2. Behavioral defects are probably the least well understood of all congenital abnormalities from any cause. Their complexity matches that of the biochemical alterations, from which they may not always be separated. The most notable behavioral defect associated with toxic exposure *in utero* is the mental deficiency among children born around Minamata Bay in Japan. This exposure was due to organic mercury that accrued in fishery products from industrial wastes containing mercurial salts, and was first noted among local cats who became ataxic after eating shellfish from the Bay. The condition in human babies manifested itself as permanent, severe mental retardation, and is still a problem more than 20 years later.

   Another more recently "discovered" behavioral defect due to intrauterine exposure is the fetal alcohol syndrome. In this case mental retardation has been found among babies born to women who were exceptionally heavy drinkers during pregnancy.[18] The implications for an effect in not-so-heavy drinkers is currently being studied and it has been suggested that binge drinking among moderate drinkers may have teratogenic effects in some cases.

   A subject of some current interest is the role of environmental toxins in producing hyperactivity in children. Naturally occurring salycilates and food additives that might induce hyperactivity have been investigated following reports by Feingold. Since this association is hypothetical, treatment with special diets was undertaken to see if a reduction of hyperactivity could be accomplished by feeding foods that did not include these presumably toxic substances. The apparent success of this treatment regimen led to the publicizing of this association in a book and was followed by many studies. These studies indicate that the presumed relationship between hyperactivity and food additives is not well enough understood. In addition, hyperactivity itself is a very difficult condition to measure with confidence.

Another association between hyperactivity and a toxic substance has been suggested from low-level, chronic lead intoxication. The question here, however, was which comes first, the hyperactivity or the lead ingestion due to pica appetite. A group of children who were classed as hyperactive were measured for blood lead concentration. Those for whom a presumed non-lead-related cause of hyperactivity could be ascertained had the same blood lead levels as normally active children. The group without known cause for hyperactivity had an increased lead burden, which implies that the hyperactivity did not cause the increased ingestion of lead.

Although both these relationships between toxic substances and hyperactivity are presumably due to ingestion of toxic substances after birth, they raise the possibility that some association may be due to exposures *in utero*.

The major problem with attributing behavioral effects to any cause is the measurement of behavioral effects, per se. Several scales have been used in addition to the Stanford-Binet IQ scale and its derivatives (which are only applicable to older children). Among the scales which have been developed for infants are the following:

Apgar Score: This score is usually computed at one minute and five minutes after birth and is a measure of the general condition of the infant. It is based on a simple assessment of the normality of the heart rate, respiratory effort, muscle tone, reflex irritability and color, and it is moderately predictive of future neurologic status.

Prechtl Neurologic Examination: This is an assessment of neurologic adequacy for newborn babies and requires special training for its administration and evaluation.

Brazelton Neonatal Behavioral Assessment: This is a well-developed scale intended as a means of measuring interactive behavior. Although it contains a few neurologic items, its main thrust is behavioral and is an attempt to score the infant's available responses to his environment. The examination is usually performed on the third or fourth day of life and scoring is based on the infant's best performance rather than on an average response. The examination has shown good reliability when repeated at one month of age.

Bayley Scale: This scale is an assessment of gross motor, fine motor, language, and social functioning in older infants. It is in many ways comparable to an IQ scale, but is not always a reliable predictor of standard IQs measured at older ages.

There are a number of other scales that could be used and special procedures have been developed for specified problems. One advantage of the Apgar Score, in spite of its very limited scope, is that it is very fast, is routinely recorded following most deliveries in the U.S., and is often included on birth certificates. All of the others require somewhat elaborate examinations and have mostly been limited to special research protocols. Thus, while of undoubted importance in evaluating risks due to intrauterine exposures to potentially toxic substances, the assessment of functional behavioral defects is a difficult task.

3. Immunologic defects due to toxic exposure are just beginning to be investigated. There is some evidence that polybrominated biphenyls may reduce immunologic response in adults and similar findings among newborns could lead to more research in this area. Laboratory procedures to test immunologic competence are complex and probably not well enough developed at present for large scale studies, however.

To complete the spectrum of functional defects which could be due to teratogenic agents, one should include physical defects. These are discussed as malformations, although there is clearly overlap between anatomic malformations and functional defects.

Pharmacologic

A final group of infant health problems that can be included under adverse reproductive outcomes are those in which the toxic substance has crossed the placenta and produced a relatively pure pharamacologic response in the fetus. A prime example of this is heroin, in which the newborn infant suffers withdrawal symptoms similar to older persons who stop using the drug. Other examples would include drugs used for medical or psychotropic purposes, but such drugs have not been included in the present discussion of environmental exposures.

The Occurrence of Adverse Reproductive Outcomes as Related to Environmental Toxic Exposures

At the present time, at least 5% of all infants born alive in the U.S. either have a serious disease, a significant anatomic malformation, or a severe functional defect or die soon after birth. Estimates of the causes of these adverse outcomes have been made as follows[19]:

Genetic factors	20%
Infections	2–3%
Maternal conditions	1–2%
Environmental exposures	2–3%
Chromosomal aberrations	3–5%
Unknown	65–70%

As can be seen, the ascribed environmental causes account for only a small fraction of these, and this includes exposure to nonchemical agents such as radiation and trauma. An undetermined portion of the unknown causes, and some of the genetic and chromosomal causes, may be the result of toxic exposure *in utero* or to the gametocytes. Considering the large number of spontaneous abortions and the undetermined amount of infertility or subfertility that may exist in the population, plus the functional behavioral deficits not detectable until children are older, the possibilities for adverse reproductive outcomes which might be caused by chemicals that are toxic to the reproductive system or its products are large.

One way of modeling the various outcomes and their relation to toxic exposure is by use of a graded dose-response relationship. Such a model would appear as in FIGURE 1, and would include most of the outcomes that have been discussed.

This model suggests that a given toxic substance could lead to several different outcomes with different incidence rates at the same dose as well as show a dose-response increase in effect for any one outcome. This is consistent with the concept that almost any chemical is toxic to almost every system at some dose. The ordering of the different adverse reproductive outcomes might vary somewhat with different chemical toxins, and the slopes of the dose-response curves would be expected to vary as well.

The advantage of viewing the relationship between toxic chemical exposures and adverse reproductive outcomes according to the above model is that for any toxic chemical, certain outcomes are likely to be more sensitive indicators of effects. Thus, it becomes important to try to select the specific outcome that is most sensitive when assessing the effects of toxic exposure. For example, comparing the rates of all malformations may not reveal differences between exposed and nonexposed populations, while rates for a specific malformation, such as cleft lip/palate or digital hypoplasia, may do so in the case of the fetal hydantoin syndrome.[20] Or, in the case of smoking during pregnancy, there is considerable controversy over the effect on fetal loss, but intrauterine growth retardation has been a consistent finding.

Another important consideration in describing the relation between exposure and outcome is the interrelationship between various factors that effect the same outcome. This is termed confounding. For those confounding co-variables that are not of interest to the investigation of the teratogenic effects of toxic substances, an adjustment can often be made during analysis of data. An example of this is in measuring the effect of a teratogenic agent on chromosomal aberrations that increase with maternal age. If exposed women have higher rates of trisomy than nonexposed women at all ages, then the relationship between these factors is said to be independent and additive, and no interaction exists. There are situations, however, when the relationship is not additive, and an interaction does exist. If the combined effect of two (or more) agents on adverse reproductive outcome is greater than would be expected when each is acting alone and independently, then the agents are said to be synergistic; an example might be diphenylhydantoin and phenobarbital, which have been suspected of teratogenic synergism.[21] If the combined effects are less than expected when either agent is acting

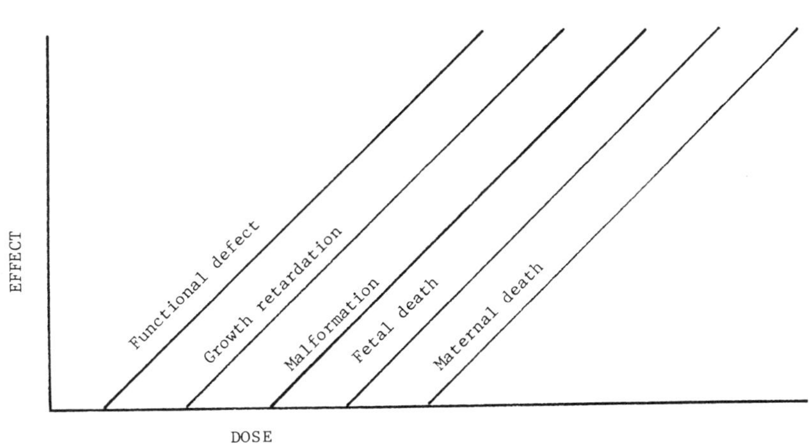

FIGURE 1.

alone, they are said to be antagonistic, as in the case of phenobarbital protecting against the teratogenicity of cyclophosphamide.[22] Thus, synergism between two or more factors may make it easier to detect teratogenic effects, but may also be misleading, while antagonistic effects may obsure the detection of teratogenic potential. Interactions between environmental agents involve more than chemical-chemical interactions, for it has been established that teratogenicity of various compounds can be modified by altering the oxygen supply to or the temperature of the fetus.[22A] And to complicate matters further, a vast literature on the great importance of gene-teratogen interaction has accumulated over the past several years.[22B]

An additional consideration in assessing teratogenicity is the detection of anamnestic response. Such a response is due to a form of sensitization that can occur *in utero* and is a potential problem when measuring adverse outcomes following exposure to substances which mimic hormones and antigens. Early exposures to the fetus may alter subsequently occurring responses to otherwise benign agents.

This brief description of the range of relationship between toxic chemicals and their effects on human reproductive outcome should serve to assist in the design of studies and the interpretation of data on teratogenic potential.

Measurement of the Association between Toxic Chemical Exposure and Adverse Reproductive Outcome

In order to establish that an association exists between an exposure to any risk factor, such as a chemical toxin, and an adverse health outcome, such as congenital malformation, one of two general methods are employed. The prospective approach is to identify an exposed population and one that has not been exposed and then to examine them for the presence of a health condition possibly attributable to the exposure. A comparison of the rates of disease between the exposed and nonexposed populations will indicate the additional risks presumably due to the exposure.

The retrospective approach is to identify a group of individuals in which some adverse health outcome of interest has occurred, and a group of "healthy" individuals and determine their status *vis-à-vis* past exposure to some agent. Both methods are equivalent for determining the relative increase (decrease) in disease that might be attributable to the exposure. In addition, in both situations a number of other factors can be taken into account and dose-response relationships can sometimes be established.

The advantage of the prospective method is that the time sequence of events, i.e., exposure *before* outcome, can be established unequivocally since exposure status is determined before knowledge of outcome status. Furthermore, during the course of such a study, the rate of occurrence of the outcome in the population being studied can be determined. If the outcome is rather rare, however, large populations may have to be followed and examined in order to make statistically reliable comparisons between exposed and nonexposed groups. The major advantage of the retrospective approach is that it is much more efficient in terms of time and resources, particularly for a rare disease outcome. A number of study designs using one or the other of these approaches have been developed, and others are possible.

The enormous number of potentially harmful chemicals and the wide variety of exposures plus the broad spectrum of adverse reproductive outcomes makes it impossible to measure all associations which might exist. Epidemiologic investigations, therefore, must attempt to concentrate on either the exposure or the outcome. The prospective method is generally used when exceptionally exposed populations can be identified, such as occupational groups or those living or working near an accidental introduction of a toxic chemical into the environment. The usual strategy is to select those chemicals having high teratogenic potential as determined from laboratory or chemical evidence or those chemicals that are structually similar to known toxins and for which some group of persons have received sufficient exposure for harmful effects to occur with a frequency high enough to be detectable. The rate of occurrence of adverse health effects among highly exposed individuals can be compared to those of the general population or to a group of unexposed controls. This has been fruitful in discovering disease-causing agents among occupational and accidentally exposed groups, but is not always relevant to those receiving ambient exposures at lower doses—often a much larger group. For example, while pesticide workers may show a decrease in reproductive efficiency due to subfertility caused by certain pesticide exposures, it is not possible to extrapolate the effects to the population as a whole, which will be exposed to much lower doses. Furthermore, occupational groups are by definition "employed," and employed persons may be "healthier" and perhaps react differently to many chemical compounds than unemployed persons.

Another and more direct way to measure exposure is through the laboratory determination of toxic chemical levels in human tissues and fluids, such as hair, blood, fat, urine, and milk. Lead and mercury exposures have been measured in this manner and individuals identified for further follow-up of health effects. These procedures are also quite useful in determining cumulative exposures to certain compounds that accumulate in body fat, such as pesticides, halogenated biphenyls, and other aromatic hydrocarbons. Both fat samples and breast milk have been subjected to laboratory analysis, and in conjunction with other information on the subjects from which these samples have been obtained, some groups of exposed individuals have been identified. Consumers of sport fish from fresh water sources in the U.S. have shown increased PCBs in breast milk and persons living in rural areas show higher levels of some pesticides than city dwellers.

The major usefulness of these studies, however, has been to identify areas of the country where exposures are high or low and to document long-term trends in general exposure. Recent data from the routine collection and analysis of fat samples and other materials from cadavers and biopsies across the U.S. by the Environmental Protection Agency has shown a decline in DDT in the 1970s following the curtailment of its domestic agricultural use. The national breast milk survey has identified areas in the East which have higher levels of PCBs in human breast milk than in the West. Both of these compounds are known or strongly suspected of being toxic to the reproductive system. The limited amount of material available for analysis makes it impossible to test for a large number of chemical compounds on a routine basis.

Thus, the prospective method of identifying and following up exposed groups has been most useful in verifying adverse reproductive or other detrimental health effects among heavily exposed occupational groups or those subjected to accidental dissemination of toxic chemicals. The method is most applicable in cases where some toxicity is known or is highly suspicious based on other evidence.

The retrospective method of epidemiologic investigation begins with the identification of cases and compares their history of previous exposure to a group of healthy controls. Most of the teratogenic agents known to produce human congenital malformations have been discovered by noting an exceptionally large number of some relatively unusual condition among a small group of children in a short span of time. How many cases among how many children and in how short a time are necessary to alert investigators that something unusual has occurred depends on the rarity of the condition and the proportion of cases which are attributable to the particular exposure. Fewer than 20 cases of scrotal cancer led Sir Percivall Pott to conclude that this was an occupational hazard to young chimney sweeps in England in the eighteenth century, while 86 cases of congenital cataract alerted Gregg to the teratogenic effects of Rubella virus infection during pregnancy in Australia. It has taken hundreds of cases to document some of the problems associated with intrauterine irradiation and literally thousands of low birthweight babies to establish the growth retarding effects of maternal cigarette smoking.

Even though most of the human teratogens have been detected through retrospective investigations of unusual congenital malformations, the reporting of these conditions remains remarkably incomplete. A number of major easily detectable anomalies such as anencephaly and spina bifida may not be recorded on birth certificates and are, thus, unavailable for investigation. The complete reporting of anomalies is elaborate and expensive, although the Center for Disease Control has established a surveillance system covering approximately one-fourth to one-third of all U.S. births and a more thorough reporting in one large city (Atlanta). Reporting of still births of at least 20 weeks gestation and of deaths to infants born alive is required in the United States, but there is no reporting of infertility, spontaneous abortions, or of childhood morbidity because it would be prohibitively expensive and extremely difficult to do so on a routine basis.

Potentially, there are routinely collected data that might be useful in assessing adverse reproductive outcome of some types among geographic areas, and by extension, to other special population groups. Data such as fetal and infant death rates could be monitored and, as recently suggested by Dobbins, birth spacing data obtainable on birth certificates in 40 of the states could be used as an indicator of fertility. Both subfertility and early abortion losses will result in longer time periods between successive live or still births. In addition, special surveys might be conducted using brief telephone interviews to identify populations with varying reproductive outcome histories.

A county study of all births and perinatal deaths has recently been initiated, and should be continued on a reasonably current basis. Areas with high perinatal loss or unusually long birth spacing could then be investigated for accuracy of reporting and for unusual local circumstances that might contribute to these outcomes. Preliminary analysis of early neonatal deaths in this manner indicates that those areas in the Northeast and Midwest where industrial pollution is noteworthy are not among the counties with significantly elevated rates. Early neonatal losses are significantly higher in the Appalachia and Rocky Mountains and in isolated areas across the country. This pattern is not similar to that observed for cancer mortality in adults, particularly for gastrointestinal cancers, which are concentrated in the Northeastern counties near industrial areas.

Thus, while retrospective methods have led to the identification of several important human teratogens, they will require a more complete reporting system to collect case material than has been available if their potential is to be fully realized. It is unfortunate that we have had to rely so heavily on the observations of alert practitioners.

Summary

The spectrum of adverse reproductive outcomes that may be the result of toxic insult to human reproduction should be expanded to include fertility problems, through structural malformation, fetal growth abnormalities, and fetal loss to childhood morbidity and behavioral functional abnormality. The large number of chemical agents potentially responsible plus the many possible means of human exposure makes it imperative that a systematic way to detect associations be considered. An attempt to describe the relationship between exposures and outcomes, and their interactions, has been presented and includes the ordering of outcomes and dose-responses leading to various possible outcomes. The methods for studying these relationships include following up observed exposures to determine outcomes and tracing back observed outcomes to determine exposures. Both methods have been useful and both have some severe limitations as well. The job is enormous, but the opportunities are ubiquitous.

References

1. Kay, K. 1977. Polybrominated biphenyls (PBB) environmental contamination in Michigan, 1973–1976. Environ. Res. **13**(1): 74–93, February.
2. Fuller, J. G. 1977. The Poison that Fell From the Sky. Random House. New York.
3. Angle, C. R. & M. S. McIntire. 1964. Lead poisoning during pregnancy. Am. J. Dis. Child **108:** 431–439.
4. Källén, B. 1978. Hexachlorophene teratogenicy in humans reputed. JAMA **240:** 1585086.
5. Hirayama, C. 1976. Clinical Aspects of PCB Poisoning. In PCB Poisoning and Pollution. Chapter 6, pp. 87–104. Academic Press. New York.
6. EPA. DDT—A review of scientific and economic aspects of the decision to ban its use as a pesticide. EPA 540/1-7-022.
7. Mes, J., D. S. Campbell, R. N. Robinson & D. J. A. Davies. 1977. Polychlorinated biphenyls and organochlorine pesticide residues in adipose tissue of Canadians. Bull. Environ. Contam. Toxicol. **17**(2): 196–203.
8. EPA National Conference on Polychlorinated Biphenyls. Conference proceedings. EPA 650/6-75-004.
9. Kuratsune, M., Y. Masuda & J. Nagayama. Some recent findings concerning Yusho. EPA 560/6-75-004, 14–29.
10. Allen, J. & D. H. Norback. Pathobiological responses of primates to polychlorinated biphenyl exposure. EPA 560/6-75-004, 43–49.
11. McKinney, J. Toxicology of selected symmetrical hexachlorobiphenyl isomers: Correlating biologic effects with chemical structures. EPA 560/6-75-004.
12. Adams, C. E., M. F. Hay & C. Lutwak-Mann. 1961. The Action of Various Agents Upon The Rabbit Embryo. J. Embryol. Exp. Morphol. **9:** 469–491.
13. Palmisona, P. A., R. C. Sneed & G. Cassady. 1969. Untaxed whisky and fetal lead exposure. J. Pediatr. **75**(5): 869–871.
14. David, O., J. Clark & K. Voeller. 1972. Lead and hyperactivity. Lancet **1:** 900–903, October 28.
15. Landrigan, P., E. L. Baker, R. G. Feldman, *et al.* 1976. Increased lead absorption with anemia and slowed nerve conduction in children near a lead smelter. J. Pediatr. **89**(6): 904–910, 1976.
16. Baker, E. L., D. S. Folland, T. A. Taylor, *et al.* 1977. Lead poisoning in children of lead workers. New Engl. J. Med. **296,** February.
17. Allen, J. R. & D. A. Barsotti. 1976. The effects of transplacental and mammary movement of PCB's on infant rhesus monkeys. Toxicology **6:** 311.
18. Clarren, S. K. & D. W. Smith. 1978. The fetal alcohol syndrome. New Engl. J. Med. **298:** 1063–1067.
19. Wilson, J. J. 1973. Environment and Birth Defects. Academic Press. New York.
20. Elshone, J. & J. H. M. VanEck. 1971. Aangeboren misvormingen met name gespleten lip met of zonder gespleten verhemlte, bijkinderen van moeders met epilepsie. Ned. Tijdochor Geneeskd. **115:** 1371.
21. Barr, M., A. K. Poznaski & R. D. Schmickel. 1974. Digital Hypoplasia and Anticohvulsants During Gestation: A teratogenic syndrome? J. Pediatr. **84:** 254–256.
22. Gibson, J. E. & B. A. Becker. 1968. Effect of phenobarbital and SKF-525A on the teratogenicity of cyclophosphamide in mice. Teratology **1:** 393–398.

22a. FRANKLIN, J. B., A. F. GOLDFARB, R. MATSOMOTO & R. L. BRENT. 1963. Modification by progestational compounds of the teratogenic effect of uterine vascular clamping. Fertil. Steril. **14:** 365–369.

BRENT, R. L., B. T. BOLDEN & A. WEISS. 1968. The use of uterine vascular clamping in the pregnant rat to modify the embryotoxic effects of anti-cancer drugs. Cancer Res. **28:** 2001–2006.

EDWARDS, M. J. & R. A. WANNER. 1977. Extremes of Temperature. *In* Handbook of Teratology. Vol. 1. Chapter 12. J. G. Wilson & F. C. Fraser, Eds. Plenum Press. New York.

22b. FRASER, F. C. 1977. Interactions and multiple causes. *In* Handbook of Teratology. Vol. 1. Chapter 13. J. G. Wilson & F. C. Fraser, Eds. Plenum Press. New York.

23. BRILLIANT, L. B., G. VAN AMBURG, J. ISBISTER, *et. al.* 1978. Breast-milk monitoring to measure Michigan's contamination with polybrominated biphenyls. Lancet: 643–646.

PART VII. IMMUNOLOGIC ABNORMALITIES

LABORATORY STUDIES ON THE IMMUNE EFFECTS OF HALOGENATED AROMATICS

Michael I. Luster, Robert E. Faith,* and George Clark

Environmental Biology and Chemistry Branch
National Institute of Environmental Health Sciences
Research Triangle Park, North Carolina 27709

INTRODUCTION

We have seen an increasing interest in the effect of environmental chemicals on the immune response, and immune parameters are now often assessed in the routine evaluation of toxicity by these chemicals. As recently reviewed by Vos,[1] chemicals of environmental concern that have clearly been shown to cause impairment of immune functions or reduce host resistance to infectious agents include organometallics such as triphenyltin and methylmercury, inorganic metals including lead and cadmium, organotin compounds and halogenated hydrocarbons such as polychlorinated biphenyls, (PCBs) and 2,3,7,8-tetrachlorodibenzo-*p*-dioxin (TCDD). Immunologic studies with 2,3,7,8-tetrachlorodibenzofuran (TCDF), an extremely toxic chemical related to TCDD, have not been reported although TCDF has been shown to cause severe thymus atrophy in chickens, guinea pigs, and mice.[2]

Most recently Firemaster® BP-6, a chemical fire retardant consisting of a mixture of polybrominated biphenyls (PBBs) has been suspect in inducing an altered immune status in many Michigan dairy farm residents who ate PBB-contaminated farm products.[3] Support for these observations have been found in chronic PBB-feeding studies with rodents.[4] These effects upon the immune system are focused upon thymus-derived lymphocytes (T-cells) although humoral immune functions associated with bursa-equivalent derived lymphocytes (B-cells) may be slightly compromised. TCDD also appears to primarily affect thymic-dependent immune functions.[5] Furthermore, recent studies in our laboratory have suggested that helper cell function is not affected in TCDD-exposed rats; thus, TCDD appears to selectively effect different functional subsets of T-cells.[6]

The mechanisms of TCDD-induced immunosuppression or for that matter any of the environmental immunosuppressive chemicals are presently unknown. Studies with TCDD have suggested that the suppression is probably not caused by an indirect action through alterations in the adrenals, pituitary[7] or production of thymic hormone, zinc deficiency or altered phagocytic activity.[8] Attempts to study the direct action of TCDD on lymphocytic functions may have been hampered because of the insolubility of TCDD in physiologic buffers.[9,10]

In this report we provide additional information on the immunosuppression induced by halogenated hydrocarbons in experimental animals. In addition to providing evidence that exposure to 2,3,7,8-TCDF can lead to immunosuppression, evidence is provided indicating that *in vitro* exposure of TCDD to splenic lymphocyte can directly alter lymphocyte functions.

*Present Address: Biological Research Center, Oral Roberts University, Tulsa, Oklahoma 74171.

0077–8923/79/0320–0473 $01.75/0 © 1979, NYAS

METHODOLOGY

Since there are a variety of functions associated with the mammalian immune system, it is necessary to study multiple parameters to properly evaluate the immune effects from chemical exposure. The numerous cell types, such as macrophages and at least three distinct types of lymphocytes, and their cooperative interactions required for normal immune responses makes this task difficult. It is helpful in preliminary studies to distinguish between alterations in humoral-mediated functions, i.e. antibody synthesis and immunoglobulin levels, and cell-mediated functions such as transplantation immunity, delayed hypersensitivity reactions, etc. Assays are available to accurately quantitate humoral immunity. It is convenient to determine serum levels of the various immunoglobulin classes by conventional radial immunodiffusion analysis

TABLE 1

BODY AND BODY-TO-THYMUS WEIGHT RATIOS IN TCDD-EXPOSED RATS

Age (Days)	TCDD Exposure	Body Weight (g) Mean	SE	Thymus/Body Weights × 10^{-3} Mean	SE
18	Controls	31.32	2.95	4.3	0.4
	Postnatal	23.63	1.29**	3.3	0.9
	Prenatal and postnatal	22.81	1.68**	2.9	0.9**
35	Controls	73.59	6.61	4.3	0.5
	Postnatal	65.92	9.61	3.3	0.4*
	Prenatal and postnatal	56.49	9.54**	3.1	0.6*
133	Controls	224.1	8.8	1.7	0.3
	Postnatal	215.6	5.8*	1.5	0.3
	Prenatal and postnatal	213.5	16.7	1.4	0.3*
270	Controls	308.8	21.2	ND	
	Postnatal	288.7	13.7***	ND	
	Prenatal and postnatal	327.8	27.9	ND	

*$p < 0.05$ *vs* controls.
**$p < 0.01$ *vs* controls.
***$p < 0.5$ prenatal/postnatal *vs* postnatal.

since heavy chain specific antisera and reference standards are commercially available or can be prepared for the particular species being studied.[11] Antibody synthesis is quantitated following immunization with a suitable antigen and determining agglutination titers or preferably the more sensitive antibody plaque assay.[12] It is sometimes advantageous to determine antibody production against a T-dependent as well as a T-independent antigen.

Unlike humoral-mediated function, it is far more difficult to satisfactorily quantitate subtle differences in cell-mediated immune responses. The most popular assay presently used is to determine lymphocyte blastogenesis by the rate of cellular DNA or protein synthesis following activation with selective B-and/or T-cell mitogens.[13] In rodents we employ splenic lymphocytes activated with the T-cell mitogens phytohemagglutin (PHA) and Conconavalin A (Con A) and the B-cell mitogens pokeweed (PWM) and *E. coli* lipopolysaccharide (LPS) although not all strains respond to LPS. Blastogenesis is determined by the amount of ^{3}H-thymidine precursor incorporated into DNA acid-insoluble cell material.[14] However, mitogens are polyclonal cell activators and the assay does not measure effector cells. Effector tests for specific cell-mediated immune response following activation with a specific antigen are available. We feel it is necessary to assay for several of the more consistent functional tests, particularly since recent studies have shown that functional subsets of T-cell exist in the mouse distinguishable by specific Ly phenotypic markers.[15] In the mouse, for example, cells involved in delayed hypersensitivity reactions possess identical Ly markers as cells exhibiting helper function, whereas cells that respond to the mitogen, PHA, are of a distinct phenotype.[16] We usually perform *in vitro* mixed lymphocyte cultures (MLC) reactions although splenic lymphocytes are not suitable in rodents because of the presence of suppressor macrophages.[17] In addition, delayed hypersensitivity reactions (DHRs), presumably an *in vivo* correlate of MLC, are performed using PPD as a sensitizing agent. DHRs are quantitated by a radiometric ear test similar to the procedure described by Lefford.[18] We also test effector mechanism by determining the ability of antigen-sensitized cells to release migration inhibition factor (MIF) which prevents macrophage migration *in vitro*. We employ an agarose plate modification of the original procedure, which has been described by Braendsturp.[19]

Although phagocytosis studies have not been routinely performed, it is nonetheless a critical area in immune surveillance studies. While determination of *in vivo* uptake is of some significance, we suggest that not only phagocytosis but also intracellular killing by polymorphonuclear and mononuclear phagocytes be evaluated *in vitro*.

RESULTS AND DISCUSSION

In Vivo *Studies with TCDD*

TCDD has previously been shown to induce atrophy of thymus and peripheral lymph nodes of rats, mice and guinea pigs.[20] In addition, TCDD has been shown to suppress thymic-dependent immune functions and resistance to bacterial infection.[21] The suppression of cell-mediated immune function includes suppression of mitogen responsiveness,[9,22] skin-graft rejection[9] and delayed hypersensitivity responses.[5] In rats the immune system is most severely effected by TCDD during the developmental phase.[5,9] Studies were performed in which nursing Wistar/Fischer hybrid or Fischer rats were dosed with TCDD (5 μg/kg body weight) on postnatal days 0, 7, and 14. A second group of pregnant rats received TCDD (5 μg/kg body weight) on gestational day 18 and at parturition these neonates were cross-fostered to nursing females that received doses of TCDD (5 μg/g body weight) on postnatal days, 0, 7, and 14. TABLE 1 depicts representative body weight and thymic weight effects at four specific time periods. In general, the weight effects between the two exposed groups were unremarkable at 18 days of age with both the postnatal and pre-/postnatal TCDD-exposed groups showing significantly decreased body weights and thymus-to-body weight ratios compared to controls. At 35 and 133 days of agc the pre-/postnatally exposed groups usually reveal more severely decreased body and relative thymic weights than the postnatal group, while at 270 days both treated groups appear normal. Thymus/body weight ratios were not determined at 270 days because of the involution of the thymus. Delayed hypersensitivity reactions (DHRs: TABLE 2) indicated that both exposed groups had significantly decreased responses

TABLE 2

DELAYED HYPERSENSITIVITY RESPONSES OF TCDD-EXPOSED FEMALE RATS TO PURIFIED PROTEIN DERIVATIVE

Age (days)	No. of rats	TCDD exposure	DHR Index ± SD†
35	10	Controls	9.60 ± 1.44
	10	Postnatal	4.81 ± 0.84*
	10	Prenatal and postnatal	3.08 ± 1.39*,**
133	10	Controls	5.64 ± 1.15
	10	Postnatal	3.17 ± 0.94*
	10	Prenatal and postnatal	2.32 ± 0.96*,***
210	10	Controls	4.24 ± 2.21
	10	Postnatal	3.95 ± 0.90
	10	Prenatal and postnatal	4.43 ± 2.48

†DHR index was calculated by dividing cpm of the challenged ear by the cpm of the control ear.

*$p < 0.01$ *vs* controls.

**$p < 0.01$ prenatal/postnatal *vs* postnatal.

***$p < 0.05$ prenatal/postnatal *vs* postnatal.

from 35–133 days of age although the pre-/postnatal exposed rats were more severely affected. Similar findings were noted when lymphocyte blastogenesis was examined using the T-cell mitogens PHA and Con A in female rats (TABLE 3). At seven months of age, lymphocyte blastogenic responses as well as DHRs in post- and pre-/postnatally exposed groups were similar to controls indicating, that the immunosuppression was reversible.

FIGURE 1 depicts serum antibody titers from control and TCDD-exposed Fischer rats injected with bovine gammaglobulin (BGG) (primary injection at 25 days of age). There was no statistical difference in response of the treated versus the controls to primary and secondary injections of BGG. Since normal responses to BGG requires cooperation of a specific subset of T-cells (helper cells), this suggested that TCDD did not effect T-helper cell function. To substantiate this observation adoptive transfer experiments were performed. Spleens were removed from 5 BGG-primed normal and postnatally TCDD-exposed rats as well as 5 nonimmunized normal and TCDD-exposed rats. Spleen cells were enriched for T-cells by passage through glass wool,[23] and one spleen cell equivalent was injected into 39 day-old neonatally thymectomized Fischer rats. All these animals were injected with BGG 24 hours after cell transfer. FIGURE 2 depicts the antibody response 7 days after transfer. Rats transferred with unprimed cells from either normal or TCDD-exposed animals had antibody titers slightly higher ($p < 0.05$) than thymectomized controls. There was essentially no difference between animals transferred with control or treated unprimed cells. Rats that received primed cells from either normal or TCDD-exposed animals also had similar antibody titers. These results indicated that helper cell function was not effected by TCDD. Recent studies have shown that various functional subsets of T-cells, characterized phenotypically by Ly markers, exist in the mouse.[15,16] Although there is no evidence, as yet, that T-cell subsets exist in the rat, these studies would suggest that DHRs and helper cell function belong to distinct T-cell subsets.

In Vitro Studies with TCDD

The mechanisms of TCDD-induced immunosuppression are presently unknown. Studies have suggested that the suppression is probably not caused by an indirect action through the adrenals, pituitary,[7] or production of thymic hormone,[8] zinc deficiency,[8] or altered phagocytic activity.[8] Attempts to study the direct effects of TCDD on lymphocytes have been hampered, probably by the insolubility of TCDD in physiologic buffers.[24] Nonetheless, aryl hydrocarbon hydroxylase activity was induced in cultured human lymphocytes following addition of as little as 0.1 ng of TCDD per ml of culture media.[25] Studies were undertaken in our laboratory to determine the direct effects of TCDD on lymphocytic function. Stock solutions of TCDD and various related chemicals were prepared in dimethyl-sulfoxide (DMSO) and diluted appropriately in DMSO. Spleens from female $B_6C_3F_1$ mice were submerged in these DMSO solutions for 10 seconds and rinsed in RPMI 1640 culture media. Spleen cell suspensions were prepared and cultured with various mitogens as described in methodology. Lymphocytic function was determined by incorporation of either radiolabeled protein, RNA or DNA precursors into cellular material. As summarized in TABLE 4 a decrease was noted in total CPM following treatment with DMSO alone as compared to controls although this was not reflected in the stimulation index (E/C). This effect apparently is due to a 10–20% loss in cell viability that occurs

TABLE 3

EFFECT OF TCDD ON ^{3}H-THYMIDINE INCORPORATION INTO DNA BY MITOGEN-STIMULATED LYMPHOCYTES

Age (days)	No. of rats	TCDD exposure	PHA S.I. ± S.D.	Con A S.I. + S.D.
25	3	Controls	9.82 ± 0.40	41.68 ± 2.28
	3	Postnatal	1.49 ± 0.71*	1.78 ± 0.63*
	3	Prenatal and postnatal	1.42 ± 0.09*	1.82 ± 0.29*
39	3	Controls	10.47 ± 2.06	24.18 ± 7.36
	3	Postnatal	6.34 ± 1.19*	6.19 ± 1.76*
	3	Prenatal and postnatal	3.89 ± 0.75*,**	5.50 ± 0.84*
59	3	Controls	13.91 ± 3.24	37.43 ± 4.46
	3	Postnatal	4.84 ± 0.63*	6.40 ± 1.06*
	3	Prenatal and postnatal	3.42 ± 0.67*	4.31 ± 0.46*
128	3	Controls	14.55 ± 2.93	19.18 ± 5.26
	3	Postnatal	10.39 ± 6.34	13.62 ± 4.04*
	3	Prenatal and postnatal	7.69 ± 9.08	7.78 ± 1.74*,**
210	3	Controls	10.84 ± 3.38	29.78 ± 10.04
	3	Postnatal	13.59 ± 2.58	23.28 ± 7.24
	3	Prenatal and postnatal	11.59 ± 1.70	28.58 ± 9.60

†S.I.: a stimulation index was calculated by dividing the mean counts per minute (cpm) of triplicate mitogen-stimulated cultures by the mean cpm of the nonstimulated cultures.

*$p < 0.01$ *vs* controls.

**$p < 0.05$ prenatal/postnatal *vs* postnatal.

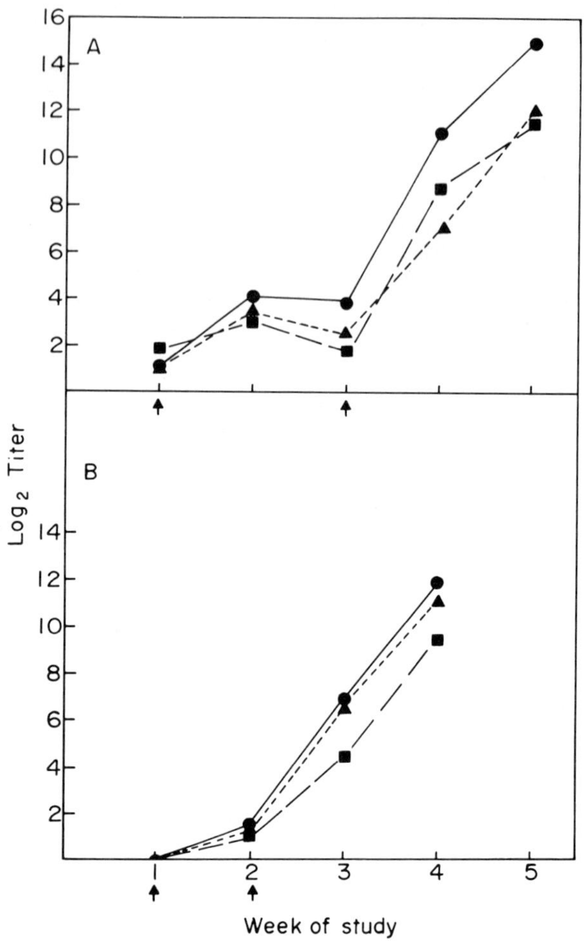

FIGURE 1. Antibody response of TCDD exposed animals. (A) Primary and secondary response to injection with bovine gamma globulin; ■——■, nonexposed animals; ○——○, animals exposed to TCDD postnatally by means of maternal dosing on days, 0, 7, and 14; and △——△, animals exposed to TCDD pre- and postnatally by means of maternal dosing on gestation day 18 and on days 0, 7, and 14. (B) Response of animals injected with bovine gamma globulin two consecutive weeks. Legend as above. In both A and B the arrows represent the days of antigen injection. At week 1 of the studies represented in both A and B the animals were 25 days of age. (From Faith & Moore.[5] By permission of *Journal of Toxicology and Environmental Health.*)

within 1 to 4 hours following treatment with DMSO. For this reason, all statistical analysis of TCDD-treated groups were compared to DMSO-treated controls rather than untreated controls. A marked depression of DNA synthesis occurred following TCDD treatment, particularly with the T-mitogens PHA and Con A at the 50 ng exposure level. The suppression was dose dependent when analyzed by the stimulation index. Similar observations were noted when protein and RNA synthesis was determined.

Uptake studies with ^{14}C-TCDD (specific activity, 148.5 Ci/mole) indicated splenic uptake was dose-dependent varying from 0.06 ng in spleens exposed to 14 ng TCDD/ml of DMSO to 0.2 ng in spleens exposed to 50 ng TCDD/ml of DMSO

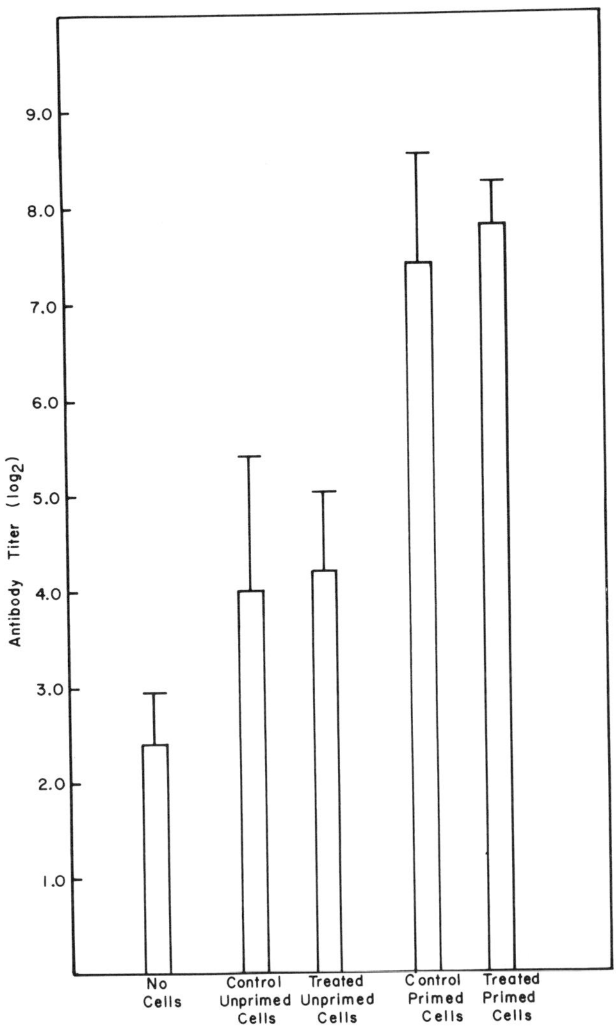

FIGURE 2. Antibody response in neonatally thymectomized rats adoptively given T-cells from Control and TCDD-exposed animals. Both groups receiving unprimed cells differed from controls (group receiving no cells) at the $p < 0.05$ level. Both groups receiving primed cells differed from those animals receiving unprimed cells and controls at the $p < 0.05$ level. (From Faith, Luster & Moore.[6] By permission of *Cellular Immunology*.)

TABLE 4

PROLIFERATIVE RESPONSE OF SPLENIC LYMPHOCYTE CULTURES TO PHA, CON A AND LPS FOLLOWING *IN VITRO* TREATMENT WITH TCDD IN DMSO†

Stimulus	Treatment	Thymidine Incorporation: Unstimulated Cultures (C) cpm ± S.E.	Stimulated Cultures (E) cpm ± S.E.	Stimulation Index (E/C) ± S.E.
PHA	None	307.5 ± 37.8	2.790 ± 243	9.68 ± 1.42
	DMSO	140.7 ± 29.0*	1,381 ± 155*	10.48 ± 1.10
	0.5 ng TCDD/ml of DMSO	132.4 ± 29.6	1,003 ± 296	7.66 ± 1.76
	5.0 ng TCDD/ml of DMSO	231.6 ± 52.0	1,354 ± 239	6.07 ± 0.66****
	50 ng TCDD/ml of DMSO	85.5 ± 15.0	461 ± 121***	5.08 ± 0.78****
Con A	None		47,770 ± 4,999	170.8 ± 37.8
	DMSO		16,689 ± 1,250**	133.0 ± 20.9
	0.5 ng TCDD/ml of DMSO		15,273 ± 4,723	113.4 ± 28.3
	5.0 ng TCDD/ml of DMSO		18,966 ± 5,180	80.7 ± 11.6
	50 ng TCDD/ml of DMSO		5,815 ± 1,490***	65.2 ± 13.0****
LPS	None		5,680 ± 435	19.25 ± 2.04
	DMSO		2,902 ± 837	19.83 ± 2.19
	0.5 ng TCDD/ml of DMSO		1,867 ± 541	13.56 ± 2.19
	5.0 ng TCDD/ml of DMSO		2,478 ± 1,158	9.08 ± 1.96****
	50 ng TCDD/ml of DMSO		895 ± 58**	12.33 ± 3.12

†Mean value ± S.E., 5 animals per group.
*$p < 0.05$ none *vs* DMSO controls.
**$p < 0.01$ none *vs* DMSO controls.
***$p < 0.01$ DMSO *vs* treated.
****$p < 0.05$ DMSO *vs* treated.

(TABLE 5A). This does not indicate whether TCDD was penetrating the entire spleen or was localized in peripheral areas such as the splenic capsule. ^{14}C-TCDD uptake was also dependent on the time spleens were immersed in the TCDD-DMSO solution (TABLE 5B). This varied from 0.293 ng in spleens immersed for 5 seconds to 0.666 ng in spleens immersed 60 seconds.

Viability studies indicated that DMSO treatment alone increased the loss of viable cells during the first 4 hours of culture by about 15%. A 10–20% loss of viable cells due to TCDD treatment occurred after 48 hours of culture but did not appear to be dose-dependent.

TABLE 6 summarizes the ability of splenic lymphoid cells to synthesize DNA after exposure to equal concentrations of either TCDD, 3,4,3′,4′-tetrachlorobiphenyl (TCB) or 2,3,7-trichlorodibenzo-*p*-dioxin-amine (TriCDD-NH_2). While TCDD severely suppressed DNA synthesis ($p < 0.01$) in Con-A-mitogen-stimulated cultures, statistically significant effects were not induced by either the TCB or

TriCDD-NH_2. The slight suppression noted with these chemicals, particularly TCB may be real since TCB induces toxicity in monkeys[26] and Aroclors, which are commercial preparations of polychlorinated biphenyls, are immunosuppressive in guinea pigs.[27]

We are uncertain of the significance of these results. Although these studies indicate that TCDD can directly effect lymphocytic function, the mechanisms are unknown. The question remains as to whether TCDD directly interacts with lymphocytes *in vivo* to alter lymphocytic function as it does *in vitro*. Tissue distribution studies with ^{3}H-TCDD given orally to rats[28] and dibenzofurans in mice[29] have shown splenic retention to be relatively higher. In our laboratory studies with ^{3}H-Con A have shown that mitogen binding is not effected following *in vitro* TCDD exposure indicating that these receptors are not being blocked. This would suggest that an active event is being affected by TCDD treatment. Quantitation of cyclic-AMP, protein phosphorylation and terminal deoxyribonucleotidy transferase activity in treated lymphocytes may be helpful in localizing the TCDD effect since these appear to be prerequisites for normal lymphocyte blastogenesis.

In Vivo *Studies with TCDF*

The similarity between TCDD and 2,3,7,8-tetrachlorodibenzofuran (TCDF) in their chemical structure, toxicologic effects and level of exposure to induce effects would suggest that TCDF is also immunosuppressive. Additionally, TCDF has been shown to cause severe thymus atrophy in chickens, guinea pigs, and mice.[2]

TABLE 5A

SPLENIC UPTAKE OF ^{14}C-TCDD*

Dose (ng TCDD/ml DMSO)	Number Tested	ng TCDD/Spleen† ± S.D.
14.0	4	0.062 ± 0.013
25.0	4	0.092 ± 0.013
50.0	4	0.205 ± 0.016

*Spleens were immersed in DMSO containing ^{14}C-TCDD for 10 sec and rinsed thoroughly in RPMI 1640 media.

†Values were obtained from a standard curve consisting of 1.0 ml NCS, Spleen, 10 ml aquasol cocktail and various concentrations of ^{14}C-TCDD adjusted to 0.04 ml in DMSO. All values were subtracted from background.

TABLE 5B

SPLENIC UPTAKE OF ^{14}C-TCDD *VS* IMMERSING TIME*

Immersing Time (sec)	Number Tested	ng TCDD/Spleen
5	3	0.293 ± 0.047
10	3	0.312 ± 0.018
30	3	0.565 ± 0.021
60	3	0.666 ± 0.110

*Spleens were immersed in DMSO containing 100 ng/ml ^{14}C-TCDD for indicated times and rinsed thoroughly in RPMI 1640 media.

TABLE 6

PROLIFERATIVE RESPONSE OF SPLENIC LYMPHOCYTE CULTURE FOLLOWING *IN VITRO* TREATMENT WITH RELATED CHEMICALS IN DMSO

Stimulus	Treatment	Thymidine Incorporation (Stimulation Index ± S.E.)*
Con A	DMSO	169.56 ± 56.02
	100 ng TCDD/ml	17.39 ± 8.96**
	100 ng 3,4,3′,4′-Tetrachlorobiphenyl/ml	79.59 ± 18.86†
	100 ng 2,3,7-Trichlorodibenzodioxin-amine	90.12 ± 24.06†

*Mean value ± S.E., 8 animals per group
**$p < 0.01$ *vs* DMSO.
†Not significant.

In these experiments, groups of female Hartley strain guinea pigs received 6 weekly doses of 0, 0.05, 0.17, 0.5 and 1.0 μg TCDF/Kg body weight. The TCDF in corn oil was administered orally. Toxicity was noted in a number of animals from the 1.0 μg group. However, those animals that survived in the 1.0 μg dose appeared overtly normal and were included in some of the immune studies. Earlier toxicity studies with TCDF in Hartley guinea pigs indicated that the single oral LD_{50} dose lies between 5 and 10 μg/kg body weight.[30]

Total leukocyte counts were normal in all groups tested. Weight gain of guinea pigs receiving TCDF were in almost all cases less than normal and was related to dosage level, although statistical analysis indicated that none of the groups were significantly reduced (TABLE 7). Thymus-to-body-weight ratios were significantly reduced in the 0.5 and 1.0 μg/kg groups while spleen to body weight ratios were normal in all dosage groups. Histologically, spleens and thymuses from all dosage groups appeared normal.

Splenic lymphocytes were harvested at 10 days after the final exposure. These cells were cultured in the presence of the T-mitogens PHA and Con A and the B-mitogen LPS. Reduced lymphocyte blastogenesis was found in the 0.5 and 1.0 μg dosage

TABLE 7

SELECTED WEIGHT EFFECTS IN TCDF-EXPOSED GUINEA PIGS

Days after Last Exposure	No. Tested	Dosage Group (μg/kg B.W.)	Body Weight ± S.E. (g)	Thymus/BW Ratio (10^{-3})	Spleen/BW Ratio (10^{-3})
0	4	0	516.5 ± 22.1	1.7 ± 0.1	1.4 ± 0.1
	4	0.05	479.8 ± 23.0	1.7 ± 0.1	1.6 ± 0.3
	4	0.17	454.8 ± 9.4	1.5 ± 0.1	1.2 ± 0.1
	4	0.5	532.2 ± 16.1	1.4 ± 0.2*	1.6 ± 0.2
	4	1.0	405.7 ± 61.8	1.2 ± 0.1**	1.6 ± 0.2
7	4	0	507.3 ± 20.4	1.6 ± 0.2	1.3 ± 0.1
	4	0.05	587.5 ± 41.7	1.7 ± 0.2	1.5 ± 0.1
	4	0.17	473.0 ± 12.5	1.4 ± 0.1	1.3 ± 0.1
	4	0.5	476.2 ± 24.1	1.3 ± 0.1*	1.5 ± 0.2
	4	1.0	544.5 ± 20.5	1.2 ± 0.1**	1.5 ± 0.1

*$p < 0.1$.
**$p < 0.05$.

TABLE 8

EFFECT OF TCDF ON ^{3}H-THYMIDINE INCORPORATION INTO DNA BY MITOGEN-STIMULATED LYMPHOCYTES

Dosage Group (μg/kg b.w.)	No. Tested	PHA S.I. ± S.D.†	Con A S.I. ± S.D.	LPS S.I. ± S.D.
0	6	27.27 ± 3.73	17.53 ± 2.40	16.65 ± 2.14
0.05	7	23.52 ± 4.91	12.08 ± 1.51	12.64 ± 2.54
0.17	8	26.60 ± 7.02	17.49 ± 5.38	11.89 ± 2.48
0.5	7	17.49 ± 3.91**	14.33 ± 3.47	9.59 ± 2.53**
1.0	8	16.28 ± 3.11*	11.86 ± 3.12	8.26 ± 1.64*

†S.I. = stimulation index; the mean cpm of triplicate mitogen-stimulated cultures divided by the mean cpm of the nonstimulated cultures.

*$p < 0.05$.

**$p < 0.1$.

groups with PHA and LPS (TABLE 8). Lymphocyte blastogenesis was not affected in the 0.17 and 0.05 μg/kg dosage groups.

TABLE 9 summarizes the effect of TCDF exposure on cell-mediated immune parameters. Dose response trends were found when both the ability to produce macrophage inhibition factor (MIF) and ability to mount delayed hypersensitivity reactions were evaluated. These function, however, were not dramatically compromised and only the decreased production of MIF in the 0.5 μg/kg dosage group was statistically significant. Animals from the 1.0 μg/kg dosage group were not available for testing.

The effect on humoral immunity was evaluated by quantitating serum IgG levels and the antibody response following immunization with bovine gammaglobulin (BGG). Serum IgG levels were normal in all treated groups. Other than a slight decrease in the antibody titer in the 0.5 μg/kg group 4 weeks after immunization the antibody response to BGG was unaffected. The 1.0 μg/kg group was not available for testing. The relatively normal humoral response appears inconsistent with the decreased mitogenic response following activation with the B-mitogen LPS. However, it has recently been shown that the antibody response to LPS (a T-independent

TABLE 9

EFFECTS OF TCDF ON CELL-MEDIATED IMMUNE PARAMETERS

Dosage Group (μg/kg B.W.)	No. Tested	Macrophage Inhibition Index ± S.E.M.†	No. Tested	Delayed Hypersensitivity Index ± S.E.M.‡
0	3	5.98 ± 1.78	5	1.90 ± 0.43
0.05	3	3.94 ± 0.50	4	1.52 ± 0.15
0.17	3	3.10 ± 1.23	5	1.53 ± 0.08
0.5	3	3.04 ± 0.63**	5	1.36 ± 0.07*
1.0		ND		ND

†Macrophage inhibition index was obtained by dividing mean migration area in antigen-stimulated cultures by mean migration area in nonstimulated cultures.

‡Delayed hypersensitivity index was calculated by dividing the cpm for the challenged ear by the cpm for the control ear.

*$p = 0.15$.

**$p < 0.05$.

antigen) is independent of the mitogenic properties of the molecule. This is supported by the fact that the mitogenicity of LPS can be inhibited without affecting the antibody response.[31] The LPS mitogenic response in mouse spleens following *in vitro* TCDD exposure is also affected (see previous section) suggesting that TCDD and TCDF induce a similar immunologic profile. Alternatively, the humoral immune parameters employed in our studies may not be sensitive enough to detect subtle alterations in humoral immunity.

In any case, it would appear that like TCDD, TCDF is also capable of suppressing cell-mediated immunity. The suppression induced by TCDF *via* this dosage regimen was by no means dramatic. However, in earlier studies in which adult guinea pigs were exposed to TCDD in a similar experimental design, the immunosuppressive effects were also not very severe.[22] It was not until TCDD was administered during the developmental phase that the severe consequences of TCDD on the immune system were recognized.

Summary

The effects of TCDD exposure on the developing immune system were investigated in Wistar/Fischer hybrid or Fischer rats. Fetal and neonatal rats were exposed to TCDD through maternal dosing (5 μk/kg) on day 18 of gestation and on days 0, 7, and 14 of postnatal life (group 1). Another group of neonatal rats was exposed to TCDD through maternal dosing on days 0, 7, and 14 of postnatal life only (group 2). Variable but significant effects on body weights and thymus/body weight ratios were found up to 133 days of age. Cell mediated immune functions were depressed up to 133 days of age in both groups but less severely in animals exposed only postnatally. Furthermore, TCDD suppressed cell-mediated immune functions without affecting humoral immune function. Adoptive cell transfer studies indicated suppression of T-cell functions was selective in that "helper" cell function was not suppressed.

In other studies, the effects on lymphocyte function following brief exposure of spleens from $B_6C_3F_1$ mice to TCDD in dimethylsulfoxide (DMSO) were investigated. DNA, RNA and protein synthesis were inhibited at concentrations less than 2×10^{-7} M TCDD in DMSO. This concentration accounted for ~0.2 ng TCDD uptake per spleen. The structurally related chemicals 3,4,3′,4′-tetrachlorobiphenyl and 1-amino, 3,7,8-trichlorodibenzop-*p*-dioxin did not show significant lymphocyte effects even at two-fold higher concentrations. The ability of lymphocyte mitogens to bind to their cell surface receptors was not affected by TCDD treatment. TCDD was slightly cytolytic to lymphocytes after 48 hours of culture. DMSO treatment alone was also slightly toxic to lymphoid cells as indicated by a 10–20% loss of cell viability, although this occurred within 4 hours after DMSO exposure.

Studies were performed to investigate the effects of 2,3,7,8-tetrachlorodibenzofuran (TCDF) on immune function in adult Hartley guinea pigs. Animals received 6 weekly doses of either 0, 0.05, 0.17, 0.5 or 1.0 μg TCDF/kg body weight. TCDF slightly depressed cell-mediated immune functions, particularly at the higher dose levels as indicated by decreased lymphocyte blastogenesis, delayed hypersensitivity reactions, and production of macrophage inhibitor factor. Additionally, thymus-to-body-weight ratios were slightly reduced in the 0.5 and 1.0 μg dosage groups. Serum IgG levels and antibody titer to BGG did not differ from controls. These results indicate that TCDF-induced immunosuppression is similar to that of TCDD.

REFERENCES

1. VOS J. G. 1977. Immune suppression as related to toxicology. Crit. Rev. Toxicol. **5:** 67–101.
2. MOORE, J. A., B. N. GUPTA & J. G. VOS. 1976. Toxicity of 2,3,7,8-tetrachlorodibenzofuran—preliminary results. *In* Proc. Nat. Conf. on Polychlorinated Biphenyls, Environmental Protection Agency, Washington, D.C., pp. 77–80.
3. BEKESI, J. G., J. F. HOLLAND, H. A. ANDERSON, A. S. FISCHBEIN, W. ROM & I. J. SELIKOFF. 1978. Lymphocyte function of Michigan dairy farmers exposed to polybrominated biphenyls. Science **199:** 1207–1209.
4. LUSTER, M. I., R. E. FAITH & J. A. MOORE. 1978. Effects of polybrominated biphenyls on immune response in rodents. Environ. Health Perspect. **23:** 227–232.
5. FAITH, R. E. & J. A. MOORE. 1977. Impairment of thymus-dependent immune functions by exposure of the developing immune system to 2,3,7,8-tetrachlorodibenzo-*p*-dioxin. J. Toxicol. Environ. Health **3:** 451–464.
6. FAITH, R. E., M. I. LUSTER & J. A. MOORE. 1978. Chemical separation of helper cell function and delayed hypersensitivity responses. Cell. Immunol. **40:** 275–284.
7. VAN LOGTEN, M. J., B. N. GUPTA & J. A. MOORE. Role of the endocrine system in the action of 2,3,7,8-tetrachlorodibenzo-*p*-dioxin on the thymus. In preparation.
8. VOS, J. G., J. G. KREEFTENBERG, H. W. ENGEL, A. MINDERHOUD & L. M. VAN NOORLE JANSEN. 1978. Studies on 2,3,7,8-TCDD-induced immune suppression and decreased resistance to infection. Endotoxin hypersensitivity, serum zinc concentrations and effects of thymosin treatment. Toxicology **9:** 75–86.
9. VOS, J. G. & J. A. MOORE. 1974. Suppression of cellular immunity in rats and mice by maternal treatment with 2,3,7,8-tetrachlorodibenzo-*p*-dioxin. Int. Arch. Allergy Appl. Immunol. **47:** 777–789.
10. BEATTY, P. W., K. J. LEMBACH, M. A. HOLSCHER & R. A. NEAL. 1975. Effects of 2,3,7,8-tetrachlorodibenzo-*p*-dioxin on mammalian cells in tissue cultures. Toxicol. Appl. Pharmacol. **31:** 309–312.
11. MANCINI, G., A. D. CABRONARA & J. E. HEREMANS. 1965. Immunochemical quantitation of antigens by single radial immunodiffusion. Immunochem. **2:** 235–243.
12. DRESSER, D. W. & M. F. GREAVES. 1973. Assays for antibody producing cells. *In* Handbook of Experimental Immunology. D. M. Weir, Ed., 27.1–27.29, Blackwell Scientific, Oxford.
13. PARK, B. H. & R. A. GOOD. 1972. A new micromethod for evaluating lymphocyte responses to phytohemagglutinin: Quantitative analysis of the function of thymus-dependent cells. Proc. Nat. Acad. Sci. USA **69:** 371–373.
14. ADLER, W. H., T. TAKEGUCHI, B. MARSH & R. T. SMITH. 1970. Cellular recognition by mouse lymphocytes *in vitro*. I. Definition of a new technique and results of stimulation by phytohemagglutinin and specific antigens. J. Exp. Med. **131:** 1049–1078.
15. KISIELOW, P., J. A. HIRST, H. SHIKU, P. C. L. BEVERLY, M. H. HOFFMAN, E. A. BOYSE & H. F. OETTGEN. 1975. Ly antigens as markers for functionally distinct subpopulations of thymus-derived lymphocytes of the mouse. Nature **253:** 219–220.
16. HIRST, J. A., P. C. L. BEVERLY, P. KISIELOW, M. K. HOFFMAN & H. F. OETTGEN. 1975. Ly antigens: Markers of T cell function on mouse spleen cells. J. Immunol. **115:** 1555–1557.
17. OEHLER, J. R., R. B. HERBERMAN, D. A. CAMPBELL & J. Y. DJEU. 1977. Inhibition of rat mixed lymphocyte cultures by suppressor macrophages. Cell. Immunol. **29:** 238–250.
18. LEFFORD, M. J. 1974. The measurement of tuberculin hypersensitivity in rats. Int. Arch. Allergy Appl. Immunol. **47:** 570–585.
19. BRAENSTRUP, O., M. MAGNUSSON, J. MAXIDAL & O. WERDELIN. 1975. Demonstration of delayed-type hypersensitivity in guinea-pigs by the agarose plate technique. Acta Path. Microbiol. Scand. Sect. C. **83:** 52–58.
20. HARRIS, M. W., J. A. MOORE, J. G. VOS & B. N. GUPTA. 1972. General biological effects of TCDD in laboratory animals. Environ. Health Perspect. **5:** 101–109.

21. Thigpen, J. E., R. E. Faith, E. E. McConnell & J. A. Moore. 1975. Increased susceptibility to bacterial infection as a sequela of exposure to 2,3,7,8-tetrachlorodibenzo-*p*-dioxin. Infect. Immun. **12:** 1319–1324.
22. Vos, J. G., J. A. Moore & J. G. Zinkl. 1973. Effect of 2,3,7,8-tetrachlorodibenzo-*p*-dioxin on the immune system of laboratory animals. Environ. Health Perspect. **5:** 149–162.
23. Handwerger, B. S. & R. H. Schwartz. 1974. Separation of murine lymphoid cells using nylon wool column. Recovery of the B cell-enriched population. Transplantation. **18:** 544–548.
24. Matsumura, F. & H. J. Benezet. 1973. Studies on the bioaccumulation and microbial degradation of 2,3,7,8-tetrachlorodibenzo-*p*-dioxin. Environ. Health Perspect. **5:** 253–258.
25. Kouri, R. E., H. Ratric, S. A. Altas, A. Niwa & D. W. Nebert. 1975. Aryl hydrocarbon hydroxylase induction in human lymphocyte cultures by 2,3,7,8-tetrachlorodibenzo-*p*-dioxin. Life Sci. **15:** 1585–1595.
26. McNulty, W. 1976. PCB poisoning in rhesus monkeys. *In* Proc. Nat. Conf. on Polychlorinated Biphenyls, Environmental Protection Agency, Washington, D.C. pp. 342–356.
27. Vos, J. G. and L. van Driel-Grootuhuis. 1972. PCB-induced suppression of the humoral and cell-mediated immunity in guinea pigs. Sci. Total Environ. **1:** 289–301.
28. Van Miller, J. P., R. J. Marlar & J. R. Allen. 1976. Tissue distribution and excretion of tritiated tetrachlorodibenzo-*p*-dioxin in non-human primates and rats. Cosmet. Toxicol. **14:** 31–34.
29. Morita, M. & S. Oishi. 1977. Clearance and tissue distribution of polychlorinated dibenzofurans in mice. Bull. Environ. Contam. Toxicol. **18:** 61–66.
30. Moore, J. A., B. N. Gupta & J. G. Vos. 1976. Toxicity of 2,3,7,8-tetrachlorodibenzofuran—Preliminary results. *In* Proc. Nat. Conf. on Polychlorinated Biphenyls, Environmental Protection Agency, Washington, D.C. pp. 77–80.
31. Poc, W. J. & J. G. Michael. 1976. Separation of the mitogenic and antigenic responses to bacterial lipopolysaccharide. Immunology **30:** 241–248.

EFFECTS OF 2,3,7,8-TETRACHLORODIBENZO-*p*-DIOXIN (TCDD) ON SPLENIC LYMPHOCYTE TRANSFORMATION IN MICE AFTER SINGLE AND REPEATED EXPOSURES

R. P. Sharma

Toxicology Program
Utah State University
Logan, Utah 84322

P. J. Gehring

Toxicology Research Laboratory
The Dow Chemical Co.
Midland, Michigan 48640

2,3,7,8-Tetrachlorodibenzo-*p*-dioxin (TCDD) is one of the most toxic man-made chemicals.[1] Manifestations of its toxicity are diffuse and include atrophy of the thymus. It had been reported to be an immunosuppressant. Buu-Hoi *et al.*[2] reported that TCDD produced a dose-related atrophy of the thymus at sublethal dose levels. More intense investigations revealed that TCDD depressed the delayed type tuberculin skin reactivity in young guinea pigs but not in rats.[3] The graft-versus-host (GVH) reactivity of mice spleen cells was also reduced.[3]

In a follow-up report, Vos and Moore[4] showed the suppression of cellular immunity in rats and mice after maternal and postnatal exposures to TCDD. Allograft (tail skin) rejection times were also prolonged in both rats and mice. The response of rat thymocytes and spleen cells to phytohemagglutinin (PHA) was reduced by TCDD exposure. No effect was noticed on the response of thymocytes to concanavalin A (Con A). Young animals were much more sensitive to the immunosuppressive effects of TCDD. Thigpen *et al.*[5] have reported that sublethal levels of TCDD increased mortality and shortened the incubation time in mice when challenged by *Salmonella* infection. On the other hand, there was no effect on the mortality of mice infected with pseudorabies virus (PRV).

The above findings suggest a possible selective depression of immune processes by TCDD. The cellular processes appear to be more sensitive than the humoral ones. In addition, there appears to be a species and age-dependent variation in the immunologic effects of TCDD. The effects of varying levels of TCDD on mice were hence investigated on the immune processes. In addition, the effects of a single exposure to TCDD were evaluated at different time intervals. Splenic lymphocytes from treated animals were cultured *in vitro* with and without the presence of mitogenic lectins that stimulate different populations of these cells and the incorporation of labeled thymidine was measured to evaluate the blast formation.

Methods

Male CD-1 mice (weighing approximately 28 g each at the beginning of the experiment) were obtained from Charles River (Willmington, Mass.) and divided into five groups after an initial one week of acclimation. These animals were housed two

0077–8923/79/0320–0487 $01.75/0

per cage, and dosed with TCDD or corn oil. The animals were kept in temperature- and humidity controlled rooms with a 12-hour light/dark cycle. The access to animal rooms was limited and kept to a minimum.

TCDD solution for dosing these animals was prepared by dissolving 5 mg of the chemical (> 99% purity) initially into 10 ml acetone (glass distilled) and mixing this solution with 190 ml of corn oil (Nutritional Biochemicals, Cleveland, Ohio). This stock solution, containing 25μg TCDD/ml was diluted with corn oil containing 5% acetone to provide appropriate dosing solutions. The control group was dosed with corn oil containing 5% acetone. The concentrations of all dosing solutions were appropriate to provide 0.1 ml/10 g to mice. The animals were weighed and dosed once a week with 0.0, 0.01, 0.10, 1.0, and 10 μg/kg of TCDD, for 8 weeks. In addition, one group of mice were given a single oral dose of TCDD at 10 μg/kg body weight.

Four animals were randomly selected from each group and sacrificed at 2,4, and 8 weeks after the treatment started. Similarly, the animals were killed 2,4, and 8 weeks after a single treatment with TCDD. At the time of sacrifice, the blood was saved for serum and selected organs were collected and weighed. The liver, kidney, and thymus were placed in formalin while the spleen was saved in sterilized saline for lymphocyte cultures.

The tissues from these animals were routinely processed and stained for histopathologic evaluation. Microscopic examination of all organs was performed in a random-blind fashion.

Splenic Lymphocyte Cultures

Immediately after their collection, the spleens were washed with sterilized isotonic saline solution and mashed with a pair of forceps in saline to release their cellular constituents. The cells were separated from the intact tissue by serially passing through 18, 21, and 26 gauge needles. The solution was allowed to stand in a glass tube for nearly 10 minutes to remove cell aggregates, after which it was transferred to a different tube leaving the sediment and glass adherent materials behind. This suspension was centrifuged at 4000 rpm for 10 minutes at room temperature and the cell pellet suspended in a small volume (3–5 ml) of culture medium. An aliquot was used for counting cells in an electronic counter after lysis of erythrocytes.

Lymphocyte transformation was determined in cultures by a method essentially modified from Alhaji *et al.*[6] The culture media used was RPMI 1640 (Grand Island Biologicals, Grand Island, N.Y.) containing 10% heat-inactivated fetal calf serum and 100 units of penicillin and 100 μg of streptomycin per ml. The media was supplemented with 300 μg/ml of L-glutamine. The cell suspension was adjusted to provide approximately 8×10^6 cells/ml and 50 μl of this layered into each well of a micro-tissue-culture plate (Microtest II, Falcon 3040, Falcon Plastic, Oxnard, Ca). To each well, 50 μl of additional culture media was added and 50 μl of media either without any mitogen or containing optimized concentrations of phytohemagglutinin (PHA, 1 μl, reconstituted, per culture well) or pokeweed mitogen (PWM, 1 μl, per culture well, both mitogens obtained from Grand Island Biologicals, Grand Island, N.Y.). The plates were incubated for 52 hours at 37° C in humidified 5% CO_2 in air atmosphere. After this incubation, 50 μl of culture media containing 0.5 μCi of ^{3}H-thymidine (New England Nuclear, Boston, Mass.) was added to each culture well and the incubation continued for an additional 16 hours. The cells were harvested and vigorously washed with physiologic saline using a model M12V Cell Harvester (Brandel, Co., Rockville, Md.) and collected on Reeve Angel 934AH glass fiber discs (Whatman, Inc., Clifton, N.J.). These discs were placed directly in regular counting vials, 10 ml of scintillation solution (Aquasol, New England Nuclear, Boston, Mass.) added and the radioactivity counted after shaking in a Mark III Scintillation spectrometer (Searle Analytics, Inc., Chicago, Ill.). The counts were converted to disintegrations per minute (DPM) using a standard quench curve, averaged and adjusted for 10^6 cells in the culture system. All cultures were run in triplicate. The stimulation index for PHA or PWM was calculated by dividing the DPM for cultures in the presence of respective mitogen by the DPM for cultures in which no mitogen was present.

To determine the effect of TCDD directly on lymphocytes in culture, the spleens were collected from control animals and cultured in media containing various concentrations of TCDD (see RESULTS section) as described above.

Estimation of Immunoglobulins

Total immunoglobulins in sera of animals were estimated by a quantitative immunoelectrophoresis (QIEF) technique modified from Bjerrum *et al.*[7] Rabbit anti-mouse Ig, globulin fraction (BioRad Laboratories, Richmond, CA), was carbamylated and incorporated into a 1% agarose gel plate (10 cm × 10 cm × 1.5 mm). Appropriately diluted serum samples were applied in duplicate in wells punched into gel plates containing respective antiglobulin and electrophoresis performed (300 volts for 3–4 hours or 100 volts for 16 hours) using a Model 1400 electrophoresis chamber (BioRad Laboratories, Richmond, CA). The plates were pressed, dried and stained with coomassie brilliant blue, and the heights of rockets compared to appropriate standard samples in the same plate. A pooled mice serum sample (pooled from juvenile male animals) was used as a reference standard and all Ig concentrations represented as a ratio to this pooled sample.

Statistical Analysis

The values are represented as means and standard deviation or standard error of means. The parameters of treatment group were compared to those in the respective control group by using a t-test. In some cases, two-way analysis of variance was performed. The body weights, organ weights and their ratios were compared using a Dunnett's 't'.[8]

RESULTS

TCDD produced toxicologic manifestations in mice at high levels of exposure. Their body and organ weights, at different periods during the experimentation, are given in TABLE 1. As indicated, the increase in the body weight of mice exposed to 10 μg/kg/wk TCDD for 8 weeks was due to the accumulation of ascitic fluid (animals with severe ascites were not sampled). The only definitive changes in organ weights were increased liver weights in mice receiving 10 μg TCDD/kg/week.

Histopathologic examination of organs showed consistent treatment-related changes only in mouse livers. The hepatic changes were similar to those described previously for rats.[9] The lesions ranged from swelling and vacuolation of hepatocytes to autolysis and necrosis (both focal and diffuse) of parenchymal tissue. The lesions were primarily in peripheral lobular regions and in some cases nearly ⅔ of the total tissue indicated extensive necrosis. Some sections also exhibited fibrosis, increased

TABLE 1

ORGAN AND BODY WEIGHTS OF MICE TREATED WITH TCDD FOR DIFFERENT INTERVALS*

Duration of Exposure	Exposure Level μg/kg/week	Body Weight (g)	Thymus (g)	Thymus (g/100g)	Spleen (g)	Spleen (g/100g)	Liver (g)	Liver (g/100)	Kidney (g)	Kidney (g/100g)
Preexposure	0	28 ± 4	0.05 ± 0.01	0.18 ± 0.02	0.09 ± 0.02	0.31 ± 0.06	1.67 ± 0.24	6.06 ± 0.35	0.48 ± 0.07	1.75 ± 0.09
2 weeks	0	33 ± 1	0.07 ± 0.01	0.20 ± 0.03	0.07 ± 0.01	0.20 ± 0.04	1.90 ± 0.17	5.79 ± 0.42	0.55 ± 0.05	1.69 ± 0.03
	0.01	36 ± 1	0.06 ± 0.01	0.17 ± 0.03	0.08 ± 0.01	0.23 ± 0.03	1.99 ± 0.11	5.53 ± 0.32	0.56 ± 0.05	1.55 ± 0.15
	0.1	36 ± 2	0.07 ± 0.01	0.18 ± 0.02	0.10 ± 0.02‡	0.28 ± 0.05	2.12 ± 0.20	5.87 ± 0.21	0.55 ± 0.06	1.52 ± 0.15
	1	35 ± 2	0.05 ± 0.01	0.13 ± 0.01‡	0.10 ± 0.01‡	0.27 ± 0.03	2.24 ± 0.15	6.36 ± 0.39	0.64 ± 0.06	1.81 ± 0.10
	10	35 ± 2	0.06 ± 0.01	0.16 ± 0.01	0.09 ± 0.02	0.26 ± 0.06	2.77 ± 0.32‡	7.83 ± 0.57‡	0.57 ± 0.08	1.61 ± 0.17
4 weeks	0†	38 ± 3	0.06 ± 0.02	0.16 ± 0.04	0.09 ± 0.03	0.25 ± 0.05	1.94 ± 0.32	5.14 ± 0.53	0.55 ± 0.07	1.45 ± 0.09
	0.01	39 ± 3	0.04 ± 0.02	0.11 ± 0.05	0.09 ± 0.02	0.24 ± 0.03	1.91 ± 0.20	4.92 ± 0.38	0.58 ± 0.07	1.48 ± 0.10
	0.1	39 ± 3	0.04 ± 0.01	0.10 ± 0.03	0.07 ± 0.02	0.19 ± 0.03	2.00 ± 0.24	5.16 ± 0.46	0.58 ± 0.07	1.48 ± 0.16
	1	38 ± 5	0.05 ± 0.01	0.13 ± 0.03	0.09 ± 0.02	0.24 ± 0.06	2.32 ± 0.34	6.11 ± 0.90	0.61 ± 0.10	1.59 ± 0.17
	10	37 ± 2	0.03 ± 0.01	0.08 ± 0.04	0.11 ± 0.04	0.30 ± 0.10	2.79 ± 0.53‡	7.54 ± 0.94‡	0.62 ± 0.11	1.68 ± 0.24
8 weeks	0	40 ± 1	0.05 ± 0.01	0.13 ± 0.01	0.09 ± 0.01	0.23 ± 0.04	1.96 ± 0.19	4.92 ± 0.43	0.57 ± 0.02	1.43 ± 0.07
	0.01	39 ± 4	0.06 ± 0.02	0.16 ± 0.05	0.13 ± 0.10	0.34 ± 0.29	2.03 ± 0.16	5.19 ± 0.27	0.54 ± 0.09	1.38 ± 0.23
	0.1	42 ± 3	0.07 ± 0.03	0.16 ± 0.07	0.11 ± 0.04	0.26 ± 0.12	2.30 ± 0.17	5.50 ± 0.08	0.60 ± 0.07	1.43 ± 0.20
	1	44 ± 4	0.07 ± 0.01	0.16 ± 0.04	0.11 ± 0.02	0.25 ± 0.05	2.83 ± 0.22‡	6.49 ± 0.41‡	0.60 ± 0.07	1.36 ± 0.13
	10	49 ± 3‡	0.05 ± 0.01	0.11 ± 0.02	0.11 ± 0.03	0.23 ± 0.06	3.28 ± 0.36‡	6.75 ± 0.46‡	0.56 ± 0.03	1.16 ± 0.11

*Mean ± S.D. of 4 animals per group, except as indicated.

†Only 3 animals in this group.

‡Indicate significant deviation from control by Dunnett's test ($p < .05$). The increased body weight in the last group in the table (8 week exposure, 10 μg/kg/week) was considered to be due to the presence of ascitic fluid.

leukocytic infiltration and perivascular degeneration and necrotic changes. The incidence of liver and thymus lesions in mice exposed to various levels of TCDD for different durations is shown in TABLE 2. The lesions observed in control and 0.01 μg/kg/wk TCDD-exposed group were minor.

Treatment-related lesions of the thymus were noted only after 4 weeks of exposure in groups given 1 or 10 μg TCDD/kg/week (TABLE 2). The lesions observed in the thymus were primarily a reduction of the cortical area and in some cases a focal depletion of thymocytes in the cortex. Cyst formation around Hassall bodies was infrequent, and in a few cases the cortical area of thymus assumed a nodular appearance. Pyknotic thymocytes were also present.

TABLE 2

INCIDENCE OF DISCERNIBLE LESIONS IN THE LIVER AND THYMUS OF MALE MICE EXPOSED TO TCDD*

Exposure Level (μg/kg/week)	Organ	Duration of Exposure 2 Weeks	4 Weeks	8 Weeks
0 (Control)	Liver†	2/4	1/3	3/4
	Thymus	1/4	0/3	0/4
0.01	Liver†	0/4	1/4	4/4
	Thymus	1/4	0/4	2/4
0.1	Liver‡	1/3	2/4	4/4
	Thymus	0/4	1/4	2/4
1	Liver‡	4/4	4/4	4/4
	Thymus§	0/4	3/4	1/4
10	Liver‡	4/4	4/4	4/4
	Thymus§	1/4	4/4	0/4

*Sections examined microscopically in a blind random fashion. Only data on liver and thymus are summarized here since kidney showed no alterations in any group. The numbers indicate the number of animals showing abnormalities/total number of animals in that group.

†The hepatic lesions in control and those receiving 0.01 μg/kg/week group involved vacuolation or slight degeneration of hepatic cells.

‡The organs showed extensive perilobular diffuse degeneration and/or necrosis.

§Marked cortical atrophy of thymus with or without the depletion of thymocytes was noticed in these groups. Other groups did not show any definite lesions.

TCDD caused a stimulation of lymphocyte transformation when no mitogens were present in the culture system (FIGURE 1). After 4 weeks of TCDD administration, there was a significant increase in the thymidine uptake of splenic cultures at all TCDD dose levels as compared to controls. This trend appeared to be reduced after 8 weeks of exposure.

The extent of phytomitogen (PHA or PWM)-induced lymphocyte blast formation was reduced by TCDD exposure (FIGURE 2). The decrease in the stimulation indices (DPM in the presence of a phytomitogen divided by the DPM in an identical culture but without the mitogen) in TCDD groups is partly because of the increased thymidine uptake in unstimulated cultures. A reduction seen in stimulation indices after 2 weeks of TCDD exposure became marked after 4 weeks and seemed to disappear at 8-week period. The decrease was observed both in PHA- and PWM-induced stimulation indices, and a significant reduction (in paired comparisons) was observed in one or more dose levels at different time intervals despite the large variation inherent in this type of assay system.[10]

The serum immunoglobulins in mice at different periods of TCDD exposures are shown in TABLE 3. The treatment with 0.01 or 0.1 μg TCDD/kg/week appeared to increase the immunoglobulin, while the higher doses (i.e. 1 and 10 μg/kg/week) of TCDD decreased these proteins.

When mice were treated orally with a single dose (10 μg/kg) of TCDD and sacrificed at 2, 4, and 8 weeks after the treatment, a significant increase in spontaneous lymphocyte transformation was seen only at 2 weeks after the challenge. The details of these results have been presented elsewhere.[11] The effects on

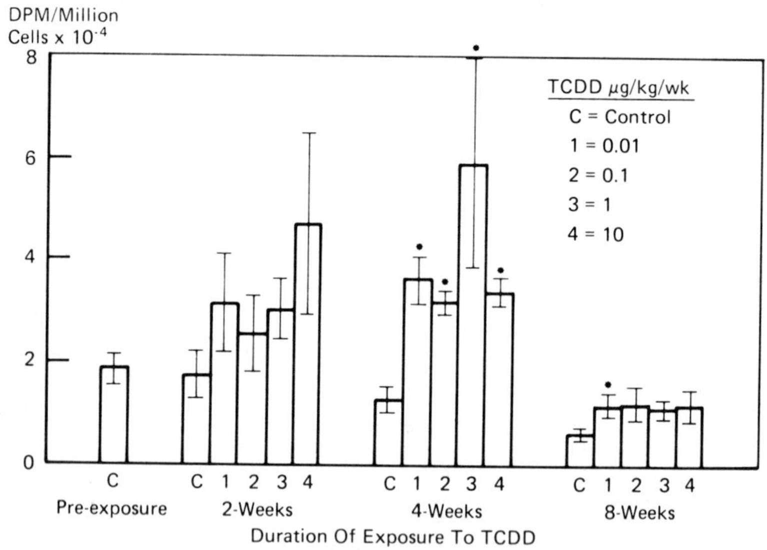

FIGURE 1. Uptake of ^{3}H-thymidine by unstimulated mouse splenic cultures at different periods after exposure to TCDD. Although a decrease with time is seen in control cultures, all TCDD-exposed groups showed an increase over respective control groups. *Indicates a significant deviation from control ($p < 0.05$).

lymphocyte transformation, apparent at 2 weeks post-administration persisted at the 4-week interval after treatment but were not significant, and no differences between control and treated animals were observed in cultures 8 weeks after doing. A similar trend was observed in the respective stimulation indices. The thymic changes and the increase in liver weights also followed a similar trend. The histologic effects in liver, however, persisted, although the degenerative lesions observed microscopically were less severe after 8 weeks.

When TCDD was directly added to the media in splenic cultures, the lymphocytes did not show any increased blastogenic activity. In fact, a dose-related decrease in ^{3}H-thymidine incorporation by lymphocytes was observed (FIGURE 3) with increasing TCDD concentrations. FIGURE 4 presents data on a similar system when the cells were

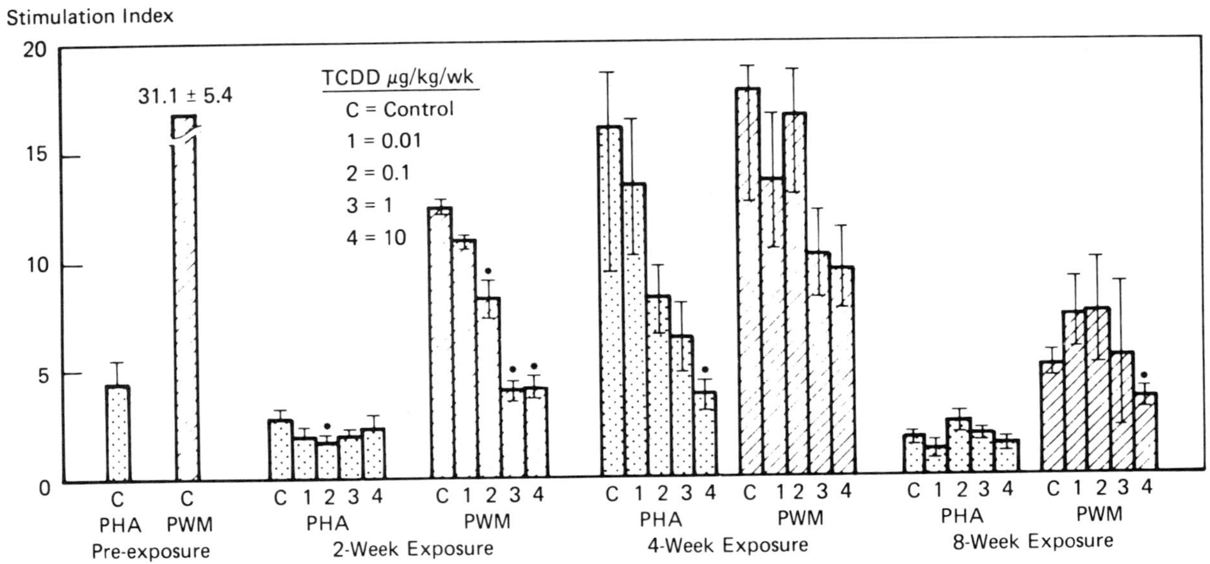

FIGURE 2. Stimulation indices by phytomitogens, phytohemagglutinin (PHA) and pokeweed mitogen (PWM) in mouse splenic cultures in relation to TCDD exposure. The cultures are similar to those in FIGURE 1. *Indicates significant difference in group when compared to respective controls ($p < 0.05$). Stimulation index = DPM from a culture in the presence of either PHA or PWM divided by the DPM in a similar culture without a mitogen.

TABLE 3

SERUM IMMUNOGLOBULINS IN MICE EXPOSED TO DIFFERENT LEVELS OF TCDD

TCDD Exposure (μg/kg/wk)	Preexposure	2 Weeks	Serum Ig at* 4 Weeks	8 Weeks
0	0.49 ± 0.06	0.94 ± 0.25	1.14 ± 0.06	0.91 ± 0.12
0.01	—	1.06 ± 0.11	1.48 ± 0.16	1.95 ± 0.99
0.1	—	1.04 ± 0.15	1.67 ± 0.39**	2.58 ± 0.47**
1	—	1.17 ± 0.27	1.43 ± 0.18	0.85 ± 0.11
10	—	0.87 ± 0.13	0.79 ± 0.08**	0.72 ± 0.18

*The values expressed are ratios of serum Ig in relation to a pooled normal mouse serum sample. Mean ± S.E. of 4 observations per group.

**Indicate significant difference from the respective control at $p < .05$.

cultured in the presence of mitogenic agents. No effect of TCDD on stimulation indices suggests that the observation in FIGURE 3 was possibly due to the direct cytotoxic effect of TCDD on lymphocytes.

DISCUSSION

Our experiments with TCDD in mice confirm the early findings that high dose levels of TCDD are immunosuppressive. It has been shown to be an immunosuppressant in guinea pigs, and rats[3] and developing rats and mice after maternal and

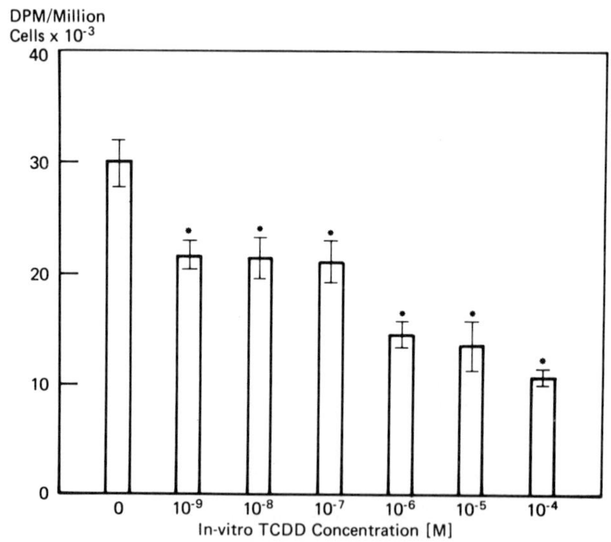

FIGURE 3. Effect of TCDD added directly to the incubation medium on lymphocyte transformation in mouse splenic lymphocyte cultures. Cells isolated from normal mice were incubated in culture media containing appropriate concentrations of TCDD. *Indicates significant difference from the control cultures at $p < 0.05$.

postnatal treatment.[4] The immunosuppression in our experiments was evident by decreased stimulation indices of splenic lymphocytes to the mitogenic agents, PHA and PWM.

The results presented in this report, however, differ from the earlier ones in several respects. The immunosuppressive effect of TCDD in mice was not necessarily dose-related in the range of concentrations that we have used. Usually the pathologic lesions were apparent, particularly in the liver, at levels of TCDD exposure lower than that were shown to be immunosuppressive. Generally the immune suppression was seen in mice at doses of 1 μg/kg/wk or more, when exposure was continued for several weeks. At lower exposure levels, the immunosuppressive effect was not apparent while liver changes were observed. The changes in thymus, described in earlier reports[3] were

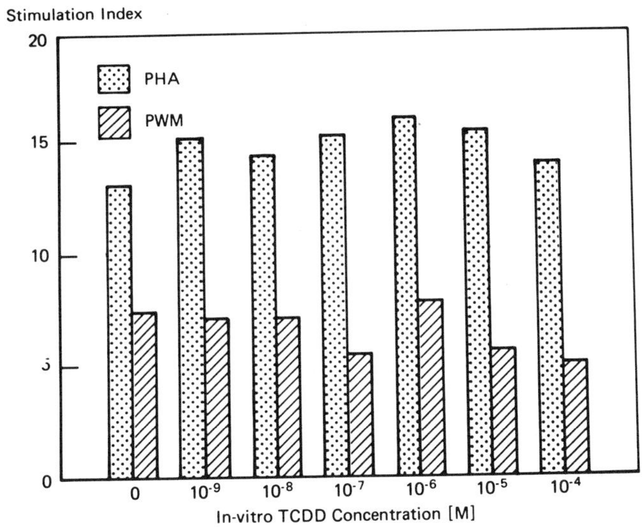

FIGURE 4. Stimulation indices in splenic lymphocyte cultures incubated in the presence of PHA or PWM and different concentrations of TCDD. The cultures are similar to those indicated in FIGURE 3. Stimulation index = DPM from a culture in the presence of either PHA or PWM divided by the DPM in an identical culture without mitogens. Addition of TCDD did not have any effect on stimulation indices by either PHA or PWM.

observed in our studies, but these were not always dose-related. One unique feature of such thymic lesions was the absence of any changes in the two highest dose levels of TCDD exposure at the end of 8 weeks in mice. The only appreciable atrophy of thymic cortex and depletion of thymocytes were seen in these treatment levels at the end of 4 weeks. In mice, even the decrease in thymus weights that was observed after 2 and 4 weeks of TCDD exposure, was not seen after 8-week treatment. Although not conclusive, this finding suggests that there may be a recovery of thymic tissue even if the exposure to TCDD continued at high levels. It may be interpreted as a sort of adaptation in this organ to the presence of TCDD. The same was not true, however, for liver lesions, which were progressive with time and also with the level of TCDD exposure.

An important observation of this study is an increase in the spontaneous blast formation in splenic lymphocytes without any mitogenic inducer. This increase was consistent at all time periods and particularly obvious in mice receiving low dose levels of TCDD. The exact implication of this finding is not clear, but it suggests a stimulation in the functional activity of lymphocytes. This can be compared to an increase when the animals are challenged with an antigen.[12] In our additional studies, such increase was seen in control groups that were injected with an antigen mixture (tetanus toxoid and Freund's adjuvant; data not given here). One possible explanation of this stimulation in spontaneous lymphocyte transformation may be that the administration of TCDD produces an antigenic response, either by combining with a body protein, or by causing tissue damage leading to the release of altered proteins that might be antigenic. Poland and Glover[13] have reported a specific and high affinity binding of TCDD with hepatic cytosol.

The decrease in the stimulation indices to mitogens by TCDD exposure may be the result of immune tolerance produced by the possible antigenic effect of this chemical. Both antigens and antibodies are known to produce immune tolerance or immune unresponsiveness.[14] In our studies, an immune unresponsiveness (or immune suppression) was seen only at relatively high exposures to TCDD, which may suggest that a saturable amount of either antigen or antibody may be present at these treatment levels.

The hypothesis of an autoimmune response is also supported by the finding that in mice TCDD produced an increased immunoglobulin synthesis at the low levels of exposure. A significant increase in total Ig appeared at 4 and 8 weeks after 0.1 μg/kg/week of TCDD treatment. At high treatment levels of TCDD, the decrease in immunoglobulin levels may be related to the damage of the organs or cells that are sites of globulin production (e.g. lymphoid tissue).

Vos and Moore[4] have reported that PHA responsiveness on splenic cells was decreased in one-month-old mice but not in 4-month-old animals. In our studies using 4–5 month old mice (data not shown) both PHA and PWM responsiveness were significantly decreased 2 weeks after a single dose of 10 μg/kg TCDD, although this decrease was not as remarkable as in growing animals. In these animals as well, the spontaneous increase in DNA synthesis was seen in splenic cultures. One of the reasons for the relative insensitivity of adult animal lymphocytes may be that the lymphocyte response to mitogens as well as spontaneous blast formation in culture is considerably decreased with age and the influence of the chemical treatment may not be remarkable. Adult animals may also bind TCDD more efficiently in selected organs (i.e. liver) than in the young ones, thus there may be less amount of TCDD available in the adults in other tissues.

Summary

Male CD-1 mice were orally treated with 0.01, 0.1, 1 and 10 μg TCDD/kg body wt./week for up to 8 weeks. Randomly selected animals were sacrificed at 2, 4, and 8 weeks of exposure. An additional group was given orally 10 μg TCDD/kg and the animals similarly sacrificed. Splenic lymphocytes from these animals were cultured *in vitro* with or without the presence of phytomitogens, phytohemagglutinin, and pokeweed mitogen. The incorporation of ^{3}H-thymidine was measured as an indication of relative blast formation. Exposure of animals to TCDD, even at the lowest level (0.01 μg/kg/wk for 2 weeks) caused a marked increase in the thymidine uptake by cultured lymphocytes. The blastogenic response of mitogens was reduced at high levels of TCDD exposure, indicating an immunosuppressive effect. Following a single treatment with 10 μg TCDD/kg, the increase in the blast formation was noted at 2 weeks, the effect reduced at 4 weeks, and no difference noticed in treated vs. control cultures 8 weeks after the treatment. Small doses of TCDD stimulate the splenic lymphocyte transformation and this effect, although somewhat dose-related, can be reversed in a relatively short period of time.

REFERENCES

1. SCHEWETZ, B. A., J. M. NORRIS, G. L. SPARSCHU, V. K. ROWE, P. J. GEHRING, J. L. EMERSON & C. G. GERBIG. 1973. Toxicology of chlorinated dibenzo-*p*-dioxins. Environ. Health Perspect. **5:** 87–99.
2. BUU-HOI, N. P., P. H. CHANN & G. SESQUE. 1972. Organs as targets of dioxin (2,3,7,8-tetrachlorodibenzo-*p*-dioxin) intoxication. Naturwissenschaften **59:** 174–175.
3. VOS, J. G., J. A. MOORE & J. G. ZINKL. 1973. Effect of 2,3,7,8-tetrachlorodibenzo-*p*-dioxin on the immune system of laboratory animals. Environ. Health Perspect. **5:** 149–162.
4. VOS, J. G. & J. A. MOORE. 1974. Suppression of cellular immunity in rats and mice by maternal treatment with 2,3,7,8-tetrachlorodibenzo-*p*-dioxin. Int. Arch. Allergy **47:** 777–794.
5. THIGPEN, J. E., R. E. FAITH, E. E. MCCONNELL & J. A. MOORE. 1975. Increased susceptibility to bacterial infection as a sequela of exposure to 2,3,7,8-tetrachlorodibenzo-*p*-dioxin. Infect. Immun. **12:** 1319–1324.
6. ALHAJI, I., C. MUSCOPLAT & D. W. JOHNSON. 1974. Deoxyribonucleic acid and early protein synthesis in caprine lymphocytes stimulated with phytohemagglutinin and pokeweed mitogen in microcultures. Am. J. Vet. Res. **35:** 441–443.
7. BJERRUM, O. J., A. INGILD, H. LOWENSTEIN & B. WEEKE. 1973. Carbamylated antibodies used for quantitation of human IgG. A routine method. *In* Quantitative Immunoelectrophoresis. N. H. Axelsen, J. Kroll & B. Weeke, Eds. Universitetsforlaget. Oslo. pp. 145–148.
8. STEEL, R. G. D. & H. H. TORRIE. 1960. Principles and Procedures of Statistics. McGraw-Hill Book Co., Inc. New York.
9. KOCIBA, R. J., P. A. KEELER, C. N. PARKS & P. J. GEHRING. 1976. 2,3,7,8-tetrachlorodibenzo-*p*-dioxin (TCDD): Results of a 13-week oral toxicity study in rats. Toxicol. Appl. Pharmacol. **35:** 553–574.
10. GRAF, P. A. & F. MATHER. 1972. A statistical model for the evaluation of antigen-stimulated blastogenic transformation as measured by scintillation spectrometry. Cell. Immunol. **4:** 143–151.
11. SHARMA, R. P. & P. J. GEHRING. 1979. Reversal of immunologic and toxicologic effects of a single exposure of 2,3,7,8-tetrachlorodibenzo-*p*-dioxin in mice. Trace Substances in Environmental Health. XII D. D. Hamphill, Ed. Univ. of Missouri. Columbia. In press.
12. LING, N. R. 1968. Lymphocyte Stimulation. North-Holland Publishing Co. Amsterdam.
13. POLAND, A. & E. GLOVER. 1976. Stereospecific, high affinity binding of 2,3,7,8-tetrachlorodibenzo-*p*-dioxin by hepatic cytosol. J. Biol. Chem. **251:** 4936–4946.
14. LACHHMANN, P. J. 1976. Immunological tolerance and unresponsiveness. *In* The Immune System. M. J. Hobart & I. McConnel, Eds. Blackwell Scientific Publishers. Oxford. pp. 152–164.

IMMUNE SUPPRESSION AS A CONSEQUENCE OF A SELECTIVE CYTOTOXIC ACTIVITY OF CERTAIN ORGANOMETALLIC COMPOUNDS ON THYMUS AND THYMUS-DEPENDENT LYMPHOCYTES

Willem Seinen and André Penninks

Working Group, Pathology-Toxicology
Pathology and Pharmacology-Toxicology Department
Faculty of Veterinary Sciences
State University
Utrecht, the Netherlands

The biologic effects of organometallic compounds have been reviewed by Barnes and Magos[1] and more recently by Thayer.[2] Organometallic compounds are used in a variety of applications. Organoarsenicals were much in use as antibiotics until about 1940, when they were largely replaced by penicillin; they also were developed as poison gas. From the organolead compounds, tetraethyllead has an especially important application as gasoline additive. The present report deals mainly with organotin compounds, which are applied as heat stabilizers, catalytic agents, and biocidal compounds.[3] Trialkyltin compounds are extensively used as preservatives for wood and paper and as paint additives especially to control marine fouling. Triaryl compounds are effective fungicides and insecticides. Dialkyltins are mainly used as heat stabilizers, especially for the stabilization of plastics. Only some dioctyltin derivates are permitted by the U.S. Food and Drug Administration as additives in polyvinyl chloride (PVC) products which come in contact with food. In European countries dimethyltins are also allowed in PVC for food packing purpose.

With the possible exception of mercurials (not included in this article) organotin compounds have received more attention for their biologic effects than organic compounds of any other metal. Their toxicity has been extensively reviewed by Barnes and Stoner[4] and more recently by Piver.[5] However, no attention was paid to the effects of organotin compounds on the lymphoid system. In recent studies, Seinen and co-authors[6–9] have shown that the main effect of some dialkyltin compounds in rats was on the thymus and thymus-dependent lymphoid structures. The present report reviews the relationship of chemical structure of organotin, organolead, organogermane, and some organoarsenic compounds and their lymphocytotoxic activity. The consequence of organotin-induced lymphocytotoxicity for the functioning of the immune system will be evaluated.

In Vivo Studies with Various Organometallic Compounds

Atrophy of Thymus and Thymus-Dependent Lymphoid Tissues in Rats Fed Di-n-Octyltindichloride

In rats fed di-*n*-octyltindichloride (DOTC) at dietary levels of 0, 50 or 150 mg/kg diet (ppm) a prominent and dose-related decrease of thymus and popliteal lymph node weight was noted.[6] Thymic atrophy progressively increased (FIGURE 1). Already after

0077–8923/79/0320–0499 $01.75/0 © 1979, NYAS

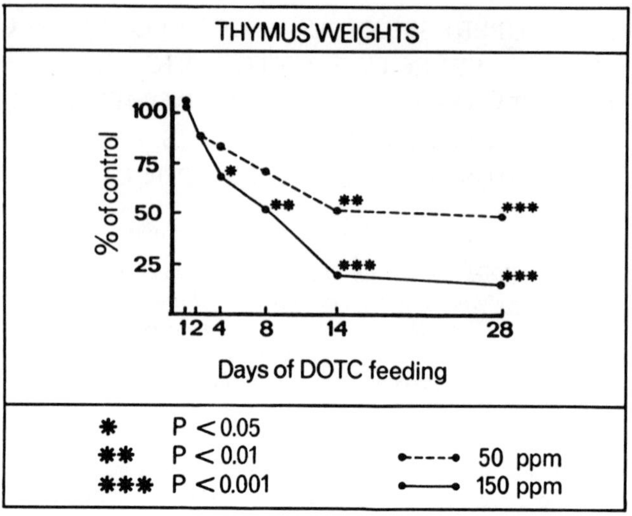

FIGURE 1. Thymus weights of rats fed 0, 50, or 150 ppm DOTC for 1, 2, 4, 8, 14 and 28 days (four animals per group).

a 4-day feeding period thymus weights were significantly lower as compared to control animals. Thymus weights of rats fed 50 or 150 ppm DOTC for 4 weeks were only 48% and 16% of the control weights. Histologically, lymphocyte depletion was observed in the thymus and in thymus-dependent areas of the spleen and lymph nodes, whereas in other organs no treatment-related histopathologic changes were noted. Although most of the lymphoid tissue of the thymus had disappeared, signs of cell destruction (karyorrhexis and "starry sky" formation) as seen after treatment with corticosteroids[10,11] antimetabolites, alkylating agents, and other antitumor compounds[12] were never observed after DOTC treatment. This does not, however, exclude lymphocytotoxicity, as the viability of nucleated cells isolated from thymus and spleen of rats fed 50 or 150 ppm DOTC for 4 weeks was significantly decreased (FIGURE 2). However, the effect on thymocyte number was much more pronounced than the effect on cell viability. Already, after 2 days feeding of DOTC the total number of thymocytes showed a dose-related decrease but cell viability was unaltered (as measured with the dye exclusion test). Total thymocyte counts progressively diminished to 33% and 6% of the control value at week 4 in animals fed 50 or 150 ppm DOTC, respectively. These data indicate that DOTC primarily induces an inhibition of the division of thymic lymphocytes and secondarily causes cell death.

The total number and viability of spleen and peripheral lymph node cells was less severely decreased by DOTC, than the number and viability of thymus cells. After 4 weeks feeding of 150 ppm DOTC, the numbers of nucleated spleen and peripheral lymph node cells were 75% and 70%, respectively, of the control counts. The total number of peripheral blood lymphocytes was generally unaltered by organotin treatment. Also, the number and viability of bone marrow cells was not decreased by DOTC (FIGURE 2) feeding. Effects of organotin compounds on bone marrow stem cells are further excluded by the spleen colony assay technique[13] with balb/c mice. This experiment was carried out with di-*n*-butyltin dichloride (DBTC), a compound with effects on lymphoid tissue similar to those observed with DOTC. The numbers of

10-day spleen colonies produced in lethally irradiated recipient mice injected with bone marrow cells of donor mice treated 1 or 3 days before cell transplantation with 0, 2, or 10 mg DBTC/kg intravenously, were essentially the same (TABLE 1). In addition, spleen weights of mice reconstituted with bone marrow cells from DBTC-treated and -untreated animals were comparable. Thus, neither the number and viability of bone marrow cells nor the number of colony-forming stem cells was decreased by organotin treatment. These results further support the selective cytotoxicity of DOTC for thymus and thymus-dependent lymphocytes.

A selective lymphocyte depletion in the thymus can also be induced by a stress-related release of glucocorticoids.[14] The selective effects of DOTC on the thymus are not related to stress since: (1) Thymus weights were as severely decreased in adrenalectomized as in sham-adrenalectomized DOTC-fed rats; (2) relative

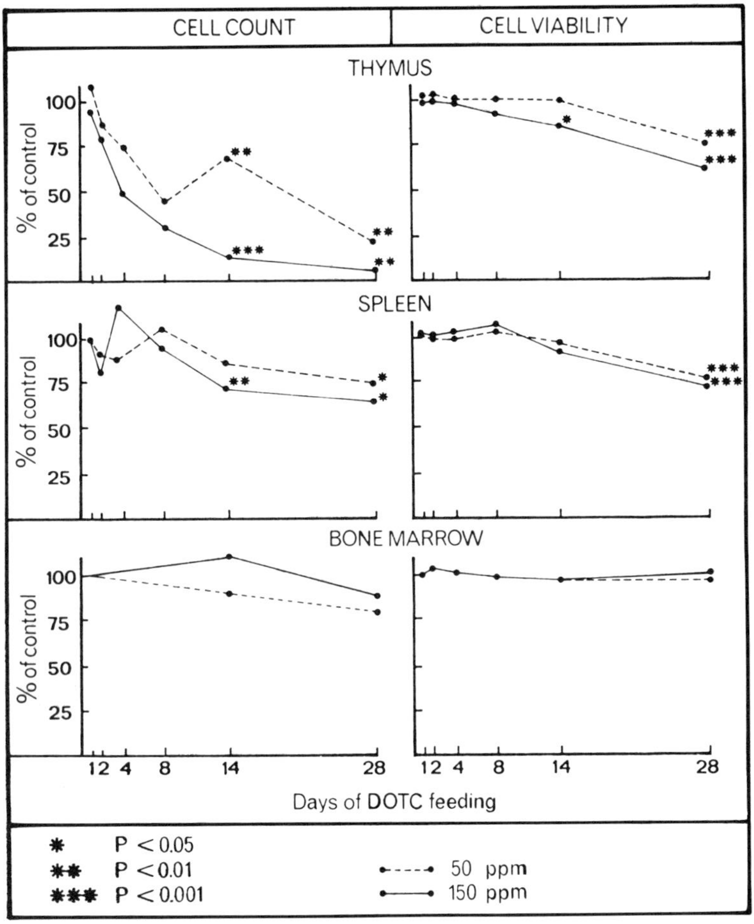

FIGURE 2. Total nucleated cell count and cell viability of thymus, spleen, and bone marrow from rats fed DOTC. Results are given as percentages of control values and are means of two animals (1, 2, 4, and 8 days) or of four animals (14 and 28 days).

TABLE 1

THE EFFECT OF DBTC ON BONE MARROW STEM CELLS OF BALB/C MICE*

Donor Mice				Recipient Mice	
i.v. Dose of DBTC (mg/kg)	Day of Treatment before Cell Transfer	Thymus Weight Reduction (%)	Number of Bone† Marrow Cells Transplanted $\times 10^4$	Number of 10-day Spleen Colonies‡	Spleen Weight‡ (mg)
0	3	0	100	n.c.§	80 ± 24
10	3	64	100	n.c.	88 ± 11
0	3	0	6	14.6 ± 4.5	42 ± 8
10	3	52	6	16.4 ± 4.0	45 ± 10
0	1	0	6	14.4 ± 3.8	44 ± 8
2	1	12	6	11.6 ± 1.8	47 ± 6
10	1	32	6	17.8 ± 3.6	40 ± 8

*Measured by the spleen colony assay according to Till and McCullough.[13]
†Bone marrow was sampled from femurs of 6 mice.
‡Mean values ± S.D. of 10 animals per group.
§n.c. = not counted, since too many colonies were present.

adrenal weights were the same in DOTC-fed and control animals; (3) histologically, no signs of hyperactivity of the adrenal cortex, nor destruction of lymphocytes in the thymic cortex were observed; and (4) *in vitro,* DOTC is cytotoxic for rat thymocytes.

Reversibility of Thymus Atrophy

Although DOTC-induced thymus atrophy occurs rapidly (FIGURES 1 and 2) and its severity mimics a "chemical thymectomy," the atrophy was completely reversible. A 78% reduction of thymus weight induced by feeding 150 ppm DOTC for 4 weeks was followed by a fast recovery, when the animals were returned to the stock diet (FIGURE 3). After a rehabilitation period of 2 weeks, thymus weights of DOTC-treated animals did not differ significantly from the controls. After 4 weeks thymus weights of treated animals were even higher than the controls, which is considered to be a compensatory reaction as is seen in many healing processes. Eight weeks after treatment thymus weights were the same as the controls. These recovery data are comparable with those of McAnulty and Dickerson[15] after "nutritional thymectomy." By restriction of food intake in weanling rats for 28 days, they induced a thymus reduction of 81%. But on rehabilitation the thymus weight returned to control values within 16 days. The similarity of the recovery of the thymus after DOTC feeding and nutritional deprivation suggests that the half-life of DOTC may be short.

Comparative Feeding Studies with Various Organometallic Compounds in Rats

From comparative short-term feeding studies with rats,[7] it appeared that not only DOTC, but also other dialkyltin compounds caused lymphoid atrophy (TABLE 2). The extent of dialkyltin-induced thymus atrophy was strongly related to the length of the alkyl chain (FIGURE 4) and is probably related to the water-lipid distribution of homologues. Dimethyl- and diethyltin dichlorides are water soluble and do not (methyl) or only slightly (ethyl) induce thymus involution, whereas the water insoluble di-*n*-butyl and di-*n*-octyltindichlorides (maximum water solubility is 0.12 mMol) caused severe thymus atrophy. Also, the dialkyllead compounds tested did not induce a selective lymphoid atrophy. The water-insoluble organogermanes compounds tested[16] did not affect lymphoid tissues, but these compounds more resemble organosilicum than organotin compounds and display inorganic properties only slightly.[17] The di-estertindichlorides were also inactive *in vivo*.[18] Probably, these compounds are hydrolyzed in the gastrointestinal tract and converted to a dicarboxylic acid. Phenylarsinedichloride, a compound which binds dithiol groups as do the dialkyltin compounds, also induced thymus atrophy but to a lesser extent than DBTC and DOTC.

From the trialkyltins tested only tri-*n*-butyltin chloride induced thymus atrophy. Both diphenyl and triphenyltin chlorides produced a selective lymphoid atrophy, but it was less pronounced than that induced by DOTC and DBTC. Mono and tetraoctyltins never affected lymphoid tissues.

Comparative Intravenous Treatment Studies with Various Organometallic Compounds in Rats

After repeated intravenous injections of DBTC and DOTC a marked thymus atrophy was observed. This atrophy occurred already at doses of 1 and 2 mg/kg, which did not affect other organ systems nor the body weight of the animals. Diethyltindi-

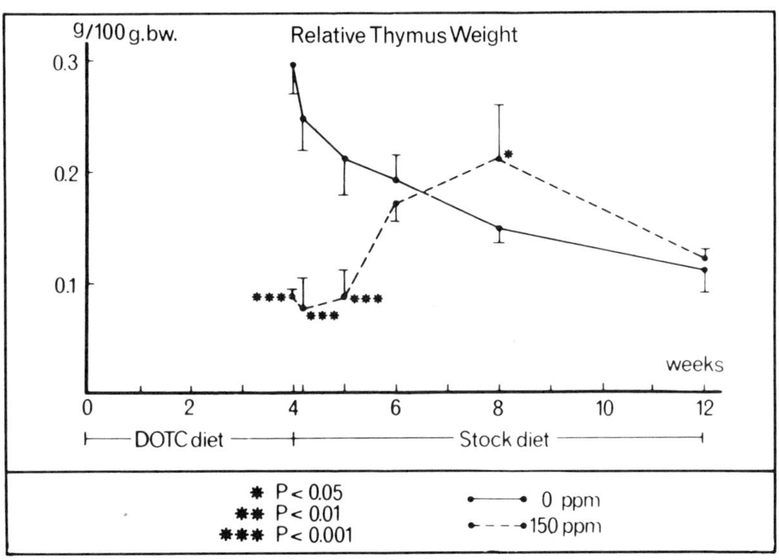

FIGURE 3. Reversibility of DOTC-induced thymus weight reduction in rats kept on a 150 ppm of DOTC diet for 4 weeks and after that period on a stock diet for 8 weeks. (means ± S.D. of five or more animals per group).

TABLE 2

EFFECTS OF VARIOUS ORGANOMETALLIC COMPOUNDS ON LYMPHOID TISSUES OF RATS

Compounds Tested	Dietary Level* (ppm)	Thymus Atrophy†
Di-alkyltins		
Dimethyltindichloride	150	−
Diethyltindichloride	150	+
Di-*n*-propyltindichloride	150	++
Di-*n*-butyltindichloride	150	+++
Di-*n*-octyltindichloride	150	+++
Di-*n*-dodecyltindibromide	150	−
Di-*n*-octadecyltindibromide	150	−
Tri-alkyltins		
Trimethyltinchloride	15	−
Triethyltinchloride	15	−
Tri-*n*-butyltinchloride	150	++
Tri-*n*-octyltinchloride	150	−
Mono-alkyltins		
Mono-*n*-octyltintrichloride	150	−
Tetra-alkyltins		
Tetra-*n*-octyltin	150	−
Aryl tins		
Diphenyltindichloride	150	+
Triphenyltinchloride	150	+
Ester tins		
β-carbobutoxyethyltintrichloride	150	−
Bis-β-carbobutoxyethyltindichloride	150	−
Bis-β-carbomethoxyethyltindichloride	150	−
Di-alkyl leads		
Di-*n*-butylleaddiacetate	150	−
Di-*n*-hexylleaddiacetate	150	−
Di-alkyl germanes		
Dimethylgermanedichloride	450	−
Di-*n*-butylgermanedichloride	450	−
Organo arsenics		
Phenylarsinedichloride	150	+

*Weanling rats were fed a diet containing the different compounds during a 2-week period.

†Quantitative effects on thymus are gradually expressed with + and − signs.

− denotes no effect; + denotes a reduction of thymus weight of less than 25%; ++ denotes a reduction of thymus weight of less than 50%; and +++ denotes a reduction of thymus weight of more than 50%.

chloride also depressed thymus weight, but only at higher levels, which also decreased growth rate and increased liver weights. Dimethyltindichloride and di-*n*-dodecyltindibromide did not induce thymus atrophy. Even a single injection of 1 mg/kg DBTC or DOTC caused thymus atrophy, which was most pronounced 4 days after treatment. Given by the intravenous route DBTC and DOTC were severely toxic. Doses of 10 mg/kg body weight produced death in about 50% of the animals (both mice and rats). The lower dialkyltin homologs are less toxic upon intravenous treatment,[7] which contrasts with the trialkyltin compounds. Both upon oral and intravenous administration trimethyl and triethyltinchlorides are much more toxic than tri-*n*-butyl and tri-*n*-octyltin chloride.[19] Upon intravenous treatment the toxicity of the organometallic compounds tested was much more pronounced as compared with oral treatment, indicating that these compounds are only slightly resorbed from the intestinal tract.

Species Differences

In contrast to rats, lymphoid atrophy did not occur in DOTC- or DBTC-fed mice, guinea pigs, Japanese quail, or chickens.[20] However, after intravenous administration of DBTC and DOTC in rats[7] and mice (TABLE 1) comparable thymus effects were found. So there may be differences in uptake and/or distribution between rats and mice after oral administration. Although a selective organotin-induced lymphoid atrophy could not be established in guinea pigs, weanling animals fed 150 ppm DOTC died within 3 weeks. These animals appeared to be emaciated, showed abdominal edema and enlarged caecums filled with a fluid. Because of the antibacterial activity of dialkyltin compounds,[21] caecal fermentation was disturbed, by which the production of free fatty acids in the caecum was almost completely inhibited.[22] Morphologically a severe villous atrophy was observed in the gastrointestinal tract of these animals, which may be related to the cytotoxic activity of DBTC and DOTC. Japanese quail appeared to be very insensitive to the toxic effects of dialkyltin compounds,[7] since feeding levels up to 600 ppm did not induce any sign of toxicity.

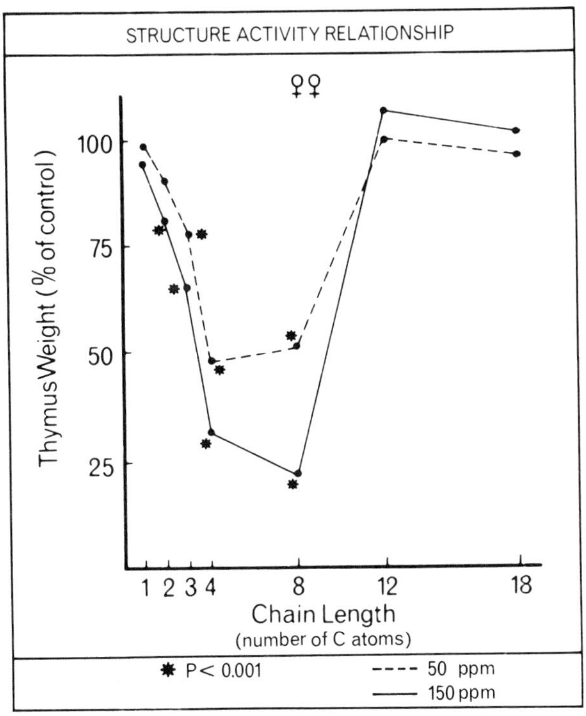

FIGURE 4. Relationship between length of the dialkyltin compounds and its effect on thymus weight determined in groups of 10 female rats fed various dialkyltin compounds at 0, 50 or 150 ppm for 2 weeks.

IMMUNE FUNCTION STUDIES AFTER *IN VIVO* EXPOSURE TO ORGANOTIN COMPOUNDS

Since the thymus and thymus-derived lymphocytes play a central role in the immune system, a number of immune function studies were performed in rats, mice, and guinea pigs. In the thymus (blood-borne) stem cells[23] migrate into the cortex where, under the stimulatory and perhaps regulatory influence of thymic epithelial cells, cell division and differentiation begin, giving rise to the forming of various subclasses of (thymus-derived) T-lymphocytes. This process starts at an early stage of ontogeny.[24] So, thymectomy of neonates results in a reduced number of lymphocytes, a diminished capacity to produce serum antibodies, and failure to reject skin grafts from foreign strains. These animals often die within 3 or 4 months from runting disease.[25,26] Thymic ablation of adult animals is not accompanied by severe disturbances in lymphopoiesis or immune functions. As DBTC and DOTC cause a severe thymic atrophy suggestive of a "chemical thymectomy," the immune function studies presented here are carried out with animals treated after weaning age as well as exposed during the developmental phase of the immune system (pre and/or postnatal exposure).

The function studies carried out and the effects of DBTC and/or DOTC[8,9,27] are summarized in TABLE 3. It appears that all immune reactions in which T-lymphocytes participate are compromised in the rat. Cell-mediated immunity was depressed in all test systems (delayed type hypersensitivity to tuberculin, skin graft rejection, graft-versus-host reaction, resistance to *Listeria monocytogenes* infection, lymphocyte transformation by T-cell mitogens). However, the humoral response was impaired with sheep red blood cells (SRBC) a thymus-dependent antigen, but the antibody

TABLE 3
IN VIVO AND *IN VITRO* FUNCTION TESTS OF THE IMMUNE SYSTEM OF DOTC- OR DBTC-TREATED RATS

Cell-Mediated Immunity	Effect*	Reference
In vivo		
—delayed type hypersensitivity to tuberculin	+	8
—allograft rejection	+	8
—graft-versus-host reaction	+	9
—resistance to *Listeria monocytogenes* infection	+	24
In vitro		
—lymphocyte transformation by Phytohemagglutinin (PHA) and Concanavalin A (Con A)	+	9
Humoral Immunity		
In vivo		
—thymus-dependent antibody synthesis to sheep red blood cells (SRBC)	+	8
—thymus-independent antibody synthesis to *E. coli* lipopolysacharide (LPS)	−	8
In vitro		
—plaque formation against SRBC	+	8
—transformation of lymphocytes by LPS	−	9
Phagocytosis by macrophages of carbon particles	−	8
Sensitivity to LPS	+	24

*+ Denotes a suppressive effect on immune parameters or an increase of susceptibility to infection or *E.coli* LPS; − = no effect.

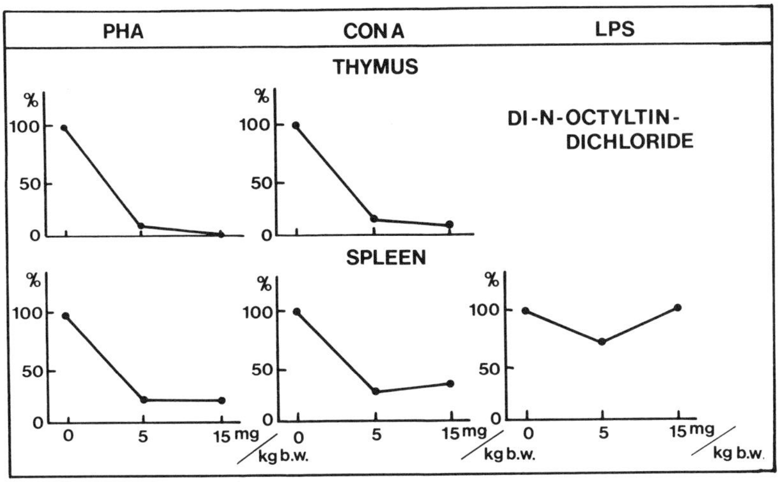

FIGURE 5. The effect of postnatal DOTC treatment of rats upon the ^{3}H-thymidine incorporation of thymus and spleen cells in response to PHA, Con A and LPS. DOTC was given by gavage, 3 times a week during 4 weeks, in amounts of 0, 5 or 15 mg/kg body weight dissolved in arachis oil. Results are expressed as percentages of the ^{3}H-thymidine incorporation in the control cultures and represent the mean of cultures from 3 or 4 animals.

response to *E. coli* lipopolysacharide (LPS) a thymus-independent antigen[28] was unaffected. The response of lymphocytes from thymus and spleen to the T-cell mitogens phytoheamagglutinin (PHA) and concanavalin A (Con A) was severely depressed, in particular in pups that were exposed postnatally by oral intubation during 3 weeks following birth (FIGURE 5), whereas the response to the B-cell mitogen LPS was not impaired. From these data, it seems likely that DOTC and DBTC suppress various subpopulations of T-lymphocytes without affecting B lymphocytes. This selectivity may be explained by a selective toxic action on lymphocytes in the thymus, resulting in a reduced number and reduced responsiveness of peripheral T-lymphocytes. Also macrophages do not seem to be compromised, since blood monocyte numbers (as precursors of macrophages) and the *in vivo* phagocytizing capacity of macrophages as measured by the clearance of carbon particles were the same for treated and control animals.

Immune suppression was most pronounced in rats exposed immediately after birth. Oral intubation with 5 mg DBTC/kg, three times a week caused growth stunting and mortality without major pathology except for severe atrophy of lymphoid organs. This is suggestive of a runting syndrome as first described by Miller[29] in neonatal thymectomized mice and also in rat pups exposed perinatally by maternal treatment with busulphan[2,3,7,8] and tetrachlorodibenzo-*p*-dioxin (TCDD).[31,32]

Immune functions of DBTC- or DOTC-fed mice and guinea pigs were not impaired,[8] which is in agreement with the absence of lymphoid atrophy observed in these species. No function studies were performed in mice that were exposed to these compounds by the intravenous route.

A remarkable effect of DOTC and DBTC, probably not related to the immune-system is the increase of endotoxin sensitivity. Weanling rats fed 150 ppm of the organotin compounds for 4 weeks died from an i.v. injection of 100 μg/*E.coli* LPS whereas the controls died at concentrations of 2500 μg LPS.[27]

In Vitro Studies with Various Organometallic Compounds

A cytotoxic effect of both DBTC and DOTC is demonstrated in suspension cultures of rat thymocytes[7] (Figure 6). Thymocyte survival decreased in direct proportion to the DBTC concentration in the nutrient medium. Thymocyte number was already decreased after incubation with a concentration as low as 0.05 μg of DBTC/ml medium for 24 hours. The thymocyte count progressively diminished to 31% and thymocyte viability (scored by a dye exclusion test) to 61% of the control value after 24 hours of incubation with 50 μg DBTC/ml. These results indicate that DBTC does not need metabolic activation, but is itself cytotoxic.

In contrast to the effect on rat thymocytes, DBTC and DOTC did not reduce rat bone marrow cell survival under the same culture conditions (Figure 7). Number and viability of rat bone marrow cells remained the same when cultured with 0, 5, or 50 μg of DBTC or DOTC/ml for periods up to 40 hr.

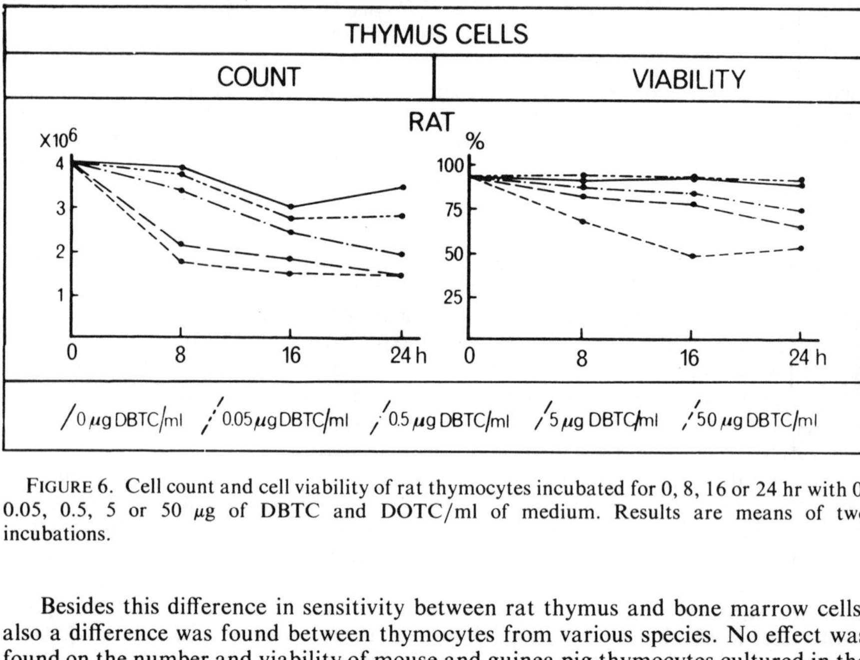

Figure 6. Cell count and cell viability of rat thymocytes incubated for 0, 8, 16 or 24 hr with 0, 0.05, 0.5, 5 or 50 μg of DBTC and DOTC/ml of medium. Results are means of two incubations.

Besides this difference in sensitivity between rat thymus and bone marrow cells, also a difference was found between thymocytes from various species. No effect was found on the number and viability of mouse and guinea pig thymocytes cultured in the presence of DBTC for 24 hours.[7] Also, the viability of human thymocytes was not distinctly decreased, but these cells aggregated in a dose-dependent fashion in the presence of 0.5 μg and more DBTC or DOTC per ml medium[7] (Figure 8). Agglutination did not occur when the cells were cultured at low concentrations in a shaking waterbath.[9] But, the ability of these thymocytes to form so-called E-rosettes with SRBC was rapidly reduced indicating that DBTC and DOTC interfere with lymphocytic receptors for SRBC that are specific for human T-lymphocytes. The number of E-rosettes decreased in a direct proportion to the dose of organotin compounds in the incubation medium (Figure 9). Concentrations, as low as 0.02 μg DBTC/ml and 0.5 μg DOTC/ml already decreased the formation of E-rosettes.

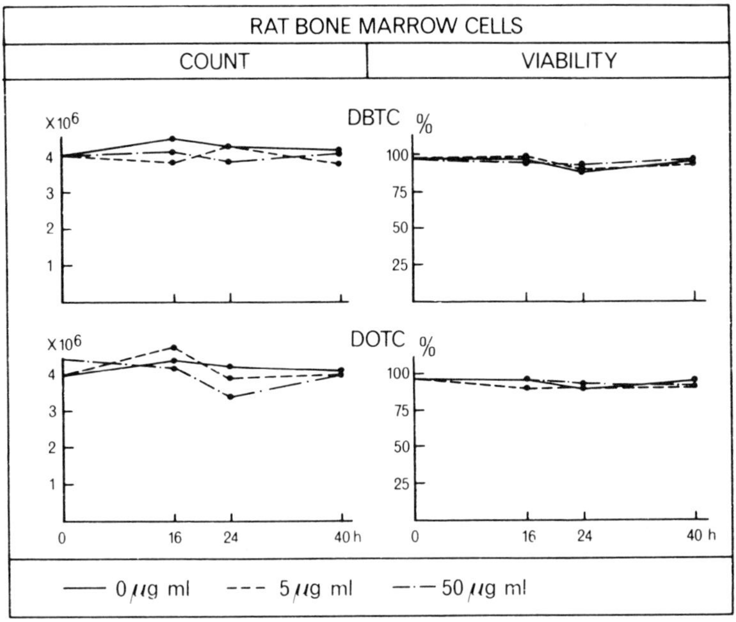

FIGURE 7. Cell count and cell viability of rat bone marrow cells incubated for 0, 16, 24, or 40 hr with 0, 5, or 50 μg of DBTC and DOTC/ml of medium. Results are means of two incubations.

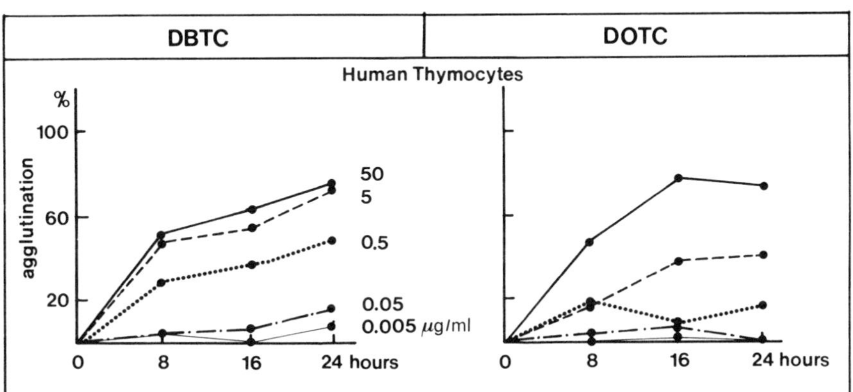

FIGURE 8. Agglutination of human thymocytes upon exposure to various concentrations of DBTC or DOTC. Two million thymocytes were cultured in duplicate in a total volume of 0.5 ml RPMI-1640 culture medium in loosely capped (12–75 mm) polystyrene culture tubes and incubated with various concentrations of the organotin compounds in a CO_2-incubator. Agglutination was calculated from the decline of cells counted electronically in the organotin exposed cultures.

Besides the dye exclusion test, which measures cell death, also the effects of various compounds on cell division in mitogen-stimulated as well as unstimulated lymphocyte cultures were studied. In the presence of DBTC and DOTC the division of thymus and spleen lymphocytes was dose-relatedly inhibited (FIGURE 10). ^{3}H-Thymidine incorporation into DNA was already decreased at a concentration of 0.02 μg DBTC/ml and was completely inhibited at a concentration of 0.1 μg DBTC per ml culture medium. At a concentration of 0.5 μg/ml medium DOTC, as well as the other

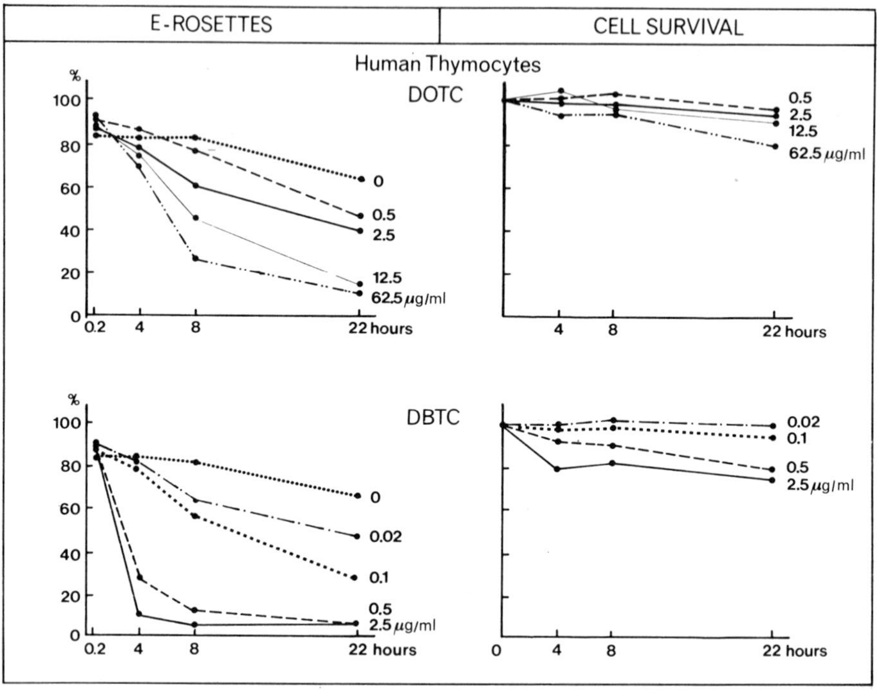

FIGURE 9. The *in vitro* effect of various concentrations of DBTC and DOTC upon the E-rosette formation and survival of human thymocytes. Three hundred thousand thymocytes were cultured in a total volume of 1 ml medium in polystyrene culture tubes incubated with various concentrations of the organotin compounds in a shaking waterbath. Cell viability was scored by a dye exclusion test. For the rosette formation 1 × 10^7 SRBC were added to the cultures, after which the cells were centrifuged and the cell pellet was incubated at room temperature for 1 hour. Rosettes were defined microscopically by the agglutination of 3 or more SRBC around a nucleated cell. Results are expressed as the mean percentage of rosette forming cells in triplicate cultures.

dialkyl compounds tested (TABLE 4), induced a complete inhibition of DNA synthesis of rat thymocytes. The viability of spleen and thymus cells was not impaired at these concentrations by the dialkyltin compounds. Therefore we conclude that dialkyltin compounds primarily induce an inhibition of cell proliferation and secondarily cause cell death.

In the lymphocyte transformation test no distinct differences in antiproliferative

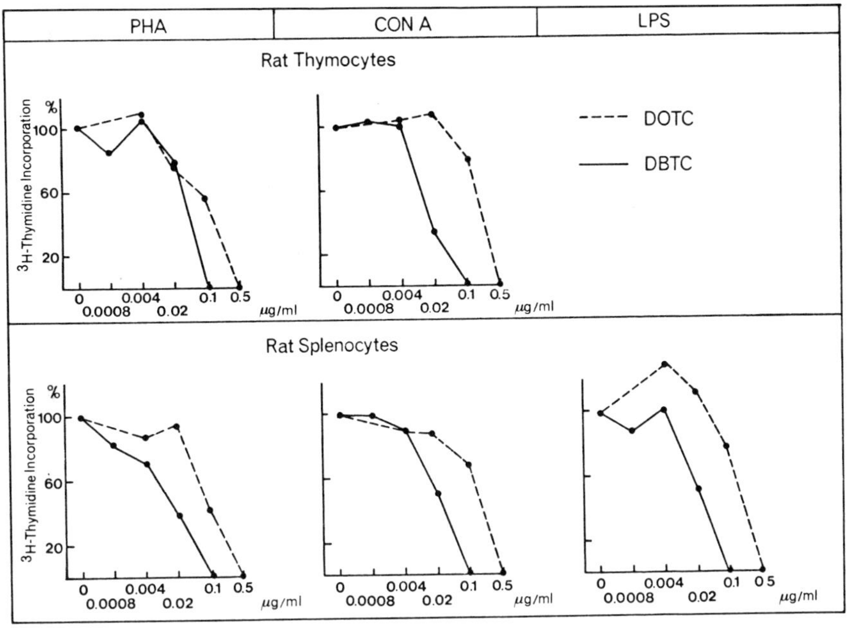

FIGURE 10. The *in vitro* effect of different concentrations of DBTC and DOTC upon ^{3}H-thymidine incorporation of rat thymus and spleen cells. Cells were cultured in triplicate during 72 hours and ^{3}H-thymidine was present during the last 24 hours of the culture period. Results are expressed as the mean percentages of the ^{3}H-thymidine incorporation of control cultures.

TABLE 4

COMPLETE INHIBITION OF DNA-REPLICATION OF HUMAN, RAT, AND MOUSE THYMOCYTES AND RAT BONE MARROW CELLS UPON INCUBATION WITH VARIOUS ORGANOTIN COMPOUNDS

	Concentration(μg/ml) Inducing Complete Inhibition of ^{3}H-Thymidine Incorporation			
Compound	Rat Thymocytes	Human Thymocytes	Rat Bone Marrow Cells	Mouse Thymocytes
Dimethyltindichloride	0.5	—	2.5	0.5
Diethyltindichloride	0.5	—	2.5	0.5
Di-*n*-butyltindichloride	0.1	0.1	0.5	0.1
Di-*n*-octyltindichloride	0.5	2.5	2.5	0.5
Di-*n*-phenyltindichloride	0.5	—	—	0.5
Tri-*n*-phenyltinchloride	0.5	—	—	0.5
Bis (β-carbobutoxyethyl) tindichloride	2.5	—	—	—
Bis (β-carboxyethyl) tindichloride	>360	—	—	—

activity of the various dialkyltins were observed (TABLE 4). So, the differences in lymphocytotoxicity observed after *in vivo* exposure between the water soluble and water insoluble dialkyltins can not be explained by differences in *in vitro* activity of these compounds.

Species differences as observed after *in vivo* exposure to organotin compounds,[7] did not occur in the lymphocyte transformation test, since comparable effects were found on DNA-replication of rat and mouse as well as human thymocytes. This indicates an absence of species differences in organotin sensitivity *in vitro*. Although, with viability testing,[7] species differences *in vitro* were noted. This discrepancy can not be explained yet, but indicates that the absence of organotin-induced cell death, as measured with a dye exclusion test, will not exclude a cytotoxic action of these compounds.

The estertin compound bis-β-carbobutoxyethyltin dichloride, which did not cause lymphoid atrophy *in vivo*, actively decreased the thymidine uptake of lymphocytes *in vitro*. However, its hydrolysis product only slightly decreases thymidine incorporation at the maximum concentration of 360 μg per ml medium. Probably, bis-β-carboxyethyltin does not enter lymphocytes. Also other dicarboxylic acids, such as succinic acid and fumaric acid are not able to penetrate cells.

A remarkable although not large difference was noted in the blast transformation test between thymus and bone marrow cells (TABLE 4). Complete inhibition of DNA-replication of bone marrow cells occurred at concentrations 5 times higher than with thymus and spleen cells. Also with tumor cell lines inhibition of DNA replication was obtained, but often at higher concentrations than with thymocytes.

In Vitro Effects of Various Organometallic Compounds on the Energy Metabolism of Rat Thymocytes

Incubation of rat thymocytes with DBTC results in an impairment of the glucose metabolism of these cells.[33] The glucose consumption of thymocytes cultured for 4 hours in the presence of DBTC was dose-relatedly stimulated until a maximum value at a concentration as low as 5 μMol (FIGURE 11). When the uptake of glucose is increased, it is completely converted to lactate and pyruvate. This indicates a complete inhibition of the entrance of pyruvate into the tri-carboxylic acid cycle. Up to concentrations including a maximum glucose uptake, the consumption of oxygen was only slightly decreased. A further increase of the DBTC dose results in a rapid decline of oxygen consumption and glucose uptake to zero levels at concentrations of 60 and 30 μMol respectively. At these concentrations also a slight decrease of thymocyte count and viability is observed.

The same effects on glucose metabolism were also found upon incubation with other dialkyltin, dialkyllead, dialkylgermane compounds and also by phenylarsinedichloride (TABLE 5). All these compounds easily hydrolyse into products with two free binding sites at the metal atom, with the exception of dialkylgermane halides, which are hydrolysed into germoxanes or germanoic anhydrides in aquous solutions.[17] But the impairment of the glucose metabolism is seen at different concentrations of the various compounds. Phenylarsinedichloride appeared to be the most active compound. From the dialkyltins especially DBTC and diethyltindichloride (DETC) were the most active. Dimethyl (DMTC) and DOTC induce the same effects but at much higher concentrations. This is at variance with the *in vivo* results. DOTC induce a severe thymus atrophy *in vivo* but is less active *in vitro*, whereas DETC impairs the glucose metabolism at very low concentrations but its *in vivo* effect on the thymus is

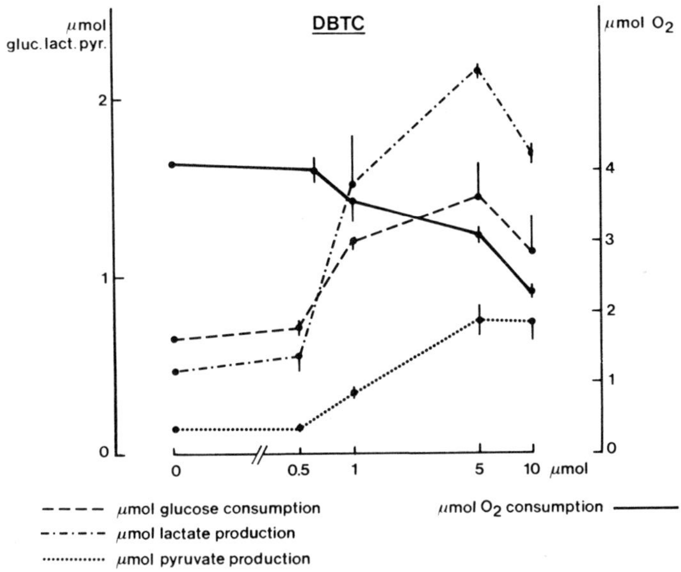

FIGURE 11. Oxygen consumption, glucose uptake and lactate and pyruvate production of 1×10^8 rat thymocytes incubated in a respirometer for 4 hours at 37°C in the presence of various concentrations of DBTC. Results are expressed as the mean ± S.E. of three incubations.

TABLE 5

STIMULATION OF GLUCOSE CONSUMPTION AND INHIBITION OF PYRUVATE DEHYDROGENASE ACTIVITY OF RAT LYMPHOCYTES CULTURED IN THE PRESENCE OF VARIOUS ORGANOMETALLIC COMPOUNDS

Compound	Maximal Glucose Consumption Concentration (μMol)	Complete Inhibition of Pyruvate Dehydrogenase Concentration (μMol)*
Dimethyltindichloride	120	30
Diethyltindichloride	10	5
Di-*n*-butyltindichloride	5	5
Di-*n*-octyltindichloride	120	10
Di-*n*-butylgermanedichloride	>120	120
Di-*n*-hexylleaddiacetate	5	1
Phenylarsinedichloride	1	0.5
Bis-β-carbobutoxyethyltindichloride	5	1
Bis-β-carbomethoxyethyltindichloride	120	30
Bis-β-carboxyethyltindichloride	—	—†

*Denotes the lowest concentration inducing a complete conversion of glucose to lactate and pyruvate.

†—Indicates no effect on glucose metabolism up to concentrations of 120 μMol.

less pronounced.[7] The *in vitro* effects of di-*n*-hexylleaddiacetate and DBTC were similar, whereas their *in vivo* effects on the thymus were completely different.[7] In comparison with the dialkyltin and dialkyllead compounds, di-*n*-butylgermanedichloride is less active *in vitro* as well as *in vivo*. Probably this is related to differences in dissociation grade.

The cytotoxic activity of bis-β-carbomethoxy and bis-β-carbobutoxyethyltindichloride, as measured by an impaired glucose metabolism, is comparable with DMTC and DBTC respectively. However, their hydrolysis product bis-β-carboxyethyltindichloride did not show cytotoxicity, which is also noted with the lymphocyte transformation test (TABLE 4). Inhibition of α-ketoaciddehydrogenase is observed in rat brain brei and rat liver mitochondria incubated with DETC and phenylarsenious acid by Aldridge and Cremer.[34] Since these compounds complexes dithiol groups, they suggested a binding with lipoic acid, a cofactor of α-ketoaciddehydrogenase systems. Also, DBTC inhibited these enzyme systems in liver mitochondria.[33] In the presence of 120 μMol DBTC the consumption of pyruvate and α-ketoglutarate is completely depressed (FIGURE 12). Incubation with L-glutamic acid results in an accumulation of α-ketoglutarate.

General Discussion

Cytotoxic effects of various organometals are demonstrated in cultures of rat thymocytes, and are related to an impairment of the energy metabolism of these cells.[33] Dialkyltins, dialkylleads, dialkylgermanes and phenylarsenites cause inhibition of the pyruvate and α-ketoglutarate dehydrogenase activity, resulting in an inhibition of the tri-carboxylic-acid cycle and therefore an energy shortage will occur in the cells. Probably these compounds complex with lipoic acid, a cofactor of α-ketoaciddehydrogenase systems, as originally suggested by Aldridge and Cremer.[34] Trialkyltins induce an energy shortage by an inhibition of the oxidative phosphorylation.[35,36] *In vitro,* all these compounds are extremely cytotoxic. A concentration of 0.5 μg/ml of the dialkyltin compounds tested completely inhibited the proliferation of thymus lymphocytes. Both in the lymphocyte transformation test and in the glucose metabolism studies only slight differences in cytotoxic activity of the various compounds were observed. However, *in vivo* marked differences between the various compounds were established regarding atrophy thymus and thymus-dependent lymphoid tissues. *In vivo* only dialkyltin compounds with chainlength of 4–8 C atoms and trialkyltin compounds with a chainlength of 4 C atoms induced a remarkable thymus atrophy, whereas the other organometals tested were inactive or only slightly decreased thymus weight. Since the fat-water partition of the various compounds are different, which is a main factor determining membrane passage and therefore the distribution over the body, the differences of thymus effects are probably related to differences in distribution pattern of various compounds. This may be the explanation of the differences between the *in vitro* and *in vivo* effects of the various compounds.

The effects of DBTC, DOTC and TBTC on lymphoid tissues *in vivo* were selective. They induce thymus atrophy without signs of myelotoxicity and without compromising other organ systems. This lymphoid atrophy is a result of a selective lymphocytotoxicity. Upon *in vivo* treatment the number as well as the viability of lymphoid cells from thymus, spleen, and lymph nodes was dose-relatedly decreased, whereas the number and viability of bone marrow cells was not changed. Possibly the selectivity of effects on lymphoid organs could be related to relatively high concentrations of this compound in thymus, as was found by Seidler *et al.*[37] for DOTC. However, this is not confirmed by Penninks and Seinen,[38] who found relatively high concentrations of ^{14}C DBTC in the liver and the kidneys and relatively low levels in the thymus of rats after oral and intravenous administration. Therefore, the selective *in vivo* thymus effects can not be explained by a difference in distribution of DBTC over the body. So, thymocytes seem to be more sensitive to DBTC-, DOTC-, and TBTC-induced energy shortage than other cells. Also, *in vitro* differences in sensitivity were noted between thymus and bone marrow cells.

As a consequence of the selective lymphocytotoxicity, an immune suppressive action of DBTC and DOTC is demonstrated in many experimental models for both cell-mediated and thymus-dependent humoral immunity in the rat. These compounds

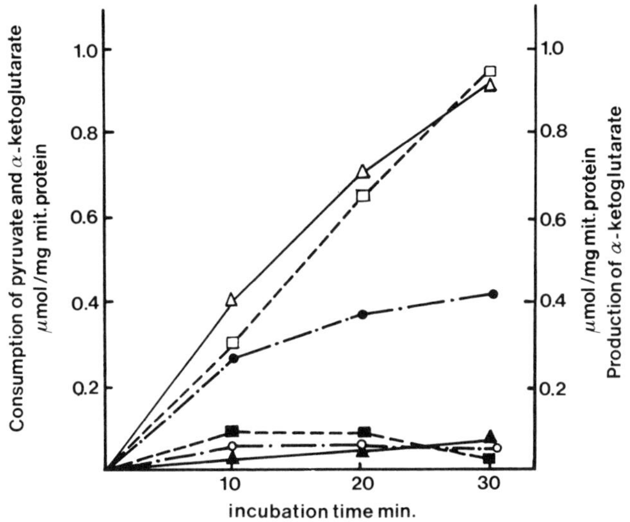

FIGURE 12. Consumption of pyruvate-□- - - ■ -, α-ketoglutarate △—▲ and the production of α-ketoglutarate- · ○ - - - ● - - by rat liver mitochondria incubated respectively with pyruvic acid/malic acid and α-ketoglutaric acid/malic acid or L-glutamic acid in the presence of 0 (open symbols) or 120 μMol DBTC (closed symbols). Mitocondria, 3.5 mg protein per ml were incubated in 10 ml medium in 25 ml Erlenmeyer flasks, vigorously shaken at 25°C. At the times indicated 1.5 ml samples were withdrawn for enzymatic assays.

seem to act selectively in those test-systems in which T-lymphocytes are predominantly involved. In other animal species no selective lymphocytotoxicity nor immune suppression is demonstrated upon oral treatment. Since *in vitro* DBTC and DOTC have comparable effects on human and rat thymocytes, possibly these compounds also produce immunosuppression in man. Therefore, the use of organotins has to be reevaluated, since toxicologic evaluation of these compounds was done at a time when the immunosuppressive properties were not known—especially since the effects on the lymphoid system are the most sensitive parameters of DBTC, DOTC, and TBTC in short-term studies.

ACKNOWLEGMENTS

The authors thank Dr. E. J. Bulten, Institute for Organic Chemistry, TNO, Utrecht, for the supply of various organometallic compounds and Dr. R. S. Talbot, AKZO Chemie, Liverpool for the supply of the estertin compounds.

REFERENCES

1. BARNES, J. M. & L. MAGOS. 1968. The toxicology of organometallic compounds. Organometal. Chem. Rev. **3:** 137–150.
2. THAYER, J. S. 1974. Organometallic compounds and living organisms. J. Organometal. Chem. **76:** 265–295.
3. ROSS, A. 1965. Industrial applications of organotin compounds. Ann. N.Y. Acad. Sci. **125:** 107–123.
4. BARNES, J. M. & H. B. STONER. 1959. The toxicology of tin compounds. Pharmacol. Rev. **11:** 211–231.
5. PIVER, W. T. 1973. Organotin compounds: Industrial application and biological investigation. Environ. Health Perspect. **4:** 61–80.
6. SEINEN, W. & M. I. WILLEMS. 1976. Toxicity of organotin compounds. I. Atrophy of thymus and thymus-dependent lymphoid tissue in rats fed di-n-octyltindichloride. Toxicol. Appl. Pharmacol. **35:** 63–75.
7. SEINEN, W., J. G. VOS, I. VAN SPANJE, M. SNOEK, R. BRANDS & H. HOOYKAAS. 1977. Toxicity of organotin compounds. II. Comparative in vivo and in vitro studies with various organotin and organolead compounds in different animal species with special emphasis on lymphocyte cytotoxicity. Toxicol. Appl. Pharmacol. **42:** 197–212.
8. SEINEN, W., J. G. VOS, R. VAN KRIEKEN, A. PENNINKS, R. BRANDS & H. HOOYKAAS. 1977. Toxicity of organotin compounds. III. Suppression of thymus-dependent immunity in rats by di-n-butyltindichloride and di-n-octyltindichloride. Toxicol. Appl. Pharmacol. **42:** 213–224.
9. SEINEN, W., J. G. VOS, & H. HOOYKAAS. 1978. Suppression of GVH reactivity, lymphocyte transformation and E. rosette formation by di-*n*-butyltindichloride and din-*n*-octyltindichloride. In preparation.
10. DOUGHERTY, T. F., M. L. BERLINER, G. L. SCHNEEBELI, & D. L. BERLINER. 1964. Hormonal control of lymphatic structure and function. Ann. N.Y. Acad. Sci. **113:** 825–843.
11. CLAESSON, M. G. & C. RØPKE. 1969. Quantitative studies on cortisol-induced decay of lymphoid cells in the thymolymphatic system. Acta Pathol. Microbiol. Scand. **76:** 367–382.
12. THOMPSON, H. A. & G. R. GALE. 1971. Cis-dichlorodiammineplatinum. II. Hematopoietic effects on rats. Toxicol. Appl. Pharmacol. **19:** 602–609.
13. TILL, J. E. & E. A. MCCULLOUCH. 1961. A direct measurement of the radiation sensitivity of normal mouse bone marrow cells. Radiation Res. **14:** 213–222.
14. SELEYE, H. 1950. The physiology and pathology of exposure to stress. First edition, p. 286. Acta Inc., Medical Publishers. Montreal.
15. MCANULTY, P. A. & J. W. T. DICKERSON. 1973. The cellular response of the weanling rat thymus gland to undernutrition and rehabilitation. Pediat. Res. **9:** 778–785.
16. SEINEN W. & V. RUTTEN. 1978. Unpublished results.
17. RIJKENS, F. & G. J. M. VAN DER KERK, 1964. Investigations in the field or organogermanium chemistry. Germanium Information Center, Kansas City, Miss.
18. PENNINKS, A & W. SEINEN. 1978. Comparative toxicity of di-estertin and di-alkyltin stabilizers. In preparation.
19. BARNES, J. M. & H. B. STONER. 1958. Toxic properties of some dialkyl-and trialkyl tin salts. Br. J. Ind. Med. **15:** 15–23.
20. HELLER, E. D. 1977. Personal communication.
21. KAARS SIJPESTEYN, A., J. G. A. LUYTEN & G. J. M. VAN DER KERK. 1969. Organometallic

fungicides. *In* Fungicides, Vol. II, pp. 331–366. D. C. Torgeson, Ed. Academic Press. New York.

22. WINCKELS, H. & W. SEINEN. 1978. Intestinal changes of guinea pigs fed di-n-octyltindichloride. In preparation.
23. MOSIER, D. & P. L. COHEN. 1975. Ontogeny of mouse T-lymphocyte function. Fed. Proc. Fed. Am. Soc. Exp. Biol. **34:** 137–140.
24. GREAVES, M. F., J. J. T. OWEN & M. C. RAFF. 1973. T and B lymphocytes. pp. 13–35. Excerpta Medica. Amsterdam.
25. MILLER, J. F. A. P. 1964. Effect of thymic ablation and replacement. *In* Thymus in Immunobiology, pp. 436–460. R. A. Good & A. E. Gabrielson, Eds. Harper and Row. New York.
26. JANKOVIĆ, B. D., B. H. WAKSMAN, & B. G. ARNASON. 1962. Role of the thymus in immune reactions in rats J. Exp. Med. **116:** 159–176.
27. SEINEN, W. Unpublished data.
28. VOS, J. G., J. BUYS, P. BEEKHOF & A. M. HAGENAARS. 1978. Quantification of total IgM and IgG and specific IgM and IgG to a thymus-independent (LPS) and a thymus dependent (tetanus toxoid) antigen in the rat by enzyme-linked immunosorbent assay (elisa). Ann. N.Y. Acad. Sci. Submitted for publication.
29. MILLER, J. F. A. P. 1961. Effect of neonatal thymectomy on the immunological responsiveness of the mouse. Proc. Roy. Soc. Ser. B. **156:** 415–423.
30. PINTO-MACHADO, J. 1970. Influence of prenatal administration of busulphan on the postnatal development of mice. Production of a syndrome including hypoplasia of the thymus. Teratology **3:**363–369.
31. VOS, J. G. & J. A. MOORE. 1974. Suppression of cellular immunity in rats and mice by maternal treatment with 2, 3, 7, 8—tetrachlorodibenzo-*p*-dioxin. Int. Arch. Allergy. Appl. Immunol. **47:** 777–790.
32. MOORE, J. A. & R. E. FAITH. 1976. Immunology response and factors affecting its assessment. Environ. Health. Perspect. **18:** 125–131.
33. PENNINKS, A. & W. SEINEN. 1978. Impairment of energy metabolism of thymocytes by various organometallic compounds. In preparation.
34. ALDRIDGE, W. N. & J. E. CREMER. 1955. The biochemistry of organotin compounds. Biochem. J. **61:** 406–418.
35. ALDRIDGE, W. N. 1958. The biochemistry of organotin compounds. Biochem. J. **69:** 367–376.
36. ALDRIDGE W. N., B. W. STREET & D. N. SKILLETER. 1977. Oxidative phosphorylation. Halide-dependent and halide-independent effects of triorganotin and trioganolead compounds on mitochondrial functions. Biochem. J. **168:** 353–364.
37. SEIDLER, H., M. HÄRTIG & H. WOGGEN. 1971. Das Verhalten von ^{14}C-markiertem Di-*n*-octylzinndithioglycolsäure-2-äthylhexylester and Di-n-octylzinnchlorid säure^{35} S-markiertem Thioglycolsäure-2-äthylhexylester in Rattenorganismus. Z. Ges. Hyg. **17:** 603–608.
38. PENNINKS, A. H. & W. SEINEN. 1978. Unpublished data.

QUANTIFICATION OF TOTAL IgM AND IgG AND SPECIFIC IgM AND IgG TO A THYMUS-INDEPENDENT (LPS) AND A THYMUS-DEPENDENT (TETANUS TOXOID) ANTIGEN IN THE RAT BY ENZYME-LINKED IMMUNOSORBENT ASSAY (ELISA)

J. G. Vos, J. Buys, P. Beekhof, and A. M. Hagenaars

National Institute of Public Health
Bilthoven, the Netherlands

After the introduction of the enzyme-linked immunosorbent assay (ELISA) by Engvall and Perlmann[1] for the determination of rabbit IgG, the ELISA has been used to measure antigens, haptens or antibodies (reviewed by Wisdom,[2] Engvall,[3] and Schuurs and van Weemen).[4] Advantages of ELISA are its reliability, its simplicity for detecting IgM and IgG antibodies, and its potential for large scale screening, since mechanization of practically all steps of the assay is feasible.[5] Recently, an on-line macro-ELISA system for screening of 4000 sera daily has been developed.[6]

One of the types of enzyme immunoassays is the "sandwich" ELISA, in which the antigen reacts with excess solid-phase antibody and after incubation followed by washing, the bound antigen is reacted with excess enzyme-labeled antibody. After further washing, the bound label is assayed, and this provides a direct measure of the amount of antigen present. The first part of this paper describes the determination of rat IgG and IgM using the "sandwich" immunoassay. Results of the IgG and IgM measurements are correlated with results obtained by the radial immunodiffusion method.[7] Details of this study will be published elsewhere.[8] Particularly in the field of toxicology, routine determination of immunoglobulin levels are useful for the detection of possible effects of chemicals on the lymphoid system.[9] To date, such determinations are not included in routine procedures for toxicity testing of chemicals, partly because of a lack of a rapid and simple but reliable laboratory method.

The second part of this paper compares the quantification of antibodies in the rat against *Escherichia coli* lipopolysaccharide (LPS) and tetanus toxoid by ELISA and the passive hemagglutination test. In the ELISA for antibodies, antigen is coupled to a solid phase, incubated with antiserum, and the amount of antibody bound to the solid phase is quantified by peroxidase-labeled antiimmunoglobulin. Details of this study will also be published elsewhere.[10]

In the mouse, the antibody response to LPS does not require cooperation of T-helper cells with B-cells,[11] whereas both the IgM and IgG responses to tetanus toxoid are thymus-dependent[12] as are responses to most protein antigens. Also in the rat, tetanus toxoid seems to be a thymus-dependent antigen, since pretreatment with antithymocyte serum prevented a humoral immune response in this species.[13] To get an answer about the possible thymus-independence of the antibody response against LPS in the rat, the primary and secondary antibody responses to LPS were determined and compared with the response to tetanus toxoid, the secondary response being identical to the primary response with a thymus-independent antigen.[14] In addition, a second parameter for thymus-independence of a humoral immune response was studied: the antibody response to LPS in congenitally athymic (*nu/nu*) rats as compared with the response in heterozygous thymus-bearing (+/*nu*)rats.

0077-8923/79/0320-0518 $01.75/0 © 1979, NYAS

Using the antibody response to a thymus-dependent and a thymus-independent antigen, insight can be obtained into the mode of action of chemicals that interfere with the humoral immune response.[9]

Materials and Methods

Animals

Wistar-derived SPF rats were randomly bred in the Institute. The animals were housed in plastic cages and received food (Muracon, Trouw Ltd, Putten, the Netherlands) and water *ad libitum*. Female homozygous *nu/nu* rats and heterozygous +/*nu* rats were obtained from Dr. R. Leyten, University of Leuven, Belgium.

Antisera

Anti-rat IgG was prepared by hyperimmunization of a sheep with purified IgG, emulsified in complete Freund's adjuvant. The antiserum to IgG (SaRaIgG) was rendered monospecific by absorption with F(ab^1)2 fragments, obtained by peptic digestion of rat IgG (courtesy Dr. J. Nagel, Laboratory for Immunochemistry). Goat anti-rat IgM (GaRaIgM) was obtained commercially from Nordic Pharmaceuticals and Diagnostics, Tilburg, the Netherlands.

Immune globulin fractions of both antisera were prepared by fractionation with caprylic acid according to the method of Steinbuch and Audran.[15] The antisera to IgG and IgM appeared to be monospecific as tested by gel precipitation (Ouchterlony technique). In addition, the sera were tested in a more sensitive ELISA system. For that purpose, wells of polystyrene microplates were coated with SaRaIgG or GaRaIgM, incubated with serially diluted normal rat serum, and finally incubated with peroxidase-labeled GaRaIgM and SaRaIgG. In this test system, the GaRaIgM appeared to contain less than 1% cross-reacting (anti-IgG) antibodies, whereas anti-IgM antibodies were virtually absent ($< 0.01\%$) in the SaRaIgG.

Enzyme-Conjugated Antiimmunoglobulins

Horse radish peroxidase (HRP), type VI, R.Z.3, was obtained commercially from Sigma Chemical Company, St. Louis, U.S.A. HRP was conjugated to the immunoglobulin fraction of SaRaIgG and GaRaIgM by the method of Nakane and Kawaoi.[16]

Antigens

Escherichia coli lipopolysaccharide 0 127:B 8 (LPS), prepared by the phenolextraction method, was obtained from Difco Laboratories (Detroit, Michigan, U.S.A.). Tetanus toxoid, plain, was prepared in the Institute, and contained 2.5 mg protein/1000 Lf.

"Sandwich" Macro-ELISA for Quantification of Total IgM and IgG

Disposable polystyrene tubes (50 by 11 mm, LKB-produkter AB, Stockholm, Sweden) were coated by adding 1 ml of antiimmunoglobulin solution (15 μg protein/ml) in sodium carbonate buffer (0.1 M; pH 9.6) with 0.02% NaN_3 as preservative. After incubation under rotation at 37°C for 18 hours, tubes were flushed on a special home-made washing machine[6] using 120 ml tap water containing 0.05% Tween 20. For the IgM and IgG analysis, sera (reference serum of pooled adult rat serum as well as the test sera) were diluted with phosphate-buffered saline (PBS; 0.01 M; pH 7.2) containing 2% bovine serum albumin (BSA) to which 0.05% Tween 20 was added. One ml quantities were added to each tube and tubes were incubated under rotation at 37°C for 1 hour. After washing, 1 ml of the conjugate (in an optimal dilution of PBS with 2% BSA and 0.05% Tween 20) was added. Tubes were incubated under rotation at 37°C for 2 hours. Excess conjugate was removed by washing. Finally, the amount of enzyme bound to the tubes was determined by adding 1 ml of the specific substrate, prepared by dissolution of 80 mg of 5-amino-salicylic acid in 100 ml of distilled water at approx. 70°C. Directly prior to use the pH of this stock solution was brought to 6.0 with 1 N NaOH. To 9 ml of this solution 1 ml of 0.05% H_2O_2 was added. Incubation was performed at room temperature and extinctions at 449 nm were measured after 1 hour on a semiautomated equipment for enzyme analysis (Olli system 3000, Olli, Kivenlahti, Finland).

Control reactions included: (1) conjugate control (i.e. antiimmunoglobulin-coated tube, to which conjugate was added and after proper incubation and washing, the substrate), and (2) replacement of the normal rat serum by normal human serum. This control reaction was only used for IgM measurements because of cross-reaction of anti-rat IgG with human serum.

Quantification of Total IgM and IgG by Single Radial Immunodiffusion (SRID)

The method of Mancini *et al.*[7] was used. Precipitates of antigen-antibody complexes were stained with amido black. Results (mean values of duplicate analyses) were expressed as percentages of the concentration present in the reference serum.

Total IgM and IgG Measurements in Dilutions of Sera of Rats of Different Ages as Measured by ELISA

In this experiment, the reliability of testing one serum dilution for the routine measurement of serum IgM and IgG was studied. For that purpose, sera of 25 rats of different ages (ranging from 10 to 55 days) as well as the reference serum, were tested in twofold serial dilutions, thus providing an indication whether the slopes of the dilution curves (^{2}log serum dilution versus extinction) of the different sera were the same. In that case, they can be considered to be dilutions of each other. Results are expressed as mean values of duplicate analyses of independent dilutions.

Correlation of Total IgM and IgG Measurements between ELISA and SRID

In order to study the correlation between the IgM and IgG concentrations measured in ELISA and SRID, sera of 25 rats of different ages were used. In the ELISA, sera were diluted as described in the previous paragraph. For the "linear" range of the dilution curve of the reference serum, linear regression analysis was used to calculate IgG and IgM concentrations in the different sera. Extinctions were corrected for the extinction value of the conjugate control, which was always below 0.025. Results of ELISA and SRID were expressed as percentages of the IgM and IgG level in the reference serum, and the correlation coefficient and the ELISA:SRID ratio was calculated.

Comparison of the Precision of IgM Measurement by ELISA and SRID

The precision of the enzyme immunoassay was determined by IgM measurements in 20 independent dilutions of pooled normal rat serum. Final dilutions of 1:5000, made in 3 steps, were made by hand using Precision pipettes (Medical Laboratory Automation, Inc., Mount Vernon, N.Y.). To determine the precision of the ELISA itself, one diluted serum sample was measured in 20-fold. Finally, the precision of the SRID was determined by IgM measurement in 20-fold of the pooled normal rat serum. All determinations were done on a single day. Results were expressed as coefficients of variation.

ELISA for Determining Antibody Titers to LPS and Tetanus Toxoid

Enzyme immunoassays were essentially performed as described by Ruitenberg *et al.*[17] Polystyrene microplates with 12×8 flat-bottomed cups were coated with antigen by adding to each cup 100 μl of tetanus toxoid solution (2 Lf/ml) in sodium carbonate buffer (0.1 M; pH 9.6) with NaN_3 (0.02%). In preliminary studies this concentration of tetanus toxoid gave optimal sensitivity. As *E. coli* LPS itself attached poorly to polystyrene, the following procedure was adopted after preliminary trials: LPS was initially coupled to methylated BSA (Miles Laboratories Inc., Slough, England). A mixture of 1 ml LPS solution containing 0.5 mg of LPS in sodium carbonate buffer, 0.1 M, pH 9.6 and 0.05 ml methylated BSA solution (10 mg/ml distilled water) was incubated for 20 minutes at room temperature on a magnetic stirrer. Before coating the solution was diluted with carbonate buffer to a final concentration of 5 μg LPS/ml (a concentration that gave optimal sensitivity in trial studies). After coating for 2 hours at 37°C, the microplates were washed with tap water of neutral pH containing 0.05% Tween 20. Serial twofold dilutions of the sera were prepared in phosphate-buffered saline (PBS, 0.01 M, pH 7.2) containing 2% bovine serum albumin (BSA) and 0.05% Tween 20. Trays were incubated with 100 μl of the serum dilutions for 1 hour at 37°C. After washing 100 μl of conjugate, diluted properly with PBS (0.01 M, pH 7.2) containing 2% BSA and 0.05% Tween 20, was added. Microplates were incubated under rotation at 37°C for 2 hours. The trays were washed again and 100 μl of substrate was added. After incubation at room temperature the reaction product was evaluated visually and expressed as ^{2}log of the highest dilution, giving a positive reaction.

As control reactions were included a negative serum control using serum of a nonimmunized rat and a conjugate control. In addition, absorption experiments were done. Sera were incubated at 37°C under rotation for 2 hours with different concentrations of LPS and tetanus toxoid, stored overnight at 4°C and centrifuged for 1 hour at 20,000 rpm.

In one experiment the antibody response to LPS was measured by the mechanized macro-ELISA.[5,6] For that purpose disposable polystyrene tubes (50 by 11 mm, LKB-produkter AB, Stockholm, Sweden) were incubated with 1 ml quantities of the different reagents described above. Specific anti-LPS IgM antibodies were measured spectrophotometrically in two-fold serially diluted sera.

Hemagglutination Method for Determining Antibody Titers to LPS and Tetanus Toxoid

The hemagglutination assay for LPS was done largely according to the method of Andersson and Blomgren[11] using sheep red blood cells (SRBC) coated with LPS. The hemagglutination method for tetanus toxoid was performed with formalinized SRBC, coated with tetanus toxoid with the aid of tannic acid. Details of the hemagglutination method used will be published elsewhere.[10] Results were expressed as ^{2}log of the highest dilution giving positive agglutination.

Comparison of ELISA and Hemagglutination Method for Determining Antibody Titers to LPS and Tetanus Toxoid

Eight week-old male rats were immunized intravenously on day 0 and 10 with 100 or 300 μg LPS (the optimal immunogenic dose of LPS when administered intravenously ranges from 0.1 to 1000 μg[10]) or with 1, 5, 10 or 25 Lf tetanus toxoid. Blood was drawn from the orbital plexus on day 5, 10 (prior to booster injection) and 21. Antibody titers were determined with ELISA and hemagglutination method.

Primary and Secondary IgM and IgG Response to LPS and Tetanus Toxoid: Primary IgM and IgG Response to LPS in Nu/Nu versus +/Nu Rats

In order to study the thymus-independence of the antibody response to LPS, 8 week-old male rats were immunized intravenously with 1 μg LPS on day 0 and 60. Using this time interval in mice, it was shown that the kinetics of the primary and secondary antibody response to type III pneumococcal polysaccharide (a thymus-independent antigen) were identical, thus showing that immunologic memory did not develop after primary immunization.[18] For comparison, rats were injected intravenously on day 0 and 60 with 5 Lf of the thymus-dependent antigen tetanus toxoid. Blood was drawn from the orbital plexus on day 0, 5, 10, 20, 60 (prior to booster injection) 65, 70 and 80. In addition, 4 weeks-old female *nu/nu* and +/*nu* rats were injected intravenously with 1 μg LPS. Blood was drawn at day 5. Specific IgM and IgG antibody titers were determined by ELISA.

RESULTS

"Sandwich" ELISA for Quantification of Total IgM and IgG

The ELISA for IgG gave low extinction values when the absorbed and monospecific SaRaIgG was used for both coating and conjugate. This problem was not encountered in the IgM analysis. The reason for the low-extinction values when IgG was measured might be the exhaustive absorption of the antiserum resulting in a specific antiserum that was too weak to be used for the quantification. Therefore, tubes were coated with unabsorbed antiserum, and as conjugate, the absorbed SaRaIgG labeled with peroxidase was used. Quantification of IgG was now possible.

Plotting the ^{2}log dilutions of the reference serum against the extinction values, a straight line was obtained between extinction values of 0.2 and 1.0 for both IgM (FIGURE 1) and IgG (FIGURE 2).

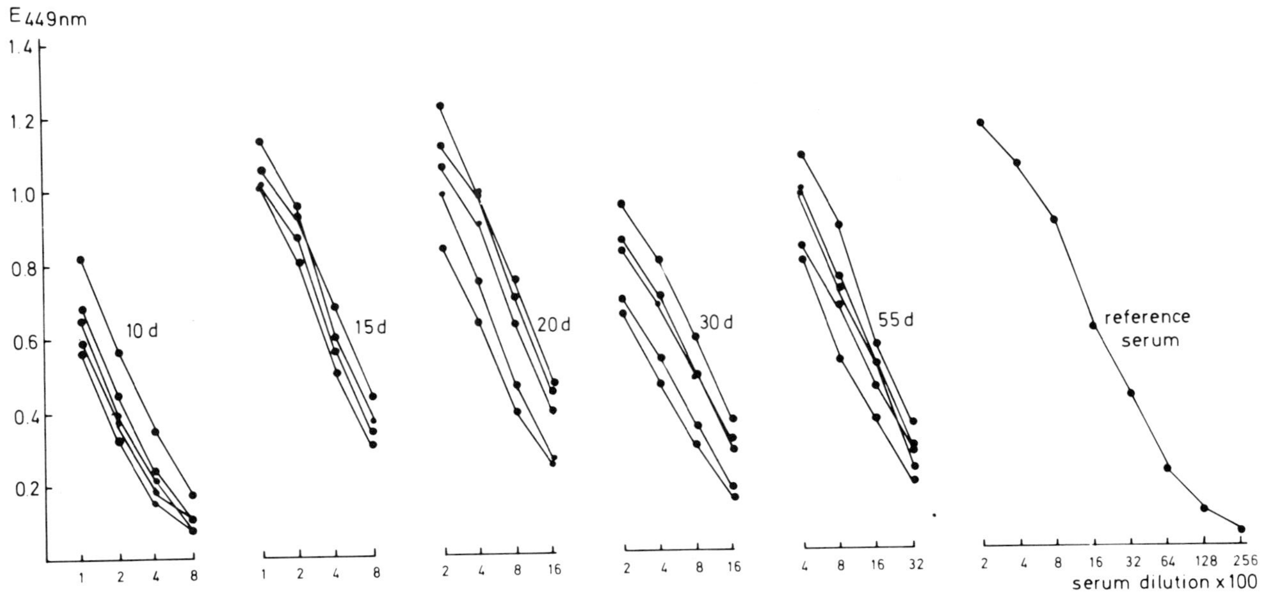

FIGURE 1. Dilution curves of IgM in a reference serum and in sera of 25 rats of different age groups as measured with the "sandwich" ELISA. Extinction values are plotted against ^{2}log of the different serum dilutions. Between extinction values of 0.2 and 1.0 straight lines with equal slopes are found.

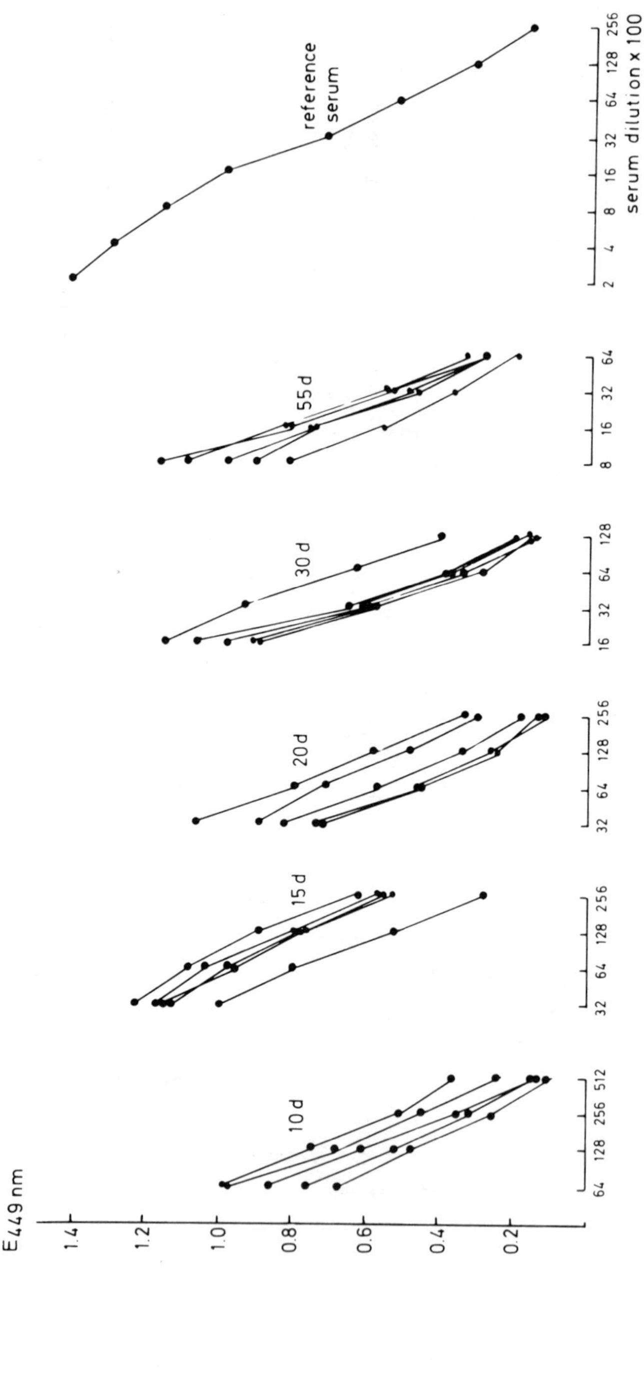

FIGURE 2. Dilution curves of IgG in a reference serum and in sera of 25 rats of various ages, showing straight lines with equal slopes between extinction values of 0.2 and 1.0.

Total IgM and IgG Measurements in Dilutions of Sera of Rats of Different Ages as Measured by ELISA

As shown in FIGURES 1 and 2, all 25 sera of rats ranging from 10 to 55 days were tested at four twofold dilutions. Between extinction values of 0.2 and 1.0, straight lines were found. The reliability of testing one serum dilution for routine measurement is demonstrated by the fact that the slopes of the lines of the 25 test sera were similar to each other and to the slope of the reference serum.

FIGURES 1 and 2 clearly demonstrate that serum concentrations of IgM gradually increase with age, whereas IgG levels are highest at day 15 as a result of intestinal absorption of IgG from milk.

Correlation of Total IgM and IgG Measurements between ELISA and SRID

As shown in FIGURE 3, using sera of 25 rats, a good correlation was found between ELISA and SRID IgM concentrations, as expressed in percentages of the IgM level in the reference serum. The correlation coefficient was 0.94. The geometric mean ratio between ELISA and SRID was 0.63.

The correlation between IgG levels obtained with ELISA and SRID is given in FIGURE 4. Data do not include IgG values of sera of rats having the highest IgG concentrations (15 day-old animals) as these sera were not tested in the SRID at a proper dilution. The correlation coefficient was 0.92 and the geometric mean ratio between ELISA and SRID was 1.25.

Comparison of the Precision of Total IgM Measurement by ELISA and SRID

The coefficient of variation of IgM measurement by ELISA in 20 independent dilutions (1:5000) of pooled rat serum was 5.4%. For the ELISA itself the coefficient

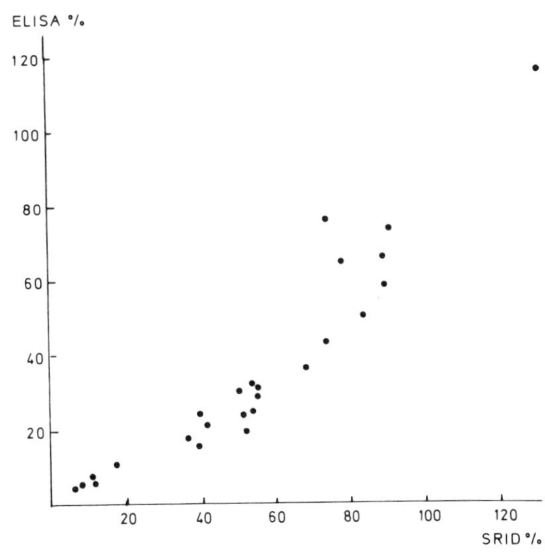

FIGURE 3. Correlation between IgM levels determined by ELISA and single radial immunodiffusion (SRID) in sera of 25 rats of different age groups (ranging from 10 to 55 days). The IgM concentrations are expressed as percentages of the IgM concentration in the reference serum. Correlation coefficient = 0.94.

of variation, determined by IgM measurement in 20-fold in a single serum sample (dilution 1:5000), was 3.2%. For the SRID, as determined by IgM measurement in 20-fold, the coefficient of variation was 4.7%.

ELISA for Determining Antibody Titers to LPS and Tetanus Toxoid

Using the ELISA, antibody titers to LPS and tetanus toxoid could be easily determined. The specificity of the assay is shown in FIGURE 5. Absorption of a positive serum with increasing amounts of LPS reduced the ^{2}log IgM antibody titer against LPS from 11 to 2. Similarly, absorption of a positive serum with tetanus toxoid reduced the IgM and IgG antibody titers against tetanus toxoid to zero.

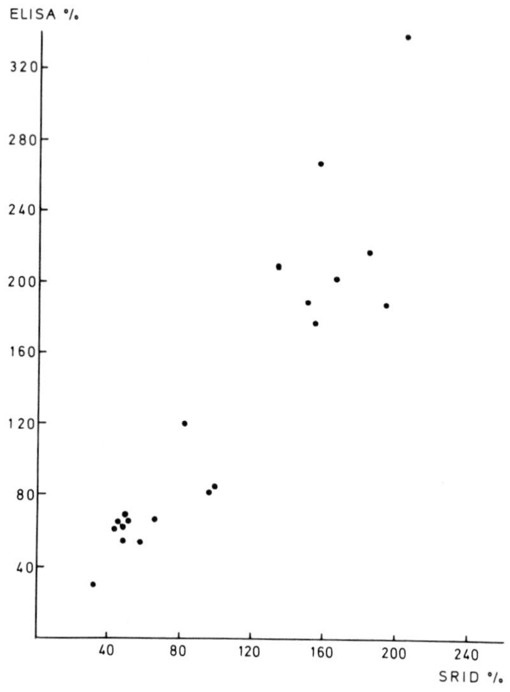

FIGURE 4. Correlation between IgG levels (as percentages of the level in the reference serum) determined by ELISA and SRID in sera of 20 rats of various ages. Correlation coefficient = 0.92.

Titrations of 3 sera containing IgM antibodies to LPS are shown in FIGURE 6. Also shown in this figure is the remarkable finding that at low serum dilutions the quantity of reaction product in the ELISA was lower than at higher dilutions. This inhibition was caused by a heat-labile serum component as heating of the serum at 56°C for 30 minutes prevented this phenomenon. No inhibition occurred at low serum dilutions when antibodies to tetanus toxoid were titrated.

In FIGURE 7, results are given of an experiment in which IgM antibodies to LPS are measured by the macro-ELISA. Twofold serial dilutions of the sera were tested and the extinction measured. In the macro-ELISA sigmoid curves were found that approached a linear function between the ^{2}log serum dilutions of 4 to 7 or 8. These preliminary data indicate that within that dilution range specific antibody concentrations can be quantified using one serum dilution.

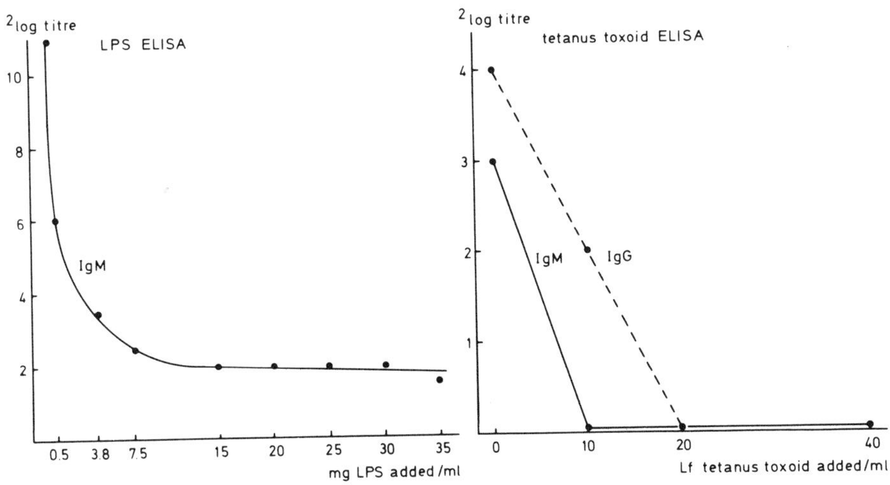

FIGURE 5. Reduction of antibody titers as measured by ELISA, after absorption of sera of immunized rats with increasing amounts of LPS or tetanus toxoid.

Comparison of ELISA and Hemagglutination Method for Determining Antibody Titers to LPS and Tetanus Toxoid

A comparison of ELISA IgM and the hemagglutination method to determine antibody titers to LPS is given in FIGURE 8. Rats were immunized on day 0 and 10, and blood was collected at days 5, 10 (prior to the second injection), and 21. In all

FIGURE 6. ELISA for the demonstration of IgM antibodies to LPS, showing a reduction of reaction product at low serum dilutions in three positive sera, day 5 after immunization (rows A, B and C). Rows D, E and F: inhibition of the reaction is not present in these same sera after heating at 56°C for 30 min. Rows G and H are serum control and conjugate control, respectively.

samples the ELISA IgM antibody titers were much higher than the hemagglutination titers. The IgM antibody response was the same after 5 and 21 days.

As shown in FIGURE 9, the optimal immunogenic dose of tetanus toxoid in the rat when administered intravenously is 5 or 10 Lf. Doses of 1 and 25 Lf did not induce detectable IgM antibodies at day 5, as measured by ELISA. At day 10, antibody titers were higher, in particular antibodies of the IgG class. At both days 5 and 10, antibodies could not be detected by the hemagglutination method. In contrast, hemagglutinating antibodies were present at day 21, and might even be slightly higher when compared with the ELISA IgG antibody response (respectively ^{2}log titers of 8 and 7). IgM antibodies were relatively low during the secondary response.

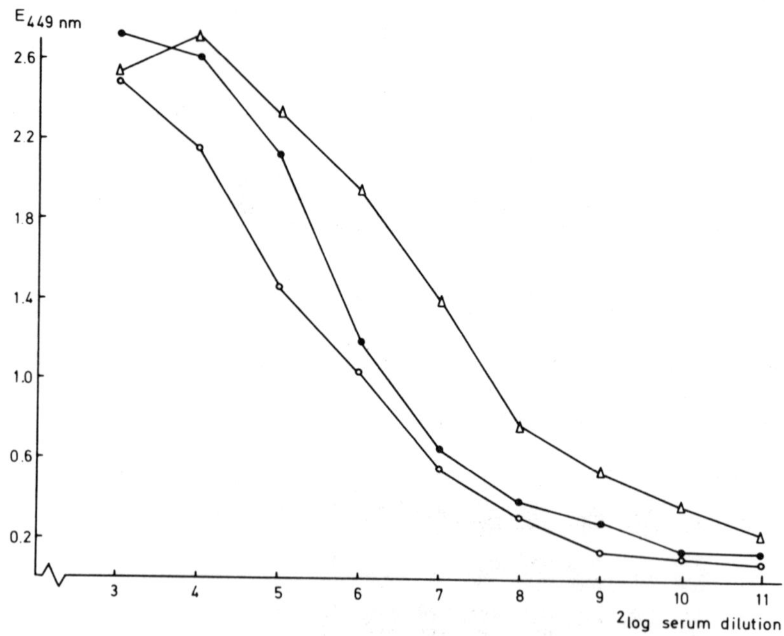

FIGURE 7. Quantification of IgM antibodies against LPS by macro-ELISA. Sera of 3 rats were twofold serially diluted and extinctions measured. The sigmoid curves approach a linear function between the ^{2}log serum dilutions of 4 to 7 or 8, thus allowing the quantification of antibody concentrations at a single reading within that region.

Primary and Secondary IgM and IgG Response to LPS and Tetanus Toxoid: Primary IgM and IgG Response to LPS in Nu/Nu versus +/Nu Rats

The primary IgM and IgG response to an optimal (1 μg) intravenous dose of LPS and the response obtained in rats given a second injection of 1 μg LPS, 60 days after primary immunization, are presented in FIGURE 10. No difference is noted in the primary and secondary IgM and IgG response, with the exception that ^{2}log titers were higher in the secondary response. The apparent reason for this phenomenon is the presence of circulating antibodies at day 60 (prior to the second injection). By contrast, the primary and secondary IgG response to the thymus-dependent antigen

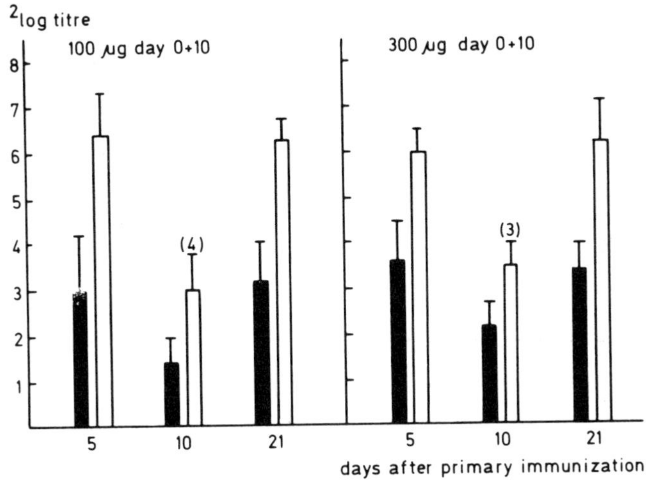

FIGURE 8. Comparison of ELISA IgM and hemagglutinating antibody titers during the primary and secondary response to LPS. Results are expressed as geometric means ± S.D. of 5 female rats per group. In brackets, number of animals when deviating from 5. ■ = hemagglutination; □ = ELISA IgM.

tetanus toxoid (intravenously injected at day 0 + 60 at a dose of 5 Lf) is quite different (FIGURE 11). Up to day 60, IgG antibody titers gradually increase. On secondary immunization with tetanus toxoid, the synthesis of IgG antibodies accelerates rapidly. The secondary IgM response resembles the primary response, although the titers are higher.

Regarding the IgM antibody titer to LPS, as measured 5 days after an intravenous

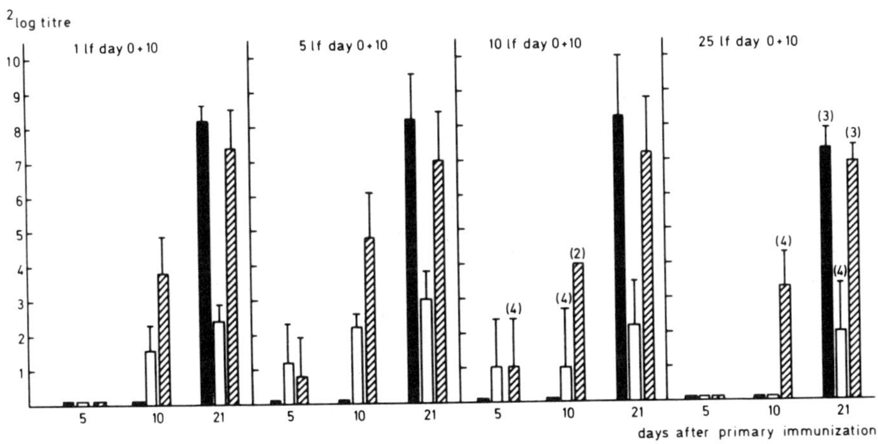

FIGURE 9. Comparison of ELISA IgM and IgG and hemagglutinating antibody titers during the primary and secondary response to tetanus toxoid. Results are expressed as geometric means ± S.D. of 5 female rats per group. In brackets, number of animals when deviating from 5. ■ = hemagglutination; □ = ELISA IgM; ▨ = ELISA IgG.

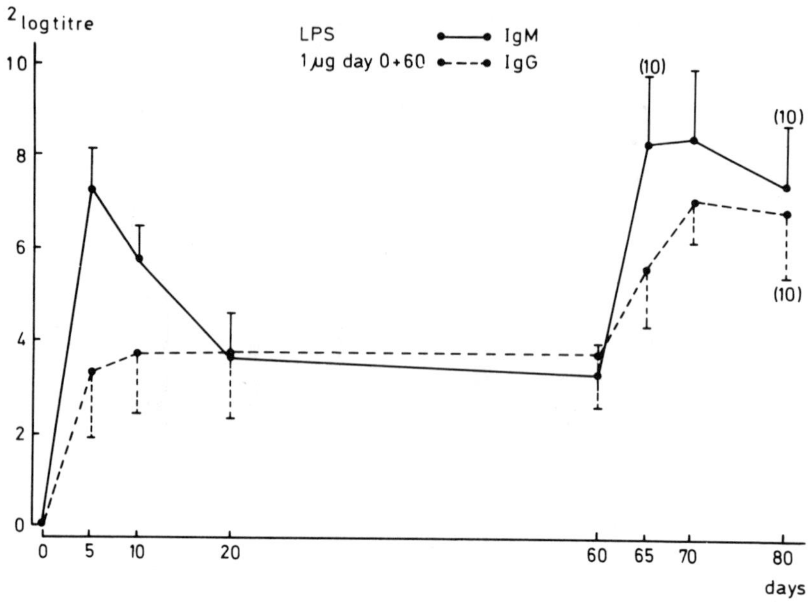

FIGURE 10. Primary and secondary IgM and IgG response to LPS as measured by ELISA. Rats were injected intravenously on day 0 and 60 with 1 μg LPS. Results are expressed as geometric means ± S.D. of 12 male rats per group. In brackets, number of animals when deviating from 12.

injection of 1 μg LPS, no significant difference was found between congenitally athymic *nu/nu* rats and heterozygous thymus bearing +/*nu* rats. The ^{2}log IgM titer in *nu/nu* rats was 3.83 ± 0.41 compared to 5.00 ± 1.41 in the +/*nu* rats (mean values ± S.D. of 6 animals per group). IgG antibodies could not be detected in either group, probably because of the low age of the animals.

DISCUSSION

In the ELISA IgM and IgG experiments, sigmoid standard curves were obtained with the reference rat serum. Between extinction values of 0.2 and 1.0, straight lines were found with the reference serum as well as with sera of 25 rats of various age. The slopes of the linear ranges were practically equal. Good correlation was obtained between the "sandwich" ELISA and single radial immunodiffusion, despite the different principles of these assays; the correlation coefficients being 0.94 and 0.92 for IgM and IgG, respectively. On the basis of these results, it can be concluded that the "sandwich" ELISA is a suitable method for determining IgM and IgG in rat serum. However, the geometric mean ratios ELISA: SRID did differ from 1 (0.63 for IgM and 1.25 for IgG). The consequence of this as yet unexplained difference is that for the time being results of total IgM and IgG measurements cannot be given in absolute quantities. In practice, this poses no real problem. For example, in the field of toxicology, results can either be presented as percentage of the values of control animals, or as percentage of the IgM or IgG concentration present in a reference serum.

As shown in this study, the precision of the "sandwich" ELISA is the same as found with the SRID, despite the dilutions that are necessary for the ELISA.

Particularly for the routine measurement of rat IgM and IgG, the ELISA has several advantages over the radial immunodiffusion test. It is less time consuming and the amount of antiserum used in the macro-ELISA is one order of magnitude lower than used in the radial immunodiffusion test, and two orders of magnitude lower in the mechanized micro-ELISA that is currently being developed in our Institute.[19]

In conclusion, the mechanized ELISA is a specific, reliable, sensitive, and economic test for routine measurements of total IgG and IgM levels in rat sera.

In the second part of this study it is shown that ELISA is a sensitive and specific method to determine antibodies against LPS and tetanus toxoid. The ^{2}log titers of antibodies against LPS were approximately two times higher in the ELISA than in the hemagglutination method. Using ELISA, IgM and IgG antibodies could be easily discriminated during the primary immune response to tetanus toxoid, whereas the hemagglutination method did not detect antibodies at this time. Only during the secondary response to tetanus toxoid, the hemagglutination method proved successful, with antibody titers that were slightly higher than those obtained by ELISA IgG. As IgM antibodies are very effective in the hemagglutination reaction one would expect that the hemagglutination test would be more sensitive when predominantly IgM antibodies are present e.g. against LPS. But, the reverse seems to be true: the hemagglutination test was only a sensitive assay during the secondary response to tetanus toxoid, when high titers of IgG antibodies are found. Thus, the difference in sensitivity between both methods cannot be explained by the class of antibodies.

The sensitivity of ELISA is also demonstrated by the results of a study in which

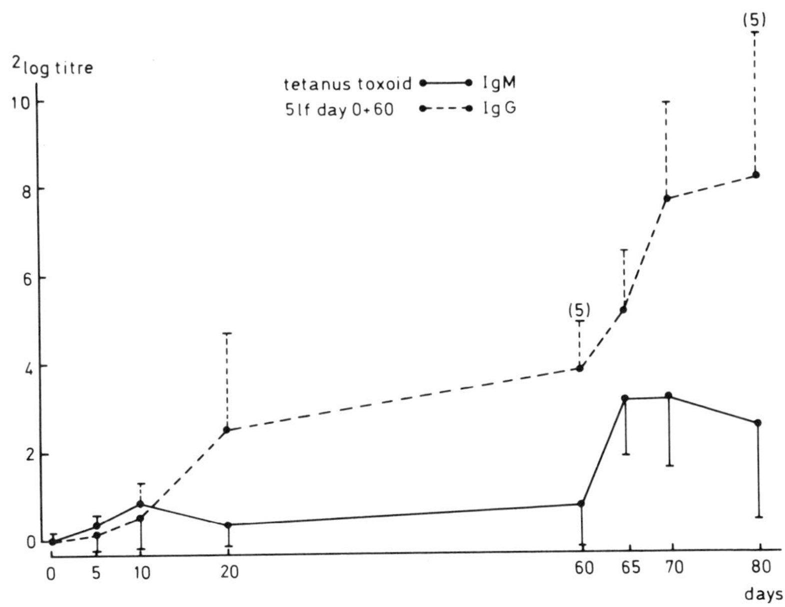

FIGURE 11. Primary and secondary IgM and IgG response to the thymus-dependent antigen tetanus toxoid. Rats were injected intravenously with 5 Lf tetanus toxoid on day 0 and 60, and antibody titers measured by ELISA. Results are expressed as geometric means ± S.D. of 6 animals per group. In brackets, number of animals when deviating from 6.

antibodies to LPS could be detected in suckling rats that were immunized as early as one day after birth.[20] This is of particular importance for the study of possible immunosuppressive effects of chemicals, in which studies rats are preferably exposed during the ontogenesis of the immune system.[9]

The discriminatory power of ELISA can possibly be further increased. This is of particular importance when subtle effects on the humoral immunity are studied. Preliminary results with the macro-ELISA showed that at certain dilutions of LPS antisera extinctions were measured that approached a linear function if plotted against the dilution. In this way, specific antibody concentrations can be measured by a single reading as is done for total IgM and IgG, in stead of endpoint titration.

In addition to the sensitivity of ELISA, a major advantage lies in the fact that antibodies of the IgM and IgG class can be easily determined separately, if high-titered specific antisera are available. Measurement of IgG antibodies does not depend on prior degradation of IgM antibodies, e.g. by treatment with 2-mercaptoethanol, as is practiced for the hemagglutination method.

Finally, evidence is presented in this study that the humoral immune response of the rat to LPS *E. coli,* is a thymus-independent phenomenon: i.e., T-helper cells are not involved in the generation of an immune response. This conclusion is based on the similarity of the IgM antibody response to LPS *E. coli* in *nu/nu* versus +/*nu* rats. In addition, the secondary immune response to LPS was practically equal to the primary immune response, suggesting that immunologic memory did not develop after the primary immunization with LPS. As such, LPS mimics the antibody response to type III pneumococcal polysaccharide (SSS-III)[14] and contrasts sharply with the response obtained with the thymus-dependent antigen, tetanus toxoid. Furthermore, the peak serum antibody response to LPS is found after 5 days,[20] which is another argument that LPS is a thymus-independent antigen.[21] The fact that LPS generates also an IgG response, is no argument that T-helper cells are involved in the antibody response.[22]

In conclusion, in performing function studies of the immune system of the rat, the antibody response to LPS *E. coli* and tetanus toxoid can be used to study the thymus-independent and thymus-dependent humoral immunity. For quantification of specific and total IgM and IgG concentrations, the ELISA is an excellent technique. Work is in progress to quantitate total IgA and IgE levels in rat sera using the "sandwich" assay described in this paper.

Summary

Rat IgM and IgG was determined by mechanized "sandwich" enzyme-linked immunosorbent assay (ELISA) using peroxidase labeled anti-rat-IgM and -IgG. Linear ranges in standard curves of a reference rat serum had a slope similar to the slopes found with sera of 25 rats of various age. IgM and IgG measurements by ELISA in these sera correlated well with results obtained by single radial immunodiffusion (SRID). In addition, the precision of the enzyme immunoassay was the same as obtained with the SRID. Compared with SRID, ELISA is less time consuming and the amount of antiserum used in the macro-ELISA is one order of magnitude lower; and again 10 times lower in the mechanized micro-ELISA that is currently being developed. In conclusion, the ELISA is a specific, reliable, sensitive, and economic method for routine measurement of rat serum IgG and IgM e.g. in toxicity studies.

In the second part of this study, ELISA and the passive hemagglutination test were compared to determine the primary and secondary antibody response to *E. coli* lipopolysaccharide (LPS) and tetanus toxoid in rats. In the ELISA, the antigens were bound to the wells of polystyrene microplates. Tetanus toxoid was coated directly, LPS after complexing with methylated bovine serum albumin. After incubation with dilutions of the rat sera, the amount of antibody bound to the solid phase was quantified by means of peroxidase-labeled antiimmunoglobulin. The specificity of the enzyme immunoassay was tested by absorption of the sera with the respective antigens. ELISA proved to be more sensitive than the hemagglution reaction, except when titers were determined during the secondary response to tetanus toxoid. Besides its specificity and sensitivity, ELISA is a convenient method for measuring both IgM and IgG antibodies.

Finally, evidence is presented that in the rat, the humoral immune response to LPS is a thymus-independent phenomenon. Thus, by using the antibody response to LPS and tetanus toxoid in function studies of the immune system of the rat, insight can be obtained in the thymus-independent and thymus-dependent humoral immune response.

Acknowledgments

The authors thank Drs. E. C. Beuvery, E. I. Krajnc, and F. X. R. van Leeuwen for valuable discussion.

References

1. Engvall, E. & P. Perlmann. 1971. Enzyme-linked immunosorbent assay (ELISA). Quantitative assay of immunoglobulin G. Immunochemistry **8:** 871–874.
2. Wisdom, G. B. 1976. Enzyme-immunoassay. Clin. Chem. **22:** 1243–1255.
3. Engvall, E. 1977. Quantitative enzyme immunoassay (ELISA) in microbiology. Med. Biol. **55:** 193–200.
4. Schuurs, A. H. W. M. & B. K. van Weemen. 1977. Enzyme-immunoassay. Clin. Chim. Acta **81:** 1–40.
5. Ruitenberg, E. J., P. A. Steerenberg, B. J. M. Brosi & J. Buys, 1976. Reliability of the Enzyme-Linked Immunosorbent Assay (ELISA) for the serodiagnosis of *Trichinella spiralis* infections. J. Immunol. Methods **10:** 67–83.
6. Ruitenberg, E. J., J. A. van Amstel, B. J. M. Brosi & P. A. Steerenberg. 1977. Mechanization of the enzyme-linked immunosorbent assay (ELISA) for large scale screening of sera. J. Immunol. Methods **16:** 351–359.
7. Mancini, G., A. O. Carbonara & J. F. Heremans. 1965. Immunochemical quantitation of antigens by single radial immunodiffusion. Immunochemistry **2:** 235–254.
8. Vos, J. G., F. M. H. Debets, J. Buys, P. Beekhof & A. M. Hagenaars. Quantification of rat immunoglobulins M and G by mechanized enzyme-linked immunosorbent assay (ELISA). In preparation.
9. Vos, J. G. 1977. Immune suppression as related to toxicology. CRC Crit. Rev. Toxicol. **5:** 67–101.
10. Vos, J. G., J. Buys, J. G. Hanstede & A. M. Hagenaars. Comparison of ELISA and passive haemagglutination method for quantification of antibodies to LPS and tetanus toxoid in the rat. Submitted for publication.
11. Andersson, B. & H. Blomgren. 1971. Evidence for thymus-independent humoral antibody production in mice against polyvinylpyrrolidone and *E. coli* lipopolysaccharide. Cell. Immunol. **2:** 411–424.
12. Willcox, H. N. A. 1975. Thymus dependence of the antibody response to tetanus toxoid in mice. Clin. Exp. Immunol. **22:** 341–347.
13. Ruitenberg, E. J. & A. M. Hagenaars. Unpublished data.
14. Baker, P. J., P. W. Stashak, D. F. Amsbaugh & B. Prescott. 1971. Characterization of the antibody response to type III pneumococcal polysaccharide at the cellular level. I.

Dose-response studies and the effect of prior immunization on the magnitude of the antibody response. Immunology **20:** 469–480.

15. STEINBUCH, M. & R. AUDRAN. 1969. The isolation of IgG from mammalian sera with the aid of caprylic acid. Arch. Biochem. Biophys. **134:** 279–284.
16. NAKANE, P. K. & A. KAWAOI. 1974. Peroxidase-labeled antibody. A new method of conjugation. J. Histochem. Cytochem. **22:** 1084–1091.
17. RUITENBERG, E. J., P. A. STEERENBERG & B. J. M. BROSI. 1975. Micro-system for the application of ELISA (Enzyme-linked Immunosorbent Assay) in the serodiagnosis of *Trichinella spiralis* infections. Medikon Nederland **4:** 30–31.
18. BRALY-MULLEN, H. 1975. Secondary IgG responses to type III pneumococcal polysaccharide. I. Kinetics and antigen requirements. J. Immunol. **115:** 1194–1198.
19. RUITENBERG, E. J. & B. J. M. BROSI. 1978. Automation in enzyme immunoassay. Scand. J. Immunol. **8** (Suppl. 7): 63–72.
20. VOS, J. G. & J. G. HANSTEDE. Unpublished data.
21. JONES, J. M., D. F. AMSBAUGH & B. PRESCOTT. 1976. Kinetics of the antibody response to type III pneumococcal polysaccharide (SSS-III). I. Use of ^{125}I-labeled SSS-III to study serum antibody levels, as well as the distribution and excretion of antigen after immunization. J. Immunol. **116:** 41–51.
22. SHARON, R., P. R. B. MCMASTER, A. M. KASK, J. D. OWENS & W. E. PAUL. 1975. DNP-Lys-Ficoll: a T-independent antigen which elicts both IgM and IgG anti-DNP antibody-secreting cells. J. Immunol. **114:** 1585–1589.

HEXACHLOROBENZENE-INDUCED STIMULATION OF THE HUMORAL IMMUNE RESPONSE IN RATS

J. G. Vos, M. J. van Logten, J. G. Kreeftenberg, and W. Kruizinga

National Institute of Public Health
Bilthoven, the Netherlands

Hexachlorobenzene (HCB, C_6Cl_6) is a stable compound used industrially as an intermediate in organic synthesis and agriculturally as a fungicide.

The present study, in which different function tests were performed of the humoral immunity, the cell-mediated immunity and the phagocytizing capacity of macrophages as reviewed earlier,[1] was initiated because of seemingly contradictory results obtained in a semi-chronic toxicity in rats.[2] At intermediate and high dose levels of HCB, rats had markedly increased spleen weights, whereas thymus weights were depressed at the high dose level. Histologically, the size of the follicles and of the marginal zone was increased in the spleen, while cortex atrophy was observed in the thymus. Increased spleen weights in rats were also reported by Kimbrough and Linder[3] and by Kuiper-Goodman *et al.*[4] In dogs, HCB administration caused hyperplasia of lymphoid tissue in the fundus and pyloric antrum of the stomach.[5] An effect on rat alveolar macrophages (decreased Fc receptor activity) has been described by Ziprin and Fowler.[6]

During the course of the present study, Loose *et al.*[7] published results of experiments in mice that show that HCB suppresses the humoral immune response. In addition, it was shown that HCB treatment of mice enhances their susceptibility to a *Plasmodium berghei* infection as well as their susceptibility to endotoxin.[8] The latter aspect is also included in the present report.

Materials and Methods

Animals and Diet

Specific pathogen-free Wistar rats were randomly bred in the Institute. For the different experiments 3–4 week old male rats were used. A semisynthetic diet (Trouw Ltd., Putten, the Netherlands) and water were provided *ad libitum.*

Hexachlorobenzene (BDH Chemicals Ltd., Poole, England; purity $\geq$ 99.5%) was mixed in the diet at concentrations of 500, 1000, and 2000 mg/kg.

Experimental

In the first experiment, groups of six weanling rats received diets containing 0, 500, 1000, and 2000 mg HCB/kg. Food consumption and body weight gain was recorded weekly. At the end of the 3-week experimental period, rats were anesthetized with ether and exsanguination was done from the aorta. Weights of thymus, spleen, mesenteric lymph node, popliteal lymph node, pituitary, adrenals, thyroid, testes, liver, and kidneys were determined. These organs were fixed in 10% buffered formalin and processed for histopathologic examination.

At the termination of the experiment, total leukocyte and differential leukocyte

0077–8923/79/0320–0535 $01.75/0 © 1979, NYAS

counts were carried out. Sera were stored at −20°C for total IgM and IgG analysis by the enzyme-linked immunosorbent assay (ELISA).[9,10] Results (mean value of triplicate analyses) were expressed as percentage of the IgM and IgG level in the control group.

On the basis of the first study, the 1000 mg HCB/kg diet level was chosen for the different function studies. Prior to these tests, rats were kept on the test diet for 3 weeks. Dosing was continued during the course of the function studies. The humoral immunity was studied by measuring antibody response against a thymus-independent antigen (LPS) and a thymus-dependent antigen (tetanus toxoid).[10] Function tests of the cell-mediated immunity included resistance to *Listeria monocytogenes* infection, rejection of skin transplants and delayed type hypersensitivity to tuberculin. The phagocytizing capacity of macrophages was studied by measuring the clearance of carbon particles. Mitogenic responsiveness was determined of lymphocytes from thymus, spleen and peripheral blood. Finally, the effect of HCB on the susceptibility of rats to endotoxin was studied.

IgM Antibody Response to LPS

Rats were immunized on day 21 in the tail vein with 1 μg of *Escherichia coli* lipopolysaccharide (LPS, *E. coli,* 0127 : B8, Difco Labs, Detroit, Michigan). The IgM antibody response to LPS was determined 5 days after immunization using the enzyme-linked immunosorbent assay (ELISA).[10,11] The test was performed in polystyrene microplates using serial twofold dilutions of the sera. Results were expressed as ^{2}log of the highest dilution giving a positive reaction.

IgM and IgG Antibody Response to Tetanus Toxoid

Rats were immunized in the tail vein with 5 Lf tetanus toxoid plain (prepared in the Institute) on day 21 and 31. Blood was drawn from the orbital plexus on day 31 (prior to the booster injection) and from the aorta in ether-anesthetized animals on day 42. The primary and secondary IgM and IgG antibody response to tetanus toxoid was determined by ELISA[10,11] and expressed as ^{2}log of the highest dilution giving a positive reaction.

Resistance to Listeria Monocytogenes Infection

Groups of rats were injected intravenously on day 21 with different (9.6×10^4, 4.8×10^5, 2.4×10^6 and 1.2×10^7) doses of *Listeria monocytogenes* bacteria that were cultured as described by Ruitenberg and van Noorle Jansen.[12] Mortality was scored until day 31.

Rejection of Skin Transplants

The method of Bailey and Usama[13] for grafting the skin of tails was used. Two donor tailskin grafts from inbred ICI-Wistar rats (reared in the Institute) were transferred to the tail graft beds of rats that were anesthetized with Fluothane (Imperial Chemical Industries Ltd., Macclesfield, England). These donor and acceptor strains apparently differ at an important histocompatibility locus, since skin allografts are rejected after approximately 10 days. To facilitate the evaluation of the rejection process one autograft was grafted. The condition of the grafts was evaluated daily to determine the allograft rejection time of the successful grafts; and the mean rejection time of the two allografts was calculated.

Delayed-Type Hypersensitivity to Tuberculin

This test was done largely following a previously published method.[14] Animals were sensitized on day 21 by a subcutaneous injection in the hind foot pad of 0.1 ml of an oil suspension containing killed *Mycobacterium tuberculosis*, strain H 37 Ra (complete H 37 Ra adjuvant, Difco Labs, Detroit, Michigan). Intradermal skin tests were done on day 35 with 10 μg tuberculin PPD (prepared in the Institute) in 0.1 ml phosphate buffered saline. The diameter and the thickness of the skin reactions were measured with calipers 24 hours after tuberculination.

Clearance of Carbon Particles

In vivo phagocytosis of carbon was measured following the technique described by Stuart *et al.*[15] Colloidal carbon suspension (C 11/143 a, Günther Wagner, Pelikan Werke, Hannover, Germany) was injected into the tail vein at a dose of 20 mg/kg body weight (a low dose of carbon was used to avoid complications such as platelet aggregations that occur at high concentrations of carbon[16]). Twenty μl of blood was drawn from the orbital plexus of lightly ether-anesthetized animals after intervals of 3 and 10 minutes and transferred to 2 ml of distilled water. The optical density was measured at 634 nm in a colorimeter. The phagocytic index K was measured using the equation of Biozzi:

$$K = [\log (OD)_1 - \log (OD)_2]/[t_2 - t_1]$$

where $(OD)_n$ is the optical density measured at time t_n. The phagocytic index α, which corrects for differences in body weight and liver and spleen weight, was not calculated as HCB induces marked increased spleen and liver weights. At histologic examination it was shown that weight increases are not due to the presence of an increased number of mononuclear phagocytes.

Mitogenic Responsiveness of Lymphocytes

Thymus and spleen of rats were removed under sterile conditions. Cell suspensions were made using a glass homogenizer in minimum essential medium (MEM, Gibco F 14) buffered with Tris (0.025 M) and supplemented with streptomycin (100 μg/ml), penicillin (100 IU/ml), and inactivated fetal calf serum (3%).

Cell suspensions were washed twice in the same medium. The viability was determined with a dye exclusion test, the number of nucleated cells was counted in chambers and the total number of cells calculated per organ. For the cultivation of peripheral lymphocytes blood was drawn from the orbital plexus, mixed with an equal volume of medium without serum, layered on a Ficoll Isopaque solution (specific density 1.078) followed by centrifugation during 15 minutes at 1000 $\times$ *g* in the interphase. The lymphocyte-rich interphase was obtained and washed twice with MEM with 3% fetal calf serum. A total of 1×10^6 thymus cells, 5×10^5 spleen cells or 1.5×10^5 peripheral blood lymphocytes were cultured in triplicate in 0.15 ml complete medium with 20% fetal calf serum in the wells of U-form microtiter plates, type M 220-24 AR (Greiner, Nürtingen, Germany) and 5 μl of reconstituted phytohaemagglutinin (PHA, Wellcome Laboratories, Beckenham, England), 1.25 μg concanavalin A (Con A, Pharmacia, Uppsala, Sweden), 10 μl reconstituted pokeweed (PWM, Gibco) or 20 μg lipopolysaccharide (LPS, *E.coli,* 0127: B8, Difco) in 10 μl medium were added. Microplates were sealed with plastic adhesive tape. After cultivation for 24 hours (spleen and thymus cells) or 48 hours (peripheral blood lymphocytes) at 37°C , 2 μCi (methyl - ^{3}H) thymidine (^{3}H - TdR, The Radiochemical Centre, Amersham, England) with a specific activity of 600 mCi/mM was added to the wells. Twenty-four hours later, the cells were harvested on glass fiber filters using a multiple cell culture harvester (Skatron, Lierbyen, Norway). Counting was done in a Packard liquid scintillation counter. Results were expressed in counts per minute (cpm) of the median value of triplicate cultures. In addition, the responsiveness per whole thymus or spleen was calculated.

Susceptibility to Endotoxin

Rats were injected intravenously on day 21 with different doses of endotoxin (LPS, *E.coli,* 0127: B8, Difco). Groups of four control animals received 32, 16, 8, 4 or 2 mg of LPS. HCB treated rats received injections of 8, 4, 2, 1, or 0.5 mg of LPS. The mortality due to endotoxin shock was scored after 48 hours.

Statistical Analysis

Student's t-test was used to calculate two-sided significances of differences between values of treated and control animals.

Results

The effect of 3-weeks feeding of HCB on body and organ : body weight ratios is given in Table 1. Body weight and food consumption was significantly lower in the 1000 mg/kg group when compared with the controls. However, no effect was present at the highest dose level, nor was there a reduction in body weight at the 1000 mg/kg level in the follow-up function studies. Most marked effects produced by HCB were significantly increased relative weights of spleen, mesenteric and popliteal lymph nodes, and the liver. Small but significant increases were noted in the relative weight of adrenals, thyroid, testes, and kidneys. Relative thymus weight was about 20% lower in the high dose group, but this difference was not statistically significant.

As shown in Table 2, HCB caused a dose-related increase in the total number of peripheral blood leukocytes. The number of neutrophilic granulocytes was increased twofold in the three test groups. A dose-related increase was found for the number of basophilic granulocytes and the number of monocytes. Lymphocyte counts were slightly, but statistically not significantly, higher in the 1000 and 2000 mg HCB/kg groups.

Results of total IgM and IgG measurements in serum are given in Table 3. Total IgM levels were dose-related increased. In the 2000 mg HCB/kg group, the increase

TABLE 1

BODY AND ORGAN : BODY WEIGHT RATIOS (‰) OF MALE RATS FED HCB FOR 3 WEEKS*

	Dietary Concentration (mg/kg)			
	0	500	1000	2000
Body	175 ± 13.6	169 ± 11.2	154 ± 15.1†	159 ± 11.3
Thymus	3.84 ± 1.1	3.41 ± 0.8	3.48 ± 0.6	3.03 ± 0.7
Spleen	3.44 ± 0.6	4.26 ± 0.8	4.85 ± 1.4†	5.20 ± 0.5§
Lymph node, mesenteric	0.57 ± 0.11	0.61 ± 0.15	0.69 ± 0.17	0.75 ± 0.09†
Lymph node, popliteal	0.032 ± 0.007	0.060 ± 0.012§	0.086 ± 0.028‡	0.091 ± 0.047†
Pituitary	0.035 ± 0.010	0.043 ± 0.004	0.042 ± 0.012	0.040 ± 0.004
Adrenals	0.219 ± 0.02	0.240 ± 0.03	0.287 ± 0.04‡	0.294 ± 0.05†
Thyroid	0.073 ± 0.02	0.095 ± 0.02	0.108 ± 0.03†	0.109 ± 0.03†
Testes	10.4 ± 0.8	10.6 ± 1.10	11.3 ± 1.0	11.5 ± 0.6†
Liver	43.7 ± 3.5	57.8 ± 7.1‡	62.0 ± 12.9‡	71.9 ± 11.5§
Kidneys	8.47 ± 0.72	9.30 ± 0.42†	9.05 ± 0.68	9.81 ± 0.20‡

*Means ± S.D. of 6 animals per group.
†$p < 0.05$.
‡$p < 0.01$.
§$p < 0.001$.

was statistically highly significant, with a mean value that was 2.6 times the level of control animals. HCB treatment did not alter serum IgG levels.

Results of the histopathologic examination are summarized in TABLE 4. Slight cortex atrophy of the thymus was observed in 2 out of 6 animals from the high dose group. Marked dose-related effects were found in the spleen (FIGURE 1). The splenic white pulp was enlarged because of an increase in size of both marginal zones and follicles, whereas the thymus-dependent areas (periarteriolar lymphocyte sheaths, PALS) apparently did not differ from the controls. In addition, extramedullary hemopoiesis occurred to a considerable degree in the red pulp of animals from all HCB groups. A remarkable effect was found in very high incidence in the mesenteric lymph

TABLE 2

PERIPHERAL LEUKOCYTE COUNTS OF MALE RATS FED HCB FOR 3 WEEKS*

	Dietary Concentration (mg/kg)			
	0	500	1000	2000
Leukocytes (10^6/l)	10,833 ± 1,502	11,100 ± 2,004	12,400 ± 959	13,083 ± 1,784†
Lymphocytes (10^6/l)	9,872 ± 1,564	9,453 ± 1,722	10,695 ± 801	11,140 ± 1,439
Neutrophils (10^6/l)	499 ± 203	991 ± 448†	967 ± 359†	1,096 ± 294‡
Eosinophils (10^6/l)	68 ± 72	95 ± 38	74 ± 76	31 ± 53
Basophils (10^6/l)	0	27 ± 46	42 ± 33†	64 ± 59†
Monocytes(10^6/l)	378 ± 161	535 ± 128	622 ± 275	752 ± 354†

*Means ± S.D. of 6 animals per group.
†$p < 0.05$.
‡$p < 0.01$.

TABLE 3

SERUM IMMUNOGLOBULIN CONCENTRATIONS OF MALE RATS FED HCB FOR 3 WEEKS*

	Dietary Concentration (mg/kg)			
	0	500	1000	2000
IgM	100 ± 41	143 ± 51	174 ± 82	262 ± 76†
IgG	100 ± 34	99 ± 47	93 ± 38	127 ± 85

*Means ± S.D. of 6 animals per group. Values are expressed as percentage of the IgM and IgG level in the control group.

†$p < 0.001$.

node, and in a somewhat lower incidence in the popliteal lymph node. The number of high-endothelial venules was increased in all treatment groups (FIGURE 2). The lengthening and arborization of high endothelial venules suggest that they are the result of growth of high endothelium into side branches that are originally lined by low endothelial cells. In addition, increased numbers of migrating lymphocytes were found infiltrating the walls of high-endothelial venules. Incidentally, a mitotic figure was seen in high-endothelial cells. In the liver of all HCB-exposed rats, liver cell hypertrophy and cytoplasmic hyalinization was observed. The localization was centrolobularly and the degree of these alterations increased with increasing HCB exposure. No lesions were observed in pituitary, adrenals, thyroid, testes and kidneys.

IgM Antibody Response to LPS

No difference was found between the day 5 IgM response to LPS of controls and HCB-treated rats. Mean ^{2}log antibody titers of 6 animals (± S.D.) were 5.83 ± 0.75 for the controls and 5.67 ± 1.03 for the treated animals.

TABLE 4

INCIDENCE OF HISTOPATHOLOGIC CHANGES IN MALE RATS FED HCB FOR 3 WEEKS*

	Dietary Concentration (mg/kg)			
	0	500	1000	2000
Thymus examined	6	6	6	6
slight cortex atrophy				2
Spleen examined	6	6	6	6
enlarged white pulp		1	4	6
increased extramedullary hemopoiesis		4	3	6
Lymph node, mesenteric examined	6	6	6	6
proliferation of high-endothelial venules		5	6	6
Lymph node, popliteal examined	6	6	6	6
proliferation of high-endothelial venules		3	1	4
Liver examined	6	6	5	6
centrolobular hypertrophy and cytoplasmic hyalinization of hepatocytes		6	5	6

*No lesions were observed in pituitary, adrenals, thyroid, testes, or kidneys.

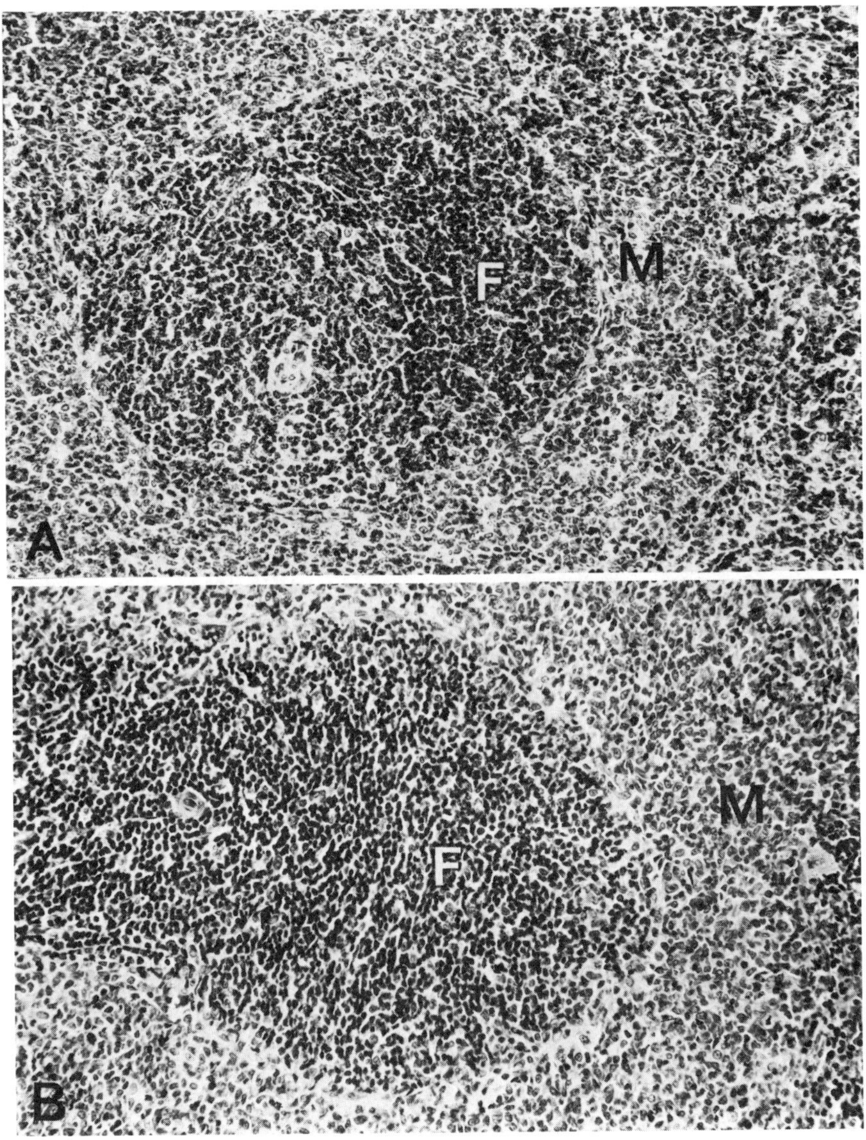

FIGURE 1. **A:** Spleen of a control rat. **B:** Spleen of a rat receiving a diet containing 1000 mg HCB/kg for a 3-week period. Note the enlarged marginal zone (M) and follicle (F). Hematoxylin-eosin, magnification $\times$ 190.

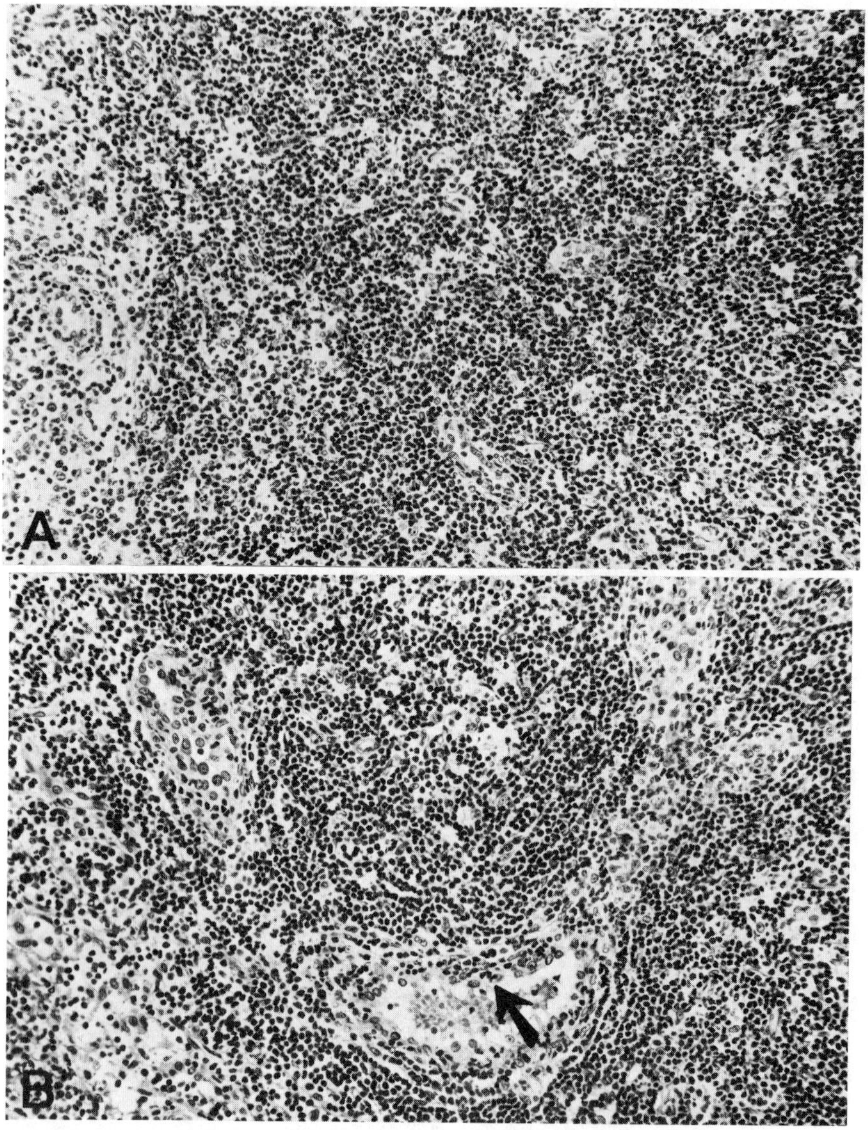

FIGURE 2. **A:** Mesenteric lymph node of a control rat. **B:** Mesenteric node of a rat fed a diet containing 500 mg HCB/kg, showing an increased number of high endothelial venules, with an increased number of recirculating lymphocytes migrating through the endothelium (arrow). Hematoxylin-eosin, magnification $\times$ 190.

IgM and IgG Antibody Response to Tetanus Toxoid

Results of this experiment are shown in FIGURE 3. Both the primary and secondary IgM response was significantly ($p < 0.10$) increased in HCB-exposed rats. Similarly, the IgG response was enhanced but the difference was only statistically significant ($p < 0.05$) during the primary response. In the average, specific antibody levels were approximately threefold higher in HCB-treated rats.

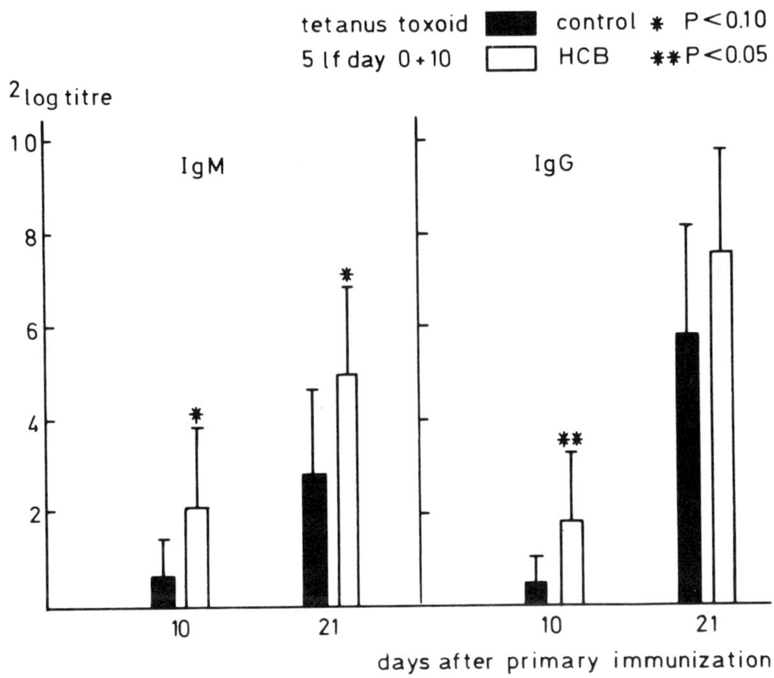

FIGURE 3. Primary and secondary IgM and IgG antibody response to tetanus toxoid in rats fed diets containing 0 or 1000 mg HCB/kg. Values represent means ± S.D. of 5–6 animals per group.

Resistance to Listeria monocytogenes *Infection*

As shown in TABLE 5, there was no clear difference in mortality between controls and treated rats. In addition, the time interval between injection and death was approximately the same.

Rejection of Skin Transplants

Rats exposed to HCB rejected skin transplants in 11.7 ± 0.5 days as compared to 11.2 ± 0.6 days in the control group (means ± S.D. of 7 animals).

TABLE 5
INCIDENCE OF MORTALITY AFTER INTRAVENOUS INOCULATION OF *Listeria monocytogenes* IN HCB-EXPOSED MALE RATS*

Dietary Concentration (mg/kg)	Number of *Listeria monocytogenes* Injected			
	1.2×10^7	2.4×10^6	4.8×10^5	9.6×10^4
0	5/5 (2.8)	3/5 (4.7)	0/5	0/5
1000	5/5 (2.8)	5/5 (4.0)	0/5	0/5

*Rats were inoculated with *Listeria monocytogenes* on day 21. Mortality was scored until day 31. The average interval inoculation-death in days is given in parentheses.

Delayed-Type Hypersensitivity to Tuberculin

The diameter of the 24-hour skin reaction (in mm) in the control group was 8.2 ± 2.9 as compared to 8.3 ± 3.1 in the experimental group (means ± S.D. of 10 animals). In addition to the diameter, no difference was present in the thickness (in mm) of the skin reaction: 1.1 ± 0.6 in the controls versus 1.2 ± 0.5 in the HCB-exposed rats.

Clearance of Carbon Particles

Results of this experiment are given in TABLE 6. The phagocytic index of rats on a HCB-diet was slightly lower when compared with the controls, but the difference was not statistically significant.

Mitogenic Responsiveness of Lymphocytes

As shown in TABLE 7, the number of nucleated cells in the thymus of control animals was the same as of HCB-exposed rats. With the exception of a reduced DNA synthesis in unstimulated cultures of the HCB group, the responsiveness of PHA-, Con A- or PWM-stimulated cultures was the same for the control and the experimental group. Similarly, DNA synthesis expressed for the whole thymus was the same.

The responsiveness of spleen cells to LPS was significantly enhanced (TABLE 8). No effect was present in PHA-, Con A- or PWM-stimulated cultures. As HCB

TABLE 6
IN VIVO PHAGOCYTOSIS (CARBON CLEARANCE) OF RATS FED HCB FOR 3 WEEKS*

Dietary Concentration (mg/kg)	Phagocytic Index (K)
0	0.127 ± 0.062
1000	0.098 ± 0.051

*Means ± S.D. of 6 animals per group. Rats were injected intravenously with 20 mg colloidal carbon per kg body weight. Blood was drawn from the orbital plexus 3 and 10 minutes after injection.

treatment markedly increased the number of nucleated cells in the spleen, all derived values per whole spleen were significantly increased.

The mitogenic response of peripheral blood lymphocytes is shown in TABLE 9. The incorporation of ^3H − TdR into DNA was significantly increased in Con A-stimulated cultures. No effect was seen in unstimulated and PHA-, PWM- or LPS-stimulated cultures.

Susceptibility to Endotoxin

Results of this experiment are given in TABLE 10. Pretreatment of rats with HCB slightly enhanced their susceptibility to endotoxin. A 100% mortality occurred in controls injected with 32 mg of *E.coli* endotoxin, whereas a similar mortality was seen in HCB-exposed rats receiving 8 mg of endotoxin. Remarkably, no clear dose-response relationship was present with regard to the endotoxin susceptibility in HCB-treated animals.

TABLE 7

MITOGENIC RESPONSE OF THYMUS CELLS FROM MALE RATS FED HCB FOR 3 WEEKS*

Dietary Concentration (mg/kg)	Number of Cells (× 10⁸)	Medium	PHA	Con A	PWM
		Responsiveness in cpm (× 10²) per culture			
0		11.8 ± 2.3	52.5 ± 11.6	165 ± 50	71.8 ± 23.2
1000		6.8 ± 2.0‡	82.6 ± 39.6	149 ± 50	57.4 ± 18.0
		Responsiveness in cpm (× 10⁵) per thymus			
0	10.1 ± 3.6	12.1 ± 4.6	51.0 ± 13.5	154 ± 34	67.2 ± 19.6
1000	10.4 ± 3.6	5.5 ± 3.3†	62.5 ± 44.9	112 ± 59	42.4 ± 20.7

*Means ± S.D. of 6 animals per group. 1 × 10^6 viable cells were cultured in triplicate with medium alone, PHA, Con A, or PWM, and responsiveness measured by the incorporation of ^3H-TdR into DNA.

†$p < 0.02$.

‡$p < 0.01$.

DISCUSSION

Regarding the effects of HCB on the immune system, considerable species differences are present. In mice, Loose *et al.*[7] reported no effect on weight and histology of thymus and spleen of HCB-exposed mice. In contrast, in rats splenomegaly has been observed in different studies.[2–4] In addition, at high HCB doses, thymus atrophy was found in this species.[2] In dogs, HCB administration resulted in hyperplasia of lymphoid tissue in the stomach.[5] In the present and in a previous study,[2] changes in the spleen consisted of an increase in size of follicles and of the marginal zone. In addition, an increase in extramedullary hemopoiesis was present in the splenic red pulp, probably as a reaction upon the HCB-induced decreased number of erythrocytes and reduced hemoglobin content.[2,4] Marked weight increases were found of the popliteal and mesenteric lymph nodes. At the low HCB dose, the relative weight of the popliteal lymph node was still two times the weight of the controls.

TABLE 8

MITOGENIC RESPONSE OF SPLEEN CELLS FROM MALE RATS FED HCB FOR 3 WEEKS*

Dietary Concentration (mg/kg)	Number of Cells ($\times 10^8$)	Medium	PHA	Con A	PWM	LPS
		Responsiveness in cpm ($\times 10^2$) per culture				
0		11.3 ± 5.8	387 ± 111	169 ± 48	64.7 ± 22.3	88 ± 25
1000		19.7 ± 9.5	415 ± 81	159 ± 45	67.0 ± 14.3	135 ± 48†
		Responsiveness in cpm ($\times 10^5$) per spleen				
0	2.4 ± 1.0	5.0 ± 2.0	176 ± 52	77 ± 25	28.3 ± 6.8	42 ± 19
1000	4.0 ± 0.8§	17.1 ± 10.0‡	332 ± 95§	127 ± 47†	56.1 ± 22.6‡	113 ± 63‡

*Means ± S.D. of 5–6 animals per group. 5×10^5 viable cells were cultured in triplicate with medium alone, PHA, Con A, PWM or LPS, and responsiveness measured by the incorporation of ^{3}H-TdR into DNA.

†$p < 0.05$.

‡$p < 0.02$.

§$p < 0.01$.

TABLE 9

MITOGENIC RESPONSE OF PERIPHERAL BLOOD LYMPHOCYTES FROM MALE RATS FED HCB FOR 3 WEEKS*

Dietary Concentration (mg/kg)	Responsiveness in CPM ($\times 10^2$) per Culture				
	Medium	PHA	Con A	PWM	LPS
0	5.1 ± 3.0	121 ± 64	97 ± 31	54.8 ± 19.6	8.6 ± 1.7
1000	4.1 ± 1.7	122 ± 99	176 ± 74†	53.8 ± 16.5	6.3 ± 3.9

*Means ± S.D. of 6 animals per group. 1.5×10^5 viable cells were cultured in triplicate with medium alone, PHA, Con A, PWM, or LPS, and responsiveness measured by the incorporation of ^{3}H-TdR into DNA.

†$p < 0.05$.

Histologically, proliferation of high-endothelial venules was observed. In addition, increased numbers of migrating lymphocytes were found infiltrating the walls of high endothelial venules. Identical changes are described by Anderson *et al.*[17] in stimulated lymph nodes (regional nodes draining skin allografts). The number of peripheral lymphocytes was slightly increased in the present 3-week study. Marked increases in absolute lymphocyte numbers were found in previous studies in which rats were exposed for 3–4 months.[2,4]

The above discussed effects of HCB on the immune system of the rat, with the exception of thymus atrophy observed at high dose levels, point towards a stimulatory effect of HCB on the immune system. As shown in the present study, a stimulation of the humoral immune response did occur. Total IgM serum levels were markedly increased. In addition, IgM and IgG antibodies to tetanus toxoid, a thymus-dependent antigen, were increased both during the primary and secondary immune response. Remarkably, no difference was found in the antibody response against the thymus-independent antigen LPS. An increased response could be expected as HCB markedly enhanced DNA synthesis of spleen cells when cultured in the presence of LPS. Although less pronounced, the incorporation of ^{3}H-TdR into DNA, as expressed on a whole organ basis, was also increased by the T-mitogens PHA and Con A and the T-B mitogen pokeweed (PWM). Finally, the response of peripheral lymphocytes to Con A was enhanced in HCB-exposed rats. Regarding the different *in vivo* function tests of the cell-mediated immunity, no significant effect of HCB was found on incidence of mortality to a *Listeria monocytogenes* infection, on rejection of skin transplants, and on the delayed type hypersensitivity to tuberculin.

The results of the present experiments in rats strongly contrast with the findings of

TABLE 10

INCIDENCE OF MORTALITY AFTER INTRAVENOUS INJECTION OF ENDOTOXIN IN HCB-EXPOSED MALE RATS*

Dietary Concentration (mg/kg)	Milligram of Endotoxin Injected						
	32	16	8	4	2	1	0.5
0	4/4	1/4	0/4	0/4	0/4		
1000			4/4	1/4	1/4	0/4	1/4

*Rats were injected with endotoxin (LPS, *E. coli,* 0127: B8) on day 21. Mortality scored after 48 hours.

Loose, *et al.*[7] in mice, in which total serum IgG, IgM, and IgA were reduced. In addition, the primary and secondary antibody response against sheep red blood cells (thymus-dependent antigen) was reduced as measured by the splenic direct plaque-forming cell (PFC) method. Moreover, the susceptibility to an infection with *Plasmodium berghei* was enhanced in mice;[8] resistance to this type of infection depends mainly on the cell-mediated immunity.[18]

In the present and in a previous study,[2] HCB increased the number of circulating basophils. In general, basophilic granulocytes (and mast cells) are related to immediate-type hypersensitivity reactions. As such, a study to determine the effect of HCB on an allergic reaction, e.g. by means of the passive cutaneous anaphylaxis (PCA) test, would be of interest. Also, it would be worthwhile to investigate if HCB itself is immunogenic.

HCB did increase the susceptibility of rats to endotoxin, but this increase was marginal. Mice exposed to HCB were 20-fold more susceptible to the lethal effects of endotoxin.[8] Regarding the mechanism of increased susceptibility to endotoxin, several possibilities exist: e.g., impairment of endotoxin inactivation by liver parenchymal cells or by macrophages.[19]

No significant decrease was found in the present study in the blood clearance of carbon particles by HCB-exposed rats. But, HCB seems to interact in some way with the mononuclear phagocyte system, as the number of circulating monocytes (as precursors of macrophages) was increased in the present and in an earlier study.[2] This probably represents recruitment to certain tissues (e.g., the lung), since increased numbers of macrophages were observed in the lung of rats.[2,3] The Fc receptor activity of lung macrophages was inhibited by HCB, possibly as a result of a shift in the lung population of macrophages, or by a reduced activity of the Fc receptor[6] (a receptor that binds IgG either free or in the form of antigen-antibody complexes, through the Fc fragment of the immunoglobulin molecule).

Summary

Rats were fed diets containing 0, 500, 1000, and 2000 mg HCB/kg during a 3-week period. Marked weight increases of spleen, popliteal and mesenteric lymph nodes and of the liver were found. Histologically, the white pulp in the spleen was enlarged because of an increase in size of marginal zones and follicles. In addition, there was an increase of extramedullary hemopoiesis. In the lymph nodes, the number of high endothelial venules was increased at all dose levels. The number of neutrophils, basophils and monocytes in the peripheral blood was significantly increased, whereas peripheral lymphocyte counts were slightly higher. Total serum IgM levels were markedly increased, but IgG concentrations were unaltered. On the basis of this experiment, the 1000 mg HCB/kg diet level was chosen for the different function studies that were carried out after a 3-weeks dietary regimen. Regarding the humoral immunity, IgM antibodies to LPS were unaltered, whereas primary and secondary IgM and IgG antibody titers to tetanus toxoid were increased approximately three-fold. HCB did not significantly alter the cell-mediated immunity, as shown by the following parameters: resistance to *Listeria monocytogenes* infection, rejection of skin transplants, and delayed-type hypersensitivity to tuberculin. The phagocytizing capacity of macrophages was studied by measuring the blood clearance of carbon particles. HCB did slightly depress the phagocytic index, but the difference with control animals was statistically not significant. The *in vitro* responsiveness of thymus cells to the mitogens PHA, Con A, and PWM was not changed by *in vivo* HCB-treatment. On a cell-for-cell basis, the responsiveness of spleen cells was increased when cultured in the presence of LPS. On a whole organ basis, the response to PHA, Con A, PWM, and LPS was markedly enhanced because of an increase in the number of nucleated spleen cells. Regarding peripheral lymphocytes, only the response to the mitogen Con A was higher. On the basis of these studies it is concluded that HCB stimulates the humoral immune response in the rat, enhances the *in vitro* responsiveness of spleen cells to the different mitogens mainly as a result of an increase in the number of splenic lymphocytes, but does not alter the cell-mediated immunity as shown with *in vivo* tests. This result contrasts with data in the literature that show that HCB suppresses the humoral and cell-mediated immunity in mice. Finally, HCB pretreatment only marginally increased the susceptibility of rats to endotoxin, whereas mice have been shown to be 20-fold more susceptible to the lethal effects of bacterial endotoxin.

REFERENCES

1. VOS, J. G. 1977. Immune suppression as related to toxicology. CRC Crit. Rev. Toxicol. **5:** 67-101
2. SPEIJERS, G. J. A., D. C. VILLENEUVE, E. M. DEN TONKELAAR, J. G. VOS, M. J. VAN LOGTEN, P. A. GREVE & G. J. VAN ESCH. Combined effect of food deprivation and hexachlorobenzene feeding in female rats. In preparation.
3. KIMBROUGH, R. D. & R. E. LINDER. 1974. The toxicity of technical hexachlorobenzene in the Sherman strain rat. A preliminary study. Res. Commun. Chem. Pathol. Pharmacol. **8:** 653–664.
4. KUIPER-GOODMAN, T., D. L. GRANT, C. A. MOODIE, G. O. KORSRUD & I. C. MUNRO, 1977. Subacute toxicity of hexachlorobenzene in the rat. Toxicol. Appl. Pharmacol. **40:** 529–549.
5. GRALLA, E. J., R. W. FLEISCHMAN, Y. K. LUTHRA, M. HAGOPIAN, J. R. BAKER, H. ESBER & W. MARCUS. 1977. Toxic effects of hexachlorobenzene after daily administration to beagle dogs for one year. Toxicol. Appl. Pharmacol. **40:** 227–239.
6. ZIPRIN, R. L. & Z. F. FOWLER. 1977. Rosette-forming ability of alveolar macrophages from rat lung: inhibition by hexachlorobenzene. Toxicol. Appl. Pharmacol. **39:** 105–109.
7. LOOSE, L. D., K. A. PITTMAN, K. F. BENITZ & J. B. SILKWORTH. 1977. Polychlorinated biphenyl and hexachlorobenzene induced humoral immunosuppression. J. Reticuloendoth. Soc. **22:** 253–271.
8. LOOSE, L. D., J. B. SILKWORTH & K. A. PITTMAN, K. F. BENITZ & W. MUELLER. 1978. Impaired host resistance to endotoxin and malaria in polychlorinated biphenyl- and hexachlorobenzene-treated mice. Infect. Immunity **20:** 30–35.
9. VOS, J. G., F. M. H. DEBETS, J. BUYS, P. BEEKHOF & A. M. HAGENAARS. Quantification of rat immunoglobulins M and G by mechanized enzyme-linked immunosorbent assay (ELISA). In preparation.
10. VOS, J. G., J. BUYS, P. BEEKHOF & A. M. HAGENAARS. 1979. Quantification of total IgM and IgG and specific IgM and IgG to a thymus-independent (LPS) and a thymus-dependent (tetanus toxoid) antigen in the rat by enzyme-linked immunosorbent assay (ELISA). Ann. N.Y. Acad. Sci. **320:** 518–534. This volume.
11. VOS, J. G., J. BUYS, J. G. HANSTEDE & A. M. HAGENAARS. Comparison of ELISA and passive haemagglutination method for quantification of antibodies to LPS and tetanus toxoid in the rat. Submitted for publication.
12. RUITENBERG, E. J. & L. M. VAN NOORLE JANSEN. 1975. Effect of *Corynebacterium parvum* on the course of a *Listeria monocytogenes* infection in normal and congenitally athymic (nude) mice. Zentralbl. Bakteriol. Parasitenkd. Infektionskr. Hyg., Abt. 1 Orig. Reihe A **231:** 197–205
13. BAILEY, D. W. & B. USAMA. 1960. A rapid method of grafting skin on tails of mice. Transplant. Bull. **7:** 424–425.
14. VOS, J. G., J. A. MOORE & J. G. ZINKL. 1973. Effect of 2,3,7,8-tetrachlorodibenzo-

p-dioxin on the immune system of laboratory animals. Environ. Health Perspect. **5:** 149–162.

15. STUART, A. E., J. A. HABESHAW & A. E. DAVIDSON. 1973. Phagocytes *in vitro*. *In* Handbook of Experimental Immunology, Chapter 24, 2nd edition. D. M. WEIR, Ed. Blackwell Scientific. Oxford.
16. DONALD, K. J. & R. J. TENNENT. 1975. The relative roles of platelets and macrophages in clearing particles from the blood: the value of carbon clearance as a measure of reticuloendothelial phagocytosis. J. Pathol. **117:** 235–245.
17. ANDERSON, N. D., A. O. ANDERSON & R. G. WYLLIE. 1975. Microvascular changes in lymph nodes draining skin allografts. Am. J. Pathol. **81:** 131–154.
18. POELS, L. G. 1977. Release of antigens and immunogenicity versus the host response during infection with a rodent malaria parasite. Thesis, University of Nijmegen, The Netherlands.
19. COOK, J. A., N. R. DILUZIO & E. O. HOFFMAN. 1975. Factors modifying susceptibility to bacterial endotoxin: the effect of lead and cadmium. CRC Crit. Rev. Toxicol. **3:** 201–229

IMMUNOLOGIC EFFECTS OF VINYL CHLORIDE IN MICE

R. P. Sharma

Toxicology Program
Utah State University
Logan, Utah 84322

P. J. Gehring

Toxicology Research Laboratory
The Dow Chemical Co.
Midland, Michigan 48640

Exposure to vinyl chloride (VC) has been associated with various disease syndromes in man and animals. In workers exposed to high levels of VC, hepatic fibrosis associated with or leading to portal hypertension and angiosarcoma of the liver has been described.[1] Vinyl chloride disease has also been associated with dermal induration, acro-osteolysis, splenomegaly and thrombocytopenia.[2,3] In animals, VC exposure has been reported to cause several types of tumors.[4]

In a recent report, Ward and coworkers[5] suggested the presence of aberrant immune complexes in workers exposed to vinyl chloride. The incidence and the level of these complexes were correlated with the level of vinyl chloride exposure as well as other manifestations of toxicity. An increase in immunoglobulins and a corresponding reduction in peripheral T-lymphocytes was suggested. In addition, the presence of circulating immune complexes (inferred by the presence of mixed cryoglobulins and decreased values of complement) and autoantibodies against various tissue components were demonstrated in exposed workers. On the basis of their findings, Ward *et al.*[5] suggested that the vinyl chloride disease may be an immune complex disorder and the immune response may be initiated by the adsorption of vinyl chloride or its metabolite on the tissue or plasma proteins.

In another related report, Page *et al.*[6] demonstrated elevated levels of carcinoembryonic antigen (CEA) in the plasma of workers exposed to vinyl chloride. Lange *et al.*[3] have reported both increases and decreases of different fractions of immunoglobulins in 10 patients suffering from VC disorders and on the basis of their findings excluded the possibility of an autoimmune phenomenon in this syndrome.

On the basis of these contradictory reports, we investigated the effects of vinyl chloride exposure on the immune system of mice. For the evaluation of immune reactivity in individuals and animals, the response of lymphocytes in culture, particularly blast formation in the presence of antigenic substances, has been suggested as a useful index.[7] Sensitized lymphocytes have been shown to play a very significant role in the production of autoimmune diseases.[8] The findings that T- and B-lymphocytes can be selectively transformed *in vitro* by different lectins[9–11] have provided a working model to evaluate the responsiveness of these lymphocytes in cell cultures. Our results, described in this report, suggest that VC exposure caused stimulation of lymphocyte transformation in cultures. Direct exposure of splenic cultures to VC *in vitro* did not have such an effect. Metabolite(s) of VC rather than the parent chemical itself appear to be involved in the process of this stimulation.

0077–8923/0320–0551 $01.75/0 © 1979, NYAS

METHODS

Vinyl chloride monomer (obtained from Matheson Gas Products, Joliet, Ill.) of 99.9% minimum purity was used for exposure of mice. The exposures were conducted in large (6′ × 6′ × 8′) inhalation chambers. The concentration of VC was controlled to provide atmospheric concentrations of 0 (control), 10, 100, and 1000 ppm, 6 hr/day, 5 days/week, for durations of up to 8 weeks. The time-weighted averages of vinyl chloride concentrations in the chamber, as determined by a gas chromatograph are described in TABLE 1.

Male, CD-1 mice (approximately 7 weeks old and weighing 35 g each at the beginning of exposure) were obtained from Charles River (Wilmington, Mass.) and acclimated for 2 weeks with the surroundings. The animals were housed 4 to a cage, identified by ear punch, and were allowed free access to food and water except during the exposure period. No food or water was allowed during the exposure. The animals were weighed once a week. Groups of mice (4 animals/group/VC level) were sacrificed at 2, 4, and 8 weeks of exposure and blood samples obtained. Liver, kidney, spleen, and thymus were weighed and these organs, except the spleen, were fixed and processed for histologic examination. The spleen was saved in sterilized saline for lymphocyte cultures.

TABLE 1

EXPOSURE CHAMBER ANALYSES FOR VINYL CHLORIDE*

	VC Concentration, ppm		
Desired Concentration	10	100	1000
Concentration determined (means ± S.D.)	9.9 ± 0.64	101.3 ± 4.8	982.8 ± 34.9
Percent of exposure days within 10% of mean analytic concentration	87	98	100
Number of analyses (days × analytic determinations per day)	45 × 4	45 × 4	45 × 4

*VC concentrations were determined on a Varian 2400 gas chromatograph equipped with a flame ionization detector. The column used was 3′ × ⅛″ stainless steel, Porpak QS, 80/100 mesh, maintained at a temperature of 120°C (injector and detector at 260°C and 225°C, respectively). Carrier gas was He, 30 ml/min, and samples of 1 ml analyzed in several replicates and compared with an appropriate standard curve prepared simultaneously.

Hematology and Serum Immunoelectrophoresis

Total erythrocyte and leukocyte counts and differential leukocyte counts were performed on all blood samples by routine methods. The relative concentrations of total Ig were determined by immunoelectrophoretic techniques modified from Bjerrum *et al.*[12]

Splenic Lymphocyte Cultures

The cultures were performed as described elsewhere in detail.[13] The spleens were washed with sterilized saline and mashed in saline with a forceps. Subsequently, the agglutinated cells and large particles were removed by passing the cells through a series of hypodermic needles and the resulting cell suspension was transferred to a glass tube. After separating the cells by centrifugation, the cells were suspended in culture medium (RPMI-1640 supplemented with glutamine, penicillin, and streptomycin), an aliquot counted electronically, and the concentration of cells adjusted to approximately 8×10^6 cells/ml. Aliquots of this suspension were transferred to microculture plates, more medium with or without mitogenic lectins was added, and the system incubated at 37°C in humidified air containing 5% carbon dioxide. Optimal concentrations of phytohemagglutinin (PHA) and pokeweed mitogen (PWM) were used. After 50–52 hours of incubation, 0.5 μCi of ^{3}H-thymidine (New England Nuclear, Boston, Mass.) was added to each culture and incubation continued for an additional 16 hours. The cells were then harvested and the radioactivity counted using a liquid scintillation spectrometer. Cells from each animal were cultured in triplicate and the three values were averaged after corrections for quenching. Stimulation indices for PHA and PWM were determined for each spleen by dividing the disintegrations in the presence of the mitogen by those in the system containing no mitogen.

When splenic lymphocyte cultures were incubated in the presence of VC, these were held in specially designed humidified glass containers which were well purged with either 5% CO_2 in air or with 1000 ppm VC in 5% CO_2 in air mixture. These containers were kept in a constant temperature incubator at 37°C for the incubation as described above.

Alteration of VC Metabolism

Groups of mice were injected i.p. with either saline, phenobarbital sodium (100 mg/kg), or ethanol (3.2 g/kg as 30% solution) 30 minutes prior to VC exposure. The concentration of phenobarbital and ethanol were adjusted so that all animals (including those receiving saline) were given 10 ml solution/kg (0.3 ml/30 g mouse). These were divided into equal subgroups and were exposed either to normal air atmosphere or to 1000 ppm vinyl chloride in 1 meter^3 inhalation chambers for 5 days. Two days after the last exposure, their spleens were obtained and cultured as mentioned above.

Statistical

The results are expressed as means and their standard errors. Comparisons were made with the respective control group using a t-test[14] and a predetermined value of $p = 0.05$.

RESULTS

In mice, vinyl chloride exposure for 8 weeks produced no consistent effects on growth, or weights of liver, kidney or thymus (TABLE 2). Spleen weights were increased at the highest level of exposure, 1000 ppm. This effect was noticed after 2 weeks of exposure and persisted throughout the exposure. A slight reduction in liver/body weight ratios after 8 weeks of exposure to 1000 ppm VC may be attributable to greater body weights in mice as compared to controls. The body weights of mice exposed to 1000 ppm VC were consistently higher than controls after

TABLE 2

ORGAN AND BODY WEIGHTS OF MICE EXPOSED TO VINYL CHLORIDE FOR DIFFERENT DURATIONS

			Organ Weights and Their Relation to Body Weight‡							
Duration of Exposure	Level of Exposure (ppm)	Body Weight (g)	Thymus (g)	Thymus (g/100g)	Spleen (g)	Spleen (g/100g)	Liver (g)	Liver (g/100g)	Kidney (g)	Kidney (g/100g)
Preexposure	035 ± 20.05 ± 0.010.14 ± 0.010.10 ± 0.01 0.27 ± 0.01 1.91 ± 0.07 5.42 ± 0.28 0.54 ± 0.051.53 ± 0.20									
2 weeks	0	36 ± 3	0.04 ± 0.02	0.12 ± 0.03	0.09 ± 0.01	0.24 ± 0.00	1.98 ± 0.30	5.46 ± 0.46	0.59 ± 0.09	1.64 ± 0.18
	10	35 ± 5	0.05 ± 0.01	0.14 ± 0.03	0.08 ± 0.02	0.24 ± 0.03	1.93 ± 0.41	5.48 ± 0.47	0.58 ± 0.09	1.67 ± 0.20
	100	36 ± 2	0.04 ± 0.01	0.12 ± 0.02	0.09 ± 0.02	0.23 ± 0.04	1.99 ± 0.23	5.49 ± 0.35	0.57 ± 0.06	1.58 ± 0.16
	1000	35 ± 4	0.04 ± 0.01	0.10 ± 0.02	0.13 ± 0.01*	0.38 ± 0.07*	2.09 ± 0.29	5.92 ± 0.29	0.64 ± 0.08	1.83 ± 0.05
4 weeks	0	34 ± 2	0.05 ± 0.01	0.15 ± 0.04	0.07 ± 0.02	0.21 ± 0.05	1.75 ± 0.15	5.15 ± 0.23	0.55 ± 0.07	1.63 ± 0.21
	10	38 ± 4	0.05 ± 0.02	0.14 ± 0.03	0.10 ± 0.03	0.25 ± 0.05	2.29 ± 0.22*	5.98 ± 0.45*	0.66 ± 0.08	1.74 ± 0.29
	100	35 ± 1	0.05 ± 0.01	0.13 ± 0.04	0.08 ± 0.01	0.22 ± 0.02	2.06 ± 0.19	5.92 ± 0.45*	0.58 ± 0.06	1.66 ± 0.15
	1000	37 ± 2	0.05 ± 0.01	0.14 ± 0.03	0.11 ± 0.01*	0.30 ± 0.04*	2.11 ± 0.11*	5.69 ± 0.29	0.64 ± 0.05	1.73 ± 0.12
8 weeks	0	38 ± 4	0.04 ± 0.01	0.11 ± 0.03	0.09 ± 0.03	0.23 ± 0.06	2.09 ± 0.33	5.57 ± 0.37	0.67 ± 0.11	1.78 ± 0.18
	10	42 ± 2	0.03 ± 0.01	0.07 ± 0.02	0.10 ± 0.01	0.23 ± 0.03	2.46 ± 0.21	5.90 ± 0.42	0.76 ± 0.04	1.83 ± 0.10
	100	42 ± 3	0.05 ± 0.01	0.13 ± 0.04	0.10 ± 0.01	0.23 ± 0.05	2.30 ± 0.08	5.51 ± 0.30	0.70 ± 0.06	1.67 ± 0.12
	1000	42 ± 1	0.05 ± 0.02	0.11 ± 0.05	0.11 ± 0.01	0.26 ± 0.03	2.02 ± 0.15	4.82 ± 0.26*	0.64 ± 0.06	1.53 ± 0.14

‡Mean ± S.D. of 4 animals per group.

*Indicates significant deviation from control ($p < .05$, Dunnett's test).

4 weeks of exposure when the body weights of all animals in that group were considered (data not shown in the table). No differences were seen in the hematologic parameters of any group exposed to VC (data not shown). At the time of sacrifice, there were no gross lesions observed in any of the organs that can be related to the treatment. Subsequent histologic examination of selected tissues revealed no differences.

Vinyl chloride exposure caused stimulation of spontaneous lymphocyte transformation in mouse splenic cultures. This effect was observed at the highest exposure level of 1000 ppm of VC after 2 weeks of exposure. After 4 and 8 weeks of exposure, an increased transformation of lymphocytes prepared from the exposed animals was observed at all levels of VC (FIGURES 1–3). The stimulation indices for PHA were

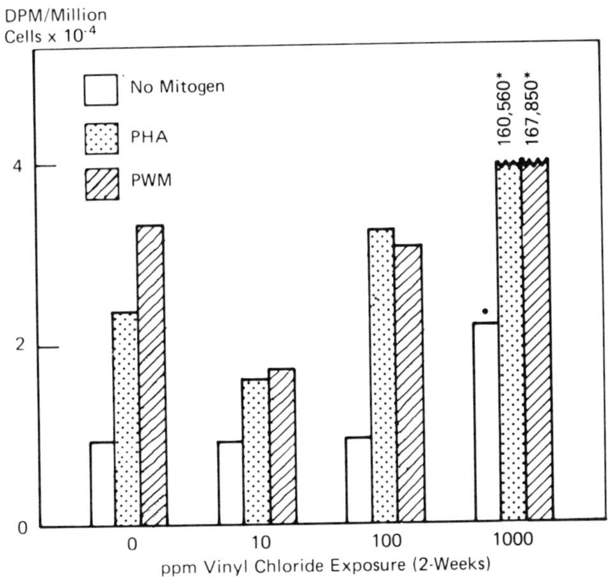

FIGURE 1. Thymidine uptake by lymphocyte cultures prepared from mice exposed for 2 weeks to vinyl chloride. Average of 4 animals, the cells cultured in triplicate. Values significantly different from controls are marked with an asterisk ($p < 0.05$).

significantly increased in mice exposed to 1000 ppm VC for 2 weeks (FIGURE 4). At subsequent time intervals, however, there was no definite dose-response, since in the mice exposed to 1000 ppm VC the indices were lower than for those exposed to the two lower levels. The stimulation indices at these points were, however, consistently greater than those observed for controls (FIGURES 5 and 6).

For all groups of mice, the DNA synthesis by lymphocytes in culture decreased with time. Such findings were consistent with those generally observed in growing animals. This was consistent in all exposure groups including control, the latter showing a considerable drop in the stimulating indices by PHA and PWM.

TABLE 3 shows the relative amounts of immunoglobulins in mouse serum at different intervals of VC exposure. After 2 and 4 weeks of exposure, an increase was

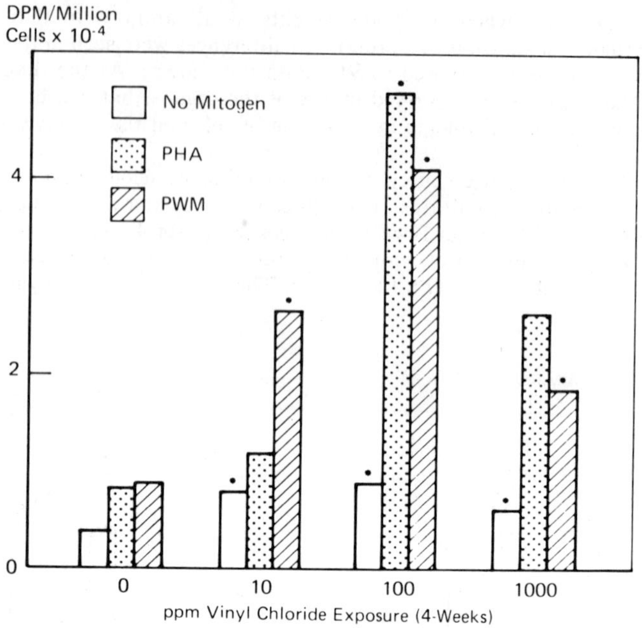

FIGURE 2. Thymidine uptake by lymphocyte cultures prepared from mice exposed for 4 weeks to vinyl chloride. Average of 4 animals, the cells cultured in triplicate. *Indicates value significantly different from respective control ($p < 0.05$).

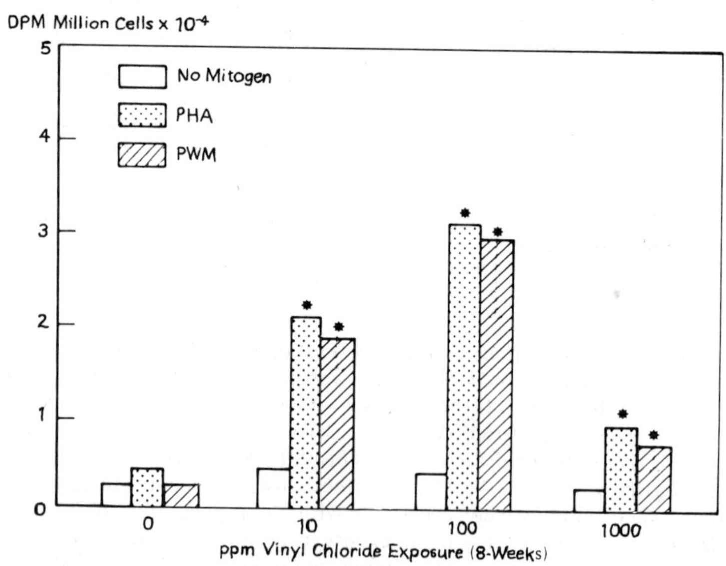

FIGURE 3. Thymidine uptake by lymphocyte cultures prepared from mice exposed for 8 weeks to vinyl chloride. Average of 4 animals, the cultures performed in triplicate. *Indicates value significantly different from control at $p < 0.05$.

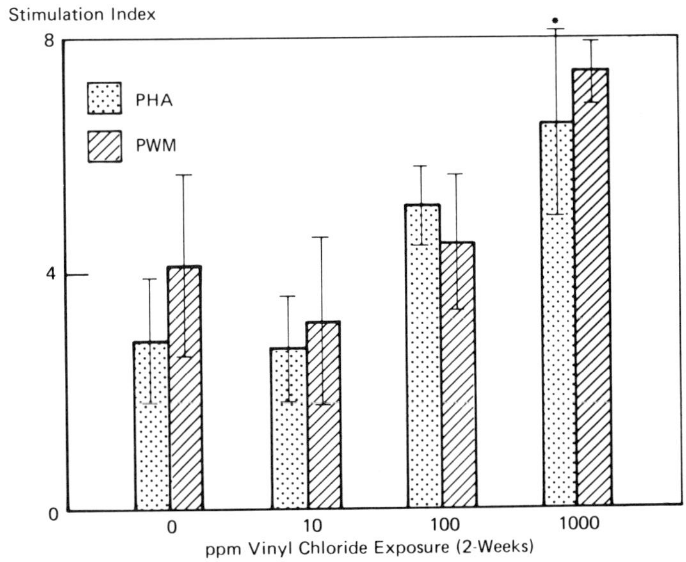

FIGURE 4. Response of mouse splenic lymphocytes to PHA and PWM after 2-week exposure to vinyl chloride. All values are average of 4 animals, the cells cultured in triplicate. Vertical bars represent standard error of mean and * indicates value significantly different from control ($p < 0.05$). Stimulation index = dpm in the presence of a mitogen divided by the dpm without any mitogen.

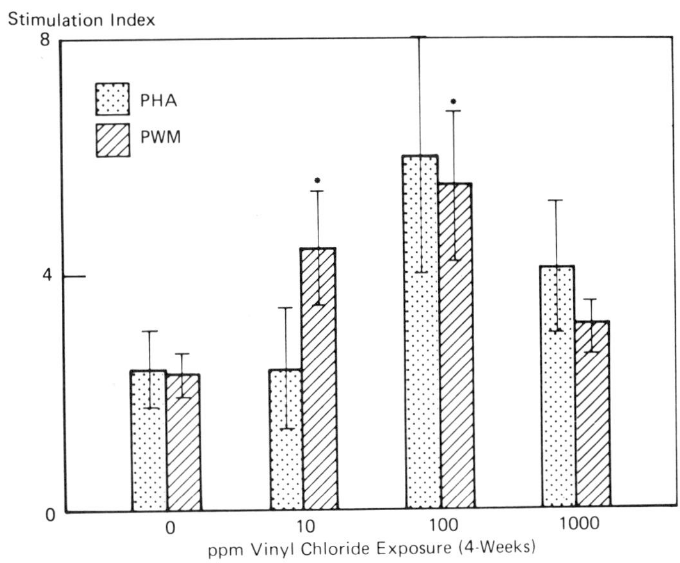

FIGURE 5. Response of mouse splenic lymphocytes to PHA and PWM after 4-week exposure to vinyl chloride. All values are average of 4 animals, the cells cultured in triplicate. Vertical bars represent standard error of mean and * indicates value significantly different from control at $p < 0.05$. Stimulation index = dpm in the presence of a mitogen divided by the dpm without any mitogen.

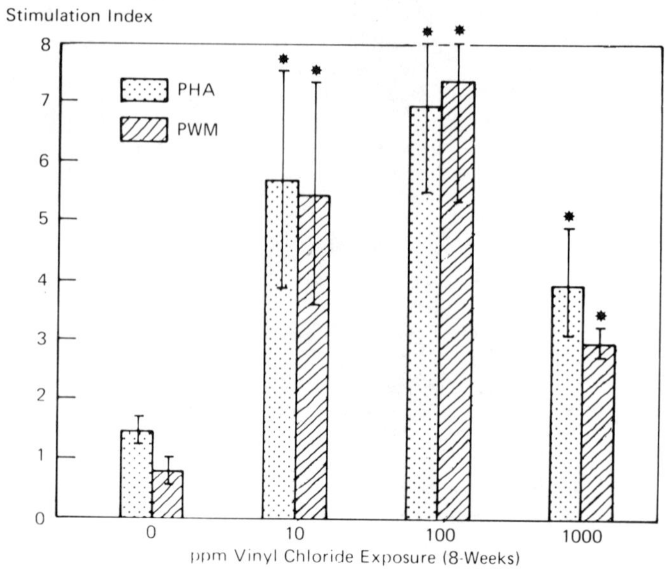

FIGURE 6. Response of mouse splenic lymphocytes to PHA and PWM after 8-week exposure to vinyl chloride. All values are average of 4 animals, the cells cultured in triplicate. Vertical bars represent standard error of mean and * indicates value significantly different from control at $p < 0.05$. Stimulation index = dpm in the presence of a mitogen divided by dpm without any mitogen.

noticed in VC-exposed animals as compared to the respective control group. A significant increase was noted in 10 ppm VC level at 4 weeks of exposure.

The presence of vinyl chloride *in vitro* in the incubating atmosphere of lymphocyte cultures had no stimulating effect on mouse splenic lymphocyte transformation. The results of one such experiment are shown in TABLE 4. When animals were exposed *in vivo* to 1000 ppm of VC for a period of 8 weeks, the lymphocytes obtained from their spleens had an increased spontaneous transformation when incubated in normal (CO_2:air) atmosphere, as compared to control animals. The presence of 1000 ppm VC in the incubation environment, in fact, slightly decreased the blast formation of lymphocytes enhanced by exposure of mice to 1000 ppm VC for 8 weeks and the

TABLE 3

SERUM IMMUNOGLOBULIN IN MICE EXPOSED TO VINYL CHLORIDE

Vinyl Chloride Concentration ppm	Preexposure	Serum Ig at* 2 weeks	4 weeks	8 weeks
0	0.8 ± 0.1	1.2 ± 0.1	0.9 ± 0.1	2.1 ± 0.2
10	—	1.2 ± 0.1	1.6 ± 0.3**	1.2 ± 0.2**
100	—	1.2 ± 0.1	1.2 ± 0.2	1.7 ± 0.3
1000	—	1.8 ± 0.5	1.4 ± 0.3	1.7 ± 0.1**

*Values indicated are the ratios of total serum Ig to that in normal pooled mouse serum. Mean ± S.E. of 4 animals per group.

**Indicates significant deviation from control at $p < .05$.

values were comparable to those obtained for lymphocytes cultures prepared from mice with no VC exposure. The stimulating indices by PHA and PWM were likewise unaltered.

Experiments similar to those shown in TABLE 4 were also conducted in the presence of hepatic metabolizing systems. Addition of liver supernatant supplemented with necessary cofactors (required for microsomal oxidation system)[15] inhibited the mitotic activity of lymphocytes considerably. Approximately 30% of normal thymidine uptake was seen in the presence of such a metabolizing system and the results were inconclusive (data not shown). The inhibitory effect of VC *in vitro*, as shown in TABLE 4, for splenic lymphocytes from VC-exposed animals, however, was not observed under such conditions.

Culturing of lymphocytes from animals exposed to vinyl chloride and treated simultaneously with chemicals to modify its metabolism provided interesting results. Administration of phenobarbital or ethanol alone to mice increased the transformation rate of their splenic lymphocytes. To handle this problem, respective controls that

TABLE 4

LYMPHOCYTE TRANSFORMATION IN MOUSE SPLENIC CULTURES FROM ANIMALS EXPOSED TO VINYL CHLORIDE AND CULTURES INCUBATED IN THE PRESENCE OF VINYL CHLORIDE (VC) ATMOSPHERE*

Animal Group	Incubation Atmosphere	DPM/10^6 Cells (without mitogen)	Stimulation Index with PHA	Stimulation Index with PWM
Control	Normal	6699 ± 1473	8.8 ± 1.5	7.6 ± 1.1
	VC	5080 ± 1804	11.3 ± 2.2	8.2 ± 1.6
VC-exposed	Normal	10262 ± 1107	8.5 ± 3.2	10.8 ± 2.0
	VC	6385 ± 822	9.1 ± 3.3	9.3 ± 0.6

*Splenic cells from either control or vinyl chloride-exposed (1000 ppm, 6 hrs/day, 5 days/wk, for 8 weeks) mice were cultured in appropriate chambers containing either 5% CO_2 in air or 1000 ppm vinyl chloride in CO_2:air atmosphere. The values are mean ± S.E. for 3 animals, each culture run in quadruplicate.

were given these chemicals were exposed to room air simultaneously in control chambers. Saline treated controls were also used in both VC-exposed and unexposed groups. Since VC exposure increases the lymphocyte transformation in culture, the differences of ^{3}H-thymidine uptake in VC-exposed animals and nonexposed animals were calculated for each drug treatment. The results are shown in TABLE 5. As established previously, both spontaneous and mitogen-induced thymidine uptake were considerably increased by exposing animals to vinyl chloride. Administration of 80 mg/kg/day phenobarbital prior to VC exposure reduced this increment in lymphocyte transformation. Administration of 3.2 g/kg of ethanol prior to VC exposure decreased markedly the stimulating effect of VC exposure to an extent that when the cultures were incubated in the presence of PHA and PWM, their transformation rate, as indicated by tritiated thymidine uptake, was lower than that for animals treated with ethanol and not exposed to VC. This lower transformation rate is the reason for the last two negative values presented in TABLE 5.

DISCUSSION

The results reported here suggest that vinyl chloride stimulates the immune responsiveness of mice. The enhanced responsiveness develops rapidly upon exposure to high levels, whereas 4 weeks of repeated exposure to low levels (10 ppm) are needed to elicit a discernible response. Blast formation of lymphocytes in splenic cell cultures, as evidenced by an increased incorporation of tritiated thymidine into cellular materials was the most sensitive indication of increased immune responsiveness. Vinyl chloride exposure at all levels increased the spontaneous lymphocyte transformation in splenic cell cultures, the effect being similar to that produced by injection of antigenic substances. The stimulating effect of two plant lectins, phytohemagglutinin and pokeweed mitogen, was also increased, further indicating an enhanced immune reactivity in these VC-exposed animals.

The stimulating effect of vinyl chloride exposure on immune responsiveness does not appear to be a nonspecific phenomenon. It is unlikely that this phenomenon is stress induced since stress is generally known to cause immunosuppression. The decrease in lymphocyte transformation in splenic cultures observed in the control group of mice with increasing time interval appears to be consistent with stress-induced immune suppression, since the exposure environment may provide some stress.

The immunostimulatory effect of VC was not always dose-related. After a 2-week

TABLE 5

CHANGE IN LYMPHOCYTE TRANSFORMATION IN MICE SPLENIC CULTURES PRODUCED BY THE INHALATION OF 1000 ppm VINYL CHLORIDE FOR ONE WEEK AND IN MICE TREATED WITH PHENOBARBITAL AND ETHANOL TO ALTER THE METABOLISM OF VINYL CHLORIDE

Animals Treated with	Δ DPM/10^6 Cells* in the presence of No mitogen	PHA	PWM
Saline	43,113	121,976	167,844
Phenobarbital	27,770	26,147	55,478
Ethanol	4,929	−147,786	−78,668

*(DPM in animals exposed to vinyl chloride)–(DPM in animals held in control chamber). Each value is obtained from an average of 4 animals and cultures of each animal run in triplicate. Negative values shown in last two columns were obtained because ethanol-treated mice had lower counts in those treated with ethanol and also exposed to vinyl chloride.

exposure period only mice exposed to 1000 ppm VC showed such an effect. After 4 and 8 weeks of exposure the effect was seen at all levels of VC exposure. However, the effect in mice exposed to 1000 ppm VC for 8 weeks was less pronounced than observed in mice exposed for shorter duration. This suggests that with continued exposure to high levels of VC, the initial increase in immune responsiveness may reverse either as a result of adaption or a direct depressive effect of high levels of VC exposure.

The metabolism of vinyl chloride has been shown to be a saturable process in the body.[16] A dose-related saturable binding of vinyl chloride has also been shown in rat liver. Also, protein alkylating metabolites of vinyl chloride in an *in vitro* liver microsomal system have been demonstrated.[15] In view of these results, it is not surprising that the immunostimulating effect of vinyl chloride is saturable since it too may possibly be mediated subsequent to biotransformation of VC in the body. It has been speculated that chloroethylene oxide is the reactive intermediate in the biotransformation of vinyl chloride. Formation of this alkylating intermediate, although not conclusively shown, has been hypothesized as a possible proximate carcinogen. Alkylation of a normal body protein by such a chemical may produce an antigenic determinant and the resultant immunostimulation. Whether such possibility exists needs to be investigated further.

The immunostimulatory effects of vinyl chloride agreed with the recent reports[5] that the vinyl chloride syndrome may involve an autoimmune phenomenon. The studies reported by these workers included patients with clinical effects of VC exposure. In our animal studies, the effects are shown after a relatively short duration of exposure. In these animal studies, an increased reactivity of lymphocytes, including those of "T" cells occurred since PHA is somewhat selective to mouse T-cells. A stimulation of T-cells has been reported to be an indication of an autoimmune phenomenon,[8] a finding also consistent with the hypothesis proposed by Ward *et al.*[5]

The immune stimulation caused by vinyl chloride exposure may be considered similar to the immune modulation caused by levamisole.[17] This chemical, which was first introduced as an animal anthelmintic, has been tried as a therapeutic adjunct in cancer patients and has been shown to prolong the disease-free interval in patients with lung and breast cancer and malignant melanoma.[18,19] At the same time, this chemical enhances survival of allogeneic tumors in animal models and also in some patients. In the case of vinyl chloride, its carcinogenic property may be distinct from its immune effects, but the latter may enhance the carcinogenic process once initiated even if not actually caused by it.

Other reports involving the immunostimulating properties of different chemicals have recently appeared. Renoux *et al.*[20] reported immunostimulating property of propylene glycol, a common solvent and vehicle for medicinal preparations. Goodman and Weigle[21] have reported a dose-related lymphocyte stimulation by 2-mercaptoethanol and α-thioglycerol. These chemicals have a thiol moiety in common in their chemical structure, a property also shared by the molecule of levamisole.[22] Two major metabolites of vinyl chloride, i.e., thiodiglycolic acid and *N*-acetyl-*S*-(hydroxyethyl)-cysteine, have been described[16] and are a result of glutathione conjugation by this chemical. It is possible that these metabolites may somehow be involved in the immunostimulatory property of vinyl chloride.

Direct evidence that metabolism of vinyl chloride is a determining factor for its lymphocyte-stimulating property is derived when its metabolism is altered. Ethanol is known to decrease the metabolism of VC,[23] and simultaneous administration of ethanol interfered with the immunostimulating property of VC in splenic cultures of exposed mice. Use of phenobarbital also exhibited a similar phenomenon. In studies reported previously, vinyl chloride metabolism was not enhanced by microsomes obtained from phenobarbital-treated animals.[15] Similarly, although the rate of VC metabolism was not altered, its binding in hepatic tissue was increased.[24] This suggests that phenobarbital may induce metabolism through another pathway leading to the formation of a product which more readily binds to macromolecules in liver and thus have little effect on lymphocytes. A slight reduction of lymphocyte-stimulating activity in VC-exposed animals by simultaneous administration of phenobarbital may be explained on this basis.

The suggestion that the epoxide metabolite of VC, chloroethylene oxide, may be responsible for stimulated lymphocyte transformation is also supported by our other studies (unreported previously) with chemicals possessing a similar chemical structure to that of vinyl chloride. A derivative of vinyl chloride, vinyl benzene (styrene monomer), has no effect on lymphocyte transformation when added directly to lymphocyte cultures *in vitro*. Addition of styrene oxide, on the other hand, in concentrations as low as 2.5×10^{-7}M caused a remarkable increase in such mitotic activity of lymphocytes in splenic cultures.

These studies, together with our other reports that vinyl chloride substantially increases lymphocyte mitosis in splenic cultures provide credence to previous reports by Ward *et al.*[5] that vinyl chloride syndrome may involve an immunologic phenomenon. A recent report[25] involving peripheral blood lymphocyte cultures from normal and VC-exposed workers has, although, failed to support this hypothesis. The response of blood lymphocytes in persons exposed occupationally to VC was essentially unaltered.

Summary

Male CD-1 mice were exposed to 10, 100 or 1000 ppm vinyl chloride (VC) for 2–8 weeks @ 6 hr/day, 5 days/week. A slight increase in the spleen weight of mice was noted at the highest exposure level. Spleens were obtained from these animals (4 mice/group after 2, 4, and 8 weeks of exposure) and their lymphocytes cultured *in vitro* with or without the presence of phytomitogens, phytohemagglutinin (PHA) and pokeweed mitogen (PWM). Relative blast formation and the DNA synthesis was measured by the incorporation of ^{3}H-thymidine in the cultured cells. The response of splenic lymphocytes to the phytomitogens was increased several-fold by VC exposure. The effects were apparent at 1000 ppm VC after 2 weeks of exposure and at all levels of VC exposure after 4–8 weeks. The effects were generally more pronounced at 100 ppm VC exposure than those at 1000 ppm. *In vitro* culture of splenic lymphocytes from control or VC-exposed mice in the VC atmosphere did not show an enhancement of blast formation. Alteration of VC metabolism during the VC exposure *in vivo* yielded results that indicated that metabolites of VC may be responsible for the stimulation of lymphocyte transformation observed in splenic cultures.

Acknowledgments

The valuable help of S. J. Gorzinski and H. O. Yakel during this study is gratefully appreciated.

References

1. Berk, P. D., J. F. Martin, R. S. Young, J. Creech, I. J. Selikoff, H. Falk, P. Watanabe, H. Popper & L. Thomas. 1976. Vinyl chloride associated liver disease. Ann. Int. Med. **84:** 717–731.
2. Harris, D. K. & W. G. F. Adams. 1967. Acroosteolysis occurring in men engaged in the polymerization of vinyl chloride. Brit. Med. J. **3:** 712–714.
3. Lange, C. E., S. Juhe, G. Stein & G. Veltman. 1974. The so-called vinyl chloride disease—an occupational systemic sclerosis? Int. Arch. Arbeitsmed. **32:** 1–32.
4. Maltoni, S. & G. Lefemine. 1974. Carcinogenicity bioassays of vinyl chloride, 1. Research plan and early results. Environ. Res. **7:** 387–405.
5. Ward, A. M., S. Udnoon, J. Watkins, A. E. Walker & C. S. Darke. 1976. Immunological mechanisms in the pathogenesis of vinyl chloride disease. Brit. Med. J. **1:** 936–938.
6. Page, M., L. Theriault & F. Delorme. 1976. Elevated CEA levels in polyvinyl chloride workers. Biomedicine **25:** 279.
7. Dutton, R. W. & J. D. Eady. 1964. An *in vitro* system for the study of the mechanism of antigenic stimulation in the secondary response. Immunology **7:** 40–53.
8. Orgad, S. & I. R. Cohen. 1974. Autoimmune encephalomylitis: Activation of thymus lymphocytes against syngeneic brain antigens *in vitro.* Science 8883: 1083–1085.
9. Janossy, G. & M. F. Greaves. 1971. Lymphocyte activation. Response of T- and B-lymphocytes to phytomitogens. Clin. Exp. Immunol. **9:** 483–498.
10. Stockman, G. D., M. T. Gallagher, L. R. Heim, M. N. South & J. J. Trentin. 1971. Differential stimulation of mouse lymphoid cells by phytohemagglutinin and pokeweed mitogen (35410). Proc. Soc. Exp. Biol. Med. **136:** 980–982.

11. ANDERSSON, J., G. MÖLLER & O. SJÖBERG. 1972. Selective induction of DNA synthesis in T- and B-lymphocytes. Cell. Immunol. **4:** 381–393.
12. BJERRUM, O. J., A. INGILD, H. LOWENSTEIN & B. WEEKE. 1973. Carbamylated antibodies used for quantitation of human IgG. A routine method. *In* Quantitative Immunoelectrophoresis N. H. Axelsen, J. Kroll & B. Weeke, Eds pp. 145–148. Universitets Forlaget. Oslo.
13. SHARMA, R. P. & P. J. GEHRING. 1979. Effects of 2, 3, 7, 8-tetrachlorodibenzo-*p*-dioxin (TCDD) on splenic lymphocyte transformation in mice after single and repeated exposures. Ann. N.Y. Acad. Sci. **320:** 487–497. This volume.
14. STEEL, R. G. D. & H. H. TORRIE. 1960. Principles and Procedures of Statistics. McGraw-Hill Book Company, Inc. New York.
15. KAPPUS, H., H. M. BOLT, A. BUCHTER & W. BOLT. 1976. Liver microsomal uptake of ^{14}C vinyl chloride and transformation to protein alkylating metabolites *in vitro*. Toxicol. Appl. Pharmacol. **87:** 461–471.
16. WATANABE, P. G., G. R. MACGOWAN & P. J. GEHRING, 1976. Fate of ^{14}C vinyl chloride after single oral administration in rats. Toxicol. Appl. Pharmacol. **36:** 339–352.
17. RENOUX, G., M. RENOUX, M. N. TELLER, S. MCMAHON & J. M. GUILLAUMIN, 1976. Potentiation of T-cell mediated immunity by levamisole. Clin. Exp. Immunol. **25:** 288–296.
18. ROJAS, A. F., E. MICKIEWICZ & J. N. FEIERSTEIN. 1976. Levamisole in advanced human breast cancer. Lancet **1:** 211–215.
19. WILKINS, S. A. & Z. L. OLKOWSKI, 1977. Immunocompetence of cancer patients treated with levamisole. Cancer **39:** 487–493.
20. RENOUX, M. L., G. DE MONTIS, A. ROCHE, & D. HEMON, 1976. Experimental study of the study of a polyvitaminic preparation on humoral antibody response. Evidence for an immunostimulating activity of propylene glycol. La Nouvelle Presse Medicale **5:** 2053–2056.
21. GOODMAN, M. G. & W. O. WEIGLE, 1977. Lymphocyte stimulation by 2-merceptoethanol (2-ME) and α-thioglycerol (αTG). Federation Proc. **36:** 1288.
22. RENOUX, B. & M. RENOUX, 1977. Roles of imidazole or thiol moiety on the immunostimulant action of levamisole. *In* Control of Neoplasia by Modulation of the Immune System. M. A. Chirigos, Ed. pp. 67–30. Raven Press. New York.
23. HEFFNER, R. E. Jr., P. G. WATANABE & P. J. GEHRING, 1975. Preliminary studies of the fate of inhaled vinyl chloride monomers (VCM) in rats. Ann. N.Y. Acad. Sci. **246:** 135–148.
24. WATANABE, P. G., J. A. ZEMPEL, D. G. PEGG & P. J. GEHRING. 1978. Hepatic macromolecular binding following exposure to vinyl chloride. Toxicol. Appl. Pharmacol. **44:** 571–579.
25. FORTWENGLER, H. P., M. E. DEVER, C. H. TAMBURRO & E. ESPINOSA, 1978. Lymphocyte transformation tests in vinyl chloride (VC) workers. Fed. Proc. **37:** 362

INVESTIGATIONS ON THE EFFECTS OF 2,3,7,8-TETRACHLORODIBENZO-*P*-DIOXIN (TCDD) ON PARAMETERS OF VARIOUS IMMUNE FUNCTIONS

Robert E. Faith* and Michael I. Luster

Environmental Biology and Chemistry Branch
National Institute of Environmental Health Sciences
Research Triangle Park, North Carolina 27709

Introduction

As interest in the biologic effects of environmental pollutants grows it becomes increasingly evident that a number of these chemicals can cause alterations in the immune function of an exposed subject. Of the chemicals of environmental concern that have been shown to induce immune alterations 2,3,7,8-tetrachlorodibenzo-*p*-dioxin (TCDD) has been investigated most extensively.

TCDD is a chlorinated hydrocarbon found to be a contaminant in biocides (trichlorophenol), herbicides (2,4,5-T), and products from chemical sensitizing processes in which polychlorinated phenols are used.[1] This compound has previously been shown to induce atrophy of thymus and peripheral lymph nodes of rats, mice, and guinea pigs.[2–4] In addition, TCDD has been shown to suppress thymic-dependent immune function[4–6] and resistance to bacterial infection.[7] The suppression of cell-mediated immune functions includes suppression of mitogen responsiveness,[4–6] suppression of skin graft rejection[5] and suppression of delayed hypersensitivity responses.[6,8] Suppression of these T-cell-dependent immune functions appears to occur without helper cell function being affected.[8]

Effect of Developmental TCDD Exposure on Immune Function in Fischer/Wistar Rats

Previous studies utilizing rats to investigate the effects of developmental TCDD exposure on immune functions have utilized Fischer rats.[5,6,8] Recently it was reported that the Fischer rat is a poor immunologic responder.[9] In order to investigate the effects of developmental TCDD exposure on a strain of rat with a good immunologic response, experiments performed previously in Fischer rats[6] were repeated in Fischer/Wistar rats.[9]

To accomplish this, female Fischer rats bred to male Wistar rats were randomly distributed into three dosage groups. (1) The control group was dosed with corn oil (0.1 ml/100 g body weight) on gestational day 18. At parturition, the neonates were cross-fostered among these females; all litters were standardized at eight pups per litter. Nursing rats were also dosed with corn oil (0.1 ml/100 g body weight) on postnatal days 0, 7 and 14. (2) Females in the postnatal group were allowed to litter, and neonates were cross-fostered among these females; litters were standardized at

*Present Address: Biomedical Research Center, Oral Roberts University, Tulsa, Oklahoma 74171.

0077–8923/79/0320–0564 $01.75/0 © 1979, NYAS

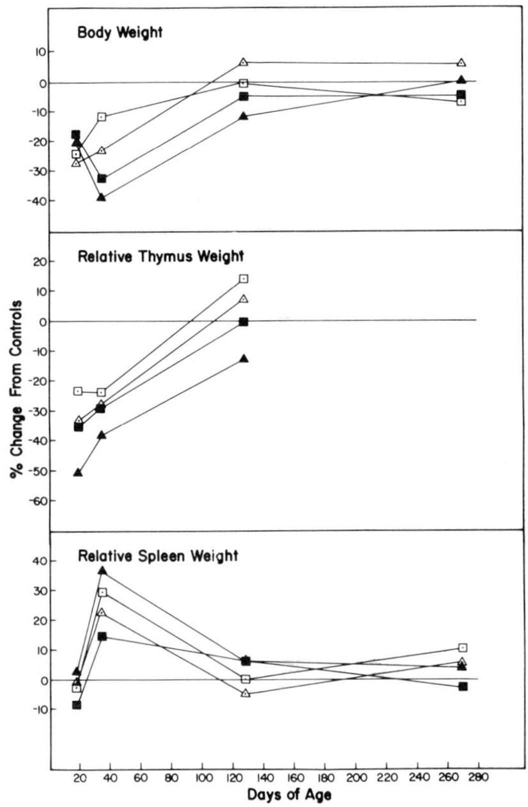

FIGURE 1. Effect of developmental TCDD exposure on body weight and selected organ weights. □, female animals exposed to TCDD postnatally via maternal dosing (5 μg/kg) on day 0, 7 and 14; ■, male animals exposed to TCDD postnatally via maternal dosing as above; △, female animals exposed to TCDD pre- and postnatally via maternal dosing (5 μg/kg) on gestational day 18 and on days 0, 7 and 14; and ▲, male animals exposed to TCDD via maternal dosing as above. Relative thymus and spleen weights represent organ-to-body-weight ratios.

Age in Days	Significant Variations from Control $p < 0.05$	$p < 0.01$
18		
body wt.		□, ■, △, ▲
rel. thy. wt.	□	■, △, ▲
rel. spleen wt.	All N.S.*	
35		
body wt.	□	■, △, ▲
rel. thy. wt.	■	□, △, ▲
rel. spleen wt.	△	□, ▲,
128		
body wt.		▲
rel. thy. wt.		▲
rel. spleen wt.	All N.S.*	
270		
body wt.	All N.S.*	
rel. thy. wt.	All N.S.*	
rel. spleen wt.	All N.S.*	

*N.S. indicates nonsignificant differences from control.

eight pups per litter. Nursing rats were dosed with 5 μg/kg TCDD (Dow Chemical Co., Midland, Michigan, lot 851-144-11, more than 99% pure) on postnatal days 0, 7, and 14. (3) Females in the prenatal-postnatal group were dosed with 5 μg/kg TCDD on gestational day 18. At parturition the neonates were cross-fostered to untreated females and the litters were standardized at eight pups per litter. The nursing females received doses of TCDD (5 μg/kg) on postnatal days 0, 7, and 14. All rat pups were weaned at 25 days of age. These animals were utilized to investigate the effects of TCDD exposure on various parameters of immune function.

FIGURE 1 illustrates the effect of TCDD exposure on body weight, relative thymus weight, and relative spleen weight. Developmental TCDD exposure resulted in depressed body and relative thymus weights in males and females of both treatment groups at 18 and 35 days of age. By 128 days of age body weights and relative thymus weights were still slightly suppressed in the prenatally treated males whereas values were normal or above normal for the other treated animals. This is similar to results previously obtained in Fischer rats.[6] At 35 days of age the relative spleen weights in both groups of TCDD-exposed animals were greater than those of nonexposed animals. This ratio reflects the change in body weight, as actual spleen weights did not vary significantly between the groups.

In vitro mitogen stimulation of lymphocyte cultures was performed using spleen and thymus cells from TCDD-exposed and nonexposed animals as previously described.[6] Results of mitogen stimulation were evaluated by calculating stimulation indices (counts per minute incorporated into stimulated cultures divided by counts per minute incorporated into nonstimulated cultures) and relating the indices of stimulation of cultures from TCDD-exposed animals to the indices of stimulation of cultures from nonexposed animals. FIGURE 2 illustrates the effect of developmental TCDD exposure on mitogen responsiveness of splenic and thymic lymphocytes. Prenatal and postnatal and postnatal only exposure to TCDD suppressed the response of splenic and thymic lymphocytes to stimulation by the mitogens phytohemagglutinin (PHA) and Concanavalin A (Con A). The response of splenic lymphocytes had returned to normal by 270 days of age. Thymic cultures were not performed at 270 days of age because of extreme thymic involution at this time. There was a trend towards recovery of mitogen responsiveness in thymic lymphocytes over the time course studied.

Delayed hypersensitivity reactions (DHR) were utilized as an *in vivo* assessment of cell-mediated immune function. The DHRs were performed according to the radiometric assay method of Lefford.[10] The effect of developmental TCDD exposure on delayed hypersensitivity responsiveness is illustrated in FIGURE 3. TCDD exposure resulted in a suppression of delayed hypersensitivity responses at 35 and 133 days of age. By 270 days of age the response of exposed animals had returned to normal.

The final assay of immune function investigated in the Fischer/Wistar animals was antibody production. Animals were immunized with primary and secondary injections of bovine gamma globulin (BGG) in complete Freund's adjuvant at various ages. Serum was obtained from the immunized animals at approximately weekly intervals for 5 weeks and antibodies to BGG assayed for by passive hemagglutinatron reactions. FIGURE 4 illustrates the antibody responses obtained in these animals. Developmental TCDD exposure had no apparent effect on antibody responses. These results indicate that at least one T-cell function, that of helper cell function, is spared from TCDD suppression, as BGG is a T-dependent antigen.

ALTERATIONS IN LYMPHOCYTE HOMING PATTERNS AS A RESULT OF DEVELOPMENTAL TCDD EXPOSURE

Inbred Fischer rats were exposed to TCDD as described for the prenatal-postnatal exposure group described above. At 45 days of age these animals were utilized to

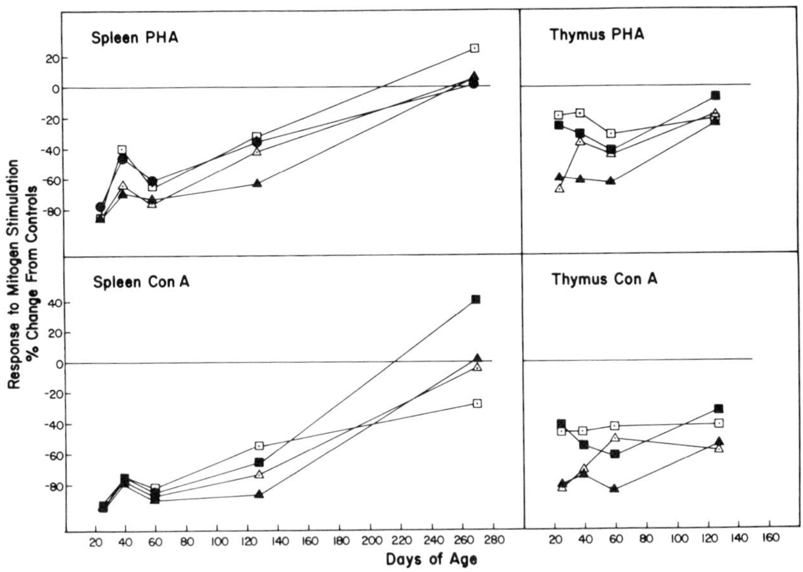

FIGURE 2. Effect of developmental TCDD exposure on *in vitro* lymphocyte responsiveness to mitogen stimulation. Legend as in FIGURE 1.

Age in Days	Significant Variations from Control $p < 0.05$	$p < 0.01$
25		
Spleen PHA		□, ■, △, ▲
Con A		□, ■, △, ▲
Thymus PHA	□	■, △, ▲
Con A	□, ■	△, ▲
35		
Spleen PHA		
Con A		
Thymus PHA	□	■, △, ▲
Con A		□, ■, △, ▲
59		
Spleen PHA		□, ■, △, ▲
Con A		□, ■, △, ▲
Thymus PHA	□	■, △, ▲
Con A	□	■, △, ▲
128		
Spleen PHA		■, ▲
Con A		□, ■, △, ▲
Thymus PHA	□, ▲	△
Con A		□, ■, △, ▲
270		
Spleen PHA	All N.S.*	
Con A	All N.S.*	

*N.S. indicates nonsignificant difference from control.

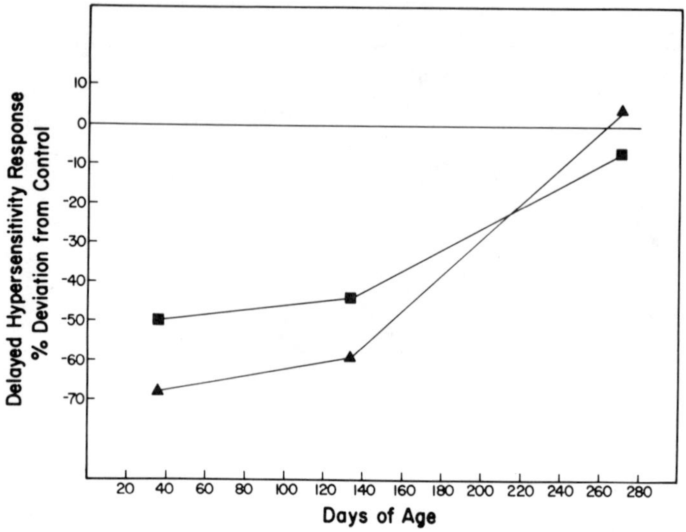

FIGURE 3. Effect of developmental TCDD exposure on delayed hypersensitivity responsiveness. ■, animals exposed to TCDD postnatally through maternal dosing (5 μg/kg) on days 0, 7 and 14; and ▲, animals exposed to TCDD pre- and postnatally by maternal dosing (5 μg/kg) on gestational day 18 and on days 0, 7, and 14. AT 35 and 133 days of age both TCDD dosage groups responded at a level significantly lower than nonexposed animals ($p < 0.01$). At 270 days of age there was no significant difference between responses of exposed and nonexposed animals.

investigate the effects of TCDD exposure on lymphocyte homing patterns. Lymphocytes from spleen, thymus, bone marrow and lymph node were isolated from TCDD exposed and nonexposed animals, labeled with ^{51}Cr,[11] and injected in TCDD-exposed and nonexposed recepients. TCDD exposure altered the homing of exposed spleen cells to the thymus in nonexposed animals. In addition, TCDD exposure altered the pattern of homing of nonexposed thymus cells in exposed animals and the pattern of homing of exposed thymus cells in nonexposed animals.

CONCLUSIONS

The highly toxic chlorinated hydrocarbon TCDD appears to be equally as immunosuppressive in a strain of rat that is a good immunologic responder (Fischer/Wistar) as it is in the Fischer rat,[5,6,8] which is a poor immunolgic responder.[9] The immunosuppression induced by developmental exposure to TCDD in the Fischer/Wistar rat is long lasting, as has previously been shown to be true in the Fischer rat.[6] The exposed animals do recover from the induced immunosuppression by 270 days of age, however.

At present the mechanism of induction of immunosuppression by TCDD is unknown. The results of these experiments indicate that whatever the mechanism the effects are reversible.

In an effort to understand more fully the types of defects in immune function induced by exposure to TCDD, studies on the effects of TCDD exposure on homing patterns of lymphocytes were performed. It was found that TCDD exposure significantly increases the thymic uptake of spleen cells from TCDD-exposed animals injected into nonexposed recipients. Thymic cells from non exposed animals showed decreased homing to spleen and increased homing to thymus in TCDD exposed recipients. At the same time thymic lymphocytes from TCDD exposed donors showed decreased homing to thymus in non-exposed recipients.

While there is controversy about the role of cell surface markers in control of homing patterns of lymphocytes,[12] the data obtained here are consistent with TCDD-induced modification of normal cell-surface markers. This modification could

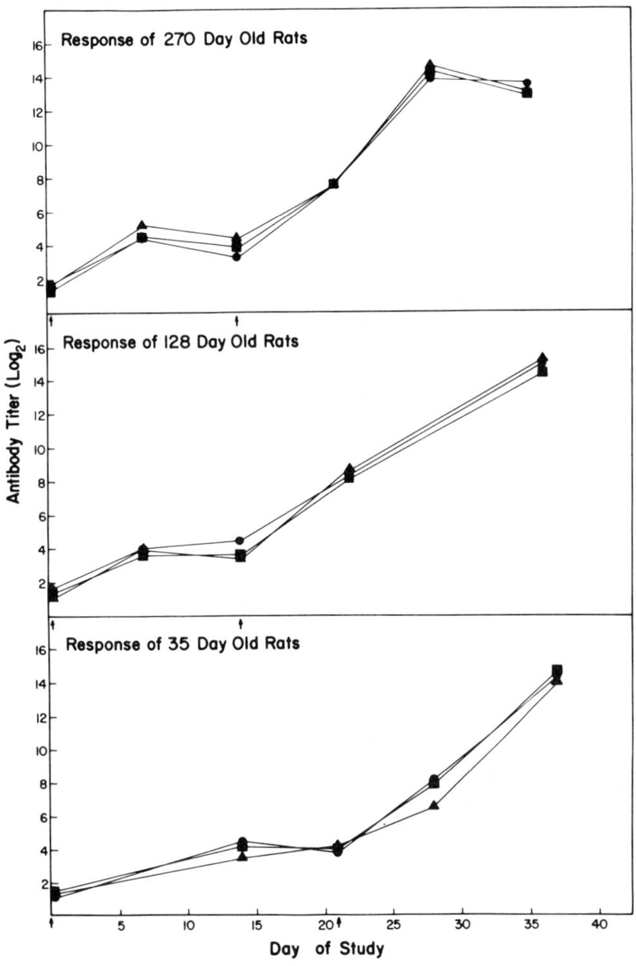

FIGURE 4. Effect of developmental TCDD exposure on antibody response to bovine gamma globulin. ●, nonexposed animals; ■ animals exposed to TCDD postnatally by means of maternal dosing (5μg/kg) on days 0, 7, and 14; and ▲, animals exposed to TCDD pre- and postnatally by means of maternal dosing (5μg/kg) on gestational day 18 and on days 0, 7, and 14. No significant differences were observed between nonexposed and TCDD-exposed animals.

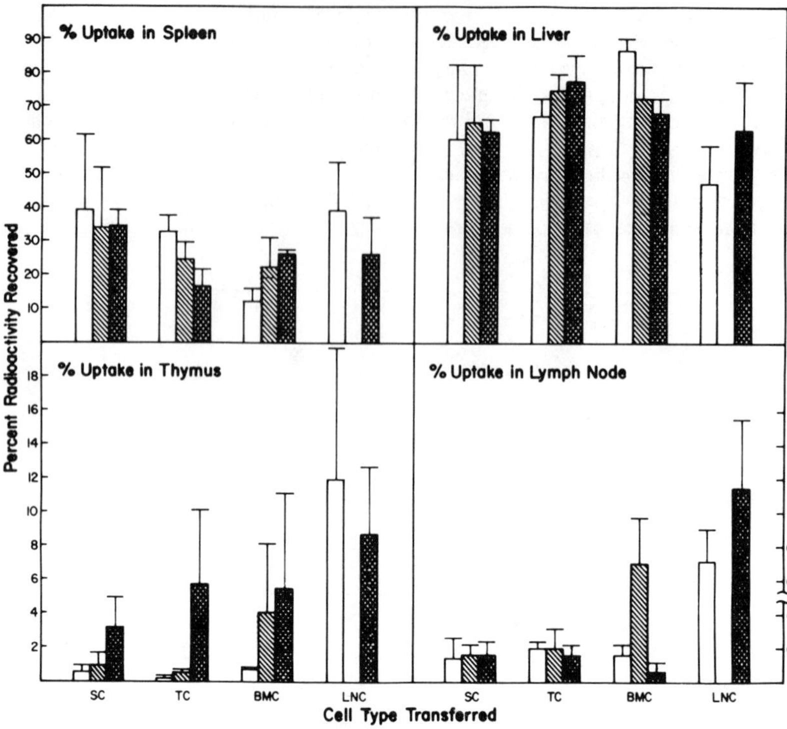

FIGURE 5. Effect of developmental TCDD exposure on lymphocyte homing patterns. The open bars represent cells from nonexposed animals injected into nonexposed animals, the shaded bars represent cells from nonexposed animals injected into TCDD-exposed animals, and the crosshatched bars represent cells from TCDD-exposed animals injected into nonexposed animals. Significant differences occurred in splenic uptake of thymus cells. Nonexposed cells homed to nonexposed spleen in greater numbers than to exposed spleen ($p = 0.06$) and in greater numbers than exposed cells homed to nonexposed spleen ($p < 0.01$). In addition, nonexposed cells homed to exposed spleen in greater numbers than exposed cells homed to nonexposed spleen ($p = 0.054$). TCDD-exposed bone marrow cells homed to nonexposed spleen in greater numbers than nonexposed bone marrow cells migrated to nonexposed spleen ($p < 0.01$). TCDD-exposed spleen cells homed to nonexposed thymus in greater numbers than nonexposed spleen cells homed to nonexposed or exposed thymus ($p < 0.01$). TCDD-exposed thymus cells homed to nonexposed thymus in greater numbers than nonexposed thymus cells homed to nonexposed thymus ($p < 0.01$). Nonexposed bone marrow cells home to nonexposed liver in greater numbers than exposed bone marrow cells homed to nonexposed liver ($p < 0.01$). All other comparisons were nonsignificant.

be accomplished by at least two mechanisms: (1) a TCDD-induced change in cellular metabolism resulting in an altered cell membrane; or (2) alteration of the cell membrane by insertion of TCDD into the membrane. In any case, TCDD exposure appears to alter homing markers in the spleen and thymus, as well as on the lymphocyte, as evidenced by alteration of homing patterns of normal thymic and splenic lymphocytes when injected into TCDD-exposed recipients.

Summary

The effects of TCDD exposure on parameters of immune function during the developmental period were investigated. Exposures were performed in Fischer/Wistar rats. Fetal and neonatal rats were exposed to TCDD through maternal dosing (5μg/Kg) on day 18 of gestation and on days 0, 7, and 14 of postnatal life (group 1). Another group of neonatal rats were exposed to TCDD through maternal dosing on days 0, 7, and 14 of postnatal life only (group 2). Body weights and relative thymus weights were found to be suppressed up to 135 days of age in group 1 but only up to 35 days of age in group 2. Parameters of cell-mediated and humoral immune function were investigated. TCDD suppressed cell-mediated immune function without affecting hummoral immune function. TCDD-exposed animals had recovered normal cell-mediated immune function by 270 days of age.

A group of inbred Fischer rats was exposed to TCDD as described for group 1 above. At 45 days of age these animals were utilized in lymphocyte homing studies. It was found that TCDD exposure alters homing patterns of lymphocytes from exposed animals when adoptively transferred to untreated animals. In addition, lymphocytes from nonexposed animals did not home normally when injected into TCDD-exposed recipients.

References

1. Moore, J. A. 1973 Prespective on chlorinated dibenzodioxins and dibenzofurans. Environ. Health Perspect. **5:** 1.
2. Gupta, B. N., J. G. Vos, J. A. Moore, J. G. Zinkl & B. C. Bullock. 1973. Pathologic effects of 2,3,7,8-tetrachlorodibenzo-*p*-dioxin in laboratory animals. Environ. Health Perspect. **5:** 125–140.
3. Harris, M. W., J. A Moore, J. G. Vos & B. N. Gupta. 1973. General biological effects of TCDD in laboratory animals. Environ. Health Perspect. **5:** 101–109.
4. Vos, J. G., J. A. Moore & J. G. Zinkl. 1973. Effect of 2,3,7,8-tetrachlorodibenzo-*p*-dioxin on the immune system of laboratory animals. Environ. Health Perspect. **5:** 149–162.
5. Vos, J. G. & J. A. Moore. 1974. Suppression of cellular immunity in rats and mice by maternal treatment with 2,3,7,8-tetrachlorodibenzo-*p*-dioxin. Int. Arch. Allergy Appl. Immunol. **47:** 777–794.
6. Faith, R. E. & J. A. Moore. 1977. Impariment of thymus-dependent immune function by exposure of the developing immune system to 2,3,7,8-tetrachlorodibenzo-p-dioxin (TCDD). J. Toxicol. Environ. Health **3:** 451–464.
7. Thigpen, J. E., R. E. Faith, E. E. McConnell & J. A. Moore. 1975. Increased susceptibility to bacterial infection as a sequela of exposure to 2,3,7,8-tetrachlorodibenzo-*p*-dioxin. Infect. Immun. **12:** 1319–1324.
8. Faith, R. E., M. I. Luster & J. A. Moore. 1978. Chemical separation of helper cell function and delayed hypersensitivity responses. Cell. Immunol. In press.
9. Tada, N., K. Itakura & M. Aizawa. 1974. Genetic control of the antibody response in inbred rats. J. Immunogenet. **1:** 265.
10. Lefford. J. J. 1974. The measurement of tuberculin hypersensitivity in rats. Int. Arch. Allergy Appl. Immunol. **47:** 570–585.
11. Morse, H. C., III & Asofsky, R. 1974. *In vivo* effects of antithymocyte serum on the homing patterns and graft-verus-host reactivity of murine splenic lymphocytes. Cell. Immunol. **11:** 19–29.
12. Schlesinger, M. 1976. Cell surface receptors and lymphocyte migration. Immunol. Comm. **5:** 775–793.

ASSESSMENT OF IMMUNOLOGIC ALTERATIONS CAUSED BY HALOGENATED AROMATIC HYDROCARBONS

Michael I. Luster and Robert E. Faith*

Environmental Biology and Chemistry Branch
National Institute of Environmental Health Sciences
Research Triangle Park, North Carolina 27709

Laboratories assessing immunologic alterations in laboratory animals resulting from exposure to environmental chemicals are confronted by the same question often asked by clinical immunologists, "What type of tests should be performed?" There is no general answer to this, thus, the methodology justifiably varies in different laboratories. The problem is aggravated by the fact that the immune effects may vary dependent upon the chemical and dosage level as well as the species and age of the animal. For example, inorganic metals may have their primary effect on humoral-mediated immune responses,[1] whereas the halogenated aromatic hydrocarbons act on cell-mediated immune responses.[2,3] Furthermore, low-level exposure to many of these chemicals such as lead,[1] tobacco smoke,[4] and polychlorinated biphenyls[5] appears to result in immunoenhancement while higher levels are immunosuppressive. Animals exposed during the developmental phase of the immune system may be more severely affected than animals exposed as adults,[2] while the natural loss of immune responsiveness in senescent animals[6] may complicate interpretation during chronic studies.[1]

In general, immune tests that are fairly reliable, i.e. minimal variation within a normal population, that are capable of detecting subtle differences in immune status, and are specific for particular aspects of the immune system are used in our laboratory. As indicated in an earlier review by Vos,[7] a suggestion of whether a particular chemical has immunosuppressive properties can often be obtained during routine toxicity studies (TABLE 1). At necropsy, central and peripheral lymphoid organs particularly the thymus, spleen and bursa de Fabricus in avian species can be obtained for gross examination, histopathology and organ-to-body weight ratios. The adrenal glands may also be of interest, since glucocorticoids have immunoregulatory potentials[8] and there is evidence that polychlorinated biphenyls affect serum corticosteroid levels.[9] *In vitro* trypan-blue exclusion can be employed to determine whether the chemical is directly cytotoxic to lymphocytes, although the water-insoluble nature of the majority of the halogenated aromatic hydrocarbons may make this difficult. Hematograms including leukocyte counts and differentials can be performed to determine the presence of lymphopenia although this is not a prerequisite for immunosuppression induced by chemicals. Serum electrophoresis and the quantitation of serum immunoglobulin classes[10,11] also is beneficial and requires no prior immune manipulation of animals in toxicology studies.

The direct involvement of at least three subpopulations of lymphocytes, macrophages and their cooperative interactions to elicit immune responses requires the assessment of multiple immune parameters to properly evaluate the effects of environmental chemicals on the immune response. However, initially one may desire

*Present Address: Biological Research Center, Oral Roberts University, Tulsa, Oklahoma 74171.

0077–8923/79/0320–0572 $01.75/0 © 1979, NYAS

TABLE 1

IMMUNOLOGIC ASSESSMENT IN ROUTINE TOXICITY STUDIES

1. Examination of lymphoid organs (thymus, spleen, lymph nodes and bursa of Fabricius).
   - Gross examination
   - Histopathology
   - Lymphoid organ to body weight ratio
   - Quantitation of viability
2. Hematology
   - Differential
   - Leukocyte count
3. Immunologic
   - Serum immunoglobulin concentrations
   - Lymphocyte transformation (PHA and LPS)
   - T-cell rosettes (species limitation)
   - Surface antigens on lymphocytes (species limitation)
4. Clinical signs—infections, tumors

to focus on general immune tests. For example, tests to determine the ability of exposed animals to synthesize antibodies following immunization as a measure of humoral-mediated immunity (HMI) or delayed hypersensitivity reactions as a correlate for cell-mediated immunity (CMI) can be used as a general screen. If an abnormal response is detected at this stage, it is desirable to localize the defective stage of the proliferative response. For example, the early stages of immune ontogeny are seldom considered in immune surveillance studies. Production of bone marrow precursors of lymphocytes can be quantitated *in vivo* by transfer of bone marrow stem cells into irradiated recipients or *in vitro* where colony formation can be observed.[12]

Lymphocyte transformation provides a simple, reproducible and semiquantitative *in vitro* correlate for CMI.[13] Transformation can be defective in the absence of lymphopenia. It does not, however, measure effector function, nor does it represent a normal response. Phytohemagglutinin (PHA) and concanavalin A (Con A) are the T-cell mitogens of choice. It is advantageous to use both mitogens since they may be selective to T-cells at different stages of maturation.[14] Mature T-cells have similar PHA/Con A ratios and possess the properties listed in TABLE 2. Immature T-cells that are found in the bone marrow have altered PHA/Con A ratios. Thus, similar mitogenic reactivity in bone-marrow lymphocytes and lymphocytes from peripheral lymphoid organs would suggest a defect in the post-bone-marrow maturation process. Lack of response in bone marrow lymphocytes would indicate defects in early bone marrow precursor cells. Mitogens that selectively stimulate B-cells are also available.[15]

TABLE 2

CHARACTERISTICS OF T-CELL SUBSETS

PHA/con A	1.0	0.3
θ Density	3+	1+
Location	Lymph node and spleen	Spleen and bone marrow
Radiation sensitivity	Relatively sensitive	Relatively resistant
Function	Mixed lymphocyte reactivity Amplifier cell in graft-versus-host reactions	Specific killer cell Precursor of effector cells in graft-versus-host reactions Required for DNA synthesis in response to antigen

TABLE 3

ASSESSMENT OF CELL-MEDIATED IMMUNITY

1. *In vitro* transformation of lymphocytes by PHA and Con A.
2. T-cell rosettes.
3. T-cell surface antigen (rabbit anti-mouse brain [C3H] adsorbed).
4. *In vitro* mixed lymphocyte interactions.
5. *In vivo* delayed hypersensitivity reactions using radiometric ear test.
6. Agarose plate technique to test ability of antigen-sensitized cells to release migration inhibition factor.

Recent evidence has indicated that several of these mitogens are selective to B-cells at different stages of maturation (personal communication). Mature B-cells are preferentially stimulated by lipopolysaccharide from *E. coli,* while dextran sulfate preferentially stimulates early B-cells. Unfortunately, Fischer rats, which are commonly used in toxicology studies, are poor responders to these mitogens.[16] Furthermore, peripheral blood lymphocytes from most species do not respond to these mitogens. In this case, pokeweed mitogen, which stimulates both B- and T-cells, can be used.

TABLE 3 enumerates the techniques commonly employed to assess cell-mediated immune responses in experimental animals. Enumeration of peripheral or splenic T-cells by spontaneous rosette formation with erythrocytes is not practical in mice or rats since less than 1.0 percent of their T-cells form rosettes. This technique, however, is quite suitable with T-cells from humans, new world monkeys, and dogs as well as guinea pigs, which form spontaneous rosettes with rabbit erythrocytes.[17] In mice, T-cells can be enumerated by a fluorescent procedure employing antiserum specific to the theta antigen present on mouse T-cells.[18] Fluorescein isothiocyanate (FITC)-conjugated anti-mouse brain (C3H) absorbed antiserum, which is specific to theta antigen in most species of mice, is commercially available (Litton Bionetics, Kensington, Md.). In mice, a series of T-cell membrane antigens termed Ly have been described that appear to represent various functional subsets of T-cells.[19] Commercial production of antiserum to Ly markers would be extremely beneficial.

The mixed-lymphocyte culture reaction (MLC) detects lymphocyte transformation induced by allogenic lymphocytes, although it does not measure effector mechanisms. In human transplantation studies, the test correlates well with graft survival and is a measure of cell-mediated immunity. It is sometimes difficult to demonstrate MLC reactivity in spleens from mice and rats because of the presence of suppressor macrophages.[20] If this is the case, blood lymphocytes obtained following hypaque-ficol density gradient centrifugation is adequate in most species.

Various criteria for evaluating dermal delayed hypersensitivity reaction (DHR) in man and experimental animals are currently in use including erythema, induration, and increase in skin thickness. In experimental animals, DHRs can be quantitated by a radiometric ear test.[21] This technique is based on radioactive labeling of bone-marrrow precursors of monocytes, which subsequently migrate into sites of inflammation. Purified protein derivative (PPD) can be used as the sensitizing antigen in rats and guinea pigs, while in mice several protein antigens including bovine serum albumin can be employed.[22] This is an extremely useful assay since it requires many cell types for a normal reaction and is, thus, useful as a general screen.

Sensitized lymphocytes, when activated *in vitro* by a specific antigen, produce soluble lymphokines, one of which is termed migratory inhibition factor (MIF) which inhibits the spontaneous migration of macrophages. Production of MIF can be semiquantitated in an agarose plate technique.[23] The assay is a correlate for CMI, which unlike lymphoblastogenesis measures the effector stage of the immune system by semiquantitating production of the lymphokine.

TABLE 4 lists the assays employed in our laboratory to detect alterations in humoral-mediated immune responses. B-cells are enumerated using FITC-conjugated anti-immunoglobulin reagents since immunoglobulin is bound to the membrane of B-cells.[24] Alternatively, one may employ "erythrocyte-antibody-complement" (EAC) rosettes since B-cell membranes also possess a receptor for the third component of complement.[25]

Radial immunodiffusion analysis offers a sensitive and reliable method of quantitating serum immunoglobulin classes.[10] The limitation in these studies is usually the availability of commercial heavy chain specific reagents. Commercial kits containing IgM^-, IgA^-, and IgG (and often IgG subclasses)- specific antisera and reference standards for cattle, mice and humans are available. Heavy-chain-specific antisera are available commercially in other species, but reference standards are difficult to obtain unless prepared in the laboratory. We suggest each laboratory reconfirm the specificity of the antisera and reference standards for their particular assay. Clinically, increased levels of IgG and IgM have been noted in patients with various liver diseases, while decreased levels of IgA are found in some patients with sinopulmonary infections[26] as well as neonatally thymectomized experimental animals.[27]

Antigen-induced antibody synthesis is examined at the cellular level by a routine antibody plaque assay which enumerates antibody-secreting plasma cells.[28] The plasma cell antibody secretion rate can be determined mathematically by direct plaque size. Recently, Archer and coworkers have adapted the Mishell-Dutton system to screen for potentially immunosuppressive compounds including water-insoluble compounds.[29] This procedure has the advantage of ascertaining the effected cell type. For convenience, sheep red blood cells (SRBC) are usually the immunizing antigen of choice in most species. However, Fischer rats and Hartley guinea pigs do not respond well to SRBC, in which case a protein antigen can be substituted. If the antibody response to a T-dependent antigen is effected, it is recommended to determine the rresponse to a T-independent antigen. This will ascertain whether the altered immune response is related to a T-cell or B-cell defect, although a defect in macrophage processing should be considered. Bacterial lipopolysaccharide from *E. coli* is the T-independent antigen of choice in most experimental animals; however, it is necessary to substitute another T-independent antigen in Fischer rats.

Alterations in the immune system may result from an indirect cause. For example, chemically induced alterations in RES function, hormones, particularly corticosteroids and estrogens, complement as well as the kinin system can effect the immune system (TABLE 5). *In vivo* uptake of a particular, e.g. colloidal carbon, offers only a

TABLE 4

ASSESSMENT OF HUMORAL MEDIATED IMMUNITY

1. *In vitro* transformation of lymphocytes by LPS and dextran sulfate or PWM
2. B-cell surface antigens
   - mIg
   - complement receptor
3. Quantitation of serum immunoglobulins
4. Antibody response to a T-dependent and T-independent antigen
5. Local production of antibody

rough screening for RES function. *In vivo* phagocytic uptake and killing of bacteria is hampered by bactericidal properties of blood and tissues and does not distinguish between the possible participation of either PMN or mononuclear phagocytes. Thus, *in vitro* quantitation of the ability of ingestion and intracellular killing (independently), although time consuming, is probably the method of choice for determining RES function.[30] Macrophage cytotoxicity tests against tumor cell and macrophage proliferation assays should also be considered.

Published studies on the effects of halogenated aromatics on the complement system have not been reported. A direct and simple assay for the entire complement system is complement-dependent hemolysis which tests the overall activity of the system.[31] If a defect is seen at this stage, it would be desirable to determine which components of the system are being affected.

Other factors may alter the immune system, such as α-fetoprotein,[32] malnutrition,[33] and zinc deficiency.[8] Assessment of these factors in immunotoxicology studies have been adequately reviewed by Vos.[7] Since the number of indirect factors that can

TABLE 5

ASSESSMENT OF INDIRECT IMMUNITY

1. Macrophage
   - *In vivo* uptake of aggregated ^{125}I-BSA or colloidal carbon
   - *In vitro* or *in vivo* determination of phagocytosis and intracellular killing of bacteria
   - Cytotoxicity and proliferation
2. Hormones
   - Cortisol/cortisone
   - Estrogens or androgens
3. Complement
4. Host resistance to infection
5. Resistance to tumor challenge
6. Endotoxin hypersensitivity

alter the immune system are almost limitless, it may be easier to ascertain whether the chemical can directly effect lymphocytic function *in vitro*.

Finally, it is critical to determine host resistance to infection following exposure to chemicals that are suspect in altering normal immune status. Not only would this represent a practical approach, but is essential since abnormal functional immune tests have never been directly correlated to host resistance. Resistance to *Listeria monocytogenes* and *Streptococcus pneumoniae* are helpful, since resistance to these organisms appear to be dependent primarily upon normal T-cell and B-cell functions, respectively. There are numerous alteratives, however, including viral resistance and parasitic expulsion.

SUMMARY

The rational for employing a particular approach and some of the difficulties that can be encountered when adapting immune surveillance to toxicology studies in experimental animals are discussed. Detailed description of methodology will not be reviewed. In general, assays are selected for their reliability, i.e. fairly standard within a normal population; that are capable of detecting subtle differences in immune status;

and are specific for particular aspects of the immune system. Since normal immune responses are directly dependet upon at least three distinct subpopulations of lymphocytes, macrophages and their cooperative interactions as well as a number of indirect factors, it is essential to study multiple parameters to properly evaluate the effects of environmental chemicals on the immune system. Immunologic assessment is further complicated by the fact that the immune effects will vary dependent upon the chemical and dosage level as well as the species and age of the experimental animals.

References

1. Koller, L. D., J. G. Roan, J. A. Brauner & J. H. Exon. 1977. Immune response in aged mice exposed to lead. J. Toxicol. Environ. Health **3:** 535–543.
2. Vos, J. G. & J. A. Moore. 1974. Suppression of cellular immunity in rats and mice by maternal treatment with 2,3,7,8-tetrachlorodibenzo-*p*-dioxin. Int. Arch. All. Appl. Immunol. **47:** 777-794.
3. Faith, R. E. & J. E. Moore. 1977. Impairment of thymus-dependent immune functions by exposure of the developing immune system to 2,3,7,8-tetrachloro-dibenzo-*p*-dioxin. J. Toxicol. Environ. Health **3:** 451–464.
4. Holt, P. G. & D. Keast. 1977. Environmentally induced changes in immunological functions: Acute and chronic effects of inhalation of tobacco smoke and other atmospheric contaminants in man and experimental animals. Bact. Rev. **41:** 205-216.
5. Bonnyns, M. & C. H. Batonsky. 1976. Cellular immune response in rats exposed to PCB. Separatum Experientia **32:** 522–523.
6. Kay, M. M. & T. Makinodan. 1976. A review: Immunobiology of aging: Evaluation of current status. Clin. Immunol. Immunopath. **6:** 394–413.
7. Vos, J. G. 1977. Immune suppression as related to toxicology. Crit. Rev. Toxicol. **5:** 67–101.
8. Bach, J. F. 1975. Corticosteroids. *In* The Mode of Action of Immunosuppressive Agents. A. Neuberger & E. L. Tatum, Eds. Amsterdam. pp. 21–70.
9. Sanders, O. T. & R. L. Kirkpatrick. 1977. Reproductive characteristics and corticoid levels of female white-footed mice fed *ad libitum* and restricted diets containing polychlorinated biphenyl. Environ. Res. **13:** 358–363.
10. Mancini, G., A. O. Carbonara & J. F. Heremans. 1965. Immunochemical quantitation of antigens by single radial immunodiffusion. Immunochem. **2:** 235–243.
11. Engvall, E. & P. Perlmann. 1971. Enzyme-linked immunosorbent assay (ELISA). Quantitative assay of IgG. Immunochemistry **8:** 871–881.
12. Metcalf, D., G. J. V. Nossal, N. L. Warner, J. Miller, T. E. Mandell, J. E. Layton & G. A. Gutman. 1975. Growth of B-lymphocyte colonies *in vitro*. J. Exp. Med. **142:** 1534-1549.
13. Catalona, W. J., J. L. Tarpley, C. Potvin & P. B. Chnetien. 1975. Correlations among cutaneous reactivity to DNCB, PHA-induced lymphocyte blastogenesis and peripheral blood E rosettes. Clin. Exp. Immunol. **19:** 327–333.
14. Katz, D. H. 1977. Ontogeny of the two major classes of lymphocytes. *In* Lymphocyte Differentiation, Recognition, and Regulation. F. J. Dixon & H. G. Kunkel, Eds. Academic Press. New York. pp. 119–151.
15. Reed, N. D., J. K. Manning & J. A. Rudbach. 1973. Immunologic responses of mice to lipopolysaccharide from Escherichia coli. J. Infect. Dis. **128:** 570–577.
16. Tada, N., K. Itakura & M. Aizawa. 1974. Genetic control of the antibody response in inbred rats. J. Immunogenet. **1:** 265–273.
17. Hoffman, T. & H. G. Kunkel. 1976. The E rosette test. *In In Vitro* Methods in Cell-Mediated and Tumor Immunity. B. R. Bloom, J. R. David, Eds. Academic Press. New York. pp. 71–82.
18. Katz, D. H. 1977. Surface antigens and markers on T-lymphocytes. *In* Lymphocyte

Differentiation, Recognition, and Regulation. F. J. Dixon & H. G. Kunkel, Eds. Academic Press. New York. pp. 10–39.

19. KISIELOW, P., J. A. HIRST, H. SHIKA, P. C. L. BEVERLY, M. K. HOFFMAN, E. A. BOYSE & H. F. OETTGEN. 1975. Ly antigens as markers for functionally distinct subpopulations of thymus-derived lymphocytes of the mouse. Nature **253:** 219–220.
20. OEHLER, J. R., R. B. HERBERMAN, D. A. CAMPBELL & J. Y. DJEU. 1977. Inhibition of rat mixed lymphocyte cultures by suppressor macrophages. Cell. Immunol. **29:** 238–250.
21. LEFFORD, M. J. 1974. The measurement of tuberculin hypersensitivity in rats. Int. Arch. All. Appl. Immunol. **47:** 570–585.
22. KATSURA, Y., K. INABA, T. IZAMI & I. UESAKA. 1977. Cell-mediated and humoral immune responses in mice. II. Sensitizing conditions for delayed-type hypersensitivity. Int. Arch. All. Appl. Immunol. **53:** 329–340.
23. MYGIND, K. & S. STENBJERG. 1974. Leucocyte migration test in agarose. Acta Path. Microbiol. Scand., Section B **82:** 277–282.
24. WINCHESTER, R. J. & G. ROSS. 1976. Methods for enumerating lymphocyte populations. *In* Manual of Clinical Immunology. N. R. Rose & H. Friedman, Eds. American Society for Microbiology. Washington, D.C.. pp. 64–76.
25. ROSS, G. D. 1976. Lymphocyte complement receptors and their relationship to other membrane receptors. *In* Clinical Evaluation of Immune Function in Man. S. D. Litwin, C. L. Christian, G. W. Siskind, Eds. Grune and Stratton. New York. pp. 27–45.
26. HOROWITZ, S. & R. HONG. 1975. Selective IgA deficiency. *In* Immunodeficiency in Man and Animals. D. Bergsma, Ed. Sinauer Associates, Sunderland, Mass. p. 129.
27. CLOUGH, J. D., L. H. MIMS & W. STROBER. 1971. Deficient IgA antibody responses to arsanilic acid bovine serum albumin in neonatally thymectomized rabbits. J. Immunol. **106:** 1624–1629.
28. DRESSER, D. W. & M. F. GREAVES. 1973. Assays for antibody-producing cells. *In* Handbook of Experimental Immunology. D. M. Weir, Ed. Chap. 27. Blackwell Scientific, Oxford.
29. ARCHER, D. L., B. G. SMITH & J. A. BUKOVIC-WESS. 1978. Use of an *in vitro* antibody-producing system for recognizing potentially immunosuppressive compounds. Int. Arch. All. Appl. Immunol. **56:** 90–93.
30. COHN, Z. A. & S. I. MORSE. 1959. Interactions between rabbit polymorphonuclear leucocytes and staphylococci. J. Exp. Med. **110:** 419–433.
31. MAYER, M. M. 1971. Complement and complement fixation. *In* Experimental Immunochemistry, E. A. Kabat, Ed. Charles C. Thomas. Springfield, Ill. pp. 133–240.
32. MURGITA, R. A. & T. B. TOMASI. 1975. Suppression of the immune response by α-fetoprotein. II. The effect of mouse α-fetoprotein on mixed lymphocyte reactivity and mitogen-induced lymphocyte transformation. J. Exp. Med. **141:** 440–452.
33. MALAVE, I. & M. LAYNISSE. 1976. Immune response in malnutrition. Differential effect of dietary protein restriction on the IgM and IgG response to alloantigens. Cell Immunol. **21:** 337–343.

APPLICATION OF IMMUNOCOMPETENCE ASSAYS FOR DEFINING IMMUNOSUPPRESSION*

Jack H. Dean, Martin L. Padarathsingh, and Thomas R. Jerrells

Department of Immunology
Biomedical Research Division
Litton Bionetics, Inc.
Kensington, Maryland 20795

INTRODUCTION

Impaired immunocompetence as indicated by depressed humoral and/or cell-mediated immune responses have been observed in a high proportion of rodents dosed with often small amounts of certain environmental chemicals including 2,3,7,8-tetrachlorodibenzo-*p*-dioxin (TCDD[1–5]); polychlorinated biphenyl (PCB[6–11]); polybrominated biphenyl[12] and metals.[13–16] Similarly, impaired immune responses have been observed in humans inadvertently exposed to polybrominated biphenyls (PBBs) through contaminated food sources.[17] Likewise, impaired immunocompetence parameters have been observed in a high proportion of cancer patients,[18–22] animals bearing tumors,[23–26] in individuals from cancer-prone families,[27,28] and individuals born with immunologic deficiencies.[29–30]

During the last few years our laboratory has been involved in developing and refining assays of humoral and cell-mediated immunocompetence. Our studies with cell-mediated immunity (CMI) assays have focused on questions of standardization and quality control; defining appropriate assay endpoints and data analysis parameters; and the application of these assays to study the modulation of general and tumor specific immunocompetence in mice bearing chemically induced transplantable tumors and during the serial monitoring of human cancer patients.

I would like to discuss several CMI assays that appear most applicable for describing and defining changes in host immunocompetence. These are the I^{125} isotopic footpad assay (IFP) for measuring cutaneous delayed-type hypersensitivity to recall antigens; Cunningham's modification of Jerne Plaque assay, which examines T-cell-dependent B-cell responses represented by the appearance of antibody-producing cells; the microculture lymphocyte proliferation (LP) assay, which utilizes T- and B-cell mitogens and alloantigens in a one-way mixed leukocyte culture (MLC) to examine proliferative function; and finally a series of surface marker assays for quantitating the percentage of T-cells, B-cells, γ Fc receptor cells and macrophages. In addition, I would like to relate our experience with standardizing these assays, methods of data analyses and more importantly, describe studies and approaches to define the mechanisms responsible for the immunosuppression, which are quite similar to the problems encountered in defining and characterizing immune suppression induced by environmental chemicals. Studies will be presented describing changes in immune response parameters in animals bearing a methylcholanthrene (MCA)-induced sarcoma and animals treated with the known cytoreductive agent cyclophosphamide (cytoxan), to illustrate these points.

*This work was supported by USPHS Contract N1H-NO1-CB-63987 from the Division of Cancer Biology and Diagnosis of the National Cancer Institute of the National Institutes of Health.

0077–8923/79/0320–0579 $01.75/0 © 1979, NYAS

Materials and Methods

Mice

Male and female inbred BALB/c mice weighing 20 to 24 g (approximately 8 to 10 weeks old) were obtained from the production colonies of the National Institutes of Health (NIH) and used throughout these studies.

Tumors

Sarcoma Meth A was kindly provided by Dr. E. A. Boyse, Sloan-Kettering Memorial Cancer Center, New York, N.Y. The tumor was developed and has been maintained in this laboratory in ascitic (ASC) form by serial passage in syngeneic BALB/c mice. The tumor dose 50 (TD_{50}) of Meth A is 10^3 cells respectively for the ASC form of the cells. Animals bearing Meth A tumors were used for the LP studies. Animals bearing a subcutaneous mineral oil-induced ADJ-PC5 plasmacytoma were also employed. This tumor was obtained from Dr. Michael Potter of the NCI and passaged weekly by s.c. inoculation of 3×10^6 viable tumor cells suspended in HBSS into syngeneic BALB/c mice.

Isotopic (^{125}I) Footpad Assay (IFP)

BALB/c mice were immunized to Keyhole limpet hemocyanin (KLH) by a subcutaneous (s.c.) injections with 100 μg of KLH in incomplete Freund's adjuvant (IFA) followed by three subsequent s.c. injections without IFA at weekly intervals Animals were immunized to *M. tuberculosis* $H_{37}R_A$ by a s.c. injection of 100 μg of viable organisms emulsified in IFA followed by three subsequent injections without IFA at weekly intervals. Two to 6 weeks following the last immunization, control and cytoxan-treated animals were challenged with test antigen to measure the delayed cutaneous hypersensitivity response to recall antigens using the IFP assay previously described.[31]

KLH (100 μg/mouse) and PPD (10 μg/mouse) were injected i.d. using a disposable 0.5 cc glass syringe (Becton, Dickinson and Co., Rutherford, NJ) fitted with a 27-gauge needle into the right hind footpad (RHFP) of immunized, cytoxan-treated or control mice (six mice per group). Physiological pyrogen-free saline was injected into the left hind footpad (LHFP) as a trauma control. This latter practice was recently discontinued since significant swelling or radioisotopic leakage was not found to be associated with injection trauma at the 20-hour sampling period.

Footpad reactions were measured by the method described by Paranjpe and Boone[32] with slight modifications.[31] Immediately following the injection of test antigen, each mouse was injected i.p. with 0.1 ml of ^{125}I-human serum albumin (Mallinchrodt Chemical Works, St. Louis, MO; specific activity 10 μCi/ml or an amount equal to 10^6 counts/minute/mouse. Twenty hours after injection of the test antigen each mouse was sacrificed, the test footpads and contralateral footpads were washed well with PBS to remove contamination by radioactive urine and the feet were excised at the junction of the lower and middle thirds of the tibia (i.e., hock). The feet were placed in gamma-counting tubes (RPI, Elk Grove Village, IL) and counted for 10 minutes each in a gamma counter (Nuclear Chicago, Inc., Chicago, IL). Results were expressed as the foot count ratio (FCR) =

$$\frac{\text{CPM test foot}}{\text{CPM contralateral control foot}}$$

and significance between positive IFP responses to test antigen in immune and normal mice was determined by Student's t-test.

Plaque-Forming Cell Assay

The plaque-forming cell (PFC) response to sheep erythrocytes (SRBC) was performed as described by Cunningham[33] and as diagrammed in FIGURE 1. Briefly, SRBC obtained in Alsever's solution (Microbiological Associates, Bethesda, MD) were washed three times in phosphate buffered saline and resuspended to a 10% (v/v) solution. Tumor-bearing, cytoxan-treated and control mice were immunized with SRBC by giving each animal an i.v. injection of 0.1 ml of the 10% suspension. After 4

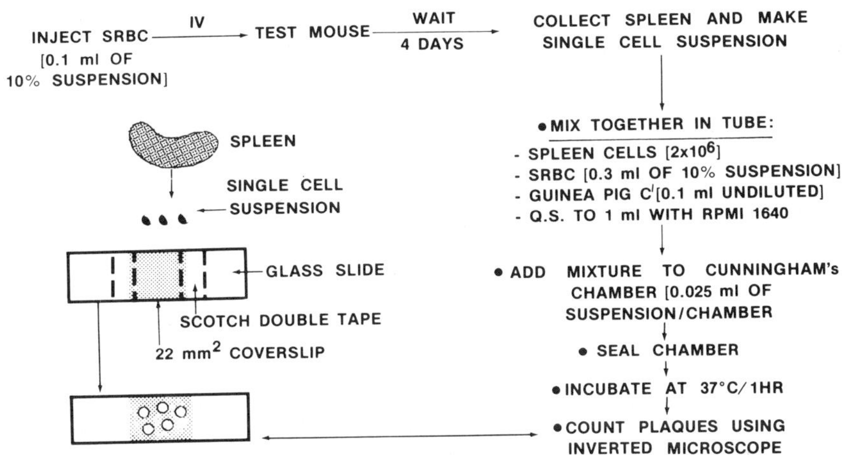

FIGURE 1. Diagram of Cunningham's modification of Jerne plaque assay.

days the animals were sacrificed and their spleens asceptically removed. The spleen cells were adjusted to 2×10^7/ml in RPMI-1640 and 0.1 ml (2×10^6) was added to 12×75 mm glass tubes. SRBC were added (0.3 ml of 10% solution) along with guinea pig complement (0.1 ml undiluted) and the volumes adjusted to 1 ml with fresh media. This mixture was added to Cunningham chambers formed by attaching 2 strips of double-stick tape to a microscope slide to which a 22 mm^2 coverslip was added. The loaded chambers were sealed with petroleum jelly (vaseline) to prevent drying and incubated at 37°C for 1 hour. Plaques in the SRBC lawn were enumerated using an inverted microscope. The average number of PFC counted in three replicate slides was taken as the number of hemolysin-producing cells in each individual spleen. Data were expressed as PFC/10^6 spleen cells and PFC/spleen.

Lymphocyte Proliferation (LP) Assay

A microculture (200 μl) LP assay that utilized general mitogens was employed to assay the proliferative ability of immunocompetent spleen cells. This assay was previously described by Dean *et al.*[34] Briefly, 2 × 10^5 viable spleen cells from tumor-bearing, cytoxan-treated or normal mice were cultured in four replicate flat-bottom wells of a microtiter plate (Falcon Plastics, Oxnard, CA). The culture medium consisted of RPMI-1640 (GIBCO, Grand Island, NY) supplemented with 25 mM HEPES buffer, 5% heat-inactivated (56°C for 30 minutes) fresh-frozen human AB serum (Irvine Scientific, Orange, CA), 1% glutamine, and 50 μg gentamicin/ml. Purified PHA (0.25 μg/well; Burroughs-Wellcome Co., Research Triangle Park, NC); Con A (0.25 μg/well; Calbiochem, San Diego, CA), T-cell mitogens and *Escherichia coli* LPS (5 μg/well; Difco Laboratories, Detroit, MI) a B-cell mitogen were added to each culture. Six hours before culture termination, 1.0 μCi of [^{3}H]TdR (New England Nuclear, Boston, MA; sp act, 6.7 Ci/mM) was added to each culture. After 72 hours of culture, the cells were harvested with a multiple sample precipitator (Biomedical Research and Development Laboratories, Inc., Rockville, MD), and the amount of [^{3}H]TdR incorporated into 5% cold trichloroacetic acid-insoluble material was determined by liquid scintillation spectroscopy. These data were expressed as the mean CPM ($\bar{x}$CPM) of the four replicates which usually had standard errors of 5% or less; as a stimulation index (SI), the CPM of mitogen-treated cultures (E) divided by the CPM of medium controls (C), or E/C; and net CPM (E-C). Mitogen data from spleen cell cultures of tumor-bearing and cytoxan-treated mice were also expressed as percent change relative to the response of normal animals.

$$\text{Percent Change} = 1 - \frac{\bar{x}\text{ CPM of treated or tumor bearing animals}}{\bar{x}\text{ CMP of normal mice}} \times 100$$

Results

Measurement of DHS Responses to Recall Antigens in Mice and Effects of Cytoxan-Induced Immunosuppression

Mice immunized with KLH and $H_{37}R_A$ developed strong and easily quantitated interdermal DHS responses (3 to 5-fold) to KLH and PPD as measured in the IFP assay (Table 1). The I^{125} albumin incorporation response peaked at 20 hours after

Table 1

Evidence of Delayed Cutaneous Hypersensitivity in BALB/c Mice to KLH and PPD Using the Isotopic Footpad Assay

Immunization	Footpad Challenge	I^{125} Control Foot CPM ± S.D.	I^{125} Antigen Challenge CPM ± S.D.	Footcount Ratio
None	KLH[100μg]	3,670 ± 871†	4,220 ± 1,011†	1.15
KLH immune‡	KLH[100μg]	3,085 ± 421	16,128 ± 2,513	5.23*
None	PPD[10μg]	3,109 ± 1,839	4,551 ± 2,328	1.51
$H_{37}R_A$ immune§	PPD[10μg]	4,048 ± 721	12,237 ± 2,895	3.06*

*Significantly different from control counts at $p < 0.01$.

†Mean counts per minute (CPM) of I^{125} incorporation ± S.D. from 6 normal and immune mice.

‡BALB/c mice were immunized by subcutaneous injection of KLH (100μg) in incomplete Freund's adjuvant followed by 3 weekly s.c. injections without adjuvant. Mice were used 3 weeks following the last immunization.

§BALB/c mice were immunized by s.c. injection of $H_{37}R_A$ (100μg) in incomplete Freund's adjuvant followed by 3 s.c. injections at weekly intervals without adjuvant.

TABLE 2

EFFECT OF CYTOXAN-INDUCED IMMUNOSUPPRESION ON RESPONSE TO KLH AND PPD IN IFP ASSAY OF BALB/c MICE

Immunization	Treatment	Footpad Challenge	CPM I^{125} Control Foot	CPM I^{125} Challenged Foot	Footcount Ratio
KLH immune	None	KLH†	3,227	13,420‡	4.16*
KLH immune	Cytoxan§	KLH	3,112	5,118	1.64
$H_{37}R_A$ immune	None	PPD§	3,052	11,270	3.69*
$H_{37}R_A$ immune	Cytoxan¶	PPD	3,468	4,200	1.21

*CPM of I^{125} incorporation in test feet were significantly different than in control feet at $p < 0.01$.

†All mice were challenged intradermally (footpad) with 100 μg of KLH.

‡Data represent mean CPM of I^{125} incorporation in 6 mice.

§Immunized mice were given 180 mg/kg of cytoxan at weekly intervals × 2 and were used in the IFP assay 24 hours following the second cytoxan treatment.

¶All mice were challenged intradermally with 10 μg of PPD and 100 μg of KLH.

interdermal challenge with antigen, which correlated well with visual changes observed in footpad thickness. The maximum footcount ratio to KLH occurred at the 100 μg challenge level in mice challenged at both higher and lower antigen concentrations, while the maximum response to PPD occurred at the 10μg level. Positive footcount responses were not observed in $H_{37}R_A$ immunized mice challenged with KLH and vice versa. The level of I^{125} albumin incorporated into antigen-challenged feet from immunized mice was significantly greater than the mean CPM in normal BALB/c mice challenged with recall antigen at $p < 0.01$.

Mice with established DHS to KLH and PPD were next treated with the cytoreductive agent cytoxan to examine the effects of this known immunosuppressant on memory response to recall antigens. Mice were treated by two i.p. injections of cytoxan (180 mg/kg) spaced at a one week interval and were challenged with antigen 24 hours following the second cytoxan treatment. These data are presented in TABLE 2. The DHS response to both KLH and PPD were significantly reduced following cytoxan treatment and was no longer considered positive using our criteria. High levels of I^{125} albumin incorporation was observed in the feet of immunized and non-treated mice following antigen challenge, but not in cytoxan-treated mice. A single injection of cytoxan (data not shown) greatly reduced the IFP response to antigen, although a weakly positive response ($p < 0.05$) was still obtained.

Effects of Tumor Bearing or Drug Treatment on the PFC Responses to SRBC

The PFC response to SRBC was found to be significantly enhanced in mice bearing small (8–10 mm) s.c. mineral oil-induced ADJ-PC5 plasmacytomas (TABLE 3) and in mice bearing the MCA-induced sarcoma Meth A (TABLE 4) as measured by the sensitive Cunningham modification of the Jerne Plaque assay for immunoglobulin M production. Mice bearing large ADJ-PC5 tumors (TABLE 3) demonstrated significantly reduced ($p < 0.05$) PFC response to SRBC which correlated with tumor burden. Small ADJ-PC5 tumor bearing had a 154% enhanced response over control mice, while mice bearing large ADJ-PC5 tumors (20–25 mm diameter) were 58% reduced in their capacity to produce PFC to SRBC. In contrast, mice bearing large SC Meth A tumors did not demonstrate (TABLE 4) depressed PFC responses to SRBC, but were actually enhanced in their PFC capacity. Thus, B-cell immunocom-

TABLE 3

THE EFFECT OF ADJ-PC5 PLASMACYTOMA ON THE PFC RESPONSE OF BALB/c MICE TO SRBC

Spleen Cell Donors (Tumor Size)	PFC/10^6 Spleen Cells† Mean (± S.D.) % Change	PFC/Spleen Mean
Normal BALB/c	102 (16) (—)	8.0×10^4
Small ADJ-PC5 tumor-bearing (8-10 mm)	260 (54)* (154%↑)	2.96×10^5
Large ADJ-PC5 tumor bearing (20–25 mm)	43 (13)* (58%↓)	3.6×10^4

*PFC response significantly different from control group at $p < 0.05$.

†All mice (6/group) received an intravenous injection of 0.1 ml of a 10% suspension of SRBC on Day −4 and were sacrificed at day 0 to measure the PFC response.

petency in the Meth A sarcoma model was suppressed, while mice bearing ADJ-PC5 plasmacytoma had quite suppressed PFC responses.

Mice given the known immunosuppressant cytoxan without or in combination with the immunopotentiator Levamisole were likewise studied for changes in their PFC responses to SRBC. The rationale for combining immunopotentiator with immunosuppressant was to determine if we could override the immunosuppressive effect of cytoxan and to examine the synergistic effect of combined drug therapy. A single injection of cytoxan (TABLE 5) 3 days prior to SRBC immunization suppressed the PFC response of BALB/c mice by 68%, which was significantly different from nontreated controls at $p < 0.01$. Levamisole therapy 13 and 6 days prior to SRBC failed to significantly potentiate the PFC response. However, treatment with Levamisole prior to cytoxan treatment even further depressed the PFC response to SRBC (94% depression) which was significantly different than cytoxan treatment alone at $p < 0.05$. Thus, these two drugs appear to have a synergistic effect when the immunopotentiator precedes the cytoreductive agent. The treatment of mice with Levamisole following cytoxan (data not shown) failed to show this synergistic immunosuppressive effect on the PFC response.

Lymphoproliferative (LP) Response to Mitogens in Tumor-Bearing and Cytoxan-Treated Mice

BALB/c mice were treated with the immunosuppressive drug cytoxan (180 mg/kg) and LP responses to mitogens were monitored at weekly intervals (TABLE 6).

TABLE 4

THE EFFECT OF METH A SARCOMA ON THE PFC RESPONSE OF BALB/c MICE TO SRBC

Spleen Cell Donor (Tumor Size)†	PFC/Spleen Mean
Normal BALB/c	5.0×10^4
Small Meth A tumor-bearing (10 mm diameter)	2.7×10^5*
Medium Meth A tumor-bearing (17 mm diameter)	1.9×10^5*
Large Meth A tumor-bearing (20–25 mm diameter)	1.8×10^5*

*PFC response significantly different from control mice at $p < 0.05$.

†All mice (6/group) received an i.v. injection of 0.1 ml of a 10% suspension of SRBC on Day −4 and were sacrificed at day 0 to measure the PFC response.

TABLE 5
THE EFFECT OF CYTOXAN AND LEVAMISOLE TREATMENT ON THE PFC RESPONSE OF BALB/c MICE TO SRBC

Pretreatment of Spleen Cell Donors	SRBC Challenge†	PFC/10^6 Spleen Cells† Mean [% Change]	PFC/Spleen Mean
None	Day −4	2,040 (—)	3.9×10^5
Cytoxan (180 mg/kg on Day −7)	Day −4	653* (68%↓)	1.0×10^5
Levamisole (0.5 mg/kg on Day −17 and −10)	Day −4	1,840 (10%↓)	4.0×10^5
Levamisole (Day −17 and −10) Cytoxan (Day −7)	Day −4	127* (94%↓)	2.0×10^4

*Significantly different from control at $p < 0.01$.
†All mice (6/group) received an intravenous injection of 0.1 ml of a 10% suspension of SRBC on Day −4 and were sacrificed at Day 0 to measure the PFC responses.
‡Cytoxan treated mice received 180 mg/kg of cytoxan.

The B-cell response to LPS (>100%) was more severely depressed than the PHA (75.5%) and Con A (70.4%) responses (T-cell) at 1 week after drug. However, the B-cell response recovered to normal levels more rapidly (3 weeks) than did the T-cell responses (5 weeks), although both T- and B-responses were comparable to nontreated control mice at 5 weeks.

Mice bearing large (20–25 mm diameter) subcutaneous Meth A sarcomas demonstrated depressed LP responses (TABLE 7) to the T-cell mitogens PHA (60%) and Con A (58%), borderline depression to the T-cell-dependent B-cell response to PWM (21%), but were not depressed to the T-cell-independent B-cell mitogens LPS (11% enhanced). Depresssion appeared to increase as a function of increasing tumor size (data not shown).

Further studies with large Meth A tumor-bearing mice will be described in order to characterize the nature of the suppression observed (TABLE 7). Passage of spleen cells from mice bearing large tumors through nylon wool columns to remove adherent cells followed by LP studies with nonadherent cells indicated that the cells suppressing T-cell responses to PHA and Con A resided in the T-cell-enriched fraction, since suppressor activity was not removed by nylon wool. The nylon wool population of cells was >90% T-cells as determined by FITC-labeled rabbit anti-mouse brain reagent. The percentage of change in each case is calculated against the appropriate treated control cell (data not shown). B-cell responses were not monitored in this population

TABLE 6
LYMPHOPROLIFERATIVE RESPONSES OF CYTOXAN-TREATED MICE TO MITOGENS

Spleen Cell Donors	PHA Response (% Change)† nCPM	SI	Con A Response (% Change) nCPM	SI	LPS Response (% Change) nCPM	SI
Normal BALB/c	248,598	94.2	260,269	98.5	10,673	5.0
Cytoxan (180 mg/kg on Day −14 and −7)	60,976 (75.5%↓)	32.2	77,064 (70.4%↓)	40.4	−432 (>100%↓)	0.8

$$\dagger\% \text{ Change} = 1 - \frac{\text{nCPM Treated}}{\text{nCPM Nontreated}} \times 100.$$

because of the known effect of NW passage on B-cell responses. Coculture experiments in which large tumor bearing spleen cells are cocultured with normal spleen cells and mitogens in LP assays revealed a consistent depression of T-cell, but not B-cell responses (lower portion of Table 7) in the normal spleen cell component. Depletion of T-cells within the tumor-bearing spleen cell suspension with anti-theta serum and guinea pig complement before coculture with normal spleen cells and mitogens abolished the ability of spleen cells from tumor-bearers to depress T-cell responses to PHA and Con A in normal spleen cells, again B-cell responses were not depressed. Surgical removal of tumor (data not shown) results in reversal of depressed responses and loss of suppressor cell activity by 10 days following surgery.

TABLE 7

EVIDENCE FOR SUPPRESSOR CELL ACTIVITY IN SPLEENS OF METH A (MCA) TUMOR-BEARING MICE

Donors	Treatment	PHA (.5 g) CPM (% Change)	Con A (.5 g) CPM (% Change)	PWM (2.5 g) CPM (% Change)	LPS (1.25 g) CPM (% Change)
Normal BALB/c	None	136,469 —	158,222 —	36,173 —	17,979 —
Large Meth A tumor-bearing	None	54,813† (60%↓)**	65,897 (58%↓)**	28,565 (21%↓)*	19,908 (11%↑)
Large Meth A tumor-bearing	Nylon Wool nonadherent	60,887 (57%↓)**	70,023 (41%↓)**	NT —	NT —
Large Meth A tumor-bearing nonadherent	Cocultured with normal BALB/c (1:1)	84,838 (53%↓)**	67,549 (42%↓)**	35,882 (8%↓)	18,392 (2%↓)
Large Meth A tumor-bearing treated with $\alpha \emptyset S + C$	Cocultured with normal BALB/c (1:1)	138,239 (1%↑)	153,195 (3%↓)	31,882 (12%↓)	19,198 (7%↑)

NT = not tested.
*= $p < 0.05$ relative to appropriate control.
**= $p < 0.01$ relative to appropriate control.
†Values represent mean CPM of quadruplicate microcultures from six mice per group with % change shown below. Percent change was calculated against appropriate control cultures which are not shown in all cases.

DISCUSSION

A series of parameters of immunocompetence were evaluated in BALB/c mice, treated with the known immunosuppressive drug cytoxan, which results in harsh and easily measurable effects; evaluations were also made in mice bearing tumors, which often produce milder immune impairment that is more difficult to detect. These studies were undertaken in order to provide insight into the applicability of these assays for measuring alteration of immunocompetence following exposure to environmental chemicals. These examples provided a spectrum of effects including immune depression resulting from reduced lymphocyte numbers following cytoreductive drugs; loss of memory cells for recall antigens; and depressed responses resulting from activation of suppressor cells of the adherent and nonadherent cell types. It is felt that by the time most environmental chemicals have been evaluated for immunologic effects this spectrum of known mechanisms of immunosuppression will be broadened and many of the mechanisms described here will be implicated.

The following CMI assays appear to be most applicable for defining immunocompetence in rodents: (1) DHS responses to recall antigens using the isotopic (I^{125}) footpad swelling response. This assay can be applied to examine the effect of chemical or tumor burden on performed immunity (i.e., memory responses) in which the animal is exposed or challenged with tumor after immunization. More commonly, animals are exposed to chemicals prior to immunization to examine their effects on the hosts' ability to be immunized. The DHS response as measured by the IFP assay is probably the best single parameter for immunologic evaluation since it requires competence at both the recognition and effector phases of the CMI response and requires the cooperation of accessory cells of the macrophage/monocyte series; (2) The plaque-forming cell (PFC) response to SRBC encompasses a T-cell-dependent antibody response, which is easily quantitated using the agar-free modification introduced by Cunningham[33] of the plaque-cell response for detecting antibody of the IgM class as described by Jerne and Nordin.[35] This response requires the assistance of macrophage/monocyte accessory cells as well as T-cell and appears to be quite sensitive for detecting immunosuppression. For the best resolution, animals are generally exposed to test compound or stressed by tumor burden prior to sensitization with SRBC for measuring impairment; and (3) The LP response of mouse lymphocytes to T- or B-cell mitogens is an easily quantitated response, which in recent years has been scaled down in volume through the use of 96 well microculture plates and semi-automated through the use of the multichannel cell harvesters. The lymphoproliferative response appears to be a sensitive indicator of immune depression resulting from cell depletion, maturation impairment or the induction of suppressor cells. These three measures of immunologic competence appear to be sensitive indicators of impairment. The only additional meritorious parameter not discussed which might be included in a battery of immunocompetence tests would be the leukocyte migration inhibition assay for the production of lymphokine. Assays of humoral immunity such as the quantitation of a specific antibody response (i.e., the antibody response to KLH) and quantitation of circulating immunoglobulin level should likewise be considered.

The DHS response to KLH and PPD in BALB/c mice were easily quantitated by the IFP assay with responses of 3 to 5-fold. Treatment of mice with two sub-lethal injections of cytoxan significantly reduced the footpad swelling response to nondetectable levels. Since mice were tested and sacrificed within 48 hours of their second cytoxan treatment no statement can be made regarding the recovery of this response. TCDD has previously been shown to "erase" memory responses to recall antigens.[3]

The PFC response to SRBC is variable depending upon the strength of the depression signal, that is in mice with both small ADJ-PC5 and Meth A tumors the PFC responses were significantly enhanced. In mice with medium and large ADJ-PC5 tumor the PFC responses were significantly depressed as shown by Padarathsingh *et al.*[26] and others[23,24] which in this model is accompanied by suppressor cells of the adherent type as shown by Padarathsingh *et al.*[36] In contrast, mice bearing medium and large Meth A sarcomas had a somewhat enhanced PFC response, which is consistent with the finding of suppressor cell activity in these animals which specifically depress T-cell, but not B-cell responses.[37]

Studies with the cytoreductive agent cytoxan revealed the strong immunosuppressive action of this drug on the PFC response in mice tested as late as 7 days after the administration of the drug. Levamisole is a drug previously shown by us[38] and other laboratories[39,40] to be a strong immunopotentiator of CMI function. In this current study, Levamisole alone had no effect on the PFC response. However, if Levamisole was given prior to treatment with cytoxan it resulted in a synergistic effect which enhanced the cytoxan-induced suppression of the PFC response. This is consistent with the findings in humans treated concomitantly with an immune activator and cytoreductive agent. These data suggest a mechanism whereby the potentiator may activate cells of the immune system and make them more susceptible to the cytolytic agent, which has a greater known effect on active cells. This observation indicates the importance of screening chemicals that may routinely be used in combination for synergistic effects on the immune system and illustrates the importance of the PFC response as a parameter and screen for immunosuppressive chemicals. The PFC response has previously been shown to be suppressed by a variety of environmental chemicals including TCDD,[1,41] PCB[9,12] and heavy metals.[14,16] In some studies, low doses of chemicals were found to enhance the PFC response while larger doses were strongly immunosuppressive, similar to our observation with the ADJ-PC5 tumor.

The lymphoproliferative response in Meth A tumor bearing BALB/c mice illustrates the application of this assay for detecting immunosuppression using selective T- and B-cell mitogens. Our laboratory, as well as others have previously described immunosuppressed LP responses to mitogens in cancer patients[18–22] and tumor bearing mice.[25,26] In previous studies with ADJ-PC5 tumor bearing or aniline mustard treated mice[26] we have observed suppression of LP responses to both T- and B-cell mitogens, which was reversible by drug removal of tumor burden. In the ADJ-PC5 tumor system suppressed LP responses were the direct effect of a suppressor cell of the adherent type which resided in the monocyte/macrophage series and was dependent on tumor burden for activity.[37,42]

Treatment of mice with aniline mustard, a known alkylating agent, gave similar LP depression of both T- and B-cell responses.[38] Cytoxan[43] as well as a variety of environmental chemicals such as TCDD,[2,3,41] polyhalogenated biphenyls,[12,17] and metals[16] have likewise been shown to depress mitogen induced LP responses. In contrast, mice bearing sarcoma Meth A had depressed LP responses to T-cell, but not B-cell mitogens as we have previously reported from preliminary studies.[37] Further studies as presented here revealed the presence of a suppressor cell in these mice that was characterized as being nonadherent to nylon wool columns; sensitive to anti-theta serum and complement; suppressed T-cell mitogen-induced LP responses of normal spleen cells in coculture experiments;[37] and which traveled in an Fc receptor bearing, T-cell enriched fraction.[37] Suppressor activity in this tumor system was reversible by surgical removal of tumor burden.

Those wishing to study and define the effects of environmental chemicals on immunocompetence must be equipped with several assay systems most of which were developed to study immune parameters as they related to cancer. These CMI assays can and have been efficiently applied to study the biologic effects of environmental chemicals on the immune system once one has given attention to the refinement and standardization of the procedure. Because of the biologic nature and sensitivity of these immunologic procedures, constant care should be taken when one changes reagents or reagent lot used in the procedure that might effect the reproducibility, accuracy or longitudinal monitoring.

Acknowledgments

We sincerely thank Miss Kathy Murphy, Miss Carolyn Fox, and Mrs. Connie Labash for their help in the preparation of this manuscript. We also acknowledge the expert technical assistance of John Northing, Dale Lewis, and Gary Richardson of the Department of Immunology at LBI.

References

1. Vos, J. G., J. A. Moore & J. G. Zinkl. 1973. Effect of 2,3,7,8-tetrachlorodibenzo-p-dioxin on the immune system of laboratory animals. Environ. Health Perspect. **5:** 149.
2. Vos, J. G. & J. A. Moore. 1974. Suppression of cellular immunity in rats and mice by maternal treatment with 2,3,7,8-tetrachlorodibenzo-p-dioxin. Int. Arch. Allergy Appl. Immunol. **47:** 777.
3. Moore, J. A. & R. E. Faith. 1976. Immunologic response and factors affecting its assessment. Environ. Health Perspect. **18:** 125.
4. Faith, R. E. & J. A. Moore 1977. Impairment of thymus-dependent immune function by exposure of the developing immune system to 2,3,7,8-tetrachlorodibenzo-*p*-dioxin (TCDD). J. Toxicol. Environ. Health **3:** 451.
5. Faith, R. E., M. I. Luster & J. A. Moore. 1978. Chemical separation of helper cell function and delayed hypersensitivity responses. Cell. Immunol. In press.
6. Friend, M. & D. O. Trainer. 1970. Polychlorinated biphenyl: interaction with duck hepatitis virus. Science **170:** 1314.
7. Vos, J. G. & R. B. Beems. 1971. Dermal toxicity studies of technical polychlorinated biphenyls and fractions thereof in rabbits. Toxicol. Appl. Pharmacol. **19:**617.
8. Vos, J. G. & T. H. De Roij. 1972. Immunosuppressive activity of a polychlorinated biphenyl preparation on the humoral immune response in guinea pigs. Toxicol. Appl. Pharmacol. **21:** 549.
9. Vos, J. G. & L. Van Driel-Grootenhuis. 1972. PCB-induced suppression of the humoral and cell-mediated immunity in guinea pigs. Sci. Total Environ. **1:** 289.
10. Koller, L. D. & J. E. Thigpen. 1973. Reduction of antibody to pseudorabies virus in polychlorinated biphenyl exposed rabbits. Am. J. Vet. Res. **34:**1605.
11. Vos, J. G. & H. Genderen. 1973. Toxicological aspects of immunosuppression. *In* Pesticides and the environment: A continuing controversy. W. B. Deichman, Ed. Intercontinental Medical Book. New York.
12. Luster, M. I., R. E. Faith & J. A. Moore. 1978. Effects of polybrominated biphenyls (PBB) on immune response in rodents. Environ. Health Perspect. In press.
13. Koller, L. D. 1973. Immunosuppression produced by lead, cadmium and mercury. Am. J. Vet. Res. **34:** 1457.
14. Koller, L. D., J. H. Exon, J. G. Roan. 1976. Humoral antibody response in mice after single dose exposure to lead or cadmium. Proc. Soc. Exp. Biol. Med. **151:** 339.
15. Koller, L. D., J. H. Exon & B. Arbogast. 1977. Methylmercury: effect on serum enzymes and humoral antibody. J. Toxicol. Environ. Health **2:** 1115.
16. Seinen, W., J. G. Vos, R. Van Krieken, A. Penninks, R. Brands & H. Hooykas. 1977. Toxicity of organotin compounds. III. Suppression of thymus-dependent immunity in rats by di-n-butyltindichloride and di-n-octyltindichloride. Toxicol. Appl. Pharmacol. **42:** 213.
17. Bekesi, J. G., J. F. Holland, H. A. Anderson, A. S. Fischbein, W. Rom, M. S. Wolff & I. J. Selikoff. 1978. Lymphocyte function of Michigan dairy farmers exposed to polybrominated biphenyls. Science **199:** 1207.
18. Hersh, E. M. & J. J. Oppenheim. 1965. Impaired *in vitro* lymphocyte transformation in Hodgkin's disease. New Engl. J. Med **273:** 1006–1012.
19. Han, T. & H. Takita, 1972. Impaired lymphocyte response to allogeneic cultured lymphoid cells in patients with lung cancer. New Engl. J. Med. **286:** 605–606.
20. Catalona, W. J., W. F. Sample & P. B. Chretien. 1973. Lymphocyte reactivity in cancer patients: correlation with tumor histology and clinical stage. Cancer **31:** 65–71.
21. Golub, S. H., T. X. O'Connell & D. L. Morton. 1974. Correlation of in vivo and in vitro assays of immunocompetence in cancer patients. Cancer Res. **34:** 1833–1837.
22. Dean, J. H., R. Connor, R. B. Herberman, J. L. McCoy & J. S. Silva. 1977. The relative proliferation index as a more sensitive parameter for evaluating lymphoproliferative responses of cancer patients to mitogens and alloantigens. Int. J. Cancer **20:** 359–370.
23. Zolla, S. 1972. The effect of plasmacytomas on the immune response of mice. J. Immunol. **108:**1039–1048.

24. TANAPATCHAIYAPONG, P. & S. ZOLLA. 1974. Humoral immunosuppressive substance in mice bearing plasmacytomas. Science **186:** 248–250.
25. KIRCHNER, H., T. M. CHUSED, R. B. HERBERMAN, H. T. HOLDEN & D. H. LAVRIN. 1974. Evidence of suppressor-cell activity in spleens of mice bearing primary tumors induced by Moloney sarcoma virus. J. Exp. Med. **139:** 1473–1487.
26. PADARATHSINGH, M. L., J. L. MCCOY, J. H. DEAN, D. D. LEWIS, J. W. NORTHING & L. W. LAW. 1977. Examination of general and tumor-specific cell-mediated immune responses in mice bearing progressively growing plasmacytomas. J. Nat. Cancer Inst. **58:** 1702.
27. CREAGAN, E. T. & J. F. FRAUMENI, JR. 1973. Familial gastric cancer and immunologic abnormalities. Cancer, **32:** 1325.
28. FRAUMENI, J. E., JR., W. WERTELECHI, W. A. BLATTNER, R. D. JENSEN & B. G. LEVENTHAL. 1975. Varied manifestations of a familial lymphoproliferative disorder. Am. J. Med. **59:** 145–150.
29. LISCHNER, H. W., H. W. PUNNETT & A. M. DIGEORGE. 1967. Lymphocytes in congenital absence of the thymus. Nature (London) **214:** 580–582.
30. OPPENHEIM, J. J., R. M. BLAESE & T. A. WALDMAN. 1970. Defective lymphocyte transformation and delayed hypersensitivity in Wiskott-Aldrich syndrome. J. Immunol. **104:** 835–844.
31. DEAN, J. H., D. D. LEWIS, M. L. PADARATHSINGH, J. L. MCCOY, J. W. NORTHING & L. W. LAW. 1977. Cellular immunity to SV40-induced tumor cells and solubilized tumor-associated antigens in immune and tumor-bearing mice using an isotopic footpad assay. Int. J. Cancer, **20:** 951–959.
32. PARANJPE, M. S. & C. W. BOONE. 1972. Delayed hypersensitivity to Simian Virus 40 tumor cells in BALB/c mice demonstrated by a radio-isotopic footpad assay. J. Nat. Cancer Inst. **48:** 563–566.
33. CUNNINGHAM, A. J. 1965. A method of increased sensitivity for detecting single antibody-forming cells. Nature **207:** 1106.
34. DEAN, J. H., J. L. MCCOY, D. D. LEWIS, E. A. APPELLA & L. W. LAW. 1975. Studies of lymphocyte stimulation by intact tumor-cell and solubilized tumor antigen. Int. J. Cancer **16:** 465–475.
35. JERNE, N. K. & A. A. NORDIN. 1963. Plaque formation in agar by single antibody-producing cells. Science **140:** 405.
36. PADARATHSINGH, M. L., J. H. DEAN, T. R. JERRELLS, J. L. MCCOY, D. D. LEWIS & J. W. NORTHING. 1979. Evidence for and characterization of suppressor cells in BALB/c mice bearing ADJ-PC5 plasmacytomas. J. Nat. Cancer Inst. In press.
37. DEAN, J. H., J. L. MCCOY, D. D. LEWIS & L. W. LAW. 1978. Perturbation of general and anti-tumor immune competence in mice bearing a chemically induced tumor. 69th Annual Meeting of the American Association for Cancer Research, Washington, D.C.
38. PADARATHSINGH, M. L., J. H. DEAN, J. L. MCCOY, D. D. LEWIS & J. W. NORTHING. 1979. Restorative effects of Levamisole on cell-mediated immune responses following chemotherapy with aniline mustard in mice bearing ADJ-PC5 plasmacytoma. Cancer Treatment Reports. **62:** 1627–1635.
39. RENOUX, G., M. RENOUX & M. N. TELLER, *et al.* 1976. Potentiation of T-cell mediated immunity by Levamisole. Clin. Exp. Immunol. **25:** 288–296.
40. SAMPSON, D. & A. LUI. 1976. The effect of Levamisole on cell-mediated immunity and suppressor cell function. Cancer Res. **36:** 952–955.
41. SHARMA. R. P., R. J. KOCIBA & P. J. GEHRING. 1978. Immunotoxicologic effects of 2,3,7,8-Tetrachloridibenzo-p-Dioxin in laboratory animals. Presented at the Society of Toxicology Meeting (March).
42. DEAN, J. H., M. L. PADARATHSINGH, J. L. MCCOY, T. R. JERRELLS, D. D. LEWIS, J. W. NORTHING & L. W. LAW. 1978. Evidence for tumor-associated antigens specific cell-mediated immunity in plasmacytoma immune and tumor bearing mice. In Proceedings of 8th International Symposium on Comparative Leukemia Research, Amsterdam, The Netherlands. In press (August).
43. WINKELSTEIN, A. 1973. Mechanisms of immunosuppression: Effects of cyclophosphamide on cellular immunity. Blood **41**(2): 273–283.

THE PROBLEM OF PCBs IN THE HUDSON RIVER SYSTEM

Edward G. Horn, Leo J. Hetling, and T. James Tofflemire

New York State Department of Environmental Conservation
Albany, New York, 12233

INTRODUCTION

On September 8, 1976, the New York State Department of Environmental Conservation (DEC), Albany, New York, and the General Electric Company (GE), Hudson Falls, New York, signed an agreement bringing to a close the action brought against GE relating to discharges of polychlorinated biphenyls (PCBs) into the Hudson River. A history of the problem, the settlement and a description of the study plan has been presented elsewhere.[1] As a result of the settlement, DEC instituted a series of major studies of the river system related to PCBs and potential actions for managing the problem of extensive in-place sediment contamination. This report summarizes the most significant results of these studies. A more detailed summary is also available.[2]

The Hudson River has been contaminated with PCBs from Ft. Edward, New York to New York City (FIGURE 1). From Spier Falls to Troy, the Upper Hudson River is actually a series of low-level dams, and south of Ft. Edward serves as part of the Champlain Canal (FIGURES 2 and 3). The Mohawk River serves as the eastern portion of the Erie Canal and joins the Hudson River just above the Federal Dam of Troy. The Lower Hudson Basin is tidal throughout its entire 241 kilometers (150 mi), but salt-water intrusion rarely extends beyond Beacon. The 50 mg/l (0.005%) salinity fluctuates from 32 km (20 mi) above the Battery (near Tappan Zee Bridge) to 112 km (70 mi) inland (south of Poughkeepsie) depending on the freshwater flow.

Although elevated levels of PCBs were first discovered in Hudson River biota late in 1969, their importance was not recognized for several years. By 1975, sampling of the river clearly implicated GE's capacitor manufacturing facilities at Ft. Edward and Hudson Falls, New York as the major source of PCBs in the Upper Hudson River. On December 18, 1972, GE had applied to the U.S. Environmental Protection Agency (EPA), Washington, D.C., for a discharge permit, stating that its two plants were discharging an average of 30 pounds per day (13.6 kg/day) of "chlorinated hydrocarbons" with a maximum of 47.6 pounds per day (21.6 kg/day).[3]

These discharges probably occurred for many years. Records of PCB usage by GE do not exist prior to 1966, but capacitor production using PCBs began as early as 1947 at Fort Edward. By 1976 PCB discharges were reduced to less than 250 grams/day and since May 1977 have been less than 1 gram/day.[4]

Until late summer of 1973, the PCBs were discharged into a river pool maintained by the Ft. Edward Dam. By October 1973 this dam was removed. The sedimentary regime of that portion of the river was changed and massive volumes of contaminated sediment that had collected behind the dam were suspended and transported downstream.[5] In April 1976 a 100-year flood in that region of the Hudson eroded more of the deposits left after the dam was removed. These deposits are still unstable and continue to erode and move downstream.

0077–8923/79/0320–0591 $1.75/0

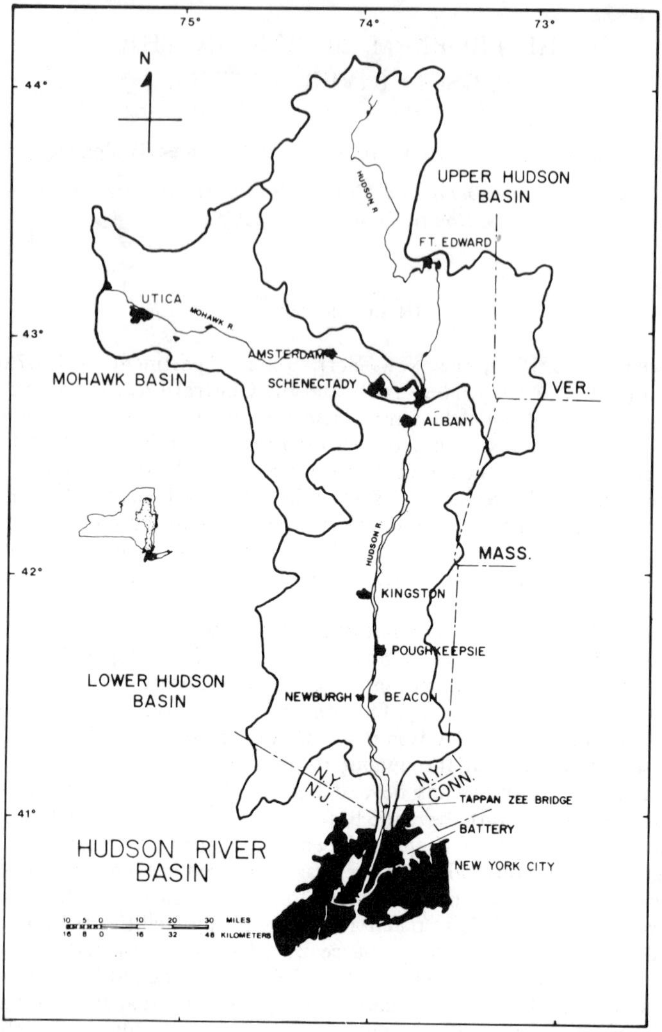

FIGURE 1. Map of Hudson River Basin.

SUMMARY OF CURRENT FINDINGS

River Sediments and Hydrology

During the spring and summer of 1977, the riverbed sediments were extensively sampled by surface grab sampling and coring.[6] The sample density was greatest in the section of the river just below Fort Edward where the PCB concentrations were previously found to be highest (FIGURE 4). Sediment sampling in the Lower River was much less extensive.

Upper River

PCB concentrations in Upper Hudson River sediments appear to be log-normally distributed (FIGURE 5). Given such a distribution, a log mean is the appropriate measure of central tendency. Although a weak trend exists for declining PCB concentration in sediments from Ft. Edward to Troy (FIGURE 4), the differences are

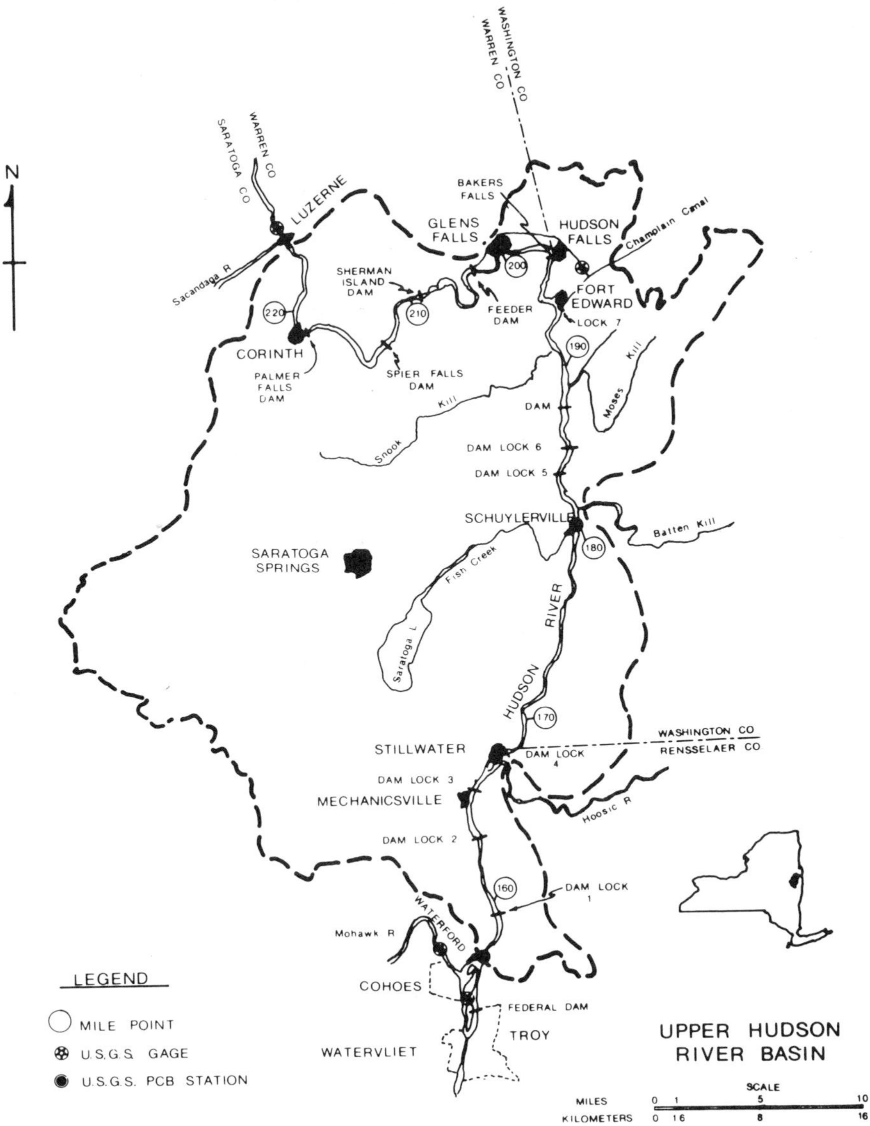

FIGURE 2. Map of Upper Hudson River Basin.

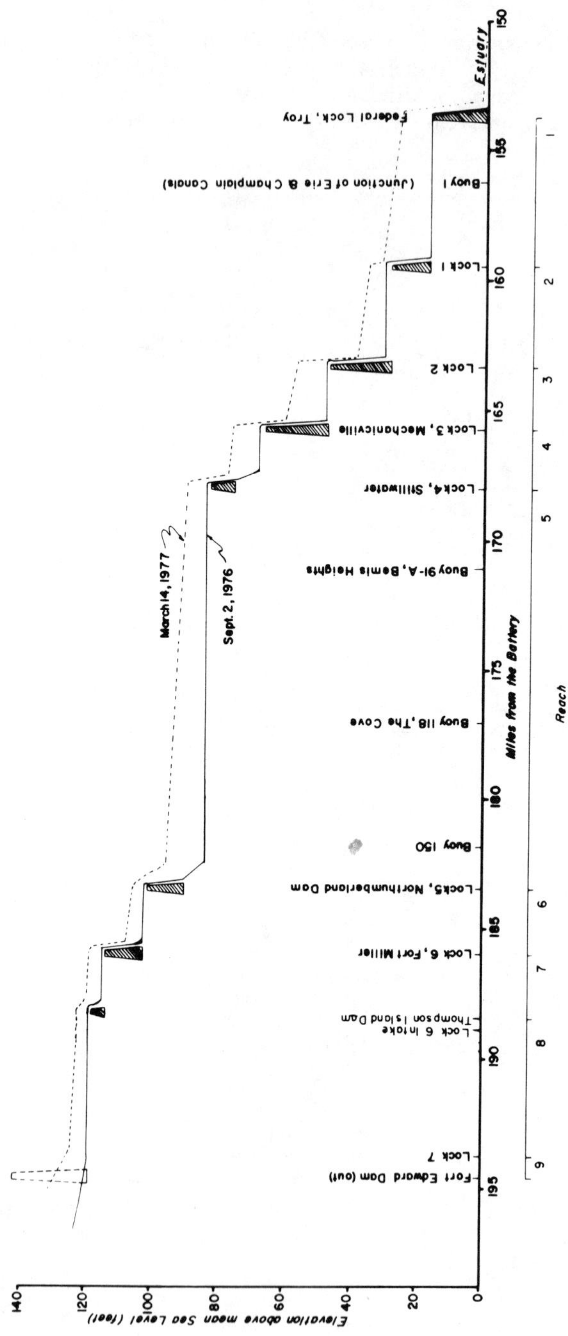

FIGURE 3. Upper Hudson River water surface profile, Fort Edward to Federal Lock. **Note:** U.S.G.S. data, dams not to scale.

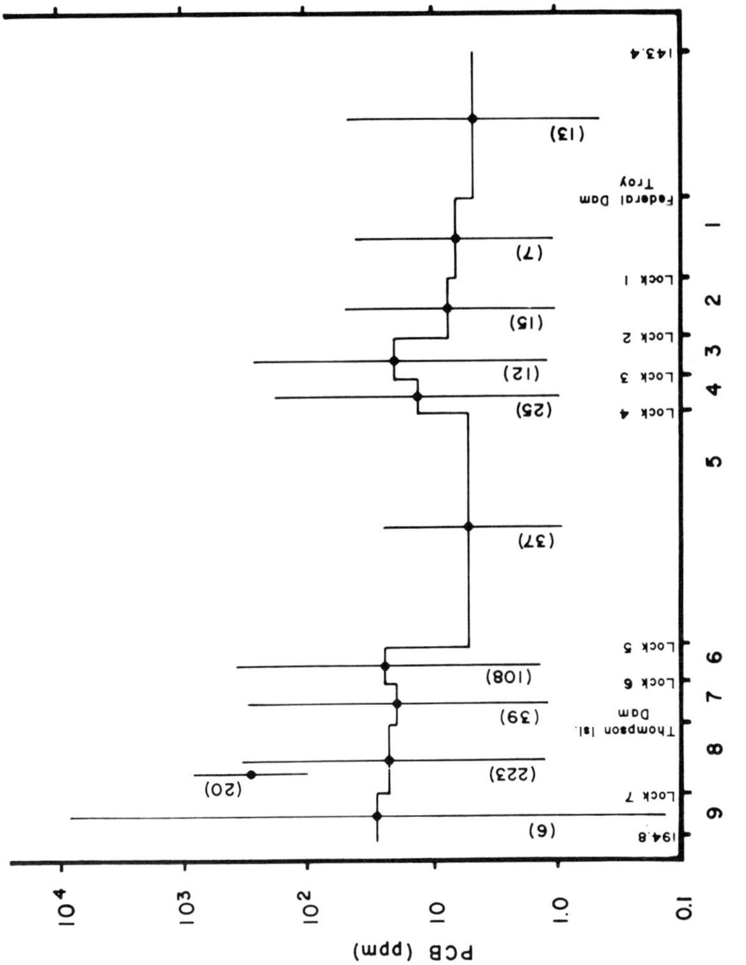

FIGURE 4. Average PCB concentration in sediments of Hudson River. Log mean and 95% confidence interval (number of samples).

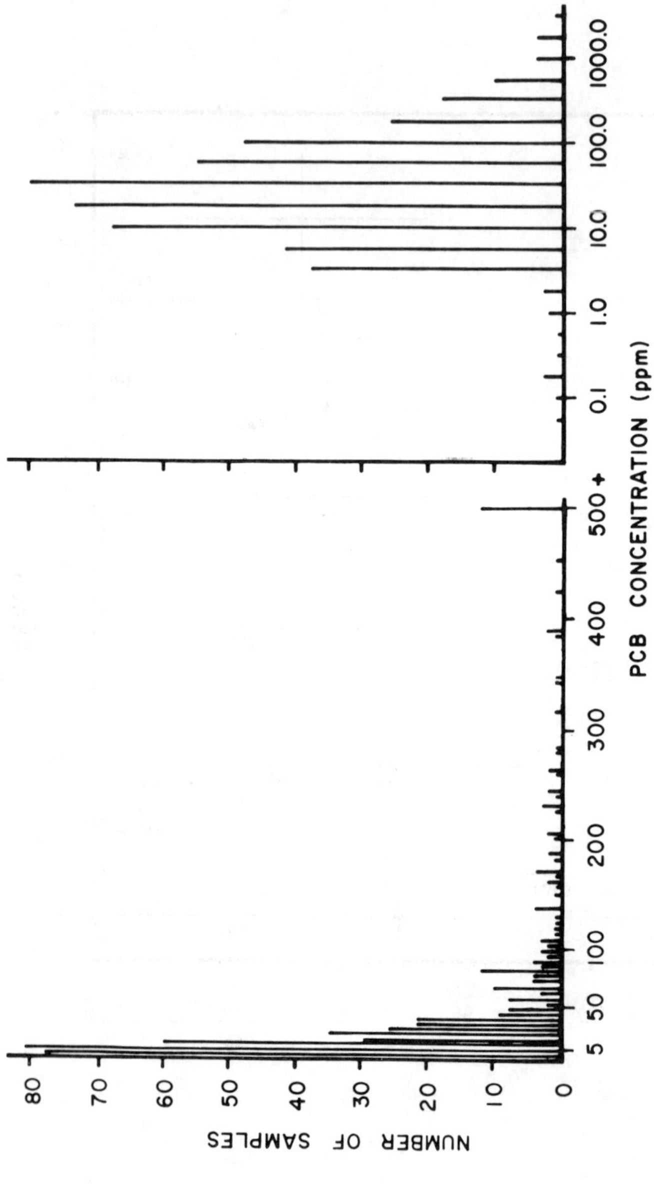

FIGURE 5. Frequency distribution of sediment PCB levels in the Upper Hudson River.

not statistically significant. Small sample sizes in the southern reaches, however, make such a comparison quite difficult. The variance, on the other hand, is reasonably constant and large in all of the reaches. This probably represents extensive, real variation of PCB concentrations within each reach. A separate set of 20 samples in reach 8 (FIGURE 4) were taken by a bucket dredge at one location. The variance of these samples is approximately one quarter the variance of the other reaches, exclusive of reach 9. Thus, although there is substantial variation in PCB concentrations at one location, larger areas exhibit even greater variation.

To provide additional insight into the measured concentrations, areas of high PCB concentrations (>50 ppm) were determined and the data stratified into two sets for each river reach; one of the measured values in high-concentration areas (hot spots), and one of measured values in low concentration areas (cold areas). The areas, average PCB concentration, and PCB mass in each of these subsets are given in TABLE 1.

TABLE 1

THE AREA, AVERAGE PCB CONCENTRATION AND MASS OF PCBS BY REACH OF HOT SPOTS (CONCENTRATION >50 ppm) AND COLD AREAS (CONCENTRATION <50 ppm) IN SEDIMENTS OF THE UPPER HUDSON RIVER

		Reaches*							
	Total	8	7	6	5	4	3	2	1
Average PCB levels (ppm)									
Hot spots	—	151	103	163	70	108	159	—	—
Cold areas	—	19.0	20.0	20.0	13.0	13.4	22.6	13.9	9.6
Overall	67.6	86.2	43.6	74.6	11.6	41.9	52.3	13.1	9.67
Area ($\times 10^5$ m^2)									
Hot spots	10.1	3.97	0.56	2.41	1.63	1.39	0.14	0	0
Cold areas	112	9.97	6.87	5.49	38.3	9.01	10.1	13.9	18.6
Total	122	13.9	7.43	7.90	39.9	10.4	10.2	13.9	18.6
PCB mass (metric tons)									
Hot spots	77	44	2.7	19	5.0	5.4	0.9	0	0
Cold areas	57	9.1	4.5	3.6	15	3.6	7.3	6.4	5.9
Total	134	54	7.3	22	20	9.1	8.2	6.4	5.9

Source: Tofflemire and Quinn.[7]
*Reaches are identified in FIGURE 3.

From studying the riverbed, cross-section maps and sediment characteristics, silt deposits are typically near the banks in the lower-velocity areas and gravel deposits are generally in the center or navigation section of the channel. The hot-spot areas occur more frequently near the banks, except for the northernmost river reaches.[7]

From the cores, it can be seen that the highest PCB concentration in the sediment is typically 6 to 12 inches (15–30 cm) below the surface (FIGURE 6). Substantially contaminated material is, however, found deeper in the sediments, particularly in the upper reaches (FIGURE 6). This may suggest that the most recent sedimentation is less contaminated.

During 1977, PCB monitoring and daily suspended-sediment stations at Glens Falls, Schuylerville, Stillwater and Waterford, New York demonstrated that PCBs continue to move from the Upper River Basin into the Lower River.[8] Although no PCBs (<0.1 ppb) were detected in any water samples from the Glens Falls station, PCBs were commonly detected at the other three sites. Throughout the year (late

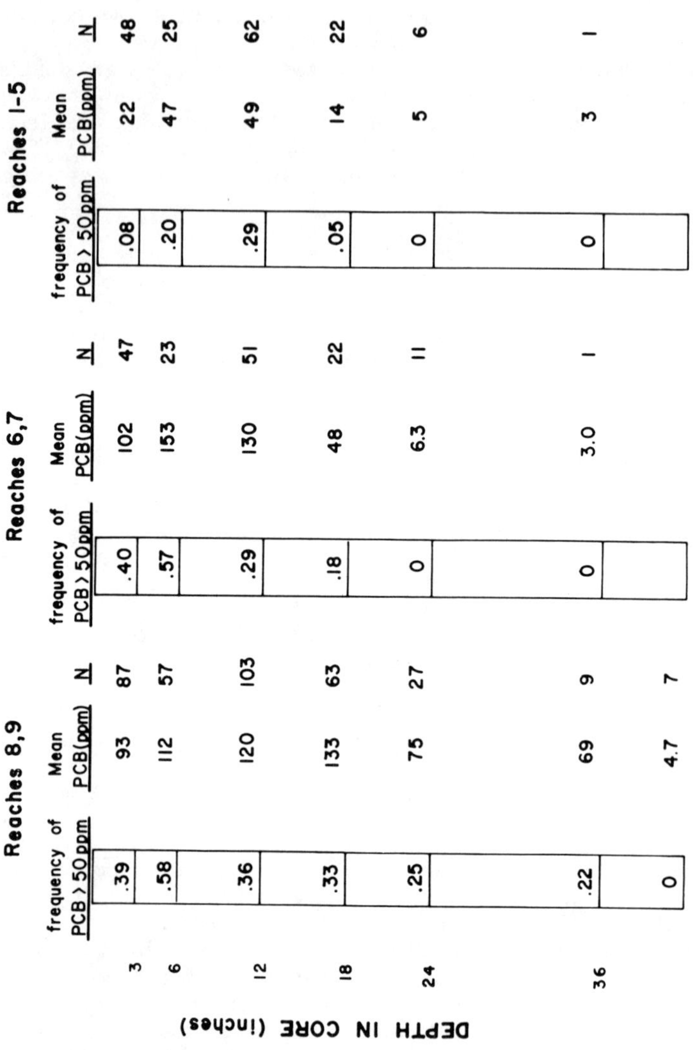

FIGURE 6. PCB penetration in cores by reach from areas where PCB > 5 ppm.

March to September) highest PCB concentrations (>1 ppb) were associated with high flows (approximately 10^6 l/sec), and the lowest PCB concentrations were associated with medium river discharges of 2 to 6 $\times$ 10^5 l/sec (FIGURE 7). With even lower flows (<10^5 l/sec or ~3500 cfs), PCB concentrations in the water at Stillwater, Schuylerville, and Wateford increase, thus indicating that a relatively constant load of PCBs was being transported except during storm events. This constant low-flow load in the Upper Hudson River was between 4 and 5 kg/day (TABLE 2). In March and April, two major storms carried approximately 750 kg (1650 pounds) of PCB into the Lower River. At peak flow approximately 112 kg/day were transported over the Federal Dam at Troy (TABLE 2). The transport of PCBs at Waterford during the entire year was approximately 2,600 kg (5,720 pounds), half of which occurred at flows less than 5.6 $\times$ 10^5 l/sec (~20,000 cfs).[8]

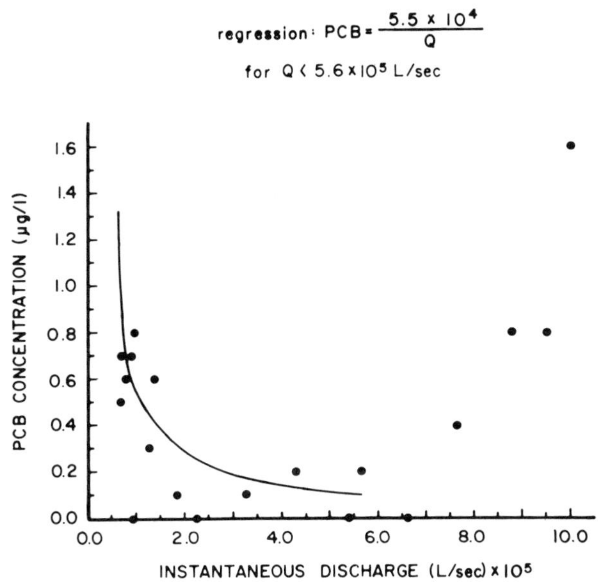

FIGURE 7. Hudson River at Schuylerville—PCB concentration vs. river discharge, March 30–September 27, 1977. (From Turk,[8] U.S. Geological Survey data.)

Lower River

Richard Bopp[9] has suggested that the Lower Hudson River can be divided into three distinct sedimentary regimes. The amounts of recent sedimentation (since the early 1950s) and the proportion of the river in each of these regimes are presented in TABLE 3. PCB contamination is confined to these recent sediments. Typical cores from these areas are shown in FIGURE 8. Here also, particularly near Albany, the surface sediments are less contaminated than those about 30 cm below the surface.

Estuary PCB concentrations in recent fine-grained sediments decrease from an average of 60 ppm in the Albany Turning Basin to an average of 6 ppm in the New York Harbor. The measurements indicate no other major historic or existing point source of PCBs to the Estuary other than the Upper River. In the core taken in the

TABLE 2

CALCULATED PCB LOAD IN THE UPPER HUDSON RIVER AT SELECTED RIVER DISCHARGES

Flow	PCB load in kg/day Schuylerville	Stillwater	Waterford
$<5.6 \times 10^5$ l/sec.	4.0	5.2	3.8
1.6×10^6 l/sec.	—	—	112

Source: data from Turk.[8]

Albany Turning Basin, changes of Aroclor 1016, 1242, and 1254 through the core are consistent with changes in the Aroclor usages at GE's Ft. Edward and Hudson Falls manufacturing facilities.

River Biota

Fish sampling from various reaches of the Hudson River and subsequent analysis of PCB levels in fish flesh have been directed at determining whether given species of fish could be considered safe for human consumption and whether fish-flesh PCB concentrations are affected by PCB discharge reductions or other remedial management actions.[10–12]

Earlier analyses[11] (1975–76) indicated that PCB levels were higher in fish taken above Troy than in fish from the Lower River (TABLE 4). In 1975–76 all species sampled from Ft. Edward to New York City had average total PCB concentrations exceeding 5 ppm (the FDA's temporary tolerance limit for PCBs in fish). Upstream of Hudson Falls fish contained only trace amounts of PCBs.

Analysis of more recent samples is not complete yet, but the general pattern, discovered in earlier samples, persists and confirms extensive contamination of Hudson River fisheries (TABLE 4).[13] Some differences do exist (e.g. yellow perch between Ft. Edward and the Federal dam at Troy), but the data are presently inadequate to assess any trends.

Sampling throughout New York State[11] from 1970–75 demonstrates that PCB contamination of fish flesh is highest in the Hudson River, but detectable levels of PCBs are present in fish from virtually any river system or large lake in the state. Fish from nonindustrialized or lightly industrialized watersheds in New York (Chemung, Susquehanna, Raquette) average less than 1 ppm in their edible flesh; although in the Black River, levels approach 4 ppm. In more heavily industrialized basins (Genesee,

TABLE 3

SEDIMENTARY REGIMES FOR THE LOWER HUDSON RIVER

Lower River Sedimentary Regime	Depth of Recent Sediment Accumulation (cm)	Proportion of River Bed (%)
Natural channel & submerged bank	<5	65
Coves and broad shallow areas	15–30	25
New York Harbor area	up to 250	10

Source: data from Bopp.[9]

Hoosic, Great Lakes), however, PCB levels in fish approach and often exceed 5 ppm.

Snapping turtles from the Lower Hudson River also exhibit elevated PCB levels. Although the muscle tissue in seventeen specimens averaged 4 ppm, a single fat sample contained 712 ppm PCBs.

In 1976 and 1977 multiple-plate samplers were tested as a method for monitoring PCBs in the water column of the Hudson River.[14] After exposing hardboard artifical substrates for periods of 4 to 5 weeks, organisms and other material that attach to the plates are removed and extracted for PCB analysis. In the Hudson River samples, macroinvertebrates comprise approximately half of the material attached to the sampling plates, the other half being detritus and attached phytoplankton. Most of the macroinvertebrates (60–70%) are midge larvae (Chironomidae) with the next most abundant organism being oligochaete worms (15–20%).[15]

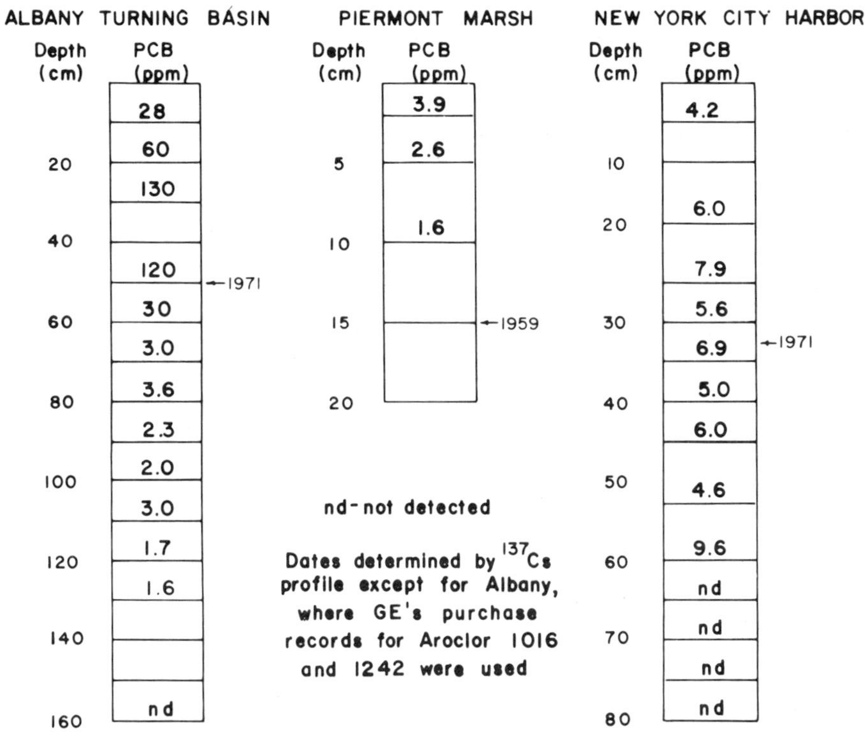

FIGURE 8. PCB levels from sediment cores in the Hudson River estuary. (Data from Bopp.[9])

As with the fish and sediments, consistently higher PCB concentrations have been found in organisms from the Upper Hudson below Ft. Edward than elsewhere in the river. Contamination of macroinvertebrates in the lower Hudson River, while less severe, was well above the background levels north of Ft. Edward (FIGURE 9).

An interesting temporal change can also be noted in FIGURE 9. In September 1977, at all the estuary stations with the exception of Troy, PCB concentrations in macroinvertebrates were higher than during September 1976 or August 1977. This suggests that PCBs are becoming more widely and evenly distributed in the Hudson River. Data from future surveys will help evaluate the significance of this trend.

In September 1977, macroinvertebrates from the Ft. Edward area had considerably elevated PCB concentrations, relative to August 1977 or the previous year. Dredging north and east of Rogers Island began in early August. Transport of PCBs from this activity undoubtedly accounts for the September increase of PCBs in the macroinvertebrates at the nearby Ft. Edward sampling station.

Enrichment cultures have been obtained from sediment core samples taken from a highly contaminated (1000 ppm total PCBs) region of the Upper River.[16] The microbial population found in the first 20 cm of core sediments ranged from 10^4 to 10^6 microorganisms per 10 g of wet sediment. Mixed cultures grown on 2% glucose as the primary carbon source were unable to degrade Aroclor 1221, 1016, or 1254 in shake-flask studies. Colonies were observed, however, on agar plates containing

TABLE 4

AVERAGE TOTAL PCB LEVELS IN EDIBLE FLESH OF FISH FROM THE HUDSON RIVER (Concentrations are Based on Wet Weight)

Location	Species	Collection date	Total PCB, (ppm) Average	(n)*	Range
Above Hudson Falls	Smallmouth bass	1975–76	Trace	(26)	
	Yellow perch	1975–76	Trace	(15)	
Fort Edward to Federal Dam Troy	Smallmouth bass	1975–76	72.6	(11)	41.5–122.9
	Largemouth bass	1975–76	61.7	(37)	12.5–164.4
		1977	70.7	(14)	6.2–235
	White sucker	1975–76	68.2	(37)	28.2–131
	Yellow perch	1975–76	134.6	(10)	79.3–299
		1977	12.2	(30)	1.57–42.7
	Goldfish	1977	576	(16)	41.8–1836
Albany/Troy	Smallmouth bass	1977	16.4	(9)	0.98–50.36
	Largemouth bass	1977	18.3	(2)	1.45–35.2
	Yellow perch	1977	9.3	(20)	1.49–27.3
	Brown bullhead	1977	45.0	(17)	3.2–110
Catskill	Largemouth bass	1977	29.5	(27)	<0.44–85.7
	Yellow perch	1977	4.21	(18)	0.84–10.51
Indian Point	Atlantic tomcod	1977	<0.96	(30)	<0.44–4.48
	American eel	1977	151	(7)	6.54–263

Source: Sloan.[13]

*Number of fish analyzed.

Aroclors 1221 and 1254 and shake-flask studies using these organisms (mixed culture) have shown them to be capable of degrading PCBs.[17] Further studies are in progress.

Terrestrial Environment

Studies have been carried out to ascertain the amount of PCB presently in landfills and dredge spoil sites in the Upper Hudson Basin and to estimate the erosion and leaching of PCBs from these sites.[18] TABLE 5 presents the results to date. Substantial quantities of PCBs exist in both types of disposal area.

Measurable quantities of PCBs are also leaching or eroding from these sites, and

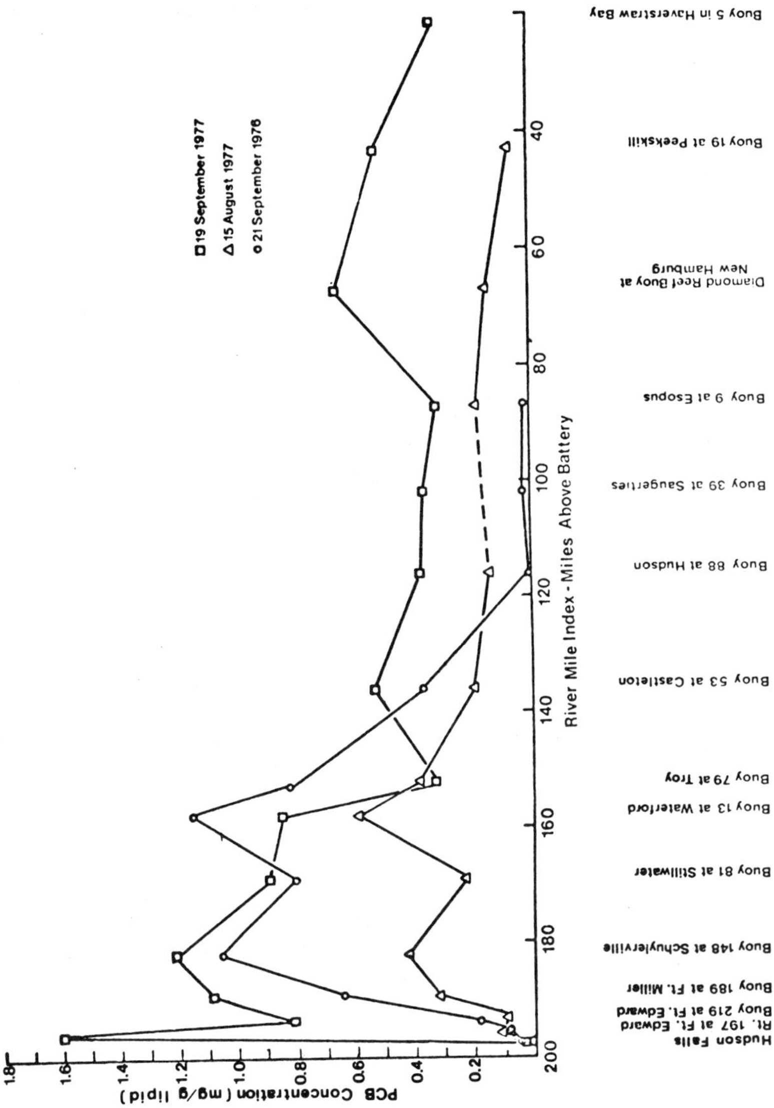

FIGURE 9. PCBs in multiplate residues. (From Simpson *et al.*[14] PCB analyses by N.Y.S. Department of Health.)

TABLE 5

ESTIMATED AMOUNT OF PCBs IN AND PROJECTED LOSS OF PCBs FROM SELECTED DREDGE SPOIL AREAS AND LANDFILLS IN THE UPPER HUDSON BASIN. (See text for important reservations about these values)

Site	Estimated PCBs in site (metric tons)	Groundwater loss (kg/yr)	Erosion loss (kg/yr)
Dredge spoil sites			
Lock 1	0.43	.08	6.8×10^{-3}
Lock 4	1.75	10.3	9.43
Site 212	1.57	5.12	11.0
Site 518	0.68	2.2×10^{-3}	7.85
S.A. 13	36.6	23.2	2.04
Moreau	20.6	5.61	1.90
New Moreau*	3.08	‡	‡
Rogers Island†	7.89	‡	‡
Total	72.6	44.4	32.2
Landfills and dumps			
Fort Miller	54.0	9×10^{-5}	168
Caputo	0.51	.16	1.18×10^{-3}
Old Ft. Edward	‡	191	0.033
Kingsbury	132	?	2.63
Ft. Edward	52.6	?	1.45
Total	240	191	172

Source: modified from Weston,[18] preliminary data.
*Estimated by Horn from NY DEC data.
†Estimated by Tofflemire from NY Department of Transportation dredging records.
‡Insufficient data.

the landfills and dumps appear to be larger contributors of PCBs to the surrounding environment than the dredge-spoil areas. One must be very cautious in interpreting this information, however, because the groundwater loss estimates do not yet incorporate local soil adsorption properties. The groundwater values are therefore probably considerable overestimates of PCB transport away from the immediate vicinity of the sites. Contaminated groundwater is not presently being used for drinking water at any of these sites.

In addition, the erosional losses from all sites should be considered "worst-case" estimates and also may not transport PCBs very far from the site. Very little of the eroded material from the Fort Miller dump, for example, probably reaches the Moses Kill or the Hudson River. On the other hand, because the dredge-spoil sites are immediately adjacent to the Hudson River, most of the eroded material from these sites probably does reach the river.

In 1977 ambient air monitoring in the Ft. Edward area revealed concentrations as high as 4000 nanograms/m^3 adjacent to the Ft. Edward manufacturing facility with average levels of approximately 1000 nanograms/m^3.[19] The levels varied with the day of the week, with higher levels detected during the work week and lower ones on weekends. After GE ceased using PCBs at their facilities in June 1977, ambient air levels decreased significantly (average-300 nanograms/m^3).

Dustfall was also sampled, and through June, the average PCB load from the atmosphere in dustfall was approximately 58 μg PCB/m^2/30 days (678 μg PCB/m^2/yr) with an average concentration of PCBs on the dust of 17 ppm.

Although the ambient air levels are considerably higher than have been previously found, the dustfall levels are not unprecedented.[20] Over Providence, Rhode Island, the maximum concentration reported was 9.4 ng/m^3; and in the Western North Atlantic over the Grand Banks, levels were usually 100 times less than that. Airborne particulates in four US cities were found to have from 27 to 230 μg PCB/g of particulates, and a continent-wide mean rate of fallout of 60 to 100 μg PCB/m^2/yr was estimated. However, dry fallout near a DDT manufacturing plant in the Los Angeles area was measured to be 600 μg PCB/m^2/yr.[20] The Ft. Edward levels are comparable to this, but understandably, above the national average.

Air samples taken at two dump sites (Caputo and Fort Miller) and the Moreau dredge-spoil site demonstrate that PCBs are clearly being emitted from the landfill sites. Dredge spoil appears to be a very weak source of PCBs to the atmosphere.

Plant samples were gathered at these same landfill and dredge disposal sites.[21] Two dozen plant species were sampled with a large number being common to all sites. Control tissues were collected several miles west of these sites. PCB concentrations in or on plant tissues from these collections varied widely from as high as 2770 ppm (dry weight) to less than detection (TABLE 6).

The extremely high levels of PCBs associated with the plant tissue suggest that surface adsorption of contaminated dust is involved rather than transport of PCBs from the soil. Additionally supporting this hypothesis is the fact that plants closest to the soil (goldenrod, aster, grasses) have the highest PCB levels at each site; staghorn sumac, the next highest, is very pubescent and somewhat sticky, making it a better dust trap than other plants. Finally, leaves and twigs of pine trees were collected at two heights (<2m and 5m) and cores of the trunks were taken. The lower tissues and the tissues with the greatest surface-to-volume ratio (leaves) have the highest concentrations of PCBs, whereas the cores (vascular tissue) have very low concentrations (1 ppm as compared to 300 ppm for the lower leaves and 90 ppm for the lower twigs).

Although root-to-shoot transport might be implicated by the presence of PCBs in xylem wood (cores), three other possibilities exist. PCBs may have diffused through the bark and be associated with the outer surface of the xylem or cambium, transport may be occurring from the leaves to the stem and roots, or possibly, inattention to contamination from handling or coring through contaminated bark could be responsible.

Even if PCBs are not transported by the plants and merely attached to the surface

TABLE 6

AVERAGE PCB LEVELS ASSOCIATED WITH PLANT TISSUES AT DUMP AND DREDGE DISPOSAL SITES ALONG THE UPPER HUDSON RIVER

(Concentrations are based on dry weight)

Site	Average PCB (ppm)	Number of Samples	Number of Species	Range
Caputo	502	12	9	2770–28.3
Fort Miller				
on dump	146	9	6	484–11.6
below dump	15.1	7	6	23.4–5.13
hill opposite	6.0	9	7	14.4–1.32
Moreau	0.58	19	10	1.34–0.25
Control	0.23	22	15	1.68–<0.2

PCB analyses by WARF Institute, Madison, Wisconsin (USA).

of the leaves and twigs, any animal that consumes these tissues will also consume the PCBs. Effects on the plant itself appear to be minimal. No gross abnormalities nor obvious signs of chlorosis that could be definitely associated with high PCB doses were noted among the contaminated plants.

Mass Balance for PCBs in the Hudson River

Mass of PCBs in Hudson River Basin

The mass of PCBs presently contaminating at least four compartments of the Hudson River Basin environment (the river sediments, its biota, dredge spoil areas and landfills and dumps) has been estimated. Insufficient data exist to quantify contamination of the general landscape, but this contamination is probably small compared to

Table 7
Summary Distribution of Residual PCBs in the Hudson River Basin

Area	Estimated Total PCBs (metric tons)	Estimated Scourable PCBs (metric tons)	Calculated Loss Rate* (metric tons/yr)
River Sediments			
Remnant deposits[23]	63.5	20.4	3.9
Upper Hudson riverbed[7,8] (Ft. Edward Dam Site to Troy)	134	44.4	2.6
Lower Hudson riverbed[9] (Estuary-troy to New York Harbor)	91	†	†
Dredge spoil areas[18] (Upper Hudson)	72.6	†	0.8
Landfills and dumps[18] (Upper Hudson)	240	—	.36
Biota	.09–.91	—	†
Lower River[27]			
Total	602	64.8	—

*Does not include volatilization.
†Insufficient data.

what has been measured. The air and water levels vary too rapidly over time to calculate a meaningful mass of PCBs.

Slightly more than 600 metric tons are estimated to contaminate the Hudson River Basin (Table 7.) A little less than one half (48%) of these PCBs is in the river sediments and a bit more than one half (52%) is in landfills or industrial dump sites. Only a very small fraction (<0.2%) of the PCBs is contained within the biota of the river. Although the Lower Hudson River sediments are far from pristine regarding PCBs, the bulk of PCBs contaminating river sediments are found in the Upper Hudson. The remnant deposits (sediment that collected behind the old Ft. Edward Dam at Ft. Edward) and the riverbed sediments between Ft. Edward and Troy contain 69% of the PCBs in the riverbed.

Approximately 57% of the PCB mass in river sediments between Ft. Edward and Troy is located in hot spots that represent only 8% of the riverbed area (Table 1). No hot spots have been found in the lower two reaches (below Lock 2), but fewer samples were taken there and small areas could have been missed.

Movement of PCBs in Hudson River

In theory, PCBs can move from riverbed sediments primarily by two mechanisms: leaching (or desorption) from the sediments into the flowing water and/or erosion of the contaminated bed material. Although these are basically physical phenomena, they may be enhanced by biologic activity. Benthic organisms pump water through the surface layer, and microbial fermentation of organic matter produces methane and other gases that can stir up the sediments in rising. Barge traffic during low water also enhances sediment scour.

Erosion (scour) probably provides the primary mechanism of PCB transport in the Upper Hudson River, but the half of PCB transport[8] that occurs at low and moderate flows may not be the result of riverbed scour. The amount of PCBs vulnerable to scour (TABLE 7) has been estimated from a computer model of river hydraulics and sediment transport.[22] Approximately one third (33%) of the total PCBs in the sediments between Ft. Edward and Troy is subject to scour (TABLE 7).

Although it is not possible to quantify the relative contribution of the hot spots to this scour, some hot spots do lie in erosional zones. The center of the river channel usually has the greatest potential for scour, but the PCB concentration in these sediments rarely exceeds 50 ppm. (N.B. High concentrations are sometimes found where the cross section of the river is greatly enlarged; the channel areas here act as transient sediment traps, which scour under extreme flows.) The river banks and shallow shoreline deposits are depositional areas where 45–90 cm of sediment have collected during the past 68 years. Sediment samples with high PCB levels are more frequently associated with these regions than the channel regions.[7]

Although the source is yet to be defined, the total mass of PCBs moving into the estuary is well documented. The Upper Hudson basin delivered approximately 2.6 metric tons of PCBs to the estuary in 1977. The sediment-transport model estimates that between 32 and 54 metric tons of PCBs will be transported into the estuary during the next 20 years if no remedial action is taken.[22]

The remnant deposits also constitute a source of readily erosible material that in some areas is highly contaminated.[23] Approximately 20 metric tons of PCBs (32% of PCBs in remnant deposits) can eventually be transported out of the area by erosion. This is occurring rapidly as the loss of PCBs from the remnant deposit is estimated at 3.9 metric tons per year.[23] Much of this will probably deposit in lower reaches for now, but some undoubtedly contributes to the present flow of PCBs over the Federal Dam at Troy (2.6 metric tons per year).

The rate of leaching of PCBs from sediments has not been investigated yet. After the scourable contaminated sediments have been swept out of the Upper Hudson Basin or to stable deposits, PCBs will move into the water column by leaching. Contamination of the water and biota would consequently continue.

Although the dumps and dredge spoil sites contain more than half of the PCB estimated to be in the basin, only a very small proportion is lost through erosion and leaching. Movement of PCB from these sites is almost an order of magnitude lower than movement in the riverbed (TABLE 7). Considering that these estimates probably exaggerate actual PCB losses and that no attempt has been made to reduce erosion or leaching at most of these sites, PCBs remain much more confined on land than in the riverbed.

Relation of Hudson River to North American PCB Use and Environmental Contamination

PCB production in the USA has rapidly declined since 1970. In that year the Monsanto plant at Sauget, Illinois (sole US manufacturer) produced about 38,000 metric tons, 33,000 metric tons of which were sold in the USA.[24] Since 1929, when PCBs were first introduced, almost 450,000 metric tons of PCBs have been sold in North America. Although the data on PCB discharges are very limited, EPA estimates that approximately 270,000 metric tons have been placed in dumps, 27,000 metric tons have been discharged to the atmosphere and 54,000 metric tons have been discharged directly to fresh and coastal waters.[24]

Between 1966 and 1974, General Electric's Ft. Edward and Hudson Falls facilities purchased 35,000 metric tons of PCBs or approximately 15% of Monsanto's domestic sales during that time.[3,24] This suggests that General Electric's discharges to the Hudson River Basin might represent close to 15% of the nationwide total discharges to the environment.

Generally speaking, contamination of the Hudson River Basin exceeds, in level and scope, any other area found to date in the USA.[25] River sediment concentrations are at least an order of magnitude higher than other contaminated systems (e.g. Lake Ontario; Lake Michigan; Escambia Bay, Florida; and soils from the Monsanto plant). The PCB level at a few other sites (e.g. Waukegan River, Illinois and Housatonic River, Connecticut) are comparable, but the resources contaminated do not appear to be as great as the Hudson River. Estimates of the quantity of PCBs in the sediments of the Great Lakes[26] are considerably lower than those for the Hudson River (e.g. Lake Ontario, 9 metric tons; Lake Erie, 35.6 metric tons; and Lake Michigan, 7 metric tons).

References

1. Hetling, Leo J. & Edward G. Horn. July 1977. Hudson River PCB Study Description and Detail Work Plan. NYS Department of Environmental Conservation. Albany, N.Y. 62 pp.
2. Hetling, L., E. G. Horn & T. J. Tofflemire. April 1978. Summary of Hudson River PCB Study Results. Technical Report #51. Bureau of Water Research. New York State Department of Environmental Conservation. Albany, N.Y. 88 pp.
3. Sofaer, A. February 1976. Interim Opinion and Order. unpublished opinion in the matter of violations of ECL by GE Company. New York State Department of Environmental Conservation. File #2833. 77 pp.
4. Data abstracted from National Pollutant Discharge Elimination System Permit Number 0007048. Files maintained in Bureau of Industrial Programs. New York Department of Environmental Conservation. Albany, N.Y.
5. Malcolm Pirnie, Inc. August 1977. Investigation of Conditions Associated with the Removal of the Fort Edward Dam. Review of 1975 Report. White Plains, N.Y. 141 pp.
6. Normandeau Associates, Inc. September 1977. Hudson River Survey 1976–1977 with Cross-section and Planimetric Maps. Bedford, New Hampshire. 351 pp.
7. Tofflemire, T. J. & S. O. Quinn. PCB in the Upper Hudson River: Mapping and sediment relationships. In preparation.
8. Turk, John. 1978. Personal communication.
9. Bopp, Richard. 1978. Personal communication.
10. Anonymous. 1977 Fish Sampling and Analysis Program—Hudson River. New York State Department of Environmental Conservation. Bureau of Environmental Protection (unpublished). 7 pp.

11. SPAGNOLI, J. & L. SKINNER. September 1977. PCBs in fish from selected waters of New York State. Pest. Monit. J. **11**(2): 69–87.
12. SLOAN, R. January 1978. Report presented to PCB Settlement Advisory Committee (unpublished).
13. SLOAN, R. June 1978. Personal communication.
14. SIMPSON, K. W., R. C. MT. PLEASANT & B. BUSH. September 1977. The Use of Artifical Substrates for Monitoring Toxic Organic Compounds: A Preliminary Evaluation Involving the PCB Problem in the Hudson River, New York. Manuscript for Kepone Seminar II. U.S. EPA. Region II. Easton, MD. 17 pp.
15. SIMPTON, K. March 1978. Personal communication.
16. GENERAL ELECTRIC CORPORATE RESEARCH & DEVELOPMENT. January 1978. Report to PCB Settlement Advisory Committee (unpublished).
17. MCFARLAND, C. March 1978. Personal communication.
18. LEIS, W. (WESTON ENVIRONMENTAL CONSULTANTS). March 1978. Personal communication.
19. ROMANO, D. January 1978. Report to PCB Settlement Advisory Committee (unpublished).
20. FULLER, B., J. GORDON & M. KORNREICH. April 1976. Environmental Assessment of PCBs in the Atmosphere. Mitre Corporation. McLean, VA. for U.S. EPA. 336 pp.
21. BENENATI, F. (WESTON ASSOCIATES). January 1978. Report to PCB Settlement Advisory Committee (unpublished).
22. LAWLER, MATUSKY & SKELLY ENGINEERS. March 1978. Upper Hudson River PCB No Action Alternative Study: Final Report. Pearl River, New York. 190 pp.
23. MALCOLM PIRNIE, INC. March 1978. Environmental Assessment of Remedial Measures at the Remnant Deposits of the Former Fort Edward Pool. White Plains, N.Y. 173 pp.
24. U.S. ENVIRONMENTAL PROTECTION AGENCY. January 1975. Scientific and Technical Assessment Report on Polychlorinated Biphenyls. Washington, D.C. EPA-600/6-75-00X. 109 pp.
25. U.S. ENVIRONMENTAL PROTECTION AGENCY. January 1976. Review of PCB Levels in the Environment. Washington, D.C. EPA-560/7-76-001. 143 pp.
26. POLLUTION FROM LAND USE ACTIVITIES REFERENCE GROUP (PLUARG). July 1978. Final Report. International Joint Commission. Windsor, Ontario. 115 pp.
27. HYDROSCIENCE ASSOCIATES, INC. April 1978. Estimation of PCB Reduction by Remedial Action on the Hudson River Ecosystem. Westwood, N.J. 107 pp.

THE FATE OF PCBs IN THE HUDSON RIVER ECOSYSTEM

Robert V. Thomann

Environmental Engineering and Science Program
Manhattan College,
Bronx, New York 10471 and
Hydroscience, Inc.
Westwood, New Jersey 07675

John P. St. John

Hydroscience, Inc.
Westwood, New Jersey 07675

Introduction and General Problem Framework

The focus of this paper is the description of the distribution and fate of polychlorinated biphenyls (PCBs) in the Hudson River with particular emphasis on the aquatic ecosystem. A settlement between the New York State Department of Environmental Conservation (NYSDEC) and the General Electric Company (GE) concerning the contamination of the Hudson River by PCBs discharged from GE's facilities at Fort Edward, New York, called for an overall study of the Hudson. The direction of the work reported on herein is to provide input into the decision making process of estimating the effects of remedial actions on PCB levels in the general biota and the larger migratory fish. This paper considers only briefly the transport of PCBs in the abiotic sector (e.g., bottom sediment, suspended particulate matter and water column); a more complete treatment is given in reference 1.

FIGURE 1 is a schematic of the Hudson River that indicates the mile points (MP) of key locations and also shows the divisions of the Hudson into reaches for the physical transport analysis and reaches for the biologic analysis. From the Federal Dam at Troy to the Ocean, the Lower Hudson is tidal and depending on the average freshwater flow, the end of the salt water intrusion oscillates approximately between the Tappan Zee Bridge (MP 25) and Poughkeepsie (MP 75) and under severe drought conditions, it may reach as far north as MP 80. The long term monthly average discharge and the 1976 monthly average discharge from the Upper Hudson to the Lower Hudson are 13,270 cfs and 22,100 cfs respectively.

The major interactions that lead to the PCB distribution in the biotic and abiotic sectors of a given aquatic environment are shown in FIGURE 2. Although the detailed dynamics of these interactions are not known at the present time, a qualitative overview is possible. PCBs exist in a body of water in two main components, the dissolved and the particulate phases. The dissolved component includes PCBs that exist in solution and the particulate component includes PCBs that are adsorbed on various forms of particulate (suspended) matter. The formation of these two components results from the characteristic low solubility and high adsorption of the PCBs. The particulate component may settle and contaminate the bottom sediment. On the other hand, the bottom sediment can be resuspended through the shearing action of the overlying water and through the actions of organisms that inhabit the upper layers of the bottom sediment. The existence of contaminated bottom sediment gives rise to dissolved PCBs in the interstitial water, which diffuses dispersively into the overlying water. At the air-water interface, there are evaporative PCB losses from

0077–8923/79/0320–0610 $1.75/0

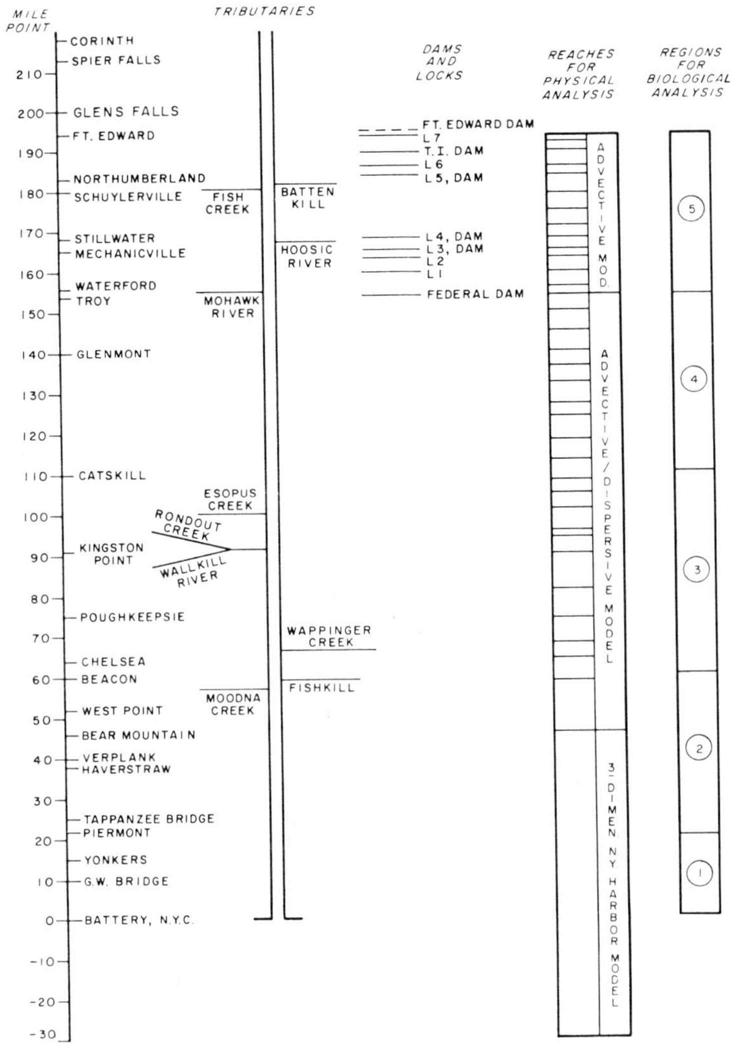

FIGURE 1. Schematic of the Hudson River and segmentation used in PCB analysis (biologic and physical).

the dissolved component into the atmosphere. In this work, for any component, only total PCBs are considered, i.e., no specific analysis of various isomers or mixtures of PCBs is attempted.

The ecosystem of any body of water is an extremely complex interaction of organisms from bacteria to higher order fish. As a simplification, one may think of the system as a food chain extending from the smaller to the larger organisms. Each segment of the food chain accumulates PCBs from two main sources, PCBs from the abiotic components and PCBs from contaminated food. The food chain also excretes

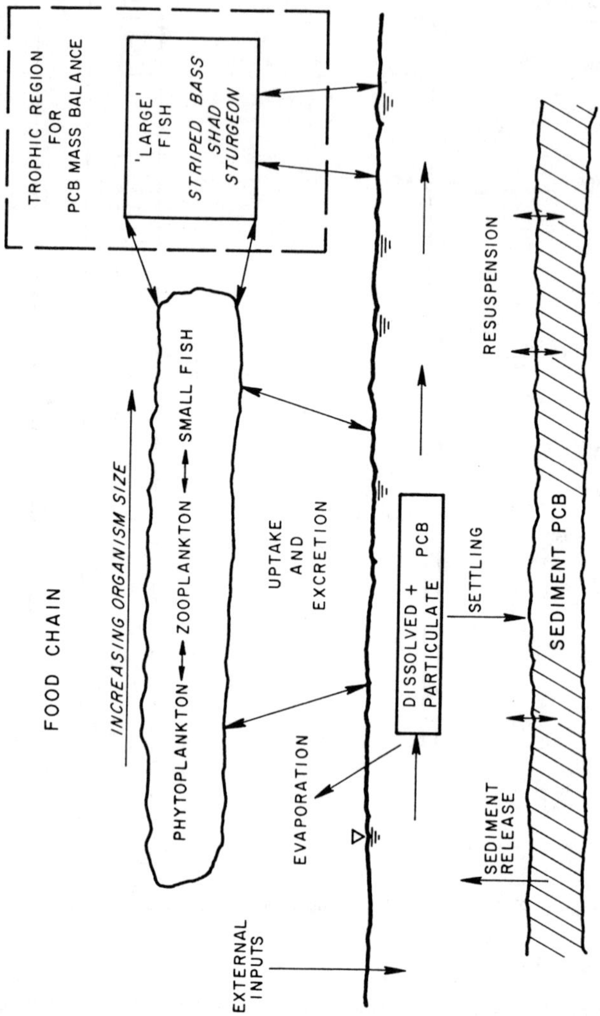

FIGURE 2. Schematic of major features in PCB distribution throughout food chain.

PCBs into the abiotic sector. Thus, the accumulation of PCBs in any segment of the food chain will depend on the PCB distribution in its ambient abiotic environment, food sources, and its physiology, which will determine its PCB clearance rates. This whole system of complex interactions is ultimately driven by the external PCB inputs.

Analysis of Water Column and Sediment PCB Data

The principle of the conservation of mass permits a first approximation to describing the spatial distribution of the various components of PCBs. The following assumptions are made: (1) the mass of PCBs in the food chain is small relative to that in the water column; (2) a local equilibrium prevails between the particulate and dissolved phases of the PCBs; (3) the adsorption phenomena is linear; (4) the river system is in temporal steady state (although longer term trends may be present); and (5) losses from the water column are principally through sedimentation, although some evaluations of losses due to evaporation and biodegradation has been made.[1] The general equation for the total PCB concentration, $c_T(\mu g/1)$ in a dimensional system can be written as:

$$\frac{udc_T}{dx} - E\frac{d^2c_T}{dx^2} + \left[\frac{am}{1+am}K_s + \frac{K_e + K_n + K_d}{1+am}\right]c_T = \left[W_{TB} + W_{TA} + W_{TE}\right] \quad \mathbf{(1)}$$

where x is distance downstream, u is the river or estuary velocity (cm/sec), E is the tidal dispersion (m^2/day), m is the mass of suspended solids (mg/l), "a" is a partition coefficient between the particulate and dissolved phases ($\mu g/g \div \mu g/1$), K_s, K_e, K_n, K_d are coefficients [day^{-1}] representing sedimentation, evaporation, photooxidation, and biodegradation, respectively; and W_{TB}, W_{TA} and W_{TE} are total input loads (kg/l-day) of PCB from bottom sediment interactions, atmospheric sources and other external sources, respectively. For the Upper Hudson area, W_{TA} can be assumed close to zero and W_{TB} can be given from the resuspended bed sediment load as $W_{TB} \approx r_b W_{ss}$ where r_b is the concentration of PCBs in the resuspended sediment (μg PCB/g sediment, a measurable quantity) and W_{ss} is the sediment input (kg sediment/liter-day). A mass balance on the suspended solids is therefore also required and is given by

$$\frac{udm}{dx} - E\frac{d^2m}{dx^2} + K_s m = W_{ss} \quad \mathbf{(2)}$$

With m from **(2)** and estimates of the coefficients for **(1)**, c_T can be calculated. Since the total concentration is the sum of the particulate and dissolved, the dissolved concentration, c_D, can be calculated from:

$$c_D = c_T/(1 + am) \quad (3)$$

This concentration then provides an estimate of the dissolved form of PCB which may be available for uptake and bioconcentration by the aquatic food chain. The approach to analyzing the PCB column data was to first fit the suspended solids data using Equation 2 which provides an estimate of K_s and W_{ss} (but not uniquely). Then using field data of r_B and the determined distribution of m and K_s, the PCB data are calibrated using Equation 1. In using this latter equation, a knowledge of the partition coefficient a is needed. The limited available data for the Upper Hudson River show a large variation from 20–500 ($\mu g/g$ per $\mu g/l$). Dexter[2] in his work examined the partitioning of PCBs in the two phases and reported values from 13–75 $\mu g/g$ per $\mu g/l$.

A value of 100 μg/g per μg/l was chosen for this work. The reaches for the physical analysis of PCBs are shown in FIGURE 1.

For the Upper Hudson (north of Federal Dam), the model was applied to PCB surveys of March 18 and March 29–31, 1976. The resulting profiles and the survey data are shown in FIGURE 3 and as shown, the agreement between the observed data and the calculation is good. However, background and tributary PCBs had to be assigned based on rather meager data.

FIGURE 4 shows the calculation profile for the entire length of the Hudson to the Battery. In the Lower Hudson, data on total PCBs were available only for the reach for MP 65–MP 90. In the top graph of FIGURE 4 are shown two profiles resulting from different assumptions on bottom PCB concentration as shown in the bottom graph. The dashed line therefore in the water column PCB calculation corresponds to an assumption of 1 μg/PCB/g sediment for the Lower Hudson. It is estimated on the basis of FIGURE 4 that the total PCB concentration from MP 0 to MP 110 ranges from 0.1–0.5 μg/l. The average dissolved PCB concentration (not measured) is estimated at about 0.1 μg/l (100 ng/l).

Various schemes have been suggested for reducing the effect of the upstream PCB sediments, including the removal of the contaminated sediment by dredging. Therefore, two dredging alternatives were considered and their impacts on the water column PCB were calculated.

Under the first alternative, only the Thompson Island pool at Fort Edward is dredged to a bottom sediment PCB concentration of 1 μg/g. Under the second dredging alternative all pools of the Upper Hudson are dredged to a bottom sediment PCB concentration of 1 μg/g.

The average PCB concentration for the Upper Hudson is calculated to be appreciably reduced from approximate present levels of 0.5 μg/l to 0.28 μg/l and 0.14 mg/l under the two alternatives. This represents an approximate 20% and 70% reduction of level from the Upper Hudson to the Lower Hudson. These results assume no change in the estuary PCB sediment concentration. However, at the mean annual flow, the estuarine sediment PCBs are the primary source of estuarine water column PCBs. Therefore, if the contaminated sediment of the Upper Hudson were not partially or totally removed or inactivated, such sediment would continue to add to the contamination of estuarine sediments by means of the naturally occurring sediment discharged over the Troy Dam.

Recognizing this indirect effect and an approximate range of 20–70% reduction in load at Troy depending on the degree of dredging, it has been assumed that the bed sediment-bed load input may drop by 50%. In this case the calculated total PCB concentration is almost halved and in the biologically active region, the average dissolved PCB concentration is estimated at 0.05 μg/l.

PCB UPTAKE AND DISTRIBUTION IN THE AQUATIC ECOSYSTEM

The available data base on PCB levels in the Hudson River ecosystem is relatively sparse and generally limited to the fish of the upper trophic levels. Further, the state of the art of understanding and predicting effects of control actions on hazardous substances such as PCBs is in a very early stage of development. As such, the rigor and quantitative nature of estimates of future scenarios is restricted by available data and understanding of mechanisms. Nevertheless, in the present problem context, some estimates of control effectiveness must be made to aid the ongoing decision-making process.

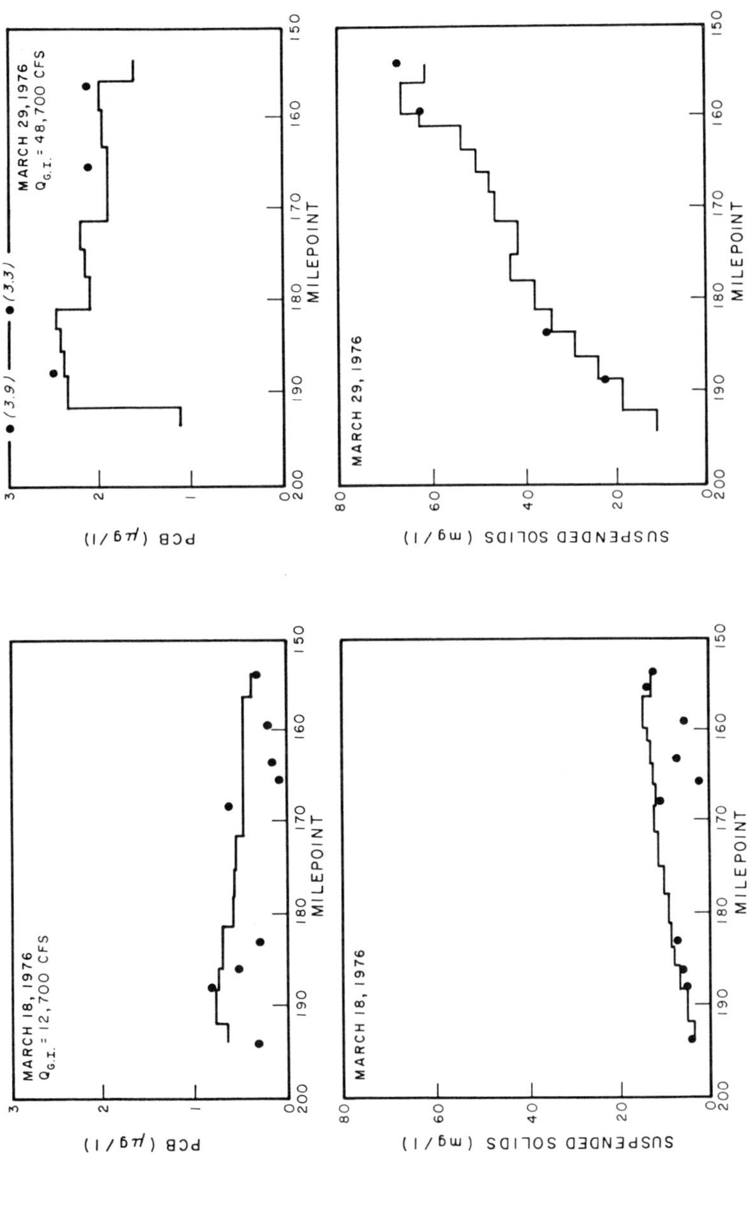

FIGURE 3. Total PCB and suspended solids calculated profiles (Upper Hudson River).

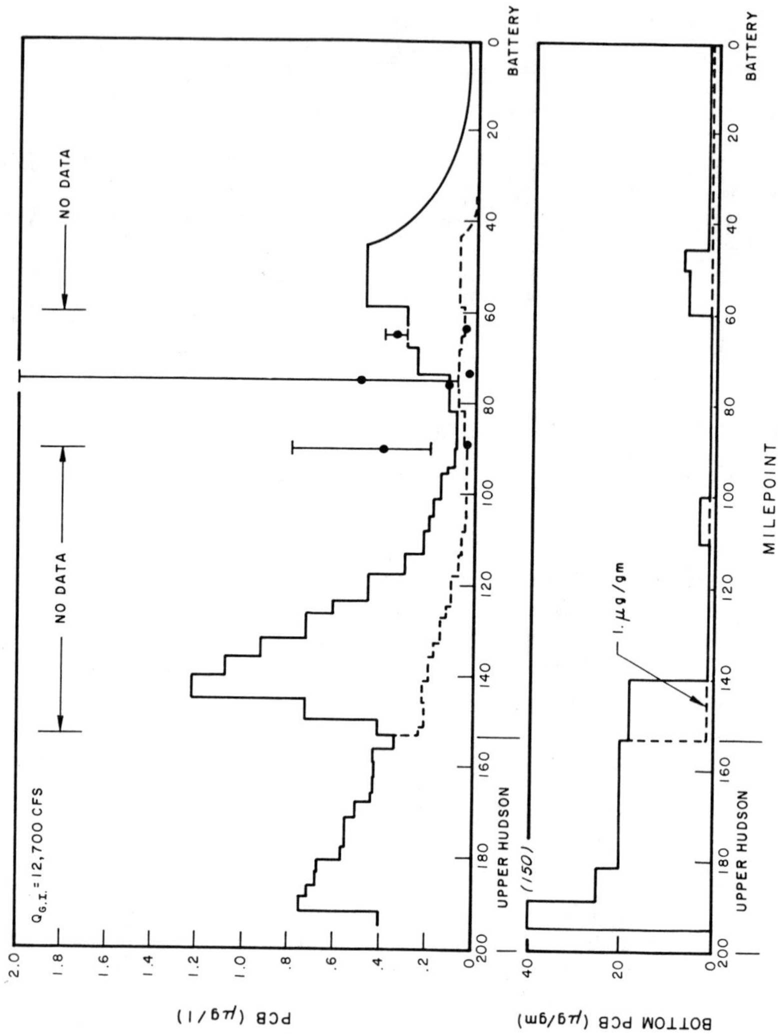

FIGURE 4. Total PCB calculated profile and bottom PCB distribution.

Recognizing this problem setting, a distinction must also be made between the *fate* of PCBs in the aquatic ecosystem and the *effect* of PCBs on the system. The former factor relates to the distribution in time and space of PCBs and throughout various sectors of the food web. The latter factor relates to the impact that a given level of PCB concentration has on a given sector of the ecosystem. Thus, an action level of 5μg PCB/g for the commercial fishery reflects a determination that such a level is detrimental in some sense. In this work, only the fate of PCBs in the Hudson is assessed as opposed to the effects of various concentrations on such factors as suitability for human consumption, reproductive behavior, and effects on growth dynamics.

In the analysis framework discussed herein, primary emphasis is placed on the upper trophic levels of the resident and migrating fish, particularly the striped bass. This emphasis is warranted because of the high degree of interest in the fishery and the recognition that the fishery represents the major present impact of high PCB levels in the Hudson.

Therefore, a "one-dimensional" food chain (not a food web) as shown in FIGURE 2 can be conceived. For this work, rather than try to describe the entire food-chain PCB distribution in detail (partly as a result of a lack of data), a mass balance is written around a defined upper trophic level representing the "large" fish such as striped bass. The following assumptions are incorporated:

1. Uptake of PCB by the food chain including top carnivores is linear with respect to water concentration and proportional to fish biomass.
2. Excretion of PCB by food chain is approximately a first order reaction.
3. Water PCB concentration, both dissolved and suspended, is not significantly influenced by biologic processes of uptake and excretion.
4. Top carnivores feed in a size range of prey from L_1 (μm) to L_2 (μm), the PCB concentration of which is known or can be calculated.
5. Spatially, carnivores are distributed equally throughout normally large (10–100 mile) "completely mixed" regions of the Hudson as shown in FIGURE 1.

The justification for assumptions 1 and 2 is discussed in Thomann[3] where data have been analyzed from laboratory experiments. It should be noted, however, that excretion rates may be quite variable among various organisms and PCBs may be incorporated in "less easily flushed" segments of the organisms (such as bone or muscle). Assumption 3 is justified on an estimate that the entire food chain throughout the lower Hudson contributes less than about 200 lbs PCB/day, which is relatively small compared to all other inputs.[1] Assumption 4 requires data on PCB concentrations in the food of the top predators or a computational model for estimating such concentrations. Very few data are available in the Hudson River on smaller organisms that are prey for the larger fish and one must fall back on simple estimation techniques. Assumption 5 is an assumption of convenience that recognizes the implausibility of detailed spatial computations of PCB in fishes given present levels of understanding and available data.

The general mass balance for PCBs in the top carnivore level for the *ith* spatial segment can be written as the following equation:

$$V_i \frac{ds_i}{dt} = (AU)_{i-1,i}s_{i-1} - (AU)_{i,i+1}s_i - K_i s_i V_i - D_i s_i V_i + (k_u McV)_i + \left(akMV \int_{L_2}^{L_1} \psi dL\right)_i + Gs_{ji}V_i \quad (4)$$

where: s_i = mass of PCB in top carnivores per liter of water for segment i (μg PCB/l) = $\gamma_i M_i$, for γ_i = mass of PCB/mass (wet) of fish biomass (μg PCB/g (wet)); M_i = biomass of top carnivores in ith segment (gram (wet)/l); V_i = volume (1) of ith segment; A = cross-sectional area of segment (m^2); U = average fish velocity (swimming speed) (m/day); K = excretion rate of PCB for top carnivores (day^{-1}); D = net loss rate of PCBs from top carnivores due to death and predation by man (day^{-1}); k_u = uptake rate of PCBs from water $(\text{l/gram wet weight-day})^{-1}$; c = concentration of PCB in water available for direct uptake (μg/l); k = feeding coefficient of top carnivores $(\text{gram (wet)-day/l})^{-1}$; a = efficiency of incorporation of prey PCBs into predator biomass; L_1, L_2 = feeding size range from top carnivores (μm); ψ = "continuous" distribution of PCBs in food chain, a density function (μg PCB/1-m); G = recruitment rate of juveniles to top carnivores (1/day); s_j = mass of PCB in juveniles per liter of water in segment i (μg PCB/l).

The full implementation of Equation 4 cannot be accomplished at the present time principally because of a lack of relevant PCB data throughout the ecosystem and detailed data and understanding of the spatial and temporal behavior of the biomass. Nevertheless, some insight can be gained in the uptake and accumulation of PCBs under a very much simplified analysis. Therefore, if steady state is assumed in a single volume and PCB intake due to food and recruitment and loss due to death rate are neglected, than Equation 4 simply becomes the standard expression for bioconcentration directly from the water.

For aquaria experiments, with no feeding, Equation 4 for the bioaccumulation of PCBs becomes:

$$\frac{ds}{dt} = k_u Mc - Ks \qquad s = 0 \text{ at } t = 0 \tag{5}$$

and for constant water concentration, c, and constant biomass M, one can show that:

$$\frac{\gamma}{c} = N = \frac{k_u}{K}(1 - e^{-Kt}) \tag{6}$$

where N is a bioconcentration factor (μg/g (wet) per μg/1). This equation indicates that the rate of buildup of PCBs in the laboratory animal provides an estimate of the excretion rate (assuming the process is reversible) and the PCB bioconcentration factor, N, at equilibrium provides an estimate of k_u/K, from which k_u can be estimated. For depuration experiments:

$$\gamma = \gamma_o e^{-Kt} \tag{7}$$

where γ_o is the body burden of the organism when placed in a tank of zero PCB concentration. It should be noted that estimates of K from Equations 6 and 7 may not always agree because of such factors as accumulation and retention of substances in the skeletal structures of the animal.

If it is assumed that the prey PCB concentration is known or can be estimated then Equation 5 can be written as:

$$\frac{ds}{dt} = k_u Mc - Ks + akMM_p\gamma_p \tag{8}$$

where M_p and γ_p are the mass of prey (g/l) and PCB concentration (μg PCB/g) of prey, respectively. The mass of prey available for food at the upper trophic levels can be estimated from field data as well as γ_p. The difficulty arises in projections of the effects of remedial action programs on γ_p and in principle requires solution of a compartment model over several levels of the food chain or a continuous approximation such as given by Thomann.[3]

Solution to Equation 8 at equilibrium and for constant food input is:

$$\gamma = \frac{k_u c + akM_p\gamma_p}{K} \tag{9}$$

Comparing Equation 9 to Equation 6, the increased body burden due to food intake can be seen. These simplified equations together with a simplified dynamic analysis form the basis of the judgments made on the fate of PCB in the ecosystem of the Hudson.

Available Data on PCBs in the Hudson Ecosystem

FIGURE 5 shows the spatial distribution of PCB in various fish species over the entire length of the Hudson from 1975–76 data. All available data on all species are not plotted. Values represent essentially whole body concentrations although some fillet data are included. The present and proposed action levels (5 μg/g and 2 μg/g) that are shown represent levels for the edible portion only . Fillet concentrations would generally be expected to be less than whole body concentrations, depending on the fat content of the fish. Fillet PCB for brook trout for example has been measured at approximately 20% of whole body PCB.[4] As shown in FIGURE 5, the spatial trend is clear. A region of very high PCB fish concentration exists in the Fort Edward area with an apparent decrease towards Troy. Values range from 20 to greater than 500 μg/g. In the estuary, values range from 3 to greater than 30 μg/g.

Considerable interest centers around the striped bass as a valuable sport and commercial species. FIGURE 6 shows the PCB distribution in the striped bass by age class over three seasons and two regions. For the spring, it is seen that for Region 2 (Mi 20–60), the region closest to the ocean, values are generally less than 7 μg/g, whereas for Region 3 (Mi 60–110), values are generally greater than 7 μg/g. This difference may reflect the variable body burden that would be associated with a resident overwintering population versus a transient migratory population entering the region with a reduced body burden because of depuration while in the coastal regions.

A summary of the distribution of available data in the food chain is indicated in FIGURE 7. This figure was compiled from NYSDEC, N.J. and Federal EPA data as well as from other sources.[5,6] The estimated phytoplankton values were calculated from partition coefficients given by Dexter[2] and the range of plankton biomass and total water PCB concentration. A general increase in PCB concentration with organism size appears to occur. The values reported for fish in the coastal-ocean regions are particularly interesting since PCB concentrations range from 0.1–10 μg/g and values of 2–5 μg/g are not uncommon for the area. This implies that ambient water concentrations of PCBs as well as food PCB levels to which the fish are exposed are at significant nonzero levels.

Estimated Effects of Remedial Action

With the estimated distribution of biomass and the distribution of PCB body burden given in FIGURE 7 the estimated mass of PCB in the biologic sector of the

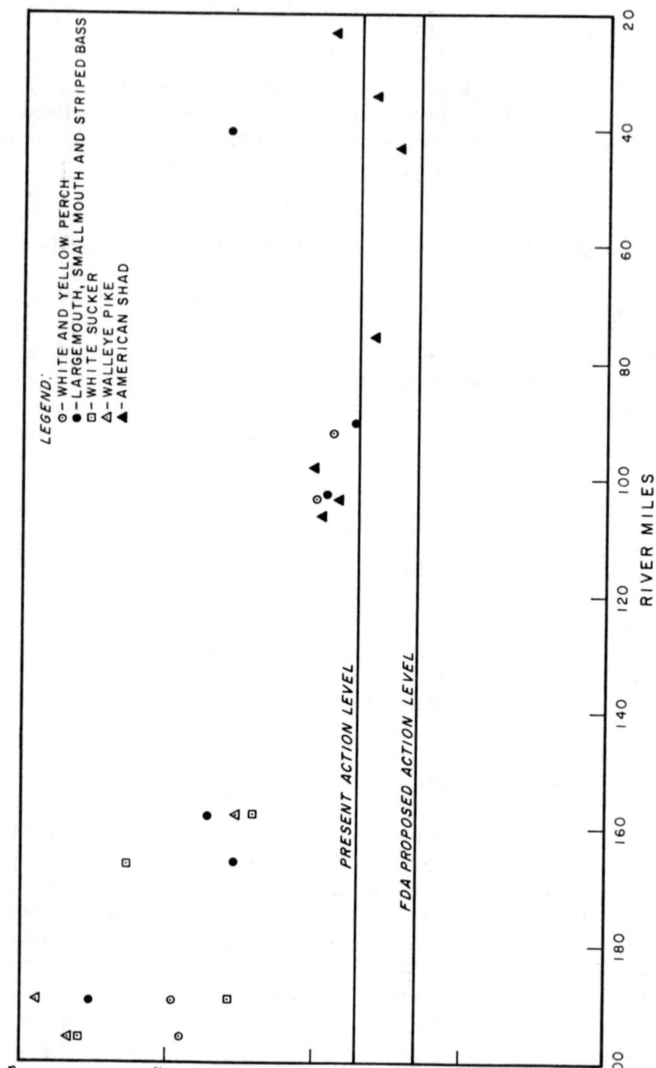

FIGURE 5. PCB body burden for fish as a function of river miles.

Hudson is shown in TABLE 1. As indicated, the total mass of PCB in certain compartments for the lower 85 miles of the estuary is about 100 kg although it should be recognized that there are several key ecological compartments, including aquatic macrophytes and juvenile fish, for which data are not available. However, even if the PCB mass were to increase by a factor of 10 to several thousand kilograms, it is still at least two orders of magnitude below the two-hundred-thousand kilograms estimated to reside in the bed sediment.

In order to estimate the expected PCB body burden in the upper trophic levels,

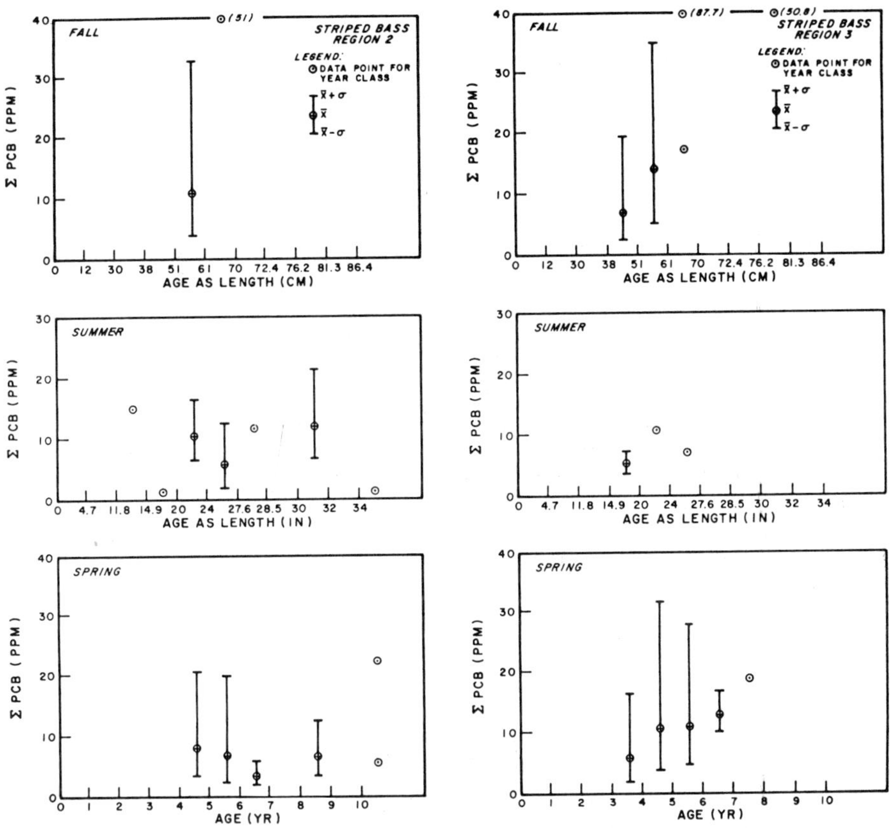

FIGURE 6. Distribution of PCB body burden for striped bass by region, size, and age class.

FIGURE 8 redrawn from Thomann[3] provides a lower bound. The data represent accumulation from the water only and do not include bioaccumulation through the food chain. The range in the bioconcentration factors is large and is from 10^{-1} to 10^{2} ($\mu g/g \div \mu g/l$) and includes physical adsorption processes in the less than 10 μm particle size to fish on the order of 0.1 meters in size. There appears to be a slight dependence on organism size although this results essentially from one point at 100 μm.

FIGURE 9 summarizes the estimated responses and indicates that at 0.1 μg PCB/l

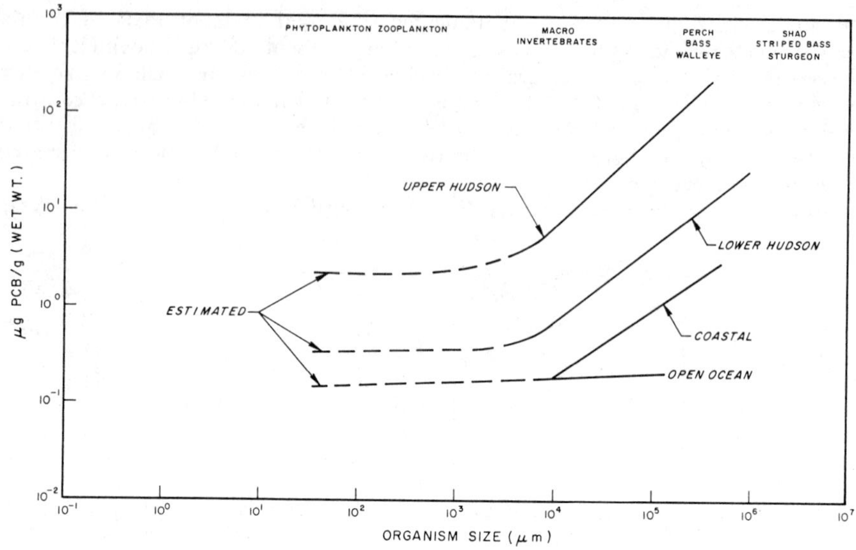

FIGURE 7. Estimated distribution of PCB body burden as a function of organism size for different regions of the Hudson River.

available for bioconcentration, and a concentration factor of 100 for the upper levels of the food chain, the body burden ranges from 5–10 µg/g or above the present action level. At .05 µg PCB/l, for the upper levels, the body burden is close to 5 µg/g and generally above 2 µg/g the proposed action level. At minimum then, available PCB concentrations would have to be reduced below 0.05 µg/l to possibly meet the 5 µg/g action level. This is especially true when it is recognized that the body burden will also increase because of PCB in the food. At .05 µg/l the action level of 2 µg/g will apparently not be met. One can extend the analysis to lower concentrations of about

TABLE 1

ESTIMATED PCB MASS IN SOME ECOLOGICAL COMPARTMENTS
LOWER HUDSON ESTUARY

Compartment	Biomass (mg wet/l)	Assigned PCB Concentration (µg PCB/g wet)	PCB Concentration (µg/l*)	PCB Mass (lbs†)
Phytoplankton	25	1	.025	120
Benthic Macro-invertebrates	50 (mg/kg wet sed)	0.5	—	1
Zooplankton	2.0	1	.002	10
0.1m Fish + 1–4 yr Str. Bass	1.5	10	.015	75
Str. Bass 4 yr.	0.3	10	.003	15
Total			.045	220

*Concentration of PCB in given compartment per liter of water.
†For lower 85 miles of estuary.

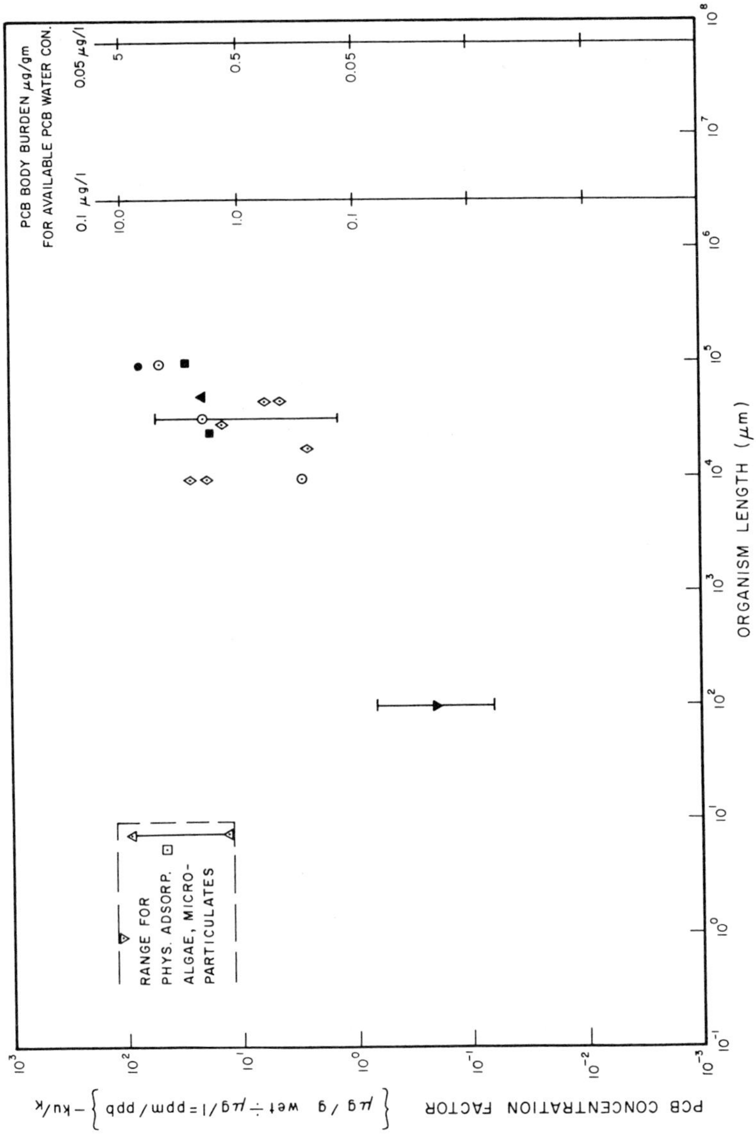

FIGURE 8. Experimental data on PCB accumulation from water only.

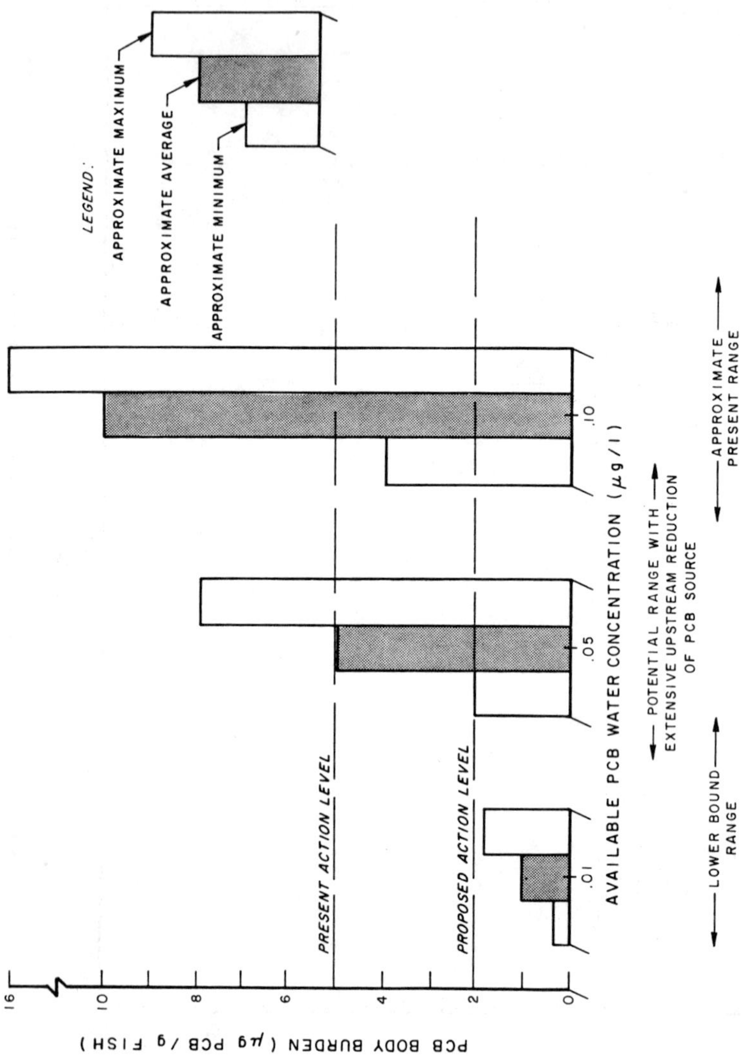

FIGURE 9. Estimated relationship between striped bass PCB level and water concentration.

.01 μg/l and conclude that only in the range of .01–.05 μm/l is there a possibility of reaching the 2 μg/g goal. At a water concentration of .01 μg/l, (10 ng/l), the fishery would probably be below the action level of 5 μg/g. However, this concentration represents a level close to conditions in the near shore regions of the Atlantic.[7] As such, it does not appear to be a feasible concentration in the Hudson at the present time.

Additional information is provided in TABLE 2, which shows some observed PCB relationships and the importance of the lipid content of the organisms. As indicated, for Lake Ontario coho salmon, a total water concentration of 0.055 μg/l (of which approximately .050 μg/l is in the dissolved state), an average concentration of 6.5 μg/g is observed. This represents an overall concentration factor of about 130. Using FIGURE 7 and a bioconcentration factor of 80–100, it is estimated then that the food consumption has increased the body burden by about 30–60%. TABLE 2 therefore indicates that, depending on lipid content, with water concentrations of .05 μg/l, average body burden may range from 1–7 μg PCB/g. This is additional evidence that the available concentration in the Hudson would have to be generally below .05 μg/l in order to reach body burdens less than 5 μg/g.

TABLE 2

SOME RELATIONSHIPS BETWEEN PCB WATER CONCENTRATION, FISH CONCENTRATION & % LIPID

Water B	Total PCB Water Concentration μg/l	Fish Concentration μg/gm	Fish	Approximate % Lipid	Reference
Lake Ontario	.055	6.5	Coho Salmon	20	8
Lake Superior	.05–1	7.1	Siscowet Trout ("Fats")	40	9
		1.0	Lean Trout	10	9
Milwaukee River	.1–.4	2–14	Suckers, Carp	2	10

DISCUSSION OF REMEDIAL ACTION RESPONSE TIMES

Analysis of net fluxes of PCB from (1) bottom sediments, (2) evaporation, and (3) biodegradation indicates a net loss from the Upper Hudson of from 4.5–14 kg PCB/day.[1] This loss under a "no action" alternative would result in a response time of several decades for the upper river to eliminate the approximately 200,000 kg PCB in the sediment.

Some insight into the response time of the fishery can be obtained by a simplified analysis using Equation 4. FIGURE 10 shows the location of the striped bass at different life stages and FIGURE 11 shows the two regions that were considered in the simplified analysis: the "estuary" region and the "coastal" region. PCB concentrations are given in each region and a striped bass cohort is allowed to spawn, mature in the estuary, and then migrate to the coastal region where generally PCB concentrations are at lower levels. Equation 4 is the basic equation for $i = 2$ regions, no food intake, no recruitment, or death and mortality are included. Uptake and excretion continues throughout the life cycle of the cohort according to specified functions that are

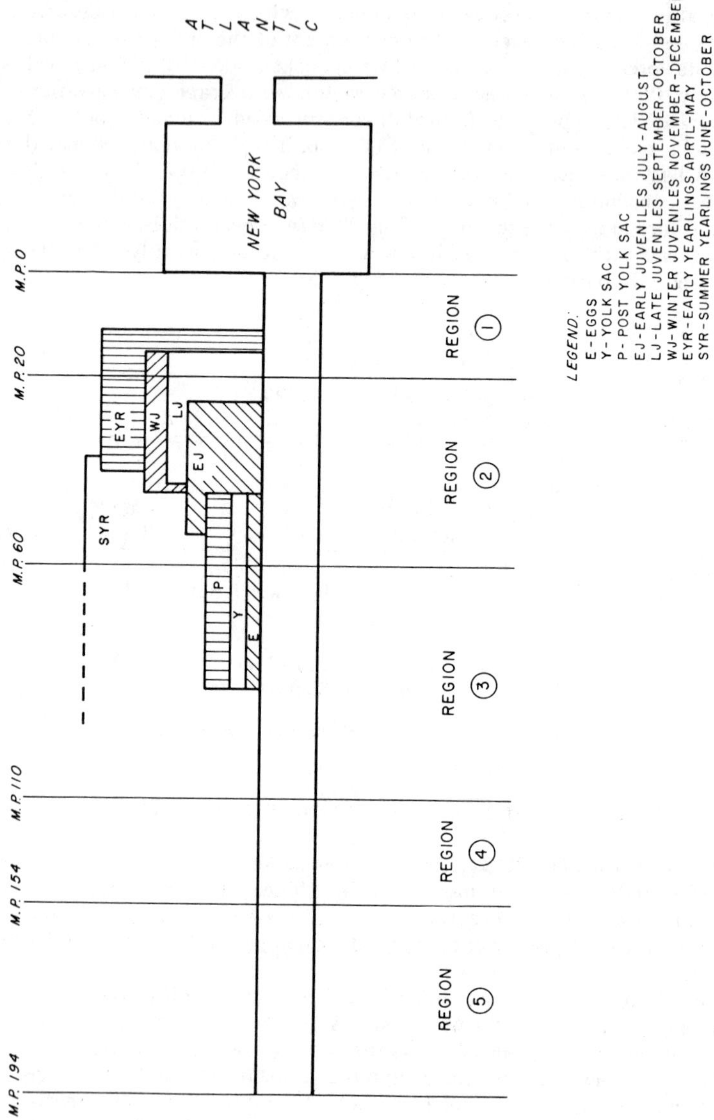

FIGURE 10. Schematic of location and life stage of striped bass in Lower Hudson River.

age-dependent. The concentration factor (k_u/K) is assumed to increase from 10 at age 1 to 100 at age 10 and the excretion rate to decrease from .04/day for the one year old class to .001/day for the 10 year old class. The fish are assumed to live to a maximum age of fourteen at uptake and excretion levels equal to that at 10 years. Further details are given in Reference 1.

FIGURE 12 shows some of the results of the simulations. FIGURE 12A is approximately representative of present conditions under an assumed coastal PCB concentration of 0.01 μg/l. The simulations indicate at least one reason for the observed wide range of PCB concentrations in the striped bass, namely, the exposure to different water concentrations at different age classes. FIGURE 12B illustrates the simulated response times if estuary concentrations were dropped to .05 μg/l. The

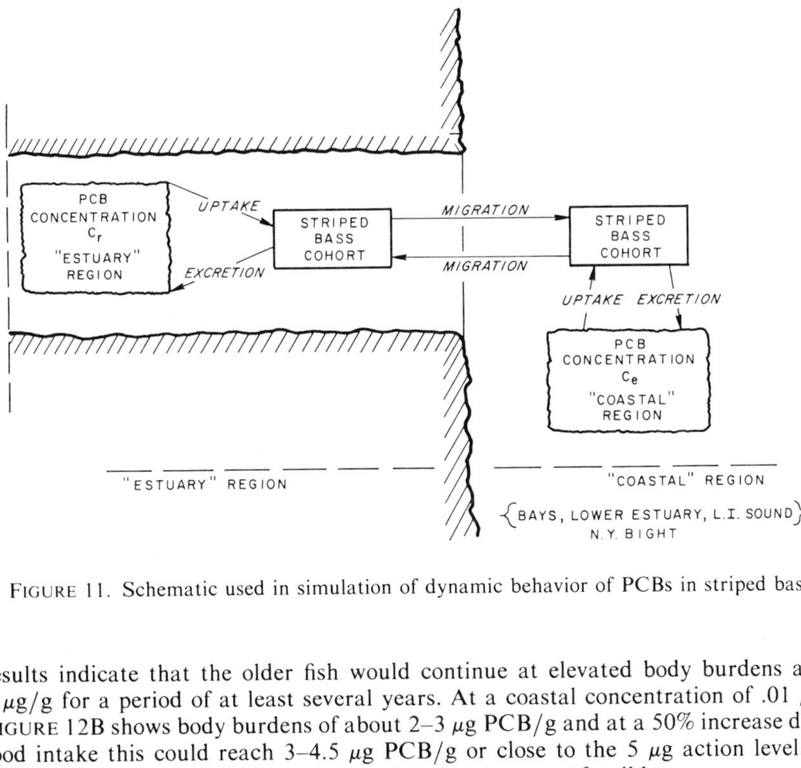

FIGURE 11. Schematic used in simulation of dynamic behavior of PCBs in striped bass.

results indicate that the older fish would continue at elevated body burdens above 5 μg/g for a period of at least several years. At a coastal concentration of .01 μg/l, FIGURE 12B shows body burdens of about 2–3 μg PCB/g and at a 50% increase due to food intake this could reach 3–4.5 μg PCB/g or close to the 5 μg action level. For either case, the 2 μg PCB/g action level does not appear feasible.

In summary, the simple analysis indicates that if the available water concentration were reduced by about one-half to .05 μg/l, response times of several years would be required to detect any substantial change in fish PCB concentrations, depending on the PCB concentration of the coastal region. Older fish may retain higher body burdens because of reduced excretion rates. In any event, a level of .05 μg/l in the estuary might still leave the fishery quite close to the 5 μg/g action level and generally above the proposed 2 μg/g action level.

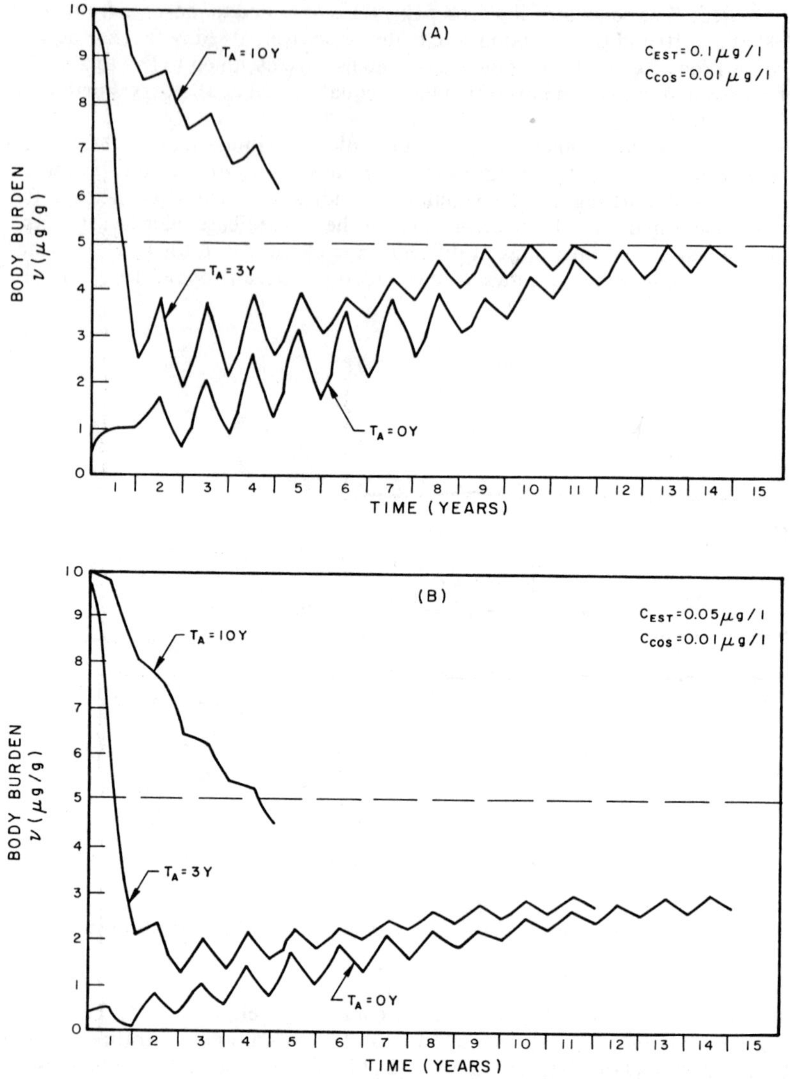

FIGURE 12. Simulated PCB body burdens for different estuary and coastal available PCB concentrations (T_A = age at t = 0).

ACKNOWLEDGMENTS

Special thanks are due to Michael Kontaxis and Thomas Gallagher of Hydroscience, Inc. for their efforts, diligence and insight in assisting in this work.

REFERENCES

1. HYDROSCIENCE, INC. 1978. Estimation of PCB reduction by remedial action on the Hudson River ecosystem. Prepared for New York State Department of Environmental Conservation by Hydroscience, Inc., Westwood, N.J. p. 107.
2. DEXTER, R. N. 1976. An application of equilibrium adsorption theory to the chemical dynamics of organic compounds in marine ecosystems, Ph.D. Dissertation, University of Washington, Seattle, Wash. p. 181.
3. THOMANN, R. V. 1978. A size dependent model of hazardous substances in aquatic food chains. EPA-600/3-78-036, ERL, Duluth, Minn., p. 40.
4. SNARSKI, V. & F. A. PUGLISI. 1976. Effects of Aroclor® 1254 on Brook Trout (Salvelimes Fontinalis). Environmental Research Laboratory, Duluth, ORD, EPA-600/3-76-112, USEPA. Duluth, Minn. p. 34.
5. DUCE, R. A., *et al.* 1974. Pollutant transfer to the marine environment, NSF/IDOE, Pollutant Transfer Workshop, NSF, Kingston, R.I. p. 55.
6. ANON. 1972. Polychlorinated biphenyls—environmental impact; a review by the panel of hazardous trace substances, Env. Res. V. **5:** 244–362.
7. HARVEY, G. R., *et al.* 1973. Polychlorinated biphenyls in North Atlantic Ocean water. Science V. **180:** 643–644.
8. HAILE, C. L. *et al.* 1975. Chlorinated hydrocarbons in the Lake Ontario ecosystem (IFYGL). EPA-660/3-75-022, NERC, EPA, Corvallis, Or. p. 28.
9. SWAIN, W. R. 1977. Persistent organic compounds in the vicinity of Isle Royale, Lake Superior. Presented at 20th Conference, Great Lakes Res. Ann Arbor, Mich. p. 29.
10. VEITH, G. 1970. Environmental chemistry of the chlorobiphenyls in the Milwaukee River. Ph.D. Dissertation University of Wisconsin, Madison, Wisc. p. 180.

THE HUDSON RIVER PCB PROBLEM: MANAGEMENT ALTERNATIVES

Leo J. Hetling,* T. James Tofflemire,* Edward G. Horn,*
Richard Thomas,† and Russ Mt. Pleasant,*

**New York State Department of Environmental Conservation*
Albany, New York 12233

†Malcolm Pirnie, Inc.,
White Plains, New York 10602

INTRODUCTION

The extent of existing environmental contamination of the Hudson River System by PCBs and some indication of environmental transfer rates has been explored in the previous papers.[1-3] A summary of the results of some of this work is presented in FIGURE 1. From this figure, it is clear the control of the PCB losses from the remnant deposit areas and the Upper Hudson riverbed are critical if further spread of PCBs to the river environment is to be prevented.

DESCRIPTION OF MAJOR CURRENT SOURCES

Remnant Deposits

During the period that GE was discharging PCBs to the river, a low-level dam was located downstream from the outfall locations (FIGURE 2). This dam was removed in 1973 and subsequent high flows scoured large volumes of the contaminated sediments downstream out of this reach. During high-flow periods, this material continues to be eroded and move downstream. The deposits reside in five distinct areas (areas 1 through 5). Areas 1,2, and 4 had PCB concentrations of 1–10 ppm, while in areas 3 and 5 PCB contamination ranged from 20-200 ppm. Area 3A had a surface (top 12 inches) contamination that averaged approximately 1,000 ppm.

Upper Riverbed Sediments

To provide additional insight into the mass of PCBs in the riverbed sediments, areas of high PCB concentration (>50 mg/l) were delineated and the data stratified into two sets for each river reach; one of the measured values in high concentration areas, and one of measured values in low concentration areas.

From studying the riverbed, cross-section maps, and sediment characteristics, it can be observed that silt deposits are typically near the banks in the lower velocity areas and gravel deposits are generally in the center or navigation section of the channel. The areas with high PCB concentrations also occurred more frequently near the banks, especially for the southern river reaches.

From an analysis of PCBs in cores, it was observed that typically, the highest PCB concentration in the sediment is approximately 30 cm below the surface (FIGURE 3).

0077–8923/79/0320–0630 $01.75/0 © 1979, NYAS

Substantially contaminated material is, however, found deeper in the sediments, particularly in the upper reaches.

As indicated above, the Upper Hudson Riverbed sediments were divided into two areas, "hot spots" (areas of PCB concentration >50 ppm) and "cold areas" (PCB concentrations <50 ppm). The mass of PCBs in each of these areas was calculated.[1] Approximately 60% of the mass is in the hot spot areas.

Management Alternatives

Overview

Early in the formulation of the plan of study, a number of potential methods of managing the PCB-contaminated sediments were suggested. Although most of these

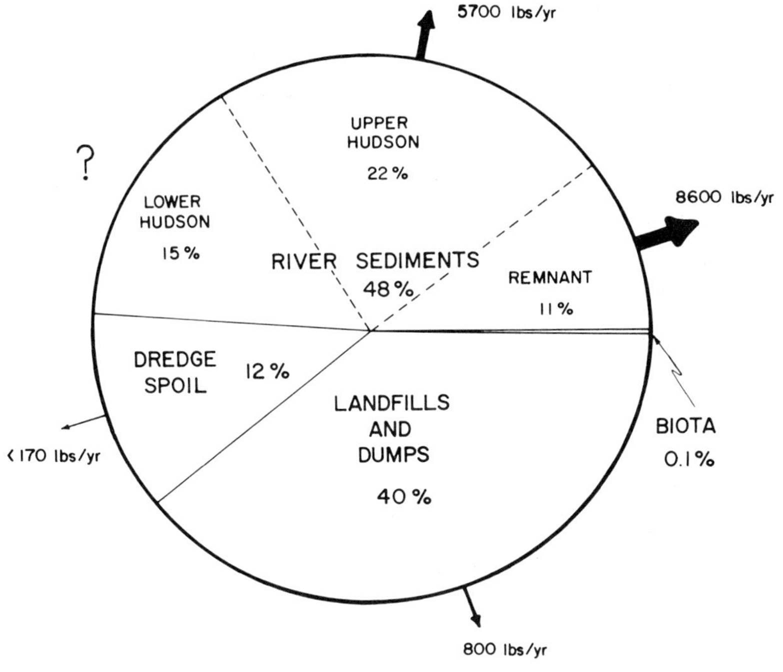

FIGURE 1. Distribution of PCBs in the Hudson River Basin.

technologic alternatives were not studied in detail, their applicability to the Hudson River problem was reviewed[5] (TABLE 1). Some of these alternatives are not applicable to the Hudson River problem because of high cost and/or impracticability, but most were judged to be insufficiently technically developed to evaluate. This judgment has since been confirmed in the studies on Kepone® in the James River.[6] Development of sufficient technology to evaluate any of these systems would take 5 to 10 years and

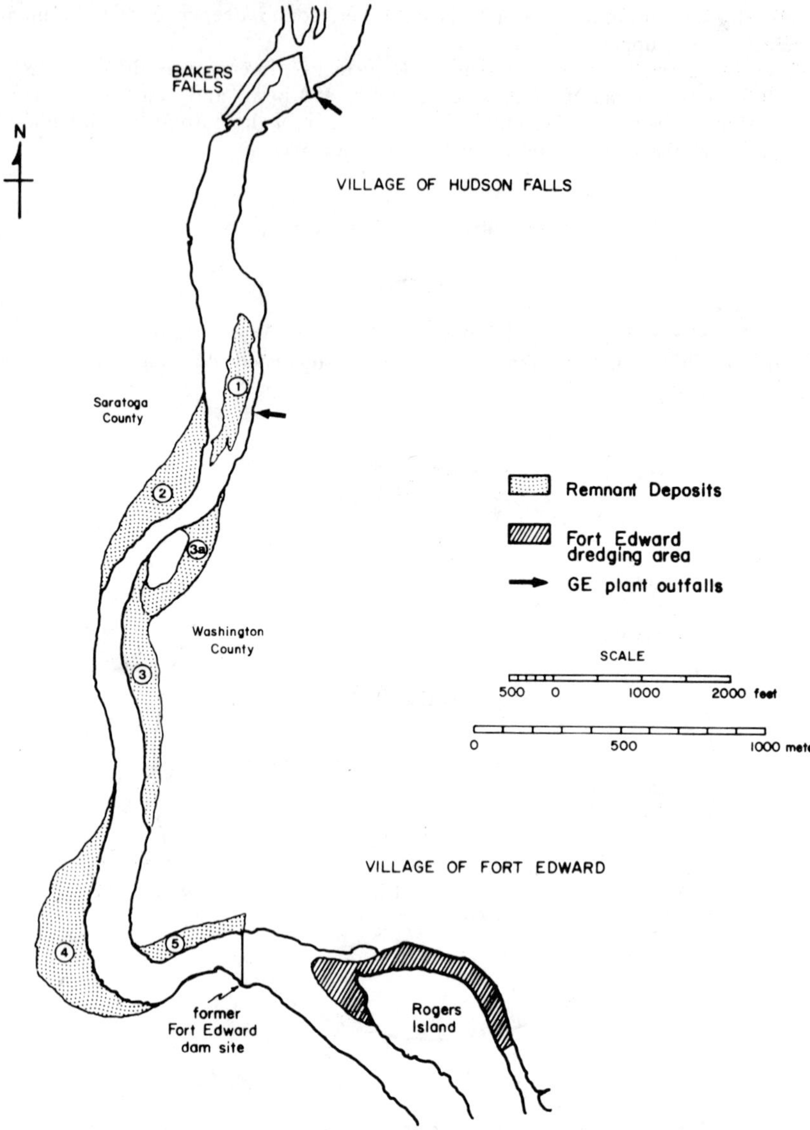

FIGURE 2. Map of remnant deposits and dredging area for maintenance dredging project of 1977–78.

cost over several million dollars. During the time period required for the development of the technology, contaminated sediments would continue to disperse through the river system and become less amenable to any remedial action. Thus, the only management alternate studied in detail was physical removal of PCB-contaminated sediments by dredging.

Dredging

Technology

Dredges in use today can be generally divided into three categories: mechanical, hydraulic, and pneumatic.[7] Mechanical dredges normally lift the dredged material above the waterline by means of buckets or scoops of various designs and deposit it in a barge or similar conveyance for transport and disposal. Hydraulic dredges generally move bottom material by means of a centrifugal pump and pipeline directly toward a disposal area, whereas pneumatic dredges move bottom material by compressed air and pipelines.

Mechanical dredges used in the U.S. include the dragline, the dipper, the grab bucket or clamshell, and the endless-chain dredge. Of these, the clamshell and dragline are the most common. The clamshell dredge consists of a derrick mounted on a barge with a "clamshell" bucket for excavating. The material is removed by forcing the opposing bucket edges into the sediment. The bucket is lifted out of the water and deposits the spoil on a barge or onshore. The dragline dredge is generally composed of a crane having a bucket suspended from a swinging boom mounted on a barge or truck. The dredge operates by scraping the material from the bottom by pulling the bucket towards the stationary crane. The spoil is then lifted and deposited on a barge or on the bank. Both the clamshell and dragline are available in a wide variety of sizes.

The hydraulic suction dredge can be fitted with various mechanisms at the suction pipe inlet to facilitate sediment removal. These mechanisms include rotary cutters or cutterheads, auger-type cutterheads, and high-pressure water jets. Mud shields or dustpans are used on some hydraulic dredges in conjunction with the water jets to reduce secondary suspension at the suction inlet. However, these dredges collect only 10% to 30% solids, cause considerable sediment agitation when mechanical cutter-

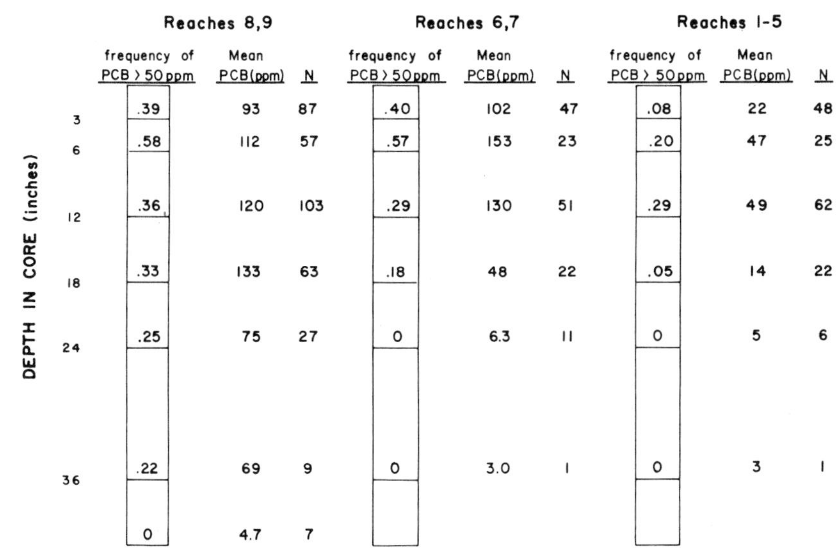

FIGURE 3. PCB penetration in cores by reach from areas where PCBs >5 ppm.

TABLE 1

SUGGESTED METHODS OF MANAGING PCB-CONTAMINATED SEDIMENTS*

Alternatives	Status of Technologic Development	Potential Constraints or Environmental Problems	Estimated Treatment Costs
Removal			
Activated carbon adsorption	Laboratory	Technology for application and retrieval has not been proposed.	\$120–1200/ha treated
Bioharvesting	Conceptual	The time and costs involved with harvesting enough fish are prohibitive. Tremendous ecological side effects.	\$1000/kg PCB removed
Adsorption by oil-soaked mats	Conceptual	Technology for application and retrieval has not been proposed. Tremendous ecological side effects.	Insufficient data
Dredging	Demonstrated on small scale	Untested on a large scale.	\$25–500/kg removed
Destruction in place			
Degradation by ultraviolet ozonation	Developed for closed system applications	Treatment requires closed reaction vessel.	Insufficient data
Biodegradation	Laboratory	Conditions for the natural microbial decompositions of PCBs must be optimized.	Potential low cost
Chemical treatment	Conceptual	Possible ecological side effects.	Insufficient data
Fixing in place			
Adsorption	Laboratory	Possible ecological side effects.	Potential low cost
Erosion control of river bottom	Conceptual	—	Insufficient data
Chemical fixation	Conceptual	Possible ecological side effects.	Insufficient data
Covering PCB-contaminated sediments	Conceptual	Massive disturbance of ecosystem. Rupture of seal or ballooning of plastic due to gas formation. Placement of cover difficult.	Particulate matter: \$1200–1600/ha Sheeting material: \$300–700/ha

*Modified from Horstman.[5]

heads are used, and can induce secondary pollution at the receiving site because of high water content in the dredged material. Consequently, without the use of sediment-control measures, conventional dredging could pose a serious threat for aggravating an existing, but possibly dormant, pollution problem. Some generic types of hydraulic dredges in the United States include the cutterhead, the plain suction, the dustpan, the hopper, the sidecaster, and the Mud Cat. The Mud Cat dredge is comparable to the cutterhead, except that, in lieu of a rotating cutter, it features an auger-type horizontal cutterhead. This dislodges the material, and the auger moves it toward the suction pipe. A mud shield surrounds the auger and thereby minimizes mixing of the disturbed bottom sediments with the surrounding water.

Dredging technology in some foreign countries surpasses that of the United States; Japan especially has well-developed technology because serious problems with in-place toxic substances prompted the development of a dredge designed to remove contaminants rather than to excavate river channels. A significant advancement in dredging technology for removing contaminants was made by improving a pneumatic dredge. The pneumatic or "Pneuma" dredge, originally developed in Italy, uses hydrostatic head pressure and compressed air to remove contaminated sediments. By applying a vacuum to a pneumatic dredge, the Japanese were able to utilize the dredge in shallow water, thereby eliminating the constraint of needing high hydrostatic head pressure. This dredge is called the Oozer dredge.

In general, mechanical dredges can remove bottom material at near-in-place density, whereas, hydraulic and pneumatic dredges have to add dilution water to form the dredged-material slurry. Effective separation of the dredged material from the water before it is discharged back into the river is necessary. Mechanical dredges, however, tend to resuspend more riverbed material than hydraulic or pneumatic dredges during the dredging operation.

The specific dredging technology applicable to a specific project is dependent on dredging objective, environmental constraints, river and dredgematerial characteristics, cost, and equipment availability.

Environmental Effects

The environmental effects of dredging are both physical and chemical. The physical effects on biota caused by the mechanical disruption of the riverbed have

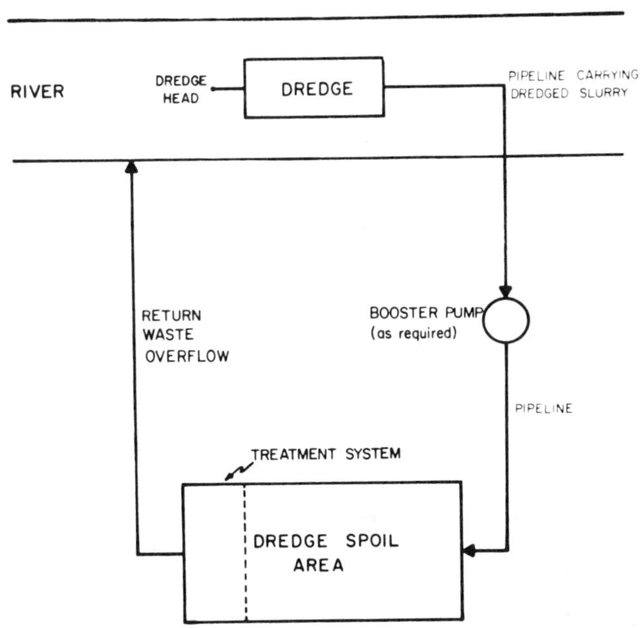

FIGURE 4. Schematic of hydraulic dredging. Modified from Malcolm Pirnie, Inc.[7]

been characterized elsewhere[8] and are not discussed here. The chemical effects occur as a result of enhancing the rate of transfer of PCBs to the water.

There are two areas where PCB transfer rates can be affected by hydraulic dredging removal efficiencies: losses at the dredge head and losses in the return waste overflow from the dredge spoil area (FIGURE 4).

Full-scale monitoring of New York Department of Transportation (DOT) hydraulic dredging in 1976 and 1977 in the Upper Hudson showed that dredge-head losses are relatively small (TABLE 2). TABLE 3 shows the return flow loss at three monitored dredging sites in the Upper Hudson. These studies show that one hour of sedimentation can provide 90% removal of the PCB-laden suspended solids in the return flow. This was further improved by the addition of a cationic polymer to promote rapid destabilization of suspended particles. Surface scums that developed during the dredging operation were found to be very high in PCBs. PCB concentrations were as high as 18,000 ppm, but the total mass was small. Their collection and removal was found to be desirable and relatively easy with proper placement of an oil boom.

New York Department of Environmental Conservation (DEC) also monitored the 1977–78 dredging near Rogers Island which involved a dragline dredge and power shovels.[11] This 150,000m³ project and its environmental effects have been evaluated in detail.[12] The adverse effects appeared to be minimal and removal efficiencies were slightly better than the original estimates. These effects were generally limited to within one river mile below the project area. TABLE 4 summarizes a portion of these data.

TABLE 2

MEASURED PCB LOSSES AT THE DREDGE HEAD CAUSED BY HYDRAULIC DREDGING IN THE UPPER HUDSON NEAR LOCK 4*

	River Cross Section	Dredge Head	Increase in River Below Dredge Head
Area	149 m^2	15 m^2	—
Water velocity	.71 kl/hr	.71 fps	—
Total PCBs	.1–.4 ppb	2.1 ppb	0.018 ppb 44 gm/day
Suspended Solids	15–35 mg/l	120 mg/l	1 mg/l

*Source: Tofflemire.[9]

The experience and data collected from the above full-scale Hudson River dredging projects clearly indicate that with proper precautions, the Upper Hudson River may be dredged with minimal adverse environmental effects, i.e., PCB and suspended-solids losses can be kept to insignificant levels.

Disposal

Three methods of final disposal of PCB-contaminated sediments removed from the Hudson River were studied and evaluated: incineration, biodegradation and engineered encapsulation.

1. Incineration

General Electric has investigated the technical, economic and environmental feasibil-

TABLE 3

SUMMARY OF HYDRAULIC DREDGING PERFORMANCE AT THREE SITES ON THE UPPER HUDSON RIVER*

Parameter	Buoy 212	Lock 4	Lock 1
Silt and clay			
(% passing #200 sieve)	5	25	26
Lagoon flow-through time			
(typical, by salt tracer)	45 min.	55 min.	15 min.
Lagoon influent			
average % solids	5	3	1
PCBs in solids (μg/g)	100	20	15
Lagoon effluent			
Without polymer			
Suspended solids (mg/l)	500	460	2000
PCBs (μg/l), composite	100	3.0	40
With polymer			
Suspended solids (mg/l)	250	230, 20†	400
PCBs (μg/l), composite	50	<1.5	8
Removal efficiencies (%)			
Without polymer			
Suspended solids	99	98	80
PCBs	98	99.5	73
With polymer			
Suspended solids	99.5	99, 99.9*	96
PCBs	99	99.8	95

*Source: amended from Tofflemire.[10]

†Polymer fed between lagoons.

TABLE 4

PERFORMANCE OF DREDGING PROJECT AT FORT EDWARD CHANNEL DURING THE SUMMER OF 1977*

(Comparison of actual dredging performance with predictions of Environmental Assessment[11] when the project was approximately 45% complete.)

Parameter	Predicted	Actual
Total suspended solids added to river by dredging (MT)†	7800 assuming 5% fines 4750 assuming 3% fines	<2730 (4% fines)
PCBs		
In river	5 ppb at low flow	0.02–1.14 ppb
Disposal site leachate	3 ppb at 5.7 m^3/day	0.47 ppb at 2.8 m^3/day‡
Heavy metals	Very low increase predicted under "worst-case" conditions.	Insignificant increases through project area.
Esthetics	Noise, appearance of work area were recognized as temporary factors.	Degraded appearance localized; environmental impact not proportional.

*Source: Mt. Pleasant.[12]

†Metric tons.

‡The disposal area could not be covered before winter. Once the site is complete, a clay cover should reduce this loss.

ity of destroying PCBs in river sediment using a gas-fired multiple hearth furnace.[13] All PCBs were destroyed when the afterburner temperature exceeded 1800°F (982°C) with 0.5 sec residency time of the exhaust gases. Although other unidentifiable chlorinated hydrocarbons were produced, they were destroyed by use of an afterburner and scrubber system. Heavy-metal losses were also monitored and not found to be a problem.

Although technically and probably environmentally possible, burning river sediments is an expensive undertaking. Fuel costs alone (current prices) would be approximately $\$13/m^3$. In order to minimize operating and investment costs, the furnace would have to have a capacity of at least 10^5 metric tons/year ($87{,}000m^3$).

Wright-Malta Corporation of Ballston Spa, New York, has devised a gasifier-gas turbine system that can generate electric power from solids and liquid wastes.[14] The process chemistry has been proven in batch and continuous benchscale equipment. Tests demonstrated that process conditions providing 1100°F, 300 psi, an alkaline catalyst, and one hour dwell time would completely destroy PCBs absorbed in paper.

A one megawatt pilot plant could be constructed with operation to commence within two years. It would handle a mixture. of 20 tons of solid waste and 30 tons of sewage sludge per day, or 15 tons of solid waste and 35 tons of dredge spoil per day. Approximately, 10,000 tons ($8{,}400m^3$) of dredged material could be treated per year at an estimated total annual operating cost of $650,000. Subtracting revenues from electricity generated, the cost would be approximately \$50/ton ($\$75/m^3$) of dredge spoil assuming no cost was assigned to concomitant solid waste disposal.

2. Biodegradation

Research has been carried out by GE to explore the possibility of using naturally occurring microorganisms to reclaim PCB-contaminated Hudson River dredge spoils. This research was predicated upon the development of PCB-degrading enrichment cultures isolated from various contaminated sources and the study of practical methods of enhancing biodegradation rates of PCB in soil.[13,15]

Twenty-two strains (6 species) of naturally occurring microorganisms that degrade PCBs have been isolated and purified. PCB degradation was confirmed by measuring the loss of PCB and the formation of yellow-colored metabolic intermediates. Measureable PCB biodegradation rates were observed in shake-flask studies using pure and mixed cultures (FIGURE 5). Satisfactory mixed-culture growth was observed in the presence of metal ion concentrations similar to those observed in dredge-spoil samples. Mixed-culture pH optima varied from about 5 to 8 and should be compatible with conditions attainable in dredge spoils. Certain surfactants have been shown to significantly enhance PCB biodegradation rates.

Biochemical Corporation of America has reported that they successfully applied an adapted strain of bacteria to degrade the PCBs in a sample of a sewage sludge received from the City of Bloomington, Indiana.[16] The sludge, containing 15% solids and 198 ppm PCB, was distributed into chemically clean Ehrlenmeyer flasks to which were added minimal salts for nutrients plus the specialized culture (except for the control flask). After two days of shaking, the PCB level was reduced to 54 ppm (74% reduction). No degradation occurred in the control flask (without the culture). A similar flask was run at the same time except that the sludge was diluted with five parts water. After five days of shaking with the culture, the PCB level was less than detectable (99.9% reduction).

They conclude from this that PCBs in sewage sludge can be degraded by bacterial cultures and that the extent of this reduction is a function of time. These tests imply that other sludges could also be handled in a similar manner given proper conditions.

Because of the rapid development in this area, it is difficult at this time to assess the actual practicability of biodegradation as a disposal alternate. Although sufficient information does not now exist to design for biodegradation, its environmental acceptability and potential low cost make it a highly desirable option, and it would be unwise to take any management action at this time that would preclude its use in the future.

3. Engineered Encapsulation

General description. Engineered encapsulation of contaminated dredged materials involves placing the contaminated material in a land burial facility in such a manner that it is permanently removed from man's normal environment. Within recent years, governments have issued detailed regulations regarding the design and operation of chemical landfills which are applicable to encapsulation of Hudson River PCB

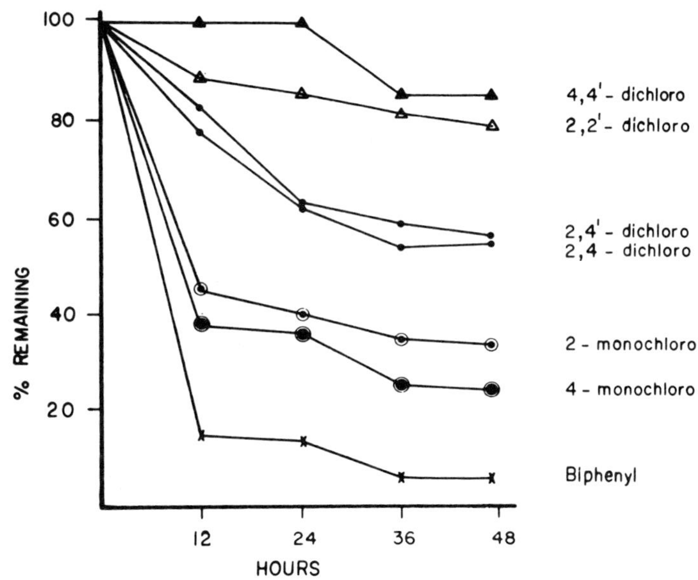

FIGURE 5. Relative biodegradability of 1221 compounds. From General Electric.[15]

contaminated dredged sediments. These criteria relate to three aspects of the design of such a facility, i.e. site selection, lining and leachate prevention, and long-term operation and maintenance.

Site selection. Potential sites for landfilling PCB-contaminated sediment from the Upper Hudson River Study area were identified on the basis of the criteria shown in TABLE 5. Using these criteria, 40 parcels of land with a total of 1300 ha were found to be generally acceptable as sites.[7] Detailed field studies would be required before final site selection can be made. Because the size and location of a disposal site depend on the actual dredging program selected, the detailed field studies and final site selection must await selection of an actual dredging program.

Lining and leachate prevention. FIGURE 6 illustrates potential paths through which PCBs can be lost to the environment from a poorly designed spoil site or landfill.

Proper design of an engineered encapsulation site secures the contaminated material so that the potential for these losses are minimized. This includes placing an inpermeable barrier beneath the contaminated materials and an impermeable cap above it to prevent excessive infiltration of precipitation from entering the contaminated spoils. The concept of limiting infiltration and percolation to the greatest possible extent is a necessary part of the design plans for any facility.

A leachate-collection system should also be installed beneath the spoils and a leachate-treatment facility designed for these sites to collect and treat any leachate

TABLE 5

CRITERIA USED TO SELECT AN ADEQUATE SITE FOR AN ENGINEERED ENCAPSULATION FACILITY TO CONTAIN PCB-CONTAMINATED DREDGE SPOIL*

Parameter	Unacceptable	Ideal
Soil	Permeability $>10^{-5}$ cm/sec, less than 1 m thick *in situ*. ($<1.4 \times 10^{-4}$ cm/sec overlayed.) Class I or II agricultural soils.	Permeability $<10^{-7}$ cm/sec, *in situ* thickness >3 m soil passing #200 sieve >30%, liquid limit >30% plasticity, index >15.
Slope	Deep gullies, slope over 15%.	Slope <10%
Surface water	Closer than 90 m to any pond or lake used for recreational or livestock purposes, or any surface water body officially classified under state law. In special flood hazard areas or recognized wetlands.	>300 m
Bedrock	Closer than 10 m to highly fractured rock or carbonates, closer than 3 m to all other rock.	More than 15 m to highly fractured rock or carbonates.
Groundwater	Closer than 3 m to groundwater, wells tapping shallow aquifers, closer than 300 m to any water supply well. Flow towards site.	>15 m to groundwater, no wells or only deep bedrock wells within 600 m radius.
Committed land	Closer than 300 m to parks, cemeteries, residential areas, historic sites, etc.	>450 m away.
Biologically sensitive areas	Endangered plant or animal habitats, unique or regionally significant environments.	No woodlands, no locally significant features.

*Source: Malcolm Pirnie, Inc.[7]

that may be produced. An extensive groundwater monitoring system should be installed prior to deposition of any wastes to detect any contamination that might leach from the site and enter the groundwater system.

FIGURE 7 shows a sketch of a typical engineered encapsulation site. Features of such a site include:

—A natural impermeable layer ($K<10^{-7}$ cm/sec) at least 1 m thick over the bottom of the site.

—Dikes (average height 4.2 m) lined with at least 60 cm of impermeable material ($K<10^{-7}$ cm/sec).

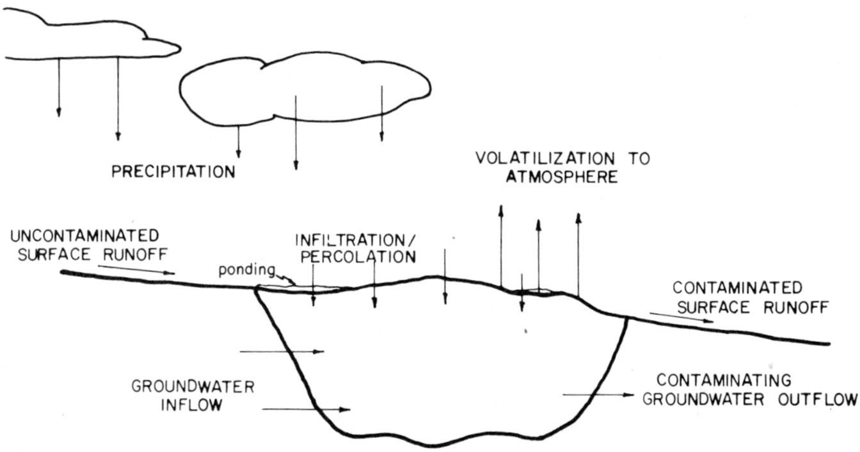

FIGURE 6. Typical landfill problems showing primary routes of contaminant loss.

—An impermeable cover ($K<10^7$ cm/sec) at least 45 cm thick over the top of the site. This, together with the 5% slope given the top of the dredged material, will facilitate runoff and minimize the amount of rainfall penetrating the fill.

—A 45 cm layer of select material for turf establishment over the top of the site. This will be seeded and graded and a vegetative cover established to stabilize the site.

—A system of monitoring wells into the site to monitor leachate generation and collect leachate for treatment.

—A system of groundwater-monitoring wells to detect contamination that might leach from the site and contaminate groundwater (not shown).

A model disposal site for containing PCB-contaminated spoils taken from the channel at Rogers Island was designed and put into operation in 1977.[11,12] This site's

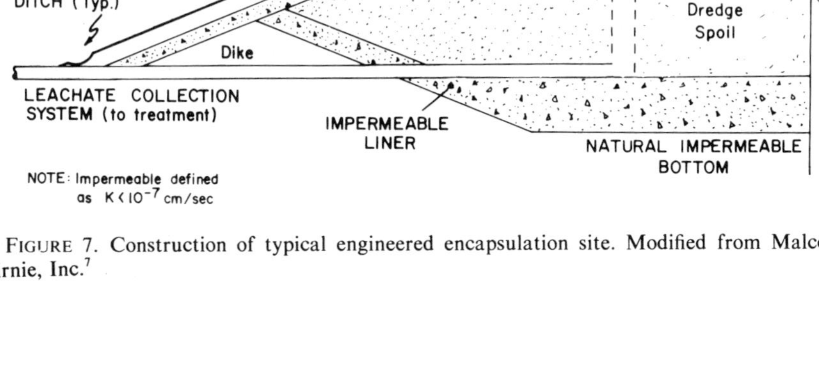

FIGURE 7. Construction of typical engineered encapsulation site. Modified from Malcolm Pirnie, Inc.[7]

design varied somewhat from the criteria described above in that it was designed before these criteria were available. Preliminary data indicate that PCB losses from the site are and will continue to be minimal. Additional monitoring of the site is planned.

Long-term operation. An engineered encapsulation site is a man-made structure and, as such, will require long-term maintenance and operation. Essential elements of a maintenance and operation program include:

—Continued compatible use of the land surface (e.g. ecological study, recreation, material storage, etc.).

—Structural maintenance sufficient to insure the integrity of the impermeable barrier.

—Proper collection and disposal of any leachate developed (either from on-site treatment or by removal to an adequate, off-site treatment facility).

—Monitoring to insure the PCB losses to the environment are not occurring (i.e. air, biota, ground and surface water).

There is little experience with long-term operation of such facilities; and thus, there are uncertainties regarding operational costs. If no major problems occur, such costs should be minimal; however, if unforseen problems occur, costs could be high.

Alternatives

Description

Although an infinite number of alternate dredging programs for removal of PCB-contaminated sediments from the Hudson River are possible, the distribution of PCBs, existing technology, and practical considerations limit the number of actual practical alternatives. Four alternatives which appear feasible are outlined in TABLE 6. Following is a description of each of these alternatives, their technologies, their cost and an assessment of their environmental impacts.

Maintenance dredging. Even if no action is taken to dredge PCB-contaminated sediments from the Upper Hudson, future river channel dredging for canal maintenance is planned. The canal is generally maintained with 3.6 m deep channel 60 m wide. Over the past 25 years, about 0.5 million m^3 were dredged from this reach of the canal and wet dumped inside designated areas. Another 1.2 million m^3 were dredged and placed in upland disposal sites.[11]

Using the past dredging experience, it is estimated that it will be necessary for the DOT to dredge and remove the river approximately 690,000 m^3 of sediment for canal maintenance over the next ten years. Assuming a sediment PCB concentration of 35 ppm and a 90% removal efficiency, this activity will remove approximately 23,000 kg of PCBs from the river system.

Remnant deposits. Because of their variable PCB concentration and erosibility, it is necessary to treat each of the identified remnant deposit areas differently.[25] No structural action is planned for areas 1, 2, and 4, because of their low PCB content. Area 5, although high in PCBs, has been previously stabilized and is not now actively contributing PCBs to the river. However, Niagara Mohawk Power Corporation is considering rebuilding the Ft. Edward Dam, and they would have to remove much or all of area 5 to construct the dam.

Because of their high PCB content, the surface sediments in area 3A would be removed immediately. Area 3 is now eroding, but a combination of further stabilization efforts and removal would reduce the input to the river from this source.

TABLE 6

SUMMARY DESCRIPTION AND EVALUATION OF POSSIBLE MANAGEMENT STRATEGIES FOR THE HUDSON RIVER*

	Dredged from River (m^3) (1978–1988)	Removed from River System (kg PCBs) (1978–1988)	Types of Technology	Status
1. Further study—No management action.	700,000	23,000	Modified DOT maintenance dredging program to assure permanent removal of >90% of PCBs dredged from river as part of maintenance program.	Continuous DOT maintenance dredging necessary for Barge Canal operation.
2. Stabilize and/or remove remnant deposits.	827,000† (680,000)‡		Bank stabilization and/or remove remnant deposits and modified DOT maintenance dredging program as outlined above.	Environmental assessment prepared; 2-year program anticipated (1978–79).
3. Remove remnant deposits and remove all river sediments with PCB concentrations greater than 50 ppm (hot spot dredging).	1,270,000	142,000	Removal of remnant deposits and hot spot dredging technology to vary from small specialized hydraulic dredges to large clam shell dredges. Maximum practical PCB containment technology would be used.	Remnant deposit environmental assessment prepared; 2-year program anticipated; hot spot environmental assessment under preparation; 3-year program anticipated (1979–81).
4. Removal of all river deposits greater than 1 ppm (<5 ppm in remnant deposit areas).	11,500,000	186,000	Massive clamshell dredging effort. Maximum practical PCB containment technology would be used.	Environmental assessment not as yet authorized; 10-year program anticipated (1979–89).

*Source: calculated from Malcolm Pirnie[7,11,18] and Tofflemire and Quinn.[4]
†Assumes areas 3, 3A, and 5 are removed from the basin.
‡Assumes areas 3A and 5 are removed and area 3 is stabilized.

TABLE 7

TOTAL TEN-YEAR COSTS OF POTENTIAL MANAGEMENT ALTERNATIVES

(Total costs are in millions of dollars; unit cost is dollars/kg of PCB removed from the river system)

	Maintenance Dredging* (Total $)	Remnant Deposits (Total $)	Hot Spot Dredging (Total $)	Complete Removal (Total $)	Study and Monitoring†	Total Cost‡ Total $	Total Cost‡ Unit $/kg
1. Further study—No management action	2.3	—	—	—	0.23	2.5	108
2. Stabilize and/or remove remnant deposits	1.9	3.8§ (2.2)¶	—	—	0.57§ (0.41)¶	6.3§ (4.5)¶	84§ 101¶
3. Remove remnant deposits and hot spot dredging	1.9	3.8	21	—	2.61	28.7	203
4. Removal of all river sediments		3.8	—	182	18.6	204	1070

Source: compiled and modified from Malcolm Pirnie, Inc.[7,18]

*Assumes DOT maintenance dredging cost of $3.25/m^3$. Maintenance dredging in alternative 4 was assumed to be minimal and does not remove substantial amounts of PCBs.

†Environmental study and monitoring will require 10% of project cost.

‡Long-term maintenance costs are not included.

§Assumes removal of areas 3, 3A, and 5.

¶Assumes removal of areas 3A and 5 and stabilization of area 3.

TABLE 8

SUMMARY OF SHORT-TERM ENVIRONMENTAL IMPACT OF VARIOUS MANAGEMENT ALTERNATIVES

	Aesthetic	Biota	Water Supply	Land
1. Further study, No management action	Minimal disturbance in sections of river dredged as part of DOT maintenance dredging program.	Localized elimination of benthic fauna in maintenance dredging areas.	Minimal increase in PCBs in vicinity of Waterford water supply intake when areas adjacent to water intake are dredged.	Land areas involved are small and land based operation will be local and of minimal environmental significance.
2. Stabilize/remove remnant deposits	Extremely localized.	Temporary increase of PCB levels in macroinvertebrates immediately downstream of dredging.	No short-term impact other than as stated above for maintenance dredging.	2–8 ha of land may be completely disrupted.
3. Remove remnant deposits plus hot spot dredging	Local water quality standard for turbidity will be violated during dredging operation (2–3 yrs).	Temporary (1–2 years) loss of benthic fauna in dredged hot spot areas. Short term rise in PCB concentration in biota immediately downstream from dredging area. Temporary increase of PCB levels in macroinvertebrates immediately downstream of dredging. Loss of wetlands.	Since there are no hot spots in vicinity of Waterford intake, no water supply impacts are expected other than the maintenance dredging impact stated above.	30–70 ha of land will be completely disrupted.
4. Removal of all river deposits with PCBs greater than 1 ppm (5 ppm in remnant deposits)	Potentially extensive. Continued local disturbance in the areas under active dredging will cause aesthetic problems for 8–10 yrs.	Temporary elimination of benthic fauna in dredged areas for period of up to 3 yrs after dredging. Temporary increase of PCB levels in macroinvertebrates immediately downstream of dredging.	Temporary increase in PCB concentrations in vicinity of Waterford water supply intake when that segment of the river is dredged.	Up to 400 ha of land will be disrupted.

TABLE 9
SUMMARY OF LONG-TERM ENVIRONMENTAL IMPACT OF VARIOUS MANAGEMENT ALTERNATIVES

	PCB Containment	Biota[11]	Water Supply	Land	Air
1. Further study—No management action	Approximately 23,000 kg of PCBs will be removed from the riverbed by maintenance dredging, leaving 175,000 kg (or 88% of the inplace upper river PCBs) to disperse and probably recycle through the environment.	PCB levels in fish and other biota in the upper and lower river will not change in the foreseeable future (upper river >10 yrs, lower river >20); lower river concentrations may rise as highly contaminated PCB sediments move from the upper river into the estuary.	Water column PCB concentrations will continue to cause public health concerns for people using the river as a source of drinking water. Any plans to use the Hudson for drinking water in the future will have to accommodate this contamination.	No long-term land impact is foreseen.	Continued probable volatilization from the river and dredge spoil from maintenance dredging.
2. Stabilize/remove remnant deposits	As much as 75,000 kg of PCBs will be permanently removed from the riverbed leaving 122,000 kg (or 63% of the upper river inplace PCBs) to disperse and continually recycle through the environment. Stabilization of the remnant deposits will substantially eliminate a known active source of PCBs to the upper river system; will reduce	Will reduce by an unknown amount the time required for PCB levels in the upper river biota to fall. Minimal effects on the lower river biota are anticipated.	May reduce water column PCB concentrations in vicinity of Waterford water intake by reducing an active source to the upper river. Will have only a minimal effect on any lower river water supplies, and planned flood skimming for NYC water supply in the foreseeable future.	2–8 ha of land will be required for spoil disposal. Permanent restrictions on use of the land will be required. Restrictions may be mitigated by using land as a nature preserve or ecological study area. (Permanent restrictions could be changed by technologic advancement in PCB bacterial, biodegradation research.)	Reduction of volatilization from river probably slight.

	size and PCB complexities of DOT maintenance dredging program.				
3. Remove remnant deposits & remove all river sediments with PCB concentra-tions >50 ppm (hot spot dredging)	Approximately 142,000 kg of PCB will be permanently removed from the riverbed leaving 56,000 kg (or 28% of the upper river inplace PCBs) to disperse and continually recycle through the environment. This option will eliminate all high concentration sediments (>50 ppm) from the river system. It will substantially reduce the size & PCB complexity of DOT maintenance dredging operations. The sediments to be removed are also generally contaminated with organic pollutants & other toxic materials. This project would rmove these materials from the river environment.	Can reduce PCB biota concentrations in upper river by 50% to fish flesh levels of 20–40 ppm. In the lower river, fish burdens may drop by about 20% to about 8–12 ppm.	Should substantially reduce water column PCB concentrations in vicinity of Waterford water supply intake. Will have a marginal effect on lower river water supplies, existing & planned in the foreseeable future.	30–70 kg of land will be required for contaminated spoils disposal. Permanent restrictions on use of land will be required. Restrictions may be mitigated by using land as a nature preserve or ecological study area. (Permanent restrictions could be changed by technologic advancement in PCB bacterial, biodegradation research).	Should substantially reduce PCB volatilization from the upper river.

TABLE 9 (continued)

	PCB Containment	Biota[11]	Water Supply	Land	Air
4. Remove all upper river deposits >1 ppm (>5 ppm in remnant deposit areas)	About 186,000 kg (~94%) of the PCBs in the upper river will be permanently removed from the riverbed and the environment. Substantial reductions in the size & PCB complexities of future DOT maintenance dredging programs will occur. Most of the sediments to be removed are also contaminated with other organic pollutants and other toxic materials. This project will remove these materials from the river environment.	Can reduce PCB biota concentrations in the upper river to fish flesh levels of 10–20 ppm. By significantly reducing PCB input into the lower river, it may reduce the body burden of larger fish by about 30–50% to levels close to 5 ppm.	Will substantially eliminate the problem of PCBs in the village of Waterford water supply intake. Will also lower level of other toxic contaminants in Waterford water supply. Can measurably reduce PCB water column problems in vicinity of estuary water supply intakes by stopping active sources of PCBs to the lower river. May reduce PCB problems associated with flood skimming project.	About 400 ha of land will be required for contaminated spoil disposal. Permanent restrictions on use of land will be required. Restrictions may be mitigated by using land as a nature preserve or ecological study area. (Permanent restrictions could be changed by technologic advancement in PCB bacterial, biodegradation research).	Will eliminate the river as a source of PCBs to atmosphere. Odors associated with extensive sludge beds should also be essentially eliminated.

Hot spot dredging. Dredging of only the high concentration areas was considered because the unit costs (\$/kg PCBs) of removal are lowest and, at the same time, the greatest PCB removal can be achieved with the least environmental disruption. Dredging can be 94% efficient in removing PCB contamination,[7] whether hydraulic or mechanical methods are utilized. The data show that 67,000 kg PCB can be removed from the Upper River by dredging hot spots, which make up only 8% of the riverbed surface area. Additionally, the remnant deposit program and normal maintenance dredging is expected to remove approximately 75,000 kg of PCBs The total, 142,000 kg of PCBs, represents 74% of the PCBs in the Upper Hudson River, and 52% of the PCBs in the entire riverbed.

This project would require more careful examination of each major hot spot so that equipment best suited for each area could be selected. The dredging could probably be completed within three years.

Complete removal of PCB-contaminated sediments. The most extensive reclamation program would involve dredging the entire 65 km section of the Upper Hudson River from Bakers Falls to Troy. The dredging yardages involved are extensive, but most (94%) of the PCBs could be removed from the riverbed above Troy, thus eliminating a major source of PCBs to the water and biota of the entire Hudson River. A dredging project of this magnitude would require approximately 10 years.

Cost

The cost of each of the management alternatives discussed above are given in TABLE 7.

Environmental Impact

Both the short-term and long-term environmental impact of the alternatives considered were evaluated. This eveluation is summarized in TABLES 8 and 9.

PRESENT STATUS

On the basis of the above results, an independent scientific Advisory Committee[19] has recommended that the partial (hot spot) dredging management program be pursued. Implementation of this aciton will involve a complex program of public education, lobbying for funds, detailed engineering, and continued monitoring and studies. A high degree of cooperation and assistance from scientists, environmentalists and local, state, and federal government will be required for success.

REFERENCES

1. HORN, E. J., L. J. HETLING & T. J. TOFFLEMIRE. 1979. The Problem of PCBs in the Hudson River System. Ann. N.Y. Acad. Sci. This volume.
2. SANDERS, J. E. Sedimentary Aspects of PCB Pollution of the Hudson River. Paper presented at the New York Academy of Sciences Conference on The Health Efforts of Halogenated Aromatic Hydrocarbons, June, 1978. Unpublished.
3. THOMANN, R. 1979. The Fate of PCBs in the Hudson River Ecosystem. Ann. N.Y. Acad. Sci. **320:** This volume.

4. Tofflemire, T, J. & S. O. Quinn. PCB in the Upper Hudson River: Mapping and sediment relationships. In preparation.
5. Horstman, K. H. 1977. Evaluation of non-dredging alternatives to the removal of PCB contamination in the Hudson River. Comp. Ed. 18 thesis, Union College, Schenectady, New York. 51 pp.
6. U.S. Environmental Protection Agency. 1978. Mitigation Feasibility for the Kepone contaminated Hopewell/James River areas. Washington, D.C. (preliminary draft).
7. Malcolm Pirnie, Inc. 1978. Dredging of PCB-contaminated riverbed materials from the Upper Hudson River, New York: Feasibility Report. White Plains, New York.
8. U.S. Army Engineers Waterways Experiment Station. 1978. Dredged material research program notes, news, reviews, etc. U.S. Army Corps of Engineers, Vicksburg, Mississippi.
9. Tofflemire, T. J. Unpublished data from October 1978.
10. Tofflemire, T. J. 1976. Preliminary report on sediment characteristics and water column interactions relative to dredging the Upper Hudson River for PCB removal. New York State Department of Environmental Conservation, Albany, New York. 82 pp.
11. Malcolm Pirnie, Inc. 1977. Environmental assessment of maintenance dredging at Fort Edward Terminal Channel, Champlain Canal. White Plains, New York. 271 pp.
12. Mt. Pleasant, R. 1978. Presentation to PCB Settlement Advisory Committee. Data summarized from Miner, W. January 1978. Fort Edward maintenance dredging project monitoring report. New York State Department of Environmental Conservation, Albany, New York.
13. Griffen, P. M., C. M. McFarland & A. R. Sears. 1978. Research on removal of treatment of PCB in liquid of sediments dredged from the Hudson River: Semi-annual progress report. General Electric Company, Corporate Research and Development, Schenectady, New York.
14. Coffman, J. A. 1978. (Wright-Malta Corporation, Ballston Spa, New York). Personal communication.
15. General Electric Corporate Research & Development. 1978. Report to PCB Settlement Advisory Committee. Unpublished.
16. Blair, J. E. 1978. (Biochemical Corporation of America, Salem, Virginia). Personal communication.
17. Leis, W. 1978. (Weston Environmental Consultants). Personal communication.
18. Malcolm Pirnie, Inc. 1978. Environmental assessment of remedial measures at the remnant deposits of the former Fort Edward Pool. White Plains, New York, 173 pp.
19. Sanders, J. A. 1978. Letter to Commissioner Peter A. A. Berle, New York State Department of Environmental Conservation, Albany, New York.

PCB REMOVAL FROM THE DUWAMISH RIVER ESTUARY: IMPLICATIONS TO THE MANAGEMENT ALTERNATIVE FOR THE HUDSON RIVER PCB CLEANUP

S. P. Pavlou and W. Hom

Aquatic Ecosystems Program
URS Company
Seattle, Washington 98121

INTRODUCTION

Because of the continuously increasing inputs of contaminants into riverine systems and the significance of their trapping potential, there has been an increased concern and interest in the possible impact on the resident biota of these systems. Therefore, concerted efforts to evaluate the effects of bottom sediments containing toxic substances have been initiated both nationally and internationally. One of the most significant toxic substance problem areas is the Upper Hudson River, which is contaminated by polychlorinated biphenyls (PCBs). The state of New York has launched a comprehensive program to evaluate the extent of the impact of high levels of PCBs and to determine the best alternative for mitigation of the problem.

In order to make the proper environmental decision and take the appropriate course of action, there is a great need to acquire relevant data. However, an extensive data bank by which to determine the best management alternative for removal of contaminated sediments is lacking.[1]

Data do exist, however, for a dredging operation on the Duwamish River, Washington, which successfully removed PCB-laden sediments from the river.[3–6] The objective of this paper is to present the cleanup of this river as a test case for comparison to the Hudson River problem.

Description and History of the Spill

The sequence of events is presented in TABLE 1. On September 13, 1974 a transformer being loaded for shipment to Alaska was dropped onto the dock and broken open, spilling the transformer fluid into the water. The fluid released was Aroclor® 1242, a PCB mixture. Since Aroclor 1242 has a density of 1.38 kg per liter, 1306 kg of the mixture quickly sank to the bottom. FIGURE 1a shows the location of the spill site at Slip 1.

The EPA, Region X, was notified shortly thereafter and resulting from an initial survey conducted 5 days later, arranged to remove the PCBs from the river bottom. For 20 days in October 1974, EPA recovered visible pools of the PCBs from the river sediment using hand dredges. However, subsequent monitoring of the residue levels in the sediments indicated that this initial cleaning was not effective. Furthermore, upriver mobilization of the residues was also observed.

The U.S. Department of Defense assumed responsibility for the cleanup in February 1975; the Army Corps of Engineers, Seattle District, was given the specific mission to clean up and remove the PCBs from the riverbed. The Corps brought in the Pneuma dredge system and dredged for 24 days in March 1976, removing approxi-

0077–8923/79/0320–0651 $01.75/0 © 1979, NYAS

TABLE 1

CHRONOLOGICAL SEQUENCE OF EVENTS

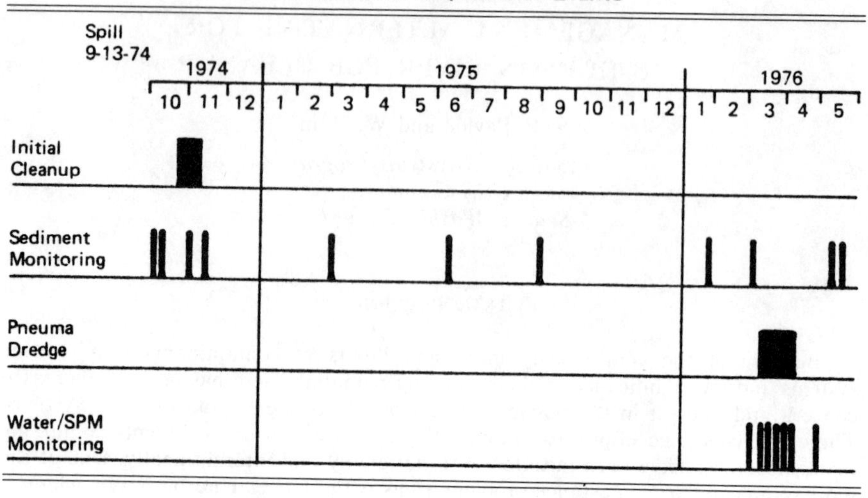

mately 5300 m^3 of sediments containing 56% of the spilled PCBs. The material was then deposited into a nearby disposal pond. The final monitoring in May 1976, showed that the PCB levels in the river were again near the previous background values.

BACKGROUND INFORMATION

The Duwamish River drains into Puget Sound, which is located in the northwestern United States (FIGURE 2). It enters the Sound via Elliott Bay, which lies at the mid-point of the Sound. In the region of the spill, the river is a well-stratified salt-wedge estuary that is influenced by the diurnal fluctuations of the tides (FIGURE 1b).

Background levels of PCBs in Puget Sound have been investigated extensively for the biota, water, and sediment phases.[7–13] TABLE 2 presents the mean total PCB values for various regions and phases in Puget Sound. A synoptic overview of the sediment levels is presented in Figure 2, which shows that the PCB levels in the sediments range from 0.02 to 1.2 ppm. FIGURE 3 shows a more detailed distribution of PCBs in Elliott Bay. Contours range from 2 ppm at the mouth of the Duwamish River to 0.1-0.2 ppm at the entrance to the Sound. The surface sediment values for navigable portions of the Duwamish River range from 0.5 ppm to 2.0 ppm. Concentrations increase from the turning basin to the mouth of the river, showing that the river is far from being pristine.

CLEANUP EFFORTS

Initial Cleanup

After establishing the limits of the problem in a pre-dredge survey, the hand dredge method was chosen in the interim period.

Scuba divers entered the water at the site to operate hand-held dredges pumping at 50 gallons per minute (gpm). By initiating the cleanup quickly after the spill, the PCBs were still found to be concentrated in pools on the bottom. This concentration facilitated the removal process by allowing the divers to visually select the PCB pools for removal by dredging. Consequently, the dredge spoils were nearly pure PCBs mixed with dilution water. The slurry was pumped into a settling pond and treated with NALC® 34 polyelectrolyte to improve the settling rates of the sludge.

During the 20-day dredge period, 215 55-gallon drums were filled with the PCB-laden sludge. The dilution water was passed through a sand and carbon filter system before its return to the river. During this period, the sediment and water were monitored for changes in the PCB level.

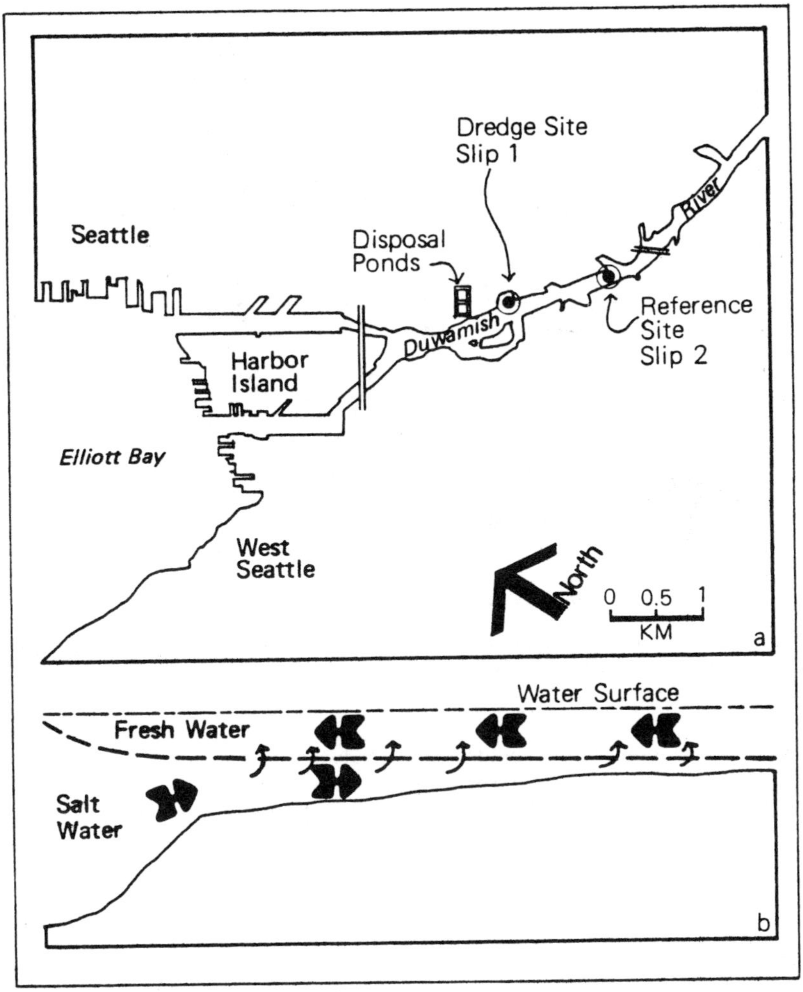

FIGURE 1. Location of dredge and reference sites in an estuarine regime.

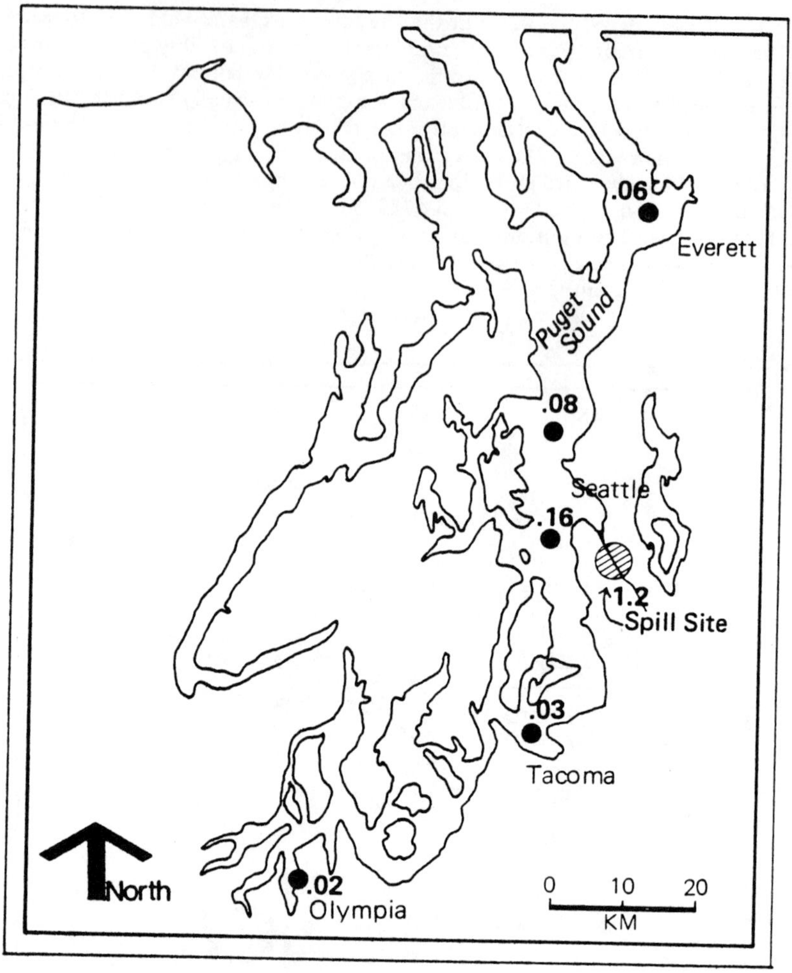

FIGURE 2. Duwamish River spill site location within Puget Sound and PCB surface sediment concentrations (ppm). See Dexter, 1976.[7]

After determining that it would require a substantially longer effort to remove the PCBs from the bottom, EPA terminated its cleanup operation after 20 days. The contaminated mud was then shipped to the hazardous waste disposal area in Idaho in the spring of 1975.

Final Cleanup

The Corps of Engineers was given the specific mission to remove the PCBs from the bottom sediments of Slip 1 after the Department of Defense assumed responsibility for the accident. The choice of removal technology was hydraulic dredging; the

TABLE 2

MEAN TOTAL POLYCHLORINATED BIPHENYL CONCENTRATIONS IN VARIOUS REGIONS OF PUGET SOUND*

Area	Water [ng l^{-1}]	SPM [ng g^{-1}]	Zooplankton [μg (g-lipid)$^{-1}$]	Sediment [ng g^{-1}]	Surface Film [ng l^{-1}]
Duwamish River	22 ± 13(60)	1770 ± 540(70)			116 ± 101(5)
Elliot Bay	11 ± 13(185)	920 ± 200(87)	6.8 ± 3.1(11)	637 ± 830(4)	98 ± 83(6)
Main Basin	4.3(1)	104 ± 31(3)	5.9 ± 4.4(4)		12(1)
Sinclair Inlet		168 ± 21(4)	16 ± 7.0(3)		
Whidbey Basin	4.4 ± 2.2(15)	82 ± 37(7)	3.7 ± 1.1(12)	30 ± 14(5)	
Northern Sound and Straits of Juan de Fuca	3.7 ± 1.7(9)		1.3 ± 1.2(3)	16 ± 13(3)	
Southern Sound				8.0 ± 8.3(6)	
Commencement Bay				28 ± 17(8)	
Hood Canal	3.0 ± 2.2(12)	88 ± 37(3)	1.8 ± 1.0(4)	12 ± 3.0(3)	

*Values in parentheses are the number of data points. See reference 15, Pavlou and Dexter, 1978.

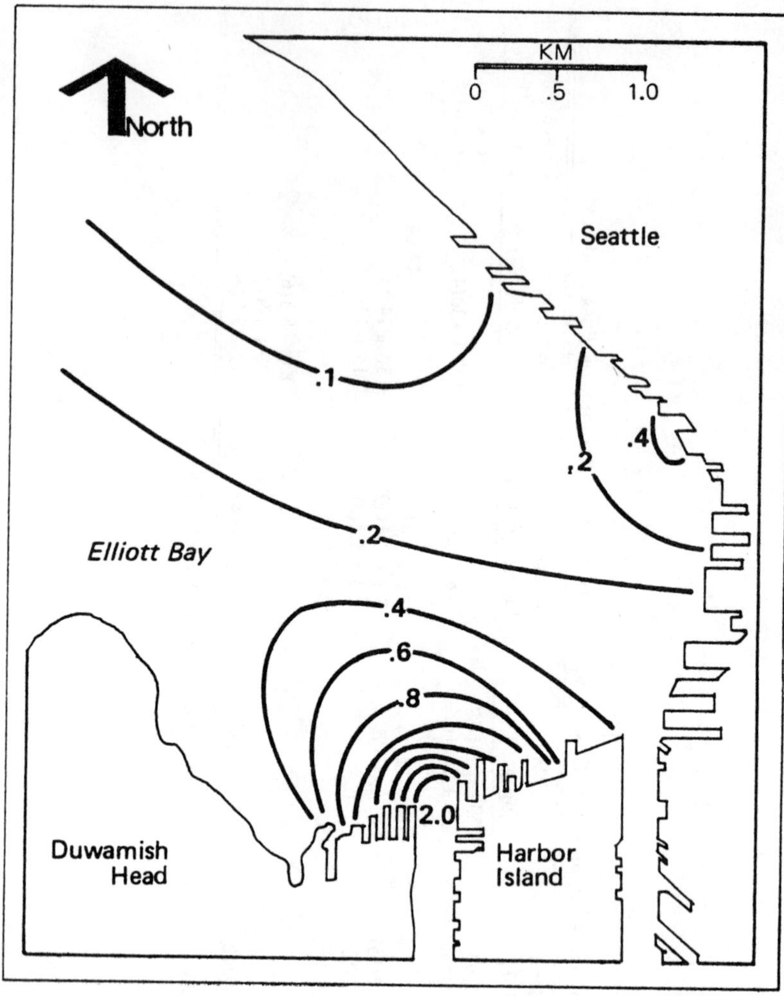

FIGURE 3. PCB sediment levels in Elliott Bay (ppm). See Dexter, 1976.[7]

patented Pneuma Dredge system, which does not extensively resuspend the neighboring sediments as it operates, was used. FIGURE 4 shows the basic operation of the Pneuma.

Operating from a barge, the area adjacent to the north side of Slip 1 was dredged (FIGURE 5). On the corner of the spill site a 30 × 40 m rectangular patch was extensively dredged. In total, approximately 5352 m^3 of sediments were dredged. For the entire dredge area, a depth of 0.6 m was removed, while at the spill site, extensive dredging removed 3.7 m of sediment. Dredging occurred down to the hard pan in order to reduce the PCB concentration to acceptable levels. The dredge spoils were carried through a 25 cm internal diameter pipe to a dual pond disposal site (FIGURE 6) approximately 650 m to the north of the spill site. These ponds were specially constructed to hold the PCB spoil materials. They were approximately 4.6 m deep with a diagonal distance of 98 m.

To aid in the treatment process, a cationic polyelectrolyte, NALCO 7134, was added to the flow about 60 meters from the entry to pond 1. This enhanced the solids' flocculation considerably, yielding very low turbidity upon entry to pond 2. The water from pond 2 was pumped through a filter system to remove algae and suspended solids into a holding pond. By pumping the water through three sand filters followed by three activated carbon filters of the EPA physical/chemical treatment truck[14] the concentrations of PCBs in the water were sufficiently reduced allowing the water to be returned to the river.

After the dredging operation had been completed, the disposal area was buried with at least 3 feet of clean soil and stablized for possible future use.

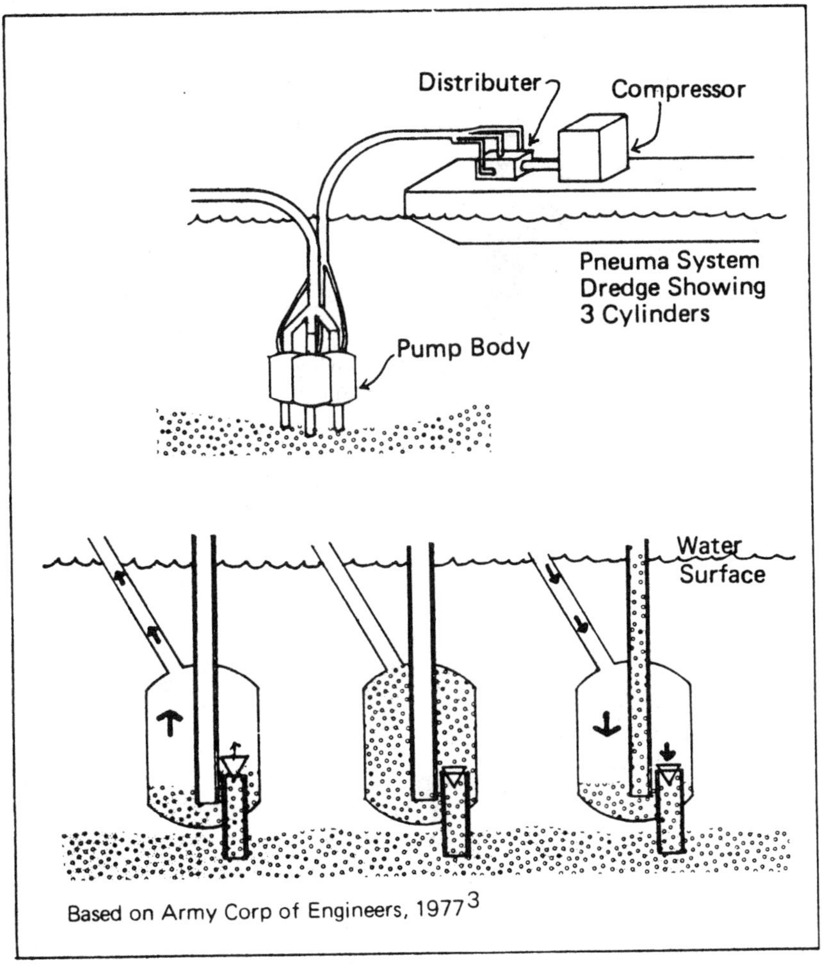

FIGURE 4. Schematic of Pneuma dredge operation.

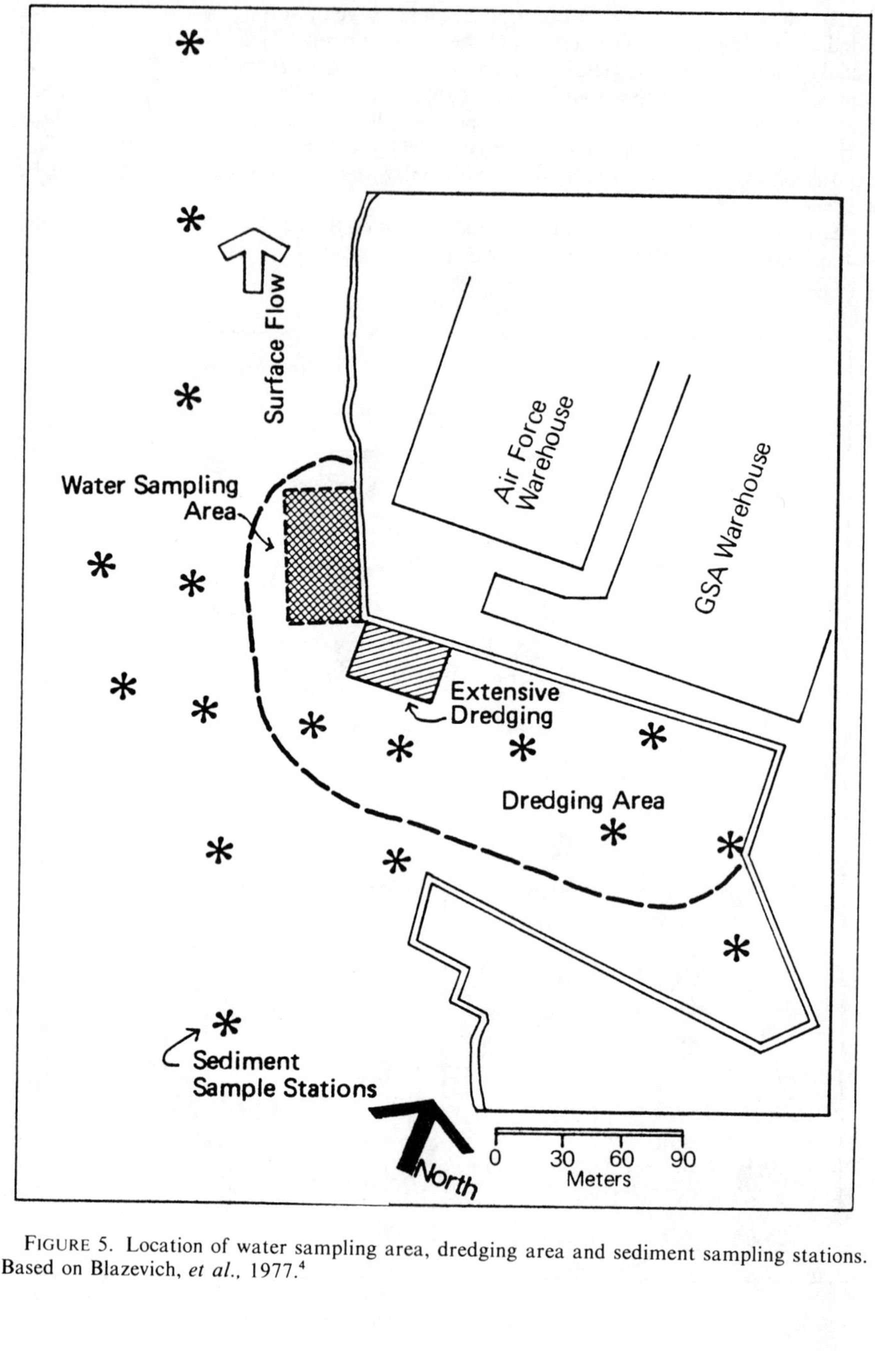

FIGURE 5. Location of water sampling area, dredging area and sediment sampling stations. Based on Blazevich, *et al.*, 1977.[4]

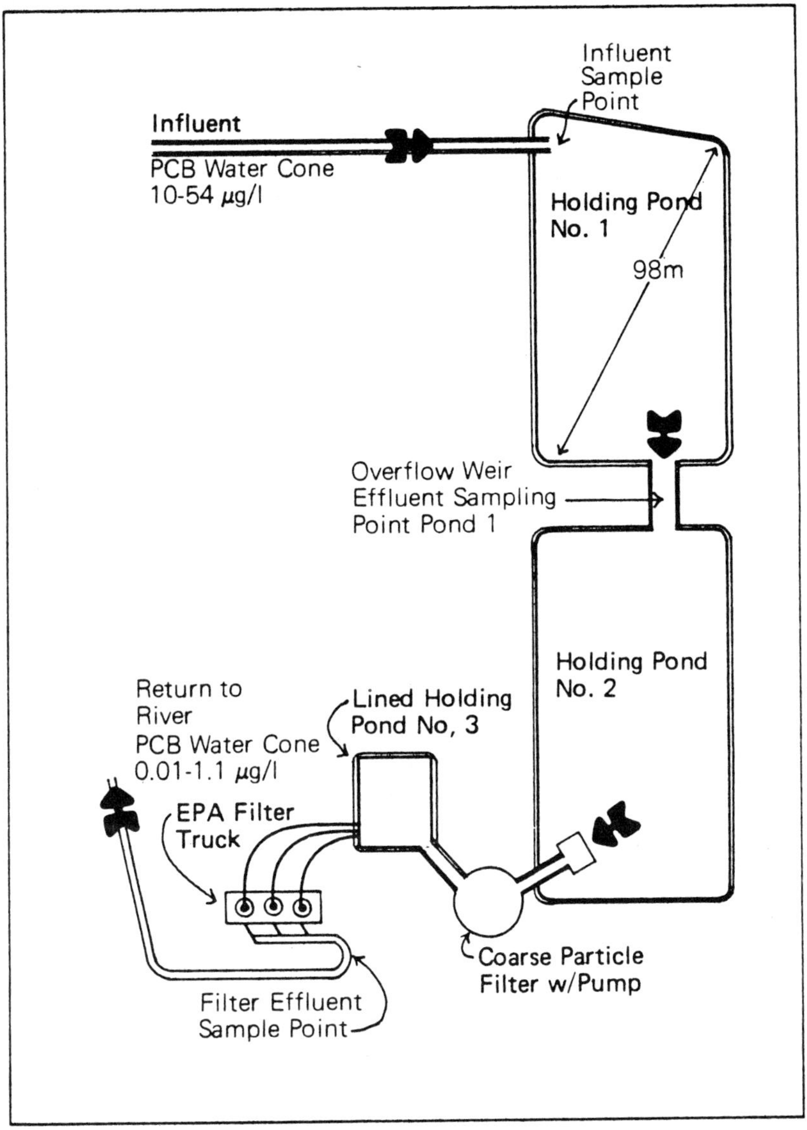

FIGURE 6. Disposal pond layout. Based on Blazevich *et al.*, 1977.[4]

MONITORING OF DREDGING AND DISPOSAL

Sediment Monitoring Efforts

The contours in FIGURE 7 are estimates of the PCB-disperson limits in the sediments and show how they changed as the dredging progressed. The stations in the monitoring effort are shown in FIGURE 4.

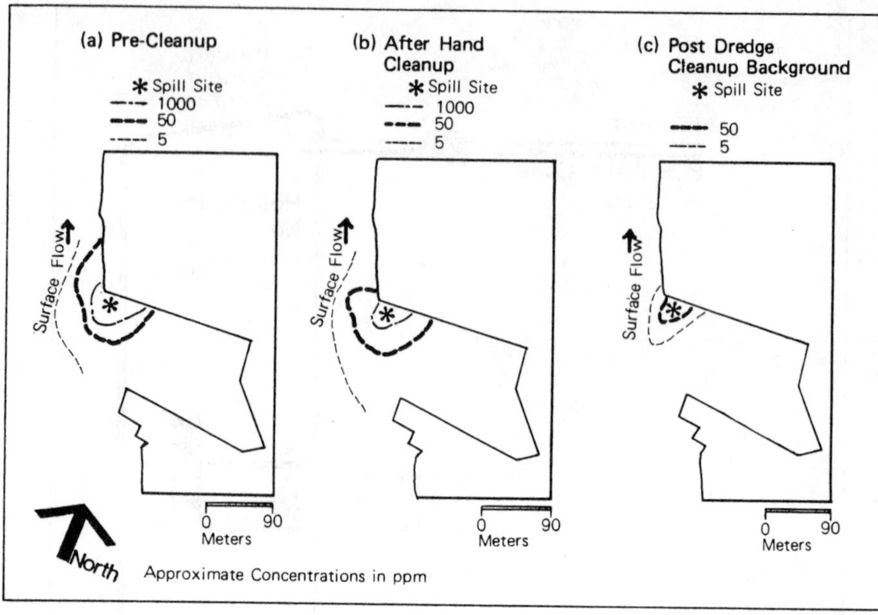

FIGURE 7. Contours of PCB sediment levels at different stages of cleanup. Based on Blazevich *et al.*, 1977.[4]

Prior to the initial effort, the PCBs had been mobilized and were found in a 10–20 m radius about the spill site (FIGURE 7A). Elevated levels in the middle of the channel were also found.

During the initial cleanup phase, the monitoring was carried out by the EPA Region X laboratory. The divers visually checked the dredged areas and reported that the liquid pools of PCB had been removed; however, analysis of the sediment showed that the PCBs had infiltrated the sediment matrix. At this juncture, it was recognized that extensive removal of the sediment would be required to effect a substantial cleanup effort.

The results of the hand-dredge cleanup are shown in FIGURE 7B. Although the extent of the PCB influence was still prominent in a 10–20 m radius, the levels had been reduced. Residual levels in the channel had decreased as well, but this was probably due to dilution or burial by more recent sediment since the channel was not dredged during the PCB-removal effort.

The second cleanup effort was greater in magnitude because of the larger area and deeper cut into the sediments. As such, the monitoring efforts were now divided into sediment, water and suspended particulate matter (SPM) segments.

The sediment monitoring had stations throughout Slip 1 and extended part way into the channel (FIGURE 4). Sampling before and after the dredging showed a dramatic drop in the PCB levels of the sediment (FIGURES 7B and 7C). In the channel region, the post-dredge results show values comparable to values observed throughout the navigable portions of the river. In the slip, the values at the spill site were smaller than the hand-dredge levels. Although they were still higher than observed background values, the major portion of the PCB-laden sediment had been removed. The surrounding regions of the spill site showed levels commensurate to previously observed values.

Water samples were collected adjacent to the dredging area. A reference station was located about 1.2 km up river. Samples of water and SPM were collected before, during and after the dredging. Because of the stratified estuarine nature of the river, samples were taken from the fresh and sea water layers at the dredge site and reference station. The results are summarized in TABLE 3. Although PCB concentrations in the water appear to be different, statistical analysis showed that the reference and dredge-site values were indistinguishable at the alpha = 0.05 level for the fresh and sea water layers.[5] For the SPM, inspection of the results show that the levels in fresh and sea water were the same regardless of whether the samples were from the dredge site or reference stations. Furthermore, the component concentration ratios were also invariant. The conclusion to be drawn from this work is that the Pneuma dredge operation did not measurably alter the water quality in either the water or suspended matter phases by injection of PCBs.

During the monitoring period, when the dredging operation was active, the normal sampling position for SPM and water was 15 m to 70 m downstream from the barge. However, one of the sampling periods brought the barge to less than 4 m from the sampling site. This particular set of samples was collected in the propeller wash from the dredge barge as the barge maneuvered near the river bank creating an increased suspended load. The propeller wash samples showed an elevated relative mass fraction (F_N) for trichlorobiphenyls when compared to previous studies and background material. This was expected, since the PCB spilled into the river consisted primarily of the 3- and 4-chlorobiphenyls (FIGURE 8). Rapid particle settling and dilution probably accounted for the transience of this impact.

FIGURE 8 also shows the background homolog distribution of water and SPM samples. When the F_N values for the surface and deep water layers at the dredge and reference sites were statistically analyzed, no difference was detected at the 95% confidence level either spatially or temporally. As such, the determination and monitoring of the relative mass fractions of PCB components provides a good technique for assessing the dispersal and identification of the source. Furthermore, the invariance of the component concentration ratios indicate that equilibrium partitioning with ambient water is a realistic index of the fractionation characteristics of PCB under dredging and normal conditions.[15–17]

TABLE 3

DUWAMISH RIVER WATER/SPM MONITORING RESULTS*

Average Total PCBs			
Phase	Station	Surface Layer	Deep Layer
Water × 10^{-12} g/g	Ref	16.9	12.8
	Dredge	24.5	22.1
SPM × 10^{-6} g/g	Ref	16.9	12.8
	Dredge	24.5	22.1
Component concentration ratio (part. coefficient × 10^4)	Ref	16.9	12.8
	Dredge	24.5	22.1

*Reference: Hafferty *et al.*, 1977.[5]

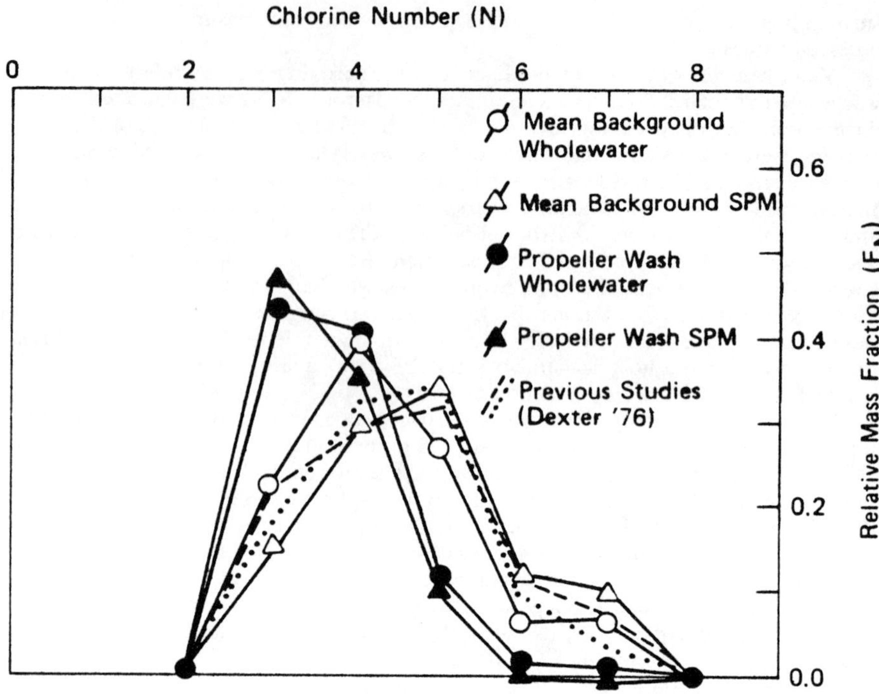

FIGURE 8. Relative mass fraction (F_N) versus chlorine number (N). See Hafferty *et al*, 1977.[5]

Monitoring of Disposal

Monitoring the disposal of the dredged spoils occurred at three locations in the spoils disposal pond. The influent, the ponded sediments, and the effluent water were each analyzed for their PCB content. As a result of erratic dredging schedule, the influent rate to pond 1 was also erratic. Consequently, water exiting pond 1 only came over the weir when a sufficient volume of water had been accumulated. During the time span of 16 March to 23 March the influent PCB slurry concentration ranged from 4.1 μg/l to 54 μg/l demonstrating that the PCB-laden sediments were being removed from the river.

After the slurry entered the pond, the solids quickly settled, forming a grade which was exposed near the entry point and covered by water at the exit. Three sampling regions were designated in an approximately concentric manner about the influent pipe. The inner most region near the slurry entry point had an average sediment concentration of 146 ppm while the two outer regions averaged 33 ppm. Based on the estimated volume of sediment in the pond, these levels show that at least 730 kg of PCB were recovered by the final cleanup effort. Monitoring of the water effluent from the ponds that had passed through the EPA physical/chemical treatment unit showed a dramatic decrease in the PCB concentration. Levels in the exiting water ranged from 0.01 μg/l to 1 μg/l, demonstrating that this technique of sediment treatment and disposal in a pond is feasible and efficient.

Summary

Using a hand-dredge, approximately 418 kg were recovered from the 170 m^3 of sediments that were dredged. Approximately 17 months later, the second, more extensive, cleanup took place. During a 24-day dredging period, 5352 m^3 of sediments were removed from the river bottom and piped to the nearby disposal pond. Treatment of the slurry with flocculent resulted in a total recovery of about 730 kg during this cleanup phase. A summary of the overall cleanup operation is shown in TABLE 4; FIGURE 9 indicates the recovery for each cleanup effort. Although the water and SPM monitoring was limited to the second cleanup effort, the results revealed several factors: (1) there was no observable change in the SPM concentration throughout the dredging operation, (2) the whole water PCB-levels were also constant in the same period and (3) both the component concentration ratio, K_N^d, and the relative mass fraction, F_N, remained constant. This suggested that under equilibrium partitioning, K_N^d, can be a fractionation index of PCBs between water and SPM, while F_N can be employed as an identifier and tracer of the PCB distribution. As such, the dredging operation did not significantly alter the PCB characteristics of the river.

In 44 days of dredging, over 90% of the PCBs were recovered and disposed of and better than 5500 m^3 of sediments were translocated in this period. Less than two years after the spill, the Duwamish River at Slip 1 had been stabilized and returned to its previous characteristics.

The total cost for the two cleanup periods was $540,000. The initial EPA cleanup cost $121,000 and the Army Corps dredging operation cost $359,000. The balance of the monies was used for monitoring of the water, sediments, and miscellaneous expenses.

TABLE 4

SUMMARY OF DUWAMISH DREDGE OPERATION

Accident
- 1306 kg PCBs, Spilled into River

Initial Cleanup by Hand Dredge
- 418 kg Recovered
- 170 m^3 Dredged
- 20 days

Pneuma Dredge
- 730 kg Recovered
- 5352 m^3 Dredged
- 24 days

Monitoring Results during Pneuma Dredging
- No change in SPM, PCB levels
- No change in water PCB levels
- Constant component concentration ratio equilibrium partitioning as fractionation index
- Constant relative mass fractions identification and tracing technique

Time in River
- Less than 2 years

Summary
- Greater than 90% of PCBs recovered
- 44 Days of dredging
- 5520 m^3 Dredged
- Cost, $540,000

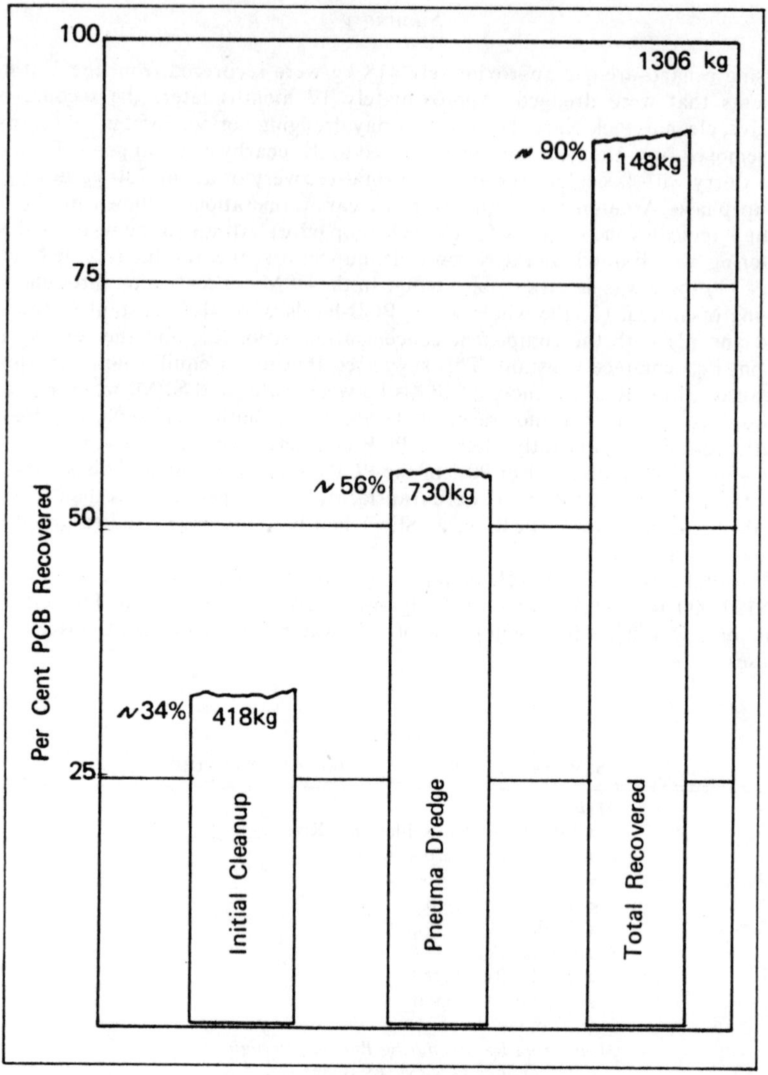

FIGURE 9. Summary of PCBs recovered.

CHARACTERIZATION OF DUWAMISH AND HUDSON RIVER PCB PROBLEMS

An obvious similarity between the Duwamish and Hudson Rivers is that both have been or are currently contaminated with PCBs in their river beds. The location of the Duwamish River contamination was in an estuarine region whereas the Hudson River problem occurs only in freshwater. Compared to the long reach of contaminated sediment in Hudson River, the Duwamish was only affected at the spill site. Estimates for PCB quantities in the Hudson show that it has 100-fold more PCBs than the

Duwamish. This is not surprising, since PCBs have been discharged into the Hudson for more than 20 years, whereas the contamination in the Duwamish resulted from an isolated incident. TABLE 5 presents a comparison of the PCB contamination of the two rivers.

Characteristics of the Rivers

As measured near the area of the spill, the Duwamish River, drains an area of 1033 km^2 and has a transport rate ranging from 19 m^3/sec to 53 m^3/sec, with a mean of 37 m^3/sec.

On the Upper Hudson River, the vicinity near river mile 200 at Glens Falls and Hudson Falls are the contaminated regions (FIGURE 10). The critical region is located just above the former Fort Edward Dam (FIGURE 11). The Upper Hudson River drains an area of 12,000 km^2 with an average transport of 217 m^3/sec, as measured at Waterford.

Applicable Results of the Duwamish to the Hudson Problem

The most significant result of the Duwamish contamination is the successful completion of an operation designed to remove PCB-laden sediment from a reach of river, which recovered greater than 90% of the injected contaminant. The Pneuma dredge is an applicable system for efficient sediment removal with low resuspension and turbidity. Monitoring efforts during the dredging operation found no differences in the PCB levels in SPM and water when the dredge site and an upriver reference sites were compared. The only problem observed was when the propeller wash of the maneuvering barge resuspended quantities of sediment. When the problem was highly localized at the beginning, hand dredging was successful in removing the identifiable pools of PCBs.

Disposal of the PCB-laden sediments to ponds and subsequent return of the water to the river was shown to be a workable system. Settling of the solids was greatly

TABLE 5

COMPARISON OF PCB-CONTAMINATED REGIONS

Characteristic	Duwamish River	Upper Hudson River
Area affected (km of River)	0.48	730
Estimated amount of PCBs in sediment (kg)	1306	205,000
Time in sediment	1 Year	20 Years
Depth in sediment	Surface-10 cm	Surface-87 cm
Type	Well-stratified Estuary	Fresh water river

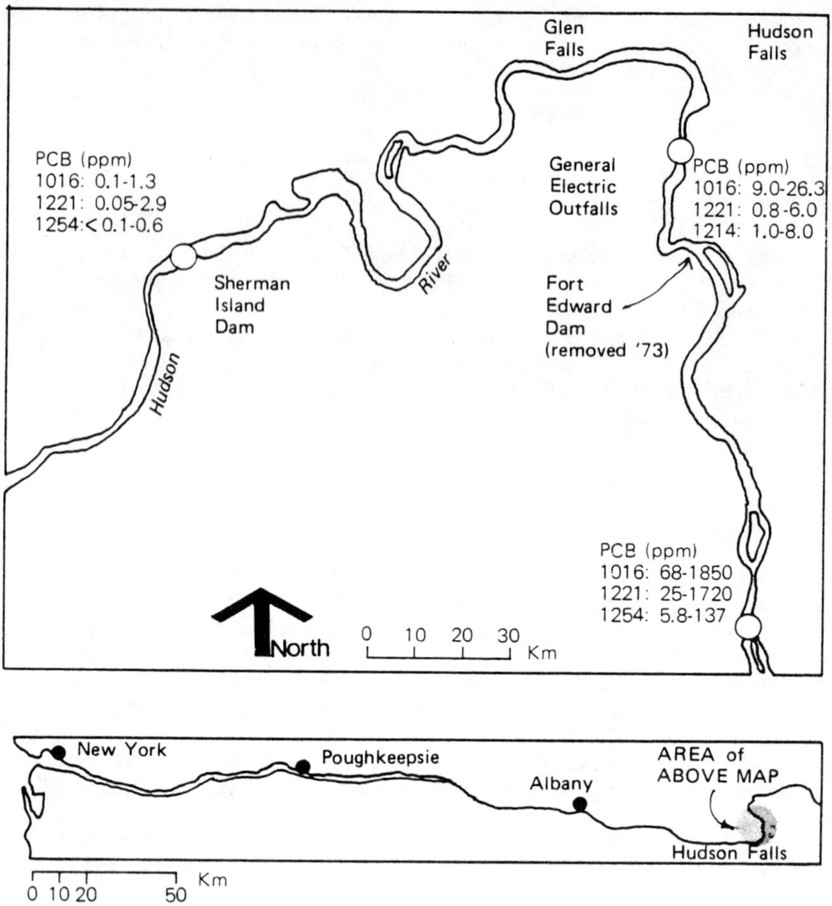

FIGURE 10. Location and levels of PCB contamination in the Upper Hudson River. See the Hudson River-PCB Study, Description and Work Plan, July, 1977.[18]

promoted by adding a flocculent to the pond inflow line. Before its return to the river, the effluent water from the ponds was filtered to remove the remainder of PCBs. This complete cycle of sediment removal, disposal, and return to the river of the dilution water, was shown to be a viable technique.

After the operation, the disposal ponds were stabilized and buried under at least three feet of clean soil for future development.

The mobilization of PCBs from the spill site into the surrounding areas was observed during the sediment monitoring efforts that followed the first cleanup. The possibility of mobilization has been a primary factor of concern in the Hudson problem. Because of this potential transport and dispersion an immediate response to cleanup and containment of the problem should be executed, or else critical time will be lost, creating a problem of larger magnitude and greater difficulty.

RECOMMENDATIONS AND REMOVAL ALTERNATIVES FOR THE HUDSON RIVER

A report has been compiled which synthesizes data from studies covering various aspects of the problem.[2] A composite presentation of the alternatives is given in TABLE 6, presenting the dredge volume, cost, technology, time, and impact of the alternatives. There are four basic potential solutions, depending on the degree of removal and cost desired.

The least demanding alternative is to take no action and study the problem further. This allows the continuous channel-maintenance dredging to remove insignificant quantities of PCBs. The second alternative is to stabilize and/or remove the remnant deposits (FIGURE 11). These are the sediments in the former depositional

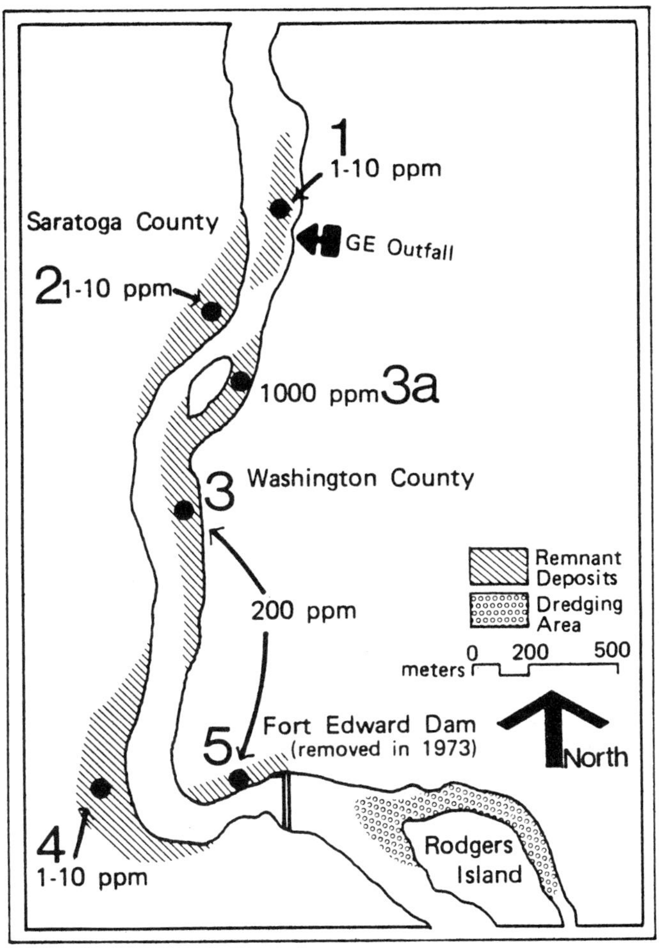

FIGURE 11. PCB-contaminated regions in the vicinity of the Fort Edward Dam. Based on Hetling *et al.*, 1978.[2]

TABLE 6

MANAGEMENT ALTERNATIVES*

	No Management Action Further Study	Stabilize and/or Remove Remnant Deposits	Remove Remnant Deposits and Sediments >50 ppm	Remove all River Sediments >1 ppm
$M^3 \times 10^6$ Dredged (1978–1988)	0.70	0.83† (0.68)‡	1.74	11.5
Percent PCB Removed§	11	37† (22)‡	74	94
Total Cost (millions)	2.5	6.3† (4.5)‡	29.4	204
Unit Cost $/kg PCB	108	84† (101)‡	192	1070
Types of Technology	Modified DOT dredging	Bank stabilization modified dredging	Hot spot dredging (Clamshell) max. PCB containment	Massive dredging max. PCB containment
Estimated Time Required	Continuous maintenance	2 yr	3 yr	10 yr
Land Impact	Minimal disturbance	0.1 – 0.3 km^2	0.3 – 0.7 km^2	4 km^2

*Based on Hetling *et al.*, 1978.[2]

†Removal of Region 3, 3A, and 5, Figure 11.

‡Removal of Region 3 and 5, Figure 11.

§Total 204,600 kg, Table 5.

areas, deposited when the Fort Edward dam was present. They are now potentially erosional since the dam has been removed. The sediments in these regions can be stabilized with riprap or removed to an encapsulated disposal site.

The third alternative is essentially similar to the second, but calls for dredging of hot spots, characterized by regions having PCB levels greater than 50 ppm. This increases the cost, the quantity of sediment dredged, and the time required, but a substantially higher portion of the PCBs would be removed from the river. The fourth alternative is to do extensive dredging to remove all sediments having PCB levels greater than 1 ppm. This effort would require a large increase in the cost, the sediment volume moved, and the time required, resulting in a small percentage increase in the total amount of PCBs recovered.

In light of the Duwamish River case, the management considerations for addressing the Hudson River problem are listed in TABLE 7. Regional, site-specific, and other impacts must be evaluated prior to the execution of a removal effort. Regionally, the distribution of PCBs, the river morphology, the water quality, and the potential impact to the overall ecosystem all need to be considered. At a specific site, similar factors must be considered, but on a smaller scale. During the review and decision stages, the impact to the industry, commerce, recreation, and social needs of the region must not be overlooked.

Several of the alternatives suggested by the Hetling *et al.*[2] report are not adequate or marginal: (1) Further study without management action is not recommended. There is no change in the level of the PCBs but the potential long-term hazard to the biota is always present and subject to the natural fluctuation and processes of the river flow. (2) The stabilization and/or removal of remnant deposits is a marginal alternative. This removes a portion of the PCB load from the river and potentially isolates the remainder from mobilizing forces. Although the regions are riprapped, the possibility for flow through and leaching still exists, creating a long-term potential hazard through continuous contamination changes in the water quality. (3) The other extreme, complete removal, is desirable but also not feasible. Aside from the high cost, the long-term disturbance would also present a long-term hazard to the ecosystem through extended exposure to elevated ambient PCB levels. In addition, the large

TABLE 7

MANAGEMENT CONSIDERATIONS CLEANUP AND MONITORING FACTORS

Regional		Site-Specific	Other Impacts
• Nonuniform/uniform distribution		• Must be localized	• Recreation
• Toxicant load		• Toxicant load	• Industry
• Dispersion/containment by natural flow		• Natural burial	• Commerce
		• Natural degration	• Social
• Deposition/erosion due to river morphology		• Dredge disposal to other locale	• Cost for cleanup
• Mechanical removal		• Onsite burial or chemical treatment	
• Water quality—O_2, temperature, nutrients		• Channelization around site	
• Overall ecological ecosystem impact on biota			
Alternatives			
No Management Action Further Study	Stabilize and/or Remove Remnant Deposits	Remove Remnant Deposits and Sediments > 50 ppm	Remove all Deposits > 1 ppm
Recommendations			
Normal maintenance not adequate to reduce problem *Not Recommended*	Will remove some PCBs but not sufficient *Marginal*	Removes significant PCB quantity *Best*	Removes significant PCB quantity *Not Recommended* Cost & Time Constraints
Tradeoffs			
No change in status, potential long-term effect on biota	Partial PCB removal, potential long-term effect on biota still exists	Majority of PCBs removed, reduced long-term biota hazard, subsequent reduction of levels due to natural processes	Total PCB removal, long project, high cost, long-term disturbance to system

volume of spoils from the dredging operation would create a secondary problem that is also difficult to solve.

The best alternative, based on the above considerations and on the results of the Duwamish case, is the removal of remnant deposits with concentrations greater than 50 ppm. This method removes a majority of the PCB load within reasonable economic limits. The PCBs are removed, decreasing the possibility of mobilization through natural processes and reducing the long-term potential hazard to the biota and other ecosystems components. The results of the Duwamish study have shown that the dredging method is not necessarily detrimental to the ambient conditions; no changes in the PCB levels of SPM and water were observed. Most importantly, the Duwamish case demonstrates that most of a toxicant load, better than 90%, can be safely recovered through the appropriate dredging technology and spoils-disposal methods. In short, the alternative of removing all the remnant deposits of greater than 50 ppm is the best.

In summary, the PCB problem in the Upper Hudson Basin must be reduced. Natural processes can mobilize and redistribute toxicant-laden sediments throughout an entire ecosystem. Current technology allows for the dredging of contaminated sediments and the proper disposal of spoils to minimize any potential long-term hazard to the ecosystem.

Summary

The results of the Duwamish River PCB cleanup and recovery project are discussed as a basis for comparison in selecting potential management alternatives for removing PCB residues in the Hudson River. Application of the Pneuma dredge system at the Duwamish River spill site removed approximately 92% of the PCBs spilled. No significant changes in the PCB load of suspended matter and water column were observed. These results suggest that a carefully planned dredge/disposal operation can be effective in minimizing environmental impacts in the Hudson River. On the basis of observations of the Duwamish case, the existing baseline data on the distribution of PCBs in the Hudson River, and the removal alternatives and associated environmental tradeoffs, hot spot dredging is recommended as the best practical alternative.

References

1. Peterson, S. A. & K. A. Randolph, Eds. 1977. Management of Bottom Sediments Containing Toxic Substances. Proceedings of the Second U.S.—Japan Experts' Meeting, October 1976—Tokyo, Japan. Ecological Research Series, EPA-600/3-77-083.
2. Hetling, L., E. Horn & J. Tofflemire. 1978. Summary of Hudson River PCB study results. Manuscript in preparation.
3. Army Corps of Engineers. 1977. PCB cleanup operations, Duwamish Waterway. Army Corps of Engineers, Seattle District.
4. Blazevich, J. N., A. R. Gahler, G. J. Vasconcelos, R. H. Rieck & S. V. W. Pope. 1977. Monitoring of trace constituents during PCB recovery dredging operations—Duwamish Waterways. EPA Report No. 910/9-77-039. EPA Region X.
5. Hafferty, A. J., S. P. Pavlou & W. Hom. 1977. Release of polychlorinated biphenyls (PCB) in a salt-wedge estuary as induced by dredging of contaminated sediments. Sci. Total Environ. **8:** 229–239.
6. EPA. 1974. Puget Sound 305-A Report. EPA Report No. 910/7-74-001, Surveillance and Analysis Division.

7. DEXTER, R. N. 1976. Ph.D. Thesis. University of Washington.
8. PAVLOU, S. P., K. A. KROGSLUND, R. N. DEXTER & J. R. CLAYTON. 1973. Data Report, R/V ONAR Cruise 434, 450, 469 and 502. Synthetic Organics in Puget Sound; Hydrographic, Chemical & Biological Measurements. Special Report No. 54, Department of Oceanography, University of Washington, Seattle, Washington.
9. DEXTER, R. N. & S. P. PAVLOU. 1976. Partitioning characteristics of trace organic compounds in Puget Sound. Proceedings of the Fifth Technical Conference on Estuaries of the Pacific Northest, April 1–2, 1976. Oregon State University, Corvallis, Oregon.
10. DEXTER, R. N. & S. P. PAVLOU. 1976. Characterization of polychlorinated biphenyl distribution in the marine environment. Bull. Environ. Contam. Toxical. **16**(4): 447–482.
11. DEXTER, R. N. & S. P. PAVLOU. 1977. Mass solubility and aqueous activity coefficients of stable organic chemical in the marine environment: polychlorinated biphenyls. Marine Chem. **2**(1978): 41–53.
12. CLAYTON, J. R., S. P. PAVLOU & N. F. BREITNER. 1977. Polychlorinated biphenyls in coastal marine zooplankton: bioaccumulation by equilibrium partitioning. Environ. Sci. Technol. **11**(July): 676.
13. PAVLOU, S. P., R. N. DEXTER & W. HOM. 1977. Polychlorinated biphenyls (PCB's) in Puget Sound: Physical/chemical aspects and biological consequences. Proceedings of a Symposium on the Use, Study and Management of Puget Sound. University of Washington. Seattle, Washington.
14. MANSON, D. G., M. K. GRUPTA & R. G. SCHOLTZ. 1972. A mobile multi-purpose treatment system for processing hazardous material contaminated waters. Proceedings of the 1972 National Conference on Control of Hazardous Spills, EPA, Houston, Texas, March, p. 153.
15. PAVLOU, S. P. & R. N. DEXTER. 1977. Physical chemical aspects of the distribution of persistent organic compounds in the aquatic environment. Proceedings of the ASTM Second Symposium on Aquatic toxicology. Cleveland, Ohio.
16. PAVLOU, S. P. & R. N. DEXTER. 1979. Distribution of polychlorinated biphenyls (PCB) in estuarine ecosystems: Testing the concept of equilibrium partitioning in the marine environment. Environmental Science and Technology. In press.
17. DEXTER, R. N. & S. P. PAVLOU. 1979. Distribution of stable organic molecules in the marine environment: Physical-chemical aspects. Marine Chemistry. In press.
18. HETLING, L. J. & E. G. HORN. 1977. Hudson River PCB study description and detailed work plan. NYS Department of Environmental Conservation. Albany, N.Y. July.

FDA'S CHEMICAL CONTAMINANTS PROGRAM: THE SEARCH FOR THE UNRECOGNIZED POLLUTANT

Pasquale Lombardo

Division of Chemical Technology
Food and Drug Administration
Washington, D.C. 20204

During the past decade, it has become increasingly evident that industrial chemicals can contaminate foods even though they are not intentionally used on agricultural products or processed foods. Extensive contamination of fish by mercury was revealed in 1969, and widespread pollution of the environment by polychlorinated biphenyls (PCBs) was uncovered at about the same time.

Government regulatory agencies were caught by surprise. Much had to be learned about these chemicals, and in order to determine the extent to which foods were contaminated, it was necessary to gain familiarity with analytic methods suitable for their detection and quantification. Considerable time and effort went into standardizing and simplifying the methods so that results would be comparable, and analyses could be run routinely.

To provide better preparation for future occurrences of this nature, FDA launched its Chemical Contaminants Program in 1971. The program covers industrial chemicals, pesticides, and metals, and its principal objectives are: (1) to develop sufficient information to establish guidelines for recognized industrial chemical contaminants of foods; (2) to anticipate and search for as yet unrecognized contaminants; (3) to evaluate their hazards if found in foods; and (4) to take steps to eliminate or minimize the problem. The basic aims, of course, are to accumulate background information in a noncrisis atmosphere, provide FDA an early warning system, and enable the agency to react rapidly in "emergency" situations.

Although the program also encompasses pesticides and metals, emphasis here will be placed on industrial chemicals. The prime responsibility for research on industrial chemicals lies with the Division of Chemical Technology of FDA's Bureau of Foods. Fundamentally, our charge is to determine which industrial chemicals have the greatest potential to become food contaminants, what areas are most likely to be affected, and to check likely target foods.

The main criteria used in the selection of industrial chemical types for study are the same as those used by most agencies concerned with the environment. They include volume of production, associated impurities or by-products, predicted environmental stability, pattern of use, oil-water partition coefficients, bioaccumulation potential, known toxicity, and means of disposal. Background information is gathered chiefly through industry and government contacts, trade journals, and the literature. One of the chief groups with which we exchange information is, of course, the Environmental Protection Agency (EPA). Among the classes of materials under investigation are flame retardants, plasticizers, electrical fluids, halogented solvents, various chemical intermediates, functional fluids, lubricating oil additives, replacements for PCBs, and chemicals present in sewage sludge and wastewaters.

Once it is decided that particular chemicals or groups of chemicals should be looked into, we determine if our traditional pesticide methodology can adequately recover them from foods. If not, our laboratory attempts to develop the necessary means for the detection of these compounds. Reference compounds are essential for

0077–8923/79/0320–0673 $01.75/0 © 1979, NYAS

this, and we have established a collection of some 500 different industrial chemicals for our use. (These materials are also made available to our field units, other government agencies and selected researchers on request.)

With analytic methodology in hand, our usual approach is to obtain food samples from geographic locations thought likely to contain the potential contaminant. The vast majority of our samples are obtained through the cooperative efforts of other federal and state agencies, whose help has proved invaluable. Our prime indicator organism is the fresh water fish, as most problem contaminants eventually find their way into this segment of the human food chain.

If we are successful in finding the suspect pollutant, our monitoring activities are expanded, up to and including nationwide surveys of various foods by our field laboratories. On the basis of the results, our toxicologists may then judge whether the chemical contaminant constitutes a hazard to the consuming public. Ultimately, regulatory limits may be set for the chemical or chemicals in question.

As a result of our investigations, we have been able to detect some of the less commonly encountered industrial chemicals in foods. These are summarized in TABLE 1. Unless otherwise indicated, the findings have been made in fish. Since almost all the samples were from inland waters near manufacturer or user sites, the residues are not necessarily reflective of what the general public is exposed to.

I will comment briefly on some of these findings.

A. *Chlorinated benzenes.* Annual production is large, totaling well over 500 million pounds. As the table illustrates, residues of these compounds have been found in fish from many different geographic locations. The chlorobenzenes are used for a variety of purposes, including chemical intermediates, dye carriers, electrical fluids, solvents, and pesticides. The findings of tri- and tetrachlorobenzenes are not surprising, as these chemicals have been used for many years in combination with PCBs in transformers. Leakage of the transformers would provide direct entry into the environment. Hexachlorobenzene, of course, is now a widely recognized environmental contaminant.

B. *Chlorinated benzyl chlorides and benzotrifluorides.* Thus far, we have found these chemical intermediates solely in fish from the lower Niagara River in New York.

C. *Other chlorinated aromatics.* We have detected the chloronitrobenzenes in fish only recently, and we intend to collect samples near other manufacturing sites to determine if these compounds are more widespread in the environment. Pentachloroaniline and pentachlorophenyl methyl sulfide are associated with the pesticide pentachloronitrobenzene (PCNB). Pentachloroanisole is derived from the pesticide pentachlorophenol, and is frequently found as a residue in fish. Octachlorostyrene is a by-product of many chlorination processes, and we have detected this compound in fish from the two locations shown.

D. *Chlorinated cycloaliphatics.* The chlorinated norbornene and norbornadiene compounds are precursors of the pesticide endrin, and have been found in Mississippi River fish below a manufacturing plant in Memphis, TN. Epoxy hexachloronorbornene is an apparent environmental conversion product of one or both of these compounds. The chlorocyclopentenes were found in fish near a plant manufacturing hexachlorocyclopentadiene (HCP), and are probably by-products. We have never detected HCP itself in fish, although the compound is an important intermediate for several pesticides and flame retardants. It has been concluded by us and others that HCP is environmentally unstable at low concentrations. (We were involved to some extent in the Louisville, KY problem, where large amounts of still bottoms from HCP production were dumped into the sewer system. In this case, we had no problem detecting HCP, octachlorocyclopentene, and a host of other chloroorganics in the sewage sludge. The concentrations, of course, were in the percent range.)

TABLE 1

INDUSTRIAL CHEMICAL RESIDUES FOUND IN FOODS*

	Range, ppm†	Location
Chlorinated Benzenes		
monochloro 1,4-dichloro 1,2-dichloro 1,3,5-trichloro 1,2,3-trichloro 1,2,4-trichloro 1,2,3,5-tetrachloro 1,2,4,5-tetrachloro 1,2,3,4-tetrachloro	tr–6.7	Tombigbee R. AL Ohio R. OH Niagara R. NY White Lake, MI Spring Creek, PA Bald Eagle Creek, PA Mississippi R. LA/MO
pentachloro	0.11	(peanut oil)
hexachloro	0.07	(peanut oil)
Chlorinated Benzyl Chlorides		
monochloro	‡	Niagara R. NY
dichloro	‡	↕
trichloro	‡	↕
3,4-dichloro	‡	↕
Chlorinated Benzotrifluorides		
4-chloro	0.3–2.0	↕
3,4-dichloro	tr–0.1	↕
2,4-dichloro	0.2–0.8	↕
trichloro	‡	↕
tetrachloro	‡	Niagara R. NY
Other Chlorinated Aromatics		
4-chloronitrobenzene	0.01–0.33	Mississippi R. MO
2-chloronitrobenzene	0.006–0.12	↕
3,4-dichloronitrobenzene	0.05	↕
2,3-dichloronitrobenzene	0.01	Mississippi R. MO
tetrachloronitrobenzene	tr	(peanut oil)
pentachloronitrobenzene	0.03	Mississippi R. MO
pentachloroaniline	0.26	↕
pentachlorophenyl methyl sulfide	0.08	Mississippi R. MO
pentachloroanisole	tr–0.15	many rivers and lakes
pentachloroanisole	0.1	(peanut oil)
pentachlorophenol	tr–0.02	Mississippi R. LA
octachlorostyrene	0.02–0.1	Lake Ontario, NY White Lake, MI
Chlorinated Cycloaliphatics		
hexachloronorbornadiene heptachloronorbornene epoxy hexachloronorbornene	0.03–16.2	Mississippi R. LA/MO
tetrachlorocyclopentene	‡	White Lake, MI
pentachlorocyclopentene	‡	↕
hexachlorocyclopentene	‡	White Lake, MI

TABLE 1 (*Continued*)

	Range, ppm†	Location
Chlorinated Aliphatics		
hexachlorobutadiene	tr–4.6	White Lake, MI; Mississippi R. LA/MO; Ohio R. OH
hexachloroethane	0.13	Ohio R. OH
pentachloroethane	‡	↕
pentachlorobutadiene	‡	Ohio R. OH
Brominated Aromatics		
polybrominated biphenyls	tr–1.1	Pine R. MI
polybrominated biphenyls		Ohio R. W.VA (animal fat)
1,4-dibromobenzene	tr	Pine R. MI
1,2,4-tribromobenzene	0.05	↕
1,2,4,5-tetrabromobenzene	0.04	↕
monobromobiphenyl	‡	↕
dibromobiphenyl	‡	Pine R. MI
Triaryl Phosphates		
triphenyl	0.06–0.15	Waukegan Harbor, IL
cumylphenyl diphenyl	0.13–0.41	↕
nonylphenyl diphenyl	0.16–0.50	Waukegan Harbor, IL
isopropylphenyl diphenyl	0.1–1.0	Saginaw R. MI; Kishacoquillas Creek, PA
trixylenyl	0.02–0.08	Columbia R. WA/OR
tricresyl	0.02–0.04	Columbia R. WA/OR
2-ethylhexyl diphenyl	1.4	(animal fat)
Aromatic Amines		
1-naphthylamine	0.01–0.12	Delaware R. DE; Buffalo R. NY
N-ethyl-*N*-phenyl benzylamine	0.001–0.17	Delaware R. DE
N-ethyl-*N*-(*m*-tolyl) benzylamine	0.001–0.01	Delaware R. DE
3,3′-dichlorobenzidine	0.1–0.3	Muskegon County, MI (waste water)
benzidine	2.5	(waste water)

tr, = Trace
*In fish unless otherwise indicated.
†Edible portion basis.
‡Not quantitated because of lack of reference material at time of analysis.

E. *Chlorinated aliphatics.* The chlorinated butadienes and ethanes are related to octachlorostyrene and hexachlorobenzene in that they are all by-products of chlorinated hydrocarbon manufacture. Hexachlorobutadiene (HCBD) is not made commercially, but is a major component of waste products derived from perchloroethylene production. HCBD has been frequently found as a contaminant of fish. Prompted by our findings of HCBD residues, FDA conducted a field surveillance program in 1975, in which five of our district laboratories collected food samples near perchloroethylene plants. Residues of HCBD were found in 8 of 9 samples of fresh water fish (0.01–1.2 ppm), 2 of 20 salt water fish (0.01–0.02 ppm), and 1 of 20 milk samples (< 1 ppm, fat basis). None was found in 15 egg and 19 leafy vegetable samples. FDA toxicologists concluded that the concentration present did not constitute a toxicologic hazard to the consumer.

F. *Brominated aromatics.* In addition to the polybrominated biphenyls, low levels of several bromobenzenes have been detected in fish collected near a bromocarbon producer in Michigan.

G. *Phosphates.* Our interest in these chemicals stemmed from the fact that they have replaced PCBs in some flame retardant plasticizer and functional fluid applications. Our first findings of residues in fish were made in Waukegan Harbor, IL. After learning that a manufacturing facility that formerly discharged large amounts of a PCB hydraulic fluid had replaced it with an aryl phosphate, we obtained fish samples from near the outfall. As shown in TABLE 1, concentrations of the major components of the fluid totaled almost 1 ppm in some fish. Subsequent findings of aryl phosphate residues have been made near other user sites.

H. *Aromatic amines.* These chemical intermediates have been found in fish near dyestuff producers. Although the concentrations are low, we have maintained interest in this class of compounds because of their carcinogenic potential.

Among other chemicals considered and temporarily shelved are chlorinated paraffins and phthalate esters. Production of chlorinated paraffins is greater than 100 million pounds annually, but they do not appear to biomagnify to any appreciable extent. We have also been unable to detect these compounds in fish and sediment collected near manufacturing facilities. The phthalates are of course widespread, but after conducting a nationwide survey, FDA concluded that the concentrations present in foods are not high enough to constitute a toxicologic threat.

We are continuing to investigate and search for other potential food contaminants. In addition to those already mentioned, we are studying lower chlorinated phenols and anisoles, volatile halocarbons (such as perchloroethylene, haloforms, carbon tetrachloride, etc.), chlorinated dibenzofurans, and aromatic nitrogen compounds. It is our hope that these activities will enable FDA to uncover contamination of foods by industrial chemicals early enough to take corrective action. At the very least, the information developed will help FDA to react quickly in emergency situations.

STUDIES ON BROMINATED CHEMICALS IN THE ENVIRONMENT

Vincent J. DeCarlo

Office of Toxic Substances
Environmental Protection Agency
Washington, D.C. 20460

Introduction

Under the Toxic Substances Control Act, data on the release of chemicals to the environment during manufacturing, processing, distribution, use and disposal are of interest. Some of our environmental measurements over the last two years have been directed to assessing the types and levels of brominated compounds released to the environment by chemical manufacturing in the states of New Jersey[1,2] and Arkansas.[3]

The studies in New Jersey were initiated to assess the environmental contamination due to the manufacturing of PBBs at facilities in Bayonne and Sayerville, N.J. Samples were taken off-property of air, soil, water, wastewater, bottom sediments, and available biota. Human hair samples were also collected in nearby communities. Some of the compounds identified in New Jersey include:

Polybrominated biphenyls, PBBs
Polybrominated biphenyl oxides, PBBOs
Bromobenzenes
Bromotoluenes

The activities in Arkansas were designed to examine environmental matrices surrounding the bromine industry in El Dorado and Magnolia, AK and to identify as much as possible, all brominated chemicals resulting from these manufacturing activities. A total of 26 brominated organic compounds were identified in environmental samples collected from a geographic area associated with the bromine industry. The major chemical facilities operated in this area include Arkansas Chemical, Great Lakes Chemical, Michigan Chemical, Ethyl and Dow Chemical. Many of the chemicals were detected off plant property, indicating that their transport into the neighboring environment was occurring.

Analytic

In both the New Jersey and Arkansas study, the principle analytic method used was gas chromatography/mass spectrometry/computer (GC/MS/COMP). In the normal operating mode, detection limits for GC/MS/COMP were inadequate for most of the compounds of interest, hence multiple ion detection was used almost exclusively. In this method the mass filter is stepped through discrete ion masses repetitively instead of scanning the entire mass range. The selected ions are detected for proportionately more of the total time thereby improving signal-to-noise for those ions. Confirmation of identity when required was accomplished either by monitoring several selected ions and comparing ion intensities with standards or by direct probe mass spectrometry.

0077–8923/79/0320–0678 $01.75/0

For PBBs, mono- through pentabromobiphenyls were analyzed using a 45 cm × 0.2 cm i.d. glass column packed with 2% OV-101 coated on Gas-Chrom Q and a flow rate of 30 cc/min helium. The column temperature was 150°C initially, raised at 25°C/min at 300°, then held constant. Hexa- through decabromobiphenyls were analyzed using the same gas chromatographic conditions as for the lower isomers except the flow rate was 20 cc/min and the column temperature was 220° initially, programmed at 12°/min to 300°, then held constant. Under these conditions, decabromobiphenyl is eluted in 6 minutes.

Samples containing mixtures of PBBs and PBBOs were successfully separated, identified, and quantified in this study. This was possible because fragmentation of the brominated biphenyl oxide by loss of an oxygen to give an ion with atomic mass exactly the same as a comparable PBB is *not* occurring. If this had occurred to any extent, the analysis of biphenyls would have been difficult since the same peaks could arise from the biphenyl oxides with corresponding numbers of bromines.

RESULTS

Selected environmental data for New Jersey are shown in TABLE 1. Concentration units for the soil, sediment, water, sludge, and biota samples are expressed in ppb; air

TABLE 1

SELECTED ENVIRONMENTAL SAMPLES

Sample Type	Bayonne, NJ	Sayerville, NJ
Soil*	1,135–4,250	40–4,600
Sediment*	<10–20	4,600§
Water*	9.8–46	<0.2–210
Sludge*	431,000	
Air†	0.06	
Biota*	<10–1,270‡	20–92,000¶

*Concentration, ppb.
†Concentration, ng/m^3.
‡Lichen.
§Marsh soil.
¶ Reeds in receiving swamp.

data is expressed in ng/m^3. A 24-hour high volume air sample taken 900 meters downwind of the facility contained 0.06 ng/m^3 of hexabromobiphenyl. A number of different biota samples were collected at each site with lichen and swamp reeds giving the highest values at Bayonne and Sayerville, respectively. The sludge sample identified as containing 431 ppm of PBBs was obtained from a discharge-treatment zone. For this sample, PBBs were confirmed by full mass spectrum matching with PBB standards in addition to bromobenzenes, bromotoluenes, bromoethylene and bromobiphenyl ethers. Specific compounds identified by GC/MS in the sludge sample are shown in TABLE 2 and include:

Di-, tetra- and hexabromobenzene
Ethylene tetrabromide
Tri- and pentabromotoluene
Tri- to decarbromobiphenyl
Tetra- to decabromobiphenyl oxide

TABLE 2

BROMINATED COMPOUNDS IN SLUDGE SAMPLES, GC/MS ANALYSIS

$C_6H_4Br_2$	$C_{12}H_4Br_6$
C_2Br_4	$C_{12}H_4Br_6O$
$C_7H_5Br_3$	$C_{12}H_3Br_7$
$C_6H_2Br_4$	$C_{12}H_3Br_7O$
$C_{12}H_7Br_3$	$C_{12}H_2Br_8$
$C_7H_3Br_5$	$C_{12}H_2Br_8O$
C_6Br_6	$C_{12}HBr_9$
$C_{12}H_6Br_4$	$C_{12}HBr_9O$
$C_{12}H_6Br_4O$	$C_{12}Br_{10}$
$C_{12}H_5Br_5$	$C_{12}Br_{10}O$
$C_{12}H_5Br_5O$	

The variability of specific PBBs found in soil and sludge samples are shown in TABLE 3. Soil samples consistently exhibited increased concentration levels for the higher substituted PBBS.

The brominated chemicals identified in Arkansas are shown in TABLE 4. Several compounds of recent interest include 1-chloro-2,3-dibromopropane (DBCP) and Tris-(2,3-dibromopropyl) phosphate (TRIS). Other compounds of interest because of potential human contamination include decabromobiphenyl oxide (Decabrom) and tetrabromobisphenol A (Tetrabrom). These were identified in samples of human hair obtained from El Dorado and Magnolia barber shops. For these measurements, in order to obtain enough material for analysis, composite samples were made and analyzed as three separate samples. One sample gave positive Tetrabrom and Decabrom values at the 2 and 5 $\mu g/kg$ level respectively. For Tetrabrom, the two ions m/e 528 and 530 were observed at the correct GC retention time with the correct intensity 1.6 vs 1.5 for the standard. Decabrom was confirmed by full scan gas chromatography-mass spectrometry. Data from the other two samples showed unconfirmed Tetrabrom and confirmed Decabrom, but at lower levels. Since the three composites surveyed a total of ~40 individuals, these findings indicate that at a minimum, 5% (2 out of 40) of the population had detectable levels of Decabrom.

These areas are still under active investigation; the presence of these compounds in the enviroment is presenting an interesting scientific challenge as well as new concerns for the health of exposed individuals.

TABLE 3

PBB LEVELS IN SELECTED ENVIRONMENTAL SAMPLES (ppb)

	Soil	Sludge
$C_{12}H_9Br$	<10	540
$C_{12}H_8Br_2$	<10	2,200
$C_{12}H_7Br_3$	<10	4,300
$C_{12}H_6Br_4$	<10	<10
$C_{12}H_5Br_5$	<10	590
$C_{12}H_4Br_6$	30	3,800
$C_{12}H_3Br_7$	30	3,300
$C_{12}H_2Br_8$	90	3,600
$C_{12}HBr_9$	2,000	22,500
$C_{12}Br_{10}$	2,100	390,000
Total	4,250	431,000

TABLE 4

BROMINATED CHEMICALS IDENTIFIED IN ARKANSAS

Substance	Matrix
Chlorodibromopropane	air
1-Chloro-2,3-dibromopropane (DBCP)	air, water, soil, sediment, front and tail brine
1,2-Dibromoethane	air, water, soil, sediment
Difluorodibromomethane	air
Methyl bromide	air
Bromobenzene	air, soil, sediment
Bromoethane	air
Bromoform	air, sediment, front and tail brine
1-Chloro-2-bromoethane	air, water, sediment
1-Chloro-3-bromopropane	air
1,3-Dichloropropane	air
Vinyl bromide	air
Bromodichloromethane	air, tail and front brine
Allyl bromide	air
1 or 2-Bromopropane	air
Bromopropene	air, water
Chlorobromopropene	air
1,1-Dibromo-2-chloropropane	air
Dibromopropane	air
Dibromoethene	water
Bromodichloropropane	water
Decabromobiphenyl oxide (Decabrom)	air particulate, soil, sediment, human hair
Tetrabromobisphenol A (Tetrabrom)	air particulate, soil, sediment, human hair
Tris-(2,3-dibromopropyl) phosphate (TRIS)	air particulate, soil
1,2-Bis(2,4,6-tribromophenoxy) ethane	air particulate, soil
Dibromochloromethane	front and tail brine
Pentabromophenol	soil

REFERENCES

1. ERICKSON, M. D., R. A. ZWEIDINGER & E. D. PELLIZZARI. 1977. Analysis of a Series of Samples for Polybrominated Biphenyls (PBBs), Contract 68-01-1978.
2. WHITLOCK, S. A. & C. L. STRATTON. 1977. Survey of Polybrominated Biphenyls (PBBs) Near Sites of Manufacture and Use in Northeastern New Jersey, Contract 68-01-3248.
3. PELLIZZARI, E. D., R. A. ZWEIDINGER & M. D. ERICKSON. 1978 Environmental Monitoring Near Industrial Sites Brominated Chemicals, Contract 68-01-1978. EPA-560/6-78-002.

SURVEILLANCE FOR FUTURE ENVIRONMENTAL CONTAMINANTS: DISCUSSION

Hans L. Falk

Office of Health Hazard Assessment
National Institute of Environmental Health Sciences
Research Triangle Park, North Carolina 27709

I was puzzled that the title of our small section is the SURVEILLANCE FOR FUTURE ENVIRONMENTAL CONTAMINANTS and yet we seem to be talking only about present environmental contaminants.

What the future will bring, we really do not know; however, after listening to these very informative talks by Dr. Lombardo and Dr. DeCarlo, I was going to make a few comments with regard to the problems which we may as yet not have realized and which therefore, may become future problems with environmental contaminants, namely, to look into some of the possible new modes of expresson of their toxicity.

We know a lot of chemicals that are present in the environment, but we do not necessarily, know what adverse effects they may have because we have not been able to test for all possible adverse effects that may be produced.

I was thinking, for example, in terms of compounds that might inhibit DNA repair mechanisms. This area has not been given much attention to date, particularly, since most testing for adverse effects of chemicals with regard to mutagenesis has deliberately preempted in the *Salmonella typhimurium* organisms, the capacity for DNA repair. On the other hand, DNA repair exists in all species, from the mammals down to the bacteria and is absolutely essential for their survival.

It would, therefore, be disastrous if some of our "new" or "old" common chemicals, that we have not tested adequately, would turn out to be DNA repair inhibitors.

We know already that there are such compounds in existence. They have been found sporadically. They have been considered useful for studying the mechanism of DNA repair and as explanation of differential susceptibility to cancer development in cocarcinogenic studies, but not as potential environmental hazards. So far the list of such inhibitors may be limited to about three dozen chemicals that belong to different chemical classes and have not yet produced any clear cut understanding of what the requirements may be for chemicals to become effective DNA repair enzyme inhibitors.

In the same train of thought, I would also mention that we should pay more attention to inhibitors of enzymes like the mixed-function oxidases produced by enzyme induction. This has been done to a much greater extent in the recent past. Research in this area has produced interesting data on enzyme kinetics and actually introduced new chemicals that were useful as pesticide synergists, which found practical application in this capacity in conjunction with pyrethroid pesticides.

Another problem I wanted to mention deals with another potential interference with some essential biologic processes. Purification of water by chlorination will allow the addition of chlorine to many chemicals, including many of natural origin, for instance some of the purines, that might produce new problems. Chlorination has been observed in the water in natural surroundings of compounds like guanine, xanthene, and caffeine. Chlorination was observed also in those positions where enzyme systems would not be able to catabolize the compounds. In the case of xanthene, this would be on the carbon in the 8 position, and the compound, therefore, would have a tendency to become persistent. These chlorinated purines may act as antimetabolites for the appropriate purines that are essential for growth of certain organisms.

I wanted to mention a few of these areas, which we are not concerned about at this time, but are "future worries." The chemicals we learned about today are really "worries of today" and I would not put them off to be only considered in the future.

SYMPTOMS AND CLINICAL ABNORMALITIES FOLLOWING INGESTION OF POLYBROMINATED-BIPHENYL-CONTAMINATED FOOD PRODUCTS

Henry A. Anderson,* Mary S. Wolff,* Ruth Lilis,*
E. C. Holstein,* José A. Valciukas,*
K. E. Anderson,† M. Petrocci,* L. Sarkozi,‡ and Irving J. Selikoff*

*Environmental Sciences Laboratory
Mount Sinai School of Medicine
The City University of New York
New York, New York 10029

†The Rockefeller University
New York, New York 10021

‡Department of Pathology
Mount Sinai School of Medicine
The City University of New York
New York, New York 10029

Unlike the now ubiquitous halogenated compounds DDT [1,1,1-tri-chloro-2, 2-bis (4-chlorophenyl) ethane] and PCBs (polychlorinated biphenyls), PBBs (polybrominated biphenyls) are not commonly present in animal or human tissues. PBBs are mixtures of brominated biphenyls, usually with 4–8 bromine atoms per molecule, which have been useful as fire retardants in thermoplastic resins.[1] One such mixture, Firemaster FF-1®, mainly hexobromobiphenyl, caused widespread contamination of livestock (dairy and beef cattle, poultry and eggs) after an accidental introduction into cattle feed in Michigan during 1973.[2] At the time the accident was detected, no guidelines existed for allowable limits of PBBs in food for human consumption. Under rapidly implemented federal and state regulations, more than 500 farms were eventually found to have cattle and/or milk over the newly established allowable limits of PBBs in food which were used as farm quarantine "action limits." Over 23,000 dairy cattle and 1.6 million chickens died or were destroyed.[3]

Unfortunately, before the cause and dimensions of the problem were fully appreciated, contamination of milk, beef, pork, eggs and poultry occurred, and ingestion of these foodstuffs resulted in PBB absorption by those who ate these products. Farm families accustomed to eating their own produce were first and perhaps most heavily exposed. In the interval of more than 9 months between the accident, the identification of its cause, the beginning of statewide testing and the establishment of quarantines, commercially marketed products entered the Michigan food chain; thus the potential existed for wide dissemination of this halogenated hydrocarbon to the general population of the state.

No previous instance of human exposure to PBBs, with possible associated health effects, was known. Possessing chemical properties which make it biologically persistent,[4] and having demonstrated toxicity to animals (liver,[2,5] endocrine,[6] skin,[7] kidney,[8] and lymphatic tissue[9]), PBBs were considered to pose a potential health hazard to those exposed. Anecdotal reports of associated disease and widespread public concern prompted health assessments of exposed individuals.[10–12]

0077-8923/79/0320-0684 $01.75/0 © 1979, NYAS

Materials and Methods

Populations

By April 1976, the Michigan Department of Agriculture (MDA) had characterized the extent of PBB contamination on 1,039 farms. Considerable information was available, including farm name and address, dates and types of sampling, serial test results, size of herd, date and length of quarantine if imposed and the number of contaminated animals dead or destroyed. These data provided a limited assessment of exposure parameters such as: PBB concentration in products consumed on the farm, time at which individual farms became aware of contamination, estimate of duration of exposure and duration from last known exposure. These lists were the most readily available source for locating participants for which some pre-1976 information was available.

Utilizing these lists, a cross-sectional clinical field study was planned to assess whether patterns of clinical or biochemical abnormalities were present among PBB exposed Michigan farm residents. Examinations of Michigan farmers were conducted during November 1976 and of Wisconsin dairy farmers (unexposed group) in March 1977. Family units were invited from four sub-groups.

Group 1: Families Randomly Chosen from the MDA List of Tested Farms

Invitations were sent to 150 farms randomly selected (using a random numbers list) from the MDA records. These were supplemented by an additional 50 invitations sent to farms which the MDA information suggested had the highest contamination and thus, presumably, the highest chance of exposure. All quarantined farms were ranked by date of quarantine and PBB level detected and number of animals dead or destroyed. The first 50 made up the supplemental "highest exposure" list. To insure a sample of the probably low exposure farms, 25 of the 242 farms on the MDA list which had never had PBBs detected by the MDA testing were randomly chosen as an additional supplement. A total of 225 unsolicited invitations so chosen were sent.

Group 2: Consumers of Produce Bought Directly from Participating Farms

Each farm family that accepted the invitation was asked to provide the names of any families who had regularly purchased produce from them since 1973 and who were not themselves livestock farmers. In one sense, these individuals ("consumers") represented an "extended family" inasmuch as they shared a common food source, with similar PBB contamination. They did differ in several aspects; they had not faced the same economic losses and disruption of usual work and livelihoods suffered by farmers from whom they had purchased food, and did not necessarily consume the same amount of contaminated food. They were considerably insulated from the psychologic impact of the PBB episode. The names of 106 families were provided, and all were sent invitations.

Group 3: Self-Selected Michigan Families

Other families sought referral to the study. In many cases, these individuals, or their personal physician, believed they had symptoms or illnesses related to PBB exposure. Some, but not all, of these families were on the MDA list of tested farms, but had not been chosen during the Group 1 selection process. It was of interest to examine these individuals to determine whether they differed significantly from the other groups by measures other than current symptoms and concern. Characteristics of this group might identify those more susceptible.

Group 4: Wisconsin Dairy Farmers Not Exposed to PBBs

The experience of the Michigan Department of Public Health (MDPH)[10] and Meester and McCoy[11] suggested that a non-PBB-exposed dairy farm population in Michigan would be extremely difficult to identify for use as a comparison population. Because of the wide dissemination of contaminated feed, as well as commercial products, a substantial farm group having no PBB exposure or body burden could not be identified in Michigan.

With the cooperation and collaboration of the Marshfield Medical Foundation and Marshfield Clinic in Marshfield, Wisconsin, dairy farms located in western Wisconsin were identified and randomly chosen to participate in a clinical field survey identical to that conducted in Michigan.

Serial Examinations

In April 1977, a subsample of the Michigan population examined in November 1976 was reexamined to investigate possible endocrine dysfunction and abnormal spermatogenesis.[13] The sample consisted of 50 males between the ages of 25 and 40. They had no medical history of cardiac, kidney, liver, or endocrine disease and reported use of less than 120 grams of alcohol per week and no current medication. These men and their wives were invited for reexamination, and each provided a 24-hour urine specimen collected during the 24-hour period prior to examination.

Examination Protocols

Clinical Examinations

Identical examination protocols were administered to all survey sub-groups by the same team of experienced field investigators. The comprehensive information gathered and laboratory tests performed are listed in Table 1.

Questionnaires were developed and laboratory tests chosen which incorporated those from the Yusho experience with PCBs,[14] our own experience in examining PCB-exposed workers,[15] the MDPH initial survey,[10] and suggestions from the clinical experience of Michigan physicians.[11,16]

Both the field investigators and the survey participants were unaware of the method of invitation of examinees. Each individual was examined and interviewed separately from the rest of his or her family. The interviews and examinations were performed in modular units, with one station gathering farm history and quarantine status and another gathering the medical information. Because of geographic constraints, the Wisconsin comparison group was examined four months after the Michigan group.

TABLE 1
PBB SURVEY PROTOCOL

Item	Information
Farm history questionnaire	Characterization of farm, chemicals used on farm, where and how much commercial feed purchased, etc.
Dietary habits	Detailed history of eating habits over past years, home or store origin
Lifetime occupational history	All jobs ever held (including part-time), detailed possible toxic exposures, years of farm residence
Past medical history	Past and present medical difficulties, medications, hospitalizations, alcohol use
Family history questionnaire	Complete pregnancy and outcome histories
Respiratory history	Medical Research Council questionnaire on chronic bronchitis, smoking history
Symptomatology questionnaire	Past and present symptoms grouped by organ systems
Physical examination (Adult)	Complete (no pelvic or rectal)
Physical examination (Children)	Complete for age
Ophthalmologic examination	Symptoms and eye-related history questionnaire, acuity, color blindness, tonometry, slit lamp, fundal examination
Growth record	Child's height and weight as on personal pediatric records
Dermatologic examination	Total body skin examination, special questionnaire
Laboratory tests	Routine urinalysis, urine for porphyrins, complete blood count, blood chemistries, zinc protoporphyrin (ZPP carcinoembyronic antigen, serum PBB, DDE, serum for special tests (farmers' lung precipitins, endocrine function, rheumatoid factor, ANA, immunocompetence testing serum for storage,) serum PBB (drawn for those volunterring to be part of MDH long-term study)
Lung function tests	Spirometry, flow-volume curves
Chest x-ray	All those over age 40, or by examining physician's request
Percutaneous needle fat biopsy	Volunteers, adults only (239 done)
Neurologic examination	Examinees with unexplained neurologic symptoms were referred to a neurologist for additional evaluation
Behavioral evaluation	Random and by physician referral

Laboratory Analyses

At the time of examination, venous blood specimens were collected using 10cc Vacutainer® tubes. After clotting, serum was separated and transferred to storage tubes which were frozen immediately and kept frozen until time of analysis.

Twenty-four hour urine specimens were collected and after volume measurement, two 100 ml aliquots were frozen in light-tight, sealed containers.

Serum PBB Analysis

After separation, serum for PBB analysis was pipetted into pesticide residue-free glass bottles with Teflon® caps. Analysis was performed at the Environmental Sciences Laboratory on a 4 ml sample using gas chromotography with electron capture detection.[17]

The limit of detection for PBB was 0.2–0.5 parts per billion. PBB levels are reported as the measure of the major hexabromobiphenyl isomer peak. In some cases ($n = 208$) a separate sample was drawn and an independent blind analysis performed by the MDPH. The correlation coefficient of compared results was 0.93 ($p < .001$).

Serum Biochemistries

Biochemical constituents of serum were performed on a Sequential Multiple Analyzer with Computer (Technicon Corp.) in the Clinical Chemistry Department at the Mount Sinai Hospital and School of Medicine. The upper limits of normal values for the laboratory tests were: alkaline phosphatase 95 IU/L, serum glutamic pyruvic transaminase (SGPT) 45 IU/L, serum glutamic oxaloacetic transaminase (SGOT) 40 IU/L, lactic acid dehydrogenase (LDH) 225 IU/L. Standard SMA quality control was used.

Urinary Porphyrins

The aliquot of urine collected was analyzed for total urinary porphyrins at The Rockefeller University of New York utilizing the methodology described by Martinez.[18]

Statistical Analyses

Prevalence data were compared by the chi-square test. Differences between means were compared using the student's t-test. P values at less than .05 were considered significant. Correlation analyses were performed using Pearson correlation coefficients and when appropriate, Spearman correlation coefficients. Statistical analyses were performed with the cooperation of The City University of New York Computer Center, using the Statistical Package for the Social Sciences.[19]

RESULTS

A total of 1,221 adults and children participated in the survey: 993 from 228 Michigan families and 228 from 67 Wisconsin families. TABLE 2 summarizes the participation rates for the four invited groups. Those invited from the list of families requesting to be examined had the highest rate of participation (86%) and the Wisconsin control group the lowest (17%).

No statistically significant differences in sex distribution were observed among the sub-groups, or all Michigan compared to Group 4 ($\chi^2 = 3.4$ df $= 1$ $p =$ n.s.), TABLE 3.

TABLE 2

SURVEY COMPOSITION

Group	Invitations Sent	Invitations Accepted	Response Rate, %	Total Participants
Group 1	225	107	48%	399
Group 2	106	59	56%	153
Group 3	72	62	86%	44*
Michigan (all)	403	228	56%	993
Group 4	400	67	17%	228

*Includes 58 "walk-ins."

Differences in mean age were observed with Group 3 significantly younger than Group 4 ($t = 5.37$ $df = 1{,}216$ $p < .01$). No differences were seen for Group 1 and Group 2 compared to Group 4.

Serum PBB Distributions

Serum PBB analysis was accomplished for 938 (95%) from Michigan Groups 1–3 and 228 (100%) from the Wisconsin comparison Group 4. TABLE 4 shows distribution of the serum PBB values by the four invitation groups.

Except for the absence of PBBs in Group 4, differences were not striking, although Group 3 had a lower prevalence of higher PBB values than Groups 1 or 2 ($\chi^2 = 25.5$ $df = 5$, $\chi^2 = 10.3$ $df = 5$). Of special interest was Group 2 (consumer group) whose PBB distribution *was not significantly different* from that of Group 1 (random farms) ($\chi^2 = 3.7$ $df = 5$ $p = ns$). TABLE 5 clearly shows the distributional differences of serum PBB with respect to quarantine status of the farms with which participants

TABLE 3

DEMOGRAPHIC CHARACTERISTICS OF THE STUDY POPULATIONS

Group	Sex*	Number Examined	Mean Age†
Group 1	Male	204	32 ± 19 yrs
	Female	203	
Group 2	Male	73	29 ± 20 yrs
	Female	80	
Group 3	Male	220	25 ± 17 yrs
	Female	213	
All Michigan	Male	497	28 ± 19 yrs
	Female	496	
Group 4	Male	128	33 ± 20 yrs
	Female	97	

*Sex distribution not significantly different between Michigan and Wisconsin farmers. $\chi^2 = 3.4$ $df = 1$ $p = n.s.$

†T-tests for mean age difference

1 vs 4 $t = 0.62$ $p = n.s.$

2 vs 4 $t = 1.90$ $p = n.s.$

3 vs 4 $t = 5.37$ $p < .01$

All Michigan vs 4 $t = 3.52$ $p < .01$

TABLE 4

DISTRIBUTION OF SERUM PBBS BY SURVEY PARTICIPANT GROUP

Serum PBBs (parts per billion)	Group 1	Group 2	Group 3	All Michigan	Group 4
Non-detectable/limit of detection	13 (3.4%)	7 (4.7%)	19 (4.7%)	39 (4.2%)	220 (99.1%)
0.2–1 ppb	50 (12.9%)	27 (18.2%)	89 (22.1%)	166 (17.7%)	2 (0.9%)
1–5 ppb	187 (48.3%)	68 (45.9%)	194 (48.1%)	449 (47.9%)	6*
5–10 ppb	51 (13.2%)	15 (10.1%)	56 (13.9%)	122 (13.0%)	0
10–100 ppb	68 (17.6%)	24 (16.2%)	38 (9.4%)	130 (13.9%)	0
100+ ppb	18 (4.7%)	7 (4.7%)	7 (1.7%)	32 (3.4%)	0
Total	387	148	403	938	228

*Six individuals (1 family) examined in Wisconsin had worked a Michigan farm until 1975. All had detectable PBB levels, highest was 1.3 ppb.

Distribution of serum PBB values significantly different among Group 1 vs Group 3 $\chi^2 = 25.5$ df = 5 $p < .01$: Group 2 vs Group 3 $\chi^2 = 10.3$ df = 5 $p < .01$
Distribution of serum PBB values not significantly different among Group 1 and Group 2 $\chi^2 = 2.7$ df = 5 p = n.s.

were associated. *Regardless of method of invitation, those associated with quarantined farms, as a group, had higher serum PBB levels* ($\chi^2 = 117.2$df = 5 $p < .001$). However, residence on a quarantined farm did not necessarily predetermine a high serum PBB level and conversely, residence on a non-quarantined farm did not always mean low or no PBBs (16% of residents of quarantined farms had levels below 1 ppb and 64% of residents of non-quarantined farms were above 1 ppb). No significant differences were observed between the consumer groups and the farm groups from which their produce came.

TABLE 5

DISTRIBUTION OF SERUM PBBS BY FARM RESIDENCE QUARANTINE STATUS

Serum PBBs (parts per billion)	Residents of Quarantined Farms	Consumers of Produce from Quarantined Farms	Residents of Non-quarantined Farms	Consumers of Produce from Non-quarantined Farms
Non-detectable/ limit of detection	12 (2.5%)	1 (1.1%)	17 (5.8%)	9 (10.8%)
0.2–1 ppb	50 (10.5%)	12 (13.6%)	82 (28.1%)	22 (26.5%)
1–5 ppb	220 (46.3%)	38 (43.2%)	149 (51.0%)	42 (50.6%)
5–10 ppb	75 (15.8%)	12 (13.6%)	31 (10.6%)	4 (4.8%)
10–100 ppb	93 (19.6%)	18 (20.5%)	13 (4.5%)	6 (7.2%)
100+ ppb	25 (5.3%)	7 (8.0%)	0	0
Total	475	88	292	83

Distribution of serum PBBs not significantly different between quarantine farm residents and consumers of their produce. $\chi^2 = 2.6$ df = 5 p = n.s.
Distribution of serum PBBs not significantly different between non-quarantine farm residents and consumers of their produce. $\chi^2 = 5.7$ df = 4 p = n.s.
Distribution of serum PBBs significantly different between quarantined-farm-associated individuals and non-quarantined-farm-associated individuals. $\chi^2 = 117.2$ df = 5 $p < .001$

TABLE 6
PREVALENCE OF SYMPTOMS AMONG SURVEY PARTICIPANT GROUPS‡

Symptom	Group 1 n = 296	Group 2 n = 100	Group 3 n = 255	Group 4 n = 150
Eye irritation	32 (11%)	13 (13%)	46* (18%)	12 (8%)
Nasal irritation	26 (9%)	12 (12%)	29 (11%)	16 (11%)
Sore throat	17 (6%)	7 (7%)	26 (10%)	9 (6%)
Cough	22 (7%)	9 (9%)	25 (10%)	16 (11%)
Wheezing	11 (4%)	6 (6%)	14 (5%)	7 (5%)
Tightness in chest	14 (5%)	8 (8%)	17 (7%)	6 (4%)
Rash	32 (11%)	7 (7%)	43* (17%)	13 (9%)
Acne	22 (7%)	15** (15%)	31* (12%)	8 (5%)
Sun sensitivity	10 (3%)	5 (5%)	16* (6%)	2 (1%)
Burning sensation	10 (3%)	5 (5%)	14* (5%)	2 (1%)
Darkening of skin	3 (1%)	4* (4%)	15** (6%)	0
Thickening of skin	3 (1%)	2 (2%)	8 (3%)	4 (3%)
Nail discoloration	10 (3%)	5* (5%)	28** (11%)	1 (1%)
Prolonged healing	16* (5%)	5 (5%)	38** (15%)	2 (1%)
Weight loss > 10 lbs.	6 (2%)	4 (4%)	16* (6%)	3 (2%)
Loss of appetite	14 (5%)	7 (7%)	16 (6%)	4 (3%)
Nausea	18** (6%)	7** (7%)	30** (12%)	1 (1%)
Vomiting	8 (3%)	4 (4%)	10 (4%)	1 (1%)
Abdominal pain	18 (6%)	9 (9%)	32 (13%)	11 (7%)
Abdominal cramps	16 (5%)	5 (5%)	16* (6%)	3 (2%)
Constipation	12 (4%)	7 (7%)	23 (9%)	8 (5%)
Diarrhea	13 (4%)	9 (9%)	38** (15%)	8 (5%)
Red urine	1 (.3%)	1 (1%)	2 (1%)	6† (4%)
Brown urine	1 (.3%)	0	6 (2%)	5† (3%)
Headache	59 (20%)	34** (34%)	104** (41%)	21 (14%)
Blurred vision	30* (10%)	4 (4%)	43** (17%)	5 (3%)
Dizziness	31** (11%)	6 (6%)	52** (20%)	5 (3%)
Depressed	30* (10%)	8 (8%)	50** (20%)	5 (3%)
Unusual fatigue	85** (29%)	24 (24%)	132** (52%)	22 (15%)
Altered perception	6 (2%)	4* (4%)	15** (6%)	0
Nervous	36** (12%)	14** (14%)	56** (22%)	3 (2%)
Difficulty sleeping	24 (8%)	12 (12%)	37** (15%)	8 (5%)
Sleepy	42** (14%)	11 (11%)	81** (32%)	9 (6%)
Weakness	25* (8%)	6 (6%)	34** (13%)	4 (3%)
Clumsiness	7 (2%)	4* (4%)	15** (6%)	0
Parasthesia	41** (14%)	12 (12%)	48** (19%)	8 (5%)
Loss of balance	21* (7%)	7* (7%)	37** (15%)	3 (2%)
Joint pain	101* (34%)	27 (27%)	110** (43%)	34 (23%)
Swelling joints	31 (11%)	7 (7%)	36* (14%)	10 (7%)
Back and leg pain	51** (17%)	18* (18%)	58* (23%)	12 (8%)
Drier skin	47* (16%)	14 (14%)	47** (18%)	11 (7%)
Voice change	6 (2%)	5 (5%)	6 (2%)	2 (1%)
Increased sweating	21 (7%)	13 (13%)	47** (18%)	9 (6%)

‡Includes only adults age 18 and over.
*Prevalance significantly greater than in Group 4. χ^2 test df = 1 p < .05.
**Prevalance significantly greater than in Group 4. χ^2 test df = 1 p < .01.
†Prevalance significantly greater in Group 4 than in Group 1 or 3. χ^2 test df = 1 p < .05.

Symptoms

Adults and children differ in their recognition and interpretation of symptoms, in current and past illnesses, habits such as cigarette smoking and alcohol use, and normal values for biochemical tests. Such differences made direct comparisons using both adults and children difficult, and the two groups were dealt with separately. One set of questionnaires and examinations were utilized for adults age 18 or over, and another for the children under age 18. This section of the report discusses the observation made of the 651 adults examined in Michigan and 153 adults in Wisconsin.

TABLE 6 summarizes the prevalence of current symptoms among the participants by method of invitation. Group 3 had the highest prevalence of symptoms (31 of 43 statistically significantly different from Group 4).

Group 1 had 14 symptoms that were statistically more prevalent than among the Wisconsin comparison Group 4. Group 2 had 10 symptoms significantly more prevalent. If Groups 1 and 2 are combined, 17 of 43 symptoms were significantly more

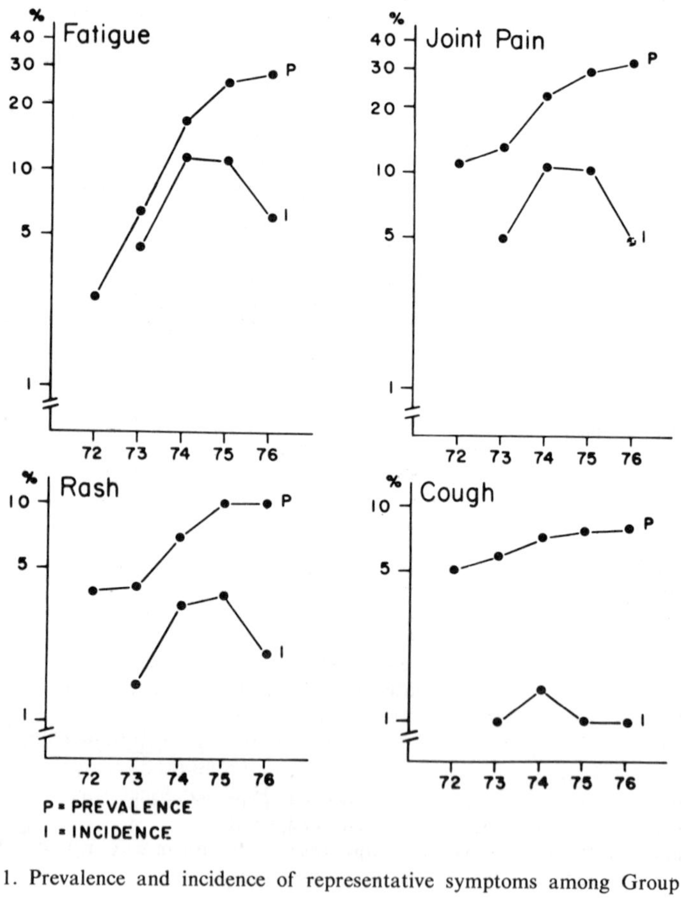

FIGURE 1. Prevalence and incidence of representative symptoms among Group 1 and 2 adults.

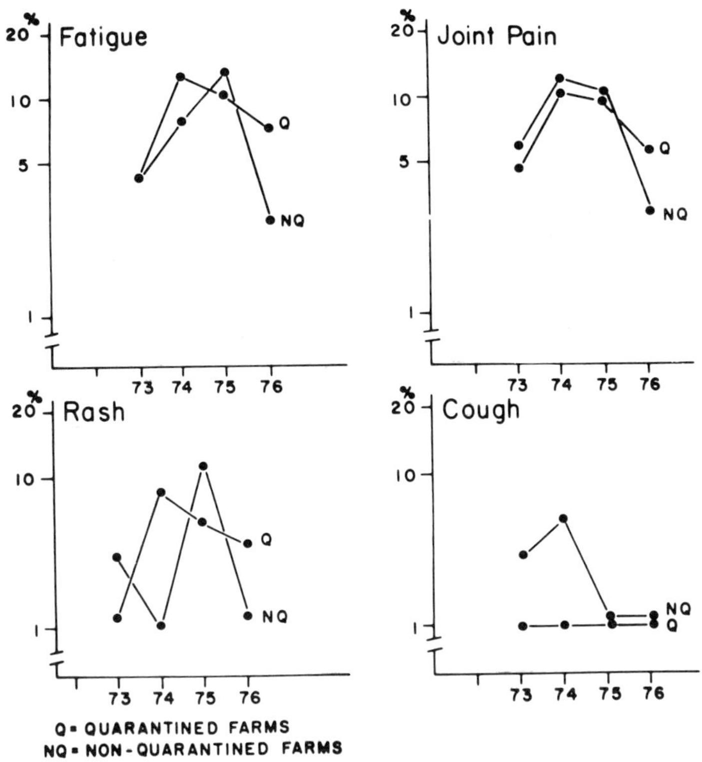

FIGURE 2. Incidence rates of representative symptoms among Group 1 and 2 adults by their associated farm status.

prevalent than among Group 4. Two symptoms were more prevalent in Wisconsin than Michigan.

In the examination protocol, the presence or absence of each symptom was recorded for each of the five years prior to examination, 1972–1976. Using 1972 as a baseline prevalence, it was possible to calculate the *incidence* of new symptoms reported in each year 1973 to 1976. It should be noted that only retrospective (recall) information was available for analysis, since no records were obtained during each incidence interval. Biases present in recall data necessitate caution in their interpretation. FIGURE 1 displays the incidence of four representative symptoms, as well as the prevalence in each year. Where the incidence of symptoms was seen to increase, peaks occurred in 1974. Symptoms associated with the skin showed peak incidence in 1975. Symptoms such as cough which had the same current (1976) prevalence in Michigan and Wisconsin showed only a slight fluctuation in incidence rate. The incidence of new symptoms in 1976, for each symptom which showed a marked incidence increase was lower than in 1974 or 1975. In most cases, the 1976 incidence was not statistically significantly higher than the 1972 baseline prevalence.

The small number of individuals reporting new symptoms in any year, precluded calculation of incidence rates for the Wisconsin population.

FIGURE 2 shows the incidence rates by quarantine and non-quarantine farm relationships. For nervous system and skin related symptoms, the quarantine farmers had incidence peaks in 1974, while the non-quarantine group peaked in 1975. For other symptoms, such as joint pain, the incidence rates in both groups were parallel.

Serum Biochemical Findings

Animal studies indicated that the liver was one of the primary target organs for PBB toxicity. Included in our survey were serum biochemical studies (alkaline phosphatase, SGPT, SGOT, LDH) which are often found elevated in liver disease. However, they are not specific for liver damage, since elevations also occur with other illnesses. TABLE 7 summarizes the prevalence of abnormalities by invitation group. SGPT, SGOT, LDH elevations were significantly more prevalent in the Michigan group (chi square test). No differences were seen for alkaline phosphatase. Although Groups 2 and 3 tended to have higher prevalences of abnormalities than Group 1, the

TABLE 7

PREVALENCE OF SERUM BIOCHEMISTRY ABNORMALITIES BY INVITATION GROUP†

Serum Biochemistry	Group 1	Group 2	Group 3	All Michigan	Group 4
Alkaline Phosphatase ≥ 96	34 (11.5%)	18 (18%)	39 (15.3%)	91 (14.0%)	25 (16.7%)
SGPT ≥ 46	30** (10.1%)	15** (15%)	38** (14.7%)	83** (12.7%)	4 (2.7%)
SGOT ≥ 41	34** (11.5%)	15** (15%)	34** (13.3%)	83** (12.7%)	3 (2.0%)
LDH ≥ 226	27* (9.1%)	7 (7%)	22* (8.6%)	56* (8.6%)	5 (3.3%)
Total examined	296	100	255	651	150

†Includes only adults age 18 and older.

*Prevalence compared to Group 4 significantly greater at $p < .05$ level using χ^2 test df = 1.

**Prevalence compared to Group 4 significantly greater at $p < .01$ level using χ^2 test df = 1.

differences did not reach statistical significance when tested with the chi square. As demonstrated in TABLE 8, differences were seen between residents of quarantined farms and non-quarantine residents only for SGOT. Non-quarantine residents had a higher prevalence of abnormal SGOT ($\chi^2 = 7.3$ df = 1 $p < .01$). Consumers of produce from quarantined farms had a higher prevalence of abnormal SGOT values than the quarantine farm residents ($\chi^2 = 3.9$ df = 1 $p < .05$). The converse was true for SGOT among participants who were not associated with a quarantined farm with the farm residents having the higher prevalence ($\chi^2 = 4.3$ df = 1 $p < .05$). In general, the non-quarantine consumers had lower, (but not statistically significant), prevalences of abnormalities than the non-quarantine farm participants and the opposite was true for quarantine farm residents and those who purchased their produce (consumers having the higher prevalence).

Among males and females, significant differences were seen for prevalence of abnormality. These prevalences are outlined in TABLE 9. Differences were investigated using the chi square test. Within the Michigan group, males had a higher prevalence of SGPT and LDH abnormality than seen among the Michigan females. In both

TABLE 8

PREVALENCE OF SERUM BIOCHEMISTRY ABNORMALITIES BY FARM RESIDENCE QUARANTINE STATUS†

Serum Biochemistry	Residents of Quarantined Farms		Consumers of Produce from Quarantined Farms		Residents of Non-quarantined Farms		Consumers of Produce from Non-quarantined Farms	
Alkaline Phosphatase ≥ 96	40	(11.6%)	13	(19.7%)	32	(16.9%)	6	(12.0%)
SGPT ≥ 46	36	(10.7%)	10	(15.2%)	29	(15.3%)	7	(14.0%)
SGOT ≥ 41	34[a,b]	(9.8%)	12[b]	(18.2%)	34[a,c]	(18 %)	3[c]	(7.0%)
LDH ≥ 226	29	(8.4%)	4	(6.1%)	19	(10.1%)	4	(8.0%)
Total examined	346		66		189		50	

†Includes only adults age 18 and over.
[a]Prevalence of pairs significantly different at $p < .01$ level using χ^2 test df = 1.
[b,c]Prevalence of pairs significantly different at $p < .05$ level using χ^2 test df = 1.

Michigan and Wisconsin, males had higher prevalences of elevated values, although statistical significance was only reached for SGPT and LDH in Michigan and LDH in Wisconsin. Comparing males in both groups, Michigan males had significantly more SGPT and SGOT abnormalities than the Wisconsin males. Michigan females had a significantly higher prevalence of elevated SGOT than the Wisconsin females.

The possible contribution of alcohol use, current medication, and past history of hepatic disease to abnormalities reported were investigated and no statistically significant associations found.

Urinary Porphyrins

Total urinary porphyrins were studied in the specimens collected by the 92 individuals seen during the follow-up studies. TABLE 10 gives the distribution of the values reported by sex. The porphyrin excretion was greater for the males than the females tested ($\chi^2 = 20$ df $= 5$ $p < .001$). However, all values observed were well below those seen in patients with clinical porphyrias.

TABLE 9

PREVALENCE OF SERUM BIOCHEMISTRY ABNORMALITIES BY SEX AND STATE†

Serum Biochemistry	Michigan Males		Michigan Females		Wisconsin Males		Wisconsin Females	
Alkaline Phosphatase ≥ 96	53	(16.0%)	38	(11.9%)	19	(22.1%)	6	(9.4%)
SGPT ≥ 46	58[a,b]	(17.5%)	25[b]	(7.8%)	3[a]	(3.5%)	1	(1.6%)
SGOT ≥ 41	50[a]	(15.1%)	33[b]	(10.3%)	2[a]	(2.3%)	1[b]	(1.6%)
LDH ≥ 226	40[b]	(12.0%)	16[b]	(5.0%)	5[c]	(5.8%)	0[c]	
Total examined	332		319		86		64	

†Includes only adults age 18 and over.
[a,b,c]Prevalence of pairs significantly different at $p < .01$ using χ^2 test df = 1.

TABLE 10

DISTRIBUTION OF TOTAL URINARY PORPHYRINS* BY SEX AMONG 92 MICHIGAN DAIRY FARMERS (1977–78)

Sex	Number Examined	Total Urinary Porphyrins (nmol/day) 0–99	100–199	200–299	300–399	400–499	500+
Males	50	1 (2%)	9 (18%)	22 (44%)	10 (20%)	6 (12%)	2 (4%)
Females	42	7 (17%)	18 (43%)	8 (19%)	8 (19%)	0 (0%)	1 (2%)
Total	92	8 (9%)	27 (29%)	30 (33%)	18 (20%)	6 (7%)	3 (3%)

*Based on 24 hour urine collection.

Mean 246 ± 126 nmol/day n = 92; mean (males) 283 ± 111 nmol/day n = 50; mean (females) 201 ± 130 nmol/day n = 42.

Distribution of total urinary porphyrin excretion significantly different between males and females. $\chi^2 = 20.0$ df = 5 $p < .001$.

INVESTIGATION OF ASSOCIATIONS

Possible associations between current symptoms and objective clinical findings were investigated. Such associations were evaluated in all examination groups, separately and as a total; however, because of the known bias in Group 3 (invited because of subjective complaints) only associations which were significant in Groups 1 and 2, either separately or combined, are presented.

Statistically, significant associations were seen between abnormal SGPT values and the presence of symptoms of excessive fatigue, sleepiness, rash and joint pain in Group 1. Significance was tested using the chi square. Except for joint pain, similar associations were seen in Group 2. TABLE 11 shows the distribution of abnormal SGPT prevalence by presence or absence of symptoms in Group 1. Individuals complaining of fatigue had nearly three times the prevalence of abnormal SGPT as those without the symptom. When one or more of the symptoms related to the nervous system were present, those with the symptoms had a significantly higher prevalence of abnormal SGPT values. No statistically significant associations were seen between these symptoms and alkaline phosphatase, SGOT or LDH when similar chi square testing was done. While individuals with symptoms tended to have the higher prevalence of

TABLE 11

DISTRIBUTION OF ELEVATED SGPT VALUES BY PRESENCE OF SYMPTOMS AMONG GROUP 1*

Symptom	Presence	Number in Group	Number with SGPT ≥ 46	Chi Square Test
Fatigue	Present	85	15 (17.6%)	$\chi^2 = 6.3$ $p < .01$ df = 1
	Absent	211	15 (7.1%)	
Sleepiness	Present	42	10 (23.8%)	$\chi^2 = 8.4$ $p < .001$ df = 1
	Absent	254	20 (7.9%)	
Rash	Present	32	7 (21.9%)	$\chi^2 = 4.1$ $p < .05$ df = 1
	Absent	264	23 (8.7%)	
Joint pain	Present	101	16 (15.8%)	$\chi^2 = 4.6$ $p < .05$ df = 1
	Absent	195	14 (7.2%)	
One or more neurologic	Present	142	20 (14.1%)	$\chi^2 = 3.9$ $p < .05$ df = 1
	Absent	154	10 (6.5%)	

*Includes only adults age 18 and older.

TABLE 12

TOTAL URINARY PORPHYRINS AND LIVER FUNCTION TESTS AMONG 92 MICHIGAN DAIRY FARMERS

Total Urinary Porphyrins	Total Number	Number with Normal Liver Function Tests	Number with Abnormal* Liver Function Tests
0–199 nmol/day	35	33 (94%)	2 (6%)
200–399 nmol/day	48	43 (90%)	5 (10%)
400+ nmol/day	9	5 (56%)	4 (44%)

*SGOT > 41 and or SGPT > 45 and or LDH > 225.
The prevalence of abnormal liver function tests is significantly greater among individuals with higher total urinary porphyrins than among those with lower total urinary porphyrins. $\chi^2 = 10.4$ df = 2 $p < .01$.

elevated values, the differences did not reach statistical significance. (No associations were seen in the Wisconsin Group 4.)

TABLE 12 shows the association of abnormal liver enzyme tests and total urinary porphyrin excretion. Individuals with higher prophyrin excretion had a higher prevalence of abnormal liver function tests ($\chi^2 = 10.4$ df = 2 $p < .01$). Biochemical values were those obtained on blood specimens collected at the time the urine was collected.

Incidence of symptoms suggested that peak symptoms may have occurred early-on, along with PBB exposures which cannot be assessed by 1976 serum-fat measures. Further, SGOT and SGPT elevations are known to be reversible in chemically related liver damage, so that persons with high PBB levels may have had acute hepatic changes resulting in enzyme elevation in the past with gradual return to normal. Such healed damage may not be reflected by 1976 biochemical testing. However, an effort to determine whether associations existed between important clinical findings and PBB serum levels was made, since the relation of PBBs with liver toxicity in animals was known.

No statistically significant positive correlations were found between serum PBB values and any individual current symptom (t-test of mean PBB values among those with and without symptoms).

Among the 613 adults in the survey, a positive correlation of current serum PBB levels with current SGPT levels was found ($\chi^2 = 4.0$ df = 1 $p < .05$). This association is outlined in TABLE 13. Pearson correlation coefficients were determined for serum

TABLE 13

DISTRIBUTION OF ABNORMAL SGPT AMONG ADULT MICHIGAN FARM RESIDENTS BY LEVEL OF SERUM PBB

PBB (ppb)	Number Examined	SGPT ≥ 45
≤ 1 ppb	143	11 (8%)
> 1 ppb	500	70 (14%)
Total	643	81 (13%)

Prevalence of abnormal SGPT values significantly higher among those with serum PBB values greater than 1 ppb.
$\chi^2 = 4.0$ df = 1 $p < .05$

PBB and serum enzyme levels. The largest coefficients were for SGPT (r = .09) and SGOT (r = .09). The correlation coefficient for serum PBB vs LDH was r = .01. SGPT and SGOT enzyme activities as performed (SMA) are quantitative in the abnormal range (SGPT > 45, SGOT > 41), but otherwise form a subjective classification of normal. Ranking of both enzyme values and PBB serum results was felt appropriate and Spearman correlation coefficients determined. They were r = .15 for both SGPT and SGOT. Using the t-test, means of PBB were compared for those with abnormal enzyme values and those with normal values. No statistically significant differences between mean PBB levels in each group were observed.

Association between serum PBB and total urinary porphyrin excretion was also observed. FIGURE 3 shows the linear regression between log PBB and total urinary porphyrin excretion. The correlation coefficient is r = 0.34.

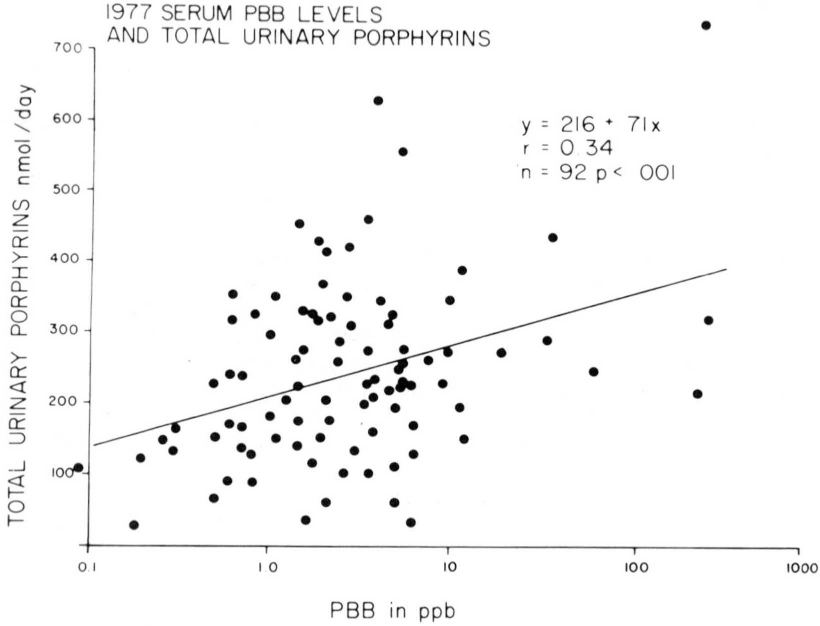

FIGURE 3. Regression analysis scattergram of total urinary porphyrin excretion and log PBB levels.

DISCUSSION

Demographically, Michigan Group 1 and 2 were similar to the Group 4 Wisconsin comparison group. No significant age or sex distributional differences were seen. Adult current and past medical histories for the Michigan Groups 1–3 and Group 4 were also similar. Of the 95 current or previously diagnosed illnesses included in the protocol, only 1 (degenerative arthritis) showed a statistically significant prevalence difference between the populations. Wisconsin had the higher prevalence of degenerative arthritis (17% vs 10%, $\chi^2 = 4.2$ df = 1 $p < .05$).[12] Alcohol consumption, medication use and tobacco use history were also similar.

Significant differences in subjective symptoms and objective findings were observed between the population of PBB-exposed farmers and the Wisconsin dairy farm comparison group. Ninety-six percent of the Michigan population had detectable levels of PBBs in their serum. As postulated, individuals living on or consuming products from farms that had been quarantined at one time had significantly higher PBB levels. No significant differences were observed between the consumer groups and the farm groups from which their produce came. This was consistent with the hypothesis that ingestion of contaminated produce was more important than place of residence. Consumers and farm residents shared the common contaminated products and thus the source of exposure.

In follow-up studies, serum PBB levels were found unchanged from 1976 to 1977, and fat-serum ratios were consistent over a wide range of concentrations.[20] These findings were interpreted as evidence of persistence of PBBs in human tissues and a virtual fat-serum equilibrium.

For current symptomatology, the self-selected group (Group 3) had the highest number of symptoms with a significantly greater prevalence than that seen in the Wisconsin Group 4. This was to be anticipated because they were invited, for the most part, because they or their physicians felt they had experienced unusual health problems which may have been related to PBB exposure. In the discussion of symptoms, this group has been excluded.

Removing Group 3, 17 of 43 symptoms were more prevalent in Michigan than in Wisconsin. Two symptoms were more prevalent in Wisconsin than in Michigan. The symptoms with the highest prevalence in both Michigan and Wisconsin were joint pain and excessive fatigue. Symptoms that are most commonly associated with the nervous system were the most prominent and severe.[21] These symptoms have been judged to have occurred earlier than others.[22] The unusual constellation of symptoms was similar to those reported after other toxic exposures: the similarity to the clinical presentation of carbon disulfide poisoning[23] and the Yusho (PCB)[14] clinical reports is of interest.

The correlation between current symptoms and current serum biochemistry abnormalities lends credence to symptomatology. The possibility of reactive depressive illness being the sole contributing cause to the excess symptoms reported is made unlikely by these associations. There is no doubt that depression played a role in some cases, since families had undergone considerable psychologic stress from economic loss and disruption of their livelihood and lifestyle. Further support for a "toxic syndrome," rather than an exclusive psychologic effect, is the observation of identical symptoms among the consumer groups, which had undergone no economic loss and were not involved in legal actions. For the most part, they were even unaware of the extent of their exposure.

There was no prior evidence to suggest that consumers of products from low-level, non-quarantined farms were at risk. *Yet the consumers had only a slightly lower prevalence of symptoms and laboratory abnormalities.*

The lack of an obvious dose-response relationship between PBB levels in serum or fat tissue and reported symptoms precludes determining a safe or non-toxic dose level. However, the lack of strong associations with this measurement is not altogether surprising.

Our examinations came three years after the initial episode and thus our observations were of chronic or cyclical effects rather than acute effects. This interpretation was supported by the calculation of incidence rates for the appearance of new symptoms. For most symptoms, incidence peaked in 1974–1975, and in 1976, incidence was only slightly higher than pre-1973.

This suggests that the major appearance of new symptoms occurred before the date of the examination and that the symptoms reported were sub-acute or chronic, rather than acute.

The dynamics and metabolism of PBBs are not fully understood with regard to the effect of a peak, intermittent dose compared with a low-level steady exposure. The later peak (1975) in incidence among lower exposure non-quarantine farmers and consumers further suggests that such exposure differences may be of clinical significance.

It has been suggested that "PBB" as estimated in the serum or adipose tissue may not necessarily be the most active toxic component in the dietary toxins ingested.[24,25] By convention, measurements reflect the hexabromobiphenyl isomer that makes up only 60% of the PBB in the original PBB, Firemaster. The Firemaster that was fed to livestock contained several other, possibly more toxic isomers.[26] Human exposure occurred after metabolism and mediation by different animals over varying periods of time. The gas chromotographic traces of PBB from directly exposed industrial workers are considerably different from those of the farmers, as well as from the original Firemaster compound.[27] Brominated napthalenes, which are more toxic in animals than PBBs, are present in Firemaster at concentrations of ±250 ppm.[24] Trace quantities of other contaminants not readily identifiable may also contribute to health effects. Recent reports on the Yusho incident now suggest that a dibenzofuran contaminant was probably an important etiologic agent, rather than the PCBs alone.[14,28] Because of the considerable time which has passed, many factors which contributed to the PBB incident may never be fully appreciated.

The observed differences in prevalence of abnormalities between sexes, with males in all cases having the higher prevalence of abnormalities as well as higher PBB levels, is of considerable interest. The explanation is not readily apparent and further investigation is needed.

A positive association, albeit a weak one, was observed between serum PBB levels and SGPT and SGOT elevation and increasing porphyrin excretion. These laboratory tests are non-specific and are affected by many variables. The combined effects of PBBs with other agents and conditions are being considered.

Finally, marked differences in clinical findings between the Michigan and Wisconsin participants cannot be fully explained without considering an etiologic role for PBB exposure.

Acknowledgments

We greatly appreciate the valuable contributions made by the over 100 professionals, technicians and field staff who participated in this survey. Special thanks go to Ms. Chris Muldoon and Mr. Steve Friedman who scheduled the Michigan participants and to the staff of the Marshfield Medical Foundation who contacted and scheduled the Wisconsin families. Their participation was invaluable. The generosity of institutions, such as the Kent Community Hospital and the Marshfield Clinic, Inc., in contributing clinic space and x-ray facilities was exemplary. The close cooperation of the institutions and departments in Michigan, New York, Wisconsin, which enabled such a comprehensive survey, was achieved through the efforts of many dedicated individuals who are, unfortunately, so numerous that they cannot be mentioned separately. All their efforts are gratefully acknowledged. We wish to thank The Computer Center of The City University of New York for their support in providing computer time and assistance and Dr. W. F. Calhoun of the Mount Sinai School of Medicine Department of Biostatistics for many helpful discussions. We wish to thank Mr. Sidney Sibel and Mrs. Valerie Josephson for the manuscript and table preparation.

This research was supported by NIEHS Center Grant No. ES 00928.

References

1. Norström, A., K. Anderson & C. Rappe. 1976. Major components of some brominated aromatics used as flame retardants. Chemosphere **4:** 255–261.
2. Jackson, T. F. & F. L. Halbert. 1974. A toxic syndrome associated with the feeding of polybrominated biphenyls-contaminated concentrate to dairy cattle. J. Am. Vet. Med. Assoc. **165:** 437–439.
3. Kay, K. 1977. Polybrominated biphenyls (PBB) environmental contamination in Michigan, 1973–1976. Environ. Res. **13:** 74–93.
4. Matthews, H. B., S. Kato, N. M. Morales & D. B. Tuey. 1977. Distribution and excretion of 2,4,5,2′, 4′,5′-hexabromobiphenyl, the major component of Firemaster BP-6. J. Toxicol. Environ. Health: **3:** 599–605.
5. Kasza, L. & M. A. Weinberger. 1978. Comparative toxicity of polychlorinated biphenyl and polybrominated biphenyl in the rat liver: light and electron microscopic alterations after subacute dietary exposure. J. Environ. Pathol. Toxicol. **1:** 241–257.
6. Corbett, T. H., A. R. Beaudoin, R. G. Cornell, M. D. Anver, R. Schumacher, J. Endres & M. Szwabowska. 1975. Toxicity of polybrominated biphenyls (Firemaster BP-6) in rodents. Environ. Res. **10:** 390–396.
7. Lambrecht, L. K., D. A. Barsotti & J. R. Allen. 1978. Responses of nonhuman primates to a polybrominated biphenyl mixture. Environ. Health Perspect. **23:** 139–146.
8. Evers, W. D., J. B. Hook & J. T. Bond. 1977. Effect of polybrominated biphenyls on renal tubular transport of organic ions. J. Toxicol. Environ. Health **3:** 759–767.
9. Luster, M., R. E. Faith & J. A. Moore. 1978. Effects of polybrominated biphenyls (PBB) on immune response in rodents. Environ. Health Perspect. **23:** 227–232.
10. Humphrey, H. E. B. & N. S. Hayner. 1975. Trace Substances in Environmental Health-IX. University of Missouri, Columbia, Mo. p. 57.
11. Meester, W. D. & D. J. McCoy. 1976. Symposium on Environmental Toxicology, Seattle, August, 1976. *In* J. Am. Med. Assoc. **236:** 1550.
12. Anderson, H. A., R. Lilis, I. J. Selikoff, K. D. Rosenman, J. A. Valciukas & S. Freedman. 1978. Unanticipated prevalence of symptoms among dairy farmers in Michigan and Wisconsin. Environ. Health Perspect. **23:** 217–226.
13. Roseman, K. D., H. A. Anderson, I. J. Selikoff, M. S. Wolff & E. Holstein. Unpublished data.
14. Kuratsune, M., Y. Masuda & J. Nagayama. 1976. Some recent findings concerning Yusho. National Conference on Polychlorinated Biphenyls, Proceedings. U.S.E.P.A. 560-6-75-004. Office of Toxic Substances, Washington, D.C.: 14–29.
15. Fischbein, S. A., *et al.* Ann. N.Y. Acad. Sci. **320.** This volume.
16. Salvati, D. T. Personal communication.
17. Wolff, M. S., B. Aubrey, F. Camper & N. Haymes. 1978. Relation of DDE and PBB serum levels in farm residents, consumers and Michigan Chemical Corporation employees. Environ. Health Perspect. **23:** 177–181.
18. Martinez, C. A. & G. C. Mills. 1971. Spectrophotofluorometric determination of porphyrins in urine. Clin. Chem. **17:** 201–205.
19. Hie, N. H., C. H. Hull, J. G. Jenkins, K. Steinbrenner & D. H. Bent. 1970. Statistical Package for the Social Sciences. McGraw-Hill, Inc. New York.
20. Wolff, M. S., H. A. Anderson, K. D. Rosenman & I. J. Selikoff. 1979. Equilibrium of polybrominated biphenyl (PBB) residues in serum and fate of Michigan residents. Bull. Environ. Contam. Toxicol. **21:** 5. In press.
21. Lilis, R., H. A. Anderson, J. A. Valciukas, S. Freedman & I. J. Selikoff. 1978. Comparison of findings among residents on Michigan dairy farms and consumers of produce purchased from these farms. Environ. Health Perspect. **23:** 105–110.

22. VALCIUKAS, J. A., R. LILIS, M. S. WOLFF & H. A. ANDERSON. 1978. Comparative neurobehavioral study of a polybrominated biphenyl-exposed population in Michigan and a nonexposed group in Wisconsin. Environ. Health Perspect. **23:** 199–210.
23. LILIS, R. 1974. Behavioral effects of occupational carbon disulfide exposure. *In* Behavioral Toxicology, Early Detection of Occupational Hazards. C. Xintaras, B. Johnson and I. deGroot, Eds.: 51–59. H.E.W. Publication No. (NIOSH) 74–126.
24. HASS, J. R., E. E. MCCONNELL & D. J. HARVAN. 1978. Chemical and toxicologic evaluation of Firemaster BP-6. J. Agric. Food Chem. **26:** 94–99.
25. KIMBROUGH, R. D., V. W. BURSE & J. A. LIDDLE. 1977. Toxicity of polybrominated biphenyl. Letter. Lancet **2:** 602–603.
26. DANNAN, G. A., R. W. MOORE & S. D. AUST. 1978. Studies on the microsomal metabolism and binding of polybrominated biphenyls (PBBs). Environ. Health Perspect. **23:** 51–61.
27. WOLFF, M. S. & B. AUBREY. 1978. PBB homologs in sera of Michigan dairy farmers and Michigan chemical workers. Environ. Health Perspect. **23:** 211–216.
28. NAGAYAMA, J., Y. MASUDA & M. KURATSUNE. 1977. Determination of polychlorinated dibenzofuranas in tissues of patients with Yusho. Food Cosmet. Toxicol. **15:** 195–198.

CLINICAL FINDINGS AMONG PCB-EXPOSED CAPACITOR MANUFACTURING WORKERS*

Alf Fischbein, Mary S. Wolff, Ruth Lilis, John Thornton,† and Irving J. Selikoff

Environmental Sciences Laboratory, Department of Community Medicine and the †Department of Biostatistics Mount Sinai School of Medicine The City University of New York New York, New York 10029

The polychlorinated biphenyls (PCBs) constitute a group of chemically stable compounds with wide use in industry since the 1930s, especially as dielectric fluids in capacitors and transformers.[1] Paints and other surface coatings such as plastics and varnishes may also contain PCBs. These are also a component of carbonless carbon paper.

Although PCBs have been used for almost half a century, and despite several experimental studies on biologic activity of PCBs,[2–5] few reports concerning the effects of occupational exposure to PCB compounds have appeared.[6–9] Increased concern with these compounds, however, followed when it was found that they were ubiquitously distributed in the environment.[10,11] Simultaneous with these findings came the reports of adverse human health effects resulting from the accidental ingestion of PCB-contaminated rice oil in Japan. The disease called Yusho (Japanese for "oil disease") became the clinical and pathologic paradigm for PCB-induced effects in humans.[12]

Present Study

In order to evaluate the health effects associated with occupational exposure to PCBs, a clinical field survey of 326 capacitor manufacturing workers was conducted. Various mixtures of PCBs (i.e. Aroclor® 1254 and 1242) had been used at the plant employing the workers since the 1940s. However, Aroclor 1016 and, to a minor degree, Aroclor 1221 had replaced the higher chlorinated PCBs in the capacitor manufacturing process for two years prior to this study (Table 1).

PCB-containing liquid was used for immersion of wound metal foil, which constitutes the principal unit of the capacitor. The intensity of exposure was estimated by industrial hygiene walk-through and on the basis of air-measurements provided by the company. Four different exposure categories were subsequently used in the analysis of the data (Table 2).

Population

The group examined were members of the United Electrical Workers Union, employed at two U.S. capacitor manufacturing plants, with total work forces of

*This work was supported by National Institute of Environmental Health Sciences Grant No. 00928.

0077–8923/79/0320–0703 $01.75/0 © 1979, NYS

TABLE 1
AROCLOR USAGE IN A CAPACITOR MANUFACTURING PLANT

Year	Aroclor 1016	Aroclor 1242	Aroclor 1221	Aroclor 1254
1975	*		*	
1974	*		*	
1973	*	*	*	
1972	*		*	
1971	*	*	*	
1970		*		*
1969		*		*
1968		*		*
1967		*		*
1966		*		*
1965		*		*
1964		*		*
1963		*		*
1962		*		*
1961		*		*
1960		*		*
1959		*		*

approximately 800 persons. Employees with more than 10 years of employment were invited to voluntarily participate in the examinations. One-hundred and sixty-eight male and 158 female workers accepted.

Invitations were extended to employees in all departments, in order to include individuals with high, low, and "no" exposure. Of those agreeing to participate, only two individuals (0.6%) failed to appear for the examination. In addition to those currently employed, 27 retirees were also examined.

MATERIALS AND METHODS

Each worker was questioned regarding past and present occupational history, and special attention was given to prior work with solvents such as carbon tetrachloride and trichloroethylene, pesticides, and metals; also, dust exposures (asbestos, talc, silica, coal, cement dust, and textile fibers). Past exposure to radiation was also recorded.

Present exposures were classified according to the information shown in TABLE 2.

TABLE 2
EXPOSURE CATEGORIES OF 326 PCB-EXPOSED WORKERS

Exposure	Air Level (8 hour time-weighted average, mg/m^3)
"None"	$0 \leq 0.07$
"Low"	$0.07 \leq 0.41$
"Medium"	$0.41 \leq 0.60$
"High"	$0.60 - 11.0$

Up to ten job categories were chronologically recorded. Both duration from onset of exposure and total length of employment at the plant were subsequently used in the analysis. Most individuals had held jobs in more than one job category.

Information concerning industrial hygiene conditions, such as use of respirators and protective clothing, smoking, and intake of food at the work site, was also recorded. A special family history questionnaire explored reproductive history.

Each worker had a careful review of past medical history and personal habits such as smoking and alcohol intake. A modified version of the MRC respiratory questionnaire was administered as was a specially designed questionnaire concerning symptoms related to various organ systems. These included acute symptoms of the upper and lower respiratory passages and skin. The latter included rashes, burning sensation of the skin, the occurrence of acne after starting work in the plant, pigmentation disturbances and discoloration and/or deformation of the nails.

Questions concerning gastrointestinal symptoms were directed at assessing weight loss, nausea and abdominal pain. Neurologic symptoms explored were headache, dizziness, general fatigue, and sleep disturbances: also, memory changes. Musculo-skeletal symptoms such as muscle and joint pain were also investigated. Each individual underwent a complete physical examination.

TABLE 3

BIOCHEMICAL PARAMETERS MEASURED IN PCB-EXPOSED WORKERS

Calcium	SGOT
Phosphorus	SGPT
BUN	Alkaline phosphatase
Creatinine	LDH
BUN/creatinine ratio	γ-Glutamyl transpeptidase
Uric acid	Triglycerides
Glucose	Total lipids
Total protein	Cholesterol
Albumin	Amylase
Globulin	Sodium
Alb/Glb ratio	Potassium
Total bilirubin	Chloride
Direct bilirubin	Iron

A broad spectrum of laboratory tests was used. These tests included complete blood count and 26 biochemical parameters tested by a Sequential Multiple Analyzer (TABLE 3). PCB levels in serum were determined by gas chromatography. Endocrinologic investigations of a subgroup included measurement of sex hormones (testosterone, progesterone, estradiol), thyroid hormones (T3 and T4) and cortisol. Endocrinologic findings will be the subject of a separate report.

Standard 14″ × 17″ chest x-rays were taken and read according to the ILO U/C classification. Pulmonary function tests included measurements of FVC, FEV_1 and FEF_{25-75}. These tests were performed by the use of a heated wire flow sensor, SRL Mod. M-10. Routine urinalyses were also included in the examination.

Statistical Methods

All statistical analyses were done using the χ^2-test and Student's t-test adjusting for unequal variances as appropriate.

TABLE 4

AGE DISTRIBUTION OF 326 CAPACITOR MANUFACTURING WORKERS BY SEX

Age	Total Number Examined	Male		Female	
		Number	Percent of Number in Age Group	Number	Percent of Number in Age Group
<30	49	33	67.3	16	32.7
30–39	61	41	67.2	20	32.8
40–49	88	48	54.5	40	45.5
50–59	93	27	29.0	66	71.0
60–69	25	13	52.0	12	48.0
70–	10	6	60.0	4	40.0
Total	326	168		158	

RESULTS

The age distribution of the examined population is shown in TABLE 4; the duration of employment is shown in TABLE 5. One-hundred and thirty-one individuals (40%) had been employed for 20 years or longer, while only 33 (10%) had been employed for less than five years. Thus, the workers generally had a long history of occupational exposure to PCBs.

Symptoms

Respiratory

To summarize, the prevalence of work-related upper respiratory irritation was 48%, and 10% of the group had at one time experienced tightness in the chest during work. A detailed report of the pulmonary findings is published elsewhere.[13]

Dermatologic

This aspect of the study is also reported in greater detail elsewhere.[14] Since it is known that chlorinated hydrocarbons seem to have a predilection for the pilosebaceous

TABLE 5

DURATION OF PCB EXPOSURE OF 326 CAPACITOR MANUFACTURING WORKERS

Duration (years)	Number of Workers	Percent
<5	33	10.1
5.0–9.9	68	20.9
10.0–14.9	57	17.5
15.0–19.9	37	11.4
20.0–24.9	95	29.1
25.0+	36	11.0
Total	326	100.0

TABLE 6

PREVALENCE OF REPORTED DERMATOLOGIC SYMPTOMS AMONG 326 CAPACITOR MANUFACTURING WORKERS

Symptom	Number	Percent
Rash	128	39.3
Burning sensation	81	24.8
Acne	35	10.7
Pigmentation (Darkening)	8	2.5
Thickening	12	3.7
Discoloration of fingernails	8	2.5

apparatus of the skin, the possible occurrence of acne in this population was of particular interest. Dermatologic findings were also prominent features in the patients with Yusho.

The overall prevalence of reported dermatologic symptoms in the present study group was high: 76 (45%) of the male workers and 87 (55%) of the female workers had a history of dermatologic complaints; skin rash, the most prevalent symptom, was reported by 128 persons (39%). Burning sensation of the skin was reported by 81 individuals (25%); a history of acne after beginning work at the plant was given by 35 workers (11%). The prevalence of reported pigmentation disturbances, thickening of the skin, and discoloration of the nails was markedly lower, approximating 3% (TABLE 6).

TABLE 7a

PREVALENCE OF GASTROINTESTINAL SYMPTOMS AMONG 326 CAPACITOR MANUFACTURING WORKERS

	Total Number Examined	Gastrointestinal Symptoms Present*	
		Number	Percent
Males	168	22	13.1
Females	158	36	22.8
Total	326	58	17.8

*Nausea, abdominal distress, weight loss, anorexia, vomiting.

TABLE 7b

PREVALENCE OF GASTROINTESTINAL SYMPTOMS BY SYMPTOM CATEGORY AND SEX AMONG 326 CAPACITOR MANUFACTURING WORKERS

	Males		Females	
Symptom	Number	Percent of Group Examined	Number	Percent of Group Examined
Anorexia	8	4.8	3	1.9
Weight loss	16	9.5	8	5.1
Nausea	7	4.2	14	8.9
Vomiting	4	2.4	2	1.3
Abdominal pain	5	3.0	15	9.5

TABLE 8
PREVALENCE OF MUSCULOSKELETAL SYMPTOMS AMONG 326 CAPACITOR MANUFACTURING WORKERS

	Males		Females	
Symptom	Number	Percent of Group Examined	Number	Percent of Group Examined
Muscle pain	5	3.0	6	3.8
Joint pain	13	7.7	24	15.2

Gastrointestinal

Of the workers, fifty eight (18%), 22 male and 36 female, gave a history of gastrointestinal symptoms. The prevalence of gastrointestinal symptoms is summarized in TABLE 7a and 7b.

Musculoskeletal

The prevalence of these symptoms is summarized in TABLE 8.

Neurologic

Sixty-four male workers (39%) and 94 female workers (58%) gave a history of neurologic symptoms. When present they were often multiple. The various neurologic symptoms reported by the workers are summarized in TABLE 9.

Physical Examination

The prevalence of abnormal dermatologic findings was marked. Sixty-eight male workers (41%) and 60 female workers (38%) had abnormalities including erythema,

TABLE 9
PREVALENCE OF NEUROLOGIC SYMPTOMS BY SYMPTOM CATEGORY AND SEX AMONG 326 CAPACITOR MANUFACTURING WORKERS

	Males		Females	
Symptom	Number	Percent of Group Examined	Number	Percent of Group Examined
Headache	33	19.6	44	27.8
Dizziness	16	9.5	22	13.9
Depression	5	3.0	19	12.0
Memory loss	2	1.2	13	8.2
Fatigue	17	10.1	36	22.8
Nervousness	21	12.5	49	31.0
Sleeplessness	10	6.0	17	10.8
Somnolence	8	4.8	18	11.4

swelling, dryness, and thickening. Sixteen individuals (5%) were found to have acneiform eruptions. Forty-eight workers (15%) showed conjunctival and palpebral abnormalities ranging from injected conjunctiva to palpebral hyperpigmentation and edema, the latter physical signs being characteristic PCB-associated effects.[8]

Seven workers (2%) had abnormal secretions from the eyes while three individuals (1%) had enlargment of the meibomian glands. Examination of the liver revealed four individuals (1.2%) with abnormal liver size, two of whom had a history of heavy alcohol intake. The other two individuals had been employed in the plant for 12 and 24 years, respectively. The former had one slightly abnormal liver function test (SGOT 55).

TABLE 10

DISTRIBUTION OF PERTINENT CLINICAL CHEMISTRY PARAMETERS IN PCB-EXPOSED WORKERS

Biochemical Parameters				
BUN	<20	20–25	*25+	
Number (percent)	250 (77.9)	54 (16.8)	17 (5.3)	Total 321
Creatinine	<1.5	1.5–1.7	*1.7+	Total 321
Number (percent)	270 (84.2)	38 (11.8)	13 (4.0)	
SGOT (i.u./1)	<50	*50–74	*75+	
Number (percent)	314 (97.8)	6 (1.9)	1 (0.3)	Total 321
SGPT (i.u./1)	<50	*50–74	*75+	
Number (percent)	297 (92.8)	20 (6.3)	3 (0.9)	Total 320
LDH (i.u./1)	<250	*250–275	*275+	
Number (percent)	313 (97.5)	7 (2.2)	1 (0.3)	Total 321
Alkaline phosphatase (i.u./1)	<50	*50–75		
Number (percent)	317 (98.8)	4 (1.2)		Total 321
γ-glutamyl transpeptidase (u./1)	<40	*40–59	*60+	
Number (percent)	315 (98.1)	5 (1.6)	1 (0.3)	Total 321
Bilirubin (mg/100 ml)	<1.0	1.0–1.3		
Number (percent)	304 (94.7)	17 (5.3)		Total 321
Cholesterol (mg/100 ml)	<250	250–299	*300+	
Number (percent)	165 (51.4)	99 (30.8)	57 (17.8)	Total 321
Triglycerides (mg/100 ml)	<199	*200–249	*250+	
Number (percent)	282 (89.5)	16 (5.1)	17 (5.4)	Total 315
Total lipids (g/100 ml)	0.5–0.8	0.8–1.0	*1.0+	
Number (percent)	283 (88.2)	27 (8.4)	11 (3.4)	Total 321

* Indicates value outside of normal range.

Routine neurologic examination did not reveal any remarkable prevalence of abnormalities; extensor weakness was observed in six individuals (1.8%), while only one worker presented tremor at physical examination.

Laboratory Findings

A summary of the pertinent laboratory findings is shown in TABLE 10. It is of interest to note that there was a paucity of abnormal results in the biochemical studies. Similar findings were noted in the results of the hematologic tests, which showed a mean hemoglobin concentration of 15.3 gm/100ml (range 12.9–17.9) for male

TABLE 11a
HEMOGLOBIN LEVELS OF 160 FEMALE PCB-EXPOSED WORKERS

Hemoglobin, g/100 ml	Number	Percent
<13	25	15.6
13.0–13.9	49	30.6
14.0+	86	53.8
Total	160	100.0

TABLE 11b
HEMOGLOBIN LEVELS OF 166 MALE PCB-EXPOSED WORKERS

Hemoglobin, gm/100 ml	Number	Percent
<13	7	4.2
13.0–13.9	6	3.6
14.0+	153	92.2
Total	166	100.0

workers and a mean hemoglobin concentration of 14.0 gm/100ml (range 11.3–16.2) for female workers. Also, there were no significant abnormalities found in erythrocyte indexes. The distribution of hemoglobin concentrations are summarized in TABLES 11a and 11b.

The distribution of plasma PCB concentrations among the examined workers is summarized in TABLE 12.

Pulmonary Function

The pulmonary function studies, reported separately,[13] showed a decrease in forced vital capacity in 34 (14%) of 243 examined workers. The pulmonary function

TABLE 12
DISTRIBUTION OF PLASMA PCB CONCENTRATIONS AMONG CAPACITOR MANUFACTURING WORKERS

	Male Workers		Female Workers	
Lower PCB Homologues, ppb	Number	Percent	Number	Percent
0 < 100	97	64.2	94	71.2
100 < 200	29	19.2	26	19.7
200+	25	16.6	12	9.1
Total	151	100.0	132	100.0
	Male Workers		Female Workers	
Higher PCB Homologues, ppb	Number	Percent	Number	Percent
0 < 25	71	47.0	83	62.9
25 < 75	55	36.4	37	28.0
75+	25	16.6	12	9.1
Total	151	100.0	132	100.0

TABLE 13

PREVALENCE OF X-RAY ABNORMALITIES AMONG 326 PCB-EXPOSED WORKERS

Abnormality	Number	Percent
Parenchymal change ≥s 1/0, ILO/UC classification	6	1.8
Pleural thickening	4	1.2
Obliteration of costophrenic angle	5	1.5
Other abnormalities (cardiac, infections, emphysema, granuloma)	14	4.3
Total abnormal	29	8.9

abnormalities were not accompanied by abnormalities on chest x-ray, as shown in TABLE 13.

Correlative Studies

Mean plasma PCB levels for the various exposure categories are summarized in TABLE 14. There was a statistically significant difference between mean plasma PCB concentration of the various exposure categories, a difference which was noticed both for lower and higher homologues of PCBs.

No association was found between reported symptoms and duration of employment. Neither was there any correlation between reported symptomatology and plasma levels of lower and higher homologues of PCBs. Regarding correlations between abnormal findings on physical examination and plasma PCB levels, there was a statistically significant association between abnormal dermatologic findings and plasma levels of higher homologues of PCBs as summarized in TABLE 15.

Despite the low prevalence of abnormal liver function tests, there was a statistically significant association between abnormal SGOT levels and plasma levels of both lower and higher homologues of PCBs (TABLES 16a and 16b). Such association was not found with any of the other biochemical tests, including SGPT, γGTP, bilirubin and LDH. Analysis of serum cholesterol and triglycerides in relation to exposure is in progress and will be reported elsewhere.

TABLE 14

PLASMA PCB CONCENTRATIONS AMONG ELECTRICAL WORKERS ACCORDING TO JOB EXPOSURE INTENSITY

Present Job Exposure Relative Intensity	Mean Concentration, ppb		
	Lower PCB Homologues*	Higher PCB Homologues*	n
High	266 ± 328	82 ± 128	43
Medium	171 ± 361†	25 ± 25	62
Low	73 ± 81	41 ± 75	158
Retired	80 ± 80	84 ± 120	26
All	124 ± 229	48 ± 85	289

*Lower PCB homologues included peaks with retention time (gas chromatography) shorter than for p, p'-DDE. Higher PCB homologues included peaks with retention times greater than p, p'-DDE.

†Including one individual with plasma PCB of 2,530 ppb.

TABLE 15

FREQUENCY OF ABNORMAL SKIN FINDINGS AND PLASMA PCB CONCENTRATION OF HIGHER HOMOLOGUES

Plasma PCBs, Higher Homologues, ppb	Total Number Examined	Number with Abnormal Skin Findings
0 < 10	65	24 (37%)
10 < 30	99	31 (31%)
30 < 50	47	12 (26%)
50 < 100	34	17 (50%)
≥ 100	26	15 (58%)
Total	271	99 37%

$\chi^2 = 11.36$ df = 4 $p < 0.05$.

DISCUSSION

In order to evaluate the health status of individuals occupationally exposed to polychlorinated biphenyls, a cross-sectional clinical field survey was conducted of 326 workers employed at two capacitor manufacturing facilities, where dielectric fluids containing PCBs and chlorinated benzenes had been extensively used for approximately 30 years.

A large portion of the study population (40%) had been employed for 20 years or longer, making it possible to attempt an evaluation of the effects of long-term occupational exposure to polychlorinated biphenyls. Such information seemed to be of value in view of the paucity of clinical studies on the effects of human PCB exposure, although experimental studies have delineated the effects of PCBs on various organ systems in many animal species.[1-5] Some observations concerning human health

TABLE 16a

SGOT LEVELS AND PLASMA LEVELS OF HIGHER HOMOLOGUES OF PCBs

HPCB (ppb)	Total Number	SGOT <50	SGOT >50	Percent Abnormal of Group
0–75	244	240	4	1.6
75+	36	33	3	8.3
Total	280	273	7	2.5

$\chi^2 = 5.77$, $0.01 < 0.025$.

TABLE 16b

SGOT LEVELS AND PLASMA LEVELS OF LOWER HOMOLOGUES OF PCBs

LPCB	Total Number	SGOT <50	SGOT >50	Percent Abnormal of Group
0–200	243	240	3	1.2
200+	37	33	4	10.8
Total	280	273	7	2.3

$\chi^2 = 12.06$ $p < 0.005$.

effects of occupational exposure to PCBs have been reported,[8,15,16] but much of both clinical and laboratory data have been derived from the "Yusho" incident in Japan, in which some 1,000 people accidentally consumed rice oil contaminated with PCBs.[12,17] This incident seemed to provide epidemiologic evidence for an association between Yusho and PCBs.

An additional possible etiologic factor was suggested when it was demonstrated in chickens that the contaminated rice oil ingested by the Yusho patients was more than twice as toxic as anticipated by the amount of PCBs alone present in the oil.[18] Therefore, the presence of other toxic substances in the contaminated rice oil was suggested. This hypothesis was subsequently confirmed by the demonstration of polychlorinated dibenzofurans (PCDFs) in the Japanese oil.[19] Other investigators have reported the presence of PCDFs in some commercial PCBs,[20,21] and studies on the comparative toxicity of PCBs and PCDFs seem to indicate that PCDFs are much more toxic, especially for the thymus, skin (more acnegenic), liver and the hematopoetic system, when administered in the diet to rats.[22] However, the exact role played

TABLE 17

COMPARISON OF SELECTED RESPONSES TO PCBs IN RATS, NON-HUMAN PRIMATES AND MAN

Response	Rat	Monkey	Man	Study Population
Susceptibility to toxicity	(+)	++	++	+
Acne	–	+	+	+
Hyperpigmentation of skin	–	(+) infants only	+	+
Hyperactive Meibomian glands	–	+	+	+
Conjunctivitis	–	+	+	+
Oedema of eyelids	–	+	+	+
Hyperplasia of hair follicle epithelium	–	+	+	NA* (comedos)
Gastric hyperplasia	–	+	NA	NA
Liver enzyme change	+	+	NA	+

Modified from IARC Monographs on the evaluation of the carcinogenic risk of chemicals to humans. Vol. 18, IARC, Lyons, France, October, 1978.

*NA = not available.

by toxic impurities or derived products, such as the PCDFs, is not known. Moreover, because of the difference in toxic effects of PCBs in humans and animals, it has been difficult to extrapolate experimental data to human experience. Comparative data on the toxic effects of PCBs in animals and humans are summarized in TABLE 17.[23–25]

The concern with potential human health effects is augmented by the fact that human exposure to PCBs seems to be widespread as a result of environmental contamination; PCBs are commonly found in human tissue.[27]

High exposure to PCBs may occur among workers manufacturing or using PCB-containing compounds, and previous reports indicate that occupational exposures may vary considerably.[9,27] Studies of workers employed in PCB-using facilities have shown blood levels of PCBs several times higher than that of control groups with no particular exposure to PCBs.[8,28]

In this study of capacitor manufacturing workers the most prevalent symptoms reported were dermatologic and central nervous system symptoms; these were also the most common features in the early phase of Yusho.[17] Dermatologic abnormalities on physical examination observed in Japan, including eye discharge and swelling of the upper eyelids, also occurred among some in this study population. The most characteristic skin changes, such as comedones and acne, occurred primarily in persons currently exposed. None of the retirees examined showed any marked current dermatologic abnormalities. It is of interest to note that the marked dermatologic and ophthalmic manifestations found in the Yusho patients gradually improved over the years following exposure.[17]

The paucity of other abnormalities on physical examination was striking; this included a very low prevalence of abnormal liver findings. It should be noted, however, that, in spite of the absence of liver abnormalities on physical examination, in a subgroup of this population, the functional capacity of the cytochrome P-450 system of the liver investigated through the plasma half life ($T_{1/2}$) of antipyrine was found to be different from that of normal non-PCB-exposed controls. Five workers studied showed a significantly lower mean antipyrine half-life than the mean half-life in non-PCB exposed normal subjects. This strongly suggests that their occupational experience and PCB exposure had resulted in an inducing effect on enzymes metabolizing antipyrine in man.[29] This had previously been shown to occur in animals.[3]

The relatively high prevalence of decreased vital capacity that was found in this group of workers is of interest in view of recent experimental data indicating accumulation of certain PCB isomers and metabolites in both bronchial mucosa and lung parenchyma.[30,31]

References

1. Penning, C. H. 1930. Physical characteristics and commercial possibilities of chlorinated diphenyl. Industr. Eng. Chem. **22:** 1180–1182.
2. Litterst, C. L., T. M. Farber, A. M. Baker & E. G. Van Loon. 1972. Effect of polychlorinated biphenyls on hepatic microsomal enzymes in the rat. Toxicol. Appl. Pharmacol. **23:** 112–122.
3. Alvares, A. P., D. R. Bickers & A. Kappas. 1973. Polychlorinated biphenyls: A new type of inducer of cytochrome P-488 in the liver. Proc. Nat. Acad. Sci. USA **70:** 1321–1325.
4. Goldstein, J. A., P. Hickman & D. L. Jue. 1974. Experimental hepatic porphyria induced by polychlorinated biphenyls. Toxicol. Appl. Pharmacol. **27:** 437–448.
5. Kimbrough, R. D. 1975/76. Toxicity of polychlorinated polycyclic compounds and related chemicals. CRC Critical Rev. Toxicol. **3:** 445–490.
6. Schwartz, L. 1936. Dermatitis from synthetic resins and waxes. Am. J. Pub. Health **26:** 586–592.
7. Schwartz, L. 1942. An outbreak of halowax acne ("Cable rash") among electricians. J. Am. Med. Assoc. **122**(3): 158–161.
8. Ouw, K. H., G. R. Simpson & D. S. Siyali. 1976. The use and health effects of Aroclor 1242, a polychlorinated biphenyl in an electric industry. Arch. Environ. Health **31**(4): 189–194.
9. Meigs, J., J. J. Albom & B. L. Kartin. 1954. Chloracne from an unusual exposure to Arochlor. J. Am. Med. Assoc. **154:** 1417–1418.
10. Jensen, S., A. G. Johnels, M. Olsson & G. Otterlind. 1969. DDT and PCB in marine animals from Swedish waters. Nature **224:** 247–250.
11. Jensen, S. 1972. The PCB story. Ambio **1:** 123–131.
12. Higuchi, K., Ed. PCB poisoning and pollution. 1976. Kidansha, Tokyo and Academic Press, New York.
13. Warshaw, R., A. Fischbein, J. Thornton, A. Miller & I. J. Selikoff. 1979. Decrease in vital capacity in PCB-exposed workers in a capacitor manufacturing facility. Ann. N.Y. Acad. Sci. **320:** 277–283. This volume.

14. FISCHBEIN, A., J. THORNTON, M. S. WOLFF & I. J. SELIKOFF. 1979. Dermatological findings in capacitor manufacturing workers exposed to dielectric fluids containing PCBs. Acta Dermato-Venereologica. In press.
15. JONES, J. W. & H. S. ALDEN. 1936. An acneform dermatosis. Arch. Dermatol. Syphilol. **33:** 1022–1034.
16. GOOD, C. K. & N. PENSKY. 1943. Halowax acne (cable rash): Cutaneous eruptions in marine electricians due to certain chlorinated naphthalenes and diphenyls. Arch. Dermatol. Syphilol. **48:** 251–257.
17. KURATSUNE, M., Y. MASUDA & J. NAGAYAMA. 1976. Some of the recent findings concerning Yusho. *In* Proceedings of a National Conference on Polychlorinated Biphenyls. Chicago, 1975. EPA-560/6-75-004, Washington, D.C. Env. Protection Agency. pp. 14–29.
18. IKEDA, Y. 1972. Toxicity of PCBs. J. Food Hyg. Soc. Japan **13:** 339–267.
19. NAGAYAMA, J., M. KURATSUNE & Y. MASUDA. 1976. Determination of chlorinated dibenzofurnas in Kanechlors and "Yusho oil." Bull. Environ. Contam. Toxicol. **15:** 9–13.
20. VOS, J. G., J. H. KOMEN, H. L. VAN DER MAAS, M. D. TEN NOEVER DE BRAUN & R. H. DEVOS. 1970. Identification and toxicological evaluation of chlorinated dibenzofurans and chlorianted naphthalene in two commercial polychlorinated biphenyls. Food Cosmet. Toxicol. **8:** 625–633.
21. ROACH, J. A. G. & I. J. POMERANTZ. 1974. The identification of chlorinated dibenzofurans in a Japanese polychlorinated biphenyl sample. Bull. Environ. Contam. Toxicol. **12:** 338–342.
22. OISHI, S., M. MORITA & H. FUKUDA. 1978. Comparative toxicity of polychlorinated biphenyls and dibenzofurans in rats. Toxicol. Appl. Pharmacol. **43:** 13–22.
23. KURATSUNE, M. 1972. An abstract of results of laboratory examinations of patients with Yusho and of animal experiments. Environ. Health Perspect. **1:** 129–136.
24. ALLEN, J. R. 1975. Response of the nonhuman primate to polychlorinated biphenyl exposure. Fed. Proc. **34:** 1965–1979.
25. ALLEN, J. R., L. A. CARSTENS & L. J. ABRAHAMSON. 1976. Responses of rats exposed to polychlorinated biphenyls for fifty-two weeks. I. Comparison of tissue levels of PCB and biological changes. Arch. Environ. Contam. Toxicol. **4:** 404–419.
26. FINKLEA, J., L. E. PRIESTER, J. R. CREASON, T. HAUSER, T. HINNERS & D. I. HAMMER. 1972. Polychlorinated biphenyl residues in human plasma expose a major urban population problem. Am. J. Pub. Health **62:** 645–651.
27. ELKINS, H. G. B. 1959. The chemistry of industrial toxicology. 2nd edit. John Wiley and Sons. New York. p. 153.
28. KARPPANEN, E. & L. KOLHO. 1972. The concentration of PCB in human blood and adipose tissue in three different research groups. *In* Proc. 2nd PCB Conference, Solna, Sweden, 1972, Stockholm. Nat. Swed. Env. Protection Board, pp. 124–128.
29. ALVARES, A. P., A. FISCHBEIN, K. E. ANDERSON & A. KAPPAS. 1977. Alterations in drug metabolism in workers exposed to polychlorinated biphenyls. Clin. Pharmacol. Therap. **22:** 140–146.
30. BRANDT, T. 1977. Tissue localization of polychlorinated byphenyls chemical structure related to pattern of distribution. Acta Pharmacol. Toxicol. **40** (Supplement II).
31. JANSON, B. 1978. Transformation of some organochlorine compounds in the environment studied by gas chromatography and mass spectrometry. Report from the National Swedish Environmental Protection Board. Special Analytical Laboratory, Wallenberg Laboratory, University of Stockholm, S-106 91 Stockholm, Sweden.

IMMUNOLOGIC DYSFUNCTION AMONG PBB-EXPOSED MICHIGAN DAIRY FARMERS*

J. George Bekesi, Henry A. Anderson,†
Julia P. Roboz, John Roboz, Alf Fischbein,†
Irving J. Selikoff,† and James F. Holland

Department of Neoplastic Diseases
The Mount Sinai School of Medicine
New York, New York 10029

†Environmental Sciences Laboratory
The Mount Sinai School of Medicine
New York, New York 10029

In 1973 a commercial preparation of polybrominated biphenyls (PBBs) was inadvertently used in place of magnesium oxide in the preparation of a special feed supplement for lactating cows, which is used to prevent hypomagnesemic tetany and to provide better buffering rumen content and stabilize fat.[1,2] The contaminating PBB preparation was Firemaster FF-1®, a flame retardant manufactured by the Michigan Chemical Company. Acute toxic effects in animals fed PBB-contaminated feed included decreased milk production, joint swelling, hyperkeratosis, persistent mastisis, cutaneous and subcutaneous infections, and abscess formation on back, legs and udder; wasting and death occurred within six months to 24 of a group of 400 exposed cows.[3,4] Similar observations were made on twelve 6–18 month old heifers and bulls. Only two animals remained alive 5 months after a short exposure to contaminated feed. In addition to general wasting among these young animals, one 18-month-old bull appeared to have atrophied testicles and abnormal semen showing no spermatozol motility with numerous headless and tailess spermatoza.[3] Positive identification of the toxic agent in the feed was not made until 9 months after its ingestion. Feed containing PBBs was used on farms in many parts of Michigan during 1973 and most of 1974. More than 500 Michigan dairy herds and poultry farms were quarantined, and over 30,000 cattle and 1.5 million chickens died or had to be destroyed.[1,3]

Dairy products containing PBBs (beef, poultry, eggs, milk, cheese, butter) were widely consumed in Michigan until 1977. PBBs were subsequently found in the serum and/or adipose tissues of dairy farm residents and workers employed by the Michigan Chemical Co. as well as in many urban Michigan residents.[5–8]

Experimental data suggest that polybrominated biphenyls, which are similar to polychlorinated biphenyls, are fat soluble and are stored in the thymus, liver, brain, and adipose tissues, and persist in those tissues for very long periods.[9,12] Because of toxicity and lack of active metabolism of polyhalogenated biphenyls, there has been concern over the short and long-term health consequences in Michigan residents.[5,6,13–18]

Data reported here represent an extension of our previously published study of immunologic dysfunctions among Michigan dairy farm residents who consumed meat and dairy products from animals which were fed food containing PBBs.[12]

*This research was supported by Center Grant for the National Institute of Environmental Health Sciences ES00928 and by National Cancer Institute Grant 5P11-CA15936.

0077-8923/79/0320-0717 $01.75/0 © 1979, NYAS

Materials and Methods

Subject Population

Three groups of subjects were examined: (1) 45 adult dairy farm residents (both quarantined and nonquarantined) who lived within 50 miles of Reed City, Michigan, and were subjected to PBB-contaminated food for a period of 3 months to 4 years; (2) 46 Wisconsin dairy farmers and members of their families who did not eat PBB-contaminated food; (3) 79 healthy subjects in the New York area. An approximately equal number of males and females participated in the study. The mean age of the PBB-exposed group was younger than that of the control populations.

Blood Samples

Peripheral blood (50 ml) was taken by venipuncture in a plastic syringe containing preservative-free heparin. The blood samples were hand-carried from Michigan or Wisconsin to New York City. In addition, control blood samples obtained from the New York City personnel participating in the field study were sent with the study samples from both Wisconsin and Michigan to ascertain whether the plane journey in some way affected the specimen.

Isolation of Lymphocytes

Peripheral blood lymphocytes (PBLs) were isolated and purified from the blood by Ficoll-Hypaque sedimentation 12–18 hours after the blood was drawn. Blood samples were diluted with equal volumes of phosphate-buffered saline, layered on a Ficoll-Hypaque gradient, and centrifuged for 30 minutes at 850 $\times$ g at 20°C. The lymphocytes were located at the interface of the gradient. After collection, lymphocytes were washed three times with RPMI-1640 medium supplemented with 20% heat-deactivated autologous plasma.

Surface Markers

Purified PBL was rosetted with sheep red blood cells according to the method of Wybran and Fudenberg.[20] Sheep red blood cells from a pretested single animal source, stored in Alsever's solution at 4°C for no more than 6 days were used. Lymphocytes with three or more E-rosettes were considered positive. All lymphocytes were counted to determine the percentage of total E-rosettes. The absolute number of total E-rosette forming cells was determined from the corresponding total and differential leukocyte counts.

The C_3 receptor on purified monocytes was determined by the EAC rosettes assay as described by Nussenzweig[21] using sheep red blood cells coated with rabbit IgM anti-sheep erythrocyte antibody in the presence of complement. Direct immunofluorescence assay for quantification of surface membrane immunoglobulins was performed as described by Preud'homme and Seligmann.[22] Fc receptors were analyzed by the EA rosettes assay and by immunofluorescence with fluorescine-conjugated aggregated gamma globulins. We have calculated percentage and number of PBLs without characteristic membrane markers by subtraction: the percentage or total number of mononuclear cells minus the percentage or absolute number of E plus EAC rosette-forming lymphocytes.

In Vitro *Lymphocyte Function Determined by Mitogens*

Lymphocyte blastogenesis for PBL was determined by two mitogens: phytohemagglutinin (PHA) for T-lymphocytes and pokeweed mitogen (PWM) for B-lymphocytes. After Ficoll-Hypaque purification, 100,000 PBLs were placed in each of five replicate wells of Falcon microplates containing RPMI-1640 medium supplemented with 20% heat-inactivated autologous or allogeneic plasma in the presence of highest-purity PHA (Burroughs Wellcome) or PWM (GIBCO). Control cultures were incubated without mitogens in the replicate wells at 37°C in a humidified atmosphere containing 5% CO_2 in air. Lymphocyte blastogenesis was determined by measuring the level of DNA synthesis upon addition of 1 μCi of [^{3}H] thymidine (New England Nuclear) to each well 18 hours prior to termination of the culture. After 90 hours, the cells were harvested with a Mash II automatic harvester and the amount of [^{3}H] thymidine incorporated was determined in a liquid scintillation spectrometer (Packard).

Mixed Lymphocyte Culture

Proliferative T-cell response was determined by one-way-mixed lymphocyte culture (MLC). Stimulating mononuclear cells were isolated and stored in liquid nitrogen from six different HLA mixmatched healthy donors. Prior to MLC assay the individual donor lymphocytes were rapidly thawed at 37°C and were washed with mixed salt and glucose media at 4°C. Next, cells were layered on 22% human serum albumin for separation of viable from nonviable cells. A viability of 95% was attained. The mixed leukocyte culture test was done as follows: stimulating cells (an equal mixture of lymphocytes from six different allogeneic donors) were incubated with mitomycin-C at 25 μg/ml of 2×10^6 cells suspension for 20 minutes at 37°C. After washing 2×10^5 stimulating cells distributed in replicate wells of Falcon microplates containing 10^5 responding lymphocytes per well in RPMI-1640 media supplemented with 20% heat-inactivated autologous plasma. After 5 days incubation 2 μCi [^{3}H] thymidine was added to each well. Cultures were harvested 18 hours later with the addition of an excess of cold thymidine. The amount of [^{3}H]-TDR incorporation was determined in a Packard liquid scintillation spectrometer.

Determination of Lymphocyte Function after Transportation

We have examined at the beginning and continued to monitor all through this investigation, the possibility that transportation of blood specimens from Wisconsin or Michigan to New York City might have an adverse effect on the functional integrity of PBL. Research personnel who participated in both the Michigan and Wisconsin field surveys, and had been previously subjected at least twice in the last 12 months to immunodiagnostic tests, served as study subjects. Base line data were established on each individual 4–7 days prior to field survey using freshly drawn blood. Lymphocytes were separated and assayed 60 minutes after specimens were drawn. For comparison, blood samples were drawn from the same individuals at the same time as those for the Wisconsin farmers and hand carried in the original plastic syringe on the Wisconsin to New York plane journey. In this case lymphocytes were separated and tested 16 hours after blood was drawn.

TABLE 1

STUDY SUBJECTS

	Control Subjects		PBB-Exposed Subjects
	New York City Residents	Wisconsin Dairy Farm Residents	Michigan Dairy Farm Residents
Number of subjects	79	46	45
Males	41	30	25
Females	38	16	20
Mean Age	43 ± 8.2*	45 ± 9.5	35 ± 12.8
Hemoglobin (GMS)	14 ± 2.1	14.4 ± 2.1	14.8 ± 1.3
WBC $\times 10^3$	6.9 ± 2.1	6.5 ± 0.2	7.1 ± 2.2
Lymphocytes (%)	30.1 ± 4.2	32.4 ± 7.9	32.7 ± 4.6

*Means ± Standard Error.

RESULTS AND DISCUSSION

Study Subjects

Routine hematologic analysis showed that not a single individual in either group differed from the established laboratory normal values for hemoglobin, white blood cells, % lymphocytes or differential counts (TABLE 1).

TABLE 2

EFFECT OF TRANSPORTATION ON THE LYMPHOCYTE FUNCTION

	Freshly Drawn Blood*				18 Hrs after Blood Samples Were Drawn—Airplane Transportation†			
	Subjects				Subjects			
	1	2	3	4	1	2	3	4
WBC $\times 10^3/\mu l$	7.21	6.85	7.62	8.09	7.30	6.71	7.78	7.94
% Lymphocytes	34	30	35	31	35	31	33	30
T-cells								
%	77	72	68	71	76	74	70	69
Absolute no./μl	1888	1479	1814	1780	1942	1539	1797	1643
B-cells								
%	18	24	21	19	17	22	20	21
Absolute no./μl	441	493	560	477	434	458	513	500
PHA								
Maximum stimulation c.p.m. $\times 10^3$	101.6	103.7	91.3	84.5	97.3	102.1	93.7	87.2
PWM								
Maximum stimulation c.p.m. $\times 10^3$	95.4	78.7	88.5	91.7	101.1	82.7	83.8	93.2

*Blood specimens were used for assays 60 minutes after blood was drawn.

†Blood samples were hand carried from Wisconsin to New York and tested 16 hours after blood was drawn.

Transportation of Lymphocytes

Data obtained from four staff members (TABLE 2) clearly indicate that transportation of blood specimens from Wisconsin to New York City did not alter the functional integrity of lymphocytes as compared to the freshly drawn and processed blood specimens.

Change of Membrane Markers of Peripheral Blood Lymphocytes in PBB-Exposed Michigan Dairy Farm Residents

For the quantification of total T and B-lymphocyte subpopulations the percentages of E and EAC rosette-forming peripheral blood lymphocytes were determined by direct surface membrane immunofluorescence. FIGURE 1 shows the values of the percentage and absolute number of T-lymphocytes in the peripheral blood of PBB-exposed Michigan dairy farm residents and of the control group of Wisconsin dairy farmers. Purified lymphocytes of Wisconsin subjects formed stable E-rosettes, at a range of 60–81% with a mean value of 74%. The absolute number of T-cells, which was determined by calculation of the corresponding total and differential leukocyte counts, was at the range 1300–2500 with a mean value of 1800 per μl. The mean value

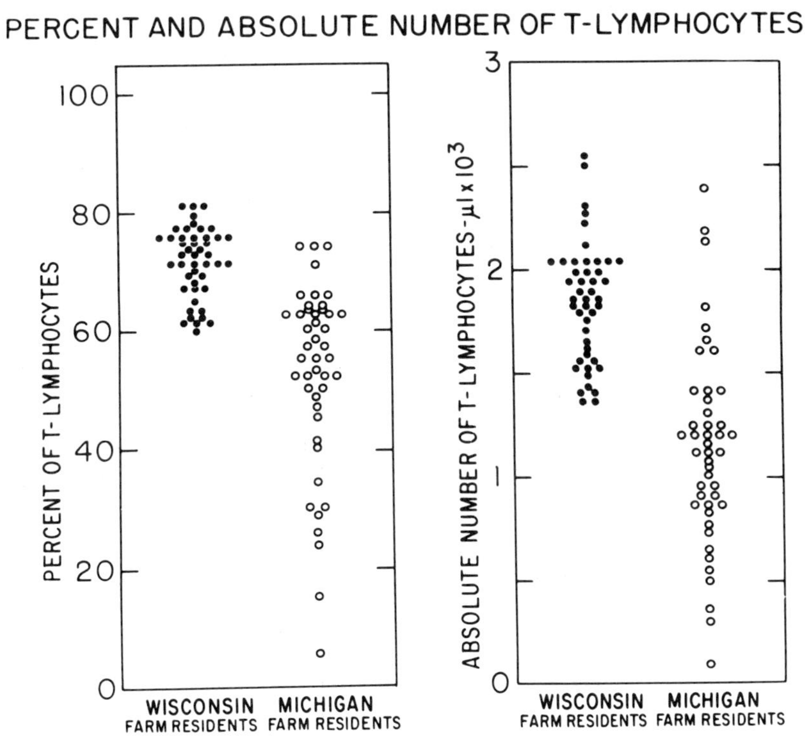

FIGURE 1. Change of T-lymphocyte subpopulation in PBB-exposed Michigan dairy farm residents.

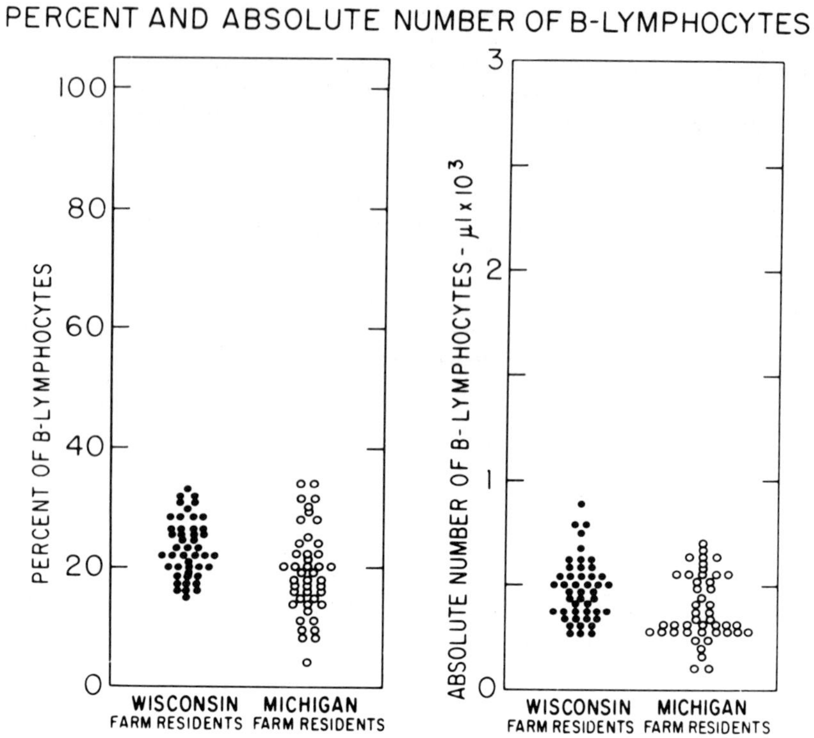

FIGURE 2. Percent and absolute number of EAC rosette-forming lymphocytes in the control group of Wisconsin dairy farmers and in PBB-exposed Michigan dairy farm residents.

of the T-cell subpopulation of the peripheral blood lymphocytes of New York City residents was 74.4 ± 2.5, which is almost the same as that of the Wisconsin Control group. Considering the normal values for T-cells obtained from the two control groups, study subjects who show a less than 50% T-lymphocytes, with an absolute number of less than 100 per/µl can be considered as exhibiting marked change from normal values. Significant deviations from the normal range in both percent and absolute number of T-cells were observed in about 35% of the examined PBB-exposed Michigan dairy farm residents.

As shown in FIGURE 2, contrary to the data obtained on T-lymphocytes, only five study subjects can be recognized in the Michigan group to have shown a marginal decrease in the B-lymphocyte subpopulation. One of the most striking changes detected in the peripheral blood lymphocytes of the PBB-exposed Michigan farm residents (FIGURE 3) is the marked increase in the number of lymphocytes without detectable membrane markers, "null cells." It is not known presently, whether this is due to immature T-lymphocytes, the effect of PBB on the thymus and/or to the selective binding of PBB on the plasma membrane masking receptor sites on T-lymphocytes.

Impaired Lymphocyte Function in PBB-Exposed Michigan Dairy Farm Residents

The *in vitro* immunocompetence of both groups of subjects was determined by selected mitogens: phytohemagglutinin (PHA) for T-cells and pokeweed (PWM) for B-cells. FIGURE 4 shows a representative stimulation pattern of Wisconsin and Michigan farm residents' PBLs by PHA and PWM. Maximum lymphocyte transformation was obtained at 0.15–0.30 μg per well for PHA and 30 μg per well for PWM for both groups. Results presented in FIGURE 5 show the response of all study subjects' PBLs to PHA and PWM. The range of maximum stimulation in the control group was 75–131 $\times$ 10^3 c.p.m. for PHA and 65–137 $\times$ 10^3 c.p.m. for PWM. Significant deviation from the normal range in lymphocyte function was apparent in at least 18 individuals in the Michigan group. In the 10 of 18 individuals who showed significantly decreased response to T-cell mitogen, this was accompanied by a proportionately lower percentage and absolute number of T-lymphocytes. Thus, it can be assumed that the decreased lymphoblastogenesis was not due to functional defects

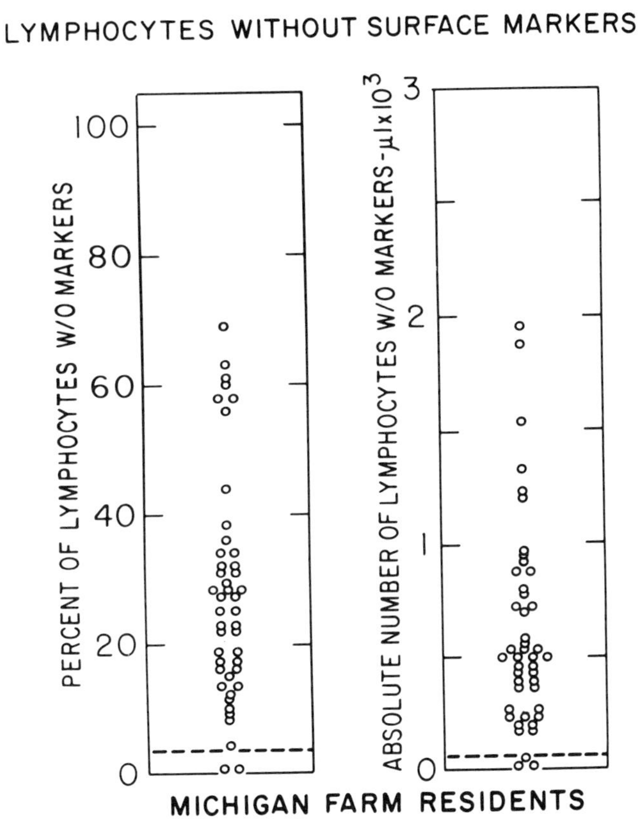

FIGURE 3. Increased percent and absolute number of "null cells" in PBB-exposed Michigan dairy farm residents.

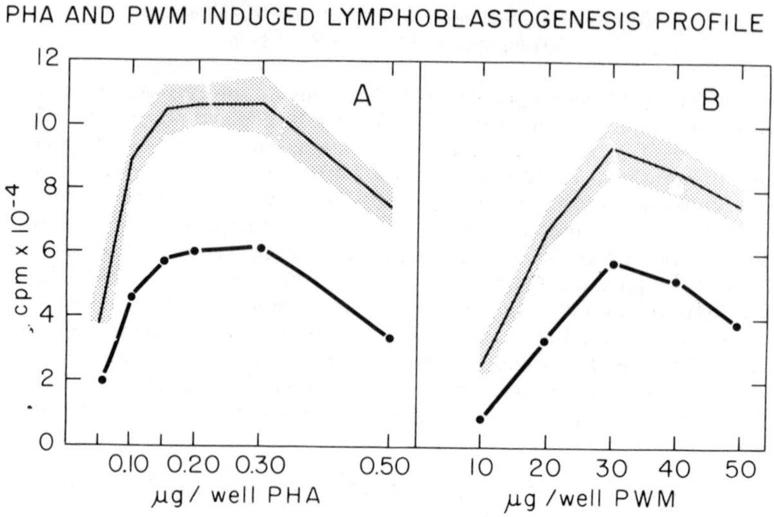

FIGURE 4. Typical PHA- and PWM-induced stimulation of peripheral blood lymphocytes from normal donor and Michigan farm residents. Normal donors———, Michigan farm residents ·—·—·

of T-lymphocytes but to the reduced number of T-cell population in the PBLs. However, in the remaining eight subjects functional impairment of T-cells was observed, since the decreased response to the mitogen occurred despite a normal range of lymphocytes in terms of percent and absolute number.

It is also important to note that while there was no significant change from normal values in the percentage and absolute number of B-lymphocytes in the PBLs of PBB-exposed Michigan farm residents, the maximum mitogenic stimulation values of 18 individuals in this group were 1/2 to 1/10 of the values observed among New York City residents and Wisconsin dairy farmers not exposed to contaminated food. This may suggest a functional defect and/or reduced helper T-cell subpopulation in PBL, which intrinsically might alter the B-cell mitogenic response (FIGURE 3).

Proliferative T-cell response as determined by one-way-mixed lymphocyte culture, confirms and further extends our observations, made with synthetic mitogens, that the immunocompetence of at least 35% of the Michigan farmers has been associated with a significant change in their lymphocyte function as compared to the control subjects (TABLE 3).

FIGURE 6 shows the PHA induced blastogenesis and percentage of T-cells in the PBLs of several family units of quarantined and nonquarantined Michigan dairy farm residents.

Results presented in FIGURE 6 show clearly that the changes in lymphocyte function detected among the PBB-exposed dairy farm residents tend to affect families as units and were independent of exposed individuals, age or sex, diminishing the possibility of genetic predisposition. Similar family clustering of plasma levels of PBB was noted by Wolff *et al.*[7,8] among the 62 examined quarantined and nonquarantined Michigan-farm families, which correlated well with the magnitude of PBB exposure from common dietary sources of the families.

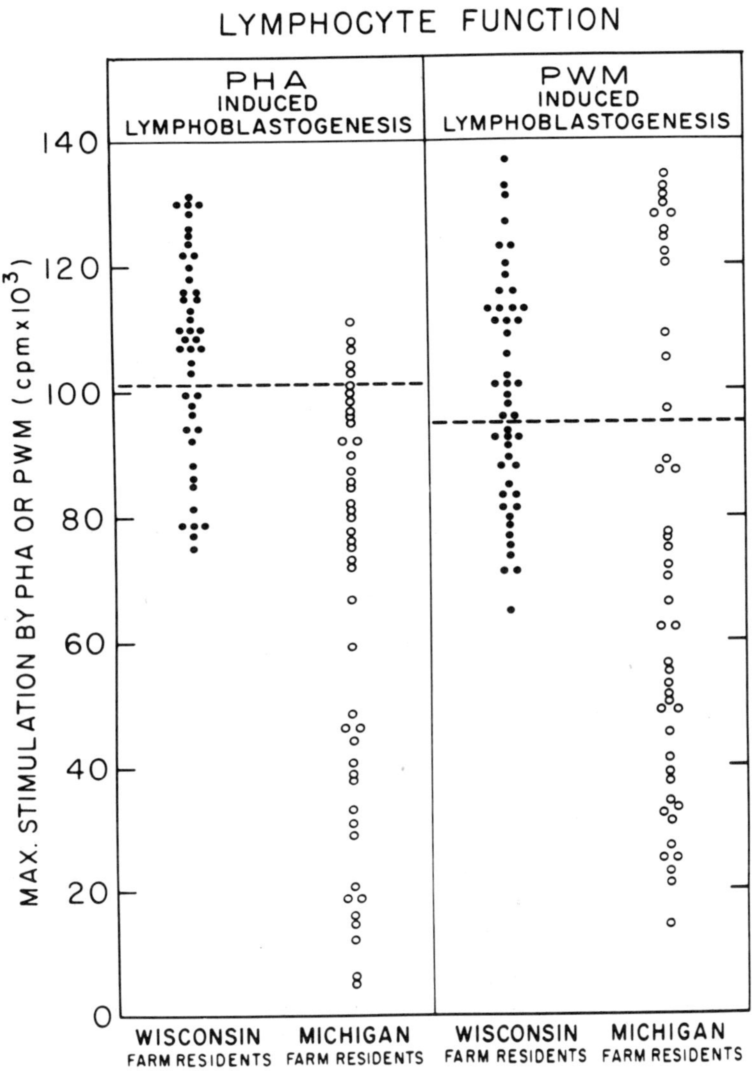

FIGURE 5. Reduced cell-mediated immunity among PBB-exposed Michigan dairy farm residents. Results are expressed as maximum stimulation c.p.m. $\times 10^3$ per 10^5 lymphocytes. The unstimulated lymphocytes gave a mean of 395 $\pm$ 24 c.p.m. and 410 $\pm$ 30 c.p.m. for Wisconsin and for Michigan groups, respectively.

Identification of PBBs in Blood Compartments

In a pilot study the presence of PBBs has been positively identified by mass spectrometry in isolated white blood cells, erythrocytes, and plasma of a small group of Michigan subjects. The major carrier in plasma was the β lipoprotein fraction. The ratios of PBB (hexa) binding to the surface membranes of white blood cells, red blood

TABLE 3
PERIPHERAL BLOOD LYMPHOCYTE FUNCTION IN PBB-EXPOSED MICHIGAN DAIRY FARM RESIDENTS

Subjects	N	Mixed Leukocyte Culture: Maximum Stimulation c.p.m.	S.I.
Wisconsin diary farm residents	46	38,172 ± 1,209	66
PBB-exposed Michigan dairy farm residents with normal PHA response	27	28,248 ± 2,072*	49
PBB-exposed Michigan dairy farm residents with abnormal PHA response	18	10,147 ± 317*	20

The data are means ± standard error. The Stimulation Index (S.I.) is calculated from the number of lymphocytes undergoing maximum stimulation divided by the number of unstimulated lymphocytes.

*Statistical significance (Student's t-test) between maximum blastogenesis of Wisconsin farm residents and that of Michigan farm residents ($p < .0001$).

cells, and β lipoproteins with respect to plasma were 86:15, 1:15, 92:15, respectively. No selective enrichment of homologs was evident. At the time of our initial report on this subject no consistent correlation was found between immune dysfunction and PBB level in plasma. Our work in progress, however, strongly suggests that the PBB distribution in various subfractions of white blood cells may be more meaningful than total plasma concentration in attempting to correlate PBB content of the blood and altered immune function observed in PBB-exposed individuals in Michigan.

SUMMARY

In 1973 inadvertent contamination occurred in a special farm feed supplement for lactating cows. Polybrominated biphenyls (PBBs) were used in place of magnesium oxide resulting in serious harm to farm animals, including cattle, chickens, geese, ducks. Farm families, accustomed to eating their own products, were most heavily exposed. To further study the impact of PBBs, 45 adult Michigan farm residents who were originally examined in a clinical field survey were further studied with respect to their immunologic status. For comparison, 46 dairy farm residents in Wisconsin, who had not eaten PBB-contaminated food, were examined, as were 79 healthy subjects in New York City. Abnormalities in the Michigan group included significant decrease in absolute numbers and percentages of T and B-lymphocytes and increased number of lymphocytes with no detectable surface markers ("null cells"). Significant reduction of *in vitro* immune function was noted in 35–40% of the Michigan farm residents who had eaten food containing PBB.

Despite the absence of any apparent numerical reduction, both T and B lymphocyte subpopulations of peripheral blood lymphocytes showed evidence of functional defect. Ten of the 45 Michigan farmers studied showed impaired PHA-induced blastogeneic response, due to the decreased number and percent of T-cells in the PBLs.

The decreased immune function detected among the PBB-exposed farm residents tended to affect families as a unit and was independent of exposed individuals' age or sex, speaking against the possibility of genetic predisposition.

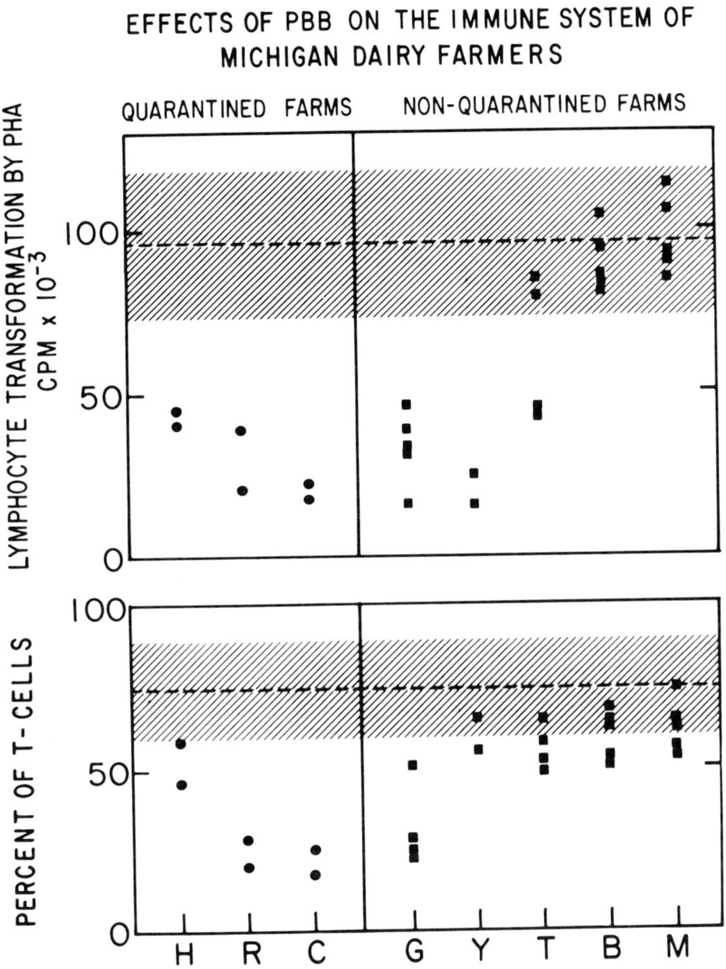

FIGURE 6. Lymphocyte function of PBL in quarantined and nonquarantined Michigan farm residents who ate food containing PBBs.

ACKNOWLEDGMENTS

The cooperation of Drs. Mary S. Wolff and William Rom is gratefully acknowledged. Thanks are also due to Rogena Brown, Frances Camper, Suzan S. Gillman, Shirley Levine, Angeles Mison, Frances Perez, and Robert Suzuki for technical assistance. We also gratefully acknowledge the excellent help and cooperation of the Marshfield Medical Foundation, Marshfield Medical Clinic in course of our Wisconsin clinical field survey.

This research was supported by Center Grant for the National Institute of Environmental Health Sciences, ES00928 and by NCI Grant 5P11-CA15936.

REFERENCES

1. CARTER, L. J. 1976. Michigan's PBB incident: Chemical mix-up leads to disaster. Science **192:** 240–243.
2. DUNCKEL, A. E. 1975. An updating on the polybrominated biphenyl disaster in Michigan, J. Am. Vet. Med. Assoc. **167:** 838–841.
3. JACKSON T. F. & F. L. HALBERT. 1974. A toxic syndrome associated with the feeding of polybrominated biphenyls-contaminated concentrate to dairy cattle, J. Am. Vet. Med. Assoc. **165:** 437–439.
4. HUMPHREY, H. E. B. & N. S. HAYNER. 1974. Polybrominated biphenyls: an agricultural incident and its consequences. An epidemiological investigation of human exposure. Michigan Department of Health, Lansing, Michigan.
5. ANDERSON, H. A., R. LILIS & I. J. SELIKOFF. 1978. Unanticipated prevalence of symptoms among dairy farmers in Michigan and Wisconsin Environ. Health Perspect. **23:** 217.
6. LILIS, R., *et al.* 1978. Comparison of findings among residents on Michigan dairy farms and consumers of produce purchased from these farms. Environ. Health Perspect. **23:** 105.
7. WOLFF, M. S., B. AUBREY, F. CAMPER & N. HAYMES. 1976 Relation of DDE and PBB serum levels in farm residents, consumers, and Michigan chemical corporation employees. Environ. Health Perspect. **23:**177–181.
8. WOLFF, M. S., N. HAYMES, H. A. ANDERSON & I. J. SELIKOFF. 1978. Family clustering of PBB and DDE values among Michigan dairy farmers. Environ. Health Perspect. **23:** 315–319.
9. FRIES & MARROW. 1975. Excretion of polybrominated biphenyls into the milk of cows. J. Dairy Science. **58:** 6.
10. GUTENMANN, W. H. & D. J. LISK. 1975. Tissue storage and excretion in milk of polybrominated biphenyls in ruminants. J. Agric. Food Chem. **23:** 1005–1007.
11. ERNEY, M. J. 1975. Confirmation of PBB residues in feeds and dairy products, using ultraviolet irradiation-gas. Liquid chromatographic technique. J. Assoc. Off. Anal. Chem. **58:** 1202–1205.
12. FEHRINGER, V. N. 1975. Determination of polybrominated biphenyls residues in dairy products. J. Assoc. Off. Anal. Chem. **58:** 978–982.
13. LEE, K. P., R. R. HEBERT, H. SHERMAN, J. F. AFTOSMIS & R. S. WARITZ. 1975. Bromine tissue residue and hepatotoxic effects of octabromobiphenyls in rats. Toxicol. Appl. Pharmacol. **34:** 115–127.
14. SLEIGHT, S. D. & V. L. SANGER. 1976. Pathologic features of polybrominated biphenyl toxicosis in the rat and guinea pig. J. Am. Vet. Med. Assoc. **169:** 1231.
15. LEE, K. P., *et al.* 1975. Octobromobiphenyl-induced ultrastructural changes in rat liver. Arch. Environ. Health **30:** 465.
16. FARBER, T. M. & A. BAKER. 1974. Microsomal enzyme induction by hexabromobiphenyl. Toxicol. Appl. Pharmacol. **29:** 102.
17. MOORE, R. W., G. DANNAN & S. D. AUST. 1976. Induction of drug metabolizing enzymes in rats nursing from mothers fed polybrominated biphenyls. Fed. Proc. **35:** 709.
18. KIMBROUGH, R. D., V. W. BURSE & J. A. LIDDLE. 1978. Persistent liver lesions in rats after a single oral dose of brominated biphenyls (FireMaster FF.1) and concomitant PBB tissue levels. Environ. Health Perspect. **23:** 265.
19. BEKESI, J. G., J. F. HOLLAND, H. A. ANDERSON, A. S. FISCHBEIN, W. ROM, M. S. WOLFF & I. J. SELIKOFF. 1978. Lymphocyte function of Michigan dairy farmers exposed to polybrominated biphenyls. Science **199:** 1207–1209.
20. WYBRAN, J., M. C. CARR & H. H. FUDENBERG. 1972. The human rosette-forming cell as a marker of a population of thymus-derived cells. J. Clin. Invest. **51:** 2537.
21. NUSSENZWEIG, V. N. 1974. Receptors for Immune Complexes on Lymphocytes. Adv. Immunol. **19:** 217.
22. PREUD'HOMME, J. L. & M. SELIGMANN, 1972. Surface bound immunoglobulins as a cell marker in human lymphoproliferative diseases. Blood **40:** 777.

Index of Contributors